# 兽医内科与临床诊疗学研究

（2012 — 2014）

中国畜牧兽医学会兽医内科与临床诊疗学分会　组编

中国农业科学技术出版社

## 编辑委员会

**主　任：**陈　越（国家自然科学基金委员会）

**副主任：**王　哲（吉林大学动物医学学院）

王俊东（山西农业大学）

韩　博（中国农业大学动物医学院）

**委　员**（以姓氏笔画为序）：

王建华（西北农林科技大学动物医学院）

文利新（湖南农业大学动物医学院）

叶俊华（公安部南昌警犬基地）

朱连勤（青岛农业大学）

刘宗平（扬州大学兽医学院）

张乃生（吉林大学动物医学学院）

张海彬（南京农业大学）

胡国良（江西农业大学）

徐世文（东北农业大学动物医学学院）

郭定宗（华中农业大学动物医学院）

唐兆新（华南农业大学兽医学院）

黄克和（南京农业大学动物医学院）

**主　编：**韩　博（中国农业大学动物医学院）

**副主编**（按姓氏笔画为序）：

曲伟杰（云南农业大学动物科学技术学院）

段　纲（云南农业大学动物科学技术学院）

黄克和（南京农业大学动物医学院）

**审　校：**王　哲（吉林大学动物医学学院）

王俊东（山西农业大学）

# 前　　言

光阴荏苒，中国畜牧兽医学会家畜内科学分会上一次在兰州举行学术研讨会至今已快两年了，现在我们又将迎来家畜内科学分会的新一次学术盛会（2014年7月19-21日，云南昆明）。两年来，我国的兽医内科与临床诊疗学科有了长足的发展，许多研究工作进入了国际前沿，研究水平和深度有了极大地提高，取得了一些重要研究成果，临床诊疗技术也取得了不少突破。

本次会议共收到147篇研究论文，其中论文全文45篇、论文摘要102篇。经初步统计，2012年8月1日至2014年6月30日，我国兽医内科诊断学培养已毕业的硕士研究生381名，培养已毕业的博士研究生50名，培养已出站的博士后11名。中国畜牧兽医学会家畜内科学分会代表获得省部级科技成果三等以上奖励5项，授权发明、实用新型专利49项，发表SCI收录论文250篇。主编兽医内科学与诊断学教材11部，主编兽医内科学与诊断学方面的专著19部，主译兽医内科学与动物诊断学著作3部。

经分会研究，本次会议将授予中国农业大学动物医学院王志教授、东北农业大学动物医学学院康世良教授和石发庆教授“中国畜牧兽医学会家畜内科学分会第四届终身成就奖”。同时，将授予SCI收录论文影响因子大于5的通讯作者：山西农业大学王俊东教授，扬州大学刘宗平教授，山东农业大学刘建柱博士、副教授，东北农业大学张志刚博士、副教授“优秀论文一等奖”；将授予SCI收录论文影响因子在4和5之间的通讯作者：山西农业大学王俊东教授，吉林大学动物医学学院张乃生教授，华中农业大学动物医学院郭定宗教授，东北农业大学动物医学学院徐世文教授，中国农业大学动物医学院韩博教授和吉林大学动物医学学院杨正涛博士、副教授“优秀论文二等奖”。

为了提高执业兽医，尤其是我国西南边陲从业人员的兽医临床诊疗水平，促进和加强执业兽医继续教育，中国畜牧兽医学会家畜内科学分会在本次会议上探索性地举办执业兽医继续教育项目培训班。经向中国兽医协会请示，同意本次培训可授予执业兽医人员继续教育学分。

随着我国兽医科技的发展，特别是兽医内科学、兽医临床诊断学和兽医临床治疗学以及小动物临床诊疗的迅速发展，家畜内科学这一名称已无法涵盖分会的内涵，况且中国畜牧兽医学会下设的兽医相关分会大都以兽医冠名，这也符合国际惯例。因此，在原有的兽医内科学基础上将兽医临床诊断学和兽医临床治疗学纳入分会，对提高分会的阵容和影响力，提升我国兽医临床诊疗的科教水平和动物疾病的临床诊疗水平，必将起到积极地推动作用。经中国畜牧兽医学会家畜内科学分会理事会研究讨论，建议将分会的名称变更为：中国畜牧兽医学会兽医内科与临床诊疗学分会。这项工作已经获得中国畜牧兽医学会的批复。

总之，上述的相关工作，展示了中国畜牧兽医学会兽医内科与临床诊疗学分会代表取得的丰硕成果，体现了中国畜牧兽医学会兽医内科与临床诊疗学分会代表的凝聚力和战斗力，表明了中国畜牧兽医学会兽医内科与临床诊疗学分会的生机和活力。希望大家在各自的岗位上继续努力奋斗，为学会的发展做出更大的贡献。

陈越

中国畜牧兽医学会兽医内科与临床诊疗学分会 理事长

2014年6月30日于北京

# 目　录

# 一、中国畜牧兽医学会兽医内科与临床诊疗学分会论文全文

# Betulinic acid prevents alcohol-induced liver damage by improving the antioxidant system in mice

YI Jin'e, XIA Wei, TAN Zhuliang

*Colleges of Veterinary Medicine, Hunan Agricultural University, Changsha* 410128, *China*

**Abstract**: Betulinic acid (BA), a pentacyclic lupane-type triterpene, has a wide range of bioactivities. The main objective of this work was to evaluate the hepatoprotective activity of BA and the potential mechanism underlying the ability of this compound to prevent liver damage induced by alcohol *in vivo*. Mice were given oral doses of BA (0.25, 0.5, and 1.0 mg/kg) daily for 14 days, and induced liver injury by feeding 50% alcohol orally at the dosage of 10 ml/kg after 1 h last administration of BA. BA pretreatment significantly reduced the serum levels of alanine transaminase, aspartate transaminase, total cholesterol, and triacylglycerides in a dose-dependent manner in the mice administered alcohol. Hepatic levels of glutathione, superoxide dismutase, glutathione peroxidase, and catalase were remarkably increased, while malondialdehyde contents and microvesicular steatosis in the liver were decreased by BA in a dose-dependent manner after alcohol-induced liver injury. These findings suggest that the mechanism underlying the hepatoprotective effects of BA might be due to increased antioxidant capacity, mainly through improvement of the tissue redox system, maintenance of the antioxidant system, and decreased lipid peroxidation in the liver.

**Keywords**: alcohol, antioxidant capacity, betulinic acid, lipid peroxidation, liver damage

## 1 Introduction

Alcohol abuse and its associated social consequences are a major health problem in many areas of the world. Alcohol abuse is a brain-centered addictive behavioral disorder that develops regardless of gender, race, age, or economic standing, and can lead to alcoholic liver disease (ALD) in many patients[5,13]. ALD presents a broad spectrum of disorders ranging from simple fatty liver to more severe forms of liver injury that include alcoholic hepatitis (AH), cirrhosis, and superimposed hepatocellular carcinoma[12]. In the United States, the Centers for Disease Control and Prevention estimate that 50% of people aged 18 or older drink alcohol[3]. Among these, 5% are classified as heavy drinkers and 15% are binge drink. The National Institute on Alcohol Abuse and Alcoholism (USA) has reported that liver cirrhosis is the 12th leading cause of death with a total of 29 925 deaths in the United States in 2007 (48% of which were alcohol-related) and is also a major contributor to liver disease-related mortality in other countries[12].

The abuse of alcohol significantly increases morbidity and mortality from infectious diseases along with the risk of cardiovascular, brain, pancreatic, renal, cerebral, and oncological diseases[13]. The overall socioeconomic cost of alcohol abuse including direct (drug and hospitalization expenses) and indirect (due to loss of work productivity, crime, and reduction in health-related quality of life) costs has been estimated to be more than $185 million annually in the United States[20,25]. Despite the profound economic and health impact of ALD, many patients continue to consume alcohol. The primary treatment recommendation for this disease is abstinence from alcohol. Other treatment options, such as pharmacological and nutritional therapies, are also available[17]. Interest in potential antioxidant therapy for treating AH has increased as a growing body of evidence indicates that oxidative stress is a key mechanism underlying alcohol-mediated. Alcohol intoxication leads to increased generation of reactive oxygen species (ROS) while suppressing the antioxidant defense system[28,31]. Antioxidants have the ability to scavenge/deactivate ROS and inhibit the oxidation of various cellular substances. Several antioxidants such as S-adenosylmethionine, N-acetylcysteine, beta-carotene, vitamin C, vitamin E, and selenium have been systematically evaluated[12,18]. It was reported that AH patients treated with vitamin E, selenium, and zinc supplements had an in-hospital mortality rate of 6.5% compared to a mortality rate of 40% for the placebo group[17]. Plants produce an extensive variety of antioxidant compounds such as phenolics, terpenoids, alkaloids, and bioactive protein or peptides. These reagents can be used as ingredients in nutritional supplements/nutraceuticals that combat oxidative stress with minimal side effects[35].

Betulinic acid (BA; 3β-hydroxy-lup-20(29)-en-28-oic acid) is a pentacyclic lupane-type triterpene (Fig. 1A) that exists widely in food, medicinal herbs, and plants, especially birch bark[9]. As an oxidation product of betulin (see Fig. 1B), BA has also been detected in extracts of white birch bark[9,32]. These reagents are of particular interest due to their wide spectrum of biological activities such as antioxidant, anti-inflammatory, anti-tumor, anti-angiogenesis, anti-viral, anti-HIV, anti-neoplastic, and anti-plasmodial properties[10,11,22,24,32]. BA was not found to be toxic at a concentration of 500 mg/kg in mice[22].

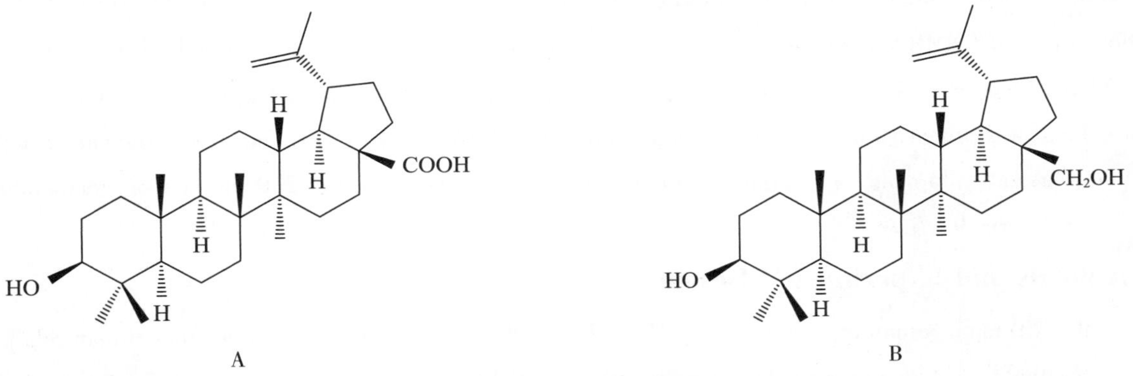


Fig. 1 Chemical structures of betulinic acid (A) and betulin (B)

Several studies have demonstrated that natural triterpenes such as lupeol, betulin, ursolic acid, and oleanolic acid effectively reduce hepatotoxicity induced by carbon tetrachloride, acetaminophen, ethanol, and cadmium *in vivo* and *in vitro* [29,38]. The protective effect of betulin, BA, and oleanolic acid against ethanol-induced cytotoxicity in HepG2 cells has also been reported[28]. The mechanisms underlying this effect include the suppression of enzymes activities that play roles in liver damage such as cytochrome P450 (CYP450), cytochrome B5 (CYB5), and cytochrome P4501A (CYP1A), or increased production of antioxidant substances including catalase (CAT), superoxide dismutase (SOD), glutathione (GSH), and metallothioneins that help protect liver mitochondria[28]. *In vitro*, BA inhibits ethanol-induced activation of hepatic stellate cells, as an antioxidant, suppressing of ROS, tumor necrosis factor-α (TNF-α) and transforming growth factor-β (TGF-β) production as well as nuclear factor-kappa B/inhibitory protein of NF-κB (NF-κB/IκB) transduction signaling[28,29]. However, ability of BA to confer *in vivo* protection against liver injury induced by alcohol has not been previously reported. The objective of this study was to investigate the possible hepatoprotective effects of BA against alcohol induced acute liver damage in mice and elucidate the underlying mechanism.

## 2 Materials and Methods

### 2.1 Extraction, Synthesis, and Identification of BA from White Birch Bark

White birch bark samples were collected during the spring of 2009 in Wroclaw, Poland. All the collected bark samples were immediately dried at 60℃ and stored in a dry, dark place. To extract betulin, about 15 g of dried bark were subjected to reflux with 200 ml methanol for 3 h at 70℃. The methanol extract was dried under negative gauge pressure and dissolved in 150 ml dichloromethane. After adding 150 ml 2 mol/L sodium hydroxide and mixing, the lower liquid layer was collected and then filtered under negative gauge pressure. The residue was dissolved in 200 ml ether. Next, 200 ml water was added and the solution was thoroughly mixed. The upper liquid layer was subsequently collected, dried, and subjected to silica gel column chromatography (Flash column; Aisimo Corporation, China)[32]. Betulin was fractionated with hexane and ethyl acetate (6 : 1, V : V). BA was synthesized in two steps: 1) the betulin was oxidized with Jones reagent (1 : 6, molar ratio) at 20℃ for 3 h to obtain betulonic acid, and 2) betulonic acid was reduced by sodium borohydride (1 : 1, molar ratio, Sigma-Aldrich, USA) in 10 ml tetrahydrofuran (Sigma-Aldrich) to produce a mixture of 3α- and 3β-hydroxyl products (5 : 95) as previously described[9,32]. Crystallization of the product from methanol resulted in the 3β-hydroxyl product (BA) with a yield of 75%.

The semi-synthetic compound was a white powder, and characterized using Fourier transform infrared spectroscopy (FI-IR), high performance liquid chromatography-mass spectrometery (HPLC-MS), and nuclear magnetic resonance (NMR) as previously described[32]. Purity of the BA was measured with HPLC using a Zorbax Eclipse XDB-C8 column (4.6 mm × 50 mm, 5 μm; Agilent Technologies, USA). The injection volume was 20 μl and the column was eluted with an ethanol : water solution (86 : 14, *V* : *V*, 4 g/L ammonium formate in water) as the mobile phase at a flow rate of 1 ml/min, and UV detection was performed at 210 nm at room temperature. The Purity of the BA was 96.5%.

### 2.2 Animals and Experimental Design

A total of 50 male Kunming mice weighing 18 – 22 g (6 weeks old) were purchased from Hunan Silaikejingda Laboratory Animal Co., Ltd. (China). The mice were acclimated to the laboratory environment for 1 week before the experiment was initiated, and fed an M02 standard mouse diet from Hunan Silaikejingda Laboratory Animal (China). The mice had access to food and water *ad libitum*, and were maintained at a constant temperature (23 ± 1)℃ and humidity (60% ± 10%) with a 12-h light/dark cycle. The present study complied with the Animal Care and Use Guidelines of China and was approved by the Animal Care Committee of Hunan Agricultural University (China)[15].

The dose of BA and duration of treatment were established based on a previous study[37]. The mice were randomly divided into five groups: group A served as the control (no alcohol, no BA), group B served as the alcoholic liver injury model (alcohol, no BA), group C received a low dose of BA (0.25 mg/kg) and alcohol, group D received a moderate dose of BA (0.5 mg/kg) and alcohol, and group E consumed a high dose of BA (1 mg/kg) and alcohol. BA was administered orally to most mice with 1% starch jelly (Sinopharm Chemical Reagent, China) every day for 14 days while animals in groups A and B were given an equivalent amount of 1% starch jelly only. The mice were fasted overnight (16 – 18 h) before receiving a single dose of 50% ethanol (Sinopharm Chemical Reagent) at the dosage of 10 ml/kg on the 14th day of the study 1 h after the last administration of BA. Group A received an equal volume of sterile saline.

9 h after alcohol administration, blood samples were collected in tubes (Eppendorf, Germany) without

anticoagulant by venous puncture with the mice under light anesthesia induced by diethyl ether (Sinopharm Chemical Reagent). The animals were then sacrificed by cervical dislocation and liver samples were collected. Serum was collected by centrifugation (Z383K Universal High Speed Centrifuge; Hermle labortechnik, Germany) at 2 000 × g for 10 min at 4℃ and frozen at −70℃ until analysis.

Livers were quickly removed, weighed, and the liver indices expressed in milligrams of liver per 10 g of body weight were calculated. The liver tissues were washed in a chilled 0.9% NaCl solution and minced into small pieces on ice. A 10% (*W/V*) homogenate (Tenbroeck tissue grinders; Wheaton, USA) was prepared in 10 mmol/L phosphate buffer (pH 7.4) and centrifuged (Z383K Universal High Speed Centrifuge; Hermle labortechnik) at 2 500 × g for 15 min at 4℃. The resulting supernatant was collected and stored at −70℃ until analysis.

### 2.3 Evaluation of Serum Biomarkers of Liver Injury

Serum alanine transaminase (ALT), aspartate transaminase (AST), total cholesterols (TC), and triacylglycerides (TG) levels were measured using Mindray commercial kits and a Mindray BS-200 automatic biochemistry analyzer (Shenzhen Mindray Bio-Medical Electronics, China). ALT and AST levels were expressed as units per liter of serum (U/L) while TC and TG concentrations were expressed as nmol per liter of serum (nmol/L).

### 2.4 Measurement of SOD, GSH-Px, and CAT Activities Along with Hepatic MDA and GSH Contents

The activities of superoxide dismutase (SOD), glutathione peroxidase (GSH-Px), and catalase (CAT) were measured along with malondialdehyde (MDA) and glutathione (GSH) levels using commercial kits (Nanjing Jiancheng Bioengineering Institute, China) according to the manufacturer's protocols. The levels of SOD, GSH-Px, and CAT activities were expressed as unit per mg of protein (U/mg protein). The MDA and GSH levels were expressed as nmol per mg protein (nmol/mg protein) and mg per g of protein (mg/g protein), respectively. Liver protein concentrations were determined with a Bradford assay using commercial kits (Nanjing Jiancheng Bioengineering Institute) according to the manufacturer's protocols.

### 2.5 Histologic Analysis of Liver Tissues

Liver tissues from the mice were removed and fixed in 10% neutral buffered formalin. The tissues were then rinsed with water, dehydrated with ethanol, subsequently embedded in paraffin, and sectioned into slices (4-μm-thick) using a Leica RM2235 rotary microtome (Leica Microsystems, Germany). The sections were mounted onto glass slides (Beyotime Institute of Biotechnology, China), dewaxed in xylene and ethanol, and stained with hematoxylin and eosin (Beyotime Institute of Biotechnology, China) for histological evaluation using a Motic BA410 microscope (Motic Incorporation, China). Image Pro-Plus Motic Med 6.0 software (Motic Incorporation) was used for image processing.

### 2.6 Statistical Analysis

The results are presented as the mean ± standard deviation (SD). The data were analyzed with a one-way analysis of variance (ANOVA) followed by the Student-Newman- Keuls test for multiple comparisons. All statistical analyses were performed using the SPSS 16.0 statistical package (SPSS, USA). A $P$ value $<0.05$ was considered statistically significant and a $P$ value $<0.01$ was considered highly significant.

## 3 Results

### 3.1 Effects of BA on Liver Indices and Serum Biomarker Expression in Mice

The effects of BA on alcohol-affected liver indices along with serum levels of ALT, AST, TC, and TG in mice were assessed (Table 1). The liver indices of group B were increased ($P<0.05$) after alcohol treatment compared to those in the control group A. Pretreatment with BA significantly restored the liver indices of groups C, D, and E in a dose-dependent manner compared to those of group B.

The activities of ALT and AST in alcohol-treated group B were increased by 45.3% and 130.5%, respectively, compared to those found in the control (group A; $P<0.01$). Pretreatment with BA reduced both serum ALT and AST levels in a dose-dependent manner. Serum ALT and AST activities were decreased by 7.6% and 9.4% ($P<0.05$), 17.8% and 17.0% ($P<0.01$), and 25.4% and 30.9% ($P<0.01$), respectively, in groups C, D, and E compared to those observed in group B. At the same time, the serum concentrations of TC and TG in group B were increased by 15.7% and 38.7%, respectively, relative to group A ($P<0.05$). BA pretreatment reduced the concentrations of TC and TG in a dose-dependent manner. Serum TC and TG levels were decreased by 6.0% and 11.1%, 11.3% and 18.2% ($P<0.05$), and 12.8% and 24.5% ($P<0.05$), respectively, in groups C, D, and E compared to group B. These findings indicate that BA pretreatment conferred significant protection against alcohol-induced fatty liver by reducing elevated serum levels of ALT, AST, TC and TG in a dose-dependent manner.

Table 1 Effects of BA on liver indices and ALT, AST, TC, and TG serum levels in mice treated with or without alcohol

| Group | Liver indices (mg/10 g of body wt) | ALT (U/L) | AST (U/L) | TC (mmol/L) | TG (mmol/L) |
|---|---|---|---|---|---|
| A | 466.79 ±40.35 | 51.92 ±4.28 | 73.08 ±5.03 | 3.45 ±0.28 | 1.94 ±0.39 |
| B | 514.57 ±52.65$^a$ | 75.43 ±8.42$^b$ | 168.45 ±18.42$^b$ | 3.99 ±0.37$^b$ | 2.69 ±0.21$^b$ |
| C | 492.93 ±47.36 | 69.72 ±7.37$^c$ | 152.58 ±12.37$^c$ | 3.75 ±0.30 | 2.39 ±0.64 |
| D | 485.37 ±42.11 | 62.04 ±5.85$^d$ | 139.76 ±14.85$^d$ | 3.54 ±0.28$^c$ | 2.20 ±0.44$^c$ |
| E | 476.14 ±34.16$^c$ | 56.28 ±5.25$^d$ | 116.34 ±13.45$^d$ | 3.48 ±0.26$^c$ | 2.03 ±0.55$^c$ |

Values are presented as the mean ± standard deviation (SD). $^aP<0.05$ and $^bP<0.01$ compared to the control group (group A). $^cP<0.05$ and $^dP<0.01$ compared to the group treated with alcohol alone (group B).

## 3.2 Effect of BA on SOD, GSH-Px, and CAT Activities in Mouse Liver

SOD, CAT, and GSH-Px are important enzymes in the antioxidant defense system. Hepatic activities of these three factors decreased significantly in alcohol-treated mice (group B) compared to the control group A (Fig. 2). Pretreatment with BA (groups C, D, and E) increased the activities of SOD, CAT, and GSH-Px in a dose-dependent manner compared to treatment with alcohol alone (group B).

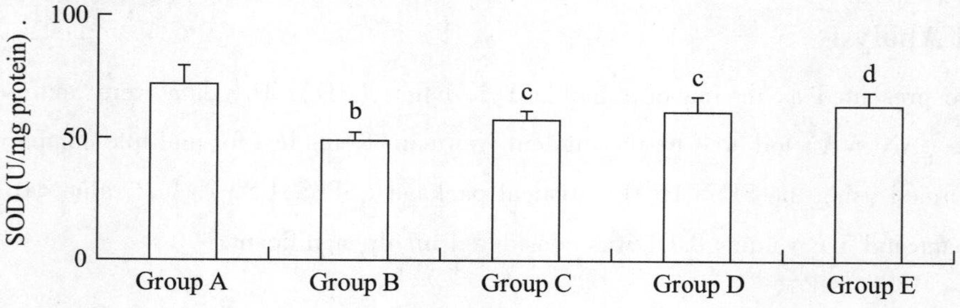


A

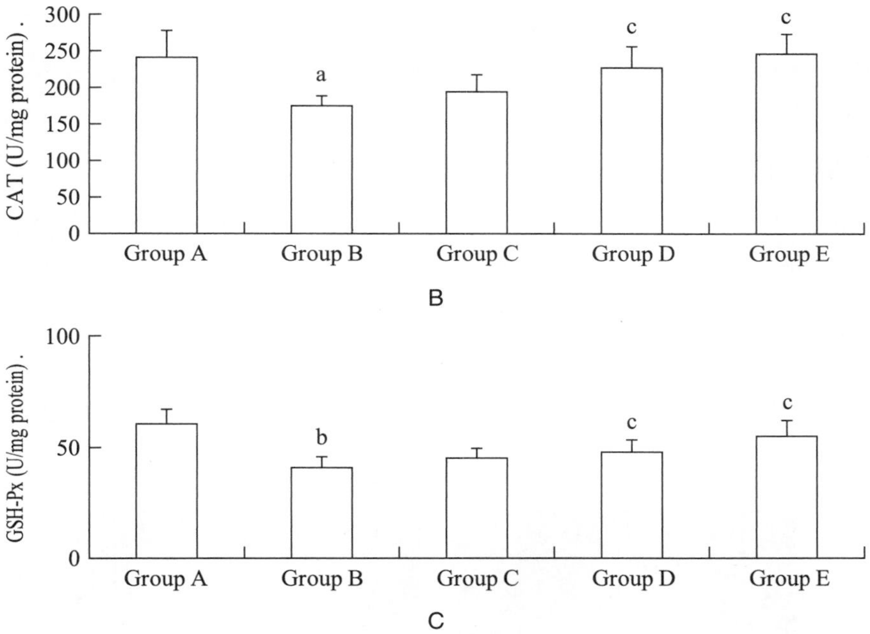


Fig. 2 Effects of BA on SOD (A), CAT (B), and GSH-Px (C) levels altered by alcohol in the liver of mice

Values are presented as the mean ± *SD*. $^{a}P<0.05$ and $^{b}P<0.01$ compared to the control group (group A). $^{c}P<0.05$ and $^{d}p<0.01$ compared to the group treated with alcohol alone (group B)

## 3.3 Effects of BA on Hepatic MDA and GSH Levels in Mice

MDA is a major biomarker of lipid peroxidation in tissues. GSH is an antioxidant that helps protect cells against ROS. As shown in Fig. 3, MDA levels was remarkably increased by 87.5% ($P<0.01$) while GSH levels were significantly decreased by 46.3% ($P<0.01$) in the alcohol-treated animals (group B) 9 h after alcohol administration compared to the control group A. BA pretreatment significantly decreased hepatic MDA contents

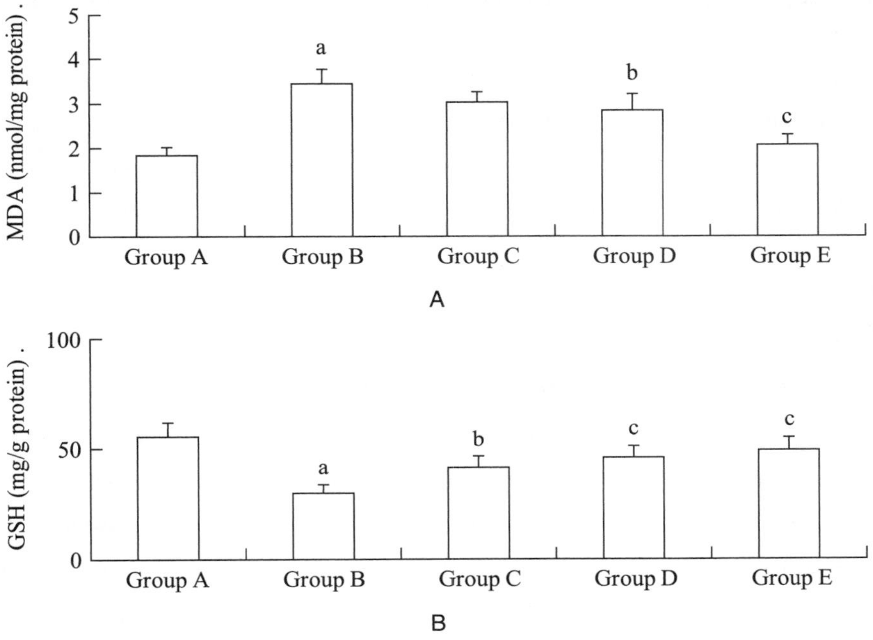


Fig. 3 Effects of BA on alcohol-induced changes of MDA (3) and GSH (3) contents in mouse liver

Values are presented as the mean ± *SD*. $^{a}P<0.01$ compared to the control group (group A). $^{b}P<0.05$ and $^{c}P<0.01$ compared to the mice treated with alcohol alone (group B)

while increased GSH levels in a dose-dependent manner. MDA contents were decreased by 12.8% ,17.4% ,and 39.7% while the GSH levels were increased by 39.4% ,54.8% ,and 66.0% in groups C,D,and E,respectively, compared to group B (Fig. 3).

### 3.4 Effects of BA on Mouse Liver Histopathology

To confirm the protective effect of BA on alcohol induced liver tissue damage, histological examinations were also performed. The control group had normal hepatic architecture (Fig. 4A) whereas the mice treated with alcohol alone exhibited diffuse cytoplasmic vacuolization (Fig. 4B) indicative of hepatocellular steatosis with no signs of necro-inflammation. In contrast, BA pretreatments reduced the degree of alcohol-induced micro-vacuolization in the liver in a dose-dependent manner (Fig 4C, D, and E). These results showed that BA protected the mouse livers against alcohol-induced hepatic steatosis, a known indicator of alcohol toxicity.

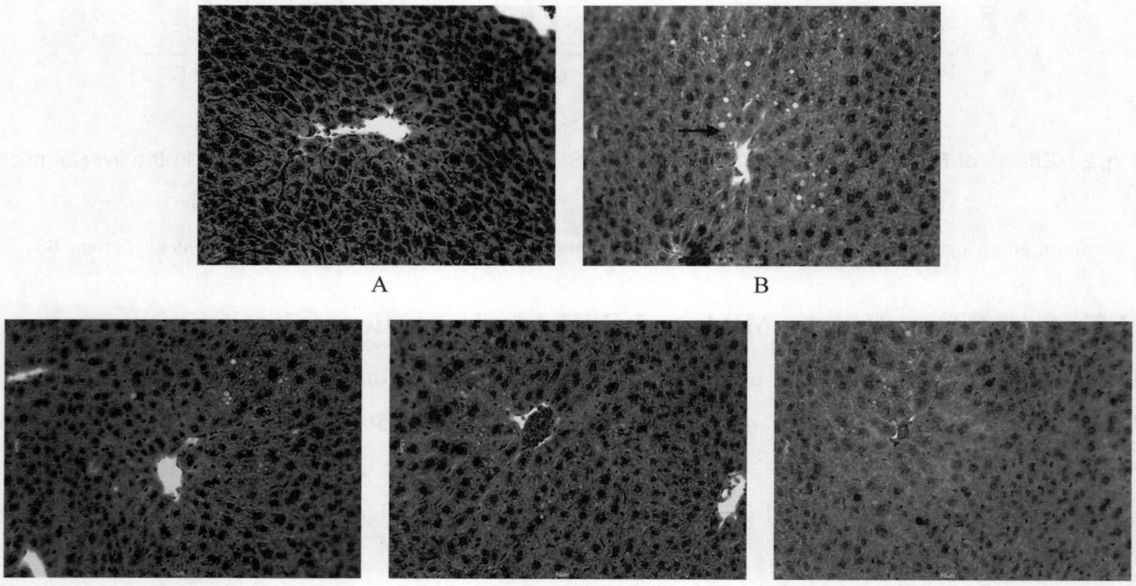

A B

C D E

Fig. 4 Histopathological changes in mouse liver (H&E staining, 400Rr × agnification)

No steatosis was observed in the control group (A). Diffuse cytoplasmic vacuolization (arrow) indicative of steatosis was observed in mice treated with alcohol alone (B). Hepatocellular steatosis was alleviated in a dose-dependent manner by the administration of 0.25 mg/kg (C), 0.5 mg/kg (D), and 1 mg/kg (E) BA for 14 days

## 4 Discussion

BA is a naturally occurring plant-derived pentacyclic triterpenoid present in many fruits and vegetables such as *Tecomella undulata*, *Coussarea paniculata*, *Caesalpinia paraguariensis*, *Vitex negundo*, and *Ilex macropoda*[14,33]. Although BA is widely available from numerous botanical sources, BA contents in these sources are extremely low. Alternatively, BA can be chemically synthesized from betulin, a closely related compound. White bark birch trees (*Betula* species), a major source of betulin, are widely distributed in the northern hemisphere. Betulin accounts for up to 30% of the dry weight of birch bark extract[9]. Therefore, a semi-synthetic method for preparing BA from betulin was used to obtain sufficient quantities of this bioactive triterpene for the current study.

Alcohol abuse affects the morphology and function of almost all organs in the body[27,31]. Liver is one of the prime target organs of alcohol-related diseases. Alcohol abuse results in hepatocytes damage, liver injury, and inflammation that lead to increased cell permeability and leakage of AST and ALT. Hence, elevated activities of AST and ALT in the plasma are hallmarks of hepatic damage[14,23]. In the present study, alcohol treatment significantly increased AST and ALT activities. These activities significantly decreased when the mice were pretreated with BA. The results were in agreement with previous findings showing that mice pretreated with BA have significantly reduced serum levels of ALT and AST elevated by D-galactosamine/lipoplysaccharide[38], indicating that the hepatocytes sustained minimal injury and their membranes were stabilized.

Alcohol alters the fatty acid composition in the liver. It has been demonstrated that increased cellular nicotinamide adenine dinucleotide hydrogen concentrations and acetaldehyde dehydrogenase activity may impair fatty acid β-oxidation and the tricarboxylic acid cycle during alcohol metabolism, leading to severe free fatty acid overload, increased TG synthesis, and subsequent steatosis[6,27]. Results from the present study showed that BA pretreatment could protect against alcohol-induced fatty liver by reducing serum levels of TC and TG as well as liver indices, which was most likely due to increased β-oxidation in the liver and energy expenditure. BA also ameliorated steatosis (Fig. 4), indicating that BA decreased lipid accumulation in the visceral adipocytes of alcohol-treated mice. This in turn could prevent toxic complications resulting from increased levels of hepatic lipids. Previously, it was also reported that BA functions as both a hypolipidemic and hypoglycemic agent when administered daily to mice fed a high-fat diet[8].

Alcohol affects the liver by inducing nutritional disturbances along with its predominant metabolism in the liver associated with oxidation-reduction (redox) reactions and oxidative stress[26]. In hepatocytes, oxidation of alcohol into water and carbon dioxide is mediated by three known enzyme systems: alcohol dehydrogenase (ADH) in the cytoplasm, the microsomal ethanol oxidizing system in the smooth endoplasmic reticulum of mitochondria, and CAT in peroxisomal membranes[13]. All of these biochemical pathways produce acetaldehyde, a highly reactive product of alcohol oxidative metabolism that damages mitochondria and promotes glutathione depletion, free radical-mediated toxicity, and lipid peroxidation[12]. Alcohol also increases liver oxidative stress via the generation of ROS and adducts. Excessive ROS or a defect of antioxidant system can directly lead to hepatocyte damage and enhance the sensitivity of hepatocytes to lipid peroxidation[27]. Endogenous enzymatic and non-enzymatic antioxidants protect the intracellular milieu against various toxic factors. Elevated lipid peroxidation results in significant depletion of cellular antioxidants[14]. Cellular GSH levels can be decreased through ROS-induced GSH oxidation or GSH export from the cells. Reduced GSH concentrations further enhance ROS production during oxidative challenge[7]. Moreover, BA significantly inhibits ROS production by hepatic stellate cells treated with ethanol[29]. All of these observations indicate that BA exerts prominent antioxidant effects against alcohol-induced oxidative damage.

Oxidative stress results in apoptosis and cell damage, and can directly disrupt the intracellular redox state or indirectly activate signal transduction pathways[37]. Apoptosis is triggered by extrinsic and intrinsic signals via two main pathways: the death receptor and mitochordria mediated cascades[16]. The death receptor pathway involves ligand-receptor binding initiated by protein-protein interactions at the cell membranes that activate initiator caspases[1]. Ligands that activate death receptors belong to the tumor necrosis factor (TNF) superfamily of cytokines that includes TNF-α, Fas ligand (FasL), and tumor necrosis factor-alpha-related apoptosis-inducing ligand (TRAIL)[2]. TNF-α, one of the most important proinflammatory mediators, induces hepatocyte apoptosis and necrosis through the activation of caspase-3, thereby leading to hepatocellular DNA fragmentation[36]. Several studies have shown that oxidative stress in cases of alcohol liver injury could result in the up-regulation of TNF-α expression[19,29,38]. TNF-α-induced intracellular ROS generation is markedly decreased by pretreatment with BA[34]. Moreover, BA decreases ethanol-induced TNF-α and TGF-β expression, probably through its ability to inhibit the

NF-κB pathway[19,29,38]. B-cell lymphoma/leukemia-2 (Bcl-2) is an anti-apoptotic protein, and Bcl-2-associated X protein (Bax) and Bcl-2 antagonist/killer (Bak) are pro-apoptotic proteins that are important regulators of apoptosis in the mitochondrial pathway[4,30]. BA promotes the expression of Bcl-2 and suppresses acute liver damage by preventing apoptosis[38]. Additionally, ROS activation of c-Jun $NH_2$-terminal kinase (JNK) can induce extrinsic or intrinsic apoptotic signaling. Previous reports have shown that BA significantly inhibits the production of ROS and down-regulates JNK activity in hepatic stellate cells treated with ethanol[21,29]. Therefore, the protective effects of BA against alcohol-induced liver injury possibly involve both exterior and interior pathways.

In conclusion, the present study demonstrated that BA exerts potent hepatoprotective effects that reduce alcohol-induced liver damage in mice. The primary mechanisms responsible for the hepatoprotective activities of BA might be due to the antioxidant capacity of this compound that improve the tissue redox system, maintain the antioxidant system, and decrease lipid peroxidation in the liver. These properties may involve both external and internal pathways. BA has a great potential for use as a beneficial nutraceuticals or dietary supplements for treating alcohol-related diseases. Further investigation is necessary to determine the exact mechanism governing the ability of BA to prevent alcohol-induced liver damage. It is also worth investigating the *in vivo* and *in vitro* roles of BA in apoptotic pathways that include signal transduction cascades and pro-apoptotic cytokines.

## Acknowledgments

Our research was supported by the National Natural Science Foundation of China (grant No. 31201964), Research Fund for the Doctoral Program of Higher Education of China (grant No. 20124320120011), Department of Education of Hunan Province, China (key project No. 12A066), and the Hunan Provincial Key Laboratory of Crop Germplasm Innovation and Utilization of China (Opening Science Fund No. 12KFXM08).

## References

[1] Ashkenazi A, Dixit VM. Apoptosis control by death and decoy receptors. *Curr Opin Cell Biol* 1999, 11, 255 – 260.

[2] Berg D, Lehne M, Muller N, et al. Enforced covalent trimerization increases the activity of the TNF ligand family members TRAIL and CD95L. *Cell Death Differ* 2007, 14, 2021 – 2034.

[3] Brandon-Warner E, Schrum LW, Schmidt CM, et al. Rodent models of alcohol liver disease: of mice and men. *Alcohol* 2012, 46, 715 – 725.

[4] Brenner C, Cadiou H, Vieira HL, et al. Bcl-2 and Bax regulate the channel activity of the mitochondrial adenine nucleotide translocator. *Oncogene* 2000, 19, 329 – 336.

[5] Cook, RT. Alcohol abuse, alcoholism, and damage to the immune system-a review. *Alcohol Clin Exp Res* 1998, 22, 1927 – 1942.

[6] Crab, DW. Recent developments in alcoholism: the liver. *Recent Dev Alcohol* 1993, 11, 207 – 230.

[7] D'Alessio M, Cerella C, De Nicola M, et al. Apoptotic GSH extrusion is associated with free radical generation. *Ann N Y Acad Sci* 2003, 1010, 449 – 452.

[8] de Melo CL, Queiroz MG, Arruda Filho AC, et al. Betulinic acid, a natural pentacyclic triterpenoid, prevents abdominal fat accumulation in mice fed a high-fat diet. *J Agric Food Chem* 2009, 57, 8776 – 8781.

[9] Flekhter OB, Nigmatullina LR, Baltina LA, et al. Synthesis of betulinic acid from betulin extract and study of the antiviral and antiulcer activity of some related terpenoids. *Pharm Chem J* 2002, 36, 484 – 487.

[10] Fujioka T, Kashiwada Y, Kilkuskie RE, et al. Anti-AIDS agents, 11. Betulinic acid and platanic acid as anti-HIV principles from Syzigium claviflorum, and the anti-HIV activity of structurally related triterpenoids. *J Nat Prod* 1994, 57, 243 – 247.

[11] Fulda S. Betulinic acid: a natural product with anticancer activity. *Mol Nutr Food Res* 2009, 53, 140 – 146.

[12] Gao B, Bataller R. Alcoholic liver disease: pathogenesis and new therapeutic target. *Gastroenterology* 2011, 141, 157 – 1585.

[13] Gramenzi A, Caputo F, Biselli M, et al. Review article: alcoholic liver disease-pathophysiological aspects and risk factors. *Aliment*

*Pharmacol Ther* 2006, 24, 1151 – 1161.

[14] Jain M, Kapadia R, Jadeja RN, et al. Hepatoprotective potential of *Tecomella undulate* stem bark is partially due to the presence of betulinic acid. *J Ethnopharmacol* 2012, 143, 194 – 200.

[15] Kong Q, Qin C. Analysis of current laboratory animal science policies and administration in China. *ILAR J* 2010, 51, e1 – e10.

[16] Mandal D, Mazumder A, Das P, et al. Fas-, Caspase 8-, and Caspase 3-dependent signaling regulates the activity of the aminophospholipid translocase and phosphatidylserine externalization in human erythrocytes. *J Biol Chem* 2005, 280, 39460 – 39467.

[17] Marsano LS, Mendez C, Hill D, et al. Diagnosis and treatment of alcohol liver disease and its complications. *Alcohol Res Health* 2003, 27, 247 – 256.

[18] Mato JM, Camara J, Fernandez de Paz J, et al. S-adenosylmethionine in alcoholic liver cirrhosis: a randomized, placebo-controlled, double-blind, multicenter clinical trial. *J Hepatol* 1999, 30, 1081 – 1089.

[19] Nanji AA, Jokelainen K, Rahemtulla A, et al. Activation of nuclear factor kappa B and cytokine imbalance in experimental alcoholic liver disease in the rat. *Hepatology* 1999, 30, 934 – 943.

[20] Neff GW, Duncan CW, Schiff ER. The current economic burden of cirrhosis. *Gastroenterol Hepatol* 2011, 7, 661 – 671.

[21] Nishitani Y, Matsumoto H. Ethanol rapidly causes activation of JNK associated with ER stress under inhibition of ADH. *FEBS Lett* 2006, 580, 9 – 14.

[22] Pisha E, Chai H, Lee IS, et al. Discovery of betulinic acid as a selective inhibitor of human melanoma that functions by induction of apoptosis. *Nat Med* 1995, 1, 1046 – 1051.

[23] Raghu R, Liu CT, Tsai MH, et al. Transcriptome analysis of garlic-induced hepatoprotection against alcohol fatty liver. *J Agric Food Chem* 2012, 60, 11104 – 11119.

[24] Recio MC, Giner RM, Manez S, et al. Investigations on the steroidal anti-inflammatory activity of triterpenoids from *Diospyros leucomelas*. *Planta Med* 1995, 61, 9 – 12.

[25] Rehm J, Mathers C, Popova S, et al. Global burden of disease and injury and economic cost attributable to alcohol use and alcohol-use disorders. *Lancet* 2009, 373, 2223 – 2233.

[26] Steinkamp-Fenske K, Bollinger L, Xu H, et al. Reciprocal regulation of endothelial nitricoxide synthase and NADPH oxidase by betulinic acid in human endothelial cells. *J Pharmacol Exp Ther* 2007, 322, 836 – 842.

[27] Sun F, Xie ML, Zhu LJ, et al. Inhibitory effect of osthole on alcohol-induced fatty liver in mice. *Dig Liver Dis* 2009, 41, 127 – 133.

[28] Szuster-Ciesielska A, Kandefer-Szerszen M. Protective effects of betulin and betulinic acid against ethanol-induced cytotoxicity in HepG2 cells. *Pharmacol Rep* 2005, 57, 588 – 595.

[29] Szuster-Ciesielska A, Plewka K, Daniluk J, et al. Betulin and betulinic acid attenuate ethanol-induced liver stellate cell activation by inhibiting reactive oxygen species (ROS), cytokine (TNF-α, TGF-β) production and by influencing intracellular signaling. *Toxicology* 2011, 280, 152 – 163.

[30] Vyssokikh MY, Zorova L, Zorov D, et al. Bax releases cytochrome c preferentially from a complex between porin and adenine nucleotide translocator Hexokinase activity suppresses this effect. *Mol Biol Rep*, 2002, 29, 93 – 96.

[31] Wang M, Zhu P, Jiang C, et al. Preliminary characterization, antioxidant activity in vitro and hepatoprotective effect on acute alcohol-induced liver injury in mice of polysaccharides from the peduncles of *Hovenia dulcis*. *Food Chem Toxicol* 2012, 50, 2964 – 2970.

[32] Yi JE, Obminska-Mrukowicz B, Yuan LY, et al. Immunomodulatory effects of betulinic acid from the bark of white birch on mice. *J Vet Sci* 2010, 11, 305 – 313.

[33] Yogeeswari P, Sriram D. Betulinic acid and its derivatives: a review on their biological properties. *Curr Med Chem* 2005, 12, 657 – 666.

[34] Yoon JJ, Lee YJ, Kim JS, et al. Protective role of betulinic acid on TNF-alpha-induced cell adhesion molecules in vascular endothelial cells. *Biochem Biophy Res Commun* 2010, 391, 96 – 101.

[35] Yu HY, Wang BL, Zhao J, et al. Protective effect of bicyclol on tetracycline-induced fatty liver in mice. *Toxicology* 2009, 261, 112 - 118.

[36] Yuan HD, Jin GZ, Piao GC. Protective effects of the supernatant of ethanol eluate from Artemisia sacrorum ledeb against acetaminophen-induced liver injury in mice. *Bio Pharm Bull* 2009, 32, 1683 - 1688.

[37] Zhang ZF, Fan SH, Zheng YL, et al. Troxerutin protects the mouse liver against oxidative stress-mediated injury induced by D-Galactose. *J Agric Food Chem* 2009, 57, 7731 - 7736.

[38] Zheng ZW, Song SZ, Wu YL, et al. Betulinic acid prevention of d-galactosamine / lipopolysaccharide liver toxicity is triggered by activation of Bcl-2 and antioxidant mechanisms. *J Pharm Pharmacol* 2011, 63, 572 - 578.

本文发表于 J Vet Sci. 2014, 15(1): 141 - 148.

# Diagnosis and treatment of periparturient metabolic diseases in dairy cow

Dr. Ottó Szenci PhD, DSc, Dipl. ECBHM

*MTA-SZIE Large Animal Clinical Research Group, Üllö-Dóra major, H-2225 Hungary*

**Abstract**: The successful genetic selection for higher milk production caused a dramatic decline in the reproductive performance of dairy cows all over the world during the last decades. Achievement of optimum herd reproductive performance (calving interval of 12 or 13 months with the first calf born at 24 months of age) requires concentrated management activities especially during the first 100 days following calving. The following management activities are needed to pursue during the early postpartum period to reach or approach the optimal reproductive performance such as careful surveillance and assistance at calving, prevention of periparturient metabolic diseases, early diagnosis and treatment of post parturient uterine diseases, accurate detection of oestrus, correct timing of insemination, reducing the effect of heat stress and early pregnancy diagnosis. Among these main activities only early diagnosis, and treatment of periparturient metabolic diseases and their effects on reproductive performance and milk production are discussed.

Fertility in dairy cows reflects the cumulative influence of metabolic, endocrine, and postpartum health components. Energy imbalance seems to be one of the most important factors, but the complex interactions of the aforementioned factors can be considered in order to be able to improve fertility, at the same timebody condition score (BCS), glucose, non-esterified fatty acid (NEFA), insulin-like growth factor-1 (IGF-1), leptin concentration from calving to AI cannot explain the low fertility rate.

Cows should be challenge-fed during the dry-off period and early lactation to prevent the incidence of metabolic disorders of the puerperal period such as milk fever, acidosis, ketosis and fat cow syndrome. These diseases can increase the incidence of reproductive diseases and reduce reproductive performance. Prevention is more preferable than treatment and requires close attention to nutrition and management. The maintenance of good condition at calving and the provision of a high-density energy diet that does not produce a fatty liver in early lactation are also very important to minimize the detrimental effect of NEB on the return of oestrous cycle after calving.

**Keywords**: dairy cow, negative energy balance, ketosis, fatty liver, fertility, milk production

The successful genetic selection for higher milk production in Holstein cows has nearly doubled the average milk production in the United States since 1960, to over 11 000 kg/year. Over the same time period, there has been a dramatic decline in the reproductive performance of dairy cows (Butler, 2000). The conception rate has decreased from 66% since 1951, to about 50% until 1975, and further more until recently to about 33.1% in Spain (López-Gatius, 2003), 33.4% in Israel (Galon, 2008), 37% in Canada (Bouchard and Du Tremblay, 2008) or 41% in Japan (Nakao, 2008). In order to decrease the longer lactations and the number of cows culled for reproductive reasons, it is very important to improve our reproductive management practices (Silva, 2003). Achievement of optimum herd reproductive performance (calving interval of 12 or 13 months with the first calf born at 24 months of

szenci. otto@ aotk. szie. hu

age) requires concentrated management activities especially during the first 100 days following calving. Early postpartum breeding of dairy cows results in more calves, and higher milk production per lactation (Britt, 1975). Poor reproductive performance can reduce the number of calves born and milk production and may increase the cost of therapy and semen.

The following management activities like careful surveillance and assistance at calving, prevention of periparturient metabolic diseases, early diagnosis and treatment of post parturient uterine diseases, accurate detection of oestrus, correct timing of insemination, reducing the effect of summer heat stress and early pregnancy diagnosis are needed to pursue during the early postpartum period to reach or approach the optimal calving interval (Szenci, 2008). Among these main factors, only the effect of periparturient metabolic diseases on fertility will be discussed.

## 1 Development of Negative Energy Balance in Postpartum Dairy Cows

Dairy cattle are usually in negative energy balance (NEB) in the first weeks of lactation because of energy intake during this period is less than half of the energy requirements for milk production. Therefore the gap between energy input and output during early lactation must be met through increased non-esterified fatty acid (NEFA) production. On the other hand the energy requirements of dairy cows is met in 60%-70% by volatile fatty acids (acetate, propionate and butyrate) fermented in the rumen therefore ruminal fluid is one of the most important sources of energy metabolism in the dairy cow. In the periparturient period feed intake is physiologically suppressed which will cause a lack of dietary energy intake results in a lack of gluconeogenesis and in turn, a lack of glucose to allow for complete oxidation of NEFA. The incomplete oxidation of fatty acid will contribute to the increased production of ketone bodies (β-hydroxybutyrate /BHB/, acetone, acetoacetate) which may cause ketosis and fatty liver. According to Oetzel (2004) two types of ketosis may develop: 'Type I ketosis' with low blood glucose due to lack of precursors for gluconeogenesis without fatty liver occurring at 3 to 6 weeks postpartum; 'Type II ketosis' associated with fatty liver just before or at calving manifesting 5 to 15 days after calving. Increasing amount of ketone bodies may also contribute to suppress feed intake. Salivation is also decreased during calving due to break of chewing or reduced period and intensity of chewing. This may contribute to the development of clinical or subclinical rumen acidosis especially when the ration of concentrate is not limited for a few days around calving. Ruminal acidosis may also negatively affect rumen motility and appetite. Similarly severe hypocalcaemia may be associated with decreased abomasal motility but it is not clear whether this can be generalized to either clinical milk fever or subclinical hypocalcaemia (Geishauser et al., 2000).

The rapid increase in energy requirements at the onset of lactation results in NEB thatmay begin a few days before calving and usually reaches its most negative level (nadir) about 2-3 weeks later and used to extend 10-12 weeks until the beginning of the usual breeding period (Bell, 1995; Butler and Smith, 1989).

The NEB that, develops spontaneously in dairy cows, represents a physiological state of undernutrition. The severity and duration of NEB is primarily related to differences in dry matter intake and its rate of increase during early lactation. Calving in moderate condition (3 to 3.75) and maintaining feed intake during the periparturient transition period are key factors to reducing NEB and avoiding metabolic disorders (milk fever, acidosis, ketosis and fat cow syndrome) that are deleterious for high milk production and fertility.

### 1.1 Endocrinology of Negative Energy Balance

As the energy requirements for milk production after calving exceed the energy available from dietary sources therefore gluconeogenesis in the liver and lipolysis (production of NEFA) in the adipose tissue will be stimulated by increased growth hormone (GH) production. Parallel with these, liver growth hormone receptors (GHR) decrease abruptly just prior to calving which reduces the insulin-like growth factor-1 (IGF-1) production in the liver, therefore IGF-1 loses its control role (negative feedback) on GH release and contributes to its elevation. GH also antagonizes the action of insulin after calving in order to support milk production by enhancing gluconeogenesis in the liver and to meet the requirement for mammary lactose synthesis (Etherton and Bauman, 1998). Peripheral insulin resistance is also specific for a high-yielding dairy cow which conserves glucose for mammary lactose synthesis by reduced glycogen synthesis and storage in the liver and a failure to suppress glucose production and release into the blood, and by reduced uptake of circulating lipids and increasedhydrolysis of stored triglycerides (Hayirli, 2006). All these endocrine changes promote the shift of the metabolism from anabolic to catabolic direction (Roberts et al., 1997). Depending on the degree of the catabolic state GnRH and/or LH, FSH secretion, granulosa cell proliferation and differentiation (follicular growth and development), oestrogen production, aromatose activity, lutheal activity and steroidogenesis, oocyte quality, embryo growth and IFN-τ production will be impaired (Zulu el al., 2002; Butler, 2005). When nutrient intake meets or exceeds the energy requirements during early lactation the catabolic state (low insulin, low IGF-1 and high GH) will be switched into an anabolic state which is characterized by a high insulin, high IGF-1 and a low GH concentration and ovarian cyclicity will be restored (Huszenicza et al., 2008). It has recently emphasized that besides IGF-1 leptin also takes part in controlling the resumption of cyclic ovarian function in dairy cow (Kadokawa et al., 2006).

### 1.2 Effect of Negative Energy Balance on Follicular Development

Following parturition, regardless of NEB due to elevated plasma FSH concentrations a wave of follicular development starts in 5 to 7 days after calving. Three types of follicular development have been described as follows (Beam and Butler, 1997):

(1) The first dominant follicle ovulatesbetween Days 16 to 20 after calving,

(2) The first dominant follicle does not ovulate, a turnover and a new follicular wave follow,

(3) The dominant follicle fails to ovulate and becomes cystic.

The development of non-ovulatory dominant or cystic follicles prolongs the interval for the first ovulation to 40-50 days after calving. Ovulation of a dominant follicle during early lactation depends on the re-establishment of pulsatile LH secretion (Butler, 2001). The NEB as physiological state of undernutrition may suppress pulsatile LH secretion and reduce ovarian responsiveness to LH stimulation and by this way deters ovulation (Butler, 2001; Jolly et al., 1995).

It is worth mentioning that prolonged anovulatory anoestrous inabout 30% of cows can be connected with reduced fertility caused by NEB (Rhodes et al., 1998; Staples et al., 1990). It seems that NEB can influence the timing of first postpartum ovulation, by which it can negatively affect fertility (Butler, 2001; Darwash et al., 2001). Cows remaining anovulatory for >50 days of lactation are less likely to become pregnant during lactation and will be culled (Frajblat, 2000).

Plasma progesterone (P4) concentrations used to elevate during the first two or three postpartum ovulatory cycles (Spicer et al., 1990; Staples et al., 1990; Villa-Godoy et al., 1988). The rate of increase in P4 is reduced or moderated by NEB (Spicer et al., 1990; Villa-Godoy et al., 1988). At the same time, high dietary intake (both energy and protein) may also increase the metabolic clearance of P4 in high yielding cows (Wiltbank et al., 2001). Through the regulation of the uterine environment, P4 plays an important role in embryonic development and growth. A slower rate of increase in P4 after ovulation may decrease embryo growth by Day 16 which may be associated with low fertility (Butler et al., 1996; Mann et al., 1996; Shelton et al., 1990). Early postpartum NEB may adversely impact quality of oocytes during the first 80-100 days after calving required for follicle development and oocyte development (blastocyst formation rate), which exert another carryover effect on fertility (Britt, 1992; Kruip et al., 2001). However, it is very difficult to reconcile the effect of NEB on follicles and oocytes with the effect of high dietary energy on oocyte quality and development to blastocysts in dairy cows (Armstrong et al., 2001; Boland et al., 2001). It appears that extremes in energy status in either direction may negatively influence fertility (Butler, 2001).

# 2 Diagnosis of Negative Energy Balance

## 2.1 Monitoring Energy Status on Site

Measuring the NEFA and ketone bodies mayallow evaluating the success of adaptation to NEB as NEFA (cut-off value: ≥0.4 mmol/L) reflects the magnitude of mobilization of fat from storage and indicates a NEB together with increased risk for fatty liver and metabolic diseases while ketone bodies reflect the completeness of oxidation of fat in the liver (Oetzel, 2004).

DVM NEFA test is a rapid, spectophotometry method to determine NEFA concentration in serum. The Pearson correlation coefficient between the DVM NEFA and the gold standard (Hitachi 911 automated analyser) NEFA determination was 0.75 (Leslie et al., 2003). Sensitivity and specificity of the test was 84% and 96%, respectively when >0.4 mmol/L was used as a cut-off level (Gooijer et al., 2006). Unfortunately, the DVM NEFA test is no longer available (Townsend, 2011).

The measurements of BHB can be carried out with a portable hand held device (Precision Xtra, Abbot Laboratories). The blood BHB concentrations measured at the farm showed a strong and significant correlation with the serum samples evaluated in the laboratory before and after freezing (Pearson correlation coefficient: >0.92) (Szelényi et al., 2011).

Several cowside tests can be used for measuring the ketone bodies such as KetoCheck powder (milk or urine), KetoStix (urine), KetoTest (milk), PortaBHB (milk), however only the Precision Xtra (blood) BHB analyser can reach high sensitivity (91%) and specificity (94%) (Townsend, 2011).

## 2.2 Metabolic Profile Test

Since the herd sizes of dairy farms gradually increase, metabolic profile test application in monitoring transition cow health and recognition of adverse consequences of peripartal diseases are widely recommended (Herdt et al., 2001). Metabolic profile tests as methods for recognizing dairy cattle nutrition and health status from blood samples, was first recorded at the Compton Institute of Animal Health, UK (Payne et al., 1970).

A metabolic profile test includes measurements of different parameters reflecting the energy balance (NEFA, BHB), protein status (urea-nitrogen, albumin), liver function (various enzymes, cholesterol, and triglycerides) and/ or macromineral homeostasis (Brydl et al., 2008; Mordak and Nicpon, 2006).

# 3 Treatment and Prevention of Negative Energy Balance

"Maximizing dry matter intake (DMI) and maintenance of a consistent intake through the last three weeks prior to calving (close-up period) is likely the hallmark of a successful transition cow program" (Duffield, 2011). If the DMI is less than 12 kg per cow per day in the close-up period the risk of subclinical ketosis after calving is significantly increased (OR: 5.7) (Osborne, 2003). Adequate pen space or stall space per cow, adequate feed bunk space, sufficient and comfortable bedding, adequate water supply and minimization of heat and management (group changes) stress are also important (Duffield, 2011). Reduction of over- condition in dairy cows in late lactation and the early dry period is also a very important tool in prophylaxis (Duffield, 2011).

Propylene glycol has been used successfully for the prevention of subclinical ketosis (Emery et al., 1964; Sauer et al., 1973). Feeding propylene glycol (Pickett et al., 2003), propylene glycol powder (Toghdory et al., 2009), or dry glycerol (Karami-Shabankareh et al., 2013) may be useful to improve negative energy balance and reproductive efficiency in dairy cow. Propylene glycol can be more effective when drenched (Duffield, 2011; Pickett et al., 2003). Glucocorticoids, vitamin $B_{12}$ and/or 50% dextrose i. v. may be administered as well (Leslie et al., 2004). Rumen protected choline may positively influence liver glycogen, triglyceride and milk production (Pipenbrink & Overton, 2003; Pinotti et al., 2010), however with the exception of cholesterol it did not affect BHB, NEFA, glucose, urea, and aspartate aminotransferase (AST) when it was used in the transition period (from 3 weeks before expected calving until Day 28 after calving) (Zahra et al., 2006).

Reducing milk harvest (about one-third of expected milk production twice a day) during the first 5 days after calving may reduce metabolic stress without compromising productivity of high-yielding dairy cows, however this hypothesis needs further confirmation (Carbonneau et al., 2012).

In summary, fertility in dairy cows reflects the cumulative influence of metabolic, endocrine, and postpartum health components. Energy imbalance seems to be one of the most important factors, but the complex interactions of the aforementioned factors can be considered in order to be able to improve fertility, at the same time BCS, glucose, NEFA, IGF-1, leptin concentration alone from calving to AI cannot explain the low fertility rate (Snijders et al., 2001; Liefers, 2004).

Cows should be challenge-fed during the dry-off period and early lactation to prevent the incidence of metabolic disorders of the puerperal period such as milk fever, acidosis, ketosis and fat cow syndrome. These diseases can increase the incidence of reproductive diseases and reduce reproductive performance. Prevention is more preferable than treatment and requires close attention to nutrition and management (Radostits et al., 1994). The maintenance of good body condition at calving and the provision of a high-density energy diet that does not produce a fatty liver in early lactation are also very important to minimize the detrimental effect of NEB on the return of oestrous cycle after calving.

## References


Armstrong DG, McEvoy TG, Baxter G, et al. Effect of dietary energy and protein on bovine follicular dynamics and embryo production in vitro: associations with the ovarian insulin-like growth factor system. *Biol. Reprod.* 2001. 64. 1624 – 1632.

Beam SW, Butler WR. Energy balance and ovarian follicle development prior to the first ovulation postpartum in dairy cows receiving three levels of dietary fat. *Biol. Reprod.* 1997, 56: 133 – 142.

Bell AW. Regulation of organic nutrient metabolism during transition from pregnancy to early lactation. *J. Anim. Sci.* 1995, 73: 2804 – 2819.

Boland MP, Lonergan P, O'Callaghan D. Effect of nutrition on endocrine parameters, ovarian physiology, and oocyte and embryo development. *Theriogenology* 2001. 55. 1323 – 1340.

Bouchard E, Du Tremblay D. Dairy herd production and reproduction in Quebec and Canada. In: Szenci, O. and Bajcsy, Á. Cs. (ed.): Factors affecting reproductive performance in the cow. *Hungarian Association for Buiatrics, Budapest, Hungary*, 2008: 60 – 66.

Britt JH. Early post partum breeding in dairy cows. A review. *J. Dairy Sci.* 1975, 58: 266 – 271.

Britt JH. Influence of nutrition and weight loss on reproduction and early embryonic death in cattle. Proceedings of the XVII World Buiatrics Congress, St. Paul, MN. 1992, 2: 143 – 149.

Brydl E, Könyves L, Tegzes Lné, et al. Incidence of subclinical metabolic disorders in Hungarian dairy herds during the last decade. *Magyar Állatorvosok Lapja* 2008, 130 (Suppl. I): 129 – 134.

Butler WR. Nutritional effects on resumption of ovarian ciclicity and conception rate in postpartum dairy cows. *British Society Animal Science Occasional Publication*,2001,26:133 - 145.

Butler WR. Nutrition. Negative energy balance and fertility in the postpartum dairy cow. *Cattle Practice* 2005,13:13 - 18.

Butler WR,Calaman JJ,Beam SW. Plasma and milk urea nitrogen in relation to pregnancy rate in lactating dairy cattle. *J. Anim. Sci.* 1996,74:858 - 865.

Butler WR,Smith RD. Interrelationship between energy balance and postpartum reproductive function in dairy cattle. *J. Dairy Sci.* 1989, 72:767 - 783.

Carbonneau E,de Passillé AM,Rushen J,et al. The effect of incomplete milking or nursing on milk production,blood metabolites,and immune functions of dairy cows. *J. Dairy Sci.* 2012,95:6503 - 6512.

Darwash AO,Lamming GE,Royal MD. A protocol for initiating oestrus and ovulation early postpartum in dairy cows. *Animal Science* 2001,72:539 - 546.

Duffield T. Managing transition cow issues. Western Dairy Management Conference,Reno,NV,Proceedings 2011:43 - 51.

Etherton TD,Bauman DE. Biology of somatotropin in growth and lactation of domestic animals. *Physiol. Rev.* 1998,78:745 - 761.

Frajblat M. Metabolic state and follicular development in the postpartum lactating dairy cow. PhD. Thesis. Cornell University,2000.

Galon N. Factors affecting reproductive performance in Israeli dairy herds. In: Szenci, O. and Bajcsy, Á. Cs. (ed.): Factors affecting reproductive performance in the cow. Hungarian Association for Buiatrics,Budapest,Hungary,2008 :28 - 36.

Geishauser T,Leslie K,Tenhag J,et al. Evaluation of eight cow-side ketone tests in milk for detection of subclinical ketosis in dairy cows. *J. Dairy Sci.* 2000,83:296 - 299.

Gooijer L,Leslie K,LeBlanc S,et al. Evaluation of rapid test for NEFA in bovine serum. In: Joshi N,Herdt Th. (eds) Production diseases in farm animals. Wageningen Academic Press,Wageningen,The Netherlands,2006:44.

Hayirli A. The role of exogenous insulin in the complex of hepatic lipidosis and ketosis associated with insulin resistance phenomenon in postpartum dairy cattle. *Vet. Res. Commun.* 2006,30:749 - 774.

Herdt TH,Dart B,Neuder L. Will large dairy herds lead to the revival of metabolic profile testing? Proceedings of the American Association of Bovine Practitioners 2001,34:27 - 34.

Huszenicza Gy,Keresztes M,Balogh O,et al. Peri-parturient changes of metabolic hormones and their clinical and reproductive relevance in dairy cows. *Magyar Állatorvosok Lapja*,2008,130 (Suppl. 1):43 - 49.

Jolly PDS,McDougall S,Fitzpatrick LA,et al. Physiological effects of undernutrition on postpartum anoestrus in cows. *J. Reprod. Fertil. Suppl.* 1995,49:477 - 492.

Kadokawa H,Blache D,Martin GB. Plasma leptin concentrations correlate with luteinizing hormone secretion in early postpartum Holstein cows. *J. Dairy Sci.* 2006,89:3020 - 3027.

Karami-Shabankareh H,Kafilzadeh F,Piri V,et al. Effects of feeding dry glycerol to primiparous Holstein dairy cows on follicular development,reproductive performance and metabolic parameters related to fertility during the early post-partum period. *Repr. Domestic Animal*,2013 (in press).

Kruip TAM,Wensing T,Vos PLAM. Characteristics of abnormal puerperium in dairy cattle and the rationale for common treatments. British Society Animal Science Occasional Publication 2001,26:63 - 79.

Leslie K,Duffield T,LeBlanc S. Monitoring and managing energy balance in the transition dairy cow. Proceedings of Minnesota Veterinary Conference,Minneapolis,Minnesota,USA,2004:101 - 107.

Liefers S. Physiology and genetics of leptin in periparturient dairy cows. Thesis,Wageningen University,Wageningen,The Netherlands, 2004.

López-Gatius F. Is fertility declining in dairy cattle? A retrospective study in northeastern Spain. *Theriogenology* 2003,60:89 - 99.

Mann GE,Mann SJ,Lamming GE. The inter-relationship between the maternal hormone environment and the embryo during the early stages of pregnancy in the cow. *J. Reprod. Fertil. Abstract Series* 1996,17:21.

Mordak R,Nicpon J. Values of some blood parameters in dairy cows before and after delivery as a diagnostic monitoring of health in herd. Electronic Journal of Polish Agricultural Universities, Veterinary Medicine, 2006. 9. Issue 2. Available Online http://

www. ejpau. media. pl/volume9/issue2/art-20. html

Nakao T. Declining fertility in dairy cows in Japan and efforts to improve the fertility. In: Szenci, O. and Bajcsy, Á. Cs. (ed.): Factors affecting reproductive performance in the cow. Hungarian Association for Buiatrics, Budapest, Hungary, 2008: 38 – 48.

Butler WR. Nutritional interactions with reproductive performance in dairy cattle. *Anim. Reprod. Sci.* 2000, 60 – 61: 449 – 457.

Emery RS, Burg N, Brown LD, et al. Detection, occurrence, and prophylactic treatment of borderline ketosis with propylene glycol feeding. *J. Dairy Sci.* 1964, 47: 1074 – 1079.

Oetzel GR. Monitoring and testing dairy herds for metabolic disease. *Vet. Clin. N. Amer. Food Anim.* 2004, 20: 651 – 674.

Osborne, T. An evaluation of metabolic function in transition dairy cows supplemented with Rumensin premix, or administered a Rumensin controlled release capsule. MSc Thesis, University of Guelph, 2003.

Payne JM, Dew SM, Manston R, et al. The use of a metabolic profile test in dairy herds. *Vet. Rec.* 1970, 87: 150 – 158.

Pickett MM, Piepenbrink MS, Overton TR. Effects of propylene glycol or fat drench on plasma metabolites, liver composition, and production of dairy cows during the periparturient period. *J. Dairy Sci.* 2003, 86: 2113 – 2121.

Pinotti L, Polidori C, Campagnoli A, et al. A meta-analyis of the effects of rumen protected choline supplementation on milk production in dairy cows. In: Crovetto GM. (ed) Energy and protein metabolism and nutrition. EAAP Publications No. 127. Wageningen Academic Publishers, Wageningen, The Netherlands, 321 – 322.

Radostits OM, Leslie KE, Fetrow J. Maintaining reproductive efficiency in dairy cattle. In: Herd Health. Food Animal Production Medicine. Second Edition. W. B. Saunders Company, Philadelphia. 1994: 141 – 158.

Rhodes FM, Clark BA, Nation DP, et al. Factors influencing the prevalence of postpartum anoestrus in New Zealand dairy cows. Proceedings of the New Zealand Society of Animal Production 1998, 58: 79 – 81.

Roberts AJ, Nugent RA, Klindt J, et al. Circulating insulin-like growth factor 1, insulin-like growth factor binding proteins, growth hormone, and resumption of estrus in postpartum cows subjected to dietary energy restriction. *J. Anim. Sci.* 1997, 75: 1909 – 1917.

Sauer FD, Erfle JD, Fisher LJ. Propylene glycol and glycerol as a feed additive for lactating dairy cows: an evaluation of blood metabolite parameters. *Can. J. Anim. Sci.* 1973, 53: 265 – 271.

Shelton K, Gayerie de Abreu MF, Hunter MG, Parkinson TJ, Lamming GE. Luteal inadequacy during the early luteal phase of subfertile cows. *J. Reprod. Fertil.* 1990, 90: 1 – 10.

Silva JW. Addressing the decline in reproductive performance of lactating dairy cows: a researcher's perspective. *Veterinary Science Tomorrow*, 2003, 3: 1 – 5.

Snijders SEM, Dillon PG, O'Farrel KJ, et al. Genetic merit for milk production and reproductive success in dairy cows. *Anim. Reprod. Sci.* 2001, 65: 17 – 31.

Spicer LJ, Tucker WB, Adams GD. Insulin-like growth factors in dairy cows: relationship among energy balance, body condition, ovarian activity and estrous behaviour. *J. Dairy Sci.* 1990, 73: 929 – 937.

Staples CR, Thatcher WW, Clark JH. Relationship between ovarian activity and energy balance during the early postpartum period of high producing dairy cows. J. *Dairy Sci.* 1990, 73: 938 – 947.

Stokes PAS SR, Goff JP. Case study: Evaluation of calcium propionate and propylene glycol administered into the esophagus of dairy cattle at calving. *The Professional Animal Scientist*, 2001, 17: 115 – 122.

Szelényi Z, Nagy K, Bérdi P, et al. A szubklinikai ketózis elöfordulásának vizsgálata magyarországi tehenészetekben (Incidence of subclinical ketosis in Hungarian dairy herds). In: Szenci O, Brydl E, Jurkovich V (szerk.) Magyar Buiatrikus Társaság XXI. Kongresszusa, Proceedings, Sümeg, 2011: 19 – 25.

Szenci O. Factors, which may affect reproductive performance in dairy cattle. *Magyar Állatorvosok Lapja* 2008, 130 (Suppl. I): 107 – 111.

Toghdory A, Torbatinejad N, Kamali R, et al. Effects of propylene glycol powder on productive performance of lactating cows. Pakistan Journal of Biological Sciences, 2009, 12: 924 – 928.

Townsend J. Cowside tests for monitoring metabolic disease. Tri-State Dairy Nutrition Conference, Fort Wayne, Indiana, USA, 2011: 55 –

60.

Villa-Godoy A, Hughes TL, Emery RS, et al. Association between energy balance and luteal function in lactating dairy cows. *J. Dairy Sci*. 1988, 71: 1063 - 1072.

Wiltbank MC, Sartori R, Sanfsritavong S, et al. Novel effects of nutrition on reproduction in lactating dairy cows. *J. Dairy Sci.* 2001, 84 (Suppl. 1): 32.

Zahra LC, Duffield TF, Leslie KE, et al. Effects of rumen-protected choline and monensin on milk production and metabolism of periparturient dairy cows. *J. Dairy Sci*. 2006, 89: 4808 - 4818

Zulu VC, Nakao T, Sawamukai Y. Insulin-like growth factor-I as a possible hormonal mediator of nutritional regulation of reproduction in cattle. *J. Vet. Med. Sci.* 2002, 64: 657 - 665.

# Diagnosis and treatment of post parturient uterine diseases in dairy cow

Dr. Ottó Szenci PhD, DSc, Dipl. ECBHM

*MTA-SZIE Large Animal Clinical Research Group, Üllö-Dóra major, H-2225 Hungary*

**Abstract**: The successful genetic selection for higher milk production caused a dramatic decline in the reproductive performance of dairy cows all over the world during the last decades. Achievement of optimum herd reproductive performance (calving interval of 12 or 13 months with the first calf born at 24 months of age) requires concentrated management activities especially during the first 100 days following calving. The following management activities are needed to pursue during the early postpartum period to reach or approach the optimal reproductive performance such as careful surveillance and assistance at calving, prevention of post parturient metabolic diseases, early diagnosis and treatment of post parturient uterine diseases, accurate detection of oestrus, correct timing of insemination, reducing the effect of heat stress and early pregnancy diagnosis. Among these main activities only early diagnosis, treatment and prevention of post parturient uterine diseases and their effects on milk production and reproductive performance are discussed.

**Keywords**: dairy cow, clinical metritis, clinical endometritis, subclinical endometritis

The successful genetic selection for higher milk production in Holstein cows has nearly doubled the average milk production in the United States since 1960, to over 11 000 kg/year. Over the same time period, there has been a dramatic decline in the reproductive performance of dairy cows. The average number of days open (interval from calving to conception) and the number of services per conception have increased substantially. In order to decrease the longer lactations and the number of cows culled for reproductive reasons it is very important to improve our reproductive management practices (Silva, 2003). Achievement of optimum herd reproductive performance (calving interval of 12 or 13 months with the first calf born at 24 months of age) requires concentrated management activities especially during the first 100 days following calving. Early postpartum breeding of dairy cows results in more calves, and higher milk production per lactation (Britt, 1975). Poor reproductive performance can reduce the number of calves born and milk production and may increase the cost of therapy and semen.

The following management activitiessuch as careful surveillance and assistance at calving, prevention of post parturient metabolic diseases, early diagnosis and treatment of post parturient uterine diseases, accurate detection of oestrus, correct timing of insemination, reducing the effect of summer heat stress and early pregnancy diagnosis are needed to pursue during the early postpartum period to reach or approach the optimal calving interval (Szenci, 2008).

Among these main activities onlyearly diagnosis and treatment of post parturient uterine diseases and their effects on milk production and reproductive performance are discussed in the present work. However, this topic has also a great importance because it is generally accepted that up to 40% of dairy cows may have clinical metritis within the first two weeks after calving and infection may persist in 10% to 15% of animals more than 3 weeks after

szenci. otto@ aotk. szie. hu

calving causing clinical or subclinical endometritis (Sheldon & Dobson,2004).

# 1 The Definition of Uterine Diseases

Sheldon et al. (2006) proposed standardized clinical definitions of uterine diseases which can be recommended to use in the field as well.

## 1.1 Clinical Metritis

Clinical metritis is an acute systemic illness due to infection of the uterus with bacteria, usually within 10 (21) days after parturition. Clinical metritis can be categorized into three grades. Grade 1 clinical metritis (CM1) can be characterized by an abnormally enlarged uterus and a purulent uterine discharge detectable in the vagina, within 21 days after calving. Grade 2 clinical metritis (CM2) or puerperal metritis can be characterized by a fetid red-brown watery uterine discharge and, usually pyrexia ( >39.5℃) (Drillich et al., 2001; Földi et al., 2006); in severe cases, reduced milk yield, dullness, inappetence or anorexia, elevated heart rate, and apparent dehydration may also be present. In some cases pyrexia even with daily monitoring of rectal temperature could not be detected (Benzaquen et al., 2007; Sheldon & Dobson, 2004) however an enlarged uterus with a thin wall and atonia is present with a fetid discharge. Puerperal metritis is often associated with retained placenta, dystocia, stillbirth or twins, and usually occurs toward the end of the first week after calving, being rare after the second week after calving (Földi et al., 2006; Könyves et al., 2009a; Marcusfeld, 1984). It is important to emphasize that puerperal metritis (CM2) may occur in around 10%-15% of cows with spontaneous calving and without retained foetal membranes (Benzaquen et al., 2007). In summary puerperal metritis should be defined as an animal with an abnormally enlarged uterus and a fetid watery red-brown uterine discharge, associated with signs of systemic illness (decreased milk yield, dullness) and fever >39.5℃, within 21 days after calving (Sheldon et al., 2006). Grade 3 clinical metritis (CM3) or toxaemic metritis can be characterized by additional signs of toxaemia (such as inappetence, cold extremities, depression and/or collapse) which has a poor prognosis (Sheldon et al., 2006).

Retained placenta  Retained placenta or retained foetal membranes may occur if the placenta has not been shed by 12 (24) h after calving. Majority of the placenta used to be expelled within 6-9 h after calving (88.7%) (Van Werven et al., 1992). The average incidence of retained placenta after normal calving used to be 7% between 3 (4)% to 11 (12)% (Eiler & Fecteau, 2007; Grunert, 1986). After abnormal delivery (e. g. twin pregnancy, Caesarean section, foetotomy, forced extraction of the foetus, abortion, premature calving) and in herds infected with brucellosis its incidence rate can range between 20% and 50% or even more. Several factors like genetic, nutritional, immunological and pathological ones may influence the separation of bovine placenta; however its aetiology is not fully understood. As retained placenta predisposes the development of uterine infections (clinical metritis, as well as clinical and subclinical endometritis (Dohmen et al., 2000), and causing a decrease in milk production (decreased milk yield, milk from treated cows withheld) (Könyves et al., 2009b; Sheldon & Dobson, 2004) and reproductive performance (increases in days open, services per conception, calving to first heat interval, days from calving to first service and culling rate) (LeBlanc et al., 2002a) therefore the aim of its therapy is to prevent its adverse side effects.

## 1.2 Clinical Endometritis

Clinical endometritis is characterized by the presence of purulent or mucopurulent uterine discharge detectable in the vagina, 21 days or more calving, and is not accompanied by systemic signs (LeBlanc et al., 2002a; Sheldon & Noakes, 1998). Due to the fact that mostly there is no endometrial inflammation (endometritis) in case of purulent vaginal discharge (PVD) therefore according to Duduc et al. (2010) the PVD terminology should be used.

Pyometra  Pyometra is characterized by the accumulation of purulent or mucopurulent material within the uterine lumen and distension of the uterus, in the presence of an active corpus luteum. There are often an increased number of pathogenic bacteria within the uterine lumen when the corpus luteum forms and pyometra occurs (Noakes et al., 1990). Although there is a functional closure of the cervix, the lumen is not always completely occluded and some pus may be discharged through the cervix into the vaginal lumen. According to Sheldon et al. (2006) pyometra is defined by the accumulation of purulent material within the uterine lumen, in the presence of a persistent corpus luteum and a closed cervix.

### 1.3 Subclinical Endometritis

Subclinical endometritis can be defined as endometrial inflammation of the uterus usually determined by cytology, in the absence of purulent exudates in the vagina (Gilbert et al., 2005). In animals without signs of clinical endometritis, subclinical disease may be diagnosed by measuring the proportion of neutrophils present in a sample collected by flushing the uterine lumen, or using a cytobrush (Kasimanickam et al., 2005a; Kasimanickam et al., 2005b). Subclinical endometritis was determined in one study by the presence of >18% neutrophils in uterine cytology samples collected 20-33 days calving or >10% neutrophils at 34-47 days calving (Kasimanickam et al., 2005b). The assessment of inflammation at Days 40 to 60 after calving corresponded approximately to >5% neutrophils (Gilbert et al., 1998). It is proposed by Sheldon et al. (Sheldon et al., 2006) that a cow with subclinical endometritis is defined by >18% neutrophils in uterine cytology samples collected at Days 21 to 33 after calving, or >10% neutrophils at Days 34 to 47 after calving, or by finding mixed echodensity fluid within the uterine lumen 21 days after calving during ultrasonographic examination, in the absence of clinical endometritis.

## 2 Diagnosis of Uterine Diseases

### 2.1 Clinical Metritis

Diagnosis of clinical metritis (CM1 to CM3) is based on the clinical signs (purulent uterine discharge within 21 days after calving (CM1), fetid watery red-brown uterine discharge, fever >39.5℃, dullness and decreased milk yield (CM2), additional signs of toxaemia such as inappetence, cold extremities, depression and/or collapse (CM3)) and abnormally enlarged uterus detected by rectal palpation mainly during the first week (10 days) after calving and rare during the second week (Drillich et al., 2001; Markusfeld, 1984).

Diagnosis of clinical metritis insmall farms is based on the presence of vaginal discharge because clinical symptoms are not characteristics. It is important to mention that vaginal discharge can be present only in 24%-33% of CM2 cases (Zhou et al., 2001) and this may explain why the farmers used to present CM2 only about 2% of calvings (Bareille et al., 2003) instead of 10% when systematic examination (e.g. measuring rectal temperature) is used (Bareille & Fourichon, 2006; Drillich et al., 2001).

In contrast, to diagnose it in large herds may be more complicated. Measuring the body temperature daily during the first 10 days after calving and performing vaginal examination at least once between Days 2 to 10 may help diagnosing clinical metritis more accurately. Measuring only rectal temperature is not enough because some cows may have no elevated rectal temperature (>39.5℃) and have CM2 puerperal metritis (Benzaquen et al., 2007; Sheldon & Dobson, 2004). The rectal measures of body temperature can be influenced "by the procedure itself (up to 0.5℃), type of thermometer (up to 0.3℃), and the penetration depth (11.5 cm or 6.0 cm in one of the experiments) into the rectum (up to 0.4℃ difference between a penetration depth of 11.5 cm and 6.0 cm in one of the experiments). Differences in rectal temperature before and after defecation are minor ( <0.1℃ ). These results may indicate that some care is required in generalizing rectal measures of body temperature" (Burfeind et al.,2010).

It is important to emphasize that this early vaginal examination must perform with great care using adequate lubricant and being as hygienic as possible. The accuracy of our diagnosis can be improved by monitoring milk yield because milk production in some cows is not increasing daily as expected after calving or there is a sudden drop in it. Reduced feeding activity during the pre-partum period can be a significant risk factor for developing puerperal metritis (Urton et al.,2005).

Retained Placenta  If the placenta has not been shed by 12 (24) h after calving the animal has to handle accordingly (Eiler & Fecteau,2007; Grunert,1986).

## 2.2 Clinical Endometritis

There are several methods (transrectal palpation /LeBlanc et al., 2002a), transrectal ultrasonography (Kamimura et al.,1993), histological examination of endometrial biopsies (Bonnet et al.,1993), manual vaginal examination (Sheldon et al.,2006), vaginoscopy (LeBlanc et al.,2002a) available in the field to diagnose clinical endometritis in the cow however each method has some limitations. In a recent study it was confirmed that vaginoscopy was a practical tool to distinguish healthy from diseased cows with clinical endometritis (Leutert et al., 2012).

In contrast, Metricheck[R] (Simcro, New Zealand) consisting of a stainless steel rod with a rubber hemisphere can be used to retrieve vaginal contents more easily and precisely. In a recent study three methods (vaginoscopy/reference method/, gloved hand and Metricheck) were compared for diagnosing clinical endometritis between Days 21 to 27 after calving and it was confirmed that somewhat more cows (47.5%) could be diagnosed by Metricheck than by the other two methods (vaginoscopy: 36.9%, gloved hand: 36.8%). At the same time this did not result in improved reproductive performance (Pleticha et al.,2009). On the other hand cytobrush cytology is also a reliable method for diagnosing clinical endometritis in cattle (Barlund et al.,2008) however it is not so practical in the field.

The character and the odour of the vaginal mucus (Sheldon & Dobson,2004) can be scored according to the followings:

Mucus character

Score 0 = clear or translucent mucus;

Score 1 = mucus containing flecks of white or off-white pus;

Score 2 = discharge containing ≤50% white or off-white mucopurulent material;

Score 3 = discharge containing ≥50% purulent material, usually white or yellow, but occasionally sanguineous.

Mucus odour

Score 0: no unpleasant odour;

Score 3: Fetid odour.

Pyometra  The diagnosis of pyometra can be based on transrectal palpationof a distended uterus and/or on transrectal ultrasonography of mixed echodensity fluid and the presence of a persistent corpus luteum, with a history of anoestrus (Sheldon et al.,2006).

## 2.3 Subclinical Endometritis

Subclinical endometritis can be diagnosed by uterine cytology in the absence of purulent discharge in the vagina. Endometrial and inflammatory cells may be collected by uterine lavage (Kasimanickam et al.,2004; Pécsi et al.,2009),or cytobrush techniques (Kasimanickam et al.,2004) to evaluate the presence of neutrophils in the uterine sample. If >5% neutrophils in uterine cytology samples collected by uterine lavage between Days 42 and 72 after calving can be found,subclinical endometritis is defined (Pécsi et al.,2009). The cytobrush technique is a consistent and reliable method for obtaining endometrial samples for cytological examination from postpartum dairy cows. If >18% neutrophils in uterine cytology samples collected beteen Days 21 and 33 after calving,or >10% neutrophils at Days 34 to 47 after calving can be found, in the absence of clinical endometritis, subclinical endometritis can be defined (Kasimanickam et al., 2005a). Subclinical endometritis can be defined by ultrasonographic examination if mixed echogenisity fluid within the uterine lumen 21 days after calving can be found,in the absence of clinical endometritis (Kasimanickam et al.,2004; Sheldon et al.,2006).

# 3 Treatment of Uterine Diseases

## 3.1 Clinical Metritis

Early treatment of clinical metritis (especially puerperal metritis) may decrease the severity of genital disorders (endometritis, cystic ovarian disease), the predisposition of metabolic disorders (left displacement abomasums,ketosis) and other complications like pyelonephritis, arthritis, endocarditis, hepatic and pulmonary abscesses (Gröhn et al.,1990; Gröhn et al.,2000; Mateus et al.,2002; Opsomer et al.,2000).

There are a great variety of treatment protocols such as intrauterine antimicrobial agents (oxytetracycline; ampicillin and cloxacillin),antiseptic chemicals (iodine solutions:500 ml of 2% Lugol's iodine immediately after calving and again 6 h later as a preventive measure), systemic antibiotics penicillin or one of its synthetic analogues:20 000 to 30 000 U/(kg · cow); ceftiofur /third generation cephalosporin/:2.2 mg/kg of body weight daily for 3 to 5 days; 2 doses of 6.6 mg ceftiofur crystalline free acid sterile suspension (CCFA-SS)/kg of body weight s. c. in the base of the ear at a 72-h interval (McLaughlin et al.,2012) or a single dose of CCFA-SS within 24 h after calving as a preventive measure (Dubuc et al.,2011; McLaughlin et al.,2013),ozone i. u. treatment (Zobel & Tkalčić, 2013), supportive therapy nonsteroidal anti-inflammatory drugs such as flunixin meglumine (Drillich et al.,2007),fluid therapy in case of dehydration,therapy with calcium and energy supplements in case of depressed appetite,and hormone therapy (oxytocin:20 to 40 U repeated every 3 to 6 h within 48 to 72 h after calving; $PGF_{2\alpha}$ or its synthetic analogues) have been introduced in the field (Risco et al.,2007). The prognosis for recovery from puerperal metritis (CM2) varies with severity of the condition.

According to ourpresent knowledge intrauterine antimicrobial and antiseptic treatments cannot be recommended because of irritating the endometrium (Risco et al.,2007). Routine use of hormone therapies ($PGF_{2\alpha}$) is also controversial and needs further confirmations. It seems that presently systemic antibiotic (ceftiofur) and supportive therapy can be recommended for the field (Drillich et al.,2001,Földi et al.,2009).

Retained placenta The aim of the treatment for retained placenta is to reduce the occurrence of puerperal metritis and subsequently clinical and subclinical endometritis, to decrease milk losses, to reduce reproductive inefficiency,and to decrease veterinary expenses (Könyves et al.,2009a; Könyves et al.,2009b). There are a great variety of treatment protocols (manual removal of retained membranes,intrauterine treatments with antibiotics or antiseptics /Lugol's iodine/ (Ahmed et al.,2013),ozone spray (Djuricic et al.,2012),hormones /oxytocin, prostaglandin/, ergot derivates, calcium, injection of collagenase into the umbilical arteries, versus no treatment) recommended for the field (Risco et al., 2007). However, all of these methods have some limited values in the treatment of retained placenta (Eiler & Fecteau, 2007). Recent findings confirm that systemic antibiotics (ceftiofur 1mg/kg) without intrauterine manipulation and treatment can be as effective as conventional treatment (Drillich et al., 2001). This was also confirmed in a later study in febrile cows (Drillich et al., 2006a, Drillich et al., 2006b). It seems that systemic antibiotic is effective if the selection of treatment based on fever which may reduce the use of antibiotics compared with intra-uterine antibiotics (Drillich et al., 2006b). Treatment with oxytocin, $PGF_{2\alpha}$, or calcium was not effective for the prevention of retained placenta (Hernandez et al., 1999) or did not hasten the passage of foetal membranes (Frazer, 2005). Treatment of acute puerperal metritis with a single dose of flunixin meglumine in addition to antibiotic treatment had no beneficial effect on clinical cure, milk yield within 6 d after the first treatment, or reproductive performance (Drillich et al., 2007, Königsson et al., 2001), while according to Amiridis et al. (2001) a single dose of flunixin meglumine (2.2 mg/kg BW) administered intravenously to cows with CM2 between Days 5 and 8 after calving accelerated the uterine involution and shortened the calving-to-first-estrous interval.

## 3.2 Clinical Endometritis

The general principle of the treatment of clinical endometritis is to reduce the load of pathogenic bacteria and enhance uterine defence and repair mechanisms and hence halt and reverse inflammatory changes that impair fertility (LeBlanc, 2008). A wide variety of therapy has been reported for clinical endometritis, including systemically or locally administered antibiotics, locally administered antiseptic solution and/or systemically injected $PGF_{2\alpha}$. Infusion of antimicrobials into the uterus is aimed at achieving high concentrations at the site of infection (Gilbert & Schwark, 1992; Gustafsson, 1984). In contrast to systemic administration, intrauterine administration achieves higher drug concentration in the endometrium, but little penetration to deeper layers of the uterus or other genital tissues.

Intrauterine treatment with 0.5g cephapirin, first-generation cephalosporin, at 24-42 days before the planned start of mating improved reproductive performance of dairy cattle, especially those that had a history of retained foetal membrane (RFM), a calf dead at calving or within 24 h of calving, or vulval discharge (McDougall et al., 2007). In an experimental study, systemic administration of cefquinome (1 mg/kg), fourth generation cephalosporin, for three consecutive days was efficient for treatment *E. coli*-induced endometritis (Amiridis et al., 2003). Intrauterine infusion of cephapirin or systemic administration of $PGF_{2\alpha}$ significantly improved the pregnancy rate of cows with clinical endometritis from which *T. pyogenes* was isolated (Pécsi et al., 2007). Meta-analysis study revealed that odd ratios for pregnancy after treatment of cows with clinical endometritis with $PGF_{2\alpha}$ or cephapirin on Days 28 to 35 postpartum were 1.5 and 1.9 ($P < 0.05$), respectively as compared to the control (Földi et al., 2009). In a large field study performed by LeBlanc et al. (2002a, 2002b), there was no benefit on time to pregnancy of treatment of endometritis before 4 wk postpartum. Moreover, administration of $PGF_{2\alpha}$ between 20 and 26 DIM to cows with endometritis that did not have a palpable CL was associated with a significant reduction in pregnancy rate. Cows with endometritis between 27 and 33 DIM, treated with cephapirin i. u. had a significantly shorter time to pregnancy than cows in the untreated groups. In cows with endometritis that had a palpable CL, there was no significant difference in time to pregnancy between those treated by intrauterine infusion of cephapirin or $PGF_{2\alpha}$. Both groups tended to have a higher pregnancy rate than in untreated cows. Numerous reviewers have concluded that $PGF_{2\alpha}$ appears to be at least as effective for clinical endometritis as any available alternative therapy

(Lugol's iodine /Pécsi et al. ,2007; polyvinylpyrrlidone-iodine solution /Nakao et al. ,1988; metacresolsulphuric acid and Lotagen /Heuwieser et al. ,2000) and presents minimal risk of harm to the uterus or presence of residues in milk or meat (Gilbert,1992; Gilbert & Schwark,1992; Paisley et al. ,1986).

In the absence of an active corpus luteum, the treatment efficacy of clinical endometritis with only prostaglandin injection is limited however such a treatment according to Lewis(2004) may bring certain advantages as well.

It is important to mention that cows with clinical endometritis treated withone or 2 $PGF_{2\alpha}$ before initiation of the timed AI program had the lowest pregnancy rate per AI and the highest pregnancy loss compared with those having no uterine diseases (Lima et al. ,2013). Similarly, 2 $PGF_{2\alpha}$ treatments between Days (37 ±3) and (51 ±3) after calving as part of an oestrous synchronization protocol and i. u. infusion of ceftiofur (125 mg) given at Day (44 ± 3) did not improve pregnancy per AI following the first postpartum insemination or the rate of pregnancy in the first 300 after calving (Galvăo et al. ,2009).

Recently, as an alternative treatment of antimicrobials for clinical endometritis, 50% Dextrose (Brick et al. , 2012) or proteolytic enzyme (containing trypsin, chymotrypsin and papain) solutions (Drillich et al. ,2005) were used however further investigations are needed to confirm the beneficial effects of such treatments (Sassi et al. , 2010).

Pyometra The best treatment protocol is to use prostaglandin ($PGF_{2\alpha}$ or its synthetic analogues) injection(s) because of the presence of a persistent corpus luteum. Due to common relapse it is recommended to repeat the prostaglandin treatment 12 to 14 days later. Intra-uterine antibiotic therapy (cephapirin) may be used as well. Complete restoration of the endometrium may need 4 to 8 weeks therefore it is very important to diagnose and treat pyometra as soon as possible after calving to decrease the destructive nature of pyometra on the endometrium (Paisley et al. ,1986).

### 3.3 Subclinical Endometritis

Subclinical endometritis can be treated with a prostaglandin i. m. injection (cloprostenol 500 mg) or/and an i. u. antibiotic therapy (cephapirin) at 20 – 33 DIM to improve the reproductive performance (Kasimanickam et al. ,2005b). Intrauterine infusion of ceftiofur hydrochloride reduced the prevalence of *T. pyogenes*, but did not affect fertility of dairy cows already receiving $PGF_{2\alpha}$ (Galvăo et al. , 2009). One or 2 treatments with $PGF_{2\alpha}$ before initiation of the timed AI program with subclinical endometritis were unable to improve uterine health, pregnancy rate per AI, and maintenance of pregnancy in lactating dairy cows (Lima et al. ,2013).

## 4 Prevention of Uterine Diseases

Cows having hypocalcaemia, dystocia, stillbirth, twins or retained placenta in the periparturient period are more likely to contract uterine infections than those cows that calve normally. Thus, management of sanitation, nutrition, population density, stress to prevent or reduce the incidence of these predisposing factors (especially dystocia) should be impeccable. Therefore prevention remains limited to general guidance on hygiene at calving (Hartigan, 1980), adequate nutrition (Ca, Se, Vit. E, etc.) and the control of infectious diseases.

Routine systemic or intra-uterine administration of ceftiofur may be beneficial for the prevention of clinical metritis, however its effect on reproductive performance is not significantly different to that of no treatment therefore it cannot be recommended for the field (Drillich et al. ,2001 and 2006b; Risco et al. ,2003; Scott et al. ,2005). Similarly controversial results were reported when a single-dose of ceftiofur crystalline free acid sterile suspension was used in dairy cows at high risk of uterine disease (twin, dystocia, or retained placenta) within 24 h after calving (Dubuc et al. ,2011; McLaughlin et al. ,2013).

One of the pharmacological approaches to the prevention and treatment of retained placenta can be the administration of prostaglandin immediately after calving (Stevens et al., 1995), however due to controversy results further studies are needed to confirm its efficacy. Repeated administration of $PGF_{2\alpha}$ to cows on Days 7 and 14 or on Days 22 and 35 after calving had no effect on the prevalence of clinical endometritis at Days 22 and 58 after calving, and there was no effect on the probability of pregnancy after insemination at oestrus among cows with a voluntary waiting period of >100 days, or at timed AI at Day 85 when Presynch was performed (Hendricks et al., 2006). Similarly preventive administration of $PGF_{2\alpha}$ at both 5 and 7 wk after calving had no positive effect on reproduction in dairy cows (Dubuc et al., 2011). Preventive ozone intrauterine (spray) treatment (Zobel & Tkalěié, 2013) or Sheng Hua Tang, a classical herbal formula consisting of Radix Angelicae sinensis, Ligustici rhizoma, Semen persicae, Zingiberis rhizoma, and Radix glycyrrhizae (Cui et al., 2014) during early puerperal period may improve the reproductive efficacy in dairy cows. In contrast, homeopathic drugs like Lachesis compositum (Lachesis), Carduus compositum (Carduus), and Traumeel LT (Traumeel) were not effective in preventing bovine endometritis or in enhancing reproductive performance in dairy cows (Arlt et al., 2009).

## References

Ahmed F. O., Elsheikh A. S. Intrauterine infusion of Lugol's iodine improves the reproductive traits of postpartum infected dairy cows. *Journal of Agriculture and Veterinary Science*, 2013, 5:89-94.

Amiridis G. S., Fthenakis G. C., Dafopolus J., et al. Use of cefquinome for prevention and treatment of bovine endometritis. *J. Vet. Pharmacol. Therap.*, 2003, 26:387-390.

Amiridis G. S., Leontides L., Tassos E., et al. Flunixin meglumine accelerates uterine involution and shortens the calving-to-first-oestrus interval in cows with puerperal metritis. *J. Vet. Pharmacol. Ther.*, 2001, 24:365-367.

Arlt S., Padberg W., Drillich M., et al. Efficacy of homeopathic remedies as prophylaxis of bovine endometritis. *J. Dairy Sci.*, 2009, 92:4945-4953.

Bareille N., Beaudeau F., Billon S., et al. Effects of health disorders on feed intake and milk production in dairy cows. *Livestock Production Science*, 2003, 83:53-62.

Bareille N., Fourichon C. Fcateurs de risqué des affections post-partum. *Point Vét.*, 2006, 37:116-121.

Barlund C. S., Carruthers T. D., Waldner C. L., et al. A comparison of diagnostic techniques for postpartum endometritis in dairy cattle. *Theriogenology*, 2008, 69:714-723.

Benzaquen M. E., Risco C. A., Archbald L. F., et al. Rectal temperature, calving-related factors, and the incidence of puerperal metritis in postpartum dairy cows. *J. Dairy Sci.*, 2007, 90:2804-2814.

Bonnet B. N., Martin S. W., Meek A. H. Associations of clinical findings, bacteriological and histological results of endometrial biopsy with reproductive performance of post partum dairy cows. *Prev. Vet. Med.*, 1993, 15:205-220.

Brick T. A., Schuenemann G. M., Bas S., et al. Effect of intrauterine dextrose or antibiotic therapy on reproductive performance of lactating dairy cows diagnosed with clinical endometritis. *J. Dairy Sci.*, 2012, 95:1894-1905.

Britt J. H. Early post partum breeding in dairy cows. A review. *J. Dairy Sci.*, 1975, 58:266-271.

Burfeind O., von Keyserlingk M. A. G., Weary D. M., et al. Repeatability of measures of rectal temperature in dairy cows. *J. Dairy Sci.*, 2010, 93:624-627.

Cui D., Wang X., Wang L., et al. The administration of *Sheng Hua Tang* immediately after delivery to reduce the incidence of retained placenta in Holstein dairy cows. *Theriogenology*, 2014, 81:645-650.

Djuricic D., Vince S., Ablondi M., et al. Effect of preventive intrauterine ozone application on reproductive efficiency in Holstein cows. Reprod. *Domes. Anim.*, 2012, 47:87-91.

Dohmen M. J. W., Joop K., Sturk A., et al. Relationship between intra-uterine bacterial contamination, endotoxin levels and the development of endometritis in postpartum cows with dystocia or retained placenta. *Theriogenology*, 2000, 54:1019-1032.

Drillich M. ,Beetz O. ,Pfützner A. ,et al. Evaluation of a systemic antibiotic treatment of toxic puerperal metritis in dairy cows. *J. Dairy Sci.* , 2001,84:2010 -2017.

Drillich M. ,Raab D. ,Wittke M. ,et al. Treatment of chronic endometritis in dairy cows with an intrauterine application of enzymes. A field trial. *Theriogenology*,2005,63:1811 -1823.

Drillich M. ,Mahlstedt M. ,ReichertU. ,et al. Strategies to improve the therapy of retained fetal membranes in dairy cows. *J. Dairy Sci.* , 2006a,89:627 -635.

Drillich M. ,Reichert U. ,Mahlstedt M. ,et al. Comparison of two strategies for systemic antibiotic treatment of dairy cows with retained fetal membranes:preventive vs. selective treatment. *J. Dairy Sci.* , 2006b,89:1502 -1508.

Drillich M. ,Voigt D. , Forderung D. , et al. Treatment of acute puerperal metritis with flunixin meglumine in addition to antibiotic treatment. *J. Dairy Sci.* , 2007,90:3758 -3763.

Dubuc J. ,Duffield T. F. ,Leslie K. E. ,et al. Definitions and diagnosis of postpartum endometritis in dairy cows. *J. Dairy Sci.* ,2010, 93:5225 -5233.

Dubuc J. ,Duffield T. F. ,Leslie K. E. ,et al. Randomized clinical trial of antibiotic and prostaglandin treatments for uterine health and reproductive performance in dairy cows. *J. Dairy Sci.* 2011,94:1325 -1338.

Eiler H. ,Fecteau K. A. Retained placenta,in:Youngquist R. S. ,Threlfall W. R. (eds),Current therapy in large animal theriogenology 2. W. B. Saunders Company,*Philadelphia*,2007:345 -354.

Földi J. ,Kulcsár M. ,Pécsi A. ,et al. Bacterial complications of postpartum uterine involution in cattle. *Anim. Reprod. Sci.* ,2006,96: 265 -281.

Földi J. ,Pécsi A. ,Szabó J. ,et al. Use of cephalosporins for the treatment of dairy cows suffering of puerperal metritis and endometritis (in Hungarian with English summary). *Magyar Állatorvosok Lapja*,2009,131:451 -455.

Frazer,G. S. A rational basis for therapy in the sick postpartum cow. Veterinary Clinics of North America:*Food Animal Practice*,2005, 21:523 -568.

Galvão K. N. ,Greco L. F. ,Vilela J. M. ,et al. Effect of intrauterine infusion of ceftiofur on uterine health and fertility in dairy cows. *J. Dairy Sci.* ,2009,92:1532 -1542.

Gilbert,R. O. Bovine endometritis:the burden of proof. *Cornell Veterinarian*,1992,82:11 -14.

Gilbert R. O. ,Schwark W. S. Pharmacologic considerations in the management of peripartum conditions in the cow. Veterinary Clinics of North America: *Food Animal Practice*,1992,8:29 -56.

Gilbert R. O. , Shin S. T. , Guard C. L. , et al. Subclinical endometritis. Prevalence of endometritis and its effects on reproductive performance of dairy cows. *Theriogenology*, 2005,64:1879 -1888.

Gilbert R. O. , Shin S. T. , Guard C. L. , et al. Incidence of endometritis and effects on reproductive performance of dairy cows. *Theriogenology*, 1998,49:251.

Gröhn Y. T. ,Erb H. N. ,McCulloch C. E. ,Saloniemi H. S. I Epidemiology of reproductive disorders in dairy cattle:associations among host characteristics,disease and production. *Preventive Veterinary Medicine*, 1990,8:25 -39.

Gröhn Y. T. ,Rajala-Schultz P. J. Epidemiology of reproductive performance in dairy cows. *Anim. Reprod. Sci.* , 2000,(60 -61):605 -614.

Grunert E. ,Etiology and pathogenesis of retained bovine placenta,in:Morrow A. D. (Ed.),Current therapy in theriogenology 2. W. B. Saunders Company,*Philadelphia*,1986:237 -250.

Gustafsson B. K. Therapeutic strategies involving antimicrobial treatment of the uterus in large animals. *JAVMA*, 1984, 185: 1194 -1198.

Hartigan P. J. Fertility management in the dairy herd:The need to control bacteriological contamination in the environment. *Irish Vet. J.* , 1980,34:43 -48.

Hendricks K. E. ,Bartolome J. A. ,Melendez P. ,et al. Effect of repeated administration of PGF2alpha in the early post partum period on the prevalence of clinical endometritis and probability of pregnancy at first insemination in lactating dairy cows. *Theriogenology*, 2006,65:1454 -1464.

Hernandez J. , Risco C. A. , Elliott J. B. Effect of oral administration of a calcium chloride gel on blood mineral concentrations, parturient disorders, reproductive performance, and milk production of dairy cows with retained fetal membranes. *JAVMA*, 1999, 215: 72 – 76.

Heuwieser W. , Tenhagen B. A. , Tischer M. , et al. Effect of three programmes for the treatment of endometritis on the reproductive performance of a dairy herd. *Vet. Rec.* , 2000, 146: 338 – 341.

Kamimura S. , Oui T. , Takahashi M. , et al. Postpartum resumption of ovarian activity and uterine involution monitored by ultrasonography in Holstein cows. *J. Vet. Med. Sci.* , 1993, 55: 643 – 647.

Kasimanickam R. , Duffield T. F. , Foster R. A. , et al. Endometrial cytology and ultrasonography for the detection of subclinical endometritis in postpartum dairy cows. *Theriogenology*, 2004, 62: 9 – 23.

Kasimanickam R. , Duffield T. F. , Foster R. A. , et al. A comparison of the cytobrush and uterine lavage techniques to evaluate endometrial cytology in clinically normal postpartum dairy cows. *Can. Vet. J.* , 2005a, 46: 255 – 259.

Kasimanickam R. , Duffield T. F. , Foster R. A. , et al. The effect of a single administration of cephapirin or cloprostenol on the reproductive performance of dairy cows with subclinical endometritis. *Theriogenology*, 2005b, 63: 818 – 830.

Königsson K. , Gustafsson H. , Gunnarsson A. , et al. Clinical and bacteriological aspects on the use of tetracycline and flunixin in primiparous cows with induced retained placenta and post-partal endometritis. *Reprod. Domest. Anim.* , 2001, 36: 247 – 256.

Könyves L, Szenci O. , Jurkovich V. , et al. Risk assessment of postpartum uterine disease and consequences of puerperal metritis for subsequent metabolic status, reproduction and milk yield in dairy cows. *Acta Vet. Hung.* , 2009a, 57: 155 – 169.

Könyves L, Szenci O, Jurkovich V, et al. Risk assessment and consequences of retained placenta for uterine health, reproduction and milk yield in dairy cows. *Acta Veterinaria Brno*, 2009b, 78: 163 – 172.

LeBlancL. J. , Duffield T. F. , Leslie K. E. , et al. Defining and diagnosing postpartum clinical endometritis and its impact on reproductive performance in dairy cows. *J. Dairy Sci.* , 2002a, 85: 2223 – 2236.

LeBlanc L. J. , Duffield T. , Leslie K. , et al. The effect of treatment of clinical endometritis on reproductive performance in dairy cows. *J. Dairy Sci.* , 2002b, 85: 2237 – 2249.

LeBlanc L. J. Postparum uterine disease and dairy herd reproductive performance: *A review. Vet. J.* , 2008, 176: 102 – 114.

Leutert C. , von Krueger X. , Plöntzke J. , et al. Evaluation of vaginoscopy for the diagnosis of clinical endometritis in dairy cows. *J. Dairy Sci.* , 2012, 95: 206 – 212.

Lewis G. S. , Steroidal regulation of uterine immune defences. *Anim. Reprod. Sci.* , 2004, 82 – 83: 281 – 294.

Lima F. S. , Bisinotto R. S. , Ribeiro E. S. , Effects of 1 or 2 treatments with prostaglandin F2α on subclinical endometritis and fertility in lactating dairy cows inseminated by timed artificial insemination. *J Dairy Sci.* , 2013, 96: 6480 – 6488.

Markusfeld O. Factors responsible for post parturient metritis in dairy cattle. *Vet. Rec.* , 1984, 114: 539 – 542.

Mateus L. , Lopes da Costa L. , Bernardo F. , et al. Influence of puerperal uterine infection on uterine involution and postpartum ovarian activity in dairy cows. *Reprod. Domest. Anim.* , 2002, 37: 31 – 35.

McDougall S. , Macaulay R. , Compton C. Association between endometritis diagnosis using a novel intravaginal device and reproductive performance in dairy cattle. *Anim. Reprod. Sci.* , 2007, 99: 9 – 23.

McLaughlin C. L. , Stanisiewski E. , Lucas M. J. , et al. Evaluation of two doses of ceftiofur crystalline free acid sterile suspension for treatment of metritis in lactating dairy cows. *J. Dairy Sci.* 2012, 95: 4363 – 4371.

McLaughlin C. L. , Stanisiewski E. P. , Risco C. A. , et al. Evaluation of ceftiofur crystalline free acid sterile suspension for control of metritis in high-risk lactating dairy cows. *Theriogenology*, 2013, 79: 725 – 734.

Nakao T. , Moriyoshi M. , Kawata, K. Effect of postpartum intrauterine treatment with 2% polyvinyl-pyrrolidone-iodine solution on reproductive efficiency in cows. *Theriogenology*, 1988, 30: 1033 – 1043.

Noakes D. E. , Wallace L. M. , Smith G. R. Pyometra in a Friesian heifer: bacteriological and endometrial changes. *Vet. Rec.* , 1990, 126: 509.

Opsomer G. , Gröhn Y. T. , Hertl J. , et al. Risk factors for post partum ovarian dysfunction in high producing dairy cows in Belgium: A field study. *Theriogenology*, 2000, 53: 841 – 857.

Paisley L. G. , Mickelsen W. D. Anderson P. B. , Mechanisms and therapy for retained fetal membranes and uterine infections of cows: A review. *Theriogenology*, 1986, 25: 353 – 381.

Pécsi A. , Földi J. , Kulcsár M. , et al. Subclinical endometritis in dairies in Hungary (in Hungarian with English summary), in Szenci O. , Brydl E. , Jurkovich V. (Eds. ), 19th International Congress of the Hungarian Association for Buiatrics, Debrecen, Hungary, *Proceedings*, 2009: 140 – 144.

Pécsi A, Földi J. , Szabó J. , Nagy P. , et al. Efficacy of different antimicrobial therapeutic protocols for treatment of puerperal metritis and endometritis in dairy cows (in Hungarian with English summary). *Magyar Állatorvosok Lapja*, 2007, 129: 590 – 599.

Pleticha S. , Drillich M. , Heuwieser W. Evaluation of the Metricheck device and the gloved hand for the diagnosis of clinical endometritis in dairy cows. *J. Dairy Sci.* , 2009, 92: 5429 – 5435.

Risco C. A. , Hernandez J. Comparison of ceftiofur hydrochloride and estradiol cypionate for metritis prevention and reproductive performance in dairy cows affected with retained fetal membranes. *Theriogenology*, 2003, 60: 47 – 58.

Risco C. A. , Youngquist R. S. , Shore M. D. : Postpartum uterine infections, in: Youngquist R. S. , Threlfall W. R. (Eds. ), Current therapy in large animal theriogenology 2. W. B. Saunders Company, *Philadelphia*, 2007: 339 – 344.

Sassi G. , Ismail S. , Bajcsy Á. Cs. , Kiss G. , et al. Evaluation of the alternatives of the intrauterine antibiotic treatments in the cow. Literature review (in Hungarian with English summary). *Magyar Állatorvosok Lapja*, 2010, 132: 516 – 527.

Scott H. M. , Schouten M. J. , Gaiser J. C. , et al. Effect of intrauterine administration of ceftiofur on fertility and risk of culling in postparturient cows with retained fetal membranes, twins, or both. *JAVMA*, 2005, 226: 2044 – 2052.

Sheldon I. M. , Dobson H. Postpartum uterine health in cattle. *Anim. Reprod. Sci.* , 2004, 82 – 83: 295 – 306.

Sheldon I. M. , Lewis G. S. , LeBlanc S. , et al. Defining postpartum uterine disease in cattle. *Theriogenology*, 2006, 65: 1516 – 1530.

Sheldon I. M. , Noakes D. E. Comparison of three treatments for bovine endometritis. *Vet. Rec.* , 1998, 142: 575 – 579.

Silva J. W. Addressing the decline in reproductive performance of lactating dairy cows: a researcher's perspective. *Veterinary Science Tomorrow*, 2003, 3: 1 – 5.

Stevens R. D. , Dinsmore R. P. , Cattell M. B. , Evaluation of the use of intrauterine infusions of oxytetracycline, subcutaneous injections of fenprostalene, or a combination of both, for the treatment of retained fetal membranes in dairy cows. *JAVMA*, 1995, 207: 1612 – 1615.

Szenci O. Factors, which may affect reproductive performance in dairy cattle. *Magyar Állatorvosok Lapja*, 2008, 130 (Suppl. I. ): 107 – 111.

Urton G. , von Keyserlingk M. A. G. Weary D. M. Feeding behavior identifies dairy cows at risk for metritis. *J. Dairy Sci.* , 2005, 88: 2843 – 2849.

Van Werven T. , Schukken Y. H. , Lloyd J. , et al. The effects of duration of retained placenta on reproduction, milk production, postpartum disease and culling rate. *Theriogenology*, 1992, 37: 1191 – 1203

Zhou C. , Boucher J. F. , Dame K. J. , et al. Multilocation trial of ceftiofur for treatment of postpartum cows with fever. *JAVMA*, 2001, 219: 805 – 808.

Zobel R. , Tkalčić S. Efficacy of ozone and other treatment modalities for retained placenta in dairy cows. *Reprod. Domest. Anim.* 2013, 48: 121 – 125.

# Effects of antibacterial peptide on antioxidant function of tissues in weaned piglets

DAN Qixiong, REN Zhihua, DENG Junliang, YUAN Wei, JIN Haitao, TIAN Chunlei, LIANG Zhen, and GAO Shuang

*Department of Veterinary Medicine, Sichuan Agricultural University, Ya'an, 625014, China*

**Abstract**: The early weaning played a role in the modern pig industry, which would caused weaned stress reactions. Therefore, how to reduced the weaned stress of weaned piglets became an important problem. This study was conducted to investigate the effects of antibacterial peptides (ABP) and astragalus polysaccharide (AP) on the antioxidant function of tissues in weaned piglets. Ninety 21-day-old healthy weaned piglets were randomly divided into five groups based on weight ($8.24 \pm 0.67$) kg. The control group was fed basal diet, AP group was fed basal diet and AP 400mg/kg, ABP group Ⅰ-Ⅲ were fed separately with basal diet and ABP at a dosage of 250, 500, 1 000mg/kg at 26 days of weaned piglets, respectively. The experiment lasted for 28 day. The liver, lung, spleen and kidney antioxidant indices were determined at 39, 53 days of age. AP and different concentrations of ABP added to standard diet significantly ($P<0.01$ or 0.05) enhanced the tissues (liver, lung, spleen, kidney) SOD、GSH-Px、CAT、T-AOC and inhibition of ·OH capacity, significantly ($P<0.01$) decreased the levels of MDA. The antioxidant function of tissues in the ABP groups were dose dependent, the optimum add dosage of ABP was found to be 1 000mg/kg, and was superior ($P<0.01$ or 0.05) to the 400mg/kg AP group. It showed that ABP could effectively improve the activity of the antioxidant enzyme of tissues in weaned piglets, act to scavenge free radicals and improve body function.

**Keywords**: Antibacterial peptide, antioxidant function, tissues, weaned piglets

## 1 Introduction

Antibacterial peptides (ABP), which are produced by several species including insects, other animals, microorganisms and synthesis, are a critical component of the natural defense system (Li et al., 2012). They can protect against a broad array of infectious agents, such as bacteria (Gordon et al., 2005), fungi (Wakabayashi et al., 2000), parasite (Dagan et al., 2002), virus (Wachinger et al., 1998) and cancer cells (Chen et al., 2003). ABP have a unique antibacterial mechanism and no residue, non-toxic side effects, will have a very good future in the application in pharmaceuticals industry and food additive (Zasloff, 2002).

Studies of ABP are mainly concentrated inantibacterial mechanism (Zasloff, 2002), immune function (Yu et al., 2010), production performance (Wu et al., 2012), disease prevention and control (Dale and Fredericks, 2005), etc.. As far as the authors' knew, there was few report on the effects of ABP on the antioxidant function in weaned piglets. However, in modern pig industry, most of piglets were weaned at 3 – 4 weeks of age, during this special period, the antioxidant function of the weaned piglets could decrease and cause high morbidity and

---

Corresponding author: Junliang Deng. E-mail address: dengjl213@126.com

YUAN Wei and Ren Zhihua contributed equally to this work and should be considered co-first authors mortality, resulting in significant economic loss (Zhu, et al., 2012). Therefore, how to improve the antioxidant function of weaned piglets become an important problem. In this study, we investigated the effects of different concentrations of ABP on the antioxidant function of tissues in weaned piglets.

# 2 Materials and Methods

## 2.1 Materials

Antibacterial peptide (synthesized by the Styela antibacterial peptide and the defensin) was provided by Rota Bioengineering Co., Ltd, Sichuan, China.

Astragalus polysaccharide (AP; net content of 65%) was bought from Centre Biology Co., Ltd, Beijing, China.

## 2.2 Animals and Experimental Design

Piglets (Landrace × Yorkshire × Duroc; weaned at 21d of age) were purchased from Xin Qiao Agricultural science and technology development co., ltd, Chengdu, Sichuan, China. 90 weanling piglets (average BW of (8.24 ± 0.67) kg) were allotted to 5 treatments in a randomized complete block, and acclimatized for 5 days. The control group was fed basal diet, AP group was fed basal diet and AP 400mg/kg, ABP group Ⅰ – Ⅲ were fed separately with basal diet and ABP at a dosage of 250, 500, 1 000mg/kg at 26 days of weaned piglets. The experiment lasted for 28 days.

Piglets were caged in elevated pens with wire flooring and fed with standard diet (Table 1) for weanling piglets (NRC 1998). Temperature (26 – 27℃) and relative humidity (65% – 70%) remained constant. Food and water were provided ad libitum during the acclimatisation period and through all parts of the study. All piglets used in this study were approved to be healthy. All study experimental manipulations were undertaken in accordance with the Institutional Guidelines for the Care and Use of Laboratory Animals.

Table 1 Ingredient composition of diets, as-fed basis[a]

| Ingredient | % |
|---|---|
| Corn | 53.15 |
| Soybean meal (44% CP) | 25.40 |
| Whey powder[b] | 6.00 |
| Fish meal (62.5% CP) | 4.50 |
| Soybean oil | 2.80 |
| Spray-dried porcine plasma (78% CP)[c] | 3.00 |
| Dicalcium phosphate | 2.0 |
| Limestone | 1.40 |
| Salt | 0.25 |
| L-Lysine HCl | 0.15 |
| DL-Methionine | 0.35 |
| Premix[d] | 1.00 |

[a]Calculated nutrient composition (as-fed basis): 3 280 kcal of DE/kg, 1.14% lysine, and 0.32% methionine. Analyzed nutrient composition (as-fed basis): 21.1% CP, 0.68% Ca, and 0.59% P. Corn was removed by 400 mg/kg, when 400 mg/kg astragalus polysaccharide was added. Corn was removed by 250, 500, and 1 000 mg/kg, respectively, when 250 500 and 1 000 mg/kg antibacterial

peptide were added.

[b]Whey powder was a product of Calva Product Inc. , Acampo, CA.

[c]Spray-dried porcine plasma was a product of Merrick's Inc. , Middleton, WI.

[d]Provided the following per kilogram of complete diet: vitamin A, 2 250 IU; vitamin $D_3$, 1 050 IU; vitamin E, 16 IU; vitamin $K_3$, 0.9 mg; vitamin $B_{12}$, 0.03 mg; riboflavin, 4.5 mg; niacin, 30 mg; pantothenic acid, 25 mg; choline chloride, 400 mg; folic acid, 0.30 mg; thiamin, 2 mg; pyridoxine, 7 mg; biotin, 0.20 mg; Zn, 80 mg; Mn, 22 mg; Fe, 80 mg; Cu, 225 mg; I, 0.50 mg; and Se, 0.25 mg

### 2.3 Assay of Tissue Antioxidant Function

At 39 and 53 days of age during the experiment, six piglets were selected randomly from each treatment and were euthanized for obtaining tissues (liver, lung, spleen, kidney). The tissues SOD, GSH-Px, CAT, T-AOC, MDA, and inhibition of · OH capacity were determined by colorimetric methods described with reagent kits (Nanjing Jiancheng Bioengineering Institute, Nanjing, China).

### 2.4 Data Analysis

The results were presented by means and the standard error of the mean (SEM). Statistical analysis was performed using one-way analysis of variance (ANOVA) test of SPSS (International Business Machines Corporation, Armonk, USA) 19.0 software. $P < 0.05$ or 0.01 were considered as significantly different.

## 3 Results

### 3.1 The Levels of SOD in Tissues

The results were showed in Table 2. During the whole experimental period, the levels of tissues (liver, lung, spleen, kidney) SOD in the AP group and ABP groups were significantly higher ($P < 0.01$) than those of the control group (except the levels of the lung of the ABP groups on 39 days of age). The values of the ABP group Ⅰ were less than those of the AP group (except the values of lung of ABP group Ⅰ on 53 days of age and the values of the kidney of ABP group Ⅰ on 39, 53 days of age). The levels of the liver and kidney SOD in the ABP group Ⅱ, Ⅲ were markedly higher ($P < 0.01$) than those of the AP group, the levels of the lung SOD in the ABP group Ⅱ, Ⅲ and the spleen of ABP group Ⅲ were markedly greater ($P < 0.01$ or 0.05) than those of the AP group on 53 days of age. The levels of SOD in ABP groups were dose dependent, the optimum add dosage of ABP was found to be 1 000mg/kg.

Table 2 Results of mensuration vigor SOD in tissues (U/mgprot) (n = 6)

| Tissues | Time | Control group | AP group | ABP group Ⅰ | ABP group Ⅱ | ABP group Ⅲ |
|---|---|---|---|---|---|---|
| Spleen | 39d | $7.81 \pm 0.27^{C}$ | $16.23 \pm 1.11^{A}$ | $13.93 \pm 0.95^{B}$ | $15.87 \pm 0.95^{A}$ | $16.84 \pm 1.10^{A}$ |
| | 53d | $12.19 \pm 0.93^{D}$ | $16.30 \pm 1.08^{BCa}$ | $15.50 \pm 0.93^{C}$ | $17.45 \pm 1.06^{ABc}$ | $18.08 \pm 0.86^{A}$ |
| Liver | 39d | $74.11 \pm 5.27^{C}$ | $93.73 \pm 3.88^{B}$ | $92.19 \pm 2.30^{B}$ | $106.39 \pm 4.89^{A}$ | $108.56 \pm 5.21^{A}$ |
| | 53d | $74.24 \pm 5.68^{D}$ | $121.27 \pm 7.89^{B}$ | $98.91 \pm 5.72^{C}$ | $171.43 \pm 7.60^{A}$ | $171.74 \pm 12.37^{A}$ |
| Lung | 39d | $10.68 \pm 0.39^{BC}$ | $14.87 \pm 1.17^{A}$ | $10.13 \pm 0.57^{C}$ | $11.57 \pm 0.31^{B}$ | $15.76 \pm 0.33^{A}$ |
| | 53d | $12.17 \pm 0.96^{D}$ | $14.48 \pm 1.01^{C}$ | $16.81 \pm 0.75^{B}$ | $16.22 \pm 0.92^{Bc}$ | $19.66 \pm 0.95^{A}$ |
| Kidney | 39d | $27.32 \pm 1.66^{E}$ | $36.63 \pm 1.81^{D}$ | $40.84 \pm 2.33^{C}$ | $51.95 \pm 2.05^{B}$ | $68.13 \pm 1.85^{A}$ |
| | 53d | $32.88 \pm 1.24^{D}$ | $42.40 \pm 1.59^{C}$ | $44.52 \pm 1.95^{C}$ | $53.42 \pm 1.72^{B}$ | $78.27 \pm 2.14^{A}$ |

ABCD means with various letters are very significant different ($P < 0.01$), and with the same letters but in different case are significant different ($P < 0.05$). The following table 3, table 4, table 5, table 6 and table 7 are the same

### 3.2 The Levels of GSH-Px in Tissues

The results were showed in Table3. During the whole experimental period, the levels of the tissues (liver, lung, spleen, kidney) GSH-Px in the AP group and ABP groups were markedly higher ($P<0.01$ or 0.05) than those of control group (except the levels of the lung and kidney of ABP group Ⅰ). The values of the tissues GSH-Px in the AP group were higher than those of the ABP group Ⅰ, in the lung and kidney (on 39 days of age) were significantly greater ($P<0.01$ or 0.05) than those of the ABP group Ⅰ. The levels of the liver, spleen, kidney GSH-Px in the ABP group Ⅱ were higher than those of the AP group, in the lung was lower than those of the AP group. During the whole experimental period, the values of the liver, spleen, kidney GSH-Px in the ABP group Ⅲ were markedly higher ($P<0.01$ or 0.05) than those of the AP group. The levels of GSH-Px in the ABP groups were dose dependent, the optimum add dosage of ABP was found to be 1 000mg/kg.

Table 3 Results of mensuration vigor GSH-Px in tissues (U/mgprot) (n=6)

| Tissues | Time | Control group | AP group | ABP group Ⅰ | ABP group Ⅱ | ABP group Ⅲ |
|---|---|---|---|---|---|---|
| Spleen | 39d | $16.31\pm1.72^{C}$ | $20.83\pm1.95^{B}$ | $20.35\pm1.89^{B}$ | $23.20\pm2.14^{Ba}$ | $26.88\pm1.96^{A}$ |
| | 53d | $24.43\pm1.79^{C}$ | $31.10\pm2.58^{B}$ | $30.69\pm3.19^{B}$ | $33.18\pm2.12^{AB}$ | $35.90\pm2.08^{Ab}$ |
| Liver | 39d | $19.89\pm1.62^{C}$ | $30.47\pm2.19^{B}$ | $30.22\pm2.29^{B}$ | $33.15\pm2.85^{AB}$ | $35.07\pm1.96^{Ab}$ |
| | 53d | $26.29\pm3.41^{C}$ | $39.82\pm4.42^{Ba}$ | $35.65\pm1.67^{B}$ | $41.07\pm2.66^{AB}$ | $45.33\pm4.25^{A}$ |
| Lung | 39d | $14.50\pm1.52^{C}$ | $23.00\pm1.82^{AB}$ | $15.76\pm1.32^{C}$ | $21.10\pm1.30^{B}$ | $24.68\pm2.31^{Ab}$ |
| | 53d | $28.71\pm2.75^{C}$ | $37.89\pm4.74^{Ab}$ | $31.26\pm2.20^{BC}$ | $36.43\pm2.99^{ABc}$ | $40.65\pm4.54^{A}$ |
| Kidney | 39d | $24.72\pm2.37^{CD}$ | $30.28\pm2.41^{Bd}$ | $20.80\pm3.17^{C}$ | $32.16\pm2.88^{B}$ | $41.44\pm3.92^{A}$ |
| | 53d | $31.93\pm3.01^{D}$ | $38.91\pm3.12^{BCa}$ | $35.50\pm2.39^{CD}$ | $41.29\pm3.92^{ABc}$ | $43.84\pm2.58^{A}$ |

### 3.3 The Levels of CAT in Tissues

The results were showed in Table 4. Compared with the control group, the levels of the tissues (liver, lung, spleen, kidney) CAT in the AP group and ABP groups were significantly superior ($P<0.01$ or 0.05) (except the levels of the spleen and kidney of AP group and ABP group Ⅰ on 39 days of age, the kidney of AP group on 53 days of age). The levels of the tissues (liver, lung, spleen, kidney) CAT in the AP group were higher than those of the ABP group Ⅰ (except the levels of the spleen on 39 days of age and the kidney on 53 days of age in the AP group). The values of the tissues (liver, lung, spleen, kidney) CAT in the ABP group Ⅱ and Ⅲ were significantly higher ($P<0.01$) than those of the AP group (except the levels of the spleen and lung of ABP group Ⅱ on 53 days of age, the liver on 39, 53 days of age). The levels of CAT in the ABP groups were dose dependent, the optimum add dosage of ABP was found to be 1,000mg/kg.

Table 4 Results of mensuration vigor CAT in tissues (U/mgprot) (n=6)

| Tissues | Time | Control group | AP group | ABP group Ⅰ | ABP group Ⅱ | ABP group Ⅲ |
|---|---|---|---|---|---|---|
| Spleen | 39d | $2.67\pm0.17^{C}$ | $2.63\pm0.15^{C}$ | $2.73\pm0.15^{C}$ | $3.94\pm0.14^{A}$ | $3.32\pm0.18^{B}$ |
| | 53d | $3.79\pm0.17^{D}$ | $5.13\pm0.14^{B}$ | $4.46\pm0.20^{C}$ | $5.24\pm0.19^{B}$ | $5.86\pm0.18^{A}$ |
| Liver | 39d | $138.16\pm14.50^{D}$ | $311.75\pm14.29^{B}$ | $242.47\pm13.85^{C}$ | $298.23\pm20.44^{B}$ | $363.30\pm13.93^{A}$ |
| | 53d | $267.02\pm10.70^{D}$ | $347.25\pm12.88^{B}$ | $307.99\pm10.09^{C}$ | $365.14\pm14.17^{B}$ | $399.71\pm13.52^{A}$ |

(continued table)

| Tissues | Time | Control group | AP group | ABP group Ⅰ | ABP group Ⅱ | ABP group Ⅲ |
|---|---|---|---|---|---|---|
| Lung | 39d | $5.73 \pm 0.20^{Dc}$ | $6.16 \pm 0.24^{C}$ | $6.05 \pm 0.17^{C}$ | $7.01 \pm 0.23^{Ba}$ | $7.35 \pm 0.18^{A}$ |
| | 53d | $6.50 \pm 0.23^{D}$ | $9.19 \pm 0.26^{B}$ | $7.89 \pm 0.17^{C}$ | $9.23 \pm 0.18^{B}$ | $10.96 \pm 0.28^{A}$ |
| Kidney | 39d | $190.96 \pm 12.37^{B}$ | $189.62 \pm 12.42^{B}$ | $180.14 \pm 9.50^{B}$ | $251.34 \pm 23.31^{A}$ | $265.89 \pm 121.88^{A}$ |
| | 53d | $221.61 \pm 12.49^{D}$ | $237.38 \pm 15.02^{D}$ | $292.89 \pm 9.53^{C}$ | $319.81 \pm 24.69^{AB}$ | $335.80 \pm 23.64^{A}$ |

## 3.4 The Levels of T-AOC in Tissues

The results were showed in Table5. Compared with the control group, the levels of the tissues (liver, lung, spleen, kidney) T-AOC in the AP group and ABP groups were significantly superior ($P<0.01$ or $0.05$) (except the levels of the spleen of AP group, the spleen and liver of ABP group Ⅰ, the kidney of ABP group Ⅰ and the spleen of ABP group Ⅱ on 39 days of age). The levels of the tissues (liver, lung, spleen, kidney) T-AOC in the AP group were higher than those of the ABP group Ⅰ, in the lung, kidney were significantly superior ($P<0.01$). The values of the liver, spleen, kidney T-AOC in the ABP group Ⅱ were higher than those in the AP group, the lung T-AOC in the AP group Ⅱ was lower than those of the AP group, there was no differences ($P>0.05$). The levels of the spleen, kidney and the liver (on 53 days of age) T-AOC in the ABP group Ⅲ were significantly higher ($P<0.01$) than those of the AP group. The levels of T-AOC in the ABP groups were dose dependent, the optimum add dosage of ABP was found to be 1 000mg/kg.

Table 5 Results of mensuration vigor T-AOC in tissues (U/mgprot) (n=6)

| Tissues | Time | Control group | AP group | ABP group Ⅰ | ABP group Ⅱ | ABP group Ⅲ |
|---|---|---|---|---|---|---|
| Spleen | 39d | $0.52 \pm 0.04^{B}$ | $0.57 \pm 0.06^{B}$ | $0.56 \pm 0.07^{B}$ | $0.60 \pm 0.07^{B}$ | $0.82 \pm 0.06^{A}$ |
| | 53d | $0.63 \pm 0.05^{C}$ | $0.72 \pm 0.05^{BC}$ | $0.63 \pm 0.05^{C}$ | $0.76 \pm 0.07^{Bc}$ | $1.39 \pm 0.11^{A}$ |
| Liver | 39d | $1.41 \pm 0.11^{C}$ | $1.66 \pm 0.10^{AB}$ | $1.51 \pm 0.12^{BC}$ | $1.71 \pm 0.12^{Ab}$ | $1.78 \pm 0.11^{A}$ |
| | 53d | $1.57 \pm 0.12^{D}$ | $1.81 \pm 0.13^{BCd}$ | $1.66 \pm 0.13^{CD}$ | $1.94 \pm 0.11^{Ba}$ | $2.18 \pm 0.14^{A}$ |
| Lung | 39d | $0.37 \pm 0.09^{C}$ | $0.66 \pm 0.08^{A}$ | $0.49 \pm 0.05^{Bc}$ | $0.61 \pm 0.05^{Ab}$ | $0.69 \pm 0.06^{A}$ |
| | 53d | $0.54 \pm 0.06^{C}$ | $0.76 \pm 0.07^{A}$ | $0.63 \pm 0.03^{Bc}$ | $0.72 \pm 0.04^{Ab}$ | $0.80 \pm 0.06^{A}$ |
| Kidney | 39d | $0.35 \pm 0.04^{C}$ | $0.65 \pm 0.05^{B}$ | $0.40 \pm 0.06^{C}$ | $0.66 \pm 0.05^{B}$ | $0.78 \pm 0.06^{A}$ |
| | 53d | $0.40 \pm 0.06^{D}$ | $0.68 \pm 0.07^{B}$ | $0.50 \pm 0.04^{Cd}$ | $0.76 \pm 0.05^{B}$ | $0.95 \pm 0.07^{A}$ |

## 3.5 The Contents of MDA in Tissues

The results were showed in Table6. During the whole experimental period, the contents of the tissues (liver, lung, spleen, kidney) MDA in the AP group and ABP groups were significantly lower ($P<0.01$) than those of control group. During the whole experimental period, the levels of the tissues (liver, lung, spleen, kidney) MDA in the AP group were lower than those of the ABP group Ⅰ, and in the ABP group Ⅱ were lower than those in the AP group (except the levels of the liver of group Ⅱ on 39 days of age), there was no differences ($P>0.05$). The contents of the liver, spleen and kidney MDA in the ABP group Ⅲ were significantly lower ($P<0.01$ or $0.05$) than those of the AP group (except the levels of the spleen of ABP group Ⅲ on 39 days of age). The contents of MDA in the ABP groups were dose dependent, the optimum add dosage of ABP was found to be 1 000mg/kg.

Table 6 Results of mensuration content MDA in tissues (nmol /mgprot) (n = 7)

| Tissues | Time | Control group | AP group | ABP group Ⅰ | ABP group Ⅱ | ABP group Ⅲ |
|---|---|---|---|---|---|---|
| Spleen | 39d | $1.43 \pm 0.16^{A}$ | $0.98 \pm 0.07^{B}$ | $1.05 \pm 0.13^{B}$ | $0.96 \pm 0.03^{B}$ | $0.89 \pm 0.06^{B}$ |
| | 53d | $0.65 \pm 0.04^{A}$ | $0.21 \pm 0.05^{BCd}$ | $0.28 \pm 0.04^{Bc}$ | $0.19 \pm 0.05^{CD}$ | $0.14 \pm 0.04^{D}$ |
| Liver | 39d | $2.02 \pm 0.12^{A}$ | $0.64 \pm 0.04^{CD}$ | $1.23 \pm 0.07^{B}$ | $0.73 \pm 0.06^{C}$ | $0.55 \pm 0.06^{D}$ |
| | 53d | $1.70 \pm 0.09^{A}$ | $0.77 \pm 0.05^{B}$ | $0.81 \pm 0.05^{B}$ | $0.68 \pm 0.06^{Cb}$ | $0.54 \pm 0.06^{D}$ |
| Lung | 39d | $0.94 \pm 0.07^{A}$ | $0.61 \pm 0.04^{BC}$ | $0.65 \pm 0.04^{B}$ | $0.59 \pm 0.03^{BC}$ | $0.55 \pm 0.04^{Cb}$ |
| | 53d | $0.83 \pm 0.08^{A}$ | $0.59 \pm 0.03^{BC}$ | $0.63 \pm 0.05^{Bc}$ | $0.55 \pm 0.05^{C}$ | $0.54 \pm 0.04^{C}$ |
| Kidney | 39d | $0.64 \pm 0.05^{A}$ | $0.50 \pm 0.07^{Bc}$ | $0.52 \pm 0.06^{B}$ | $0.49 \pm 0.04^{Bc}$ | $0.39 \pm 0.05^{C}$ |
| | 53d | $0.59 \pm 0.08^{A}$ | $0.46 \pm 0.03^{Bc}$ | $0.50 \pm 0.03^{Ba}$ | $0.46 \pm 0.05^{Bc}$ | $0.36 \pm 0.05^{C}$ |

## 3.6 The Levels of Inhibition of · OH Capacity in Tissues

The results were showed in Table7. Compared with the control group, the levels of the tissues (liver, lung, spleen, kidney) inhibition of · OH capacity in the AP group and ABP groups were significantly superior ($P < 0.01$ or 0.05) (except the values of the liver of AP group and the spleen of ABP group Ⅰ, the spleen of AP group and the liver of ABP groups Ⅰ - Ⅱ on 39 days of age). The levels of the tissues (liver, lung, spleen, kidney) inhibition of · OH capacity in the AP group were higher than those of the ABP group Ⅰ (except the levels of the liver and kidney of AP group on 53 days of age). The values of the liver, spleen and kidney inhibition of · OH capacity in the ABP group Ⅲ were significantly superior ($P < 0.01$) than those of the AP group (except the levels of the kidney of ABP group Ⅲ on 39 days of age). The levels of inhibition of · OH capacity in the ABP groups were dose dependent, the optimum add dosage of ABP was found to be 1 000mg/kg.

Table 7 Results of inhibition of · OH capacity in tissues (U/mgprot) (n = 6)

| Tissues | Time | Control group | AP group | ABP group Ⅰ | ABP group Ⅱ | ABP group Ⅲ |
|---|---|---|---|---|---|---|
| Spleen | 39d | $397.80 \pm 12.09^{C}$ | $430.19 \pm 17.00^{BC}$ | $411.34 \pm 15.07^{Cb}$ | $460.30 \pm 23.84^{B}$ | $587.72 \pm 36.80^{A}$ |
| | 53d | $456.26 \pm 26.59^{D}$ | $529.18 \pm 28.28^{BC}$ | $494.12 \pm 26.54^{CD}$ | $539.37 \pm 22.47^{Bc}$ | $692.33 \pm 31.26^{A}$ |
| Liver | 39d | $897.22 \pm 51.73^{B}$ | $886.21 \pm 41.19^{B}$ | $719.36 \pm 31.26^{C}$ | $900.01 \pm 49.75^{B}$ | $1096.71 \pm 69.18^{A}$ |
| | 53d | $977.31 \pm 61.50^{D}$ | $1047.66 \pm 48.90^{Dc}$ | $1144.79 \pm 41.23^{C}$ | $1237.63 \pm 47.87^{Bc}$ | $1353.20 \pm 47.32^{A}$ |
| Lung | 39d | $557.23 \pm 56.04^{C}$ | $851.23 \pm 72.44^{AB}$ | $758.99 \pm 41.54^{B}$ | $856.67 \pm 57.86^{AB}$ | $886.46 \pm 81.14^{Ab}$ |
| | 53d | $611.23 \pm 29.27^{B}$ | $897.31 \pm 78.46^{A}$ | $883.24 \pm 62.12^{A}$ | $914.48 \pm 97.71^{A}$ | $939.85 \pm 77.97^{A}$ |
| Kidney | 39d | $843.79 \pm 37.53^{C}$ | $1189.30 \pm 74.83^{AB}$ | $1124.16 \pm 35.66^{B}$ | $1238.94 \pm 47.84^{Ab}$ | $1131.63 \pm 106.58^{B}$ |
| | 53d | $1177.31 \pm 66.75^{D}$ | $1269.04 \pm 44.13^{Cd}$ | $1334.42 \pm 36.93^{Cb}$ | $1423.35 \pm 41.84^{B}$ | $1540.07 \pm 43.43^{A}$ |

# 4 Discussion

Oxidative stress is defined as the imbalance between reactive species such as free radicals and oxidants and the antioxidant defenses (Bowling et al., 1993). It may be reflected in an elevated metabolic rate and possibly also increased production of reactive oxygen species (ROS), e. g. hydrogen peroxide ($H_2O_2$), hydroxyl radicals (· OH) and superoxide anion radicals (· $O_2^-$), which cause lipid peroxidation (Mclaren et al., 2000). The antioxidant defense system have enzyme promotion system and non-enzyme promotion system, called the total antioxidant capacity (T-AOC). Antioxidase includes superoxide dismutase (SOD), glutathione peroxidase (GSH-Px), catalase (CAT), nitric oxide (NO), etc.. Non-enzyme promotion system includes vitamin E, vitamin C, b-carotene, and glutathione, etc. (Droge, 2002). SOD is a metal-containing enzyme that catalyze the dismutation of $\cdot O_2^-$ into $O_2$ and $H_2O_2$ (Buettner, 2011). Subsequently, $H_2O_2$ is reduced to $H_2O$ by GSH-Px in the cytosol (Olsvik et al., 2005), or by CAT in the peroxisomes (Naziroglu, 2012)[1]. SOD, CAT and GSH-Px are easily induced by oxidative stress, and the activity levels of these enzymes have been used to quantify oxidative stress in cells (Van Der Oost et al., 2003).

Xia et al. (2012) reported thatdietary supplementation with ABP 300 mg/kg in lateolabrax japonica, the hepatopancreas, head kidney and gill CAT, SOD were significantly higher ($P<0.05$) than the control group on the 20 day. Tang et al. (2009) reported that an expressed fusion peptide bovine lactoferricin-lactoferrampin could enhance serum and liver GSH-Px, POD, SOD and T-AOC in piglets weaned at age 21d ($P<0.05$). Chen et al. (2010) reported that dietary supplementation with ABP 2 000 – 3 000mg/kg, the shrimps had a higher level in activities of PO, POD, AKP, LZM and T-AOC in serum ($P<0.05$). In our study (Table 2,3,4,5), addition of AP and ABP to diet in weaned piglets, the tissues (liver, lung, spleen, kidney) SOD, GSH-Px, CAT, T-AOC activities were markedly higher ($P<0.01$) than those of control group, and ABP groups were dose dependent. The optimum add dosage of ABP was found to be 1 000mg/kg, and was superior ($P<0.01$ or $0.05$) to the 400mg/kg AP group. Our study results were the same with other people, showed that ABP could improve the tissues antioxidant capacity, protect the cell from ROS damage in weaned piglets.

Although endogenous antioxidants may protect the cell from ROS damage, excess ROS can overwhelm the cellular defense mechanisms and function as lethal oxidants that can damage DNA as well as oxidize polyunsaturated fatty acids (PUFA), amino acids, and specific enzymes (Balaban et al., 2005)[2]. The peroxidation of PUFA is the main source of Malondialdehyde (MDA) (Winston and Di Giulio, 1991). Moreover, an increased MDA concentration further attenuated the activity of antioxidant enzymes (Patel et al., 2006). Hydroxyl radical ($\cdot$OH) that known to be the most biologically active free radical, is formed *in vivo* under hypoxic conditions (Michiels, 2004).

Su et al. (2013) reported that dietary supplementation with ABP (0.4%) in weaned piglets, the serum MDA content was significantly lower ($P<0.05$), the T-AOC, SOD was significantly higher ($P<0.05$) than that of control. Zhang et al. (2011) reported that ABP from Bacillus subtilis Y-6 had certain activities on scavenging free radicals and inhibiting lipid oxidation *in vitro*, and the effects were better when the ABP concentration was higher. Xue et al. (2012) reported that pBD-1 and pBD-2 showed radicals scavenging activities and reducing power, the antioxidative activities of them were stronger when the concentrations were higher within 0 ~ 256 μg/ml. In our study (Table. 6,7), addition of AP and ABP to diet in weaned piglets, the tissues (liver, lung, spleen, kidney) MDA content were markedly lower ($P<0.01$) than those of control group, the tissues inhibition of $\cdot$OH capacity were markedly higher($P<0.01$) than those of control group, and ABP groups were dose dependent. The optimum add dosage of ABP was found to be 1 000mg/kg, and was superior ($P<0.01$ or $0.05$) to the 400mg/kg AP group. Our study results were the same with other people, showed that ABP could scavenge free radicals and improve function in weaned piglets.

## 5 Conclusion

Under the experimental condition, dietary supplementation with different concentrations of ABP in weaned piglets, significantly enhanced the tissues (liver, lung, spleen, kidney) SOD, GSH-Px, CAT, T-AOC and inhibition of · OH capacity, decreased the content of MDA. With the increase of the adding dose, the effects were more obvious and were superior to the control group and the 400mg/kg AP group. The 1 000 mg/kg of feed group was the most obvious effect. It showed that ABP could effectively improve the activity of the antioxidant, protect cells from damage in weaned piglets.

## Acknowledgments

This work was financially supported by the Changjiang Scholars & Innovative Research Team of Ministry of Education of China Funds (Grant no. IRT0848).

## References

Balaban RS, Nemoto S, Finkel T. 2005. Mitochondria, oxidants, and aging. Cell, 120:483 – 495.

Bowling AC, Schulz JB, Brown RH, et al. 1993. Superoxide dismutase activity, oxidative damage, and mitochondrial energy metabolism in familial and sporadic amyotrophic lateral sclerosis. *Journal of Neurochemistry*, 61:2322 – 2325.

Buettner GR. 2011. Superoxide dismutase in redox biology: the roles of superoxide and hydrogen peroxide. *Anti-cancer agents in medicinal chemistry*, 11:341 – 346.

Chen HM, Leung KW, Thakur NN, et al. 2003. Distinguishing between different pathways of bilayer disruption by the related antimicrobial peptides cecropin B, B1 and B3. *European Journal of Biochemistry*, 270:911 – 920.

Chen B, Cao JM, Chen PJ, et al. 2010. Effects of antibacterial peptides of musca domestica on growth performance and immune-related indicators in Litopenaeus vannamei. *Journal of Fishery Sciences of China*, 17:258 – 266.

Dagan A, Efron L, Gaidukov L, et al. 2002. In vitro antiplasmodium effects of dermaseptin S4 derivatives. *Antimicrobial Agents and Chemotherapy*, 46:1059 – 1066.

Dale BA, Fredericks LP. 2005. Antimicrobial peptides in the oral environment: expression and function in health and disease. *Current Issues in Molecular Biology*, 7:119 – 133.

Dröge W. 2002. Free radicals in the physiological control of cell function. Physiological Reviews, 82:47 – 95.

Gordon YJ, Romanowski EG, Mcdermott AM. 2005. A review of antimicrobial peptides and their therapeutic potential as anti-infective drugs. *Current Eye Research*, 30:505 – 515.

Li YM, Xiang Q, Zhang QH, et al. 2012. Overview on the recent study of antimicrobial peptides: origins, functions, relative mechanisms and application. *Peptides*, 37:207 – 215.

Mclaren JS, Himanka MJ, Speakman JR. 2000. Effect of long-term cold exposure on antioxidant enzyme activities in a small mammal. *Free Radical Biology and Medicine*, 28:1279 – 1285.

Michiels C. 2004. Physiological and pathological responses to hypoxia. *The American journal of pathology*, 164:1875 – 1882.

Naziroglu M. 2012. Molecular role of catalase on oxidative stress-induced $Ca^{2+}$ signaling and TRP cation channel activation in nervous system. *Journal of Receptors and Signal Transduction*, 32:134 – 141.

Olsvik PA, Kristensen T, Waagbo R, et al. 2005. mRNA expression of antioxidant enzymes (SOD, CAT and GSH-Px) and lipid peroxidative stress in liver of Atlantic salmon (Salmo salar) exposed to hyperoxic water during smoltification. *Comparative Biochemistry and Physiology Part C: Toxicology & Pharmacology*, 141:314 – 323.

Patel S, Singh V, Kumar A, et al. 2006. Status of antioxidant defense system and expression of toxicant responsive genes in striatum of maneb-and paraquat-induced Parkinson's disease phenotype in mouse: mechanism of neurodegeneration. *Brain research*, 1081: 9 – 18.

Su K, Wang K, Sun LJ, et al. 2013. Effects of antimicrobial peptide on growth performance and health of weaning piglets. *Journal of Henan Agricultural Sciences*, 42:112 – 115, 127.

Tang ZR, Yin YL, Zhang YM, et al. 2009. Effects of dietary supplementation with an expressed fusion peptide bovine lactoferricin-lactoferrampin on performance, immune function and intestinal mucosal morphology in piglets weaned at age 21 d. British journal of nutrition, 101:998 – 1005.

Van Der Oost R, Beyer J, Vermeulen NP. 2003. Fish bioaccumulation and biomarkers in environmental risk assessment: a review. *Environmental Toxicology and Pharmacology*, 13: 57 – 149.

Wachinger M, Kleinschmidt A, Winder D, et al. 1998. Antimicrobial peptides melittin and cecropin inhibit replication of human immunodeficiency virus 1 by suppressing viral gene expression. *Journal of General Virology*, 79: 731 – 740.

Wakabayashi H, Uchida K, Yamauchi K, et al. 2000. Lactoferrin given in food facilitates dermatophytosis cure in guinea pig models. *Journal of Antimicrobial Chemotherapy*, 46: 595 – 602.

Winston GW, Di Giulio RT. 1991. Prooxidant and antioxidant mechanisms in aquatic organisms. *Aquatic Toxicology*, 19: 137 – 161.

Wu SD, Zhang FR, Huang ZM, et al. 2012. Effects of the antimicrobial peptide cecropin AD on performance and intestinal health in weaned piglets challenged with *Escherichia coli*. *Peptides*, 35: 225 – 230.

Xia Y, Yu DG, Yu EM, et al. 2012. Efects of dietary antimicrobial peptide and Bacillus subtilis on growth performance and non-specific immunity of Lateolabrax japonicu. *Freshwater Fisherie*, 42: 52 – 57.

Xue XF, Han FF, Gao YH, et al. 2012. *In vitro* detections of antimicrobial and antioxidant activities of porcine β-defensins. *Journal of Agricultural Biotechnology*, 20: 1291 – 1299.

Yu FS, Cornicelli MD, Kovach MA, et al. 2010. Flagellin stimulates protective lung mucosal immunity: role of cathelicidin-related antimicrobial peptide. *The Journal of Immunology*, 185: 1142 – 1149.

Zasloff, M. 2002. Antimicrobial peptides of multicellular organisms. Nature, 415: 389 – 395.

Zhang J, Ding HY, Qi XY, et al. 2011. Antioxidant activities in vitro of antimicrobial peptide produced by bacillus subtilis Y-6. *Journal of Nuclear Agricultural Sciences*, 25: 518 – 522.

Zhu LH, Zhao KL, Chen XL, et al. 2012. Impact of weaning and an antioxidant blend on intestinal barrier function and antioxidant status in pigs. *Journal of Animal Science*, 90: 2581 – 2589.

# Effects of antibacterial peptide on cellular immunity in weaned piglets

YUAN Wei, REN Zhihua, DENG Junliang*, DAN Qixiong, JIN Haitao, TIAN Chunlei, LIANG Zhen, and GAO Shuang

*Department of Veterinary Medicine, Sichuan Agricultural University, Ya'an* 625014, *China.*

**Abstract**: As far as the authors' knew, there was few report on the effects of antibacterial peptide on the cellular immunity in weaned piglets. The aim of this study was to evaluate the effects of dietary supplementation of antibacterial peptide (ABP) and astragalus polysaccharide (AP) on T lymphocyte cells proliferation, T cell subset of peripheral blood, lymphocyte cycle, and apoptosis of spleen in weaned piglets. A total of 90 piglets (Duroc × Landrace × Yorkshire) were randomly allotted to five groups, including the control group, the antibacterial peptide groups containing 250, 500 and 1 000 mg/kg antibacterial peptide, respectively, and the astragalus polysaccharide group with 400 mg/kg astragalus polysaccharide. The whole experimental period was 28 days. Six piglets in each treatment group were selected randomly for obtaining blood samples from the jugular vein for T lymphocyte cells proliferation and subpopulation analysis at 32, 39, 46 and 53 days of age. Six piglets were selected from each treatment group and euthanized for cell cycle and apoptosis of spleen at 39 and 53 days of age. Supplementation with ABP increased ($P < 0.01$) the $G_0/G_1$ phase in the cycle of spleen, and decreased ($P < 0.01$) the S phases, the $G_2 + M$ phases and the PI in the cycle of spleen. The percentages of apoptotic spleen cells were significantly decreased ($P < 0.01$). Supple-mental ABP increased the value of lymphocyte proliferation ($P < 0.01$). The percentage of $CD_3^+$, $CD_3^+CD_4^+$, $CD_3^+CD_8^+$ and $CD_4^+CD_8^+$ ratio of AP group and ABP groups were increased ($P < 0.05$ or 0.01). In conclusion, the results showed that ABP intake could improve the T-cell proliferation, the relative proportion of the subsets of T-cells, and the cycle of spleen cells, and decreased percentage of apoptotic cells. The cellular immune function was finally improved in weaned piglets. The optimum add dosages were found to be 500 mg/kg for 4-week adding and 1 000 mg/kg for 2-week adding, individually.

**Keywords**: antimicrobial peptide, apoptosis, lymphocyte cycle, lymphocyte proliferation, T cell subset, weaned piglets

## Introduction

Antimicrobial peptide (ABP), are bioactive substances which are extracted, separated and purified from a variety of plants, animals, human tissues and cells *in vivo* (Wang et al., 2004). They have a broad range of functions like anti-bacteria (Koczulla et al., 2003), anti-viral (Huang et al., 2013), anti-fungal (Rossignol et al., 2011), anti-tumor (Yan et al., 2012), anti-parasitic (Torrent et al., 2012), and enhance immune function (Yu et al., 2010).

In modern pig industry, most of piglets were weaned at 3 – 4 weeks of age, while the immune functions of the weaned piglets always matured at 7 weeks of age. During this special period, some viral disease (such as CSF,

* Corresponding author: Junliang Deng. E-mail address: dengjl213@126.com.

YUAN Wei and Ren Zhihua contributed equally to this work and should be considered co-first authors.

PRRS, etc.) could cause high morbidity and mortality in weaned piglets, resulting in significant economic loss. Therefore, how to improve the immunity of weaned piglets become an important problem.

Previous studies found that antibacterial peptide could improve the immune function of mice, chicken, rabbit, and piglet. Geng et al. (2011) reported that the treatment groups fed with the concentrations of 1.5, 7.5 and 15 g/Lantimicrobial peptides, the E-rosette ratios of the treatment groups of mice were significantly improved ($P < 0.05$). Hisex hens, Hy-Line Brown young roosters for egg, and rabbits were fed the basal diet supplemented with different concentrations of antibacterial peptide, the levels of IgG, IgM, IgA and alexin $C_3$ of the treatment groups were significantly improved ($P < 0.05$) (Lv et al., 2011; Liu et al., 2012; Guo et al., 2012). Wu et al. (2012) reported that the treatment group fed with 400 mg/kg antimicrobial peptide cecropin AD improved the levels of IgG ($P < 0.05$) and IgA ($P < 0.05$) in weaned piglets.

However, as far as the authors' knew, there was few report on the effects of antibacterial peptide on the cellular immunity inweaned piglets. In this study, we investigated the effects of different concentrations of antibacterial peptide on cellular immunity in weaned piglets.

# Materials and Methods

## Materials

All chemicals were of the highest grade of purity available. Antibacterial peptide (synthesized by the Styela antibacterial peptide and the defensin) was provided by Rota Bioengineering Co., Ltd, Sichuan, China. Astragalus polysaccharide (AP; net content of 65%) was bought from Centre Biology Co., Ltd, Beijing, China.

## Animals and Experimental Design

Piglets (Landrace × Yorkshire × Duroc; (21 ± 2) d of age) were purchased from Xin Qiao Agricultural science and technology development co., ltd, Chengdu, Sichuan, China. Piglets (average BW of (8.24 ± 0.67) kg) were acclimatized for 5 days pioneer to the use. Piglets were caged in elevated pens with wire flooring and fed with standard diet (Table 1) for weanling piglets (NRC 1998). Temperature (26 – 27℃) and relative humidity (65% – 70%) remained constant. Food and water were provided ad libitum during the acclimatisation period and through all parts of the study. All piglets used in this study were approved to be healthy. All study experimental manipulations were undertaken in accordance with the Institutional Guidelines for the Care and Use of Laboratory Animals.

Table 1 Ingredient composition of diets, as-fed basis[a]

| Ingredient | % |
|---|---|
| Corn | 53.15 |
| Soybean meal (44% CP) | 25.40 |
| Whey powder[b] | 6.00 |
| Fish meal (62.5% CP) | 4.50 |
| Soybean oil | 2.80 |
| Spray-dried porcine plasma (78% CP)[c] | 3.00 |
| Dicalcium phosphate | 2.0 |
| Limestone | 1.40 |
| Salt | 0.25 |
| L-Lysine HCl | 0.15 |

(continued table)

| Ingredient | % |
|---|---|
| DL-Methionine | 0.35 |
| Premix[d] | 1.00 |

[a]Calculated nutrient composition (as-fed basis):3 280 kcal of DE/kg,1.14% lysine,and 0.32% methionine. Analyzed nutrient composition (as-fed basis):21.1% CP,0.68% Ca,and 0.59% P. Corn was removed by 400 mg/kg, when 400 mg/kg astragalus polysaccharide was added. Corn was removed by 250,500 and 1 000 mg/kg, respectively, when 250,500 and 1 000 mg/kg antibacterial peptide were added.

[b]Whey powder was a product of Calva Product Inc. ,Acampo,CA.

[c]Spray-dried porcine plasma was a product of Merrick's Inc. ,Middleton,WI.

[d]Provided the following per kilogram of complete diet: vitamin A,2 250 IU; vitamin $D_3$,1 050 IU; vitamin E,16 IU; vitamin $K_3$, 0.9 mg; vitamin $B_{12}$,0.03 mg; riboflavin,4.5 mg; niacin,30 mg; pantothenic acid,25 mg; choline chloride,400 mg; folic acid,0.30 mg; thiamin,2 mg; pyridoxine,7 mg; biotin,0.20 mg; Zn,80 mg; Mn,22 mg; Fe,80 mg; Cu,225 mg; I,0.50 mg; and Se,0.25 mg.

Ninety weanling piglets were allotted to 5 treatments in a randomized complete block design for 28 d. Dietary treatments were as follows: (1) control; (2) control + 400 mg AP/kg; (3) control + 250 mg ABP/kg; (4) control + 500 mg ABP/kg; (5) control + 1 000 mg ABP/kg. Six piglets in each treatment group were selected randomly for obtaining blood samples from the jugular vein for T lymphocyte cells proliferation and subpopulation analysis at 32, 39, 46 and 53 days of age. At 39 and 53 days of age during the experiment, six piglets were selected from each treatment were euthanized for cell cycle and apoptosis of spleen by the flow cytometry method (Beckman Coulter Corp, Fullerton, CA).

## Lymphocyte Proliferation

Lymphocyte proliferation was measured as described by Fang (2012). Two milliliter peripheral blood were collected into 5 ml heparinized vacuum tubes (Becton Dickinson Vacutainer System; Franklin Lake, NJ), then added an equal volume of Hanks' solution (Thermo Scientific Hy Clone, Logan, UT) and layered on the surface of lymphocyte separation medium (density: (1.077 ± 0.001) g/ml; Jingyang Co. , Tianjin, China) carefully, after centrifugation at 3 000 × g for 20 min at room temperature. The mononuclear cells were collected and washed thrice with RPMI 1640 (Gibco BRL, Grand Island, NY) media without fetal bovine serum. The resulting pellet was re-suspended to $2 \times 10^6$ cells/ml with complete RPMI 1640 media for proliferation assay.

Suspensions of the mononuclear cells ($2 \times 10^6$/well) were incubated into 96-well culture plates, 100 μl per well, each sample seeded 6 wells. Then another 100 μl of ConA (Con A: 10 μg/ml; Sigma Chemical Co. , St. Louis, MO) was added into per well. The plates were incubated in a humid atmosphere of 5% $CO_2$ for 44 h at 37℃. Then, add 10 μl MTT [3-(4,5-dimethylthiazol-2-yl)-2,5-diphenyltetrazolium bromide; 5 mg/ml; Sigma Chemical Co. , St. Louis, MO] into each well, and the plates were re-incubated for 4 h. After the incubation period, 100 μl of DMSO (dimethyl sulfoxide; Sigma Chemical Co. , St. Louis, MO) was added into each well. The plates were shaken for 10 min to dissolve the precipitation completely. Finally, the plates were placed in an automated ELISA reader (MQX200; BioTek Instruments, Inc. , Winooski, VT) for measurement of absorbance at 570 nm. Stimulation Indices (SI) which indicated the lymphocyte proliferation activity was calculated as following:

SI = OD value of Con A stimulating cells/OD value of Con A-free cells

## T Cell Subsets

At 32, 39, 46 and 53 days of age during the experiment, six piglets of each treatment group were selected and blood samples were obtained by puncturing the vena cava. The blood of the percentage of $CD_3^+$, $CD_3^+CD_4^+$ and $CD_4^+CD_8^+$ T cells were determined by the flow cytometry method (Beckman Coulter Corp, Fullerton, CA), as described by Chen (2009).

One milliliter peripheral blood was collected into 5 ml heparinized vacuum tubes, an equal volume of PBS (0.01 M and pH 7.4) was added and layered on the surface of lymphocyte separation medium carefully, after centrifugation at 200 × g for 20 min at room temperature. The lymphocyte was collected and transferred to another centrifuge tube, and washed with PBS. The resulting pellet was re-suspended to $1 \times 10^6$ cells/ml with PBS. One milliliter cell suspension was transferred to another centrifuge tube and centrifuged at 200 × g for 5 min. The supernatant was discarded. At room temperature, the cells were stained with 10 μl mouse anti-pig $CD_3$-phyto-erythrin (Southern Biotechnology Associates, Birmingham, AL), mouse anti-pig $CD_4$-phyto-erythrin (Southern Biotechnology Associates, Birmingham, AL) and mouse anti-pig $CD_8$a-FITC (Southern Biotechnology Associates, Birmingham, AL) for 20 min respectively, and then washed with PBS. The supernatant was discarded. The cells were resuspended in 0.5 ml PBS and determined by flow cytometry.

## Cycle of Spleen

At 39 and 53 day of age, six piglets were euthanized in each group for the determination of the cell cycle stages in the spleen by FCM, as described by Wu (2012). Immediately after death, about 0.2 $cm^3$ spleen of each piglet was put on the ring glass plate (6 cm In diameter) containing 0.5 ml normal saline, and broken into a single-cell suspension with ophthalmic scissors, and then transferred to a centrifuge tube containing 1.5 ml normal saline. The single-cell suspension was filtered through a 300-mesh nylon screen. The cells were washed twice and diluted to $1.0 \times 10^6$ cells/ml with PBS. One milliliter of the solution was transferred to another centrifuge at 200 × g for 5 min. The supernatant was discarded. Then 1 ml PI (5 μl/ml propidium iodide, 0.5% Triton X-100, 0.5% RNase, PBS) was added and stained for 20 min at room temperature, and washed with PBS. The supernatant was discarded. The cells were resuspended in 0.5 ml PBS, and the cell phases were analyzed by flow cytometry.

The proliferating index was calculated as: Proliferrating index (PI) $= \dfrac{S + (G_2 + M)}{G_0/G_1 + S + (G_2 + M)}$

## Annexin V Apoptosis Detection by Flow Cytometry

At 39 and 53 days of age, six piglets were euthanized in each group for the determination of percentage of apoptotic cells in the spleen, as described by Chen (2013). The cells were consistent with lymphocyte cycle of spleen. one hundred microliter cells suspension were transferred to a centrifuge tube, 5 μl V-FITC (BD Pharmingen, USA) and 5 μl PI (5 μl/ml propidium iodide, 0.5% Triton X-100, 0.5% RNase, PBS) were added and stained at room temperature for 15 min in the dark. Then 400 μl 1 × binding buffer was added to each centrifuge tube, and assayed by flow cytometry within 1 h.

## Data Analysis

The results were shown as means ± standard deviation ($M \pm SD$). Statistical analysis was performed using one-way analysis of variance (ANOVA) test of SPSS (InternationalBusiness Machines Corporation, Armonk, USA) 19.0 software. $P < 0.05$ or 0.01 were considered as significantly different.

# Results

## Cell cycle of Spleen

The results were showed in Table 2. At 39 and 53 days of age, the $G_0/G_1$ phases of AP group and ABP groups were significantly lower ($P < 0.01$) than that of control group, and the S phases, the $G_2 + M$ phases and the proliferation index of AP group and ABP groups were significantly greater ($P < 0.01$) than those of control group. At 39 days of age, the values of the ABP group Ⅲ were superior ($P < 0.01$) to those of AP group (except the value of $G_2 + M$). At 53 days of age, the four values of all of the ABP treatment group were significantly better ($P < 0.01$) than those of AP group. But the value of PI of the ABP group Ⅲ was significantly lower than those of another two ABP groups ($P < 0.01$) on day 53.

Table 2　Changes of cycle spleen cells, %

| Time | Items | Control group | AP group | ABP group Ⅰ | ABP group Ⅱ | ABP group Ⅲ |
|---|---|---|---|---|---|---|
| $G_0/G_1$ | 39 | $93.35 \pm 0.38^A$ | $87.17 \pm 0.33^c$ | $90.48 \pm 0.62^B$ | $87.94 \pm 0.50^C$ | $84.74 \pm 0.58^D$ |
| | 53 | $94.16 \pm 0.60^A$ | $90.06 \pm 0.61^B$ | $82.26 \pm 0.36^D$ | $81.81 \pm 0.28^D$ | $84.57 \pm 0.53^C$ |
| S | 39 | $4.00 \pm 0.38^E$ | $6.02 \pm 0.42^C$ | $5.20 \pm 0.49^D$ | $7.87 \pm 0.22^B$ | $9.22 \pm 0.63^A$ |
| | 53 | $3.83 \pm 0.40^D$ | $6.21 \pm 0.33^C$ | $9.48 \pm 0.28^a$ | $10.09 \pm 0.29^A$ | $8.41 \pm 0.66^B$ |
| $G_2 + M$ | 39 | $2.65 \pm 0.13^C$ | $6.81 \pm 0.31^A$ | $4.32 \pm 0.70^B$ | $4.19 \pm 0.69^B$ | $6.05 \pm 0.50^a$ |
| | 53 | $2.00 \pm 0.76^D$ | $3.73 \pm 0.39^C$ | $8.26 \pm 0.36^A$ | $8.10 \pm 0.14^A$ | $7.02 \pm 0.45^B$ |
| PI | 39 | $6.65 \pm 0.38^D$ | $12.83 \pm 0.33^B$ | $9.52 \pm 0.62^C$ | $12.05 \pm 0.50^b$ | $15.26 \pm 0.58^A$ |
| | 53 | $5.83 \pm 0.62^D$ | $9.94 \pm 0.61^C$ | $17.74 \pm 0.36^A$ | $18.19 \pm 0.28^A$ | $15.43 \pm 0.53^B$ |

abc means with various letters are very significant different ($P < 0.01$), and with the same letters but in different case are significant different ($P < 0.05$). The following table 3, table 4 and table 5 are the same

## Apoptosis of Spleen

As we could see from Table 3, the percentage of apoptotic spleen cells of AP group and ABP groups were markedly lower ($P < 0.01$) than that of control group at 39 and 53 days of age. At 39 days of age, except the percentage of apoptotic spleen cells of the ABP group Ⅲ were superior ($P < 0.01$) to that of AP group, there was no differences ($P > 0.05$) between the percentage of apoptotic spleen cells of AP group and another two ABP treatment groups. And the percentage of apoptotic spleen cells of ABP groups were significantly lower ($P < 0.01$) than that of AP group at 53 days of age. But the percentage of apoptotic spleen cells of the ABP group Ⅲ was significantly greater than those of another two ABP groups ($P < 0.01$) on day 53.

Table 3　Changes of apoptotic spleen cells, %

| Time | Control group | AP group | ABP group Ⅰ | ABP group Ⅱ | ABP group Ⅲ |
|---|---|---|---|---|---|
| 39 | $21.83 \pm 0.12^A$ | $18.01 \pm 0.17^D$ | $20.45 \pm 0.26^B$ | $18.41 \pm 0.25^C$ | $16.16 \pm 0.14^E$ |
| 53 | $23.13 \pm 0.16^A$ | $17.23 \pm 0.17^B$ | $14.50 \pm 0.24^D$ | $13.97 \pm 0.26^E$ | $15.68 \pm 0.30^C$ |

## Peripheral Lymphocyte Proliferation Assay

The results were showed in Table 4. During the whole experimental period, the SI values of AP group and ABP groups were markedly greater ($P < 0.01$) than that of control group, and the SI values of ABP treatment groups were dose dependent. And the levels of SI of all three ABP treatment groups were markedly greater ($P<0.01$) than those of AP group (except the valuse of ABP group Ⅰ on 32 and 39 days of age). The optimum add dosage of ABP was found to be 1 000mg/kg.

Table 4 Changes of the stimulation index of peripheral blood T-cells

| Time | Control group | AP group | ABP group Ⅰ | ABP group Ⅱ | ABP group Ⅲ |
|---|---|---|---|---|---|
| 32 | $1.23\pm0.06^{D}$ | $1.35\pm0.04^{C}$ | $1.31\pm0.07^{C}$ | $1.51\pm0.03^{B}$ | $1.62\pm0.04^{A}$ |
| 39 | $1.25\pm0.05^{D}$ | $1.43\pm0.04^{C}$ | $1.40\pm0.07^{C}$ | $1.62\pm0.03^{B}$ | $1.73\pm0.05^{A}$ |
| 46 | $1.30\pm0.09^{D}$ | $1.48\pm0.05^{c}$ | $1.54\pm0.01^{C}$ | $1.65\pm0.04^{B}$ | $1.81\pm0.05^{A}$ |
| 53 | $1.27\pm0.04^{E}$ | $1.51\pm0.02^{D}$ | $1.58\pm0.03^{C}$ | $1.71\pm0.06^{B}$ | $1.88\pm0.06^{A}$ |

## Lymphocyte Subpopulation

The results were showed in Table5. During the whole experimental period, the percentages of $CD_3^+$, $CD_3^+CD_4^+$, $CD_3^+CD_8^+$, and $CD_4^+CD_8^+$ ratios of AP group and ABP groups were significantly greater ($P<0.01$) than those of control group, and the values of ABP treatment groups were dose dependent. Also, among the whole experimental period, the percentages of $CD_3^+$, $CD_3^+CD_4^+$, $CD_3^+CD_8^+$, and $CD_4^+CD_8^+$ ratios of all three ABP treatment groups were markedly greater ($P<0.01$) than those of AP group (except the valuse of ABP group Ⅰ on 32 and 39 days of age). The optimum add dosage of ABP was found to be 1 000 mg/kg.

Table 5 Changes of peripheral blood T-cell subsets in piglets

| Items | Time | Control group | AP group | ABP group Ⅰ | ABP group Ⅱ | ABP group Ⅲ |
|---|---|---|---|---|---|---|
| $CD_3^+$ (%) | 32 | $59.27\pm0.98^{Ba}$ | $61.71\pm2.21^{A}$ | $60.28\pm1.96^{AB}$ | $61.95\pm1.95^{A}$ | $62.40\pm2.03^{A}$ |
| | 39 | $60.22\pm1.78^{D}$ | $70.86\pm1.88^{BC}$ | $68.55\pm1.55^{c}$ | $71.70\pm1.60^{Ba}$ | $73.85\pm2.00^{A}$ |
| | 46 | $59.21\pm2.04^{B}$ | $71.74\pm2.82^{A}$ | $71.77\pm2.85^{A}$ | $72.48\pm2.31^{A}$ | $74.07\pm2.23^{A}$ |
| | 53 | $61.25\pm2.57^{B}$ | $71.16\pm1.98^{A}$ | $71.66\pm2.27^{A}$ | $73.49\pm2.24^{A}$ | $72.25\pm2.01^{A}$ |
| $CD_3^+CD_4^+$ (%) | 32 | $27.86\pm1.61^{B}$ | $31.64\pm1.77^{A}$ | $29.13\pm1.37^{B}$ | $31.85\pm1.56^{A}$ | $32.31\pm1.4^{A}$ |
| | 39 | $26.61\pm1.32^{D}$ | $32.86\pm1.97^{B}$ | $29.50\pm1.47^{C}$ | $34.37\pm1.56^{Ba}$ | $36.58\pm1.44^{A}$ |
| | 46 | $25.71\pm1.61^{C}$ | $31.92\pm2.08^{B}$ | $31.60\pm1.61^{B}$ | $32.88\pm1.48^{B}$ | $35.96\pm2.03^{A}$ |
| | 53 | $25.64\pm1.37^{C}$ | $29.81\pm1.75^{B}$ | $31.18\pm0.87^{AB}$ | $32.54\pm1.18^{A}$ | $32.25\pm1.14^{A}$ |
| $CD_3^+CD_8^+$ (%) | 32 | $42.84\pm1.20^{AB}$ | $44.03\pm1.17^{A}$ | $42.15\pm0.92^{Ba}$ | $43.19\pm1.32^{AB}$ | $42.07\pm1.14^{Ba}$ |
| | 39 | $41.66\pm1.19^{Bc}$ | $43.31\pm1.18^{A}$ | $40.10\pm1.39^{C}$ | $42.51\pm1.10^{AB}$ | $43.68\pm1.07^{A}$ |
| | 46 | $41.42\pm1.31^{B}$ | $42.62\pm1.38^{B}$ | $41.70\pm1.10^{B}$ | $42.73\pm1.26^{B}$ | $45.56\pm1.29^{A}$ |
| | 53 | $40.89\pm1.47^{C}$ | $41.29\pm0.87^{BCa}$ | $41.51\pm1.59^{BCa}$ | $42.71\pm1.43^{ABc}$ | $43.28\pm1.57^{A}$ |
| $CD_4^+/CD_8^+$ | 32 | $0.65\pm0.04^{D}$ | $0.72\pm0.04^{BC}$ | $0.69\pm0.03^{Cd}$ | $0.75\pm0.03^{AB}$ | $0.77\pm0.03^{A}$ |
| | 39 | $0.64\pm0.02^{D}$ | $0.78\pm0.02^{Bc}$ | $0.74\pm0.03^{C}$ | $0.81\pm0.04^{AB}$ | $0.84\pm0.03^{A}$ |
| | 46 | $0.62\pm0.03^{B}$ | $0.75\pm0.02^{A}$ | $0.76\pm0.05^{A}$ | $0.77\pm0.04^{A}$ | $0.79\pm0.04^{A}$ |
| | 53 | $0.64\pm0.02^{C}$ | $0.72\pm0.03^{Ba}$ | $0.75\pm0.03^{A}$ | $0.76\pm0.04^{A}$ | $0.75\pm0.03^{A}$ |

# Discussion

There are four major phases of the eukaryotic cell cycle: theG$_1$ phase before DNA replication, the periods of

DNA synthesis (S phase), the $G_2$ phase before cell division, and the cell division (M phase) (Pines et al., 1995). In our study, addition of AP and ABP to diet, the $G_0/G_1$ phase was markedly decreased in the cycle of spleen and the S phases, the $G_2 + M$ phases and the PI were significantly increased in the cycle of spleen. The results showed that the lymphocyte proliferation were promoted by AP and ABP intake. ABP showed better effect than AP. From the results we could concluded that antibacterial peptide increased the number of lymphocytes of the spleen, and enhanced the celluar immunity function of the body futher. The mechanism of antibacterial peptide on lymphocyte proliferation has not been clear, and this need further studies to confirme.

Apoptosis, is a highly regulated process used to eliminate dysplastic or damaged cells from multicellular organisms (King et al., 1995; Bortner et al., 2004). In our study, addition of AP and ABP to diet, the percentages of apoptotic spleen cells were significantlydecreased. ABP showed better effect than AP. Our results suggested that antibacterial peptide could reduce apoptosis of spleen cells. Antibacterial peptide could improve the values of GSH-Px, SOD, T-AOC, CAT and OH of spleen, and decreased the values of MDA of spleen (data were not showed in this paper). We supposed that through improving the antioxidant function of spleen, ABP reduced the apoptosis of spleen cells.

Lymphocyte proliferation is an indicator reflecting the state of cellular immunity. T and B lymphocytes play an important role in enhancing immune function of organism (Minato et al., 2004). According to Wang et al. (2007) reported, after the broilers oral with ABP at 300 and 600 mg/kg, the peripheral blood T lymphocytes proliferation were promoted 20.43% ($P > 0.05$) and 22.94% ($P > 0.05$), respectively, and the effects of ABP on broilers' T lymphocytes proliferation were significantly promoted 60.95% ($P < 0.05$) and 68.96% ($P < 0.05$) in vitro. Yang et al. (2009) reported that the percentage of $ANAE^+$ of immune organs T lymphocytes was significantly increased ($P < 0.05$) at 7 days of age in chicken. In our study, addition of AP and ABP to diet, the value of lymphocyte proliferation were markedly higher, and ABP groups were dose dependent. ABP showed better effect than AP. Our study was the same with their studies.

T-cells were an indicator reflecting the cellular immunity in animals. The percentages of T-cell subsets were important indicators which represented the composition of matured T-cells in the body. The biological function of the T-cells were dependented on the composition of matured T-cells. The cellular immune function of the body were also dependented on the composition of matured T-cells. As the surface marker of matured T-cells, the $CD_3^+$ molecules reflected the matured T-cell population. Similar to other mammalian species, the matured T-cells of piglets are further classified according to the presence of $CD_4^+$ and $CD_8^+$ proteins (Janeway et al., 2001). In piglets, most $CD_4^+$ T cells are helper/inflammatory T cells responding to exogenous antigen in association with major histocompatibility complex (MHC) class Ⅱ molecules. $CD_8^+$ T cells respond to endogenous antigen in association with MHC class I molecules and generally function as cytotoxic T cells (Chan et al; 1988). The $CD_4^+CD_8^+$ ratio has been used as an end point in assessing the state of an individual's immune system. Jin et al. (2009) reported that by means of gavage in a daily of 0.2 ml at a dosage of 7.5 g/L antimicrobial bovine neutrophil, the percentage of $CD_3^+$, $CD_4^+$ and $CD_8^+$ T cells were significantly ($P < 0.05$) improved in mice. Geng et al. (2011) reported that gavage at a daily dose of 0.2 ml with the concentrations of 15 g/L antimicrobial Peptides from duck leukocytes, the percentage of $CD_3^+$, $CD_4^+$ and the $CD_4^+CD_8^+$ ratio were significantly ($P < 0.05$) improved. In our study, addition of AP and ABP to diet, the percentage of $CD_3^+$, $CD_3^+CD_4^+$, $CD_3^+CD_8^+$ and $CD_4^+CD_8^+$ ratio of AP group and ABP groups were increased. ABP showed better effect than AP. The results of our study were consistent with their studies.

According to the results of our study, ABP and AP intake could improve the T-cell proliferation, the relative proportion of the subsets of T-cells, and the cycle of spleen cells, and decreased percentage of apoptotic cells. The cellular immune function was finally improved in weaned piglets, ABP showed better effect than AP. However, the value of PI of thechanges of cycle spleen cells and the percentage of apoptotic spleen cells in the ABP group Ⅲ were significantly worse than those of another two ABP groups ( $P < 0.01$ ) on day 53. These suggested that high dose of ABP (1 000 mg/kg) shouldn't be long time used. So, the optimum add dosages were found to be 500 mg/kg for 4-week adding and 1 000 mg/kg for 2-week adding, individually.

## Conclusion

This study showed that ABP intake could improve the T-cell proliferation, the relative proportion of the subsets of T-cells, and the cycle of spleen cells, and decreased percentage of apoptotic cells. The cellular immune function was finally improved in weaned piglets. The optimum add dosages were found to be 500 mg/kg for 4-week adding and 1 000 mg/kg for 2-week adding, individually.

## Acknowledgments

The present work was supported by the Changjiang Scholars & Innovative Research Team of Ministry of Education of China Funds (Grant no. IRTO848).

## References

Bortner CD and Cidlowski JA. 2004. The role of apoptotic volume decrease and ionic homeostasis in the activation and repression of apoptosis. *Pflugers Archiv European Journal of Physiology*, 448: 313 – 318.

Chan MM, Chen CL, Ager LL, et al. 1988. Identification of the avian homologues of mammalian CD4 and CD8 antigens. *Journal of Immunology*, 140: 2133 – 2138.

Chen KJ, Shu G, Peng X, et al. 2013. Protective role of sodium selenite on histopathological lesions, decreased T-cell subsets and increased apoptosis of thymus in broilers intoxicated with aflatoxin B1. *Food and Chemical Toxicology*, 59: 446 – 454.

Chen T, Cui Y, Bai CM, et al. 2009. Decreased percentages of the peripheral blood T-cell subsets and the serum IL-2 contents in chickens fed on diets excess in fluorine. *Biological Trace Element Research*, 132: 122 – 128.

Fan YP, Hu YL, Wang DY, et al. 2012. Effects of Astragalus polysaccharide liposome on lymphocyte proliferation *in vitro* and adjuvanticity *in vivo*. *Carbohydrate Polymers*, 88: 68 – 74.

Geng J, Wang YC and Chen LY. 2011. Effects of antimicrobial peptides from duck leukocytes on immunity of mice. *Journal of Henan Agricultural Science (China)*, 40: 141 – 145.

Guo ZQ, Yang FZ, Lei M, et al. 2012. Effects of Antibacterial peptides on intestinal mucosal morphology, caecal microflora and immune function of meat rabbits. *Chinese Journal of Animal Nutrition*, 24: 1778 – 1784.

Huang ZJ, Kingsolver MB, Avadhanula V, et al. 2013. An antiviral role for antimicrobial peptides during the arthropod response to alphavirus replication. *Journal of Virology*, 87: 4272 – 4280.

Janeway CA, Travers P, Walport M, et al. 1999. Immunobiology: The Immune System in Health and Disease, fourth ed. Elsevier/Garland, London, pp. 363 – 415.

Jin Y, Liu HM, Geng J, et al. 2009. Effect of antimicrobial bovine neutrophil extract on immune functions in mice. *Journal of Henan Agricultural University*, 43: 630 – 633.

King KL and Cidlowski JA. 1995. Cell cycle and apoptosis: common pathways to life and death. *Journal of Cellular Biochemistry*, 58: 175 – 180.

Koczulla AR and Bals R. 2003. Antimicrobial peptides. *Drugs*, 63: 389 – 406.

Liu LR, Hua J, Wang XX, et al. 2012. Antimicrobial peptides: effects on blood immune indices and intestinal microflora of young roosters for egg production. *Chinese Journal of Animal Nutrition (China)*, 24: 1812 – 1818.

Lv ZZ. , Yuan XX, Cai ZW, et al. 2011. Effects of antimicrobial peptides on serum immune indices and IL-2 mRNA expression in spleen of laying hens. *Chinese Journal of Animal Nutrition*, 23:2183 – 2189.

Minato KI, Kawakami S, Nomura K, et al. 2004. An exo β-1, 3-glucanase synthesized de novo degrades lentinan during storage of lentinule edodes and diminishes immunomodulating activity of the mushroom. *Carbohydrate Polymers*, 56:279 – 286.

Pines J. 1995. Cyclins, CDKs and cancer. In: Biology Sicence, 6:63 – 72.

Rossignol T, Kelly B, Dobson C, et al. 2011. Endocytosis-mediated vacuolar accumulation of the human ApoE apolipoprotein-derived ApoEdpL-W antimicrobial peptide contributes to its antifungal activity in candida albicans. *Antimicrobial Agents and Chemotherapy*, 55:4670 – 4681.

Torrent M, Pulido D, Rivas L, et al. 2012. Antimicrobial peptide action on parasites. *Current Drug Targets*, 13:1138 – 1147.

Wang J and Li TS. 2007. The Effects of antibacterial peptides on peripheral lymphocyte transformation in broilers. *Acta Ecologiae Animalis Domastic*, 28:45 – 48.

Wang Z and Wang G S. 2004. APD: the antimicrobial peptide database. *Nucleic Acids Research*, 32: D590 – D592.

Wu BY, Cui HM, Peng X, et al. 2012. Effect of methionine deficiency on the thymus and the subsets and proliferation of peripheral blood T-Cell, and serum IL-2 contents in broilers. *Biological Trace Element Research*, 11:1009 – 1019.

Wu SD, Zhang FR, Huang ZM, et al. 2012. Effects of the antimicrobial peptide cecropin AD on performance and intestinal health in weaned piglets challenged with *Escherichia coli*. *Peptides*, 35:225 – 230.

Yan JX, Wang KR, Chen R, et al. 2012. Membrane active antitumor activity of NK-18, a mammalian NK-lysin-derived cationic antimicrobial peptide. *Biochimie*, 94:184 – 191.

Yu FS, Cornicelli MD, Kovach MA, et al. 2010. Flagellin stimulates protective lung mucosal immunity: role of cathelicidin-related antimicrobial peptide. *Journal of Immunology*, 185:1142 – 1149.

Yang YR, Liang HD and Wei HL. 2009. The preliminary study of antimicrobial peptides extracted from african ostrich skin on the immune organs indexes and the number of T lymphocytes in immune organs of chickens. *Chinese Agricultural Science Bulletin*, 25:46 – 48.

# Effects of antibacterial peptide on humoral immunity in weaned piglets

JIN Haitao, YUAN Wei, REN Zhihua, DENG Junliang*, ZUO Zhicai, WANG Ya, DAN Qixiong, JIN Haitao, TIAN Chunlei, LIANG Zhen, and GAO Shuang

*Department of Veterinary Medicine, Sichuan Agricultural University, Ya'an 625014, China*

**Abstract**: As far as the authors' knew, there was few report on the effects of composite antibacterial peptide on humoral immunity in weaned piglets. The aim of this study was to investigate the effects of dietary antibacterial peptide (ABP) supplementation on the serum IgG, IgM, IgA, CSF-Ab, and $CH_{50}$ levels in piglets. Ninety weaned piglets (Landrace × Yorkshire × Duroc; (21 ± 2) d of age; average weight of (8.24 ± 0.67) kg) were randomly allotted to five groups, including the control group, the antibacterial peptide groups containing 250, 500 and 1 000 mg/kg antibacterial peptide, respectively, and the astragalus polysaccharide group with 400 mg/kg astragalus polysaccharide. All Piglets were vaccinated with CSF vaccine at 21days of age. After 5 days adaptation to dietary daidzein, the whole experimental period was 28 days. Blood samples were collected at 32, 39, 46, and 53 days of age. The results showed that ABP as dietary additive increased the levels of CSF-Ab, $CH_{50}$, immunoglobulins, and the effects of ABP supplementation were dose-dependent. Piglets fed with 250 mg/kg ABP had significantly higher ($P<0.01$ or 0.05) the level of IgM at 39, 53 days of age, and the value of $CH_{50}$ among the whole experimental period. Supplementation with 500 mg/kg markedly increased ($P<0.01$ or 0.05) the value of CSF-Ab on 39, 46 days, the levels of IgG, IgA on 39, 46, 53days, and the values of IgM, $CH_{50}$ among the whole experimental period. High dosage (1 000 mg/kg) of ABP supplementation significantly improved ($P<0.01$ or 0.05) the levels of IgG, IgA at 39, 46, 53 days of age, and the values of CSF-Ab, IgM, $CH_{50}$ among the whole experimental period. Supplementing weaned piglets diets with ABP enhanced the humoral immune responses of weaned piglets by improving the levels of CSF-Ab, $CH_{50}$, and immunoglobulins. The optimum add dosage of ABP was found to be 1 000 mg/kg for 4-week adding.

**Keywords**: antibacterial peptide, classical swine fever antibody, total complement, immunoglobulin, weaned piglets

## Introduction

Previously, piglets are weaned occurs over several weeks or months. However, in modern pig industry, piglets are weaned early between 3 and 4 weeks of age to prevent sow-originated infectious diseases and maximize the whole herd production (Kim et al., 2004; Dong and Pluske, 2007). This practice presents a tremendous challenge to neonatal piglets, whose immune functions of the weaned piglets always matured at 7 weeks of age (Yang and Schultz, 1986). During this special period, some viral disease (such as CSF, PRRS, etc.) can cause high morbidity and mortality in weaned piglets, resulting in significant economic loss. Antibiotics were frequently used for the prophylaxis of infections in past decades (Bosi et al., 2011). However, there has been increasing pressure on the livestock industry to decrease or discontinue these additions because of the potential development of antibiotic resistance (Davis et al., 2004).

* Corresponding author: Junliang Deng. E-mail address: dengjl213@126.com.

JIN Haitao, YUAN Wei, and REN Zhihua contributed equally to this work and should be considered co-first authors.

Therefore, how to improve the immunity of weaned piglets become an important problem.

Antimicrobial peptide (ABP), are bioactive substances which are extracted, separated and purified from a variety of plants, animals, human tissues and cells in vivo (Wang et al., 2004). They have a broad range of functions like anti-bacteria (Koczulla et al., 2003), anti-viral (Huang et al., 2013), anti-fungal (Rossignol et al., 2011), anti-tumor (Yan et al., 2012), anti-parasitic (Torrent et al., 2012), and enhance immune function (Yu et al., 2010).

Previous studies found that antibacterial peptide could improve theimmune function of chicken, rabbit, and piglet. Hy-Line Brown young roosters for egg were fed the basal diet supplemented with 200, 250, 300 and 350 mg/kg antimicrobial peptide, respectively. After 42 days, the IgA contents of the treatment groups with 300 and 350 mg/kg antimicrobial peptide were significantly higher ($P < 0.05$) than that of the control group, and the IgG contents of the treatment groups were markedly higher ($P < 0.05$) than that of the control group (Liu et al., 2012). The New Zealand rabbits were fed basal diet supplemented with 150, 200 and 250 mg/kg antimicrobial peptide, respectively. Eight weeks later, the IgM contents of the the treatment group fed with 200 mg/kg antimicrobial peptide were significantly increased ($P < 0.05$) (Guo et al., 2012). Wu et al. (2012) reported that the treatment group fed with 400 mg/kg antimicrobial peptide cecropin AD markedly improved ($P < 0.05$) the levels of IgG and IgA in serum in weaned piglets which were challenged with *Escherichia coli*.

However, as far as the authors' knew, there was few report on the effects of composite antibacterial peptide on humoral immunity in weaned piglets. In this study, we investigated the effects of different concentrations of composite antibacterial peptide on the serum IgG, IgM, IgA, classical swine fever antibody (CSF-Ab), and total complement activity ($CH_{50}$) levels in weaned piglets.

# Materials and Methods

## Materials

All chemicals were of the highest grade of purity available. Antibacterial peptide (synthesized by the Styela antibacterial peptide and the defensin) was provided by Rota Bioengineering Co., Ltd, Sichuan, China. Astragalus polysaccharide (AP; net content of 65%) was bought from Centre Biology Co., Ltd, Beijing, China. CSF vaccine was purchased from Shanghai Haili Biological Technology Co., Ltd, Shanghai, China.

## Animals and Experimental Design

Piglets (Landrace × Yorkshire × Duroc; (21 ± 2) d of age) were purchased from Xin Qiao Agricultural science and technology development co., ltd, Chengdu, Sichuan, China. All Piglets were vaccinated with CSF vaccine at 21days of age. Piglets were acclimatized for 5 days pioneer to the use. Piglets (average BW of (8.24 ± 0.67) kg) were caged in elevated pens with wire flooring and fed with standard diet (Table 1) for weanling piglets (NRC 1998). Temperature (26 – 27℃) and relative humidity (65% – 70%) remained constant. Food and water were provided ad libitum during the acclimatisation period and through all parts of the study. All piglets used in this study were approved to be healthy. All study experimental manipulations were undertaken in accordance with the Institutional Guidelines for the Care and Use of Laboratory Animals.

Ninety weanling piglets were allotted to 5 treatments in a randomized complete block design for 28 d. Dietary treatments were as follows: (1) control; (2) control + 400 mg AP/kg; (3) control + 250 mg ABP/kg; (4) control + 500 mg ABP/kg; (5) control + 1 000 mg ABP/kg. Six piglets from each group were selected randomly to take blood samples respectively at 32, 39, 46, and 53 days of age.

Table 1 Ingredient composition of diets, as-fed basis[a]

| Ingredient | % |
|---|---|
| Corn | 53.15 |
| Soybean meal (44% CP) | 25.40 |
| Whey powder[b] | 6.00 |
| Fish meal (62.5% CP) | 4.50 |
| Soybean oil | 2.80 |
| Spray-dried porcine plasma (78% CP)[c] | 3.00 |
| Dicalcium phosphate | 2.0 |
| Limestone | 1.40 |
| Salt | 0.25 |
| L-Lysine HCl | 0.15 |
| DL-Methionine | 0.35 |
| Premix[d] | 1.00 |

[a]Calculated nutrient composition (as-fed basis): 3 280 kcal of DE/kg, 1.14% lysine, and 0.32% methionine. Analyzed nutrient composition (as-fed basis): 21.1% CP, 0.68% Ca, and 0.59% P. Corn was removed by 400 mg/kg, when 400 mg/kg astragalus polysaccharide was added. Corn was removed by 250, 500 and 1 000 mg/kg, respectively, when 250, 500 and 1 000 mg/kg antibacterial peptide were added.

[b]Whey powder was a product of Calva Product Inc., Acampo, CA.

[c]Spray-dried porcine plasma was a product of Merrick's Inc., Middleton, WI.

[d]Provided the following per kilogram of complete diet: vitamin A, 2 250 IU; vitamin $D_3$, 1 050 IU; vitamin E, 16 IU; vitamin $K_3$, 0.9 mg; vitamin $B_{12}$, 0.03 mg; riboflavin, 4.5 mg; niacin, 30 mg; pantothenic acid, 25 mg; choline chloride, 400 mg; folic acid, 0.30 mg; thiamin, 2 mg; pyridoxine, 7 mg; biotin, 0.20 mg; Zn, 80 mg; Mn, 22 mg; Fe, 80 mg; Cu, 225 mg; I, 0.50 mg; and Se, 0.25 mg

## Sampling Procedureand Determination of Serum Parameters

Blood samples were collected into microcentrifuge tubes by orbital vein puncture. Serum was obtained after blood centrifugation at 4 000 rpm at 4℃ for 20 min and stored at -20℃. Total IgG, IgM, IgA, CSF-Ab, and $CH_{50}$ were determined in serum, using enzyme-linked immunosorbent assay (ELISA) kits purchased from America RD Biosciences Co., Ltd. All operations were in accordance with the instructions of the kits.

## Data Analysis

The results were shown as means ± standard deviation (M ± SD). Statistical analysis was performed using one-way analysis of variance (ANOVA) test of SPSS 19.0 software. $P < 0.05$ or 0.01 were considered as significantly different.

# Results

## Effects on CSF-Ab of Piglets

As we could see from Table 2, during the whole experiment period, the levels of CSF-Ab of the treatment groups were higher than that of the control group, and the levels of CSF-Ab of the ABP groups were higher than that of AP group. The level of CSF-Ab of theABP group Ⅱ was markedly higher ($P < 0.05$) than that of the control group at 39 and 53 days of age. The level of CSF-Ab of the ABP group Ⅲ was significantly higher ($P < 0.01$ or 0.05) than that of the control group among the whole experiment period, and superior ($P < 0.05$) to that of the AP group at 32 and 53 days of age. The levels of CSF-Ab of the ABP were dose dependent, and the optimum add dosage of ABP was found to be 1 000mg/kg.

Table2 Effects of dietary supplementation with the antibacterial peptide on CSF-Ab levels in piglets (OD ratio)

| Time | Control group | AP group | ABP group Ⅰ | ABP group Ⅱ | ABP group Ⅲ |
|---|---|---|---|---|---|
| 32d | $0.25 \pm 0.02^{B}$ | $0.28 \pm 0.02^{Ba}$ | $0.27 \pm 0.03^{AB}$ | $0.31 \pm 0.04^{AB}$ | $0.33 \pm 0.04^{A}$ |
| 39d | $0.31 \pm 0.03^{B}$ | $0.34 \pm 0.02^{AB}$ | $0.35 \pm 0.03^{AB}$ | $0.36 \pm 0.01^{Ab}$ | $0.37 \pm 0.03^{A}$ |
| 46d | $0.34 \pm 0.04^{Ba}$ | $0.36 \pm 0.05^{AB}$ | $0.38 \pm 0.047^{AB}$ | $0.41 \pm 0.05^{AB}$ | $0.44 \pm 0.07^{A}$ |
| 53d | $0.30 \pm 0.07^{Cb}$ | $0.38 \pm 0.03^{BC}$ | $0.39 \pm 0.03^{ABC}$ | $0.42 \pm 0.04^{AB}$ | $0.46 \pm 0.04^{A}$ |

ABC means with various letters are very significant different ($P < 0.01$), and with the same letters but in different case are significant different ($P < 0.05$). The following table 3 and table 4 are the same

## Effects on $CH_{50}$ of Piglets

The results were showed in Table3. During the whole experiment period, the levels of $CH_{50}$ of the treatment groups were significantly higher ($P < 0.01$) than that of the control group, and the values of $CH_{50}$ of the ABP groups were markedly higher ($P < 0.01$ or 0.05) than that of the AP group (except the level of $CH_{50}$ of the ABP group Ⅰ on 32, 39 days, and the level of $CH_{50}$ of the ABP group Ⅱ on 32 days were lower than that of the AP group). The values of $CH_{50}$ of the ABP were dose dependent, and the optimum add dosage of ABP was found to be 1 000mg/kg.

Table 3 Effects of dietary supplementation with the antibacterial peptide on $CH_{50}$ levels in piglets (U/ml)

| Time | Control group | AP group | ABP group Ⅰ | ABP group Ⅱ | ABP group Ⅲ |
|---|---|---|---|---|---|
| 32d | $345.91 \pm 10.61^{C}$ | $435.65 \pm 15.86^{AB}$ | $410.34 \pm 14.13^{b}$ | $434.44 \pm 14.22^{AB}$ | $445.40 \pm 17.48^{A}$ |
| 39d | $361.03 \pm 22.06^{C}$ | $459.98 \pm 16.60^{B}$ | $453.43 \pm 10.80^{b}$ | $485.78 \pm 17.20^{AB}$ | $503.23 \pm 11.32^{A}$ |
| 46d | $370.52 \pm 14.50^{E}$ | $487.02 \pm 4.38^{D}$ | $508.70 \pm 7.34^{C}$ | $547.33 \pm 11.17^{B}$ | $571.53 \pm 10.16^{A}$ |
| 53d | $380.30 \pm 20.82^{D}$ | $503.06 \pm 8.55^{C}$ | $572.47 \pm 11.74^{B}$ | $605.58 \pm 21.00^{A}$ | $612.43 \pm 7.48^{A}$ |

## Effects on Immunoglobulin of Piglets

As we could see from Table 4, at 39, 46, and 53 days of age, the levelsof IgG of the ABP group Ⅱ, the ABP group Ⅲ, and the AP group were significantly higher ($P < 0.01$ or 0.05) than those of the control group and the ABP group Ⅰ, and there is no differences ($P > 0.05$) between the value of of IgG of the ABP group Ⅰ and that of the control group. The levels of IgG of the ABP group Ⅱ and the ABP group Ⅲ were higher than that of the AP group among the whole experiment period, but there were no differences ($P > 0.05$). The values of IgG of the ABP were dose dependent, and the optimum add dosage of ABP was found to be 1 000mg/kg.

The results were showed in Table 4. During the whole experiment period, the levels of IgM of the treatment groups were higher than that of the control group. Except there was no differences ($P > 0.05$) between the value of IgM of the ABP group Ⅰ on 32, 46 days, the value of IgM of the AP group on 32 days and that of the control group, the values of the treatment groups were significantly higher ($P < 0.01$ or 0.05) than that of the control group among the whole experiment period. The level of IgM of the ABP group Ⅰ was markedly lower ($P < 0.05$) than that of AP group at 39 and 53days of age. The values of IgM of the ABP group Ⅱ and the ABP group Ⅲ were significantly higher ($P<0.01$ or 0.05) than that of AP group among the whole experiment period (except the value of IgM of the ABP group Ⅱ on 53dys). The values of IgM of the ABP were dose dependent, and the optimum add dosage of ABP was found to be 1 000mg/kg.

Table 4 Effects of dietary supplementation with the antibacterial peptide on serum immunoglobulin levels in piglets

| Items | Time | Control group | AP group | ABP group Ⅰ | ABP group Ⅱ | ABP group Ⅲ |
|---|---|---|---|---|---|---|
| IgG (g/L) | 32d | $3.44 \pm 0.14^{A}$ | $3.49 \pm 0.19^{A}$ | $3.46 \pm 0.11^{A}$ | $3.58 \pm 0.54^{A}$ | $3.66 \pm 0.23^{A}$ |
| | 39d | $3.63 \pm 0.28^{B}$ | $4.37 \pm 0.41^{A}$ | $3.69 \pm 0.14^{B}$ | $4.50 \pm 0.32^{Ab}$ | $4.77 \pm 0.46^{A}$ |
| | 46d | $3.69 \pm 0.33^{C}$ | $4.90 \pm 0.33^{A}$ | $3.95 \pm 0.21^{BC}$ | $4.94 \pm 0.42^{Ab}$ | $5.20 \pm 0.35^{A}$ |
| | 53d | $3.47 \pm 0.59^{B}$ | $4.79 \pm 0.35^{A}$ | $3.74 \pm 0.53^{Ba}$ | $4.89 \pm 0.42^{A}$ | $5.13 \pm 0.32^{A}$ |
| IgM (g/L) | 32d | $0.71 \pm 0.02^{C}$ | $0.79 \pm 0.02^{C}$ | $0.78 \pm 0.03^{C}$ | $1.01 \pm 0.08^{Ba}$ | $1.13 \pm 0.09^{A}$ |
| | 39d | $0.76 \pm 0.04^{D}$ | $0.93 \pm 0.05^{C}$ | $0.90 \pm 0.05^{Cd}$ | $1.15 \pm 0.08^{Ba}$ | $1.27 \pm 0.09^{A}$ |
| | 46d | $0.90 \pm 0.06^{C}$ | $1.11 \pm 0.05^{Bc}$ | $1.00 \pm 0.04^{C}$ | $1.19 \pm 0.07^{A}$ | $1.28 \pm 0.010^{A}$ |
| | 53d | $0.88 \pm 0.04^{D}$ | $1.10 \pm 0.06^{Bc}$ | $0.99 \pm 0.05^{C}$ | $1.18 \pm 0.06^{Ba}$ | $1.25 \pm 0.06^{A}$ |
| IgA (g/L) | 32d | $0.52 \pm 0.02^{A}$ | $0.53 \pm 0.02^{A}$ | $0.52 \pm 0.01^{A}$ | $0.53 \pm 0.02^{A}$ | $0.54 \pm 0.05^{A}$ |
| | 39d | $0.53 \pm 0.09^{Ba}$ | $0.59 \pm 0.08^{BA}$ | $0.58 \pm 0.07^{Ba}$ | $0.64 \pm 0.04^{A}$ | $0.68 \pm 0.04^{A}$ |
| | 46d | $0.50 \pm 0.02^{C}$ | $0.65 \pm 0.04^{B}$ | $0.64 \pm 0.03^{B}$ | $0.70 \pm 0.02^{Ba}$ | $0.76 \pm 0.05^{A}$ |
| | 53d | $0.45 \pm 0.04^{B}$ | $0.67 \pm 0.06^{A}$ | $0.52 \pm 0.03^{B}$ | $0.68 \pm 0.04^{A}$ | $0.73 \pm 0.04^{A}$ |

As we could see from Table4, the levels of IgA of the treatment groups were higher than that of the control group among the whole experiment period (except the value of IgA of the ABP group Ⅰ on 32 days). At 46, 53 days of age, the value of IgA of the AP group was markedly higher ($P<0.01$) than that of the control group. At 39, 46, and 53 days of age, the levels of IgA of the ABP groups were significantly higher ($P<0.01$ or 0.05) than that of the control group (except the level of IgA of the ABP group Ⅰ on 39 and 53 days). The values of IgA of the ABP were dose dependent, and the optimum add dosage of ABP was found to be 1 000mg/kg.

# Discussion

## Effects on CSF-Ab of Piglets

The level of antibody reflects the humoral immunity of organism (Waters., 1986). Antibody usually appears after vaccination not only in the blood but also locally about one week. Injection with ABP of rabbits sacculus rotundus could markedly improve ($P<0.01$) the serum antibody titers of NDV and AIV of chickens (Wang et al., 2007). Piglets fed with 160 mg/kg antibacterial peptide cecropins could significantly increased ($P<0.05$) the blocking rate, positive rate, and qualified rate of CSFV antibody (Jiang et al., 2013). In our study, supplementation with ABP and AP might increased the levels of CSF-Ab. Our study was the same with their studies. The effects of ABP were dose dependent, and the effects of middle dosage (500 mg/kg) and high dosage (1 000 mg/kg) of ABP were better than that of AP.

## Effects on $CH_{50}$ of Piglets

Complement system plays an important role in specific immune and nonspecific immune of organism. $CH_{50}$ can reflects the humoral immune function (Janeway., 2001). 200 mg/kg antimicrobial peptide could significantly increased ($P<0.05$) the alexin $C_3$ contents (Guo et al., 2012). In our study, Piglets fed with ABP and AP markedly improved $CH_{50}$. ABP showed better effect than AP. Our results suggested that antibacterial peptide could enhance humoral immune responses of weaned piglets by improving complement system.

### Effects on Immunoglobulin of Piglets

The serum immunoglobulin titer is an indicator of humoral immunity (Kong. ,2007). The results of our study showed that ABP and AP increased the levels of IgG, IgM, and IgA. While, the effects of IgG, IgM, and IgA of middle dosage (500 mg/kg) and high dosage (1 000 mg/kg) of ABP were more obvious, and better than that of AP. In contrast to results of present experiment, Yoon et al. (2014) reported no effect on serum immunoglobulins concentrations in weanling piglets fed diet supplemented with antimicrobial peptide-A3 and P5. However, similar to results of the present study, Wang et al. (2011) reported positive effect of dietary supplementation of antimicrobial peptide on concentrations of serum IgG in weanling pigs, and Lv et al. (2011) positive effect of dietary supplementation of antimicrobial peptide on concentrations of serum IgM and IgA in hisex hens. This indicated that the effect of antimicrobial peptide on immunoglobulins was connected with different antimicrobial peptide and different experimental animals.

## Conclusion

The results showed that supplementing weaned piglets diets with ABP enhanced the humoral immune responses of weaned piglets by improving the levels of CSF-Ab, $CH_{50}$, and immunoglobulins. The optimum add dosage of ABP was found to be 1 000 mg/kg for 4-week adding.

## Acknowledgments

The present work was supported by the Changjiang Scholars & Innovative Research Team of Ministry of Education of China Funds (Grant no. IRT0848).

## References

Bosi P, Merialdi G, Scandurra S, et al. 2011. Feed supplemented with 3 different antibiotics improved food intake and decreased the activation of the humoral immune response in healthy weaned pigs but had differing effects on intestinal microbiota. *Journal of Animal Science*, 89: 4043 – 4053.

Davis ME, Maxwell CV, Erf GF, et al. 2004. Dietary supplementation with phosphorylated mannans improves growth response and modulates immune function of weanling pigs. *Journal of Animal Science*, 82: 1882 – 1891.

Dong GZ, Pluske JR. 2007. The low feed intakc in newly-weaned pigs: problems and possible solutions. *Asian Australasian Journal of Animal Sciences*, 20, 440 – 452.

Guo ZQ, Yang FZ, Lei M, et al. 2012. Effects of Antibacterial peptides on intestinal mucosal morphology, caecal microflora and immune function of meat rabbits. *Chinese Journal of Animal Nutrition*, 24: 1778 – 1784.

Huang ZJ, Kingsolver MB, Avadhanula V, et al. 2013. An antiviral role for antimicrobial peptides during the arthropod response to alphavirus replication. *Journal of Virology*, 87: 4272 – 4280.

Janeway CA, Travers P, Walport M, et al. 1999. Immunobiology: The Immune System in Health and Disease, fourth ed. Elsevier/Garland, London: 363 – 415.

Jiang GT, Yang J, Qiu MZ, et al. 2013. Effects of antibacterial peptide cecropins on growth performance and antibody level of CSFV and PRV in weaned piglets. *Acta Ecologiae Animalis Domastici*, 34: 25 – 29.

Kim SW, McPherson RL, Wu GY. 2004. Dietary arginine supplementation enhances the growth of milk-fed young pigs. *The Journal of Nutrition*, 134: 625 – 630.

Koczulla AR, Bals R. 2003. Antimicrobial peptides. Drugs, 63: 389 – 406.

Kong XF, Yin YL, Wu GY, et al. 2007. Dietary supplementation with Acanthopanax senticosus extract modulates cellular and humoral immunity in weaned piglets. *Asian Australasian Journal of Animal Sciences*, 20: 1453 – 1461.

Liu LR, Yang KL, Hua J, et al. 2012. Antimicrobial Peptides: effects on growth performance, immune indices and mrna expression of related cytokine genes in jejunum of young roosters for egg production. *Chinese Journal of Animal Nutrition*, 24: 1345 – 1351.

Lv ZZ, Yuan XX, Cai ZW, et al. 2011. Effects of antimicrobial peptides on serum immune indices and IL-2 mRNA expression in spleen of laying hens. *Chinese Journal of Animal Nutrition*, 23, 2183 – 2189.

Rossignol T, Kelly B, Dobson C, et al. 2011. Endocytosis-mediated vacuolar accumulation of the human ApoE apolipoprotein-derived ApoEdpL-W antimicrobial peptide contributes to its antifungal activity in candida albicans. *Antimicrobial Agents and Chemotherapy*, 55: 4670 – 4681.

Torrent M, Pulido D, Rivas L, et al. 2012. Antimicrobial peptide action on parasites. *Current drug targets*, 13: 1138 – 1147.

Wang JH, Wu CC, Feng J. 2011. Effect of dietary antibacterial peptide and zinc-methionine on performance and serum biochemical parameters in piglets. *Czech Journal of Animal Sciences*, 56: 30 – 36.

Wang KZ, Yu RP, Hu YX, et al. 2007. Effect of antibacterial peptide of the rabbits sacculus rotundus on serum antibody titers of AIV and chicken NDV. *Science Technology and Engineering*, 1: 108 – 111.

Wang Z, Wang GS. 2004. APD: the antimicrobial peptide database. *Nucleic Acids Research*, 32: D590 – D592.

Waters RV, Terrell TG, Jones GH. 1986. Uveitis induction in the rabbit by muramyl dipeptides. *Infection and Immunity*, 51: 816 – 825.

Wu SD, Zhang FR, Huang ZM, et al. 2012. Effects of the antimicrobial peptide cecropin AD on performance and intestinal health in weaned piglets challenged with *Escherichia coli*. *Peptides*, 35: 225 – 230.

Yan JX, Wang KR, Chen R, et al. 2012. Membrane active antitumor activity of NK-18, a mammalian NK-lysin-derived cationic antimicrobial peptide. *Biochimie*, 94: 184 – 191.

Yang WC, Schultz RD. 1986. Ontogeny of natural killer cell activity and antibody dependent cell mediated cytotoxicity in pigs. *Developmental & Comparative Immunology*, 10: 405 – 418.

Yoon JH, Ingale SL, Kim JS, et al. 2014. Effects of dietary supplementation of synthetic antimicrobial peptide-A3 and P5 on growth performance, apparent total tract digestibility of nutrients, fecal and intestinal microflora and intestinal morphology in weanling pigs. *Livestock Science*, 159: 53 – 60.

Yu FS, Cornicelli MD, Kovach MA, et al. 2010. Flagellin stimulates protective lung mucosal immunity: role of cathelicidin-related antimicrobial peptide. *Journal of Immunology*, 185: 1142 – 1149.

# Effect of cadmium on the concentration of ceruloplasmin and its mRNA expression in goats under molybdenum stress

ZHUANG Yu*, HU Guoliang*, ZHANG Caiying, CAO Huabin, GUO Xiaoquan

*Institute of Animal Population Health, College of Animal Science and Technology, JiangXi Agriculture University Nanchang* 330045, *P. R. China*

**Abstract**: The study was conducted to explore the effect of cadmium onthe concentration of ceruloplasmin and mRNA expression in goats under molybdenum Stress, and $CdCl_2$ and $[(NH_4)_6Mo_7O_{24} \cdot 4H_2O]$ were selected as the source of cadmium and molybdenum. 36 healthy goats were divided into four groups randomly: Control group (0mg/kg Mo +0 mg/kg Cd), Mo exposure group (30mg/kg Mo), co-exposure group A (30mg/kg Mo +0.5mg/kg Cd) and co-exposure group B (30 mg/kg Mo + 1 mg/kg Cd). After 50 days, blood and liver samples were collected. The concentrations of molybdenum (Mo), cadmium (Cd) and copper (Cu), and the levels of the ceruloplasmin (CP) protein and its mRNA expression were then analyzed from these samples. Results indicated that all experimental groups had shown some visible clinical signs of copper deficiency. The body weight and concentration of CP protein in these study subjects decreased significantly as well. Further, contents of Mo, Cd and Cu in the liver were found to have significantly accumulated in all treatment groups, and contents of Mo, Cd and Cu in serum or liver of the co-exposure (Mo + Cd) groups were significantly higher than those of the Mo group. In addition, the accumulated level of cadmium appeared in a dose- and time-dependent fashion. However, the expression of ceruloplasmin of the Mo group increased significantly ($P < 0.05$), while in the co-exposure groups, the level of expression increased before day 25 and then decreased rapidly. This study suggests that Cd has a promotional effect on molybdenosis, and it can aggravate the deposition of heavy metals (Mo, Cd and Cu), as well as interfere the metabolism of copper in the liver.

**Keywords**: Goat, Molybdenum, Cadmium, Ceruloplasmin, mRNA

## Introduction

Molybdenum (Mo) has been acknowledged as an essential trace element for animals and humans since 1953[1], but its requirement in animals is extremely low[2]. However, high concentrations of Mo could poison animals by its interruptive interaction with other trace elements. Previous studies revealed that ruminants were often harmed by high concentrations of Mo, especially in areas of mining and other industrial businesses[3,4]. Low doses of 3 mg/kg and 50 mg/kg of Mo could induce clinical signs of molybdenosis in cattle and sheep respectively[6]. Likewise, Cadmium (Cd) is a nonessential trace element and widespread metal contaminant. It has been classified as the sixth toxic substance that endangered human health by the Agency for Toxic Substances and Disease Registry (ATSDR), due to its protracted biological high-life, low rate of excretion, predominant storage in soft tissues, and diverse toxic effects[7].

---

The two authors contributed equally to this work

Corresponding author: Cao Huabin, E-mail: chbin20020804@163.com

Jiangxi province of China has one of the largest rare earths mines around the world, and contributes to more than 90% of the world rare earth production[8]. Nevertheless, exploitation of rare earth mine resources could cause environmental problems such as heavy metal dissemination. Recent epidemiological studies showed that many heavy metals such as Mo and Cd had severely contaminated the soil and forage in the area. Additionally, Fan (1983) investigated and reported Mo toxicity in the first ruminant species-the cattle. Significant clinical signs were detected, of which they called "red skin combined with white fleece syndrome (RSCWFS)" [9]. Their subsequent research revealed that the syndrome (RSCWFS) was induced by a combination of Mo and Cd intake[10]. Recently, researches still mainly focused on the toxicity of Mo or Cd alone, whereas very few concentrated on their combined effects. However, heavy metal diseases are always caused by two or more elements. Therefore we tested the combined effects of Mo and Cd in goats in order to explore the mechanism of interaction of these two elements. Moreover, the concentrations of cuprum enzyme (Ceruloplasmin) and its mRNA expression were also detected.

## Materials and Methods

### Experimental Animals and Treatments

36 healthy purebred bore goats weighing approximately 20kg each were procured from a commercial rearing farm. Animal care was approved by the Department of Animal Science of Jiangxi Agriculture University. All the experiments on goats followed the National Research Council's guide for the care and use of laboratory animal. Animals were individually housed in a well-ventilated shed and allowed 15 days of adjustment on the basal diet. The basal diet was composed of corn and wheat bran. The basal diet (DM basis) contained 6.2 mg Cu/kg, 144 mg Fe/kg, 39.5 mg Zn/kg, 43.9 mg Mn/kg, 1.13 mg I/kg, 0.11 mg Se/kg, 1.65 mg Mo/kg and 0.23% S. And the basal diet was formulated to meet or exceed all nutrient requirements for goats (NRC, 1997) (Table 1). The goats were then randomly distributed to 4 groups with 9 goats in each group. Subjects in the control group was orally administered corresponding quantitative deionized water, meanwhile the animals of all exposure groups were orally administered 30 mg Mo/kg BW, and co-exposure group A was orally administered 0.5 mg Cd/kg BW and co-exposure group B was given 1 mg Cd $\cdot kg^{-1} \cdot$BW per day. $CdCl_2$ and $[(NH_4)_6Mo_7O_{24} \cdot 4H_2O]$ were the source of cadmium and molybdenum. Experimental period was 50 days. All goats in the experimental period were fed with the basal diet. To avoid waste incurred by selective food intake of the goats, animals were fed restrictedly with 0.70 kg per goat feed (DM) at 07:00 and 18:00 in equal allotments per day, but were allowed to drink water (with less than 0.01mg Cu/kg, 0.01mg Mo/kg, 0.0001 Cd/kg after analysis) ad libitum.

Table 1 Composition and nutrient levels in the basal diet for the dairy goats (DM basis, %)

| Items | Content (%) |
|---|---|
| Ingredients[a] | |
| Corn | 52.5 |
| Deoiled rice | 19.00 |
| Soybean meal | 10.0 |
| Rapeseed meal | 7.0 |
| Cottonseed meal | 7.0 |
| $CaHPO_4$ | 1.0 |

(continued table)

| Items | Content(%) |
|---|---|
| Limestone | 1.5 |
| Salt | 1.0 |
| Additives[b] | 1.0 |
| Total | 100 |
| Nutrient levels | |
| DM(%) | 87.84 |
| ME/(MJ/kg) | 8.89 |
| CP (%) | 11.90 |
| EE(%) | 2.71 |
| Calcium | 0.90 |
| AP (%) | 0.78 |

[a] As fed basis.

[b] Additives provided as follows per kilogram of additives: nicotinic acid, 2 000mg; vitamin A, 1 000 000 IU; vitamin $D_3$, 2 500 000 IU; vitamin E, 24 000 mg; iron($FeSO_4 \cdot H_2O$), 2 000 mg; zinc($ZnSO_4 \cdot H_2O$), 140 000 mg; manganese($MnSO_4 \cdot H_2O$), 3 000 mg; iodine(KI, 3%), 180 mg; selenium($NaSe_3O_4 \cdot H_2O$), 100 mg

## Recording Keeping

The body weight of individual groups was recorded at thebeginning of the experiment and thereafter in the morning of every tenth day before feeding. The animal's clinical signs were also recorded.

## Collection and Preservation of Samples

Jugular blood samples were collected from each goat in non-heparinized Vacutainer tubes on day 0, 10, 20, 30, 40 and 50 for determining serum ceruloplasmin (CP) concentration. Blood samples were transported in ice packs to the laboratory, then centrifuged at 1 000g at 4℃ for 10 min and the serum was separated and storeed at −20℃ in Eppendorf tubes until the analyses were performed.

Liver samples were collected on day 0, 25 and 50by slaughtering three goats randomly chosen from each group. After recording the body weight, each goat was sacrificed with an overdose intravenous injection of sodium pentobarbital. After the abdominal cavity was opened, the liver was immediately removed and washed in distilled water after careful removal of the gall bladder. Two pieces of liver samples were collected and thoroughly washed in running deionized water, and were frozen in liquid nitrogen and stored at −80℃ in individual sample bags.

## Biochemical Assays

Concentration of CPBlood was collected via Jugular vein, and then the serum was obtained by centrifugation at 1 000g at 4℃ for 10 min. The activity of CP in serum was detected using the CP kit (Nanjing Jiancheng Bioengineering Institute, PR, China), and the process followed the protocol of the manufacturer.

Molybdenum, cadmium and copper in liver and serum The contents of Mo, Cd and Cu in liver and serum were analyzed via graphite furnace atomic absorption spectrometry (Model 5100, HGA-600 Graphite Furnace; Perkin-Elmer, USA)[11,12]. The working parameters (air, acetylene, optics, and electronics) of the device were adjusted by following the protocol of the manufacturer. All analyses were triplicated and the mean values are reported.

The samples of frozen liver were cut into small pieces with a stainless knife and were baked at 110℃ for 12 h in a baking oven; then, 0.5 g samples were transferred into beakers. For digestion, 25 ml of concentrated $HNO_3$/HCl (4:1) was added to each beaker and warmed on a low temperature electric hot plate until solution was transparent. The samples were adjusted to 10 ml using a volumetric flask by 0.5% $HNO_3$[13], and samples of serum were diluted 1 : 4 with 0.2% nitric acid solution and then analyzed by graphite furnace atomic absorption spectrometry, which is an established technique commonly used for the determination of trace-element concentrations.

## Expression of CP mRNA in the Liver

Hepatic total RNA was isolatedby using the RNesay mini kit (QIAGEN) from 20-50 mg tissue and then quantified by spectrophotometry. Aliquots of total RNA sample (5 μg) were resolved in 1.0% denaturing agarose gels and stained with ethidium bromide. If the band was clear with no trailing phenomenon, and the ratio of gray level from the 28s and 18s band was about 2.0, it can be concluded that RNA had no degradation and its quality was reliable. Otherwise, the RNA sample wasn't useful and extraction had to be done for a second time.

Thesequences of the CP gene and the actin (β-Actin) gene of goats were searched from the GenBank. Primers and probes were designed in the relatively conservative area by Primer Express3.0. The 5' end of the probe was marked FAM, the 3' end of the probe was marked TEMRA, eventually Shanghai Ying Jun technology company was entrusted for synthesis. The primer sequences were shown in Table 2.

Table 2 Primer and Probe Sequences

| Gene Name | Gene Serial Number | Primer and Probe Sequence (5' to 3') | Amplified fragment length |
|---|---|---|---|
| *CP* | *NM_001009733.1* | Upstream: TGCTATTAATGGAAGGATGTTTGG<br>Downstream: ACCGAGTGCAAGTCTACTTCATTG<br>Probe: AACCTGCAAGGCCTCAC | 78bp |
| *β-Actin* | *U39357.1.1* | Upstream TCACGGAGCGTGGCTACAG | 63bp |

TaqMan Probe method was used for Real Time (qPCR) reaction, and operation was conducted according to TaKaRa Premix Ex *Taq*™ (Probe qPCR) kit provided by the Po TaKaRa Biotechnology (Dalian) Co., LTD. The amplification cycle for gene consisted of 40 cycles of 95℃ for 30 sec, 94℃ for 10 sec, and 67℃ for 37 sec. Results were analyzed by the Softwares SDS2.4 and 1.2.1 RQManager. Ct was the cycle threshold. $\Delta Ct$(sample) = $Ct$(target gene) − $Ct$(reference gene); $\Delta Ct$(calibrator) = $Ct$(target gene) − $Ct$(reference gene); $\Delta\Delta Ct = \Delta Ct$(sample) − $\Delta Ct$(calibrator); $2^{-\Delta\Delta CT}$ represented the relative expression of the initial cDNA of the target gene. RQ generated automatically by using softwares was the relative expression of cDNA, and then data was exported.

## Statistical Analysis

Statistical analysis was done using SPSS 17.0 (Chicago, IL), and all parameters were presented as mean ± standard deviation. A significant value ($P < 0.05$) was obtained by one-way ANOVA. Differences between means were assessed using Tukey's honestly significant difference test for post hoc multiple comparisons.

# Results

## Clinical Signs

Animals ofthe control group were under a healthy condition through clinical observation. However, a shuffling gait, watery diarrhea, achromotrichia, anorexia, and emaciation were observed in the Mo exposure and co-exposure groups. Meanwhile, onset time of abnormal clinical signs in co-exposure groups were earlier than the Mo exposure group. Abnormal clinical signs in co-exposure group B (30mg/kg Mo + 1mg/kg Cd), co-exposure group A (30mg/kg Mo + 0.5 mg/kg Cd) and Mo exposure group were observed firstly on day 36, 43, 47, respectively. Clinical signs of watery diarrhea, anorexia, emaciation and achromotrichia in co-exposure group B (30mg/kg Mo + 1mg/kg Cd) were more serious than those of co-exposure group A (30mg/kg Mo + 0.5 mg/kg Cd). Screams and red skin were observed at midnight since day 41 in both co-exposure groups A and B. Red skin was observed in both co-exposure groups. Neither was present in the Mo exposure group.

## Body Weight Change

Record of body weight was presented in Table 3. Body weight of the goats in the control group had a significant increase ($P < 0.05$) during the experimental period. However, body weight in the Mo exposure and co-exposure (Mo + Cd) groups decreased significantly ($P < 0.05$). Body weight was significantly lower ($P < 0.05$) in the Mo exposure and co-exposure (Mo + Cd) groups than the control group. Body weight of the co-exposure group B (30mg/kg Mo + 1mg/kg Cd) decreased significantly ($P < 0.05$) compared with the Mo exposure or co-exposure group B (30mg/kg Mo + 0.5 mg/kg Cd) on day 40 and 50.

Table 3 Effect of Cadmium on average body weight in goats under Molybdeum Stress (kg/Goat)

| Dose (mg/kg BW) | | Experimental period (Day) | | | | | |
|---|---|---|---|---|---|---|---|
| Cd | Mo | 0 | 10 | 20 | 30 | 40 | 50 |
| 0 | 0 | $21.38 \pm 1.10^{Aa}$ | $22.27 \pm 0.41^{Aa}$ | $22.60 \pm 0.29^{Aab}$ | $23.19 \pm 0.08^{Aab}$ | $23.83 \pm 0.27^{Ab}$ | $24.38 \pm 0.14^{Ab}$ |
| 0 | 30 | $21.35 \pm 1.29^{Aac}$ | $21.72 \pm 1.06^{Ac}$ | $19.97 \pm 0.52^{Bab}$ | $19.02 \pm 0.28^{Bbd}$ | $18.62 \pm 0.13^{Bbd}$ | $16.92 \pm 0.39^{Bd}$ |
| 0.5 | 30 | $21.88 \pm 0.51^{Aa}$ | $21.94 \pm 0.93^{Aa}$ | $20.69 \pm 0.76^{Bab}$ | $18.97 \pm 1.03^{Bbc}$ | $17.75 \pm 0.51^{Bcd}$ | $16.25 \pm 0.36^{Bd}$ |
| 1 | 30 | $21.43 \pm 1.12^{Aa}$ | $21.23 \pm 0.54^{Aa}$ | $20.74 \pm 0.49^{Ba}$ | $18.59 \pm 1.34^{Bb}$ | $16.77 \pm 0.91^{Bbc}$ | $14.77 \pm 0.62^{Bc}$ |

## Analyses of the Concentrations of Mo, Cd and Cu in Serum

Concentrations of Mo, Cd and Cu in serum were presented in Figure 1. No significant differences ($P > 0.05$) were observed on concentrations of Mo, Cd and Cu in the control group. Concentration of Mo in co-exposure groups was significantly higher than the Mo exposure group (Figure 1A). However, concentration of Cd in co-exposure groups significantly increased ($P < 0.05$) while in the Mo group it decreased significantly (Figure 1B). On day 50, concentrations of Mo and Cd in co-exposure group B (30mg/kg Mo + 1mg/kg Cd) were significantly ($P < 0.05$) higher than those of co-exposure group A (30mg/kg Mo + 0.5 mg/kg Cd). Concentration of Cu in the Mo group and co-exposure (Mo + Cd) groups increased compared with the control group. However, concentration of Cu in Mo group increased significantly compared with the co-exposure (Mo + Cd) groups on day 50 (Figure 1C).

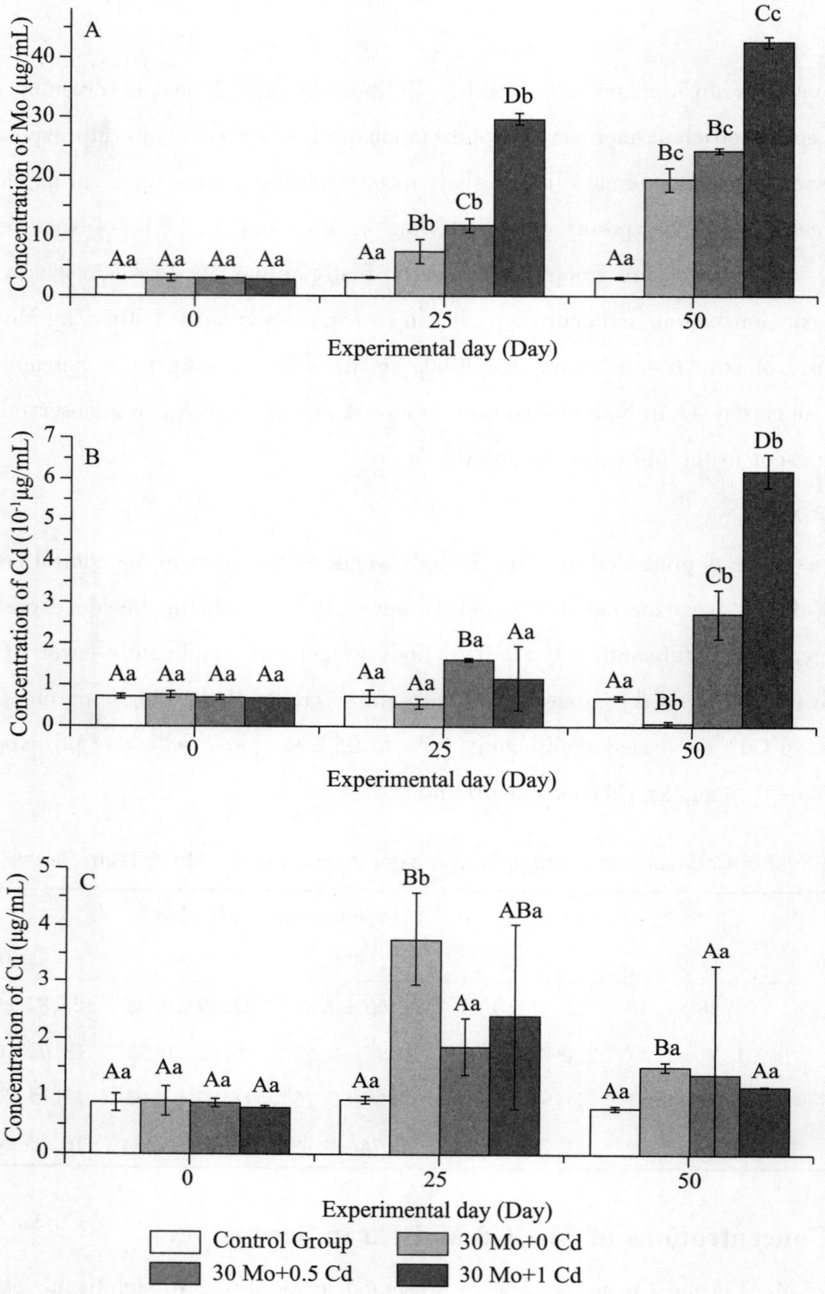


Figure 1 The effect of cadmium on concentration of Mo, Cd and Cu in serum under molybdenum stress ( n = 3 )

Means ± SEM. Different letters indicate statistical significance at $P<0.05$ among a group (a, b, c) or inter-groups (A, B, C)

## Analyses of the Contents of Mo, Cd and Cu in the Liver

Contents of Mo, Cd and Cu in the liver were presented in Figure 2. No significant differences $(P>0.05)$ were observed on contents of Cu, Cd and Cu in the control group. However, Contents of Mo and Cu in the Mo exposure and co-exposure groups increased significantly $(P<0.05)$. And contents of Cd in the co-exposure groups increased significantly $(P<0.05)$ while the contents of the Mo exposure group decreased. Both metallic elements (Cd and Cu) in the co-exposure groups (Mo + Cd) increased than the Mo exposure group.

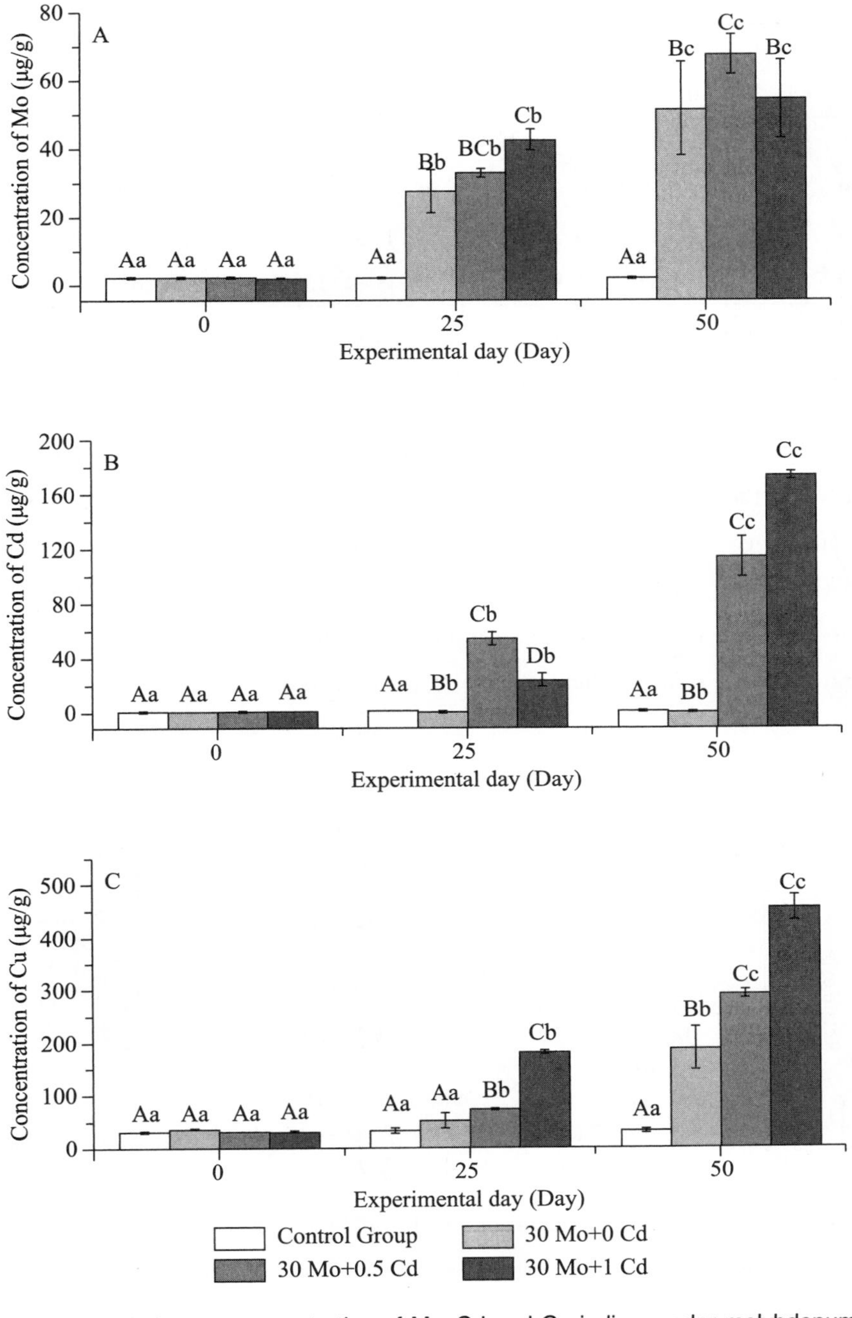


Figure 2 The effect of cadmium on concentration of Mo, Cd and Cu in liver under molybdenum stress (n = 3)

Means ± SEM. Different letters indicate statistical significance at $P < 0.05$ among a group (a, b, c) or inter-groups (A, B, C)

## Analyses of the activity of CP in Serum and expression of CP mRNA in Liver

Activity of CP in serum was presented in Table 4. No significant differences ($P > 0.05$) were observed on the activity of CP during the experimental period in the control group. Activity of CP in the Mo exposure and co-exposure (Mo + Cd) groups decreased significantly ($P < 0.05$).

Table 4 Effect of cadmium on the activity of CP in serum under molybdenum stress(μmol/L)

| Dose(mg/kg BW) | | Experimental period (Day) | | | | | |
|---|---|---|---|---|---|---|---|
| Cd | Mo | 0 | 10 | 20 | 30 | 40 | 50 |
| 0 | 0 | 42.24 ± 1.93[Aa] | 42.45 ± 1.34[Aa] | 40.73 ± 1.49[Aa] | 39.33 ± 0.46[Aa] | 41.91 ± 1.37[Aa] | 39.66 ± 0.23[Aa] |
| 0 | 30 | 44.71 ± 1.59[Aa] | 39.69 ± 0.97[Ab] | 38.61 ± 1.94[ABb] | 25.31 ± 7.98[Bc] | 20.96 ± 6.84[Bd] | 19.18 ± 1.14[Bd] |
| 0.5 | 30 | 39.82 ± 0.23[Aa] | 40.84 ± 2.78[Aa] | 36.11 ± 1.37[Ba] | 24.50 ± 0.46[Bb] | 22.73 ± 0.23[Bb] | 21.28 ± 2.28[Bb] |
| 1 | 30 | 40.09 ± 0.37[Aa] | 38.21 ± 0.68[Aab] | 36.43 ± 1.12[Bab] | 33.53 ± 12.31[Bb] | 24.18 ± 2.74[Bc] | 18.38 ± 2.74[Bd] |

Data are mean ± *SD* (n = 3). Mean with different lower case letters within a line are statistically different ($P < 0.05$), Mean with different capital case within a column are statistically different ($P < 0.05$)

Expression of CP mRNA was presented in Figure 3. No significant differences ($P > 0.05$) were observed of the expression of CP mRNA in the control group during the experimental period. On day 50, expression of CP mRNA in the Mo exposure group and co-exposure (Mo + Cd) groups increased significantly ($P < 0.05$) compared with the control group. However, on day 25, expression of CP mRNA in the co-exposure (Mo + Cd) groups increased significantly compared with the Mo exposure group. And on day 50, expression of CP mRNA in the Mo exposure group increased markedly while it had a slight decline in the co-exposure groups.

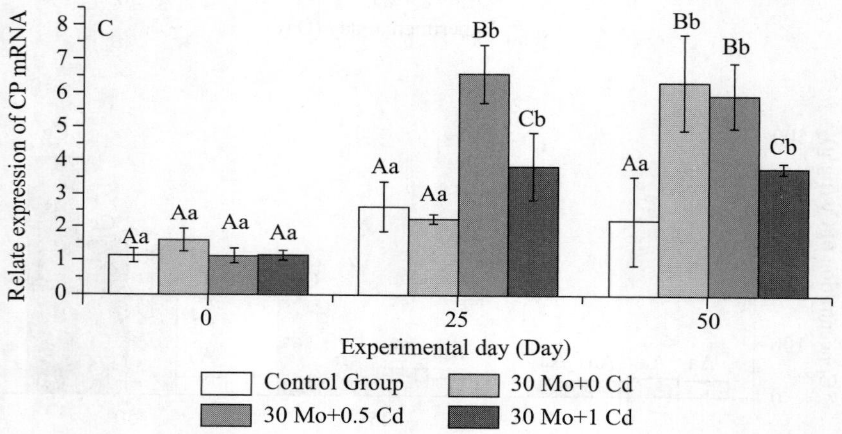


Figure 3 The effect of cadmium on expression of CP mRNA in liver under molybdenum stress (n = 3)

Means ± SEM. Different letters indicate statistical significance at $P < 0.05$ among a group (a, b, c) or inter-groups (A, B, C)

## Discussion

The toxicity and adverse effects of Mo and Cd in animals have been widely acknowledged in the pastfifty years. Even in practical feeding conditions, some ruminants can be exposed to heavy metal in the environment and develop relative diseases[10]. Previous studies revealed that high levels of Mo intake could induce copper deficiency in all livestocks[16,17], and the toxic effects of Mo can be reflected by testing the enzyme activity and gene expression of ceruloplasmin (CP), which is a copper containing metalloenzyme found in blood that is synthesized in liver and carries approximately 95% of the total plasma copper[18]. However, the element Cd was reported to have some effects on copper deficiency. Therefore, analyzing the enzyme activity and gene expression of CP allows us to explore the combined effect of Mo and Cd by adding a set level of Mo with differentiated levels of Cd.

In this study, the combined effect of Mo and Cd has been detected and the results of clinical signsof shuffling gait, watery diarrhea, achromotrichia and anorexia were observed in all treatment groups and the onset among these groups were distinct, especially in body weight of co-exposure group B, which significantly decreased ($P < 0.05$) compared with the Mo group. The result, which was consistent with previous studies about molybdenosis or copper deficiency[14,15], suggests that the level of Cd had aggravated molybdenosis in our study. Through observation of the levels of Mo, Cd and Cu in the liver, the increase of supplemental cadmium induced the deposition of Mo and Cu in the co-exposure groups compared with the Mo exposure group, and this interaction appeared in a dose- and time dependent fashion. On one hand, cadmium might have accelerated the absorption of Mo according to the analysis of serum elements (Figure 1A). However, numerous studies revealed that high level of Mo can reduce the bioavailability of Cu by forming insoluble complexes of Cu-Mo in ruminants[19,20], On the other hand, Cd have a stronger connection with MT proteins[21]. Metallothionein (MT) is a cysteine-rich, low molecular weight metal sequestering protein that has been shown to be involved in essential metal homeostasis and in the detoxification of heavy metals[22]. Cd and Cu have a high affinity of MTs.[23] Cd will lead to increased MT expression, which traps Cu in the liver, consequently disturbing the metabolism of copper. In addition, cadmium exposure was associated with alteration in oxidative stress[24,25], inhibited CP[26], and was closely associated with microflora species[27], The microflora species in lumen play an essential role in the development and health of the host by improving the intestinal tract microbial balance as well as detoxification and elimination of harmful compounds from the body[28,29]. Nevertheless, concentration of Cu in serum increased before day 25 and then reduced (Figure 1C), combined with the results of contents of Cu and copper deficiency, this phenomenon suggested that high level of Mo supplementation induced systemic copper deficiency. Copper reserves were mobilized initially, and then induced the animal body copper re-distribution. Nevertheless, the serum in Cu decreased in the later period as the result of depletion of available Cu. In addition, as a result of copper deficiency, the expression of CP in the treatment groups was significantly higher in compensation than that of the control group. Preceruloplasmin in liver has no sufficient copper to form mature ceruloplasmin. Preceruloplasmin released into the blood then degraded rapidly[30]. Nevertheless, expression of CP mRNA in liver decreased on day 50 in co-exposure groups. High heavy metal (Mo, Cd and Cu) deposition in liver caused oxidative stress, ultimately lead to disturbance of the liver's function.

In conclusion, Cd has a promotional effect on molybdenosis, and it can aggravate the deposition of heavy metals (Mo, Cd and Cu) in liver, as well as interfere the metabolism of copper in liver.

## Acknowledgment

The study was supported by the National Natural Science Foundation of China (grant number: 31101863), Provincial Natural Science Foundation of Jiangxi (grant number: 2010GQN0052). The authors also want to thank Ruowei Yang for her help in editing.

## References

[1] Nordberg G. Handbook on the toxicology of metals. 3rd edn. Academic Press, Amsterdam; Boston, 2007.

[2] Barceloux DG. Molybdenum. *Journal of Toxicology Clinical Toxicology*, 1999, 37 (2): 231 – 237.

[3] Gardner W, Broersma K, Popp J, et al. Copper and health status of cattle grazing high-molybdenum forage from a reclaimed mine tailing site. *Canadian Journal of Animal Science*, 2003, 83 (3): 479 – 485

[4] Suttle NF. Copper imbalances in ruminants and humans: unexpected common ground. *Advances in Nutrition*, 2012, 3 (5): 666 – 674. doi: 10.3945/an.112.002220.

[5] Pandey R, Singh S. Effects of molybdenum on fertility of male rats. Biometals: an international journal on the role of metal ions in biology, biochemistry, and medicine, 2002, 15 (1): 65 – 72.

[6] Mills C F, Davis G K, Molybdenum W M. Trace elements in human and animal nutrition. Ed. Mertz, W, 1987.

[7] Marettová E, Maretta M, Legath J. Changes in the peritubular tissue of rat testis after cadmium treatment. *Biological Trace Element Research*, 2010, 134 (3): 288 – 295.

[8] *Zhenggui W, Ming Y, Xun Z, et al. Rare earth elements in naturally grown fern < i > Dicranopteris linearis < /i > in relation to their variation in soils in South-Jiangxi region (Southern China)*. Environmental Pollution, 2001, 114 (3): 345 – 355.

[9] *Pu Fan, Zhili Wu, Jiyu Wang, et al. Initial research of molybdenum induced toxicity in cattle.* Acta Agriculturae Universitatis Jiangxiensis. 1981, 8(3): 1 – 9.

[10] *Pu Fan, , Xin Liu, Zhili Wu, et al. Initial research of molybdenum induced toxicity in cattle* (Ⅴ). Acta Agriculturae Universitatis Jiangxiensis. 1985, 04: 11 – 18.

[11] *Şahin ç A, Tokgöz Ī, Bektaş S. Preconcentration and determination of iron and copper in spice samples by cloud point extraction and flow injection flame atomic absorption spectrometry.* Journal of Hazardous Materials, 2010, 181(1): 359 – 365.

[12] *Lasagna-Reeves C, Gonzalez-Romero D, Barria M A, et al. Bioaccumulation and toxicity of gold nanoparticles after repeated administration in mice.* Biochemical and Biophysical Research Communications, 2010, 393(4): 649 – 655.

[13] *Liu X, Li Z, Han C, et al. Effects of dietary manganese on Cu, Fe, Zn, Ca, Se, IL-1β, and IL-2 changes of immune organs in cocks.* Biological Trace Element Research, 2012, 148(3): 336 – 344.

[14] *Kessler K L, Olson K C, Wright C L, et al. Effects of supplemental molybdenum on animal performance, liver copper concentrations, ruminal hydrogen sulfide concentrations, and the appearance of sulfur and molybdenum toxicity in steers receiving fiber-based diets.* Journal of Animal Science, 2012, 90(13): 5005 – 5012.

[15] *Yang F, Cui H, Xiao J, et al. Increased apoptotic lymphocyte population in the spleen of young chickens fed on diets high in molybdenum.* Biological Trace Element Research, 2011, 140 (3): 308 – 316.

[16] *Raisbeck MF, Siemion RS, Smith MA. Modest copper supplementation blocks molybdenosis in cattle.* Journal of Veterinary Diagnostic Investigation, 2006, 18 (6): 566 – 572.

[17] *Zhang W, Zhang YS, Zhu XP, et al. Effect of Different Levels of Copper and Molybdenum Supplements on Performance, Nutrient Digestibility, and Follicle Characteristics in Cashmere Goats.* Biological Trace Element Research, 2011, 143 (3): 1470 – 1479.

[18] *Rombach E P, Barboza P S, Blake J E. Costs of gestation in an Arctic ruminant: copper reserves in muskoxen[J]. Comparative Biochemistry and Physiology Part C: Toxicology & Pharmacology*, 2003, 134(1): 157 – 168.

[19] *Meschy F. Recent progress in the assessment of mineral requirements of goats.* Livestock Production Science, 2000, 64 (1): 9 – 14.

[20] *Gooneratne S R, Buckley W T, Christensen D A. Review of copper deficiency and metabolism in ruminants.* Canadian Journal of Animal Science, 1998, 69(4): 819 – 845.

[21] *Sarkar S, Yadov P, Bhatnagar D. Lipid peroxidative damage on cadmium exposure and alterations in antioxidant defence system in rat erythrocytes: a study with relation to time.* Biological Metals, 1998, 11: 153 ~ 157

[22] *Cobbett C, Goldsbrough P. Phytochelatins and metallothioneins: roles in heavy metal detoxification and homeostasis.* Annual review of plant biology, 2002, 53(1): 159 – 182.

[23] *Sabolić I, Breljak D, Škarica M, et al. Role of metallothionein in cadmium traffic and toxicity in kidneys and other mammalian organs.* Biometals, 2010, 23(5): 897 – 926.

[24] *Klaassen CD, Liu J, Choudhuri S. Metallothionein: an intracellular protein to protect against cadmium toxicity.* Annual Review of Pharmacology and Toxicology, 1999, 39 (1): 267 – 294.

[25] *Moulis JM. Cellular mechanisms of cadmium toxicity related to the homeostasis of essential metals. Biometals: an international journal on the role of metal ions in biology, biochemistry, and medicine*, 2010, 23 (5): 877 – 896.

[26] *Samsam Shariat S Z, Alinejad N. Inhibition of human ceruloplasmin (ferroxidase) by cadmium.* Research in Pharmaceutical Sciences, 2009, 3(1): 47 – 52.

[27] *Fazeli M, Hassanzadeh P, Alaei S. Cadmium chloride exhibits a profound toxic effect on bacterial microflora of the mice gastrointestinal tract[J].* Human & Experimental Toxicology, 2011, 30(2): 152 – 159.

[28] *Tancrede C. Role of human microflora in health and disease.* European Journal of Clinical Microbiology and Infectious Diseases,

1992,11(11):1012 - 1015.

[29] *Ravikumar S, Williams G P, Shanthy S, et al. Effect of heavy metals (Hg and Zn) on the growth and phosphate solubilising activity in halophilic phosphobacteria isolated from Manakudi mangrove. Journal of Environmental Biology*, 2007, 28(1): 109.

[30] *Sato M, Gitlin J D. Mechanisms of copper incorporation during the biosynthesis of human ceruloplasmin[J]*. Journal of Biological Chemistry, 1991, 266(8): 5128 - 5134.

# Effects of dietary antibacterial peptide supplementation on contents of serum IL-2, IL-4, IL-6, IFN-γ, and TNF-α in weaned piglets

YUAN Wei, REN Zhihua, DENG Junliang*, ZUO Zhicai, WANG Ya, DAN Qixiong, JIN Haitao, TIAN Chunlei, LIANG Zhen, and GAO Shuang

*Department of Veterinary Medicine, Sichuan Agricultural University, Ya'an* 625014, *China*

**Abstract**: As far as the authors' knew, there were limited studies evaluating the effects of antibacterial peptide on the serum cytokines. The aim of this study was to investigate the effects of dietary antimicrobial peptide (ABP) supplementation on contents of serum IL-2, IL-4, IL-6, IFN-γ and TNF-α in weaned piglets. Ninety weaned piglets (Landrace × Yorkshire × Duroc; (21 ± 2) d of age; average weight of (8.24 ± 0.67) kg) were randomly allotted to five groups, including the control group, the antibacterial peptide groups containing 250, 500 and 1 000 mg/kg antibacterial peptide, respectively, and the astragalus polysaccharide group with 400 mg/kg astragalus polysaccharide. After 5 days adaptation to dietary daidzein, the whole experimental period was 28 days. Blood samples were collected at 32, 39, 46, and 53 day of age. The results showed that ABP as dietary additive increased the levels of IL-2, IL-4, IL-6, IFN-γ, and TNF-α, and the effects of ABP supplementation were dose-dependent. Piglets fed with 250 mg/kg ABP had significantly higher ($P < 0.01$ or 0.05) the level of IL-2 at 46, 53 days of age, the levels of IL-4 and IL-6 at 39, 46, 53 days of age, the level of IFN-γ among the whole experimental period, and the level of TNF-α at 39 and 53 days of age. Supplementation with 500 mg/kg markedly increased ($P < 0.01$ or 0.05) the values of IL-2 and IL-4 at 39, 46, 53 days of age, and the values of IL-6, IFN-γ, and TNF-α among the whole experimental period. High dosage (1 000 mg/kg) of ABP supplementation significantly improved ($P < 0.01$) the the level of IL-4 at 39, 46, 53 days of age, and the values of IL-2, IL-6, IFN-γ, and TNF-α among the whole experimental period. The results indicated that dietary addition of ABP enhance the cellular and humoral immune responses of weaned piglets by improving the levels of IL-2, IL-4, IL-6, IFN-γ, and TNF-α. The optimum add dosages was found to be 1 000 mg/kg for 4-week adding.

**Keywords**: antibacterial peptide, cytokines, weaned piglets.

## Introduction

Antimicrobial peptide, are bioactive substances which are extracted, separated and purified from a variety of plants, animals, human tissues and cells *in vivo* (Wang et al., 2004). They have a broad range of functions like anti-bacteria (Koczulla et al., 2003), anti-viral (Huang et al., 2013), anti-fungal (Rossignol et al., 2011), anti-tumor (Yan et al., 2012), anti-parasitic (Torrent et al., 2012), and enhance immune function (Yu et al., 2010).

Cytokines is a kind of small peptide molecule which widely distributed in each organization of the whole body, and they are derived from epithelial cells, endothelial cells, and fibroblasts. Cytokines, mainly including interleukins (IL-1 ~ 18), interferons (IFN-α/β/γ), tumor necrosis factors (TNF-α/β) and so on. They not only participate in the maintenance of tissue integrity, but also play a central role in the cell-mediated immune response (Pié et al.,

* Corresponding author: Junliang Deng. E-mail address: dengjl213@126.com

YUAN Wei and Ren Zhihua contributed equally to this work and should be considered co-first authors

2004).

There are a few studies about the effects of cytokines of antibacterial peptide *in vitro*. Tani et al. (2000) reported that the cells of spleen were stimulated with α-defensins *in vitro*, and the levels of IL-4 and IFN-γ were increased. Hancock et al. (2000) reported that antimicrobial peptide CEMA from insecte could inhibit the expression of IL-15 of RAW cells. The innate immune cells released IL-1β by regulating of the human cathelicidin-derived peptide LL-37 (Elssner., 2004). While, LL-37 synergistically enhanced the IL-1β-induced production of IL-6, IL-10, and chemokines of human PBMC (Yu et al., 2007). Hy-Line Brown young roosters for egg were fed the basal diet supplemented with 150, 200, 250, 300 and 350 mg/kg antimicrobial peptide, individually. After 42 days, the relative expression levels of IL-6 gene of of jejunum of the treatment groups were significantly decreased ($P < 0.05$) than those of the control group (except the level of the treatment group fed with 350 mg/kg antimicrobial peptide); the relative expression levels of IFN-γ gene of jejunum of the treatment groups fed with 250, 300 and 350 mg/kg antimicrobial peptide were markedly lower ($P < 0.05$) than those of the control group; and the relative expression levels of TNF-α gene of of jejunum of the treatment groups fed with 300 and 350 mg/kg antimicrobial peptide were significantly lower ($P < 0.05$) than those of the control group (Liu et al., 2012). Lv et al. (2011) reported that the treatment groups fed with 200 and 400 mg/kg antimicrobial peptide improved the expression levels of IL-2 mRNA of spleen of Laying Hens.

However, as far as the authors' knew, there were limited studies evaluating the effects of antibacterial peptide on the serum cytokines. Therefore, our study was conducted to determine the effects of antibacterial peptide on serum cytokines and the appropriate antibacterial peptide supplemental level as an immuno-modulating agent.

## Materials and Methods

### Materials

All chemicals were of the highest grade of purity available. Antibacterial peptide (synthesized by the Styela antibacterial peptide and the defensin) was provided by Rota Bioengineering Co., Ltd, Sichuan, China. Astragalus polysaccharide (AP; net content of 65%) was bought from Centre Biology Co., Ltd, Beijing, China.

### Animals and Experimental Design

Piglets (Landraceoc × Yorkshireoc × Duroc; (21 ± 2) d of age) were purchased from Xin Qiao Agricultural science and technology development co., ltd, Chengdu, Sichuan, China. Piglets were acclimatized for 5 days pioneer to the use. Piglets (average BW of (8.24 ± 0.67) kg) were caged in elevated pens with wire flooring and fed with standard diet (Table 1) for weanling piglets (NRC 1998). Temperature (26 – 27℃) and relative humidity (65% – 70%) remained constant. Food and water were provided ad libitum during the acclimatisation period and through all parts of the study. All piglets used in this study were approved to be healthy. All study experimental manipulations were undertaken in accordance with the Institutional Guidelines for the Care and Use of Laboratory Animals.

Ninety weanling piglets were allotted to 5 treatments in a randomized complete block design for 28 d. Dietary treatments were as follows: (1) control; (2) control + 400 mg AP/kg; (3) control + 250 mg ABP/kg; (4) control + 500 mg ABP/kg; (5) control + 1 000 mg ABP/kg. Six piglets from each group were selected randomly to take blood samples respectively at 32, 39, 46, and 53 days of age.

Table 1 Composition of experimental diets

| Ingredient | % |
|---|---|
| Corn | 53.15 |
| Soybean meal (44% CP) | 25.40 |
| Whey powder | 6.00 |
| Fish meal (62.5% CP) | 4.50 |
| Soybean oil | 2.80 |
| Spray-dried porcine plasma (78% CP) | 3.00 |
| Dicalcium phosphate | 2.0 |
| Limestone | 1.40 |
| Salt | 0.25 |
| L-Lysine HCl | 0.15 |
| DL-Methionine | 0.35 |
| Premix[a,b] | 1.00 |

[a]Provided the following per kilogram of complete diet: vitamin A, 2 250 IU; vitamin $D_3$, 1 050 IU; vitamin E, 16 IU; vitamin $K_3$, 0.9 mg; vitamin $B_{12}$, 0.03 mg; riboflavin, 4.5 mg; niacin, 30 mg; pantothenic acid, 25 mg; choline chloride, 400 mg; folic acid, 0.30 mg; thiamin, 2 mg; pyridoxine, 7 mg; biotin, 0.20 mg; Zn, 80 mg; Mn, 22 mg; Fe, 80 mg; Cu, 225 mg; I, 0.50 mg; and Se, 0.25 mg.

[b]Calculated nutrient composition (as-fed basis): 3 280 kcal of DE/kg, 1.14% lysine, and 0.32% methionine. Analyzed nutrient composition (as-fed basis): 21.1% CP, 0.68% Ca, and 0.59% P

## Quantification of Cytokines

Bloodsamples were collected into microcentrifuge tubes by orbital vein puncture. Serum was obtained after blood centrifugation at 4 000 rpm at 4℃ for 20 min and stored at -20℃. Total IL-2, IL-4, IL-6, IFN-γ, and TNF-α were determined in serum, using enzyme-linked immunosorbent assay (ELISA) kits purchased from America RD Biosciences Co., Ltd. All operations were in accordance with the instructions of the kits.

## Data Analysis

The results were shown as means ± standard deviation ($M \pm SD$). Statistical analysis was performed using one-way analysis of variance (ANOVA) test of SPSS 19.0 software. $P < 0.05$ or 0.01 were considered as significantly different.

# Results

Dietary ABP and AP supplementation increased the levels of IL-2 in serum. At 32 days of age, the value of the ABP group Ⅲ was significantly greater ($P < 0.01$) than that of the control group. The values of the treatment groups were markedly greater ($P < 0.01$) than that of the control group (except the value of ABP group Ⅰ on 39 days) at 39, 46, 53 days of age. The levels of the ABP groups (except the value of ABP group Ⅰ) were superior ($P < 0.01$) to that of the AP group. ABP groups were dose dependent, and the optimum add dosage of ABP was found to be 1 000mg/kg (Table 2).

As we could see from Table 2, during the whole experimentperiod (except on 39 days), the levels of IL-4 of the treatment groups were markedly higher ($P < 0.01$) than that of the control group; the value of ABP group Ⅰ was significantly lower ($P < 0.01$) than that of the AP group; and the values of the other two groups were markedly higher ($P<0.01$ or 0.05) than that of the AP group (except the valuse of ABP group Ⅱ on 32,39,and 46 days of age). ABP groups were dose dependent, and the optimum add dosage of ABP was found to be 1 000mg/kg.

The results were showed in Table 2. During the whole experiment, the levels of IL-6 of the treatment groups were markedly higher ($P<0.01$) than that of the control group (except the value of the ABP group Ⅰ on 32 days); the value of the ABP Ⅰ group was significantly lower ($P<0.01$ or 0.05) than that of the AP group; the values of the other ABP groups were markedly higher ($P<0.01$ or 0.05) than that of the AP group. ABP groups were dose dependent, and the optimum add dosage of ABP was found to be 1 000mg/kg.

As we could see from Table 2, during the whole experimentperiod, the levels of IL-6 of the treatment groups were markedly higher ($P<0.01$) than that of the control group. The value of the ABP group Ⅰ was significantly lower ($P<0.01$) than that of the AP group (except on 32 days). Except there was no differences ($P>0.05$) between the value of the ABP group Ⅱ and that of the AP group at 53 days of age, the values of the other two ABP groups were significantly higher ($P<0.01$ or 0.05) than that of the AP group. ABP groups were dose dependent, and the optimum add dosage of ABP was found to be 1,000mg/kg.

Table 2 Effects of dietary supplementation with the antibacterial peptide on serum cytokines of piglets

| Items | Time | Control group | AP$^{a}$ group | ABP$^{b}$ group Ⅰ | ABP$^{b}$ group Ⅱ | ABP$^{b}$ group Ⅲ |
|---|---|---|---|---|---|---|
| | 32d | 227.72 ±7.36$^{B}$ | 236.04 ±8.66$^{Ba}$ | 228.34 ±11.54$^{B}$ | 233.06 ±10.23$^{Ba}$ | 251.08 ±9.92$^{A}$ |
| IL-2 | 39d | 238.61 ±9.33$^{C}$ | 260.79 ±7.68$^{B}$ | 243.49 ±7.18$^{C}$ | 269.34 ±7.85$^{Ba}$ | 282.59 ±6.15$^{A}$ |
| (ng/L) | 46d | 240.77 ±6.79$^{D}$ | 276.52 ±6.51$^{B}$ | 256.06 ±6.46$^{C}$ | 281.44 ±5.46$^{B}$ | 305.36 ±5.58$^{A}$ |
| | 53d | 228.84 ±11.62$^{D}$ | 276.89 ±6.26$^{B}$ | 251.22 ±5.35$^{C}$ | 280.38 ±6.25$^{B}$ | 304.20 ±5.44$^{A}$ |
| | 32d | 52.48 ±1.86$^{A}$ | 53.63 ±1.94$^{A}$ | 53.75 ±1.52$^{A}$ | 52.63 ±1.67$^{A}$ | 53.82 ±1.55$^{A}$ |
| IL-4 | 39d | 54.16 ±0.85$^{D}$ | 63.22 ±1.21$^{Ba}$ | 58.11 ±1.92$^{C}$ | 65.1 ±1.66$^{AB}$ | 66.67 ±2.05$^{A}$ |
| (ng/L) | 46d | 53.09 ±1.16$^{D}$ | 68.62 ±2.05$^{B}$ | 60.27 ±2.37$^{C}$ | 70.84 ±1.46$^{AB}$ | 72.83 ±1.94$^{A}$ |
| | 53d | 51.17 ±1.32$^{E}$ | 67.01 ±1.54$^{Cb}$ | 57.05 ±1.60$^{D}$ | 69.78 ±2.09$^{Ba}$ | 72.54 ±1.22$^{A}$ |
| | 32d | 463.95 ±8.55$^{B}$ | 486.51 ±9.40$^{A}$ | 471.64 ±9.39$^{Ba}$ | 489.66 ±8.47$^{A}$ | 498.58 ±7.91$^{A}$ |
| IL-6 | 39d | 468.10 ±10.08$^{D}$ | 517.84 ±10.28$^{Cb}$ | 504.09 ±9.62$^{C}$ | 537.22 ±10.48$^{B}$ | 560.42 ±12.32$^{A}$ |
| (ng/L) | 46d | 443.88 ±11.91$^{D}$ | 533.47 ±13.99$^{C}$ | 513.92 ±11.87$^{C}$ | 577.38 ±17.79$^{Ba}$ | 602.06 ±12.48$^{A}$ |
| | 53d | 378.59 ±17.34$^{E}$ | 536.01 ±11.76$^{Cb}$ | 461.85 ±14.65$^{D}$ | 560.61 ±11.85$^{Ba}$ | 583.77 ±12.31$^{A}$ |
| | 32d | 537.45 ±10.81$^{C}$ | 573.86 ±19.00$^{Ba}$ | 569.99 ±12.33$^{B}$ | 598.96 ±10.49$^{Ab}$ | 600.53 ±12.32$^{A}$ |
| IFN-γ | 39d | 547.20 ±11.90$^{E}$ | 625.43 ±11.36$^{C}$ | 592.95 ±12.78$^{D}$ | 666.37 ±11.85$^{B}$ | 692.59 ±12.47$^{A}$ |
| (ng/L) | 46d | 525.01 ±13.41$^{E}$ | 658.16 ±13.47$^{C}$ | 586.81 ±13.65$^{D}$ | 690.61 ±12.47$^{Ba}$ | 716.43 ±10.18$^{A}$ |
| | 53d | 469.57 ±12.41$^{D}$ | 647.42 ±13.01$^{B}$ | 570.29 ±13.91$^{C}$ | 664.16 ±14.05$^{B}$ | 692.10 ±11.18$^{A}$ |
| | 32d | 53.59 ±3.70$^{Db}$ | 60.86 ±4.98$^{BCa}$ | 54.95 ±4.22$^{CD}$ | 62.65 ±4.18$^{ABd}$ | 68.40 ±5.20$^{A}$ |
| TNF-α | 39d | 63.15 ±6.11$^{D}$ | 72.35 ±6.50$^{BCa}$ | 68.34 ±6.99$^{BC}$ | 78.08 ±6.73$^{ABd}$ | 86.20 ±7.23$^{A}$ |
| (ng/L) | 46d | 69.55 ±6.83$^{C}$ | 87.89 ±6.77$^{Ba}$ | 75.40 ±6.25$^{C}$ | 95.33 ±6.11$^{AB}$ | 100.84 ±6.82$^{A}$ |
| | 53d | 64.62 ±6.60$^{D}$ | 91.84 ±6.37$^{B}$ | 76.68 ±6.86$^{Cd}$ | 101.34 ±6.68$^{AB}$ | 108.8 ±6.37$^{A}$ |

abc means with various letters are very significant different ($P<0.01$), and with the same letters but in different case are significant different ($P<0.05$)

The results were showed in Table 2. During the whole experiment, the levels of TNF-αof the treatment groups were higher than that of the control group. Except there was no differences ($P > 0.05$) between the value of TNF-α of the ABP group Ⅰ and that of the control group at 32 and 53 days of age, the values of the treatment groups were significantly higher ($P < 0.01$ or 0.05) than that of the control group. The value of TNF-α of the ABP Ⅰ group was significantly lower ($P < 0.01$ or 0.05) than that of the AP group at 46 and 53 days of age; the levels of the other two ABP groups were significantly higher ($P < 0.01$) than that of the AP group (except the value of ABP group Ⅱ). ABP groups were dose dependent, and the optimum add dosage of ABP was found to be 1 000mg/kg.

## Discussion

Cytokines are crucial protein mediators in immunity (Ma et al., 2007). IL-2 is a key cytokine involved in the regulation of the immune. The functions of T cells, T suppressor cells, macrophages, and natural killer cells are enhanced by this regulation, thereby inducing T cells to generate interferons and activating Th cells (Yin., 2008). During the whole experiment period, ABP and AP given orally had been shown to increase the levels of IL-2, which suggest that the activation of Th (Th1 and Th2) cells and natural killer cells are enhanced. IFN-γ, a major Th1 cytokine produced by Th cells and natural killer cells, and mediated the cell immune response (Okamura., 2004); IL-4 is mainly secreted by Th2 cells, which can promote the proliferation of activated B cells, and mainly mediated humoral immune response. The values of IFN-γ and IL-4 were increased among the whole experiment period, which indicated that ABP and AP could improve the cellular immunity and humoral immunity function of weaned piglets. IFN-γ could synergistic induce tumor necrosis factor, thereby the levels of TNF-α of the piglets fed with ABP and AP were increased among the whole experiment period. At the same time, IFN-γ synergistic IL-2 induced the activation of LAK, and promote the expression of IL-2R of T cells (Billiau., 2000). IL-6 can promote B cells proliferation, differentiation, and produce more immunoglobulin, induction T cells to generate IL-2 and the expression of IL-2R (Davis et al., 2002). Therefore, the levels of IL-2 of the piglets were improved among the whole experiment period. TNF-α induction of cytokine and chemokine synthesis, thereby regulates leukocyte recruitment through both upregulation of adhesion molecules on vascular endothelial cells (Kips., 2001). Collectively, supplementation with ABP and AP increased the values of IL-2, IL-4, IL-6, IFN-γ, and TNF-α among the whole experiment period, those showed that ABP and AP could improve the cellular immunity and humoral immunity function of weaned piglets. The effects of 500 and 1 000mg/kg ABP were better than those of AP. The mechanism of antibacterial peptide on serum cytokines has not been clear, and this this need further studies to confirme.

## Conclusion

The results showed that supplementing weaned piglets diets with ABP enhanced the cellular and humoral immune responses of weaned piglets by improving the levels of IL-2, IL-4, IL-6, IFN-γ, and TNF-α, and the optimum add dosages of ABP was found to be 1 000 mg/kg for 4-week adding.

## Acknowledgments

The present work was supported by the Changjiang Scholars & Innovative Research Team of Ministry of Education of China Funds (Grant No. IRT0848).

## References

Billiau A, Vandenbroeck K, 2000. IFN-γ. W: Oppenheim, J. J, Feldman M. Cytokine Reference. A compendium of cytokines and other mediators of host defence. Academic Press.

Davis ME, Maxwell CV, Brown DC, et al. 2002. Effect of dietary mannan oligosaccharides and (or) harmacological additions of copper sulfate on growth performance and immunocompetence of weanlingand growing/finishing pigs. *Journal of Animal Science*, 80: 2887 – 2894.

Elssner A, Duncan M, Gavrilin M, et al. 2004. A novel P2X7 receptor activator, the human cathelicidin- derived peptide LL37, induces IL-1β processing and release. *Journal of Immunology*, 172: 4987 – 4994.

Hancock REW, Scott MG. 2000. The role of anti-microbial peptides in animal defenses. Proceedings of the National Academy of Sciences of the United States of America, 2000, 97: 8856 – 8861.

Huang ZJ, Kingsolver MB, Avadhanula V, et al. An antiviral role for antimicrobial peptides during the arthropod response to alphavirus replication. Journal of Virology, 87: 4272 – 4280.

Koczulla AR, Bals R. Antimicrobial peptides. *Drugs*, 2003, 63: 389 – 406.

Kips JC. Cytokines in asthma. *European Respiratory Journal*, 2001, 18: 24s – 33s.

Liu LR, Yang KL, Hua J, et al. Antimicrobial Peptides: effects on growth performance, immune indices and mrna expression of related cytokine genes in jejunum of young roosters for egg production. *Chinese Journal of Animal Nutrition*, 2012, 24: 1345 – 1351.

Lv ZZ, Yuan XX, Cai ZW, et al. Effects of antimicrobial peptides on serum immune indices and IL-2 mRNA expression in spleen of laying hens. *Chinese Journal of Animal Nutrition*, 2011, 23: 2183 – 2189.

Ma CS, Nichols KE, Tangye SG. Regulation of cellular and humoral immune responses by the SLAM and SAP families of molecules. *Annual Review of Immunology*, 2007, 25: 337 – 379.

Okamura M, Lillehoj HS, Raybourne RB, et al. Cell-mediated immune responses to a killed Salmonella enteritidis vaccine: lymphocyte proliferation, T-cell changes and interleukin-6 (IL-6), IL-1, IL-2, and IFN-γ production. Comparative Immunology *Microbiology and Infectious Diseases*, 2004, 27: 255 – 272.

Pié S, Lallès JP, Blazy F, et al.. Weaning is associated with an upregulation of expression of inflammatory cytokines in the intestine of piglets. *Journal of Nutrition*, 2004, 134: 641 – 647.

Rossignol T, Kelly B, Dobson C, et al. Endocytosis-mediated vacuolar accumulation of the human ApoE apolipoprotein-derived ApoEdpL-W antimicrobial peptide contributes to its antifungal activity in candida albicans. *Antimicrobial Agents and Chemotherapy*, 2011, 55: 4670 – 4681.

Tani KJ, Murphy WJ, Chertov O, et al. Defensins act as potent adjuvants that promote cellular and humoral immune responses in mice to a lymphoma idiotype and carrier antigens. *International Immunology*, 2000, 12: 691 – 700.

Torrent M, Pulido D, Rivas L, et al. Antimicrobial peptide action on parasites. Current Drug Targets, 2012, 13: 1138 – 1147.

Wang Z, Wang GS. APD: the antimicrobial peptide database. *Nucleic Acids Research*, 2004, 32: D590 – D592.

Yan JX, Wang KR, Chen R, et al. 2012. Membrane active antitumor activity of NK-18, a mammalian NK-lysin-derived cationic antimicrobial peptide. *Biochimie*, 2012, 94: 184 – 191.

Yin Y, Tang ZR, Sun, ZH, et al. Effect of Galacto-mannan-oligosaccharides or Chitosan Supplementation on Cell-mediated immune responses to a killed Salmonella enteritidis vaccine: lymphocyte proliferation, T-cell changes and interleukin-6 (IL-6), IL-1, IL-2, and IFN-γ production. *Asian-Australasian Journal of Animal Sciences*, 2008, 21: 723 – 731.

Yu FS, Cornicelli MD, Kovach MA, et al. Flagellin stimulates protective lung mucosal immunity: role of cathelicidin-related antimicrobial peptide. *Journal of Immunology*, 2010, 185: 1142 – 1149.

Yu J, Mookherjee N, Wee K, et al. Host defense peptide LL-37, in synergy with inflammatory mediator IL-1β, augments immune responses by multiple pathways. *Journal of Immunology*, 2007, 179: 7684 – 7691.

# Effect of molybdenum fed along with cadmium on spleen-related gene expression in goat

GU Xiaolong, CHEN Rongrong, HU Guoliang, ZHUANG Yu, LUO Junrong, ZHANG Caiying, GUO Xiaoquan, HUANG Aiming, CAO Huabin *

*Institute of Animal Population Health, College of Animal Science and Technology, Jiangxi Agricultural University, Nanchang* 330045, *China*

**Abstract**: In this study, the effect of molybdenum fed along with cadmium on spleen-related gene expression in goat was evaluated. Thirty-six goats weighing nearly 20kg each were randomly divided into four groups of nine goats each. Control group was drenched corresponding quantitative deionized water. Three experiment groups were drenched the same amount $CdCl_2$ (0.5mgCd/kg BW) firstly, and the different amount [$(NH_4)_6Mo_7O_{24} \cdot 4H_2O$] (15mgMo/kg BW, 30mgMo/kg BW, 45mgMo/kg BW) secondly, and the three groups were represented as group Ⅰ, Ⅱ and Ⅲ, respectively. Each group was set by three replicates. Spleen tissue was collected from 3 goats of each group after they were slaughtered on 0th, 25th and 50th day to measure expression of apoptosis gene including *Bcl*-2, *Bax*, *Cyt C*, *Caspase*-3, *Smac*and ceruloplasmin. The expression of *Bcl*-2, *Cyt C* and ceruloplasmin in each group was down-regulated significantly ($P < 0.05$), and *Caspase*-3 from each group was up-regulated significantly with the exception of group Ⅰ ($P < 0.05$). The *Bax* in group Ⅱ was up-regulated ($P > 0.05$), but that submitted a descending in group Ⅲ, and no statistical significance was observed on *Smac*. In addition, the changes in ultrastructure of spleen were detected by transmission electron microscope. Results revealed vacuoles in the lymphomacytes and apoptosis bodies of group Ⅲ. Taken together, molybdenum and cadmium could induce goat splenic cell apoptosis by involving in mitochondrial intrinsic pathway, and molybdenum and cadmium showed possible synergistic relationship, meanwhile there exists certain mechanism to repair damage from molybdenum and cadmium.

**Keywords**: Molybdenum, cadmium, spleen, apoptosis, ultrastructure, goat

## 1 Introduction

Molybdenum (Mo) is as an essential trace element for many living species since it is the important component of xanthine oxidase (XOD), aldehyde (AO), sulfite oxidase (SO) (Richert et al., 1953; Mahler et al., 1954; Cohen et al., 1971). The intake of a high concentration molybdenum in animal body produces a state of conditional copper deficiency (Suttle, 1991). Molybdenum toxicity has been reported in different species (Fairhall et al., 1945; Arrington et al., 1953; Dick, 1956; Hogan et al., 1971). The effect of high dietary molybdenum intake on animals depends on species, age, the amount and chemical form of the ingested molybdenum, the copper status and

---

Xiaolong Gu, Rong-rong Chen and Guoliang Hu contributed equally to this work.

* Corresponding author, E-mail address: chbin20020804@163.com

copper intake of the animal (i. e. the copper to molybdenum ratio), and the dietary content of inorganic sulphate and of other sulphur sources such as proteins and sulphur-containing amino acids (Berseny et al., 2008). In 1981, a suspect of Mo toxicity of cattle was initially investigated in China. The disease was "Red Skin and White Hair" syndrome by local famers since it caused persistent diarrhoea, skin redness and hair colour loss in cattle. The incidence of Mo toxicity of livestock and poultry has increased with the deterioration of the environment and industrialization. Cadmium (Cd) is a toxic metal emanated from the natural environment, and particularly due to industrial pollution. Currently, the Agency for Toxic Substances and Disease Registry (*ATSDR*) regards cadmium as the sixth toxic substance which endangers human health. The rates of cadmium exposure in the Jiangxi Province of China are much higher than other territories due to mining. Cadmium is a non biodegradable metal and has a long half-life in the animal's body, approximately 10-35 years. It can be accumulated continuously in living organisms and cause damage to the kidney, liver, lung, bone, spleen, reproductive and immune system. Waisberg et al. (2003) found that cadmium induces oxidative deterioration of lipids, proteins, and DNA. Another report indicated that the induction of cadmium triggers the body to produce large amounts of free radicals, which attack biological macromolecules (Ognjanovic et al., 2010). Although their individual effect of cadmium or molybdenum exposure has been reported in the literature, the effects of the combination of Mo and Cd have not been reported. There have been no reports on the relationship between the combination of molybdenum and cadmium and immunotoxicity of animals and humans. In this study, the spleen-related gene expression and examination of the immunotoxicity of Mo and Cd in the goat are investigated with methods of RT-qPCR and transmission electron microscopy. Furthermore, we tried to characterize the relationship between Mo and Cd *in vivo*.

# 2 Materials and Methods

## 2.1 Treatments

Thirty-six Boer goats were procured from local famer near Jiangxi Agricultural University which were 20kg to 30kg in weighing approximately, 1.5 years to 2 years of age, and randomly separated intofour groups of nine goats. They were acclimatized for two weeks in the college animal farm shed under hygienic conditions before the commencement of the experiment. They were maintained on free water and green fodder of the season. Goats in the control group were drenched corresponding quantitative deionized water. Goats in the experiment group were firstly drenched the same amount $CdCl_2$ (0.5mgCd/kg BW), and the different amount [$(NH_4)_6Mo_7O_{24} \cdot 4H_2O$] (15mgMo/kg BW, 30mgMo/kg BW, 45mgMo/kg BW) secondly. In this case, groups consisted of one control group in one side, and experiment groups (i. e. experiment group Ⅰ, experiment group Ⅱ, and experiment group Ⅲ) in the other side. The daily administration of salts was made between 8:00 and 10:00 a. m. after dissolving them in adequate amount of deionized water, according to goats' body weights. All goats were closely observed for clinical signs and mortality. The time span for experiment persisted was 50 d. All the experiments on goats follow the National Research Council's guide for the care and use of laboratory animal.

## 2.2 Biohydrogenation

Basic biohydrogenation for experimental goats were prepared according tothe standard nutritional requirement for goat breeding by NRC (2007) and the composition of basic biohydrogenation. The levels were shown in Table 1 (omitted) and Table 2 (omitted).

## 2.3 Sample Collection

Three goats from each group were randomly sacrificed with an overdose intravenous injection of sodium pentobarbital. Their spleen tissues were collected and analyzed for expression of apoptosis gene *Bcl*-2, *Bax*, *Cyt C*, *Caspase*-3, *Smac* and ceruloplasmin ( *Cp*) on 0, 25th, 50th day. The changes in ultrastructure of spleen tissues were detected by transmission electron-microscope.

## 2.4 Transmission Electron Microscopy

TEM studies were performed according to the protocol described by Correa Antúnez MI ( Correa Antúnez et al., 2011). The samples of spleen were maintained in 3.5% glutaraldehyde solution until they were moved to a sodium cacodylate buffer solution. Post-fixation was done with 1% osmium tetroxide, followed by dehydration with acetone and uranyl acetate, clarifying with propylene oxide, and embedding in Araldite. Ultrathin sections were taken with an ultramicrotome, loaded onto a grid, Reynolds-stained, and viewed under Transmission Electron Microscope Zeiss 900.

## 2.5 Determination of Spleen-related Gene Expression

Total RNA was extracted using TransZol one-step method. The concentration and purity of each sample were measured by Nucleic acid protein analyzer. Total RNA (1 to 2 μl) was separated by 1.0% denaturing agarose gel electrophoresis. Photos were taken by gel imaging system after the observation of EB dyeing RNA bands was performed. The ratio of gray level from 28s and 18s band was kept about 2.0. Primer was designed using software Primer Express3.0 according to the sequence of Bcl-2, Bax, Cyt C, Caspase-3, Smac, Cp and Actin (β-Actin) of goat published in GenBank. Primers and probes were designed in the area of relatively conservative, respectively. The 5′ end of probe was labeled with *FAM* and the 3' end with TEMRA. Eventually they were synthesized by Shanghai Ying Jun Technology Company. The primer sequences were shown in Table 3. Reverse transcription was performed via a kit provided by Quan Shi Jin Biological Technology Company. The TaqMan Probe method was used for Real Time qPCR, and it was operated using a TaKaRa Premix Ex *Taq*™ (Probe qPCR) kit (TaKaRa Bio Inc).

## 2.6 Statistical Analysis

Results were analyzed by software: SDS 2.4 and RQ Manager 1.2.1. Ct referred to cycle threshold. $\Delta Ct$ (sample) = $Ct$ (target gene) − $Ct$ (reference gene); $\Delta Ct$ (calibrator) = $Ct$ (target gene) − $Ct$ (reference gene); $\Delta\Delta Ct = \Delta Ct$ (sample) − $\Delta Ct$ (calibrator); $2^{-\Delta\Delta CT}$ represented relative expression of initial cDNA of target gene. RQ generated automatically using software was relative expression of cDNA, and data was exported. We conducted the analyses using SPSS 17.0. All data were presented in mean ± *S. D.* All experimental data were tested by analysis of variance with significant differences between means determined by the method of Duncan. Values were considered significant statistically when $P < 0.05$ and $P < 0.01$.

# 3 Results

## 3.1 Effect of Molybdenum Fed Along With Cadmium on Spleen Pro-apoptosis and Anti-apoptosis Gene Expression in Goat

Fig 1 − 4 shows the effect of molybdenum fed along with cadmium on spleen gene *Caspase*-3, *Cyt C*, *Smac* and *Bax* expression in goat, respectively. Expression of *Caspase*-3 from each experiment group increased on 50th day ($P <$ 0.01) when comparing with control group, excepting for which is from experiment group Ⅰ, expression of *Bax* from experiment group Ⅰ ascended on 50th day ($P > 0.05$) when comparing with control group, which was from experiment group Ⅱ rise on 25 − 50th day ($P > 0.05$), whereas that submitted a decreasing in group Ⅲ on 50th day ($P < 0.01$). Expression of *Cyt C* from group Ⅲ was also found a decline on 25th and 50th day ($P < 0.01$)

when comparing with control group, and no statistical significance was observed on *Smac*. Fig. 5 shows variation of *Bcl*-2 expression, which from experiment group Ⅲ decreased on 25th day ($P<0.05$) when comparing with control group, which was from each experiment group declined on 50th day ($P<0.01$).

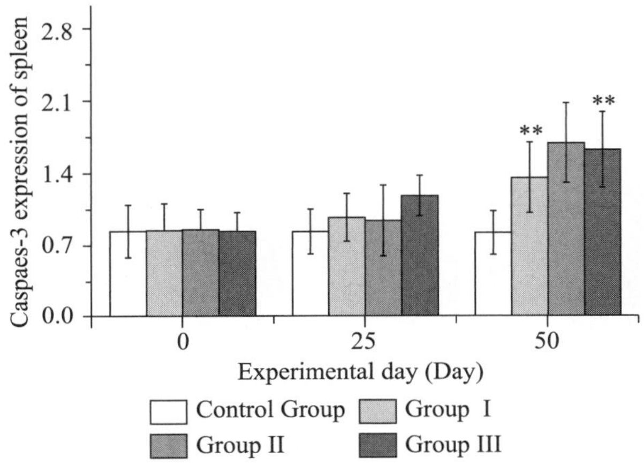


Fig. 1

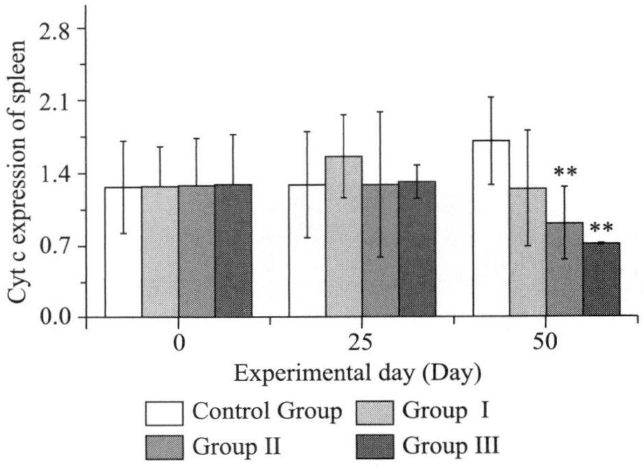


Fig. 2

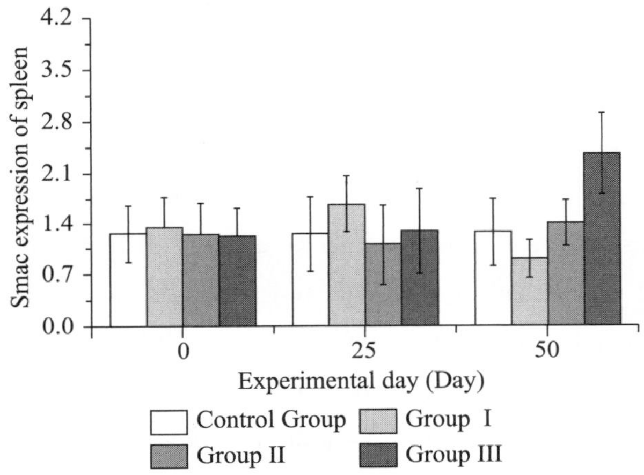


Fig. 3

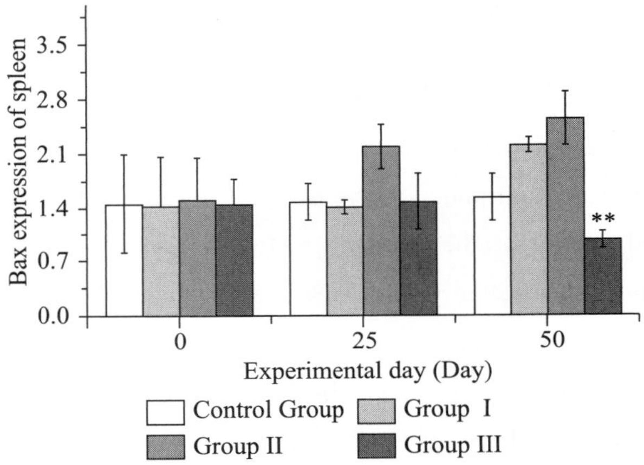


Fig. 4

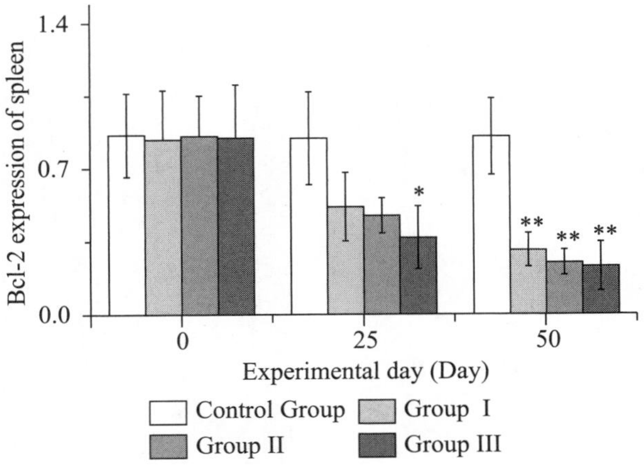


Fig. 5

Fig. 1-5 Effect of molybdenum fed along with cadmium on spleen pro-apoptosis and anti-apoptosis gene expression

## 3.2 Changes of Spleen *Cp* Expression

The effect of molybdenum fed along with cadmium on spleen *Cp* expression in goat was shown in Fig. 6. Expression of *Cp* from each experiment group descended on 25th day ($P<0.05$) when comparing with control group, excepting for experiment group Ⅱ, which was from each experiment group decreased on 50th day ($P<0.01$).

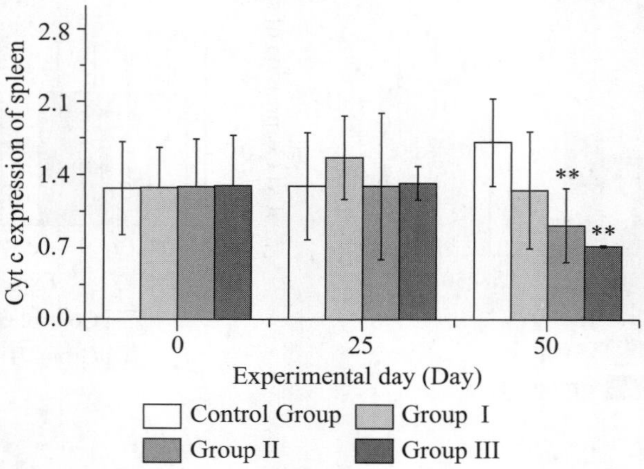


Fig. 6 Effect of molybdenum fed along with cadmium on spleen ceruloplasmin gene expression

## 3.3 Pathological Lesions

The apoptosis was also evaluated in ultrastructure of spleen in goat. As shown in Fig. 7, rough endoplasmic reticulum and nucleus were obvious in the cytoplasm and had normal structures in splenic lymphocyte cells from the red pulp area. The appearance of cells from the white area was consistent with which shown in Fig. 8. In contrast, as Fig. 9 shown, the lymphocytes decreased significantly, even dissolved and disappeared. And, apoptosis phenomenon and apoptotic body was displayed, perinuclear space of individual lymph cell nucleus became wider, its endoplasmic reticulum fractured and dissolved moreover. Swelling and vacuoles of mitochondria was observed. Fig. 10 demonstrated the effect of cadmium and molybdenum on lymphocytes of spleen in the white plup area. It illustrated apoptosis of lymphocytes, pyknosis and agglomerated lump of nuclear, unclear structure. And, vacuoles and apoptotic body was observed additionally.

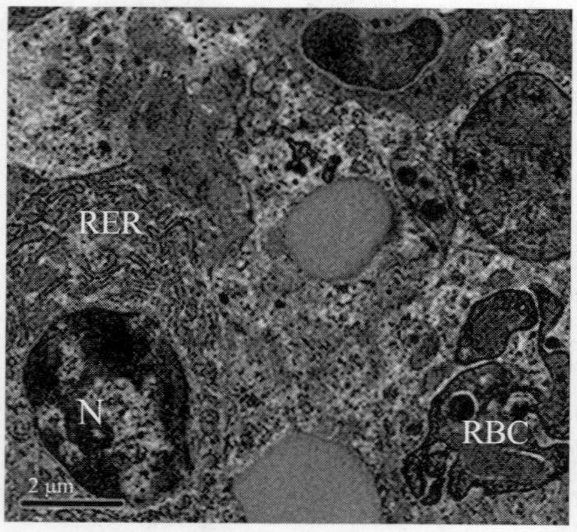


Fig. 7

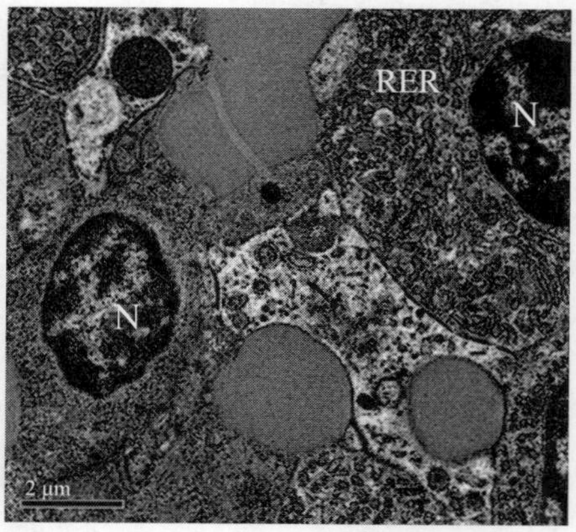


Fig. 8

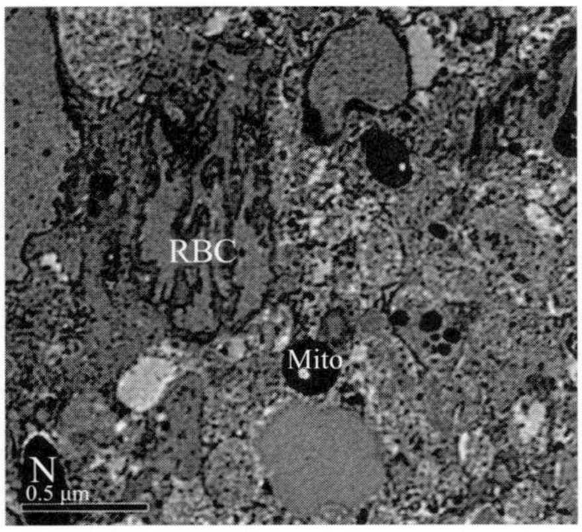


Fig. 9

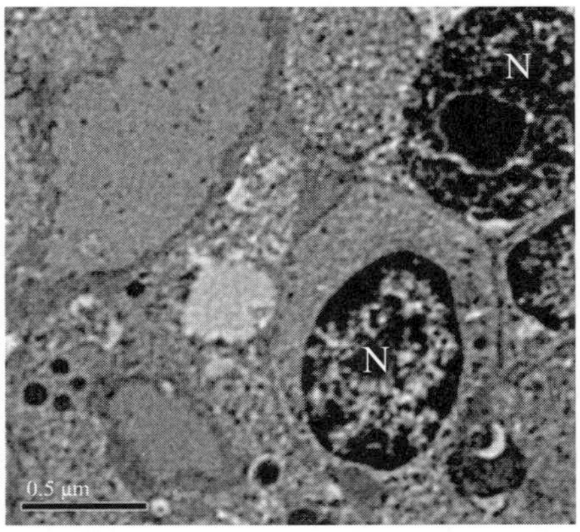


Fig. 10

Fig. 7-10 Effects of molybdenum fed along with cadmium on ultrastructure of spleen in goat 7splenic red pulp from control group, round nuclear lymphocyte, abundant organelle, erythrocyte and mononuclear cell are visible (6 800 ×); 8 splenic white pulp from control group, many lymphocytes, obvious organelle (6 800 ×); 9 splenic red pulp from high cadmium high molybdenum group, lymphomcytes decrease, serious apoptosis, broad perinuclear space, mitochondrial vacuolation (6 800 ×); 10 splenic white pulp from high cadmium high molybdenum group, apoptosis in lymphomcytes, condensed nuclear, less organelle, apoptosis body are visible (6 800 ×)

## 4 Discussion

Molybdenum is well known for its antagonistic effect with copper. However, the molecular mechanisms linking copper deficiency induced by molybdenum to apoptosis have not been identified. Cadmium could alter mitochondrial function and structure, and then induce cell apoptosis by endogenous apoptosis pathway. It could be mainly separated as three aspects: (1) There are many protein contained sulfydryl (-SH) on the mitochondrial outer membrane. Adjacent -SH may interreact after the bond of cadmium with it since high affinity between cadmium and -SH. They promote the formation of disulfide bond and the increase of mitochondrial membrane permeability. Eventually, matter which induces apoptosis within plasma membrane such as Cyt C, Smac release. (2) There are a lot of metallothionein out of the mitochondrial membrane. Cadmium is easy to combine it. The combination will stimulate a lot of reactive oxygen species (ROS) which induce apoptosis of cells by attacking DNA. (3) Cadmium is inclined to compete with calcium on mitochondrial inner membrane since their similar atom size, and the mitochondrial membrane permeability increase when -SH is influenced by cadmium. And the activity of $Ca^{2+}$, $Mg^{2+}$-ATPase on plasma membrane may also be affected. As a result, transport of intracellular $Ca^{2+}$ is inhibited and intracellular endonuclease is activated (Jingming et al., 2005).

Unlike previously identified oncogene, *Bcl*-2, which is belonged to anti-apoptotic members, is found that it may inhibit cell death rather than promoting cell proliferation (Vaux et al., 1988). *Bax* as a monomer exists in the cytoplasm in healthy cells, or adheres loosely on the plasma membrane. *Bax* experiences a conformational change under death stimuli, and infuse into mitochondrial outer membrane when oligomerization happens (Cory et al., 2003; Gross et al., 1999). *Bax* oligomerization is thought to increase the mitochondrial outer membrane permeability. It allows the release of apoptotic proteins, like *Cyt C* into the cytoplasm. As shown in Fig. 5, the toxicity from molybdenum fed along with cadmium significantly inhibit the expression of *Bcl*-2. The expression of *Bcl*-2 from each experiment group aparting from experiment group Ⅱ present a decreasing trend with the extension of experiment time. It indicates that the toxicity from molybdenum fed along with cadmium could induce the apoptosis of spleen cells through decreasing expression of *Bcl*-2. And Fig. 4 shows, expression of *Bax* from experiment group Ⅰ and Ⅱ increase on 50th day ($P > 0.05$) when comparing with control group, whereas which is from experiment group Ⅲ decrease on 50th day ($P < 0.05$) significantly. By and large, the results above mentioned is coincidence with the observation in (Li et al., 2003), and Mo facilitates the toxicity of Cd according to time × dose-depended effect. But we also observe the down-regulation of Bax when treated with high level molybdenum fed along with cadmium, it indicates that organism try to repair the apoptosis induced by the external environment under the effect of cadmium and molybdenum. In meanwhile, as shown in Fig. 3, compared with control group, the expression of *Smac* from each group is either decreasing or increasing. However, there is no statistical significance. *Smac* promote apoptosis through mediating the interaction between release of Caspases and IAPs (Inhibitor of Apoptosis Proteins). N terminal of *Smac* could induce the activation of *Caspase*-3 mediated by *Cyt C* (Kashkar et al., 2003). The mechanism of restoration is further certified by which is shown in Fig. 2, expression of *Cyt C* from each experiment group decline on 50th day ($P < 0.01$) when comparing with control group excluding for which from experiment group Ⅰ. *Cyt C* generally exists in cristae and narrow cristae joint of the inner mitochondrial membrane (Pellegrini et al., 2007). *Cyt C* is mainly mediation of apoptosis (Newmeyer et al., 1994; Kluck et al., 1997). The release of *Cyt C* could cause DNA damage (Glazunova et al., 2008; Ravi et al., 2004). However, the data in this paper is contradictory with expression of other apoptosis factor or former research. The phenomenon may be due to the interaction of pro-apoptosis and anti-apoptosis or the compensatory mechanism of cells, or some unknown reason.

Furthermore, the common role of molybdenum and cadmium should be noticed. Fig. 1 shows that expression of *Caspase*-3 from each experiment group ascend on 50th day ($P < 0.01$) except for which from experiment group Ⅰ when comparing with control group. The effect of *Caspase*-3 display a time-dependent increasing. Caspases of apoptosis in mammalian has been divided into initial caspases (*Caspases*-2, -8, -9, -10) and effect caspases (*Caspases*-3, -6, -7) according to their structure and function. Effect caspases as executor of apoptosis are the final effect protein of cell apoptosis (Sprick et al., 2004). Some researchers also indicated that caspases played an important role on cell apoptosis induced by cadmium. It indicates that toxicity from effect by molybdenum fed along with cadmium could induce the apoptosis of spleen cells. In this case, the expression of *Cp* in this experiment declined since damage of spleen. It has been shown in Fig. 6, the expression of *Cp* from each experiment group decrease on 50th day ($P < 0.01$) when Comparing with the control group. *Cp* is a copper glycoprotein primarily synthesized by the liver, and it binds 90% – 95% of blood plasma copper (Takahashi et al., 1986). There is a small amount of synthesis in the spleen, lung, brain and testis (Healy et al., 2007). In addition, *Cp* focuses on transport of copper, it has activity of superoxide dismutase and amino oxidase (Koschinsky et al., 1986). It has also been reported that *Cp* has antioxidant activity, it may scavenge ROS such as superoxide ($O_{2-}$) and hydroxyl radicals ($OH^-$) (Takahashi et al., 1986). Lamb was short of copper had been found its *Cp* in plasma, SOD, catalase and glutathione peroxidase in erythrocyte decreased. And, MDA in erythrocyte increased, resulting in inhibition of antioxidant capacity and oxidative damage (Saleha et al., 2008). Moreover, the observation in ultrastructure of spleen facilitates the integrality of this research. The vacuolation of lymphocyte mitochondria and apoptosis body in spleen from experiment group Ⅲ are visible learned from electron microscopy observation of spleen. Mitochondrial apoptosis protein was released into the cytoplasm due to the mitochondrial damage, and finally cell apoptosis was induced (Roset et al., 2007; Palmer et al., 2000). This result suggests that cadmium and molybdenum induces the apoptosis of spleen cells via multiple pathways.

## 5 Conclusion

In summary, a long intake to cadmium and molybdenum produces a predominant apoptotic mechanism of splenic cell death. It involves the mitochondrial intrinsic pathway, as evidenced by nuclear condensation, apoptosis body, activation of *caspases*-3, increase of *bax* and decline of *bcl*-2. However, the decrease of *Cyt C* and irresolution of *Smac* suggests that spleen exists certain mechanism to repair damage from Cd and Mo.

## Conflict of Interest Statement

The authors declare that they have no conflict of interest.


## Acknowledgements

The research was supported by the Program of the National Natural Science Foundation of China (project No. 31101863, Nanchang, P. R. China), National Natural Science Foundation of Jiangxi Province (project No. 2010GQN0052, Nanchang, P. R. China).


## References


Arrington, L. R., Davis, G. K., 1953. Molybdenum toxicity in the rabbit. *Journal of Nutrition* 51, 295 – 304.

Bersényi, A, Berta, E, Kádár, I., et al., 2008 Effect of high dietary molybdenum in rabbits. *Acta Veterinaria Hungarica*, 56, 41 – 45.

Cohen, H. J., Fridovich, I. and Rajagopalan, K. V., 1971. Hepatic sulphite oxidase. A functional role for molybdenum. *The Journal of Biological Chemistry*, 246, 367 – 373.

Correa Antúnez, M. I., Morán Penco, J. M., Amaya Lozano, J. L., et al., 2011. Influence of intestinal resections on biliary composition and liver ultrastructure. *Clinical Nutrition*, 30, 247 – 251.

Cory, S., Huang, D. C., Adams, J. M.. The Bcl-2 family: roles in cell survival and oncogenesis. *Oncogene*, 2003, 22, 8590 – 8607.

Dick, A. T. Molybdenum and copper relationships in animal nutrition. In: McElory, W. D., Glass, B. (eds.), A Symposium on inorganic Nitrogen Metabolism: Functions of Metallo-Flavoproteins. Baltimore. Baltimore, MA, 1956, pp. 445 – 473.

Fairhall, L. T., Dunn, R. C., Sharpless, N. E., et al. Toxicity of molybdenum. U. S. Public Health Service, Public Health Bulletin, 1945: 293, 1 – 36.

Glazunova, V. A., Shtil, A. A.. Mitochondrial mechanisms of apoptosis in response to DNA damage. Molecular Biology (Moscow), 2008, 42, 765 – 771.

Gross, A., Yin, X. M., Wang, K., et al. Caspase cleaved BID targets mitochondria and is required for cytochrome c release while Bcl-XL prevents this release but not tumor necrosis factor-R1/Fas death. *The Journal of Biological Chemistry*, 1999, 274, 1156 – 1163.

Healy, J., Tipton, K.. Ceruloplasmin and what it might do. *Journal of Neural Transmission*, 2007, 114, 777 – 781.

Hogan, K. G., Money, D. F., White, D. A., et al. Weight responses of young sheep to copper and connective tissue lesions associated with the grazing of pastures of high Mo content. *New Zealand Journal of Agricultural Research*, 1971, 14, 687 – 701.

Jingming, Z., Zhanqi, L., Fenfang, G.. The Bcl-2, Bax and renal cell apoptosis induced by cadmium. *Foreign Medical Hygiene Pathology*, 2005, 32, 201 – 205.

Kashkar, H., Haefs, C., Shin, H., et al. XIAP-mediated caspase inhibition in Hodgkin's lymphoma-derived B cells. *The Journal of Experimental Medicine*, 2003, 198, 341 – 347.

Kluck, R. M., Bossy-Wetzel, E., Green, D. R., et al. The release of cytochrome c from mitochondria: a primary site for Bcl-2 regulation of apoptosis. *Science*, 1997, 275, 1132 – 1136

Koschinsky, M. L., Funk, W. D., Van Oost, B. A., et al. Complete cDNA sequence of human preceruloplasmin. *Proceeding of the National Academy of Sciences*, 1986, 83: 5086-5090.

Li, M., Xia, T., Jiang, C. S., et al. Cadmium directly induced the opening of membrane permeability pore of mitochondria. Which possibly involved in cadmium triggered apoptosis. *Toxicology*, 2003, 194, 19 – 33

Mahler, H. R., Mackler, B., Green, D. E., et al. ldehyde oxidase. A molybdoflavoprotein. *The Journal of Biological Chemistry*, 1954, 210, 465 - 480.

Newmeyer, D. D., Farschon, D. M., Reed, J. C., Cell-free apoptosis in Xenopus egg extracts: inhibition by Bcl-2 and requirement for an organelle fraction enriched in mitochondria. *Cell*, 1994, 79, 353 - 364.

Ognjanovi, B. I., Marković, S. D., Ethordević, N. Z., et al. Cadmium-induced lipid peroxidation and changes in antioxidant defense system in the rat testes: protective role of coenzyme Q10 and Vitamin E. *Reproductive Toxicology*, 2010, 29, 191 - 197.

Palmer, A. M., Greengrass, P. M., Cavali, A. D.. The role of mitochondria in apoptosis. *Drug News and Perspectives*, 2000, 13, 378 - 384.

Pellegrini, L., Scorrano, L.. A cut short to death: Parl and Opa1 in the regulation of mitochondrial morphology and apoptosis. *Cell Death and Difference*, 2007, 14, 1275 - 1284.

Ravi, D., Das, K. C.. Redox-cycling of anthracyclines by thioredoxin system: increased superoxide generation and DNA damage. *Cancer Chemotherapy and Pharmacology*, 2004, 54, 449 - 458.

Richert, D. A., Westerfeld, W. W.. Isolation and identification of the xanthine oxidase factor as molybdenum. *The Journal of Biological Chemistry*, 1953, 203, 915 - 923.

Roset, R., Ortet, L., Gil-Gomez, G.. Role of Bcl-2 family members on apoptosis: what we have learned from knock-out mice. *Frontiers in Bioscience*, 2007, 12, 4722 - 4730.

Saleha, M. A., Al-Salahyb, M. B., Sanousic, S. A.. Corpuscular oxidative stress in desert sheep naturally deficient in copper. *Small Ruminant Research*, 2008, 80, 33 - 38.

Sprick, M. R., Walczak, H.. The interplay between the Bcl-2 family and death receptor-mediated apoptosis. *Acta Biochimica et Biophysica Sinica*, 2004, 1644, 125 - 132.

Suttle, N. F.. The interactions between copper, molybdenum, and sulphur in ruminant nutrition. *Annual Review of Nutrition*, 1991, 11, 121-140.

Takahashi, N., Ortel, T. L., Putnam, F. W.. Single-chain structure of human ceruloplasmin: The complete amino acid sequence of the whole molecule. *Proceeding of the National Academy of Sciences*, 1986, 81, 390-394.

Vaux, D. L., Cory, S., Adams, J. M.. Bcl-2 gene promotes haemopoietic cell survival and cooperates with c-myc to immortalize pre-B cells. *Nature*, 1988, 335, 440 - 442.

Waisberg, M., Joseph, P., Hale, B., etal. Molecular and cellular mechanisms of cadmium carcinogenesis: a review. *Toxicology*, 2003, 192, 95 - 117.

# Hepatoprotective effects of tocotrienols on lipid peroxidation, antioxidant scavenging enzymes, and *PPAR* expression in the liver of high-fat diet-fed mice

Froilan Bernard R. Matias[1,2], Qiong WEN[1], Zijiu WANG[1], Haibin HUANG[1], Jing WU[1], Lixin WEN[1,3 *]

[1]*Department of Veterinary Clinical Medicine, College of Veterinary Medicine, Hunan Agricultural University, Changsha* 410000, *P. R. China*, [2]*Department of Pathobiology, College of Veterinary Science and Medicine, Central Luzon State University, Science City of Muñoz* 3120 *Nueva Ecija, Philippines*, [3]*Hunan Sevoc Ecological Agriculture and Husbandry Technology Co. , LTD, Changsha* 410000, *P. R. China*; *Equal contributors*

**Abstract**: Scope Prolonged consumption of high fat diet primarilyaffects the liver thru the production of ROS that causes oxidative stress. Tocotrienols are antioxidants belonging to the vitamin E group. However, evaluation of tocotrienols physiological effects is very few compared to tocopherols. This study evaluated the hepatoprotective effects of tocotrienols on the liver of high fat diet-fed mice. Specifically, its effect on lipid peroxidation, antioxidant scavenging enzymes and expression of PPAR α and ɤ were analysed.

Methods and Results Kunming mice wereuniformly divided into 6 groups. The first group served as normal control while the second group served as high-fat group. Group 3 served as 2$^{nd}$ normal control given with tocotrienol (200mg/(kg · day)). Groups 4, 5, and 6 served as treatment groups given with high fat diet and different doses of tocotrienols (200, 100, and 50 mg/(kg · day), respectively). Tocotrienols greatly reduced lipid peroxidation while influences the total antioxidant capacity of the liver. It also increased antioxidant scavenging enzymes and increased the expression of PPAR α and ɤ.

Conclusion Fat accumulation in the liver can initiate chains of events the negatively affects its integrity. However, tocotrienols, as a hepatoprotective compound, acts in more ways than one. It not only counteracts oxidants but also helps in the expression of other oxidant-neutralizing compounds and genes that help regulate the cells metabolism.

**Keywords**: Tocotrienol, lipid peroxidation, antioxidant scavenging enzymes, PPAR, hypercholesterolemia

## Introduction

Liver is a multi-tasking organ that is primarily involved in metabolic homeostasis of the body but is also involved in detoxifying the blood by filtering-out harmful toxin. However, the liver can be vulnerable while performing its function, thus with continuous assault can gradually reduce its function. Hypercholesterolemia is a condition characterized by high level of cholesterol in the blood. This condition has been associated with diets containing high amount of saturated fats and cholesterol that can be aggravated by the presence of physiological or

* Corresponding author, E-mail add: sfwlx8015@ sina. com

pathological disorders. Liver is the primary organ that regulates the amount of cholesterol in the blood. However, in the presence of overwhelming amount of fats can significantly affect the liver's functions which can further lead to chains of physiological events. Fatty liver is the most common metabolically linked disease of the liver (reference) where there is an excessive build-up of fats in the liver cells. Fatty liver can be due to too much caloric intake and/or inability of the liver to break-down fats, such as cholesterol.

Reduction-oxidation (redox) is a fundamental reaction that occurs inside the cell during cellular metabolism produces reactive oxygen species (ROS) even in normal biological conditions[1]. ROS are highly reactive molecules that can damage or alter cell structures at high concentration[2]. ROS or oxidants are counterbalanced by antioxidant systems, thus blocks its harmful effects, however, if the oxidant level is overwhelming causing a shift in equilibrium can cause a condition called "oxidative stress"[2].

Vitamin E is a group of fat soluble vitamins, popularly known to be antioxidants, divided into 2 majors groups, namely tocopherols and tocotrienols. These groups can each be further classified into 4 compounds based on their phenol attachments and given with Greek initials ($\alpha$, $\beta$, $\gamma$, $\delta$). Between the 2 major groups, researches involving tocotrienols are very few compared to tocopherols[3].

In this study, the hepatoprotective effects of tocotrienol on the liver of high-fat diet fed mice were evaluated. Specifically, this study measured and analysed; malondialdehyde (MDA) for lipid peroxidation and total antioxidant capacity (TAOC), antioxidant scavenging enzymes superoxide dismutase (SOD), catalase (CAT) and glutathione peroxidase (GSH-Px), and nuclear receptors peroxisome proliferator activated receptors (PPAR) alpha ($\alpha$) and gamma ($\gamma$) were evaluated.

## Materials and Methods

### Reagents and Kits

The MDA, SOD, CAT, GSH-PX and T-AOC test kits were bought from the Nanjing Jiancheng Bioengineering Institute (Nanjing, P. R. China 210000), while the BCA protein assay kit was purchased from Beyotime Institute of Biotechnology Co., Ltd. (Shanghai 1500).

The TRIzon Reagent, diethyl procarbonate (DEPC), tris acetate EDTA (TAE), *Taq*MasterMix and DNA Marker 1000 were purchased from Beijing CoWin Bioscience Co., Ltd (Bejing 102206). The Ethylene bromide (EB) (10mg/ml) and RevertAid First Stand cDNA Synthesis Kit (K1621) were purchased from Thermo Fisher Scientific Inc. The tocotrienol used in the study was provided by the Hunan Sevoc Ecological Agriculture and Husbandry Technology Co., Ltd, (Changsha, P. R. China 410000). The following instruments were used in the experiment; Infinite® M2100 Pro NanoQuant™ (TECAN Group Ltd. Switzerland), Veriti® Thermal Cycler (Applied Biosystems®), (DYY-6C) Electrophoresis Power Supply (Beijing Liuyi Instrument Factory).

### Feed Formulation and Experimental Animals

The customized high-fat feeds were commissioned in Hunan SJA Laboratory Animal Co., Ltd. with the following contents; 1% cholesterol, 10% egg yolk powder, 10% lard, 0.2% bile acid sodium and 78.8% basal feeds.

Table 1　Diet and treatment dose per group

| Group | 1 | 2 | 3 | 4 | 5 | 6 |
|---|---|---|---|---|---|---|
| Feeds | NBD | HFD | NBD | HFD | HFD | HFD |
| Tocotrienol | A | A | B | B | B | B |
| dose * | - | - | 200 | 200 | 100 | 50 |

NBD - normal basal diet, HFD - high-fat diet

* given based on mg/kg bodyweight/day

A - given with physiologic saline, B- given with tocotrienol

Forty eight (48) 3 – 4 weeks old male Kunming mice were purchased from Hunan SJA Laboratory Animal Co., Ltd. The mice were acclimatized for 5 days before subjecting to the experiment. The mice were divided uniformly into 6 groups with 8 mice per group based on gain in weight having no significant differences ($P < 0.05$) in means between groups. Normal basal diet was given to group 1 and while high-fat diet was given to groups 2, 4, 5 and 6 (shown in Table I) for 2 months. Drinking water was given *ad libitum* while the room temperature was kept at 21 – 25℃. Oral administration of the treatments was given once a day based on the dose shown in Table 1. The weight was measured once a week. This research has been approved by the ethics committee and all experimental mice were treated in accordance with the approved protocol for experimental animal handling and experimentation of the College of Veterinary Medicine, Hunan Agricultural University.

## Sample Preparation

The mice were fasted for 12 hours before sacrificing. Each liver sample was rinsed with physiologic saline to remove blood and dried up using filter paper. About 100 grams of liver tissue from each sample was homogenized in 1ml PBS using a handheld homogenizer. The homogenates were centrifuged at 3 000 rpm for 10-15 minutes. The supernatant was collected and stored at -20℃ until used. Another 100 grams of liver tissue were collected from each sample for RNA extraction.

## Oxidative Stress Markers

Liver total antioxidant capacity (TAOC), malondialdehyde (MDA), superoxide dismutase (SOD), catalase (CAT) and glutathione peroxidase (GSH-Px) levels were measured in accordance with the instruction provided in the test kits. Protein content of each sample was measures using BCA protein assay kit following the protocol provided in the kit. The values of each marker were normalized based on the protein content (data not shown) of the samples.

## RNA Extraction and Reverse Transcription-polymerase Chain Reaction (RT- PCR) Analysis

RNA extraction was performed using TRizon reagent following the manufacturer's instruction. The RNA quality was checked by electrophoresis and analysed using Gel-Pro analyser. Synthesis of first strand cDNA from RNA templates was performed using RevertAid First Strand cDNA Synthesis Kit following the manufacturer's procedures. The primers used for peroxisome proliferation-activated receptor (PPAR) $\alpha$ and $\gamma$ and $\beta$-actin are provided in Table 2.

Table 2 Primer sequences for PCR

| Gene | | Primer Sequence | Length | Annealing temperature | Cycles |
|---|---|---|---|---|---|
| PPAR-α | Forward | ACGCTGGGTCCTCTGGTT | 147bp | 55℃ | 30 |
| | Reverse | GTCTTGGCTCGCCTCTAA | | | |
| PPAR-ɣ | Forward | TCTCTCCGTAATGGAAGACC | 474bp | 50℃ | 30 |
| | Reverse | GCATTATGAGACATCCCCAC | | | |
| β-actin | Forward | CATCCTGAGTCTGGACCT | 499bp | 55℃ | 30 |
| | Reverse | GTACTTGCGCTCAGGAGGAG | | | |

## Statistical Analysis

The data wereanalysed using SPSS 17.0 software. Statistical analyses of the difference between each group were performed using one-way ANOVA. The data were presented as the mean ± standard deviation. A $P$ value of $<0.05$ was considered statistically significant while $P<0.001$ was considered highly significant.

## Results

### Oxidative Stress Markers

### Lipid Peroxidation and Total Antioxidant Capacity

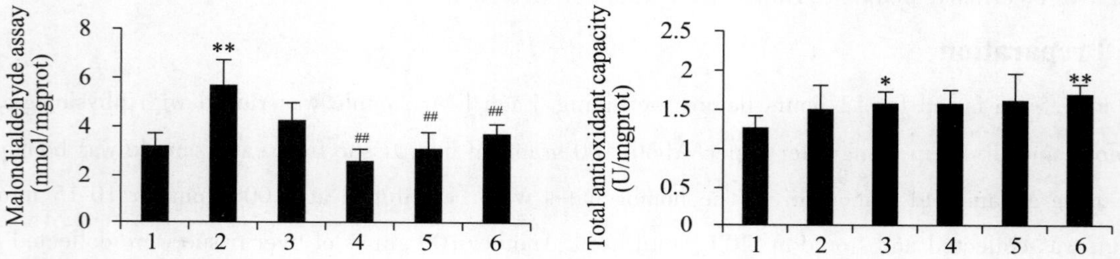


Figure 1 Liver malondialdehyde (MDA) and total antioxidant capacity of high-fat diet fed mice given with oral tocotrienols

Superscript * or ** is significant ($P<0.05$) or highly significant ($P<0.01$) compared to Group 1 while superscript# or## is significant ($P<0.05$) or highly significant ($P<0.01$) compared to Group 2

High-fat diet fedgroups given with tocotrienol showed lower MDA, specifically, group 6 given with 50mg/kg tocotrienol showed significant difference ($P<0.05$) while groups 4 and 5, given with 200 and 100 mg/kg tocotrienol, respectively showed highly significant difference. MDA results were found to have inverse relationship with the amount of tocotrienol administered. On the other hand, high total antioxidant capacity results were observed in all groups given with high fat alone or with tocotrienols. However, only group 3 showed significant result ($P<0.05$) and group 6 showed highly significant result ($P<0.01$) as compared to the control group (group 1) given with normal basal diet.

### Antioxidant Scavenging Enzymes

Table 3 Effects of tocotrienols on antioxidant scavenging enzymes of mice after 2 months

| Group | TSOD (U/mgprot) | CAT | GSH-PX (U/mgprot) |
|---|---|---|---|
| 1 | 63.66 ± 20.75 | 8.68 ± 0.91 | 374.39 ± 31.11 |
| 2 | 71.66 ± 11.75 | 7.99 ± 1.11 | 404.81 ± 25.08* |

(continued table)

| Group | TSOD (U/mgprot) | CAT | GSH-PX (U/mgprot) |
|---|---|---|---|
| 3 | 85.38 ± 14.06* | 10.02 ± 1.42 | 519.96 ± 19.20**## |
| 4 | 71.14 ± 15.27 | 9.52 ± 1.46 | 450.96 ± 23.51**## |
| 5 | 74.31 ± 13.40 | 7.42 ± 0.55 | 337.23 ± 12.93**## |
| 6 | 69.32 ± 24.12 | 6.92 ± 0.45* | 312.35 ± 25.26**## |

* or ** is significant ($P<0.05$) or highly significant ($P<0.01$) compared to Group 1

\# or## is significant ($P<0.05$) or highly significant ($P<0.01$) compared to Group 2

Significant high SOD result ($P<0.05$) was only found in group 3 which was given with normal basal diet and administered with 200 mg/kg tocotrienol, while significantly low CAT result ($P<0.05$) was found in group 6, given with high fat diet and 50 mg/kg tocotrienol. Compared to group 1, significant result was found in group 2, while highly significant results were found in groups 3, 4, 5 and 6. Compared to group 2, highly significant results were found in groups 3, 4, 5 and 6. GSH was also found to have a direct relation with the amount of tocotrienol administered. In general, group 3 fed with normal basal diet and given with 200 mg/kg tocotrienols showed to have highest antioxidant namely SOD, CAT and GSH-Px.

## Peroxisome Proliferator Activated Receptors

As shown in Figure 2 A, no significant difference was found in the PPAR-α expression between group 1 and 2. Also no significant difference was found between 1 and 3. However, highly significant results were found in groups 5 and 6, given with 100 and 50 mg/kg tocotrienol, respectively, while group 4, given with 200 mg/kg tocotrienol, showed no significant difference.

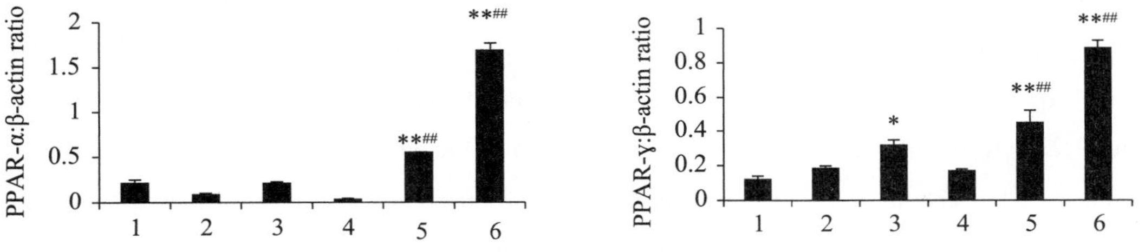


Figure 2 Effect of tocotrienols on the (A) PPAR-α and (B) PPAR-γ in mice.

The superscript * or ** is significant ($P<0.05$) or highly significant ($P<0.01$) compared to Group 1, while superscript# or## is significant ($P<0.05$) or highly significant ($P<0.01$) compared to Group 2

As shown in Figure 2 B, significant result ($P<0.05$) was shown in group 3 as compared to group 1 while no significant result was found between group 1 and 2. Highly significant results were found in groups 5 and 6 as compared while no significant difference in group 4.

## Discussion

High fat diet has been found to negatively affect the antioxidant enzymes system of liver causing oxidative stress[4,5]. Prolonged and severe oxidative stress to the liver has been suggested as a major pathophysiological mechanism in the molecular pathogenesis of non-alcoholic fatty liver disease (NAFLD)[6]. Free radicals cause oxidation of lipids, particularly the polyunsaturated fatty acids (PUFA) found in cell membrane, leading to lipid peroxidation[7]. Malondialdehyde, chief by-product of lipid peroxidation, is widely used as marker in evaluating lipid peroxidation[8]. In the present study, highest MDA result was found in the control group fed with high fat diet indicating higher lipid peroxidation. While, the high-fat diet fed groups given with different doses of tocotrienols showed lower MDA results indicating lower lipid peroxidation. Lower MDA in tocotrienols treated groups either suggests decrease in the production of free radicals or increase production of antioxidants. In a recent study[9], gamma- tocotrienol reduces lipid peroxidation induced by $H_2O_2$ by preventing MDA elevation in osteocytes. Young (2001)[10] mentioned that a reduction in total antioxidant capacity could either indicate presence of oxidative stress condition or susceptibility to oxidative damage. To further understand the tocotrienols effect, TAOC were used to evaluate the antioxidant capacity of all antioxidant in the liver. In the present study, comparable results were found between high-fat diet fed control group and groups treated with tocotrienols. Noteworthy results were only found upon comparing the normal basal diet fed groups, treated and non-treated with tocotrienols. Niki et al (2005)[11] stated that oxidative stress greater than the antioxidant capacity level may induce oxidative damage, however, a low level stress may enhance the defence ability. This may suggest that tocotrienols can do increase overall antioxidants, but tocotrienols itself is an antioxidant where vitamin E is considered as a low-molecular weight antioxidant compound[7] that can be measured together with other antioxidant using TAOC assay. In a study[12] using high dose tocotrienols showed that tocotrienol can accumulate in the liver but are primarily deposited in white adipose tissue. Fortunately, in the present study, high-fat diet fed group given with the lowest dose of tocotrienols (50mg/kg) showed highly significant difference with the control group fed with normal basal diet. With this, we can speculate that tocotrienols action is mainly focused on decreasing free radical production in a compromised lived. Lower MDA can also be link to tocotrienols ability to reduce triglyceride accumulation in vivo and in vitro thus reducing oxidative stress[13].

Within the antioxidant system, some are enzymes that function by scavenging oxidants produce during oxidative stress, particularly: superoxide dismutase, catalase and glutathione peroxidase[7]. In the present study, the sod high-fat diet fed groups given with tocotrienols did not show any significant difference compared with the untreated groups. However, a significant result was found in normal basal diet fed group given with tocotrienols compared with the untreated group. This shows that tocotrienols can improve the SOD expression in liver. SOD acts on superoxide ($O_2^-$) through a process of simultaneous oxidation and reduction, called dismutation, producing hydrogen peroxide ($H_2O_2$) and $O_2$. $H_2O_2$, unfortunately, is a free radical that even though considered as a weak oxidant and a weak reducing agent can become a highly reactive free radical in the presence of transitional metal ions[14]. This undesirable result of dismutation is removed by catalase, glutathione peroxidase and other certain peroxidases[14]. In the present study, even though no significant results were found, CAT was found to be highest in the normal basal diet fed group given with tocotrienols. On the other hand, the high production of GSH-Px showed to be highly influence by high dose of tocotrienols, as shown by the tocotrienols treated groups, especially in the normal basal diet fed group. In recent studies using streptozotocin (STZ)-induced diabetic rats, tocotrienol-rich fraction (TRF) showed to reduce oxidative stress markers on RBC membrane[15] and renal tissue[16]. Also, tocotrienols was shown to increase cardiac GSH-Px enzyme activity in rats fed with high methionine diet[17]. Tocotrienols also protects pancreatic damage by restoring antioxidant status in fenitrothion induced oxidative damage in rats[18]. Gamma tocotrienol showed intestinal cell protection thru upregulation of anti-apoptotic and downregulation of pro-apoptotic factors after radiation exposure[19]. Also, in previous studies[20,21], tocotrienol showed gastric mucosal protection in acute stress not by its anti-oxidant mechanism but by increase gastric prostaglandin $E_2$ (PGE2) thru increase COX-1 mRNA.

Within the cells are nuclear receptors that regulate the expression of specific genes involve in the development, homeostasis, and metabolism of an organism. An example of this nuclear receptor is PPAR, Peroxisome proliferator-activated receptors (PPARs) are members of the nuclear hormone receptor superfamily that heterodimerize with the retinoid X receptor (RXR) and generally function as transcriptional regulator of lipid metabolism-linked genes[22]. In the present study, PPAR $\alpha$ and $\gamma$ was found to be highly expressed in moderately and low dose of tocotrienols. PPAR $\alpha$ and $\gamma$ are molecular targets of some class of drugs which treats glucose and lipid metabolism related disorders[22]. PPAR-$\alpha$ has protective role in alcoholic liver disease[23]. Li et al (2013) showed that high fat diet significantly reduce the expression of PPAR$\alpha$ in liver tissue[24]. Activation of PPAR$\alpha$ result in increased production of enzymes involve in the $\beta$-oxidation of fatty acids thus regulate and prevent abnormal lipid deposition[22,24]. In a recent study, tocotrienols was shown to regulate the expression of PPAR $\gamma$ through enhancement of interaction between the purified ligand-binding protein domains of PPAR$\alpha$ with the receptor motif of co-activator PPAR $\gamma$ co-activator-1 alpha[25]. Also, $\gamma$ tocotrienol together with PPAR $\gamma$ agonists caused increased transcription of PPAR $\gamma$ together with increased expression of PPAR gamma and RXR[26].

## Acknowledgement

This research was financially supported by the Hunan Sevoc Ecological Agriculture and Husbandry Technology Co., Ltd.

## Conflict of Interest

The authors declare that there no conflicts of interest.

## References

[1] McCord J. The evolution of free radicals and oxidative stress. *American Journal of Medicine*. 2000, 108: 652 – 659.

[2] Birhen E, Sahiner U M, Sackesen C, et al. Oxidative stress and antioxidant defense. *World Allergy Organization*. 2012, pages 9 – 19.

[3] SenC, Khanna S, Roy S. Tocotrienols: Vitamin E beyond tocopheros. *Life Science*. 2006, 78(18): 2088 – 2098.

[4] Marczuk-Krynicka D, Hryniewieck T, Paluszak J, et al. High fat content in diets and oxidative stress in livers of non-diabetic and diabetic rats. Polish J. of Environ. Stud. 2009, 2: 249 – 253.

[5] NoemanS, Hamoda H, Baalash A. Biochemical study of oxidative stress markers in the liver, kidney and heart of high fat diet induced obesity in rats. *Diabetology & Metabolic Syndrome*. 2011, 3: 17.

[6] Videla L. Oxidative stress signaling underlying liver disease and hepatoprotective mechanisms. *World Journal of Hepatology*. 2009, 1(1): 72 – 78.

[7] SureshD, Annam V. Lipid peroxidation and total antioxidant capacity in health and disease -Pathophysiology and markers: an overview. *International Journal of Medical Science and Public Health*. 2013, 2(2): 478 – 488.

[8] Das S, Vasisht S, Das N, et al. Correlation between total antioxidant status and lipid peroxidation in hypercholesterolemia. *Current Science*. 2000, 78(4): 486 – 487.

[9] Manan N, Mohamed N, Shuid A. Effects of low dose versus high dose gamma-Tocotrienol on the bone cells exposed to the hydrogen peroxide-induced oxidative stress and apoptosis. *Evidence Based Complementary and Alternative Medicine*. 2012, 2012: 1 – 10.

[10] YoungI. Measurement of total antioxidant capacity. *Journal of Clinical Pathology*. 2001, 54: 339.

[11] Niki E, Yoshida Y, Saito Y, Noguchi N. Lipid peroxidation: mechanisms, inhibition, and biological effects. *Biochemical and Biophysical Research Communication*. 2005, 338: 668 – 676.

[12] Shibata A, Nakagawa K, Shirakawa H, et al. Physiological effects and tissue distribution from large doses of tocotrienol in rats. *Biosci. Biotechnol. Biochem*. 2012, 76(9): 1805 – 1808.

[13] Burdeos G, Nakagawa K. KimuraF, Miyazawa T. Tocotrienol attenuates triglyceride accumulation in HepG2 cells and F344 rats. *Lipids*. 2012, 47: 471 – 481.

[14] Gutteridge J. Lipid peroxidation and antioxidants as biomarkers of tissue damage. *Clinical Chemistry*. 1995,41(12):1819-1828.

[15] MatoughF,Budin S,Hamid Z,et al. Tocotrienol-rich fraction from palm oil prevents oxidative damage in diabetic rats. *SQU Med J*. 2014,14(1):95-103.

[16] Siddiqui S,Ahsan H,Khan M,Siddiqui W. Protective effects of tocotrienols against lipid-induced nephropathy in experimental type2 diabetic rats by modulation in TGF-B expression. *Toxicology and Applied Pharmacology*. 2013,273(2):314-324.

[17] NorsidahK,Asmadi A,Azizi A,et al. Palm tocotrienol-rich fraction reduced plasma homocysteine and heart oxidative stress in rats fed with a high-methionine diet. *Journal of Physio Biochem*. 2013,69:441-449.

[18] BudinS,Han C,Jayusman P,Taib I. Tocotrienol rich fraction prevents fenitrothion induced pancreatic damage by restoring antioxidant status (Abstract). *Pak J. Biol Sci*. 2012,15(11):517-523.

[19] SumanS,Datta K,Chakraborty et al. Gamma tocotrienol,a potent radioprotector,preferentially upregulates expression of anti-apoptotic genes to improve intestinal cell survival. *Food and Chemical Toxicology*. 2013,60:488-496.

[20] RodzianM,Ibrahim I,Fahami N,Ismail N. Pure tocotrienol concentrate protected rat gastric mucosa from acute stress-induced injury by non-antioxidant mechanism. *Pol J. Pathology*. 2013,1:32-38.

[21] AzlinaM,Kamisah Y,Chua K,Qodriyah H. Tocotrienol attenuayes stress induced gastric lesions via activation of prostaglandin and upregulation of COx-1 mRNA. *Evidence-Based Complementary and Alternative Medicine*. 2013,1:1-8.

[22] Ricote M,Huang J,Welch J,Glass C. The peroxisome proliferator-activated receptor gamma (PPARγ) as a regulator of monocyte/macrophage function. *Journal of Leukocyte Biology*. 1999,6:733-739.

[23] Nakajima T,KamijoY,Tanaka N,et al. Peroxisome proliferator-activated receptor alpha protects against alcohol-induced liver damage. *Hepatology*. 2004,40(4):972-980.

[24] LiH,Xu S,Li X,Zhang X,Liu W. PPAR alpha and UCP-2 expression in non-alcoholic fatty liver in mice. *Chinese Public Health*. 2013,29(9):1313-1315.

[25] FangF,Kang Z,Wong C. Vitamin E tocotrienols improve insulin sensitivity through activiating peroxisome proliferator-activated receptors. *Molecular Nutrition and Food Research*. 2010,54(3):345-52.

[26] MalaviyaA,Sylvester P. Mechanisms mediating the effects of gamma tocotrienol when used in combination with PPAR gamma agonists or antagonists on MCF-7 and MDA-MB-231 breast cancer cells. *International Journal of Breast Cancer*. 2013,1:1-17.

# Induction and mechanism of HeLa cell apoptosis by 9-Oxo-10, 11-dehydroageraphorone from *Eupatorium adenophorum*

LIAO Fei[a,b], HU Yanchun[a*], WU Lei[a], TAN Hui[a], LUO Biao[a], LUO Lijuan[a], HE Yajun[a], QIAO Yan[a], MO Quan[a], ZHOU Yancheng[a], ZUO Zhicai[a], REN Zhihua[a], DENG Junliang[a], WEI Yahui[c]

[a]*Key laboratory of Animal Disease and Human Health of Sichuan Province, College of Veterinary Medicine, Sichuan Agricultural University, Sichuan Province, Ya an 625014, China.* [b]*Qiandongnan Prefectural Center for Animal Disease Control and Prevention of Guizhou province, Kaili 556000, China.* [c] *Key Laboratory of Resource Biology and Biotechnology in Western China, School of Life Science, Northwest University, Xi'an 710069, China*

**Abstract**: 9-Oxo-10, 11-dehydroageraphorone (euptox A), a cadenine sesquiterpene, is the main toxin from *Eupatorium adenophorum*. The present study aims to explore the induction and mechanism of HeLa cell apoptosis by 9-Oxo-10, 11-dehydroageraphorone. The apoptosis inducing effect of the euptox A on HeLa cells was examined by MTT assay. Meanwhile, the mechanism was analyzed by flow cytometry and real-time PCR. Flow cytometry results suggest that euptox A could effectively inhibit the proliferation of HeLa cells, arrest the cell cycle transition from S to G2/M phase, can't continue to complete the cell cycle activity (mainly form four times and mitosis), and make the cell proliferation, the QRT-PCR detection results showed that euptox A could induce apoptosis through improving the gene expression level of apoptosis proteases such as *caspase*-10 in HeLa cells. Its mechanism of action was probably associated with up-regulating apoptotic gene expression and arresting the cell cycle.

**Keywords**: 9-Oxo-10, 11-dehydroageraphorone, *Eupatorium adenophorum*, Apoptosis, Mechanism

## 1 Introduction

*Eupatorium adenophorum* native to Mexico and CostaRica of Central America, is a worldwide noxious invasive weed[1]. After the introduction as a ornamental plant to USA in 1960s, it has spread worldwide[2], long present as a non native species in India, New Zealand, and Australia. In China, it first invaded southern areas of Yunnan Province from Burma around the 1940s[3]. Presently, *E. adenophorum* can be found in Chongqing, Yunnan, Sichuan, Guizhou, Tibet, Guangxi, Taiwan and Hubei Provinces. A rough estimate of the a nnual spreading rate of *E. adenophorum* is about 10-60 km from south to north and from west to east in China[4]. It is considered a threat to local economy and biodiversity. But as reported, several compounds have been separated and characterized from *E. adenophorum* stem, flowers and leaves, including hemiterpenes, sterides, triterpenes, flavonoid and phenylpropanoids phenol etc., and have extensive biological activity, such as anti-Inflammatory potential[5], acaricidal activity[6,7], antioxidant activity[8]. *E. adenophorum* can be used as a feed resource[9], medical resource[10], chemical material resource[11].

9-Oxo-10, 11-dehydroageraphorone (euptox A), a cadenine sesquiterpene, is the main toxin from *E. adenophorum*.

* Corresponding author, Dr. Yanchun HU, Professor, E-mail: yanchunhu@126.com

Previous studies have found that the Euptox A from *E. adenophorum* had hepatotoxicity[12] and allelopathy[13], our laboratory have proved that Euptox A had highly acaricidal activity for *S. scabiei* and *P. cuniculi* in vitro[14] and Euptox A presented significantly antitumor activity against the three tumor cell line (human lung cancer A549 cell line, Hela cell line and Hep-2 cell line) *in vitro* in a dose-dependent manner[15].

In this study we used MTT, flow cytometry and RT-PCR to detect the change of indexes before and after drugs action, then compare the availability of different methods.

# 2 Materials and Methods

## 2.1 Materials

Euptox A was provided by the laboratory of biotoxin and molecular toxicology of Sichun Agriculture University, China. The purity of the toxin we had extracted was over 96%[16].

## 2.2 Cell Cultures

Hela cell lines was cultured at 37 ℃ in a humidified atmosphere of 5% $CO_2$ in DMEM medium supplemented with 10% FCS, glutamine, and 100 U/ml penicillin, 100μg/ml streptomycin. Cell cultures were split every 3 days.

## 2.3 Colorimetric MTT Assay For Cell Proliferation

Thecytotoxicity of euptox A against Hela cells was measured by the MTT assay[17]. Hela cell line was maintained in DMEM supplemented with 10% fetal bovine serum, 100 mU/L streptomycin, and 100mU/L penicillin at 37℃ in a humidified atmosphere of 5% $CO_2$. Cells ($3\times10^3$/well) in their exponential growth phase were seeded into each well of a 96-well flat-bottomed culture plate and incubated for 24 h. Then Hela cells was incubated with the samples (euptox A) at concentrations of 0, 0.25, 0.5, 1 and 2 mg/ml. After 48 h, each well were added 20 μl of 5 mg/ml of MTT and incubated for another 4 h. After the culture media were removed, 150 μl of DMSO were added to each well. Absorbance at 490 nm was detected by a microplate ELISA reader. 5-Fu was treated as positive control. The inhibition rate was calculated according to the formula below:

$$\text{Growth inhibition rate }(\%) = \left(1 - \frac{\textit{Absorbance of experimental group}}{\textit{Absorbance of control group}}\right)\times 100\%$$

## 2.4 Flow Cytometry

Hela cells were plated at a density of 75 000/well in 24-well plates and grown with DMEM containing 10% FBS. The day after, euptox A was added at a concentration of 0, 10, 30 and 50 μg/ml (respectively) and cells were incubated for 24 h. There was a positive control (5-Fu was treated as positive control) for each incubation period. Cell cycle of Hela cells was assessed by flow cytometry analyzing DNA content in propidium iodide (PI) stained cells. Briefly, cells were harvested with trypsin/EDTA, washed, then incubated for 15 min with PI, and finally analyzed by flow cytometry.

## 2.5 QRT-PCR Detection HeLa Apoptosis Gene Expression

Logarithm period cells were divided into three groups, i. e. vehicle control, experimental control (The concentration of euptox A was 0, 10, 30 and 50 μg/ml, respectively) and positive control (5-Fu was treated as positive control). The cells were extracted after 48 h, then used for measuring gene expression by QRT-PCR.

Total RNA was isolated from $10^6$ to $10^7$ cells in a growing phase using Trizol (Aidlab, China) according to the manual. The RNA was then chloroform-extracted and precipitated with isopropanol. Synthesis of single-stranded cDNA from 5 μg of RNA was performed according to the "TUREscript 1st strand cDNA Synthesis Kit" from Aidlab

(China). Pairs of 5' and 3' primers to amplify each type of cDNA were detailed in Table 1. The reaction conditions were as follows: 94 ℃ for 1 min, 55 ℃ for 1 min, 72 ℃ for 1min (40 amplification cycles was necessary to achieve exponential amplification in which product formation was proportional to starting cDNA). Relative gene expression was then defined as a ratio of target gene (*caspase*-10 and *caspase*-3) expression versus β-actin gene expression. Each ratio was determined three times in dependent.

Table 1 The primers of PCR

| Primer species | Primer sequence | Size of PCR product (bp) |
|---|---|---|
| caspase-10 (F) | GTATCAGGCTACCCAGTCC | |
| caspase-10 (R) | CAGATCAAGCTCCACCAA | 192 |
| caspase-3 (F) | GGTTCATCCAGTCGCTTTG | |
| caspase-3 (R) | CTCTGTTGCCACCTTTCG | 97 |
| β-actin (F) | GTTGCGTTACACCCTTTC | |
| β-actin (R) | CTGTCACCTTCACCGTTC | 151 |

### 2.6 Statistical Analysis

All data are expressed as mean ± *SE* and/or confidence interval. Statistical analyses were performed to compare the treated groups with the respective control group using a one-way analysis of variance (ANOVA) complemented with the Tukey-Kramer multiple comparison test with equal sample size. All computations were done by employing the statistical software (SPSS, version 20.0)[18].

## 3 Results

### 3.1 *In Vitro* Cytotoxicity Assay

To investigate the effects of euptox A treatment on cell viability, the endpoint of cytotoxicity (MTT) assays Hela cell line the results was presented in Table 2. The results showed euptox A had significant antitumor activity against the Hela cell line *in vitro* in a dose-dependent manner. When the concentration of euptox A was at 500μg/ml, the percent inhibition of Hela cells was 68.30%.

### 3.2 Flow Cytometry

Morphological observation of HeLa cells treated with euptox A for 24 h (Fig. 1). The euptox A could effectively inhibit the proliferation of HeLa cells. The shape of Hela cell in the negative control group showed polygon, arranged in dense, firmly adherent (Fig. 1 a). Compared with the negative control group, the treated with euptox A (the concentration of euptox A was 10, 30 and 50 μg/ml), and in a certain concentration range, with the increase of the concentration of euptox A administration, suspension cells gradually increased, significantly reduced the number of adherent cells, sparsely arranged, fusiform, appeared long and irregular protrusions, membrane integrity but there apoptosis foaming phenomena characteristic changes (Fig. 1 b, c and d).

Table 2 Activation of Hela cell lines exposed to different concentrations of euptox A

| Treatment | Concentration (mg/ml) | OD (490 nm) Mean ± *SE* | Inhibition ratio (%) |
|---|---|---|---|
| Hela | | | |
| Control | — | 2.790 ± 0.226$^{A}$ | 0 |

(continued table)

| Treatment | Concentration (mg/ml) | OD (490 nm) Mean ± SE | Inhibition ratio (%) |
|---|---|---|---|
| 5-Fu | 1 | $0.487 \pm 0.064^{BC}$ | 82.54 |
| Euptox A | 0.25 | $2.201 \pm 0.393^{a}$ | 21.09 |
| | 0.5 | $0.884 \pm 0.226^{B}$ | 68.30 |
| | 1 | $0.218 \pm 0.024^{C}$ | 92.16 |

Note: The difference between data with the different capital letter within a column is significant ($P < 0.01$), and the difference between data with the different small letters within a column is significant ($P < 0.05$)

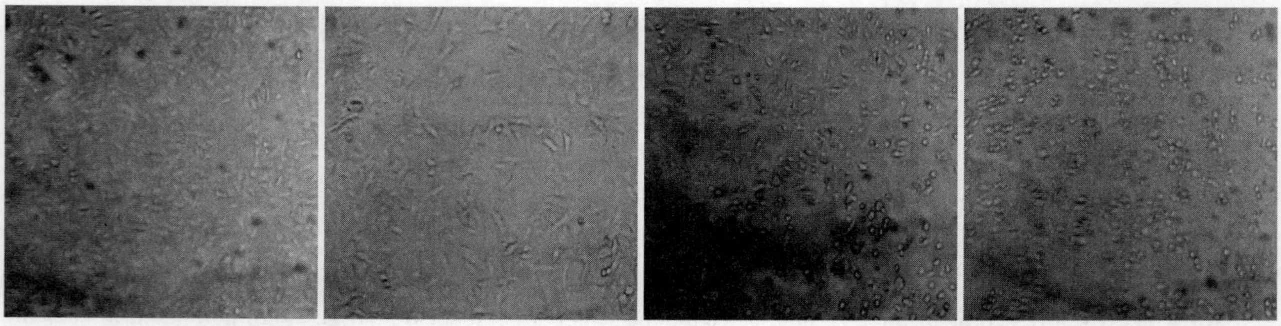


Fig 1 Morphological observation of HeLa cells treated with euptoxA for 24 h (10 ×20)

a: 0 μ/ml euptox A; b: 10 μ/ml euptox A; c: 30 μ/ml euptoxA; d: 50μ/ml euptox A

The cell cycle was analyzed by flow cytometry in the various concentration of euptox A. There were significant difference among the distribution of cell cycle after the various concentration of euptox A were given for 24 h (Fig. 3 and Table 3), and in a dose-dependent manner. Compared with the negative control group, the number of S phase form 31% to 40.936%, 48.583% and 56.950% at 10, 30 and 50 mg/ml concentration of euptox A in the experimental control group, respectively. However, the positive control group, the number of G1/M phase from 47.533% to 54.553%.

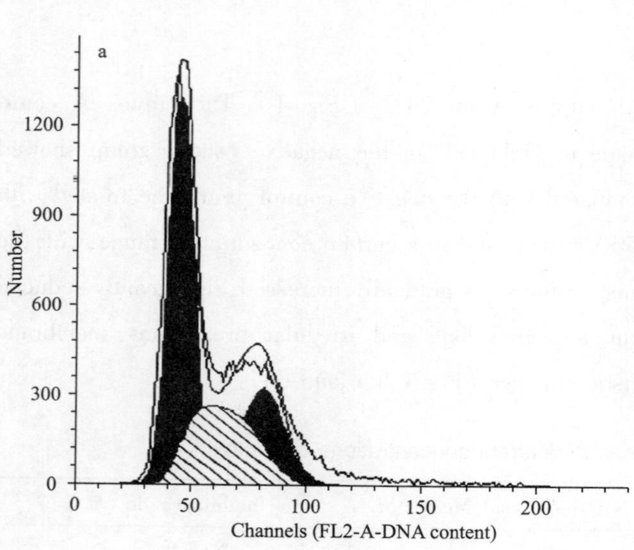


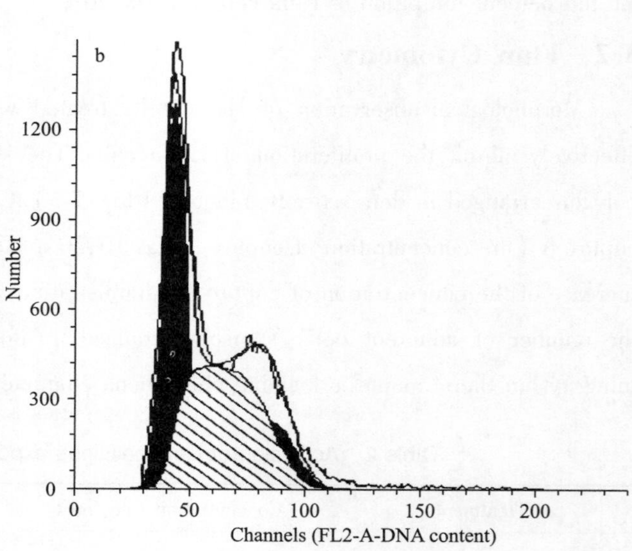

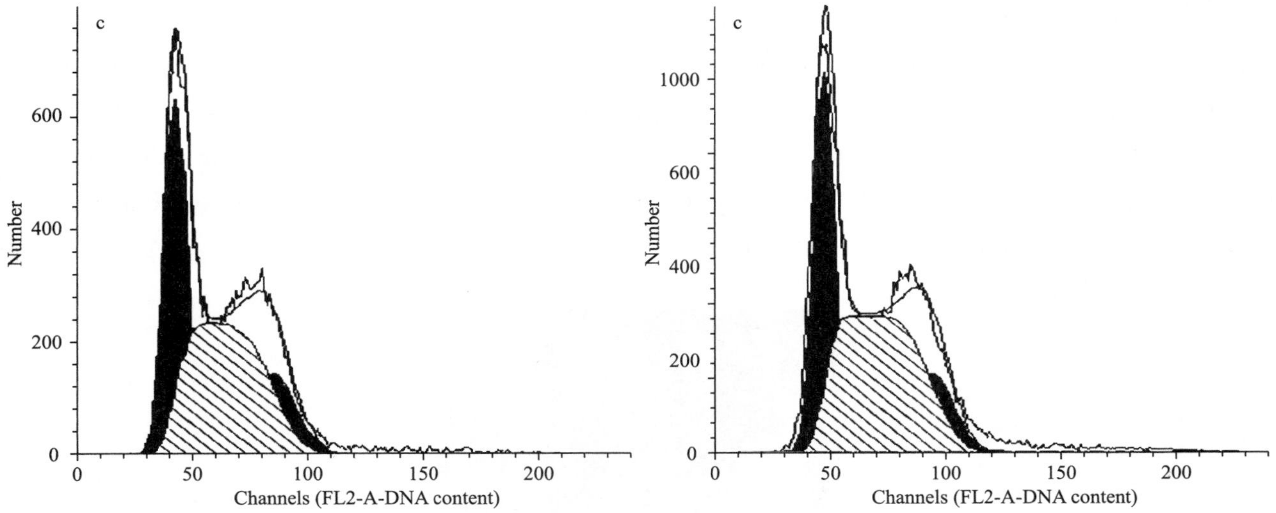


Fig 2 Effect ofeuptox A on the cell cycle of Hela cells

a:0 μ/ml euptox A; b:10 μ/ml euptox A; c:30 μ/ml euptoxA; d:50μ/ml euptox A

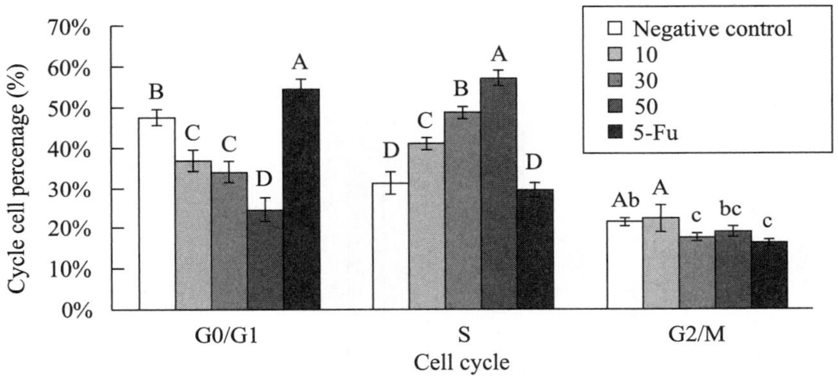


Fig. 3 Effect ofeuptox A on the cell cycle of Hela cells (Mean ± *SD*, n = 3)

5-Fu: Positive control group; Negative control group: 0 μ/ml euptox A; 10: 10μg/ml of euptox A group;

30: 30μg/ml of euptox A group group; 50: 50μg/ml of euptox A group

Table 3 Effect ofeuptox A on the cell cycle of Hela cells (Mean ± *SD*, n = 3)

| | $G_0/G_1$ | S | $G_2/M$ |
|---|---|---|---|
| 0μ/ml | $47.533 \pm 1.783^{B}$ | $31.280 \pm 2.779^{D}$ | $21.190 \pm 1.008^{Ab}$ |
| 10μ/ml | $36.876 \pm 2.392^{C}$ | $40.936 \pm 1.518^{C}$ | $22.190 \pm 3.419^{A}$ |
| 30μ/ml | $34.093 \pm 2.494^{C}$ | $48.583 \pm 1.519^{B}$ | $17.320 \pm 0.979^{c}$ |
| 50μ/ml | $24.366 \pm 3.170^{D}$ | $56.950 \pm 1.870^{A}$ | $18.683 \pm 1.363^{bc}$ |
| 5-Fu | $54.553 \pm 2.201^{A}$ | $29.603 \pm 1.745^{D}$ | $15.843 \pm 0.813^{c}$ |

Note: The difference between data with the different capital letter within a column is significant ($P < 0.01$), and the difference between data with the different small letters within a column is significant ($P < 0.05$)

## 3.3 Expression of the Caspase-10 and Caspase-3 Genes

Total RNA was extracted from Hela cell. The total RNA was integrity, purity and yield of RNA meet the requirement (Fig. 4). The expression levels of *caspase*-10 *and caspase*-3 genes in every group were illustrated in

Fig. 5. The expression levels of *caspase*-10 gene in Hela cells were increased after incubation with euptox A for 24 h, and not showed dependence with euptox A dose. Firstly increased and then decreased with the inecrease of euptox A dose. In addition, the expression levels *caspase*-10 gene in treated with 30μg/ml of euptox A reach highest. However, in Hela cells the *caspase*-3 gene was showed only minor signs decreased after incubation with euptox A for 24 h. But in group positive control (5-Fu was treated as positive control), the increased of *caspase*-10 *and caspase*-3 genes expression levels after incubation for 24 h.

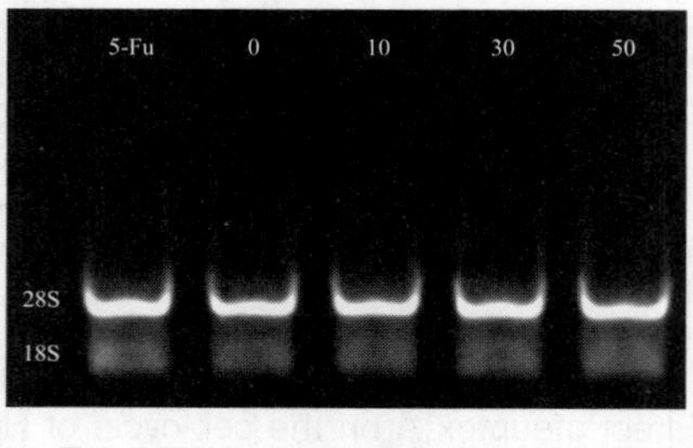


Fig 4 Total RNA was extracted from Hela cell

5-Fu: Positive control group; 0: Negative control group; 10: 10μg/ml of euptox A group; 30: 30μg/ml of euptox A group group; 50: 50μg/ml of euptox A group.

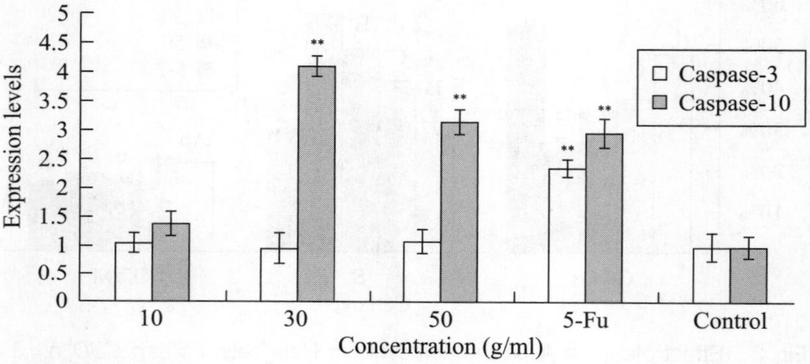


Fig 5 Effect ofeuptox A blend on mRNA expression of *caspase*-10 and *caspase*-3 in Hela cells

5-Fu: Positive control group; Control: 0 μ/mleuptox A; 10: 10μg/ml of euptox A group; 30: 30μg/ml of euptox A group group; 50: 50μg/ml of euptox A group

# 4 Discussion

Cancer is a common and frequently-occurring disease that is a serious threat to human and animal life, its mortality rate is second only to cardiovascular disease. At present, natural antineoplastic drugs have become the subject of much research. Natural products such as paclitaxol[19,20], camptothecine[21,22], podophyllotoxin[23], matrine[24], vincristine[25] have been shown to have anticancer activity. As recently demonstrated, euptox A extracted from *E. Adenophorum* markedly inhibited the growth of cancer cells directly. Euptox A was found highly active against the fast growing Hela, and its activity was concentration-dependent. A direct comparison with 5-FU in thes cell lines showed a clear superiority of euptox A, 5-FU is an antimetabolite that is used as a chemotherapeutic agent for a wide variety of cancers over 40 years[26]. However, the antitumor activity of euptox A was obviously stronger than that of 5-FU at the same concentration. The reason is most likely that the test tumors had resistance to the 5-FU. Our findings are consistent with previous studies have shown that the *E. Adenophorum* had antitumor activity[27].

In this study, we found that euptox A could effectively inhibit the proliferation of HeLa cells, arrest the cell cycle transition from S to G2/M phase, can't continue to complete the cell cycle activity (mainly form four times and mitosis), and make the cell proliferation. The findings are consistent with previous studies 10-hydroxycamptothecine (HCPT) showed a remarkable cell cycle specificity in inducing death and apoptosis in G1 phase, blocking the S phase[28]. Caspase-10 is believed to play an obligatory role in apoptosis initiation by death receptors[29]. The expression levels of *caspase*-10 gene in Hela cells were increased after treat with euptox A. The is suggested that euptox A lead to apoptosis might be regulated through influencing the expressions of *caspase*-10 gene in Hela cells.

This study provides a new way for utilization of *E. adenophorum*. Euptox A has the potential to be developed as an antitumor drug. Further studies are warranted for clinical trials, animal acute toxicity test and safety evaluation.

## Acknowledgements

This research was supported by Special Fund for Agroscientific Research in the Public Interest (Grant No. 201203062) and Chang-jiang Scholars and the Innovative Research Team in University (IRT0848).

## References (Omitted)

# RNA sequencing characterization of chicken cardiomyocyte apoptosis induced by hydrogen peroxide

WAN Chunyun[1,2], XIANG Jinmei[1], XIA Tian[1], WU Liming[1], ZHANG Guoxing[1], Muhammad Awais[1], LI Youwen[1], ZHANG Wei[1], GUO Dingzong[1*]

1. *College of Veterinary Medicine, Huazhong Agricultural University, Wuhan, 430070, China*,

2. *College of Animal Science, Yangtze University, Jingzhou* 434025, *China*

**Abstract**: Myocardial cells are terminally differentiated and have no ability to regenerate. Apoptosis is the main mechanism for myocardial cell loss and can cause heart failure or other serious consequences. Mammalian myocardial cells have been intensively studied; however, few studies have explored chicken myocardial cells and the apoptosis pathways in these cells are unclear and some key apoptosis genes are still unknown. To investigate these issues, primary cultures of embryonic chicken cardiomyocytes were maintained and treated with hydrogen peroxide to induce cell apoptosis. RNA was then extracted and RNA sequencing was performed to determine differential gene expression, identify novel transcripts and to construct a protein-protein interaction network. After sequencing, 19, 268 known genes and 2, 160 novel genes were investigated. Differential gene expression analysis identified 4 650 genes that showed significant differential expression between the apoptosis-induced and control groups. Among them, 10.4% were up-regulated while 18.9% were down-regulated. In response to apoptosis induced by hydrogen peroxide, apoptosis initiation was mainly dependent on caspase-8, 9, which activated caspase-3 and caspase-6. The signaling pathway is very similar to the mammalian myocardial cell apoptosis pathway.

## Introduction

Cell apoptosis was firstdescribed in 1972 by Kerr (Kerr J F., 1972). Soon thereafter, apoptosis was shown to occur in myocardial cells and to be involved in heart failure. Myocardial cells can undergo apoptosis and do not regenerate and remaining cells can gradually lose their function, leading to heart failure (Gaballa M A et al., 2002). Other studies have also shown that in the process of heart failure, myocardial cell apoptosis plays an important role (Sharov V G., 1996; Kang PM et al., 2003). We have found that in highly productive chickens there is a high rate of myocardial cell apoptosis, which may quickly effect production performance and lead to heart failure and death. Therefore, myocardial cell apoptosis is significant in chicken production performance and also in many other diseases.

To date, many studies have been conducted on mammalian myocardial cell apoptosis, and apoptosis pathways have been well delineated, such as the Death receptor pathway (Ashkenazi A., 1998) and the mitochondrial signaling pathway (Kong J Y., 2000; Danial N N., 2003). Many important signaling molecules have been found, such as the caspase family (Fan T., 2005), which includes cardiomyocyte-specific molecules such as ARC (apoptosis repressor with CARD domain) (Geertman R., 1996). However, limited research on chicken myocardial cells has been performed compared with that in mammals. The chicken myocardial apoptosis pathways have not been

* Corresponding author. E-mail: hlgdz@163.com

reported and some key orthologs of mammalian apoptosis genes have not been found. Hence, this study aimed to investigate the apoptosis pathways in chicken myocardial cells.

The sequencing of mRNA transcripts (termed RNA sequencing or RNAseq) is a maturing technology that is now widely used. It allows the identification of differentially expressed genes without prior annotations and can identify all genes expressed in a sample (Hampton M. ,2013; Tamim S. et al. ,2014; Schunter C. ,2014). RNAseq has only rarely been used in chickens, for example, to investigate resistance to *Campylobacter jejuni* colonization (Connell S. ,2012); however, this technology has not been applied to apoptosis of chicken myocardial cells.

To better understand signaling pathways and to find novel genes involved in apoptosis of chicken myocardial cells, apoptosis was induced using hydrogen peroxide (Gruss-Fischer T. , 2002; Suzuki T et al. , 2014). RNA samples were sequenced by 100-bp paired-end reads on the Illumina Hiseq 2000 platform.

## Materials and Methods

### Ethics Statement

The study was approved by the Animal Care and Use Committee of Hubei Province, China. All animal procedures were performed according to guidelines developed by the Chinese Council on Animal Care and protocols were approved by the Animal Care and Use Committee of Hubei Province, China.

### Pretreatment and Induction of Apoptosis in Chicken Primary Embryonic Cardiomyocytes

Monolayer cultures of embryonic chicken cardiomyocytes were prepared by the methods of DeHaan (Dong S et al. ,1967) with some modifications. Briefly, we used the differential attachment technique to gain high purity cells after 0.5 h incubation. Cells were washed three times at 8, 24, and 48 h to remove dead and non-viable cells. The serum concentration in the medium was then changed from growth (10%) to maintenance (2%) and incubation continued for 36 h. After this the cells were divided into two groups, one was kept as a control group while the other group was treated to induce apoptosis (H group). Each group had two replicates, named _1, _2 respectively (i. e. , H_1 and H_2). For RNA sequencing, all samples were bi-directionally sequenced, the direction being named as H_1_1, H_1_2. Apoptosis was induced in cells by incubation with 0.2 mmol/L hydrogen peroxide for 10 h (the dosage and time was determined by prior testing and with reference to previous studies) (Keon-Jae Park et al. ,2012; Ruotian Li et al. ,2012; Bimei Jiang et al. ,2005). The degree of apoptosis was identified by DAPI staining. The control group was treated in the same way but with the omission of the hydrogen peroxide.

### RNA Sample Collection and Preparation

Total RNA was extracted fromcells using standard protocols (Trizol) and then treated with DNase to remove potential genomic DNA contamination. RNA degradation and contamination was monitored by electrophoresis on 1% agarose gels. RNA purity was checked using a NanoPhotometer spectrophotometer (Implen, CA, USA). RNA concentration was measured using a Qubit RNA Assay Kit in a Qubit 2.0 Flurometer (Life Technologies, CA, USA). RNA integrity was assessed using the RNA Nano 6000 Assay Kit of the Bioanalyzer 2100 system (Agilent Technologies, CA, USA).

### Library Preparation for Transcriptome Sequencing

mRNA was purified from total RNA using poly-T oligo-attached magnetic beads (Fox et al. ,2010). Three micrograms of RNA was used as input material for each RNA sample preparation. Sequencing libraries were generated using a NEBNext Ultra RNA Library Prep Kit for Illumina (NEB, USA) following the manufacturer's recommendations. Index codes were assigned to attribute sequences to each sample.

The clustering of the index-coded samples was performed on a cBot Cluster Generation System using a TruSeq PE Cluster Kit v3-cBot-HS (Illumia) according to the manufacturer's instructions. After cluster generation, the library preparations were sequenced on an Illumina Hiseq 2000 platform and 100-bp paired-end reads were generated (Pomraning et al., 2012).

## Sequencing Quality Control

Raw data (raw reads) in fastq format were first processed through in-house perl scripts. In this step, clean data (clean reads) were obtained by removing reads containing adapter sequence, reads containing ploy-N, and low-quality reads. At the same time, Q20, Q30, and GC content of the clean data were calculated. All the downstream analyses were based on the clean, high-quality data.

## Reads Mapping to the Reference Genome

Reference genome and gene model annotation files weredirectly downloaded from the genome website. An index of the reference genome was built using Bowtie v2.0.6 and paired-end clean reads were aligned to the reference genome using TopHat v2.0.9. We selected TopHat as the mapping tool because TopHat can generate a database of splice junctions based on the gene model annotation file and can thus produce a better mapping result compared with other non-splice mapping tools.

## Quantification of Gene Expression Level

HTSeq v0.5.4p3 was used to count the reads mapped to each gene. The RPKM (reads per kilobase of exon model per million mapped reads) of each gene was then calculated based on the length of the gene and reads count mapped. Because this method considers the effect of sequencing depth and gene length for the reads count at the same time, it is currently the most commonly used method for estimating gene expression levels (Mortazavi et al., 2008).

## Alternative Splicing and Differential Expression Analysis

Alternative splicing was determined by software cufflinks 2.1.1 and ASprofile 1.0. Differential expression analysis was performed using the DESeq R package (1.10.1). DESeq provides statistical methods for determining differential expression in digital gene expression data using a model based on the negative binomial distribution. The resulting $P$-values were adjusted using Benjamini and Hochberg's approach for controlling the false discovery rate. Genes with an adjusted $P$-value $< 0.05$ found by DESeq were assigned as differentially expressed genes. Corrected $P$-values of 0.005 and a log2 (fold change) of 1 were set as the threshold for significant differential expression.

## GO and KEGG Enrichment Analysis of Differentially Expressed Genes

After functional annotation of unigenes (data not shown), the predicted genes were further classified by Gene Ontology (GO) assignments. GO enrichment analysis of differentially expressed genes was performed using the GOseq R package, in which gene length bias was corrected. GO terms with corrected $P$ values less than 0.05 were considered significantly enriched for differentially expressed genes (Renfro et al., 2012).

KEGG (Kyoto Encyclopedia of Genes and Genomes) is a database resource for understanding high-level functions and utilities of biological systems, such as the cell, the organism, and the ecosystem from molecular-level information, especially large-scale molecular datasets generated by genome sequencing and other high-throughput experimental technologies (http://www.genome.jp/kegg/). We used KOBAS software to test the statistical enrichment of differentially expressed genes in KEGG pathways (Mao et al,2005; Kanehisa,2008).

## PPI Analysis of Differentially Expressed Genes

PPI(Protein-Protein Interaction) analysis of differentially expressed genes was based on the STRING database, which predicts protein-protein interactions for the species existing in the database.

We constructed the networks by extracting the target gene list from the database. Blastx (v2.2.28) was used to align the target gene sequences to the selected reference protein sequences and then the networks were built according to known interactions of selected reference species.

## Novel Transcript Prediction and Alternative Splicing Analysis

The Cufflinks v2.1.1 Reference Annotation Based Transcript (RABT) assembly method was used to construct and identify both known and novel transcripts from TopHat alignment results. Alternative splicing events were classified into 12 basic types by the software Asprofile v1.0. The number of AS events in each sample was estimated separately.

## SNP Analysis

Picard-tools v1.96 and samtools v0.1.18 were used to sort, mark duplicated reads, and reorder the bam alignment results of each sample. GATK2 software was used to perform SNP calling.

# Results

## Sequencing Quality Control

AfterRNA sequencing, we assessed the quality of the data. Q20, Q30, and GC content in the clean data were calculated and are presented in Table 1. Alignments between reads and the reference genome are presented in Supplemental Table 1. Replicates of each sample were sequenced and the correlation between replicates is shown in Figure 1.

Table 1 Major characteristics ofRNAseq data between induced and control groups

| Sample name | Raw reads | Clean reads | Clean bases | Error rate (%) | Q20 (%) | Q30 (%) | GC content (%) |
|---|---|---|---|---|---|---|---|
| Con_1_1 | 61286178 | 57687170 | 5.77G | 0.03 | 97.15 | 91.50 | 53.26 |
| Con_1_2 | 61286178 | 57687170 | 5.77G | 0.04 | 96.03 | 89.60 | 53.32 |
| Con_2_1 | 57289438 | 53642113 | 5.36G | 0.04 | 97.03 | 91.19 | 54.04 |
| Con_2_2 | 57289438 | 53642113 | 5.36G | 0.04 | 95.75 | 88.93 | 54.09 |
| IA_1_1 | 59514696 | 56406015 | 5.64G | 0.03 | 97.35 | 91.95 | 52.67 |
| IA _1_2 | 59514696 | 56406015 | 5.64G | 0.04 | 96.30 | 90.12 | 52.73 |
| IA _2_1 | 62578501 | 59247377 | 5.92G | 0.03 | 97.34 | 92.00 | 53.19 |
| IA _2_2 | 62578501 | 59247377 | 5.92G | 0.04 | 96.39 | 90.34 | 53.24 |

Q20 and Q30 indicate the percentage of Phred bigger than 20 and 30 times, respectively, in total reads. GC content means the percentage of G + C bases in total bases

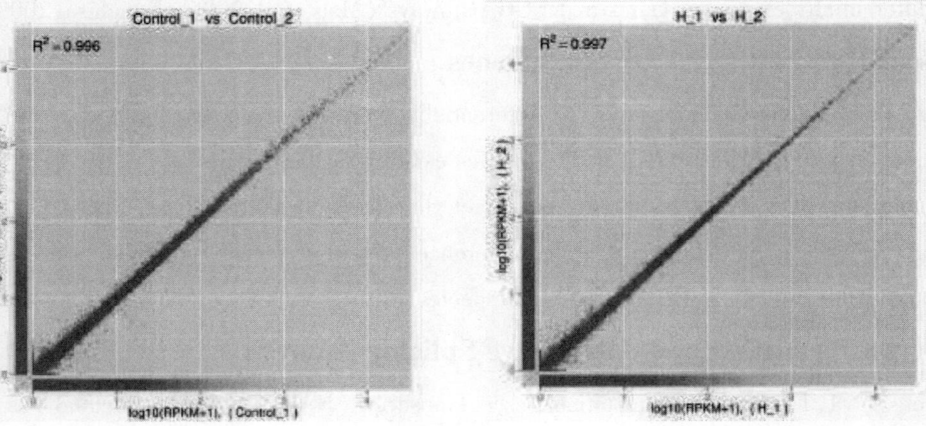


Figure 1. Correlation of biological replicates for each group

$R^2$: square of the Pearson correlation coefficient.

## Alternative Splicing and Differential Expression Analysis

Alternative splicing (AS) analysis was performed using cufflinks 2.1.1 and ASprofile 1.0. The statistics of alternative splicing events are shown in Figure 2. The most common AS types were TSS and TTS (TSS: alternative 5′ first exon, TTS: alternative 3′ last exon). Interestingly, we found some genes with more than five AS types, and in different states, it has a different position in the same AS type.

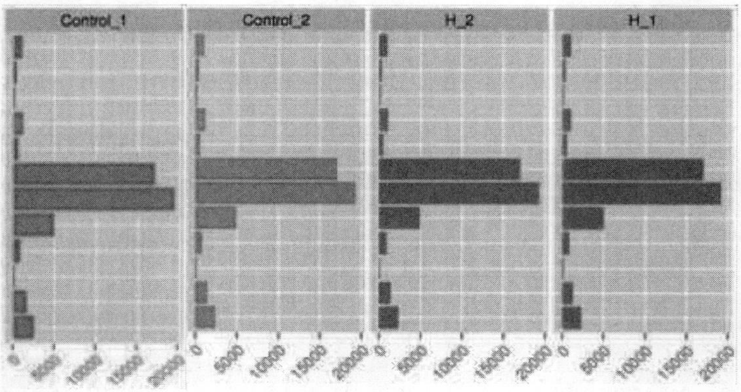


Figure 2 Statistics of alternative splicing events

Different types of alternative splicing events are represented as follows: SKIP: Skipped exon, XSKIP: Approximate SKIP, MSKIP: Multi-exon SKIP, XMSKIP: Approximate MSKIP, IR: Intron retention, MIR: Multi-IR, XMIR: Approximate MIR, XIR: Approximate IR, AE: Alternative exon ends, XAE: Approximate AE, TSS: alternative 5′ first exon, TTS: Alternative 3′ last exon

To assess global transcriptional changes after induction, we applied the former described method to identify differentially expressed genes from the normalized data. Results showed that 4 650 genes were significantly differentially expressed. A volcano plot of differential gene expression is shown in Figure 4. It showed that apoptosis induced by hydrogen peroxide causes 10.4% of genes to be up-regulated and 18.9% of genes to be down-regulated.

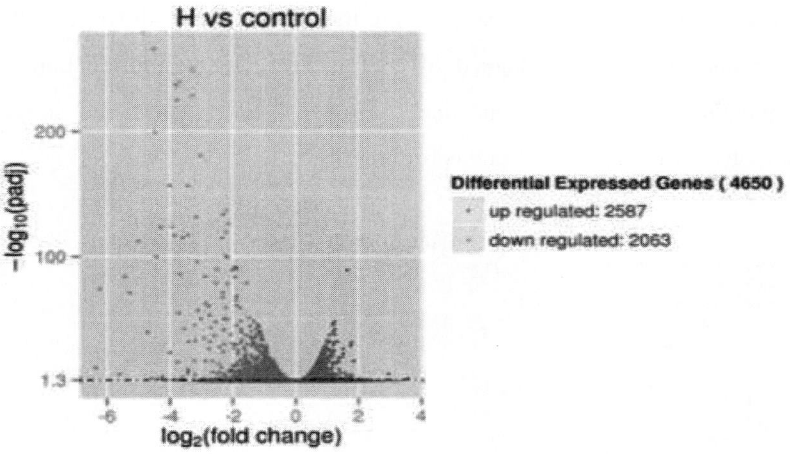


Figure 3　Volcano plot of differentially expressed genes. Each point represents one gene

Red indicates up-regulation and green indicates down-regulation.

## Validation of Differential Gene Expression Data by qPCR

To validatedifferentially expressed genes identified by sequencing, we selected 19 genes for qPCR confirmation, including nine down-regulated genes (AMPN, EGLN, DDX4, FGF10, FOXO3, GHOX7, TGFBR, NF-Kb, and MAPK) and 10 up-regulated genes (CASP8, CASP9, CASP3, Bcl, Bcl2, Cytc1A4, BAK1, TNFRSF, P53, and XIAP). The primers for qPCR assays are shown in Table S2. Pearson's correlation coefficient (r) showed that both the RNASeq differential gene expression and qPCR data were highly correlated, $r^2 = 0.97$ (Figure 3). qPCR analysis confirmed the reliability of the RNASeq differential gene expression results.

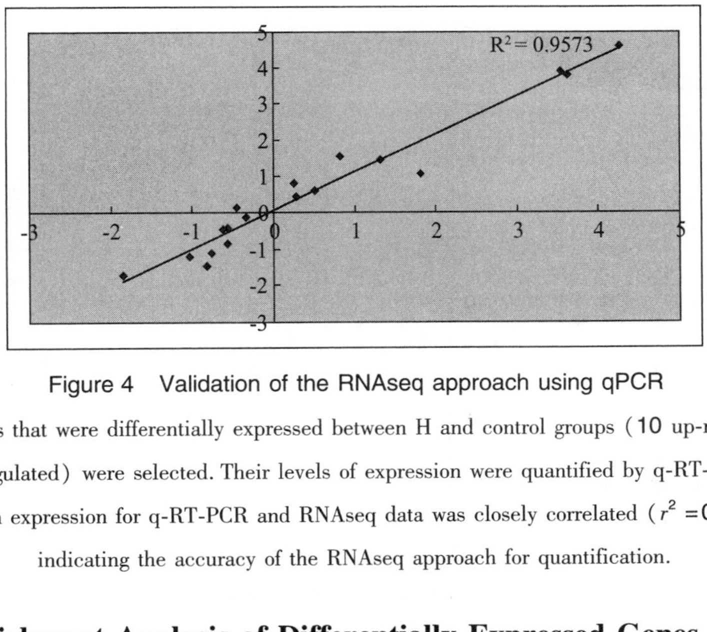


Figure 4　Validation of the RNAseq approach using qPCR

Nineteen genes that were differentially expressed between H and control groups (10 up-regulated and 9 down-regulated) were selected. Their levels of expression were quantified by q-RT-PCR and the log2 change in expression for q-RT-PCR and RNAseq data was closely correlated ($r^2 = 0.97$; $P < 0.01$), indicating the accuracy of the RNAseq approach for quantification.

## GO and KEGG Enrichment Analysis of Differentially Expressed Genes

GO enrichment analysis of differentially expressed genes was performed using the GOseq R package, in which gene length bias was corrected. GO terms with corrected $P$ values less than 0.05 were considered significantly enriched for differentially expressed genes. Genes were categorized into 10 563 GO terms consisting of three domains: biological process, cellular component, and molecular function (Figure 4). The dominant distributions in the biological process domain were for 'regulation of response to stimulus', 'response to stress', and 'macromolecule localization'. In the cell component domain they were 'intracellular', 'cell', and 'cell part'. We also observed a high percentage of genes assigned to 'intracellular signal transduction', 'regulation of signal transduction', and 'regulation of cell communication'. However, few genes were assigned to terms such as 'establishment of nucleus localization' and 'fibronectin binding'.

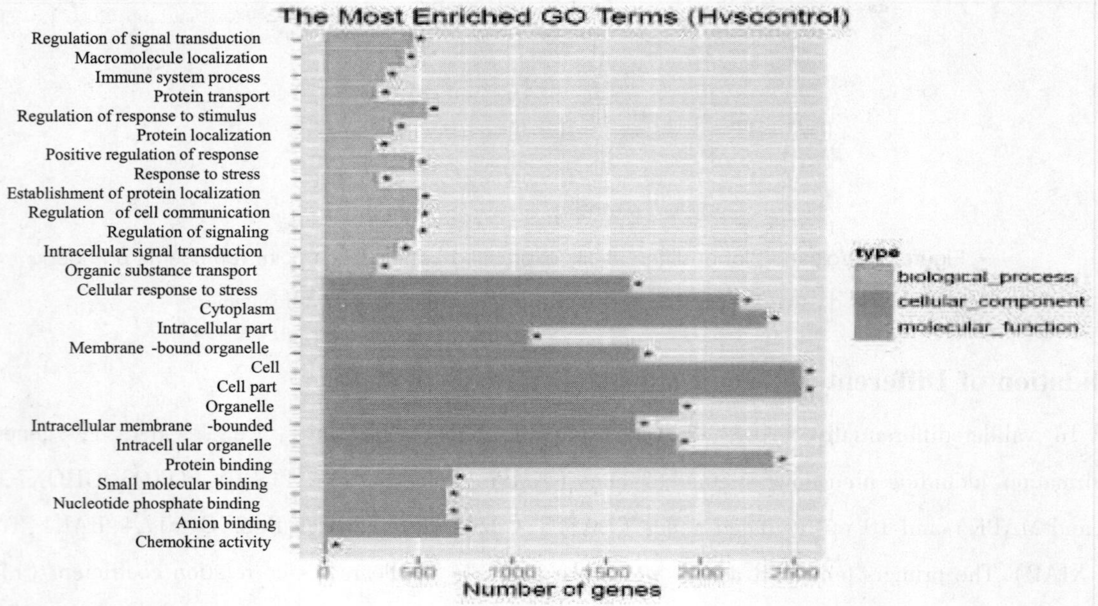


Figure 5 Bar chart of GO enrichment analyses, including biological process, cellular component, and molecular function domains.

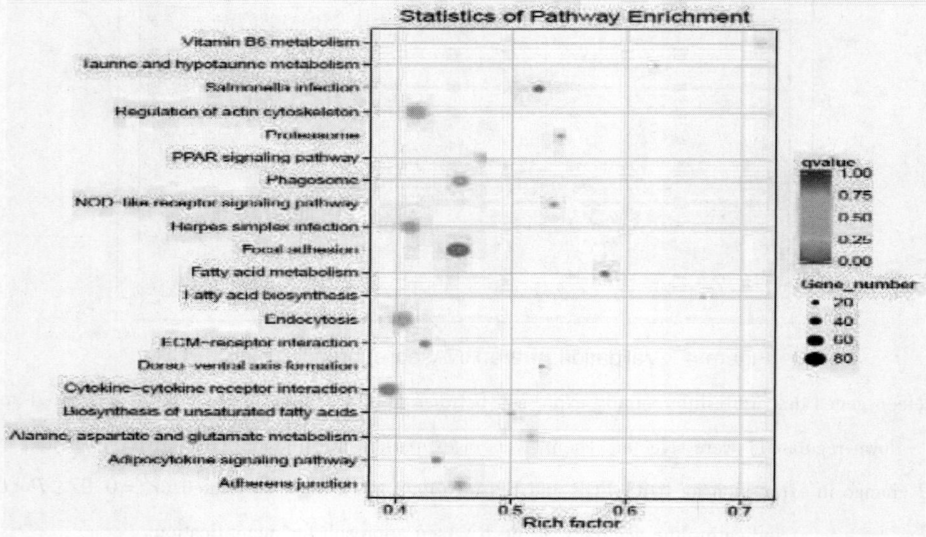


Figure 6 Statistics of KEGG enrichment.

The size of a point indicates the number of differentially expressed genes in a pathway, and the color of a point corresponds to different q values (corrected *P* values). The greater the Rich factor indicates a greater degree of enrichment

We analyzed the biological pathways that were active in our samples. Genes were mapped to the reference pathways in the KEGG; they were assigned to 152 KEGG pathways. These pathways, such as cytokine-cytokine receptor interaction, MAPK signaling pathway, metabolic pathways and regulation of actin cytoskeleton, have large numbers of differentially expressed genes. In contrast, we also found only a few genes in any one pathway, such as sulfur metabolism, illustrated in Figure 5.

## PPI Analysis of Differentially Expressed Genes

PPI analysis of differentially expressed genes was based on the STRING database, which contains known and predicted protein-protein interactions.

Networks were constructed by extracting the target gene list from the database; Additonally, Blastx (v2.2.28) was used to align the target gene sequences to the selected reference protein sequences and then the networks were built according to the known interaction of selected reference species by Cytoscape2.8.3 (Figure 6).

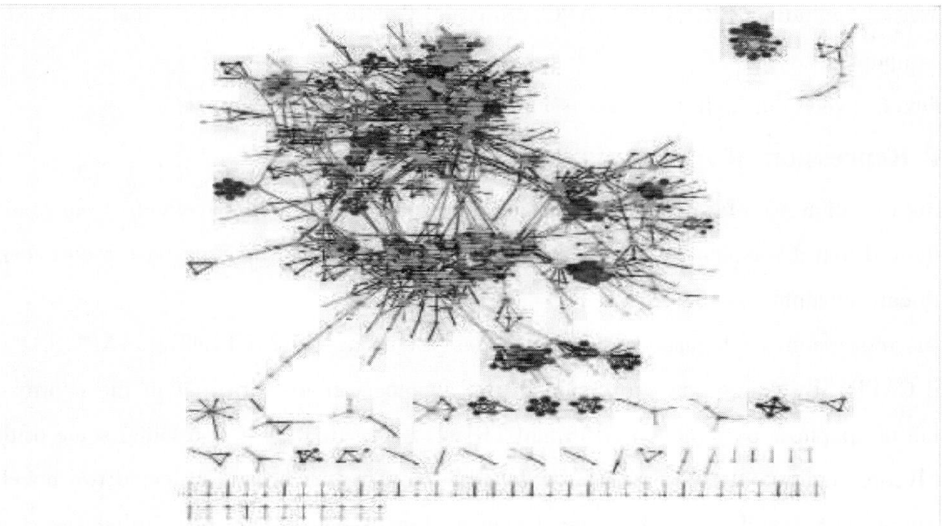

Figure 7 Protein interaction network analyses of differentially expressed genes.

The node size corresponds to the

degree of interaction. Color gradient from green to red corresponds to clustering coefficient values from low to high.

## Discussion

Myocardial cells are terminally differentiated. In the late embryonic period, the number of myocardial cells no longer increases and if any cells are lost, usually through apoptosis, they are not replaced. Furthermore, remaining myocardial cells gradually lose their function, which can lead to heart failure. This has serious consequences, including ascites, growth retardation, decrease in production performance, and even death. Little research into chicken myocardial cells has been reported and apoptosis pathways, and even key apoptosis genes, are poorly understood in these cells.

To our knowledge, this is the firstreport of an RNAseq study in chicken myocardial cells. This study identified genes that are differentially expressed in response to apoptosis induction. The assembled annotated transcriptomes provide a very valuable resource for further understanding of the molecular basis of chicken myocardial cell apoptosis.

## Screening Apoptosis Signaling Pathways

We monitored certain differentially expressed genes involved in numerous signaling pathways; for example, we compared the expression of PERK, IRE-1, and AIF-6 (endoplasmic reticulum stress response proteins) to determine whether the endoplasmic reticulum stress response is involved in $H_2O_2$-induced apoptosis. We found that the endoplasmic reticulum stress response is indeed involved.

We selected AIF, Cytc1A4, and bax to estimate if the mitochondrial pathway participated in the apoptosis induced by hydrogen peroxide. Differential gene expression analysis showed that the index for all three of these genes was increased in the H group relative to that in the control group, indicating that the mitochondrial pathway may be involved in this process.

The JNK signaling pathway was reported to play a minimal role in myocardial cell apoptosis. We compared expression of MKK4, MKK7, and JNK, and results showed that the JNK signaling pathway was also involved in apoptosis.

The WNT signaling pathway was assessed by KEGG enrichment analysis. Of 117 genes in this pathway, 38 were significantly increased, including, LRP, GSK3, APC, Axin, and β-catenin. We presume that the WNT pathway is also important for this apoptosis.

Overall, chicken myocardial cell apoptosis is highly similar to that in mammals.

### Initiation and Repression of Apoptosis

For apoptosis initiation, we elevated the expression of caspase-2, 8, 9 and 10, effective elevated by caspase-3, 6 and 7. Results showed that it was mainly rely on caspase-8, 9, effective mainly on caspase-3 and caspase-6. Caspase-3 has more significant advantages.

For apoptosis repression, we focused on several genes, including bcl-2, CFLAR, CAAP1, IAP, BFAR, TRIA1, API5, SIVA, and CAPP1. By comparing expression in the apoptosis group with that in the control group we found that the repression of apoptosis may rely on API5 and TRIA1. Interestingly, we also found some of these genes, such as IAP and BFAR, are significantly down-regulated. In addition, GO enrichment indicated two novel genes that may be involved in apoptosis repression but these need confirmation. We found no myocardial specificity for the anti-apoptotic gene, ARC.

### Others Findings

We noticed another interesting phenomenon; apoptosis and inflammation tended to have consistent gene expression patterns. Can we regard apoptosis as a special kind of inflammation induced by certain biochemical substances? Further investigation is required to test this suggestion.

## Acknowledgments

We thank Zhang X L and Cui X at NOVO gene (Beijing, China) for assistance with Illumina sequencing. We also thank Dr Gao J F for assistance with cell treatments. We thank anonymous reviewers for their very helpful comments.

### Funding

This work was supported by the National Institutes of Education [20100146110001]. The funder had no role in the study design, data collection or analysis, decision to publish, or in the preparation of the manuscript.

## Author contributions

Conceived and designed the experiments: W C-Y, X J-M, X T; performed the experiments: W C-Y, X T, W L-M, Z G X; analyzed the data: X J M, Z W, L Y W; contributed regents/materials/analysis tools: W L-M, Z G-X, Z W; wrote the paper: W C-Y.

## Supporting Information

Table S1 Alignment between reads and reference genome. Table S2 Primers used in q-RT-PCR.

# References


[1] Kerr J F, Wyllie A H, Currie AR. Apoptosis: a basic biological phenomenon with wide-ranging implications in tissue kinetics. *Br J Cancer*. 1972, 26(4): 239 - 257.

[2] Gaballa M A, Goldman S. Ventricular remodeling in heart failure. *Card Fai*, l 2002, 8(6): 476 - 485.

[3] Kang PM, Izumo S. Apoptosis in heart: basic mechanisms and implications in cardiovascular diseases. *Trends Mol Med*, 2003, 9(4): 177 - 182.

[4] Fan T J, Han L H, Cong R S. Caspase Family Proteases and Apoptosis. *Acta Biochim Biophys Sin*, 2005, 37(11): 719 - 727.

[5] SharovVG, Sabbah HN, Shimoyama H. Evidence of cardiocyte apoptosis in myocardium of dogs with chronic heart failure. *Am J Patho*, 1996, 148(1): 141 - 149.

[6] Ashkenazi A, DixitVM. Death receptors: signaling and modulation. *Science*, 1998, 281(5381): 1305 - 1308.

[7] Kong JY, Rabkin SW. Palmitate-induced apoptosis in cardiomyocytes ismediated through alterations in mitochonria: prevention by cyclosporineA. *Biocim Biopphys Acta*. 2000, 1485(1): 45 - 55.

[8] Danial NN, Gramm CF, Scorrano L. BAD and gluco-kinase reside in amitochondrial complex that integrates glycolysis and apoptosis. *Nature*, 2003, 424(6951): 952 - 956.

[9] Mao X, Cai T, Olyarchuk JG, Wei L. Automated genome annotation and pathway identification using the KEGG Orthology (KO) as a controlled vocabulary. Bioinformatics. 2005, 21(19): 3787 - 3793.

[10] Kanehisa M, Araki M, Goto S, et al. KEGG for linking genomes to life and the environment. *Nucleic Acids Res*. 2008, 36 (Database issue): D480 - 484.

[11] Geertman R, McMahon A, Sabban EL. Cloning and characterization of cDNAs for novel proteins with glutamic acid-proline dipeptide tandem repeats. *Biochim Biophys Acta*, 1996, 1306(2): 147 - 152.

[12] Connell S, Meade KG, Allan B, et al. Avian resistance to Campylobacter jejuni colonization is associated with an intestinal immunogene expression signature identified by mRNA sequencing. *PLoS One*. 2012; 7(8): e40409. doi: 10. 1371/journal. pone. 0040409.

[13] Tamim S, Vo DT, Uren PJ, et al. Genomic Analyses Reveal Broad Impact of miR-137 on Genes Associated with Malignant Transformation and Neuronal Differentiation in Glioblastoma Cells. *PLoS One*. 2014 Jan 22; 9(1): e85591. doi: 10. 1371/journal. pone. 0085591.

[14] Hampton M, Melvin RG, Andrews M T. Transcriptomic analysis of brown adipose tissue across the physiological extremes of natural hibernation. *PLoS One*. 2013 Dec 30; 8(12): e85157. doi: 10. 1371/journal. pone. 0085157.

[15] Schunter C, Vollmer SV, Macpherson E, Pascual M. Transcriptome analyses and differential gene expression in a non-model fish species with alternative mating tactics. *BMC Genomics*. 2014 Feb 28; 15(1): 167. doi: 10. 1186/1471 - 2164 - 15 - 167.

[16] Suzuki T, Yang J. Hydrogen peroxide activation of ERK5 confers resistance to Jurkat cells against apoptosis induced by the extrinsic pathway. *Biochem Biophys Res Commun*. 2014, 444(2): 248 - 253.

[17] DeHaan RL. Development of form in the embryonic heart, An experimental approach. *Circulation*. 1967, 35(5): 821 - 833.

[18] Kaestner L, Scholz A, Hammer K, et al. Isolation and genetic manipulation of adult cardiac myocytes for confocal imaging. *J Vis Exp* 2009. in press.

[19] Volz A, Piper HM, Siegmund B, *Schwartz P*. Longevity of adult ventricular rat heart muscle cells in serum-free primary culture. *J Mol Cell Cardiol*, 23 (2) (1991), pp. 161 - 173.

[20] Gruss-Fischer T, Fabian I. Protection by ascorbic acid from denaturation and release of cytochrome c, alteration of mitochondrial membrane potential and activation of multiple caspases induced by H(2)O(2), in human leukemia cells. *Biochem Pharmacol*. 2002 Apr 1; 63(7): 1325 - 1335.

[21] Soohyun Lee, Chae Hwa Seo, Byungho Lim, et al. Accurate quantification of transcriptome from RNAseq data by effective length normalization. *Nucleic Acids Research*, 2011, Vol. 39, No. 2e9 doi: 10. 1093/nar/gkq1015.

[22] KeonJae Park, YeonJeong Kim, Jeongeun Kim, et al. Protective Effects of Peroxiredoxin on Hydrogen Peroxide Induced Oxidative

Stress and Apoptosis in Cardiomyocytes. Korean Circ J. 2012 January; 42(1):23 - 32. Published online 2012 January 31. doi: 10.4070/kcj. 2012. 42. 1. 23

[23] Ruotian Li, Guijun Yan, Qiaoling Li, et al. MicroRNA-145 Protects Cardiomyocytes against Hydrogen Peroxide (H2O2)-Induced Apoptosis through Targeting the Mitochondria Apoptotic Pathway. *PLoS One*. 2012; 7(9): e44907. Published online 2012 September 18. doi: 10. 1371/journal. pone. 0044907.

[24] Baohua Wang, Jayant Shravah, Honglin Luo, et al. Propofol protects against hydrogen peroxide-induced injury in cardiac H9c2 cells via Akt activation and Bcl-2 up-regulation. *Biochem Biophys Res Commun*. available in PMC 2009, 389(1): 105 - 111.

[25] Bimei Jiang, Weimin Xiao, Yongzhong Shi, Meidong Liu, Xianzhong Xiao. Heat shock pretreatment inhibited the release of Smac/DIABLO from mitochondria and apoptosis induced by hydrogen peroxide in cardiomyocytes and C2C12 myogenic cells. Cell Stress Chaperones. 2005, 10(3): 252 - 262.

[26] Fox S, Sirigei F, Mockler TC. Application of ultra-high-throughput sequencing. Plant system biology. 2010, NewYork: humana press: 79 - 108.

[27] Pomraning KR, Smith KM, Bredeweg EL, et al. Library preparation and data analysis packages for rapid genome sequencing. *Methods Mol Biol*. 2012, 944: 1 - 22.

[28] Daniel P. Renfro, Brenley K. McIntosh, Anand Venkatraman, et al. GONUTS: the Gene Ontology Normal Usage Tracking System. *Nucleic Acids Res*. 2012, 40(D1): D1262 - D1269.

# Reflections on clinical practice: an oral history by a Chinese veterinarian

Hungchang WANG

## About Dr. H. C. Wang[1]

Dr. H. C. Wang (1915 – 2002) was born in the city of Zhenjiang, Jiansu Province, China.[2] He had the unusual opportunity of starting to learn English in fourth grade, and after sixth grade he went on to a school in Shanghai Nanyang Middle school, where most lessons were in English. In 1938 he graduated from Zhejiang University with a bachelor of veterinary science degree. Between 1938 and 1945, Dr. Wang worked on animal disease prevention in Sichuan Province and as the veterinarian at a station dedicated to the improvement of farm cattle.

In 1945, near the end of World War II, the Chinese government sent him and T. Y. Hsia to the United States to study veterinary medicine further. Through efforts of Dean W. A. Hagan[3], they had been accepted by the New York State Veterinary College at Cornell University, from which they graduated in June 1949. Dr. Wang promptly returned to China and was assigned to the Department of Animal and Veterinary Science at Beijing Agricultural University (BAU), only a few months before his country became a People's Republic. Also, until the Beijing Zoo developed its own clinic, it would send a jeep for Dr. Wang day or night, beginning in 1949.

Apart from his work, Dr. Wang was a founder of the Chinese Association of Internal Medicine in 1982 and served as its President from 1982 to 1995. Also, as a member of the editorial board, he edited the publication *Acta Veterinaria et Zootechnica Sinica*.[4]

Beginning in 1966, China went through its tumultuous Cultural Revolution. Dr. Wang burned his veterinary diploma from Cornell University to avoid punishment because of his connection with the United States.[5] The BAU faculty was moved twice during the Cultural Revolution: first, a relatively short distance to Zhuozhou farm in Heibei Province in 1969, and then further west to Yanan in 1971. There he practiced veterinary medicine for the peasants. After Mao Tzedong died in 1976, the faculty returned to Beijing, where Dr. Wang continued to work until his retirement in 1989.

During his teaching years, Dr. Wang was sent by the Chinese Ministry of Education to the Soviet Union (USSR) in 1957 to see whether he could find ways to improve Chinese veterinary education and practice. In addition to evaluating the College of Veterinary Medicine in Moscow, he visited many parts of the Soviet Union. Due to a rupture in relations between the USSR and China in 1959, he had to return after two years. Later, he led a Chinese delegation to Canada for three weeks. They visited governmental bodies in Ottawa and Vancouver, and observed veterinary education at Ontario Veterinary College as well as at the newly opened Western Veterinary College in Saskatoon, Saskatchewan.

In 1989, Dr. Wang returned to Cornell University for the fortieth reunion of his veterinary class. On behalf of himself and Dr. T. Y. Hsia, he gave Dean Robert Phemister a very large hanging scroll depicting an eagle. He stated that "The great [Chinese] character chosen for this scroll is the EAGLE because like a first class veterinary clinician the eagle has keen eyes for recognizing problems and a tenacious grasp until they are solved." Then Dr. Wang presented his reflections.

## Reflections on Clinical Practice

About at the end of World War II, I was Lend-Leased from China to study veterinary medicine in the United

States of America. Now it has been 40 years since I graduated from the New York State College of Veterinary Medicine at Cornell University with a DVM degree. I would like to tell you a little about how veterinary medicine looks to me now after such a long time as a clinician and university professor in China.

I believe strongly that practice is very important in a veterinary medical education. I do not underestimate the value of listening to lectures, doing exercises in laboratories, or reading textbooks and other materials in the library. But, Chinese students have always gotten plenty of those lessons. What they do not get enough of is clinical practice. Such clinical practice is very important for understanding how to work with sick animals. It also helps us to understand people, which is necessary because a veterinarian has to get cooperation from the people who take care of the animals in order to do correct treatment.

Before I had the good luck to attend Cornell, I had already earned a Bachelor of Veterinary Science (BVSc) degree at Zhejiang University, which is a coastal province in southeast China. Also, I had worked four and one-half years in a veterinary clinic in Sichuan Province, which is farther west in the central part of the country, something like Missouri or Arkansas in the U. S. A. At that clinic I had observed and done preventive work on such diseases as anthrax, hog cholera and rinderpest. I had also served as the chief veterinarian at a farm improvement center for two and one half years. So, when I came to America to study under great teachers like Drs. Denny Udall and Myron Fincher, and under great future teachers like Drs. Steve Roberts and Francis Fox, I had already had useful experiences with clinical problems in my own country. Let me tell you a few of my later experiences in China to show you why I think good clinical training is so necessary.

After I graduated from Cornell in 1949, only a few months before China became a People's Republic, I joined the Department of Veterinary and Animal Science as a professor of veterinary medicine at Beijing Agricultural University. Besides teaching, I set up a veterinary diagnostic laboratory there, and made some disease surveys of Beijing dairy farms. I also did surveys at the Beijing Zoo. That is how, around 1956, I got a call to see a giraffe with a bent neck.

Up to then, giraffes had been imported many times, but they had not stayed alive very long. Now this one was sick, and the first thing I could see was an unnatural bend in its long neck. The people at the zoo said that they thought its neck was bent because its feed trough was too low. I said that perhaps it had rickets, but they said no, it could not be what we call "soft bone disease" because the keepers were feeding it sufficient minerals in the correct proportions. And they also reported that they were giving it vitamins.

Well, I knew we had to examine it more closely, but restraint of wild animals often causes sudden death. Still, we restrained it... and it died. Of course, then we had to do a postmortem on it. We looked first at the ribs for evidence of rickets. They were as soft as potatoes; we could peel them with a knife, and so with other bones. So, here was a clinical question: why did giraffes get rickets at the Beijing Zoo? I had to think about the natural ecology of giraffes. They come from Africa where the sun is intense during many months of the year. The giraffe needs its thick skin to protect it from sunburn. In Beijing, which is about like Chicago in America, the sunshine is weak, especially in the winter. The animal didn't get enough ultra-violet rays to make use of what it was being fed. I suggested that the problem was related to poor vitamin D metabolism and recommended that they put some ultra-violet lights in the cages for giraffes and some other African animals. The result was pretty good. Not only did it keep these animals alive, but they also gave offspring. Those people at the zoo had been doing what the books said, but they had not been thinking about the whole life of the kind of animal that was sick. They had not been thinking like veterinary clinicians.

Earlier, about 1952, there had been a big problem with dairy cattle dying on a special farm near Beijing. The farm was set up to produce milk for high level officials and their children, especially any that were weak. First a bull had died with very fast respiration. The people in charge of the farm worried that it might be anthrax, and they had buried the body right away, so there were no good clinical or postmortem examinations. Next they had asked army medical men what was wrong. The medical men answered that animals were not their work; the farm managers would have to find a veterinarian. Well, the Chinese army had had a veterinary college for many, many years, and its clinicians had very good experience with horses, but they didn't know much about cattle. One of my students worked at the farm with the problem, and he called me to see what was going on. I was glad that I had had some good clinical experiences with dairy cattle when I was a student at Cornell.

The most important common sign tha tall of the cattle had was a very fast respiration accompanied by normal or subnormal temperatures. The respiration rates were about 130/minute, but the pulse rates were not correspondingly fast. We suspected that it might be some kind of poisoning from roughage, especially cyanide. So, we injected sodium thiosulfate, intra-venously. We also collected blood to take to the lab. We told the people at the farm to keep any animals that died, so we could make further observations. One did die that night, and we did a thorough postmortem examination. We collected materials to check for bacteria, viruses, and parasites, including blood parasites, but we couldn't find any cause.

Sodium thiosulfate had done no good, so we thought that the problem was not cyanide poisoning. Still, from the clinical signs, I suspected some kind of intoxication. We collected all kinds of their feed samples: corn, soybean cake, and so forth. All were good, healthy feeds except the sweet potatoes. On those we found some black spots. I had had experience with soft, moldy sweet potatoes in southern China, but these sweet potatoes were very firm, so I took them to some plant pathologists.

As soon as one professor saw them, he asked, "Where did you get this pure culture of *Ceratostomella fimbriata*? I have been looking for some of this. Will you give these specimens to me?"

"Sure, I'll give you them, but first can you tell me anything about whether they can harm cattle, or not?"

He had no idea if they could cause disease in animals, but he had given me the name of the fungus. That was very important; we could look for references to it in the library. There was nothing in the Chinese, English, or Russian literature about its causing disease, but we found an answer in some Japanese papers. They described signs in cattle just like the ones we had found.

This was strange because Chinese people had been planting sweet potatoes for a very long time, but we had never seen this disease before. Then we learned that this fungus disease was associated with a variety of sweet potato which had only recently come to China from Japan. China had begun importing it after the Japanese surrender because of its high productivity. It was so good the Chinese government had given it the name "Victory", and had encouraged people to plant it all over. But it was not a victory to spread such a mycosis widely. Later the government urged scientists to work hard to find resistant varieties of sweet potato because the only way to control this disease in cattle is not to let them eat sweet potatoes with the fungus.

The outbreak of this new disease in dairy cattle at Beijing made two big impressions on me. First, I learned that a clinician must not let go when the diagnosis isn't clear and quick. Second, it got me interested in mycotoxins for the rest of my life.

Of course most of the time in practice, we see the same kinds of cases over and over again, but thestudent needs to learn to look at all sorts of things when something new is presented. With the giraffe we looked at its natural home in Africa. With the dairy cattle, we kept looking at all its feeds for something different. We must also look out for new diseases when we change management practices, and that is going on rapidly in some parts of China now.

For example, Beijing is not located in an area of selenium deficiency, but China has many selenium deficient regions. Quite a few teachers from Beijing Agricultural University developed cardiac white muscle disease when our university was moved to the Yenan region during the Cultural Revolution. That is easy to understand because selenium is very low in Yenan, but why were we now seeing selenium deficiencies in chickens and swine which have been raised in the Beijing area? Well, recently, we have begun to raise chickens up in cages, so they cannot get selenium from the soil. Also, we bring them grains from selenium deficient parts of the country. So, it is not surprising, really, that this disease has come to Beijing's chickens, as well as to some swine being raised intensively. Fortunately selenium deficiency disease is not hard to control when we have found the cause. It is easy to put trace minerals in the feed when we know that management is really the problem.

As I said at the beginning, I do not underestimate the value of lectures, reading, and laboratory studies in veterinary medical education. But students need also to train their eyes to have the keen observation of an eagle. For this, they need clinical practice under the guidance of experienced eagle eyes. And, they need to be trained not to give up if the first good answer doesn't fit. They need to have teachers like mine at Cornell, who showed me how to keep looking until a sound solution could be found.

## Notes

1. The introduction to this article is by Dr. Phyllis Hickney Larsen. Dr. Wang originally dictated his talk to her in Ithaca, New York in 1989. She then worked directly with him until he was satisfied with it for saying during his 40$^{th}$ New York State College of Veterinary Medicine reunion at Cornell University June 9 – 12, 1989. Dr. Larsen attended his talk and his presentation of the eagle scroll.

2. Dr. Wang's daughter, Damin Wang, provided valuable information for the introduction after consulting her three siblings. Topics they provided included: his birth city; his earlyexperience with English; his relationship with the Beijing Zoo; the moves of BAU during the Cultural Revolution; some information about Dr. Wang's two years in the USSR; his visit to Canada; the year of Dr. Wang's retirement; and when he died. Damin Wang sent this material in E-mails to Dr. Larsen in mid-to-late October 2012.

3. Notes by P. H. Larsen in 1989 record that H. C. Wang told her then about Dean Hagan's help. Earlier, during a visit to his home in China in September of 1986, Dr. Wang had told Dr. Larsen how he had burned his original diploma, and he showed her the replacement he had received from Cornell.

4. Dr. Bo Han, editor of *Chinese Association of Veterinary Internal Medicine* 30$^{th}$ *Anniversary* provided Dr. Larsen with information about Dr. Wang's roles in that association and as editor of *Acta Veterinaria et Zootechnica Sinica*. E-mail, from Dr. Han, October 23, 2012.

5. During her visit to his home in China on September 1, 1986, Dr. Wang told Dr. Larsen how he had burned his original diploma, and he showed her the replacement he had received from Cornell in the early 1980s.

This paper published on Veterinary Heritage: Bulletin of the American Veterinary Medical History Society 2012, 35(2): 54 – 58.

# 走近中国兽医界先驱——王洪章
## ——回顾中国兽医临床实践

Phyllis Larsen
中国农业大学 陈微 博士 译

**编者注**:王洪章教授为中国知名的兽医学家,几十年如一日致力于兽医临床实践和教学工作,曾任中国畜牧兽医学会家畜内科学分会理事长,为中国的兽医临床实践和兽医教学做出卓越贡献。其先后在浙江大学和美国康奈尔大学获得兽医学士学位和兽医博士学位,主张优秀的兽医需要极其敏锐的眼光来发现问题并解决问题,临床实践是至关重要的一点。2014 年是王洪章先生诞辰 100 周年,谨以此文将王先生的兽医临床实践思想理论与大家共勉。

## 1 王洪章博士简介

王洪章博士(1915—2002 年)出生于中国江苏省镇江市,由于特殊的机遇,他在小学四年级时开始学习英语,六年级后他升入上海南洋模范中学,该中学的大部分课程用英语授课。1938 年,他毕业于浙江大学获得了兽医学士学位。1938—1945 年他在四川省动物疾病防治所任兽医,以发展耕牛业。

1945 年末第二次世界大战结束后,中国政府派遣他和夏定有赴美国进一步学习兽医学。通过院长 W. A. Hagan 的推荐,他进入了纽约州康奈尔大学兽医学院学习,1949 年 6 月毕业。在新中国成立前几个月,他应北京农业大学兽医学系的邀请毅然回国。1949 年时,北京动物园还没有自己的诊所,无论是白天还是夜间,遇到问题都会派车去请王博士会诊。

除了他的本职工作以外,王博士在 1982 年创建了中国兽医学会内科学分会。1982—1995 年出任理事长。并担任《畜牧兽医报》的编委。

1966 年初,中国进入了“文化大革命”。王博士烧掉了康奈尔大学的兽医博士学位证书,以避免因与美国的联系而受批斗。北京农业大学在“文化大革命”期间搬迁了两次。第一次于 1969 年搬迁到离北京不远的河北省涿州县农场,而后往西搬迁到延安,在那里,他的兽医工作就是服务于农民。1976 年迁回涿州,同年毛泽东去世后,该系于 1978 年又迁回北京,他继续工作,直到 1989 年退休。

1957 年,在他从事教学工作期间,为了改进中国兽医的教学和实践,王博士被中国教育部派送到苏联学习。他考察了莫斯科兽医学院并访问了苏联许多地方。两年后的 1959 年,由于中苏关系破裂他又回国了。他还率中国代表团赴加拿大进行了为期三周的访问,他访问了在渥太华的港口政府部门,考察了安大略兽医学院的兽医教育以及萨斯堪彻瓦新开办的西部兽医学院。

1989 年他回康奈尔大学参加兽医班四十周年纪念活动。他代表他和夏定有博士向院长罗伯特・费姆斯特赠送了画着一只老鹰的巨大画卷。他说:“之所以选择画着老鹰的画卷,是因为一个最出色的临床兽医应该像老鹰一样有种敏锐的眼光去发现问题,并紧紧地抓住不放直到解决问题”。

## 2 王洪章的兽医临床实践回顾

大约在第二次世界大战末,我被中国国民政府派遣来到美国学习兽医。我自纽约康奈尔大学兽医学院获得兽医博士学位至今已经四十年了。我很想告诉你们,作为中国的兽医和大学教授,在这么漫长的时间里,我是如何看待兽医学的。

在兽医教学中,临床实践是非常重要的。我并不是低估课堂教育,在实验室做实验或阅读教科书,看各种资料的作用。但是中国的学生总是花大量的时间听课,他们没有获得足够的临床实践。临床实践是诊断的重要基础。它也帮助我们了解人们,因为兽医需获得照看动物者的合作才能正确处置。

在我有幸步入康奈尔大学之前,我已经浙江大学获得了兽医学士学位。同时我在四川省一个兽医诊所工作了四年半。四川在中国西南,类似于美国的密苏里和阿拉斯加。在诊疗过程中,我对如炭疽病、猪霍乱和牛瘟疫进行了观察、了解和防治工作。之后我作为首席兽医又在农场发展中心工作了两年半。所以在我来到美国跟随大师如 Denny Udall 博士、Myron Fincher 博士、Steve Robert 博士和 Francis Fox 博士学习时,我已经拥有了与我国临床问题相关的经验。下面我谈谈我的经验,以及临床培训的重要性。

1949 年新中国成立前几个月,我从康奈尔大学毕业后在北京农业大学畜牧兽医系任兽医教授。除教学外,我创建了兽医诊断实验室,还对北京奶牛场做了一些疾病调查。同时也对北京动物园做了调查。

大概在 1956 年,动物园请我去看一只弯脖子的长颈鹿。在这以前动物园已经几次引进过长颈鹿,但它们都不能存活很久。现在这只又病了,我第一眼看到的,就是它那长长的脖子很不自然地弯着。动物园的人说脖子弯曲是由于喂食斗挂得太低引起的。我说它可能患了佝偻病,但他们说不可能。它不可能是我们所说的“软骨病”,因为饲养员按正确的比例喂给它足够的矿物质,同时喂给它各种维生素。

我们需要进一步确诊,圈养的野生动物时常会发生突然死亡。我们继续圈喂它,最后它死了。当然,我们对它进行了尸体剖检。我们首先看佝偻病的典型部位肋骨。它们像土豆一样柔软,我们用小刀就可以削动,其他部位的骨头也同样如此。所以,这儿就有一个临床问题:为什么北京动物园的长颈鹿会患佝偻病呢?我们必须考虑长颈鹿生长的自然生态环境。它们来自非洲,那里一年的很多月份有阳光照射。长颈鹿的皮肤很厚以防阳光灼伤,而北京的气候与美国芝加哥类似,阳光弱,特别是在冬天,动物不能获得足够的紫外线来利用它所获得的物质。我认为问题的所在是维生素 D 代谢水平低,建议在长颈鹿和其他非洲动物的笼子里放置一些紫外灯。效果相当好,他们不仅活了下来,而且还繁衍了后代。在动物园工作的人员只是按照书本上所说的去做,而没有考虑生病的动物所处的整个生活环境,他们没有像兽医那样思考问题。

早前大约在 1952 年,北京附近的一个专供农场的奶牛出现了大问题。该厂生产的牛奶专供高干及其子女们和特别体弱的人。第一头奶牛因呼吸急促而死亡,农场的负责人担心是不是炭疽病就立即将尸体焚烧,没有进行很好的临床诊断和尸体剖检。第二天他们请来了一位军医想了解问题出在哪儿,但军医说动物不在他们的工作范围。农场的负责人不得不去找兽医。那时中国有一个办学多年的解放军兽医学院,他们对马有丰富的临床经验,但是对牛了解不多。我的一个学生在农场处理这个问题,他就打电话请我去看看该怎么办。我很乐意去,因为我在康奈尔大学当学生时已具有一些不错的奶牛临床经验了。

所有的牛表现出的重要体征是呼吸急促,体温正常或略偏高。每分钟呼吸 130 次而脉搏并不快。我们推测可能是草料中某种物质中毒特别是氰化物。这样我们给牛注射了硫代硫酸钠及静脉输液。我们还采集了牛血送去化验室化验。我们告诉农场的人要保留死亡的动物以便我们进一步观察。那天夜里又有一头牛死亡,我们进行了剖检。我们收集材料用于细菌、病毒、寄生虫和血液寄生虫的检验,但是并没有找到原因。

注射硫代硫酸钠没有起作用,所以我们认为不是草料中毒。但是,从临床症状看,我推测是某种中毒。我们搜集了所有的饲料样本:玉米、黄豆饼等。除了甘薯外所有的饲料都很好。在甘薯上我们发现了一些黑斑。以往的经验告诉我在南方发霉的甘薯是软的,而这些甘薯却非常坚硬。于是我就带着这些甘薯去找植物病理学家。

当一位教授看到这些甘薯立刻就问:“你从哪儿搞到的 *Ceratoseomella fimbriata* 纯培养物? 我一直在找它们,你能把这些样本给我吗?”

“当然可以,我会把它们给你,但是首先你得告诉我这对奶牛是否有害?”

他也不知道是否会引起动物发病,但是他给了我这种真菌的名字。我们在图书馆寻找相关的资料。在中文、英文和俄文的文献中没有找到任何答案,但是在一些日文的报告中我们找到了答案。他们在牛上描述的症状与我们的类似。

需要强调的是，中国人种植甘薯已有悠久的历史，但是之前我们从未见过这种疾病。而后我们了解到这种真菌病与刚刚从日本进入中国的一种甘薯有关。中国政府在日本投降后开始引进这种高产的甘薯。因为这种甘薯非常好，中国政府就将它命名为“胜利”，还鼓励全国人民种植。但是它广泛传播真菌病。所以并不是一个“胜利”。之后政府要求科学家努力寻找针对这种甘薯的预防方法。控制奶牛疾病的唯一方法就是不让奶牛食用这种带霉菌的甘薯。

这次在北京暴发的奶牛疾病给我留下了两点深刻的印象。首先，我认为当不能快而准确地做出诊断时，作为一位临床工作者针对这个问题不能就让它过去。其次，这让我在业余生活中对霉菌毒素产生了兴趣。

当然在实践过程中，大部分时间我们一遍又一遍地看到同样的疾病。但是当新东西出现的时候，学生需要全面了解事物。长颈鹿我们要看在非洲的自然环境。奶牛我们要看它们的饲料改变的地方。当我们改变管理方式时，需要警惕新的疾病，目前这点在中国的一些地方已很快实施。例如，北京不位于硒缺乏的地区，但中国有许多硒缺乏地区。当我们的大学在“文化大革命”搬到延安地区时，北京农业大学极少数的教师患上了心脏白肌病。这很容易理解，因为延安地区硒含量很低。但为什么在北京地区鸡和猪的硒缺乏症也上升呢？近来，我们开始对鸡进行笼养，使它们不能从土壤里获得硒。另外，我们饲喂来自于硒缺乏地区的谷物。所以北京的鸡患上这种病，猪的病例持续上升并没什么可惊讶的。庆幸的是，在发现硒缺乏病例时，控制起来并不困难。如果我们知道管理上确实存在这种问题时，只要在饲喂时加入一些微量元素即可。

正如我在开头时所说的，我并不是低估讲课、阅读和实验在兽医教育中的价值。但是也需要培养学生拥有老鹰一样的具有敏锐观察力的眼光。他们需要在有经验有眼光的指导下进行临床实践。他们需要像我在康奈尔时的那些老师一样，执着探索直至找到正确的结论。

# TFF3 抗内毒素致新生仔猪肠黏膜上皮细胞损伤研究

邓　荣,周东海*,郭定宗,杨世锦,李朝阳

(华中农业大学动物医学院,武汉 430070)

**摘　要**:为探讨肠三叶因子(TFF3)在肠道性疾病中的作用机理,本文采用酶消化法分离培养仔猪肠黏膜上皮细胞,添加不同浓度的内毒素及重组人小肠三叶因子(rhTFF3),通过检测细胞活力来观察 rhTFF3 抗内毒素致细胞损伤,并采用荧光定量 PCR 技术检测内毒素及 rhTFF3 对肠黏膜上皮细胞中 TFF3 基因表达的影响。结果,利用酶消化法成功培养仔猪肠黏膜上皮细胞,得到的细胞具有较强的活性;rhTFF3 可以有效预防内毒素对肠黏膜上皮细胞的损伤($P<0.05$),且对不同浓度内毒素的预防效果不一($P<0.05$);低浓度(1、10 μg/ml)内毒素可抑制 TFF3 基因的表达;高浓度(100 μg/ml)的内毒素可促进 TFF3 基因的表达;而加入 rhTFF3 时,细胞中 TFF3 基因的表达也会相对降低。结果表明,酶消化法可成功培养仔猪肠黏膜上皮细胞;TFF3 具有抗内毒素致新生仔猪肠黏膜上皮细胞损伤的作用;内毒素及 rhTFF3 可以调节细胞中 TFF3 基因的表达。

**关键词**:肠三叶因子;内毒素;肠黏膜上皮细胞;重组人小肠三叶因子

# The effects of TFF3 on prevention of intestinal mucous epithelial cells of newborn piglet from injury by the colotoxi

DENG Rong, ZHOU Donghai, WU Wei

*College of Veterinary Medicine, Huazhong Agricultural University, Wuhan* 430070

**Abstract**: In order to study the mechanism of the intestinal disease by TFF3, the intestinal mucous epithelial cells (IMEC) of newborn piglet were cultured by using enzymatic digestion. The MTT measure was used to evaluate the effect of the protectin in cell from different concentration LPS by rhTFF3. The SYBR green chemistry on a real-time PCR cycler were used to analysis the expression of TFF3 gene observed which effected by LPS and rhTFF3. We found that the IMEC were cultured successfully by using enzymatic digestion, and the cells have a good liveness. The injury of LPS can be prevented by rhTFF3 in IMEC ($P<0.05$), and there were different effect in different group which was added different concentration of LPS ($P<0.05$). The inhibition expression of TFF3 can be detect in light concentration (1, 10 μg/ml) LPS group, and the higher concentration (100 μg/ml) LPS was associated with a distinct response of increased TFF3 expression. The expression of TFF3 was lower than the normal group when the rhTFF3 were existence yet. The above results indicated that the IMEC could be cultured successfully by using enzymatic digestion. The injury of LPS can be prevented by rhTFF3 in IMEC; The expression of TFF3 can be regulated by LPS and rhTFF3.

**Keywords**: TFF3, LPS, IMEC, rhTFF3

肠黏膜上皮细胞是构成肠道屏障的重要成分,它不仅可以保持黏膜屏障的完整性,更具有吸收营养物

---

**基金项目**:中央高校基本科研业务费专项资助 2011PY077

* **通讯作者**:E-mail:bigdefoot@163.com

质,阻止有害物质侵入的重要功能肠黏膜上皮细胞是体内更新较快的一类细胞,可利用其特性研究细胞的增值、信号通路、凋亡等重要的生物学信息。TFF3 在肠道主要受早期炎症因子和抗炎症因子调节,转化生长因子(如在黏膜修复进程中调节 TFF3)是一个重要的细胞因子,和表皮生长因子一样,体外已有很多试验证明 TFF3 在肠道疾病中具有重要的作用,如 TFF3 能够通过运动活性来积极参与上皮细胞的修复。早期炎症因子—肿瘤坏死因子 a(TNF)是许多黏膜损伤的参与者,在体外可被 TFF3 通过激活 NFκB 途径而抑制其转录(Tan X D et al. ,2000)。探讨这些因素可以与三叶肽的调节途径相关联,研究这些因素的可以为分析三叶肽提供更加深远的功能背景,以更好地了解三叶肽的作用方式。细菌内毒素是所有数革兰氏阴性菌外膜的主要组成部分,并且对免疫系统及炎症的发生具有重要的刺激作用,当其进入血液循环将产生如发热、肠道性疾病、感染性疾病等常见疾病,此外内毒素还可以具有调节黏蛋白及细胞因子表达的能力。

# 1 材料与方法

## 1.1 材料

初生未吃初乳的仔猪,静脉注射巴比妥麻醉后,无菌打开腹腔并无菌采集小肠段迅速分离培养。

10%胎牛血清 DMEM-F12(Hyclone)培养基,细胞角蛋白 18 单克隆抗体(武汉博士德生物工程有限公司);内毒素(sigma);rhTFF3(大连保税区联合博泰生物技术有限公司);抗兔/羊即用型 SABC 免疫组化试剂盒(武汉博士德生物工程公司;PCR mix(TaKaRa);RT-PCR 试剂盒及荧光定量 PCR 试剂盒(TaKaRa);Trizol 提取试剂(TaKaRa)。

## 1.2 小肠黏膜上皮细胞的原代培养

常规培养。

## 1.3 小肠黏膜上皮细胞的传代培养

待原代细胞长满细胞瓶的 80%以上后,弃掉原培养基,用 PBS 清洗两遍后,加入胰酶 1 ml/瓶,37℃消化 2 min后弃掉胰酶,加入完全培养基吹打细胞,待细胞完全悬浮后调整细胞密度并接种于两个新的培养瓶中。

## 1.4 仔猪肠黏膜上皮细胞的鉴定

细胞角蛋白是上皮细胞的特异性抗原蛋白,本试验通过选用角蛋白 18 单抗对原代培养的细胞进行免疫组织化学鉴定。待细胞长满时,将细胞传至有无菌盖玻片的六孔板中,培养至盖玻片上长满细胞后,进行角蛋白($CK_{18}$)免疫组化鉴定肠黏膜上皮细胞。弃掉六孔板中的培养液,加入蒸馏水清洗两遍后弃掉液体;加入冰的丙酮室温固定 10 min 左右,弃掉丙酮并用蒸馏水清洗两遍;加入 30%双氧水 1 份与 50 份纯甲醇的混合液,室温 20 min,以灭活内源性过氧化物酶,弃掉后加入蒸馏水清洗 2 遍;滴加 5% BSA 封闭液,室温 20 min,甩去多余液体,不洗;滴加适当稀释的兔抗人 $CK_{18}$的一抗,阴性对照滴加 PBS,37℃培养箱中 1 h 左右,PBS 清洗 2 min×3 次;滴加羊抗兔 IgG,37℃培养箱中 20 min,PBS 清洗 2 min×3 次;滴加试剂 SABC 室温 20 min,PBS 清洗 5 min×3 次;DAB 显色,取 1 ml 蒸馏水,加 DAB 试剂盒中的 A、B、C 试剂各一滴,混匀后加至载玻片上,反应 5~30 min;封片后显微镜下观察拍照。

## 1.5 rhTFF3 防治细胞损伤的作用

1.5.1 试验分组 待原代细胞长满后,传代并调整细胞密度后传至 96 孔板中的 35 个孔中,分为对照组,TFF3 治疗组(A 组、B 组、C 组),TFF3 预防组(D 组、E 组、F 组)七个组,每组 5 个重复。治疗组中先分别添加终浓度为 1 μg/ml(A 组)、10 μg/ml(B 组)、100 μg/ml(C 组)的内毒素作用 30 min,再加入 50 μg/ml 的 rhTFF3 作用 30 min;预防组则先添加 50 μg/ml 的 rhTFF3 作用 30 min,再加入终浓度为 1 μg/ml(D 组)、10 μg/ml(E 组)、100 μg/ml(F 组)的内毒素作用 30 min,对照组添加 PBS。

1.5.2 细胞活力的检测(MTT 法) 细胞按以上的分组处理之后,吸弃原培养液并用 PBS 清洗后,加入 180

μl 完全培养基和 20 μl MTT,4 h 后吸弃培养液并用 PBS 清洗后,加入 150 μl DMSO,振荡 10 min,用酶标仪 570 nm 波长检测细胞的吸收值。

## 1.6 内毒素对细胞中 TFF3 的影响

1.6.1 试验分组 以原代生长良好的 5 瓶细胞为研究对象,待细胞长满 80% 后,用 PBS 冲洗两遍细胞,吸弃瓶内残留的液体,其中 3 瓶细胞分别添加内毒素,并使得内毒素的终浓度分别为 1、10、100 μg/ml,另取一瓶加入 rhTFF3 至终浓度为 50 μg/ml,余下的一瓶添加生理盐水作为正常对照,细胞于 37℃,5% $CO_2$ 培养箱培养 24 h 后 Trizol 法提取总 RNA,立即反转录或保存于-70℃超低温冰箱。

1.6.2 总 RNA 的提取 用试剂盒提取。

1.6.3 反转录 反转录采用反转录试剂盒(TAKARA DRR047S),操作步骤参考说明书。

1.6.4 PCR 及荧光定量 PCR 根据 NCBI 主页 GenBank 中已公布的猪的肠三叶因子基因序列,采用 Oligo6.0 软件设计 TFF3 及 18S rRNA 的引物,并由英俊(武汉)合成。普通 PCR 及荧光定量步骤参考 TAKARA 说明书(D334A、DROX01)。

表 1 荧光定量 PCR 引物设计,S 为正义引物,AS 为反义引物

Tab. 1 Synthetic oligonucleotides designed for quantitative real-time PCR (qRT-PCR);

S:sense primer; AS:antisense primer

| 基因 | GenBank 序列号 | 引物 | 引物序列(5'-3') | 产物长度 | Tm(℃) |
|---|---|---|---|---|---|
| TFF3 | F14493 | S | GGGAGTATGTGGGCCTGTC | 174 | 58 |
| | | AS | AGGTGCATTCTGTTTCCTG | | |
| 18S rRNA | AY265350 | S | AATCGGTAGTAGCGACGG | 275 | 59 |
| | | AS | AGAGGGACAAGTGGCGTTC | | |

1.6.5 琼脂糖凝胶电泳 加完样后应立即通电电泳,电压 60 ~ 100V,样品从负极(黑色)向正极(红色)的方向移动,琼脂糖凝胶的有效分离范围会随电压的升高而降低,当看见溴酚蓝移动到距胶板下沿大约 1 cm 处时,可停止电泳,取出凝胶。

观察照相:在紫外灯下观察并于凝胶成像系统拍照保存。

# 2 结果与分析

## 2.1 肠黏膜上皮细胞培养结果

仔猪肠黏膜上皮细胞的培养,小肠组织被剪碎后在显微镜下可见大量的隐窝结构及小肠绒毛组织(图 1 A),而且组织剪得越小越容易见到绒毛组织;剪碎后的组织经 2.5%(mg:ml)的胶原酶消化 2 h 后,在消化酶的上清液中可见大量隐窝结构及肠上皮细胞存在(图 1 B)。

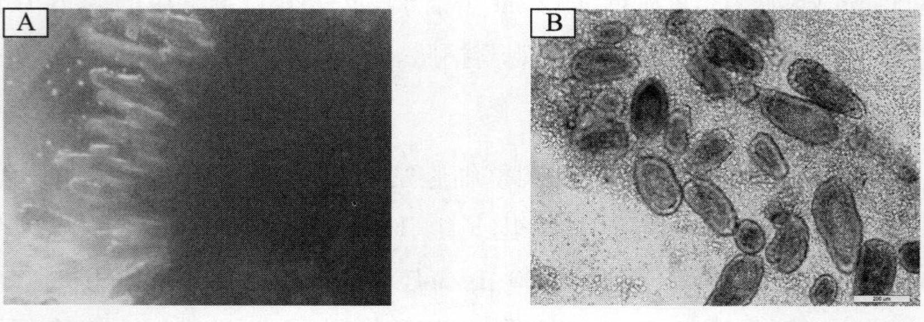


图 1 显微镜下可见小肠绒毛结构、隐窝单位及大量单个的细胞,标尺:100 μm

Fig. 1 Intestinal villus, crypt units and lots of cells were detected by microscope, scale bars represent 100 μm

在细胞原代培养 12 h 后就可见少量的细胞已经贴壁;24 h 可见大量的贴壁细胞;48 h 可见细胞或以残留的组织块辐射状生长,说明细胞具有成团生长的特性,并且已经具有一定的形态(图 2 a),成三角形或多边形,也有大量的细胞呈圆形,并可见细胞内具有两个细胞核;72 h 可见细胞紧贴瓶底,细胞体积变大,细胞分散比较均匀,大多呈典型的三角形状(图 2 b);细胞培养 96 h 以上显微镜下可见细胞长成单层,细胞状态良好,并成典型的铺路石状,此时细胞贴壁十分紧(图 2 c),细胞长满后可见细胞紧密贴壁于细胞瓶底(图 2 d、e),此外在细胞培养过程中,肠黏膜上皮细胞的体积及细胞密度会随着培养时间的增加而增大。

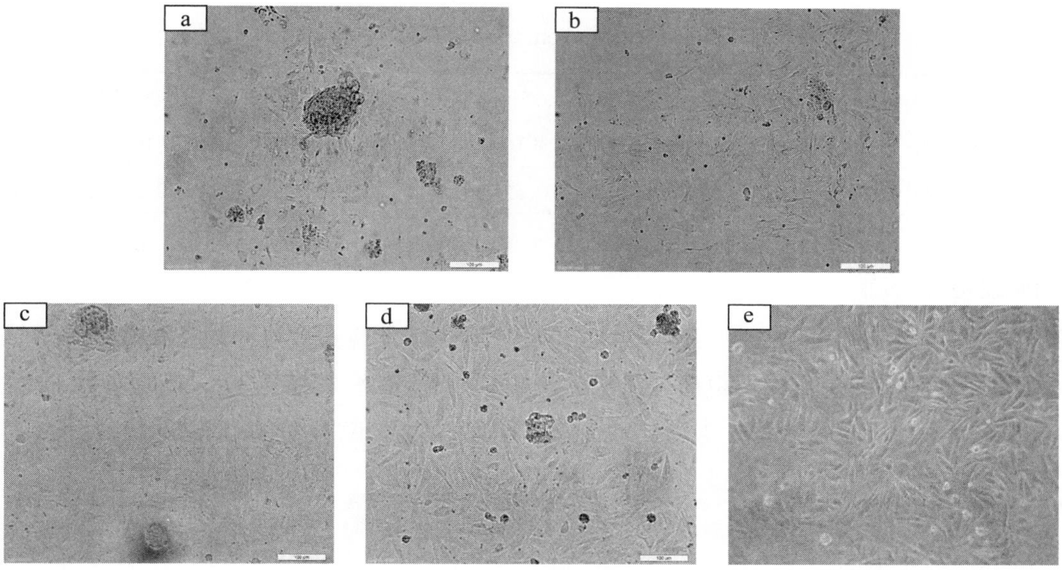


**图 2　猪小肠黏膜上皮细胞培养结果**

Fig. 2 The culture of neonatal swine intestinal epithelial cell

a: after 48 h; b: after 72 h; c: after 96 h; d, e: overgrow; scale bars represent 100 μm

a:培养 48 h, b:培养 72 h, c:培养 96h, d、e:细胞长满后,标尺:100 μm

## 2.2　肠黏膜上皮细胞鉴定

细胞角蛋白是上皮细胞的特异性抗原蛋白,本试验通过对肠黏膜上皮细胞中 $CK_{18}$ 免疫组织化学鉴定,结果发现阳性细胞中出现棕黄色,而阴性对照则不存在该现象(图 3),证明该细胞是肠黏膜上皮细胞。

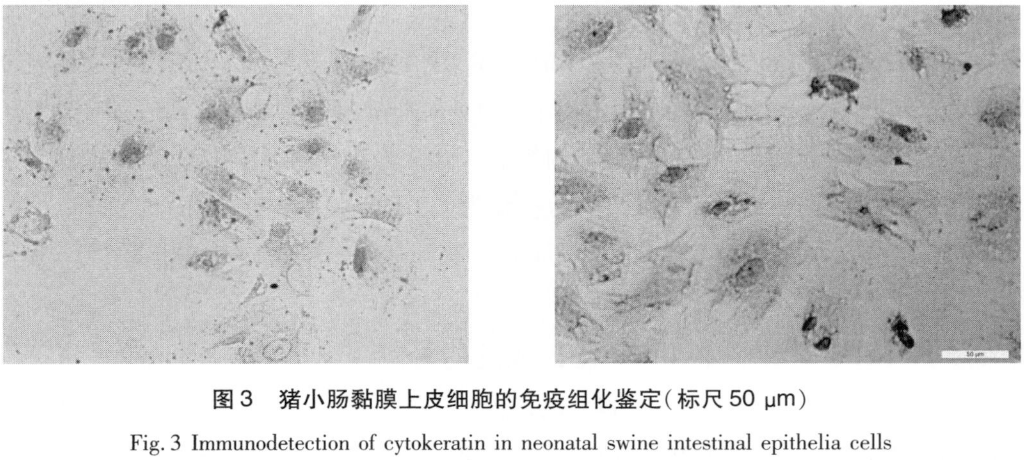

**图 3　猪小肠黏膜上皮细胞的免疫组化鉴定(标尺 50 μm)**

Fig. 3 Immunodetection of cytokeratin in neonatal swine intestinal epithelia cells

**左边:阴性对照,右边:$CK_{18}$ 鉴定上皮细胞**

Left: negative control, Right: epithelial cells assayed by Immunodetection of cytokeratin, scale bars represent 50 μm

## 2.3 rhTFF3 防治细胞损伤的结果

由表 2 可以看出,与对照组相比,细胞培养液中先添加终浓度为 50 μg/ml rhTFF3 后加入 10 μg/ml 的内毒素,即预防组 2 具有很好的预防效果($P<0.05$);预防组之间,先添加 50 μg/ml rhTFF3 后加入终浓度为 1 μg/ml和 10 μg/ml 内毒素的预防组 1 及预防组 3 之间也存在明显不同的预防效果($P<0.05$);预防组 1 与预防组 2 相比,预防效果较差($P<0.01$),说明 rhTFF3 可以有效预防肠黏膜上皮细胞免受内毒素的损伤。

表 2 rhTFF3 防治细胞损伤的结果

Tale.2 The result of prevention and cure cell damage by rhTFF3

| 数据 / 组别 | 对照组 (0 μg/ml) | 治疗组 A (1 μg/ml) | 治疗组 B (10 μg/ml) | 治疗组 C (100 μg/ml) | 预防组 D (1 μg/ml) | 预防组 E (10 μg/ml) | 预防组 F (100 μg/ml) |
|---|---|---|---|---|---|---|---|
| 吸光度 | $0.628 \pm 0.044^{a}$ | $0.668 \pm 0.028^{a}$ | $0.622 \pm 0.056^{a}$ | $0.656 \pm 0.060^{a}$ | $0.594 \pm 0.038^{ac}$ | $0.692 \pm 0.013^{b}$ | $0.688 \pm 0.074^{ad}$ |

## 2.4 1%凝胶电泳结果

2.4.1 总 RNA 凝胶电泳结果 Trizol 法提取的总 RNA 经 1% 琼脂糖凝胶电泳后在凝胶成像系统下观察结果,结果显示 RNA 经 1% 琼脂糖凝胶电泳后出现三条条带,分别为 28 S、18 S 和 5 S,其中 28 S 与 18 S 条带较亮,5 S 条带较暗,且 28 S 条带的宽度大于 18 S 条带宽度的两倍,说明 RNA 提取效果良好(图 4)。

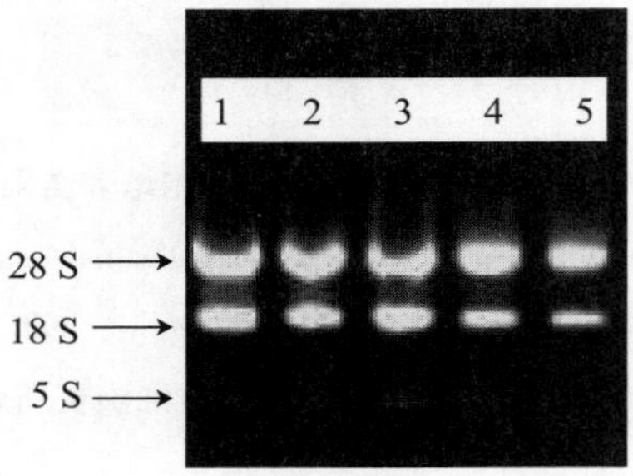


图 4 RNA 1% 琼脂糖凝胶电泳结果

Fig. 4 The result of RNA 1% Agarose Gel Electrophoresis

泳道 1-5 分别为对照、内毒素-1、内毒素-2、内毒素-3、TFF3 诱导组

Lanel 1 to 5: control group, LPS-1 group, LPS-2 group, LPS-3 group, induced by TFF3 group

2.4.2 PCR 产物凝胶电泳结果 总 RNA 经反转录、PCR 后,将 PCR 产物经 1% 琼脂糖凝胶电泳,电泳完成后在凝胶成像系统下观察结果。结果显示,PCR 产物经 1% 琼脂糖凝胶电泳后出现的条带清晰(图 5),左边的图为内参基因 18 S rRNA 扩增结果,扩增片段长度为 275 bp,与设计的引物相符;右边的图为 TFF3 目的基因,扩增片段长度为 174bp,与设计的引物长度相符。泳道 1 ~ 4 分别为内毒素终浓度为 0、1、10、100 μg/ml 组,泳道 5 为 TFF3 诱导组,泳道 6 为 DL-2000 的 Marker(图 5),结果表明,内参基因及 TFF3 基因的 PCR 扩增效果良好,达到预期的效果。

## 2.5 内毒素及 rhTFF3 对细胞中 TFF3 基因表达的影响

2.5.1 荧光定量 PCR 将提取的总 RNA 经反转录后进行荧光定量分析(图 6),荧光定量 PCR 的扩增曲线良好,CT 值都在正常的范围内,溶解曲线中,内参基因和目的基因分别出现一个单一的峰,说明设计的引物良好,不存在二聚体等结构。

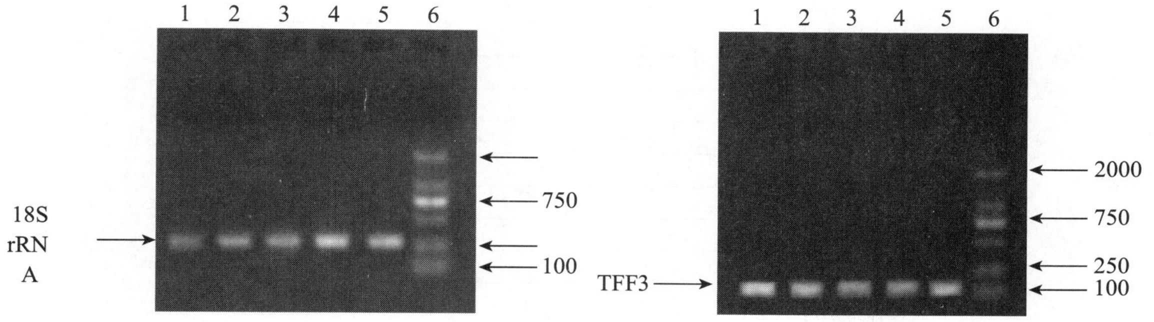


**图 5　PCR 产物 1% 琼脂糖凝胶电泳结果**

**左：18 S rRNA PCR 产物检测结果　　右：TFF3 PCR 产物检测结果**

Fig. 5 The result of PCR product 1% Agarose Gel Electrophoresis

**（泳道 1-6 分别为对照、内毒素-1、内毒素-2、内毒素-3、TFF3 诱导组、DL200 0 Marker）**

Left：The PCR result of 18 SrRNA.　　Right：The PCR result of TFF3

Lanel 1to 6：control group，LPS-1，LPS-2，LPS-3，induced by TFF3 and Marker

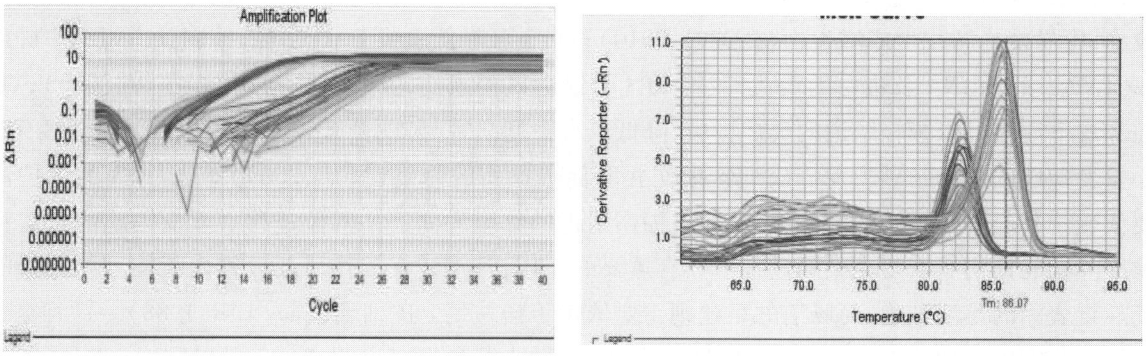


**图 6　荧光定量 PCR 结果图，左边：扩增曲线，右边：溶解曲线**

Fig. 6 The result of qRT-PCR，Left：Amplification Plot，Right：Melt Curve

2.5.2　内毒素及 rhTFF3 对细胞中 TFF3 基因表达的影响　细胞经终浓度分别为 1、10、100 μg/ml 的内毒素及 50 μg/ml rhTFF3 组中 TFF3 基因表达量分别正常组的 0.11、0.58、1.48 及 0.71 倍（图 7），表明浓度为 100 μg/ml 的内毒素对细胞中 TFF3 基因的表达具有增强作用，而浓度为 1 μg/ml、10 μg/ml 及 50 μg/ml 的 rhTFF3 则对细胞中 TFF3 基因的表达具有不同程度的抑制作用。

以上结果表明，本试验成功培养了仔猪肠黏膜上皮细胞，并通过细胞的免疫组织化学方法对细胞角蛋白进行了鉴定，证明该细胞是肠黏膜上皮细胞，并且发现细胞中存在 TFF3 蛋白的分布及表达，且 rhTFF3 可以有效预防内毒素对细胞的损伤。细胞分别受到终浓度为 1、10、100 μg/ml 的内毒素及 50 μg/ml rhTFF3 作用 24h 后，经荧光定量 PCR 检测发现，与正常组相比，在高浓度（100 μg/ml）的内毒素存在时，细胞中 TFF3 基因的表达量相对增加，可能因为该浓度的内毒素可以引起细胞出现损伤，即表现为与细胞损伤修复相关的 TFF3 基因的表达量为正常组的 1.48 倍的现象；而低浓度（1、10 μg/ml）的内毒素可抑制 TFF3 基因的表达，可而当培养液中存在 rhTFF3 时，细胞中 TFF3 基因的表达量也会相对降低，为正常组的 0.71 倍，可能是因为培养液中 rhTFF3 的存在，对细胞中 TFF3 的生成起到抑制作用。

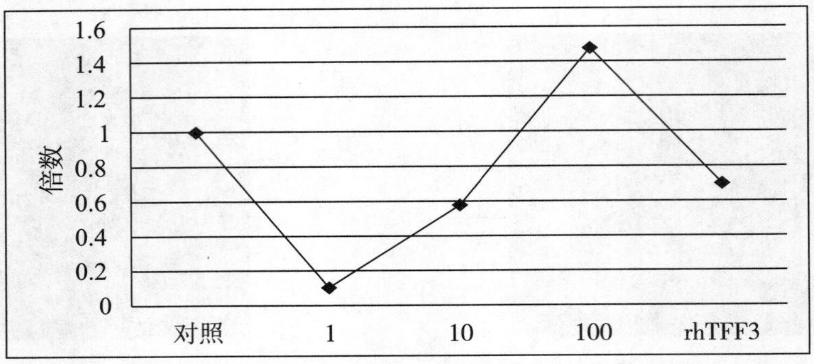


图7 内毒素及 rhTFF3 蛋白对细胞中 TFF3 基因表达的影响

Fig.7 The expression of TFF3 genes in cells effect by LPS and rhTFF3

## 3 讨 论

大量研究结果表明,TFF3 与肠道疾病间具有重要的关系,如 TFF3 分泌不足可能是导致早产儿和窒息儿更容易发生肠道黏膜损害的原因之一(田鸾英等,2011);TFF3 可能与胃癌细侵袭转移有关,其表达增多则预后不良(冯丽英等,2009);TFF3 可能参与胃黏膜早期重建(聂时南等,2002);脾胃湿热型慢性胃炎 TFF3 水平较高,更易发生肠化生或增生(曹蕊芸等,2010);TFF3 可降低血浆 DAO 活性,抑制肠道组织中肿瘤坏死因子 α(TNFα)的 mRNA 及蛋白的表达,减轻 LPS 所致幼鼠肠损伤,发挥保护作用(李军等,2006);TFF3 通过腹腔和皮下注射可以减轻肠道的炎症反应,降低氧自由基损害(陈丽萍等,2003);TFF3 可以减轻 TNBS 诱导的小鼠炎症性肠病及肠黏膜炎症病理性损伤并能促进损伤的修复(滕旭,2009)等。TFF3 的保护治疗作用主要依赖于其促进黏膜修复的能力,如促进损伤附近的细胞的向损伤部位迁移。上皮细胞的重建在维持肠黏膜屏障功能损伤后恢复完整的过程中具有重要的作用,且重建在损伤后几分钟内就能开始。胃的重建主要依靠胃表面的黏液细胞,而肠道的重建则主要依赖于肠黏膜上皮细胞(Lacy E R,1988)。这种重建方式至少包括调节细胞间的往来、细胞迁移及凋亡等过程,体外很多研究表明,TFF3 可以同时调节这 3 种过程(Hoffmann W,2005)。此外 TFF3 还可以下调钙连接蛋白的表达而影响细胞连接的功能(Meyer zum et al.,2006)。细胞迁移等具有高效的修复功能只有在细胞修复过程中细胞不出现死亡才能实现,因此细胞迁移与细胞生存之间则存在重要的联系,换言之也就是能否修复肠黏膜屏障损伤取决细胞是否死亡,也就是细胞的活力。本研究通过加入 rhTFF3 预防和治疗内毒素作用于肠黏膜上皮细胞,再采用 MTT 法检测细胞的活力发现 rhTFF3 确实可以有效预防细胞受损,对细胞起到一定的保护作用,即可以有效预防细胞因内毒素等不良因素导致肠黏膜上皮细胞的死亡。且其保护作用的程度会随着内毒素添加量的变化而变化,可能原因是肠黏膜上皮细胞对外界不良因素(内毒素)等具有一定的耐受能力,只有当不良因素达到一定的量才可能导致细胞的损伤,导致细胞的死亡,而当细胞培养液中存在 rhTFF3,细胞的活力可得到一定的保证。

在鼠的胃溃疡模型中,TFF3 和 TFF2 在损伤后的表达比经典的修复肽类如 EGF 和 TGFα 等表达得更快(Alison M R et al.,1995),在人的胃肠道细胞连接中也可见(Taupin D et al.,1999)。提示,TFF 对黏膜损伤的立即修复有重要的作用。由醋酸引发小鼠的结肠炎的过程中,急性期 TFF3 的表达处于下调状态,可能与急性期杯状细胞数量的减少有关(Tomita M et al.,1995),小鼠在甲氨蝶呤(即化疗)引起的小肠黏膜炎时,TFF3 mRNA 的表达略有增加,但蛋白的表达在急性期急剧减少,当暴露于甲氨蝶呤以及照射时,小肠和大肠中 TFF3 的总 mRNA 显著减少(Xian C J.,1999),相比之下在急性葡聚糖硫酸钠(DSS)诱导的结肠炎大鼠和小鼠中发现 TFF3 蛋白的表达增加,这可能是由于不同的模型中黏膜表面损伤程度不一样而导致的。内毒素是构成革兰氏阴性细菌外膜中内毒素的组成本质,而内毒素是引起脓毒症、感染性休克的重要因素,内毒素可直接对肠黏膜上皮细胞造成损伤,并刺激黏膜层分泌大量的细胞因子,并进一步影响细胞的通透性,对肠黏膜上皮细胞的正常功能影响很大。有证据表明,内毒素刺激血管内皮细胞后,对 P38 分裂原激活的蛋白激酶 MAPK 磷酸化程度明显,并与内毒素的剂量存在依赖关系(罗向东等,1999)。而 TFF3 具有能抑制细胞外的相关蛋白激酶和 MAPK 的部分通道的能力,TFF 家族与表皮生长因子受体(EGFR)相互之间的关系在黏膜修复的作用有 MAP 蛋白激酶信号传导途径的参与(聂时南等,2002)。这提示我们当内毒素作用于上皮细胞并对其造成损伤的过程中,有可能激活 TFF3 的信号通路,并生成 TFF3 蛋白以起到抗黏膜上皮细胞的损伤及促进黏膜损伤修复的作用,从而在内毒素与 TFF3 之间出现一定的数量关系。本研究通过查阅文献后,采用添加不同浓度的内毒素及一定量的 rhTFF3,再通过荧光定量 PCR 技术检测细胞中 TFF3 基因的表达量,来分析 TFF3 与内毒素之间的一个关系。结果证明,高浓度(100 μg/ml)的内毒素可促进 TFF3 基因表达,而低浓度(1、10 μg/ml)的内毒素对肠黏膜上皮细胞中 TFF3 基因的表达起到抑制的作用,这可能与 TFF3 蛋白具有预防细胞损伤的作用有关;当培养液中存在 rhTFF3 时,细胞本身 TFF3 基因的表达量也会相对降低,为正常组的 0.71 倍。总之,在培养的初生仔猪肠黏膜上皮细胞中,内毒素及 rhTFF3 对 TFF3 基因的表达都有不同的调节,但关于调节 TFF3 基因表达的信号通路到目前为止还不完全清楚,有待进一步的研究。

## 参考文献


[1] 曹蕊芸,毛承飞,王国耀,等. 脾胃湿热、脾虚气滞型慢性胃炎与三叶因子家族表达的相关性研究;[J]. 临床研究,2010,42(3):16 – 18.

[2] 陈丽萍,张丙宏,李艳,等. 肠三叶因子对新生鼠缺氧肠损伤模型白细胞介素 8、丙二醛的影响及意义;[J]. 中国围产医学杂志,2003,6(5):306 – 309.

[3] 冯丽英,秦玉彩,尹希,等. 三叶因子家族在胃癌及癌前病变组织中的表达及其意义;[J]. 中国全科医学,2009,12(24):2197 – 2203.

[4] 罗向东,张宗梁,杨宗城,等. 细菌内毒素对血管内皮细胞 ERK1/ERK2 和 P38 分裂原激活的蛋白激酶磷酸化的影响;[J]. 解放军医学杂志,1999,4(24):91 – 93.

[5] 李军,许玲芬,孙梅,等. 肠三叶因子在内毒素致幼鼠肠损伤中的作用及意义;[J]. 中国当代儿科杂志,2006,8(5):425 – 428.

[6] 聂时南,李兆申,湛先保,等. 乳癌相关肽及肠三叶因子在应激胃黏膜损伤的早期修复作用;[J]. 解放军医学杂志,2002,27(3):214 – 216.

[7] 裴小英. 猪小肠黏膜上皮蛋白酶激活受体 2 在早期断奶仔猪胃肠黏膜屏障中作用研究. ;[D]. 武汉:华中农业大学,2010.

[8] 滕旭. 肠三叶因子对小鼠炎症性肠病保护作用机制的研究;[D]. 北京:中国医科大学,2009.

[9] 田鸾英,易烈致,张静,等. 新生儿胎粪中肠三叶因子的意义探讨;[J]. 中国优生与遗传杂志,2011,19(1):75 – 77.

[10] 王静. 猪小肠黏膜上皮细胞培养方法的建立;[D]. 陕西:西北农林科技大学,2009.

[11] Alison M, Chinery R, Poulsom R, et al. , Experimental ulceration leads to sequential expression of SML, ITF, EGF and TGF a mRNAs in rat stomach. *J Pathol*, 1995, 175:405 – 414.

[12] Bertrand Kaeffer. Mammalian intestinal eoithelial cells primary culture:A mini-review. *In vitro* Cell Dev Bio-Animal, 2002, 38:123 – 134.

[13] Hoffmann W. TFF-triggered signals promoting restitution of mucous epithelia. *Cell Mol Life Sci*, 2005, 62:2932 – 2938.

[14] Lacy E R. Epithelial restitution in the gastrointestinal tract:Clin Gastroenterol, 1988, 10 (Suppl. 1):S72 – S77.

[15] Meyer zum, Büschenfelde D, Tauber R, et al. TFF3-peptide increases transepithelial resistance in epithelial cells by modulating claudin-1 and -2 expression. *Peptides*, 2006, 27:3383 – 3390.

[16] Tan X D, Chen Y H, Liu Q P, et al. Prostanoids mediate the protective effect of trefoil factor 3 in oxidant-induced intestinal epithelial cell injury:role of cyclooxygenase-2. *J Cell Sci*, 2000, 113 (12):2149 – 2155.

[17] Taupin D, Wu D C, Jeon W K, et al. The trefoil gene family are coordinately expressed immediate-early genes:EGF receptor-and MAP kinase-dependent interregulation. *J. Clin. Invest*, 1999, 103:R31 – R38.

[18] Tomita M, Itoh H, Ishikawa N, et al. Molecular cloning of mouse intestinal trefoil factor and its expression during goblet cell changes. *Biochem. J*, 1995, 311 (1):293 – 297.

[19] Xian C J, Howarth G S, Mardell C E, et al. Temporal changes in TFF3 expression and jejunal morphology during methotrexate-induced damage and repair. *Am. J. Physiol*, 1999, 277:G785 – G795.

# Th17 细胞的分化、调节及功能研究进展

王　婵,万涛梅,苟丽萍,任毅,袁贵强,李浪,左之才
(四川农业大学,雅安 625014)

**摘　要**:近年研究发现了效应辅助性 T 细胞的新亚群 Th17 细胞,它在机体免疫反应中起到重要作用。Th17 细胞能分泌 IL-17 等细胞因子,并介导自身免疫病、器官移植排斥反应、肿瘤及感染性疾病等的发生。本文从 Th17 细胞的发现、分化、调节以及与疾病相关性的研究进展进行综述。

**关键词**:T h17 细胞;IL-17;分化;疾病

$CD4^+$ T 淋巴细胞在受到刺激后分化成不同的 T 淋巴细胞亚群,主要包括 Th1 型、Th2 型、Th17 型效应性细胞亚群及调节性 T 细胞(regulatory T cell,Treg)亚群。各类 $CD4^+$ T 细胞亚群执行不同的生物学功能,相互间构成一个复杂的免疫调节网络[1]。$CD4^+$ T 细胞的两个新亚型 Th17 细胞和 Treg 是近年来的研究热点,Th17 细胞主要分泌 IL-17,并与自身免疫性疾病等疾病的发生发展息息相关。Treg 最早由由 Sakaguchi 提出,它是一类具有免疫抑制作用的 T 细胞新亚群,能够控制免疫应答的强度,调节机体免疫系统[2]。虽然上述细胞亚群来源于同一前体细胞,但功能却各不相同,下面就 Th17 细胞的发现、分化、调控及其相关疾病研究进展做一综述。

## 1　Th17 细胞的提出

起初认为能分泌 IL-17 的 T 细胞属于 Th1 细胞,并通过 IL-12/IFN-γ 通路介导实验性自身免疫性脑脊髓炎(experimental autoimmune encephalomyelitis,EAE)和胶原诱导的关节炎(collagen-induced arthritis,CIA)等自身免疫性疾病的发生。但经研究证实:IL-23/IL-17 通路才是自身免疫性疾病发展的关键。在研究 CIA 时发现,IL-23 基因敲除后的小鼠完全没有表现出关节炎的临床症状[3],而敲除小鼠 IL-12 或 IFN-γ 基因后,其 EAE 炎症反应非但没减轻反而更加严重[4];随后研究证实,IL-23 能促进 $CD4^+$ T 细胞产生大量 IL-l7[5],IL-17 缺失的小鼠不发生 EAE 疾病[6]。由此,证实自身免疫性疾病是由产生 IL-17 的 $CD4^+$ T 细胞介导,称为 Th17 细胞亚群。

## 2　Th17 细胞的分化

人类和小鼠的 Th17 细胞分化存在一定区别。小鼠 Th17 细胞的分化通过 TGF-β 和 IL-6 共同介导。研究表明,TGF-β 基因激活受阻或用 TGF-β 抗体阻断 TGF-β 以及 IL-6 缺失的小鼠,Th17 细胞都无法正常分化,小鼠 EAE 病情减轻[7,8,9]。人类 Th17 细胞的分化则依靠 IL-1β 和 IL-23,并刺激 Th17 细胞大量分泌 IL-17、IL-22 等细胞因子[10]。

Th17 细胞特异性的转录调节因子是维甲酸相关孤儿受体 γt(PAR-related orphan receptor gamma t,RORγt)。Ivanov 等人[11]发现,在肠道固有层内存在着大量 Th17 细胞,其 RORγt 高度表达。RORγt 通过介导 TGF-β 及 IL-6,激活信号传导及转录激活因子 3(signal transducer and activator of transcription 3,STAT3)促使 Th17 细胞进行分化。然而,在 RORγt 缺失小鼠体内依然有较低水平的 IL-17,说明 RORγt 并不是 Th17 细胞唯一的转录因子。Yang 等人[12]发现,RORα 也参与了 Th17 细胞的分化。缺失 RORγt 或 RORα 都只能部

**作者简介**:王婵(1989 -),女,硕士研究生,贵州六盘水人。E-mail:wangchan89@163.com
**通讯作者**:左之才(1974 -),男,教授,四川高县人。E-mail:ZZCJL@126.com

分抑制 IL-17 的表达,只有同时敲除这两个基因才能完全阻断 Th17 的分化。但是这两个转录因子如何相互作用调控 IL-17 的表达,还需进一步研究。

IL-23 由活化的树突状细胞产生,是 p19 亚基和 p40 亚基组成的异源性二聚体,属于 IL-12 细胞因子家族。IL-23 作用于记忆 T 细胞,是 Th17 细胞存活、增殖、功能维持的重要因子[6]。Th17 细胞能分泌 IL-21,属于 IL-2 家族成员。IL-21 通过激活 STAT3 信号通路以自分泌的形式促进 Th17 细胞的分化,缺失 IL-21 受体的 T 细胞无法正常分化 Th17 细胞。另外,依赖 IL-21 的 Th17 细胞分化途径也得到证实:在 IL-21 和 TGF-β 都存在时,缺失 IL-6 的 T 细胞也能进行 Th17 细胞的分化[13-14]。

IL-2 对 Th17 的分化起负调节作用。在 TGF-β 和 IL-6 存在时,加入外源性 IL-2,原始 T 细胞向 Th17 细胞分化减少,加入 IL-2 抗体中和后,Th17 细胞的分化增强;此外,IL-2 缺失小鼠的 Th17 细胞数量占 Th 细胞比例更大。IL-2 对 Th17 细胞负调节作用可能是通过 STAT5 信号传导通路抑制 RORγt 的活性来发挥的[15]。STAT5 结合位点位于 IL-17 启动子,STAT5 可能通过直接结合 IL-17 启动子削弱 IL-17 的产生。IFN-γ 和 IL-4 也对 Th17 细胞的分化起到抑制作用[16]。

IL-17 某些家族成员如 IL-27、IL-25,也能抑制 Th17 细胞的分化。IL-27 由活化的单核细胞及树突状细胞产生,为 P28 和 EB-13 构成的异源二聚体[17],对 Th1 应答的启动和维护起重要作用。研究显示,IL-27 依赖 STAT1 途径抑制由 IL-6、TGF-β 介导的 T 细胞增殖,抑制 Th17 细胞的分化。由于 IL-27 受体与 IL-6 受体有一个共同的蛋白结构 gp130,故推测 IL-27 可能通过与 IL-6 竞争受体来抑制 Th17 细胞的分化。但在体外 IL-27 是通过抑制蛋白 SOCS3 抑制 IL-6 信号来发挥其作用[18]。IL-27 对 Th1、Th2 细胞分化及中性粒细胞活化的也起抑制作用[19]。IL-25 通过增强 Th2 型细胞因子如 IL-13 的分泌,抑制 Th17 细胞的分化[20],且 IL-25 基因敲除的小鼠极易发生 EAE[21],说明 IL-25 对 Th17 细胞分化的调节起重要作用。

## 3 Th17 细胞及 IL-17 与相关疾病

目前发现,Th17 细胞可以介导多种自身免疫病的发生,如类风湿性关节炎(Rheumatoid arthritis,RA)、系统性红斑狼疮(systemic lupus erythematosus,SLE)[22]、银屑病等[23]。对 RA 疾病研究指出:RA 患者的滑液培养液及关节滑膜中的 IL-17 水平升高,IL-17 能促进促炎因子如 IL-6、IL-1β、TNF-α[24] 及 MMP[25] 的分泌,并且抑制基质修复组分如蛋白聚糖和胶原蛋白的产生,加剧对细胞基质的损伤。在患有 RA 的小鼠软骨组织中,IL-17 亦能加剧蛋白多糖的损失,并抑制其合成。使用 IL-17 抗体能有效减少炎症因子的产生[26],并增加 MMP 抑制剂(TIMP)的产生[27],说明 IL-17 在 RA 的发生中扮演了重要角色。

Th17 细胞及 IL-17 与哮喘的发生密不可分。在哮喘患者痰液中 IL-17mRNA 和蛋白水平明显升高[28]。IL-17 通过促进支气管上皮细胞 CXC 及 G-CSF 的释放,增加巨噬细胞炎性蛋白(macrophage inflammatory protein,MIP-2)的浓度,特异性的募集气道的中性粒细胞[29-30],参与炎症反应;同时,IL-17 可能通过诱导成纤维细胞产生多种细胞因子如 IL-6、IL-11、IL-8,放大和调节气道炎症反应,对哮喘气道重构起到间接作用[31]。龚雨新指出[32],哮喘小鼠的 Th17 细胞可能通过胸腺释放入外周血,促进中性粒细胞及嗜酸性粒细胞在气道内聚集诱导炎症发生。

研究发现 IL-17 与器官移植排斥反应相关。IL-17 在体外可使人近端肾小管上皮细胞产生炎症介质如 IL-6、IL-8 以及单核细胞趋化蛋白-1(MCP-1),$C_3$ 补体,有效刺激早期同种免疫反应的发生,提示 IL-17 与肾急性移植排斥反应的发生有关[33-34]。在发生肺脏移植急性排斥反应患者的支气管肺泡灌洗液中发现 IL-17mRNA 和其蛋白水平高度表达[35],并证实 V 型胶原蛋白-特异性淋巴细胞能够介导肺脏异体移植排斥反应的发生及 IL-17 的表达[36]。在大鼠异体心脏移植模型中,Li 等人[37]构建了一种 IL-17 抑制剂,它的表达能削弱 Th1 类细胞因子的分泌以及白细胞在移植心脏的浸润,提示使用 IL-17 抑制剂可作为器官移植的辅助治疗。但 Tang 等人[38]指出 IL-17 抑制剂仅对预防急性排斥反应有效,而对慢性排斥反应无效。

研究指出,IL-17 与某些癌症如卵巢癌、皮肤癌、前列腺癌等的发生及发展有着紧密的联系。虽然 IL-17

对血管内皮细胞或某些癌细胞的生长没有直接影响(如 IL-17 在体外对非小细胞肺癌细胞的生长率无影响),但 IL-17 能通过诱导 IL-6 的产生,激活癌基因信号转导通路及促进 STAT3 的活化,上调促血管生成因子(vascular endothelial growth factor,VEGF)的基因表达[39],使成纤维细胞、肿瘤细胞中 VEGF 及趋化因子大量分泌,促进肿瘤的生长和侵袭[40-41]。但是,IL-17 缺失小鼠的肿瘤生长更加迅速,提示了内源性 IL-17 在肿瘤免疫反应中的积极影响[42]。另一研究也表明,IL-17 能有效抑制造肥大细胞瘤 P815 和浆细胞瘤 J558L 的生长[43]。IL-17 在肿瘤生物学中的作用尚不清楚,仍需进一步探索与研究。

虽然 Th17 细胞的发现主要与自身免疫性疾病相关联,但最新数据表明,Th17 细胞也与某些细菌性感染有关,在幽门螺旋杆菌[44]、肺炎克雷伯菌[45]、结核分枝杆菌[46]、牙龈卟啉单胞菌[47]等感染中均出现 IL-17 分泌增多的现象。也有报道 Th17 细胞参与某些真菌(如白色念珠菌)感染、寄生虫病感染、莱姆病[48]、炎性肠病[49]等疾病的发生。

## 4 小 结

机体的免疫因子、免疫细胞和免疫组织构成一个复杂的免疫系统。Th17 细胞的分化、调节、功能无不证明它在机体免疫中的重要性。但 Th17 细胞及 IL-17 的作用及功能等还需进一步研究与探索。由 Th17 细胞及相关细胞因子介导的疾病可以通过阻断关键因子的表达或清除其功能进行预防和治疗。

## 参考文献


[1] Sallusto F,Lanzavecchia A. Heterogeneity of CD4+ memory T cells:functional modules for tailored immunity . *European Journal of Immunology*,2009,39(8):2076 - 2082.

[2] Sakaguchi S,Sakaguchi N,Asano M,et al. Immunologic self-tolerance maintained by activated T cells expressing IL-2 receptor alpha-chains (CD25). Breakdown of a single mechanism of self-tolerance causes various autoimmune diseases . *The Journal of Immunology*,1995,155(3):1151 - 1164.

[3] Murphy C A,Langrish C L,Chen Y,et al. Divergent pro-and antiinflammatory roles for IL-23 and IL-12 in joint autoimmune inflammation . *The Journal of Experimental Medicine*,2003,198(12):1951 - 1957.

[4] Cua D J, Sherlock J, Chen Y I, et al. Interleukin-23 rather than interleukin-12 is the critical cytokine for autoimmune inflammation of the brain . *Nature*,2003,421(6924):744 - 748.

[5] Aggarwal S,Ghilardi N,Xie M H,et al. Interleukin-23 promotes a distinct CD4 T cell activation state characterized by the production of interleukin-17 . *Journal of Biological Chemistry*,2003,278(3):1910 - 1914.

[6] Komiyama Y, Nakae S, Matsuki T, et al. IL-17 plays an important role in the development of experimental autoimmune encephalomyelitis . *The Journal of Immunology*,2006,177(1):566 - 573.

[7] Veldhoen M, Hocking R J, Flavell R A, et al. Signals mediated by transforming growth factor-β initiate autoimmune encephalomyelitis,but chronic inflammation is needed to sustain disease . *Nature Immunology*,2006,7(11):1151 - 1156.

[8] Li M O,Wan Y Y,Flavell R A. T cell-produced transforming growth factor-β1 controls T cell tolerance and regulates Th1-and Th17-cell differentiation . *Immunity*,2007,26(5):579 - 591.

[9] Bettelli E,Carrier Y,Gao W,et al. Reciprocal developmental pathways for the generation of pathogenic effector TH17 and regulatory T cells . *Nature*,2006,441(7090):235 - 238.

[10] Wilson N J,Boniface K,Chan J R,et al. Development,cytokine profile and function of human interleukin 17-producing helper T cells . *Nature Immunology*,2007,8(9):950 - 957.

[11] Ivanov I I, McKenzie B S, Zhou L, et al. The Orphan Nuclear Receptor RORγt Directs the Differentiation Program of Proinflammatory IL-17 + T Helper Cells . *Cell*,2006,126(6):1121 - 1133.

[12] Yang X O,Pappu B P,Nurieva R,et al. T helper 17 lineage differentiation is programmed by orphan nuclear receptors RORα and RORγ . *Immunity*,2008,28(1):29 - 39.

[13] Nurieva R,Yang X O,Martinez G,et al. Essential autocrine regulation by IL-21 in the generation of inflammatory T cells . *Nature*,2007,448(7152):480 - 483.

[14] Korn T, Bettelli E, Gao W, et al. IL-21 initiates an alternative pathway to induce proinflammatory TH17 cells. *Nature*, 2007, 448 (7152): 484 – 487.

[15] Laurence A, Tato C M, Davidson T S, et al. Interleukin-2 signaling via STAT5 constrains T helper 17 cell generation. *Immunity*, 2007, 26(3): 371 – 381.

[16] Iwakura Y, Ishigame H. The IL-23/IL-17 axis in inflammation. *Journal of Clinical Investigation*, S2006, 116(5): 1218 – 1222.

[17] Hunter C A. New IL-12-family members: IL-23 and IL-27, cytokines with divergent functions. *Nature Reviews Immunology*, 2005, 5(7): 521 – 531.

[18] Stumhofer J S, Laurence A, Wilson E H, et al. Interleukin 27 negatively regulates the development of interleukin 17-producing T helper cells during chronic inflammation of the central nervous system. *Nature immunology*, 2006, 7(9): 937 – 945.

[19] Tizard I R. Immunologı́a veterinaria. Netherland: Elsevier Health Sciences, 2009.

[20] Hurst S D, Muchamuel T, Gorman D M, et al. New IL-17 family members promote Th1 or Th2 responses in the lung: in vivo function of the novel cytokine IL-25. *The Journal of Immunology*, 2002, 169(1): 443 – 453.

[21] Kleinschek M A, Owyang A M, Joyce-Shaikh B, et al. IL-25 regulates Th17 function in autoimmune inflammation. *The Journal of experimental medicine*, 2007, 204(1): 161 – 170.

[22] Wong C K, Lit L C W, Tam L S, et al. Hyperproduction of IL-23 and IL-17 in patients with systemic lupus erythematosus: implications for Th17-mediated inflammation in auto-immunity. *Clinical Immunology*, 2008, 127(3): 385 – 393.

[23] Arican O, Aral M, Sasmaz S, et al. Serum levels of TNF-α, IFN-γ, IL-6, IL-8, IL-12, IL-17, and IL-18 in patients with active psoriasis and correlation with disease severity. *Mediators of Inflammation*, 2005, 2005(5): 273 – 279.

[24] Katz Y, Nadiv O, Beer Y. Interleukin-17 enhances tumor necrosis factor α-induced synthesis of interleukins 1, 6, and 8 in skin and synovial fibroblasts: A possible role as a "fine-tuning cytokine" in inflammation processes. *Arthritis & Rheumatism*, 2001, 44(9): 2176 – 2184.

[25] Jovanovic D V, Di Battista J A, Martel-Pelletier J, et al. Modulation of TIMP-1 synthesis by antiinflammatory cytokines and prostaglandin E2 in interleukin 17 stimulated human monocytes/macrophages. *The Journal of Rheumatology*, 2001, 28(4): 712 – 718.

[26] Chabaud M, Page G, Miossec P. Enhancing effect of IL-1, IL-17, and TNF-α on macrophage inflammatory protein-3α production in rheumatoid arthritis: regulation by soluble receptors and Th2 cytokines. *The Journal of Immunology*, 2001, 167 (10): 6015 – 6020.

[27] Chabaud M, Lubberts E, Joosten L, et al. IL-17 derived from juxta-articular bone and synovium contributes to joint degradation in rheumatoid arthritis. *Arthritis research*, 2001, 3(3): 168 – 177.

[28] Bullens D M, Truyen E, Coteur L, et al. IL-17 mRNA in sputum of asthmatic patients: linking T cell driven inflammation and granulocytic influx. *Respir Res*, 2006, 7(1): 135.

[29] Laan M, Cui Z H, Hoshino H, et al. Neutrophil recruitment by human IL-17 via CXC chemokine release in the airways. *The Journal of Immunology*, 1999, 162(4): 2347 – 2352.

[30] Ye P, Rodriguez F H, Kanaly S, et al. Requirement of interleukin 17 receptor signaling for lung CXC chemokine and granulocyte colony-stimulating factor expression, neutrophil recruitment, and host defense. *The Journal of Experimental Medicine*, 2001, 194 (4): 519 – 528.

[31] Molet S, Hamid Q, Davoineb F, et al. IL-17 is increased in asthmatic airways and induces human bronchial fibroblasts to produce cytokines. *Journal of Allergy and Clinical Immunology*, 2001, 108(3): 430 – 438.

[32] 龚雨新，于化鹏，邓火金，等. Th17 细胞在哮喘小鼠发病中的作用机制研究. 解放军医学杂志，2009 (1): 72 – 75.

[33] Van Kooten C, Boonstra J G, Paape M E, et al. Interleukin-17 activates human renal epithelial cells in vitro and is expressed during renal allograft rejection. *Journal of the American Society of Nephrology*, 1998, 9(8): 1526 – 1534.

[34] Loong C C, Hsieh H G, Lui W Y, et al. Retracted: Evidence for the early involvement of interleukin 17 in human and experimental renal allograft rejection. *The Journal of Pathology*, 2002, 197(3): 322 – 332.

[35] Vanaudenaerde B M, Dupont L J, Wuyts W A, et al. The role of interleukin-17 during acute rejection after lung transplantation.

*European Respiratory Journal*,2006,27(4):779 - 787.

[36] Yoshida S,Haque A,Mizobuchi T,et al. Anti-Type V Collagen Lymphocytes that Express IL-17 and IL-23 Induce Rejection Pathology in Fresh and Well-Healed Lung Transplants . *American Journal of Transplantation*,2006,6(4):724 - 735.

[37] Li J,Simeoni E,Fleury S,et al. Gene transfer of soluble interleukin-17 receptor prolongs cardiac allograft survival in a rat model . *European Journal of Cardio-thoracic Surgery*,2006,29(5):779 - 783.

[38] Tang J L,Subbotin V M,Antonysamy M A,et al. Interleukin-17 antagonism inhibits acute but not chronic vascular rejectionl . *Transplantation*,2001,72(2):348 - 350.

[39] Wang L,Yi T,Kortylewski M,et al. IL-17 can promote tumor growth through an IL-6-Stat3 signaling pathway . *The Journal of Experimental Medicine*,S2009,206(7):1457 - 1464.

[40] Numasaki M,Fukushi J,Ono M,et al. Interleukin-17 promotes angiogenesis and tumor growth . *Blood*,2003,101(7):2620 - 2627.

[41] Numasaki M,Watanabe M,Suzuki T,et al. IL-17 enhances the net angiogenic activity and in vivo growth of human non-small cell lung cancer in SCID mice through promoting CXCR-2-dependent angiogenesis . *The Journal of Immunology*,2005,175(9):6177 - 6189.

[42] Kryczek I,Wei S,Szeliga W,et al. Endogenous IL-17 contributes to reduced tumor growth and metastasis . *Blood*,2009,114(2):357 - 359.

[43] Benchetrit F,Ciree A,Vives V,et al. Interleukin-17 inhibits tumor cell growth by means of a T-cell-dependent mechanism . *Blood*,2002,99(6):2114 - 2121.

[44] Luzza F,Parrello T,Monteleone G,et al. Up-regulation of IL-17 is associated with bioactive IL-8 expression in Helicobacter pylori-infected human gastric mucosa . *The Journal of Immunology*,2000,165(9):5332 - 5337.

[45] Chung D R,Kasper D L,Panzo R J,et al. CD4 + T cells mediate abscess formation in intra-abdominal sepsis by an IL-17-dependent mechanism . *The Journal of Immunology*,2003,170(4):1958 - 1963.

[46] Aujla S J,Chan Y R,Zheng M,et al. IL-22 mediates mucosal host defense against Gram-negative bacterial pneumonia . *Nature Medicine*,2008,14(3):275 - 281.

[47] Johnson R B, Wood N, Serio F G. Interleukin-11 and IL-17 and the pathogenesis of periodontal disease . *Journal of Periodontology*,2004,75(1):37 - 43.

[48] Codolo G,Amedei A,Steere A C,et al. Borrelia burgdorferi NapA-driven Th17 cell inflammation in Lyme arthritis . *Arthritis & Rheumatism*,2008,58(11):3609 - 3617.

[49] Fujino S,Andoh A,Bamba S,et al. Increased expression of interleukin 17 in inflammatory bowel disease . *Gut*,2003,52(1):65 - 70.

# 丹参酮ⅡA 对 LPS 诱导的牛肾细胞 β-防御素、TLR4、TNF-α基因表达的影响

付开强[1],陶　爽[2],吕晓珮[1],李伟诗[1],田文儒[1],曹荣峰[1*]

(1. 青岛农业大学动物科技学院,青岛 266109;2. 信阳市浉河区畜牧局,信阳 464000)

**摘　要**:研究丹参酮ⅡA 对 LPS 诱导的牛肾细胞 β-防御素、TLR4、TNF-α 基因表达的影响,揭示丹参酮ⅡA 抗炎的作用机制。在细胞体外培养的条件下,采用 MTT 法筛选丹参酮ⅡA 对牛肾细胞的安全剂量和 LPS 诱导牛肾细胞炎症的最佳条件。在 LPS 刺激牛肾细胞 24h 后,再加入丹参酮ⅡA 作用 24h,用 MTT 法检测细胞的活性,用 RT-PCR 法检测 β-防御素、TLR4、TNF-α mRNA 的表达。结果显示,LPS 处理牛肾细胞 24h 后,TLR4 和 TNF-α 的 mRNA 表达水平显著高于对照组。丹参酮ⅡA 干预牛肾细胞炎症模型 24h 后,牛肾细胞的活性与炎症组相比显著上升,且 β-防御素、TLR4、TNF-α mRNA 表达水平与炎症组相比显著下降。丹参酮ⅡA 可抑制由 LPS 诱导的牛肾细胞 β-防御素、TLR4、TNF-α 基因的表达,对体外培养的牛肾细胞起抗炎作用。

**关键词**:丹参酮ⅡA;脂多糖;牛肾细胞;细胞因子;β-防御素

# The effect of tanshinone ⅡA on the gene expression of β-defensins、TLR4 and TNF-α in LPS induced inflammation of MDBK (NBL-1) cells

FU Kai-qiang[1], TAO Shuang[2], LV Xiao-pei[1], LI Wei-shi[1], TIAN Wenru[1], CAO Rong-feng[1*]

1. *College of Animal Science and Technology*, *Qingdao Agricultural University*, *Qingdao*, 266109

2. *Shihe Animal Husbandry Bureau*, *Xinyang*, 464000

**Abstract**: In order to study the effect of tanshinone ⅡA on the gene expression of β-defensins、TLR4 and TNF-αin LPS induced MDBK(NBL-1) cells, to reveal the anti-inflammatory mechanism of tanshinone ⅡA. MDBK cells were cultured in vitro, the safe doses of tanshinone ⅡA and the best condition of LPS induced inflammation in MDBK cells were filtrated. At first, the MDBK cells were conducted by LPS for 24h, then tanshinone ⅡA was added to cells for 24h, after that, the activity of cells were tested by MTT and the mRNA level of β-defensins、TLR4 and TNF-αwere tested by RT-PCR. Results show that the mRNA level of TLR4 and TNF-α in MDBK cells were significantly higher than the control group after were conducted by LPS for 24h. The models of LPS induced inflammation in MDBK cells were conducted by tanshinone ⅡA for 24h, the activity of cells were significantly increased than the LPS group, and the mRNA level of β-defensins、TLR4 and TNF-αwere significantly declined than the LPS group. The results indicate that tanshinone ⅡA can attenuates inflammatory responses by suppressing the mRNA level of β-defensins、TLR4 and TNF-α in LPS induced inflammation of MDBK cells.

**Keywords**: tanshinone ⅡA, LPS; MDBK(NBL-1) cells, cytokines, β-defensins

---

**基金项目**:山东省自然科学基金资助项目(ZR2011CM005)和 SDAIT-12-011-03

**作者简介**:付开强(1987-),男,山东潍坊人,硕士,主要从事动物生殖免疫研究,E-mail:xueshukaiqiangfu@163.com

* **通讯作者**:曹荣峰(1973-),博士,教授,主要从事动物生殖免疫与疾病防治方面的研究,E-mail:rfcao@126.com

脂多糖(LPS)是从革兰氏阴性菌细胞壁提取的内毒素,易使动物或细胞产生炎症反应,是试验研究中常用的炎症反应诱导物[1]。LPS 可以刺激多种细胞 TLR4、TNF-α 的高表达,引发细胞的剧烈炎症反应[2-4]。丹参酮ⅡA 是从中药丹参中提取的脂溶性物质,可以通过下调多种炎症因子的表达有效抑制 LPS 诱导的细胞炎症反应[5-6]。而最近有研究报道,LPS 在诱发炎症的同时可以上调 β-防御素的表达[7-8]。那么,丹参酮ⅡA 抑制炎症反应与 LPS 诱导 β-防御素表达的关系如何,尚未见报道。本试验通过观察丹参酮ⅡA 对 LPS 诱导的牛肾细胞表达 TLR4、TNF-α 和 β-防御素 mRNA 的影响,初步探讨丹参酮ⅡA 对牛肾细胞的保护机制和对 β-防御素的调控机制,为丹参酮药物的开发提供依据。

## 1 材料与方法

### 1.1 试验细胞

牛肾细胞系 MDBK(NBL-1)购自中国科学院细胞库。

### 1.2 主要试剂及仪器

脂多糖(LPS)购自 Sigma;丹参酮ⅡA 购自南京奥多福尼生物科技有限公司;Taq DNA 聚合酶和 DNA Marker 购自 Takara 公司;EASY spin 组织/细胞 RNA 快速抽提试剂盒购自原平皓生物技术有限公司;BioRT cDNA 第一链合成试剂盒购自杭州博日科技有限公司;噻唑蓝(MTT)和二甲基亚砜(DMSO)购自上海生工生物工程有限公司; RMPI 1640 购自 Gibco;胎牛血清(FBS)购自 Hyclone; PCR 仪(TaKaRa),GDS-8000 凝胶成像分析系统(UVP),酶标仪(Thermo)。GAPDH、TLR4、TNF-α、TAP、LAP、BNBD4、BNBD5 引物均由上海生工生物工程有限公司合成。序列如下:

表 1 引物设计

Table 1 Primer Design

| 引物 Primer | 5’-3’序列 Sequence5’-3’ | 大小 Size(bp) |
|---|---|---|
| Sense | AATTCTCAAAGCTGCCGT | 164 |
| Anti-sense | CACAGTTTCTGACTCCGC | |
| Sense | TCTTCCTGGTCCTGTCTGCT | 183 |
| Anti-sense | GCTGTGTCTTGGCCTTCTTT | |
| Sense | GCCAGCATGAGGCTCCATC | 278 |
| Anti-sense | CGTTTAAATTTTAGACGGTGT | |
| Sense | CGGGATCCGGATTTACTCAAGTAGTAAGAAATC | 143 |
| Anti-sense | CCCAAGCTTTTACCACCTCCTGCAGCAT | |
| Sense | TAACAAGCCGGTAGCCCACG | 590 |
| Anti-sense | GCAAGGGCTCTTGATGGCAGA | |
| Sense | AACCACCTCTCCACCTTGATACTG | 452 |
| Anti-sense | CCAGCCAGACCTTGAATACAGG | |
| Sense | ACCACTGTCCACGCCATCAC | 452 |
| Anti-sense | TCCACCACCCTGTTTGCTGTA | |

### 1.3 试验方法

1.3.1 牛肾细胞的培养 解冻复苏牛肾细胞,用含 10% FBS 和青、链霉素各 100 U/ml 的 RPMI-1640 培养基，在37℃、5% $CO_2$细胞培养箱中常规培养，每24～48h换液1次，3～5d传代1次。

1.3.2　LPS和丹参酮ⅡA的细胞毒性　将96孔培养板中生长良好且处于对数生长期的牛肾细胞，随机分为对照组、LPS处理组和丹参酮ⅡA处理组，LPS浓度分别为：0.1、0.5、1.0、1.5μg/ml，丹参酮ⅡA浓度分别为：10、20、30、40、50μmol/L，每组设3个重复。对照组不添加LPS和丹参酮ⅡA，另外两个处理组分别加入含不同浓度LPS、丹参酮ⅡA的RPMI-1640，上述各组在37℃、5% $CO_2$细胞培养箱中培养12h、24h、48h后做MTT细胞毒性检测。

1.3.3　LPS诱导牛肾细胞炎症反应模型的建立　取培养皿中生长良好且处于对数生长期的牛肾细胞，随机分成对照组和LPS处理组，每组3个重复。根据上述试验确定的浓度给予刺激，24h后提取细胞总RNA，RT-PCR法检测TLR4和TNF-α基因的表达，确定LPS诱导牛肾细胞炎症反应情况。

1.3.4　丹参酮ⅡA对LPS诱导牛肾细胞炎症作用的试验分组　取培养板中生长良好且处于对数生长期的牛肾细胞，随机分成对照组、LPS处理组、LPS+丹参酮ⅡA处理组，每组3个重复。LPS处理组和LPS+丹参酮ⅡA处理组添加含1μg/ml LPS的培养基，对照组添加常规培养基。培养24h后弃去培养基，LPS+丹参酮ⅡA组再加浓度为30 μmol/L丹参酮ⅡA的培养基，其他各组均添加常规培养基，再培养24h。

1.3.5　丹参酮ⅡA对LPS诱导的牛肾细胞活性影响　按照1.3.4方法处理后，弃去培养基，PBS冲洗3次，每孔加5 mg/ml的MTT 20 μl，37℃温育4h。弃去MTT溶液后，每孔加入100μl的DMSO，震荡5～10 min。用酶标仪在490nm波长下测定吸光度（A），计算抑制率（IR），确定各组细胞的活性。

1.3.6　丹参酮ⅡA对LPS诱导的牛肾细胞β-防御素、TLR4、TNF-αmRNA表达的影响　按照1.3.4方法处理后，弃去培养基，PBS冲洗3次，提取牛肾细胞总RNA，进行RT反应。PCR反应条件为：94℃预变性5min，94℃变性30s，退火（LAP 54℃、TAP 53.8℃、BNBD4 58.1℃、BNBD5 55.5℃、TNF-α 54.0℃、TLR4 54℃、NF-κB 57℃）30s，72℃延伸30s，共进行30个循环，最后72℃延伸10min。PCR产物经2%琼脂糖凝胶电泳（含0.5g/L溴化乙啶），由凝胶成像分析系统拍照鉴定。用Glyko BandScan 5.0进行灰度分析，计算mRNA的相对表达量。

### 1.4　统计分析

使用SPSS13.0软件，单因素方差分析进行统计学处理，$P<0.05$为差异具有统计学意义。

## 2　结　果

### 2.1　LPS、丹参酮ⅡA对牛肾细胞的毒性检测

MTT法筛选LPS剂量试验结果见图1。试验结果表明，LPS对细胞的抑制率均随LPS浓度的升高和刺激时间的延长而增强。LPS浓度为1.0 μg/ml、刺激时间24h时，牛肾细胞的生长抑制率小于10%，所以，确定该条件是细胞炎症反应模型的最佳条件。

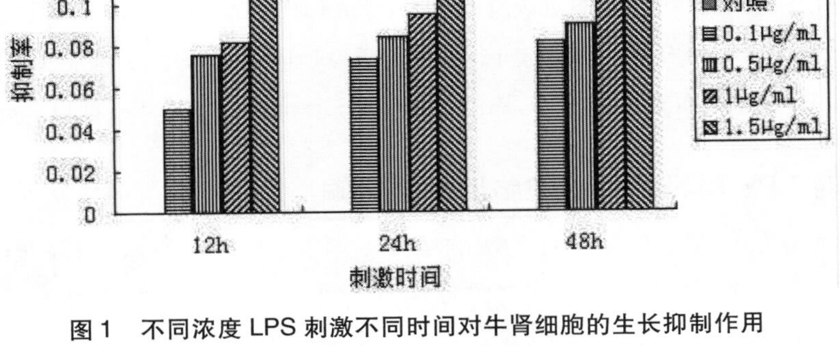


图1　不同浓度LPS刺激不同时间对牛肾细胞的生长抑制作用

Fig.1　The inhibition ratio in different concentration of LPS and different time period

MTT 筛选丹参酮ⅡA 剂量试验结果见图 2。结果表明,细胞培养 12h,各浓度丹参酮ⅡA 对细胞的 IR 值均小于 10%;培养 24h,10、20、30μmol/L 丹参酮ⅡA 对细胞的 IR 值均小于 10%,而 40、50 μmol/L 丹参酮ⅡA 的 IR 值大于 10%;培养 48h,10、20 μmol/L 丹参酮ⅡA 对细胞的 IR 值均小于 10%,而 30、40、50μmol/L 丹参酮ⅡA 的 IR 值大于 10%。所以,本试验确定丹参酮ⅡA 最佳的干预浓度为 30μmol/L,处理 24h。

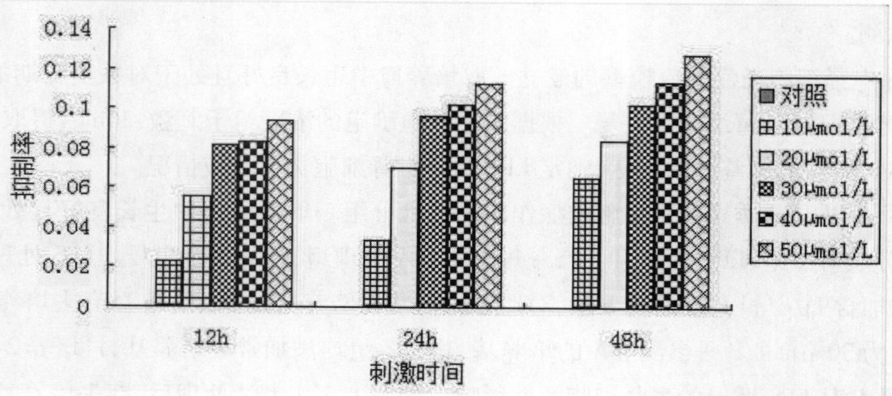


图 2　不同浓度丹参酮ⅡA 刺激不同时间对牛肾细胞的生长抑制作用

Fig. 2　The inhibition ratio in different concentration of tanshinone ⅡA and different time period

## 2.2　LPS 诱导牛肾细胞炎症模型的建立

LPS 处理组和对照组 TLR4 和 TNF-α 的 mRNA 表达结果见图 3,相对含量见图 4。结果发现,LPS 处理组与对照组中均有 TLR4 和 TNF-α 基因的表达;LPS 处理组与对照组相比 TLR4 和 TNF-α 的 mRNA 表达水平均显著上升($P<0.05$)。说明牛肾细胞炎症模型成功建立。

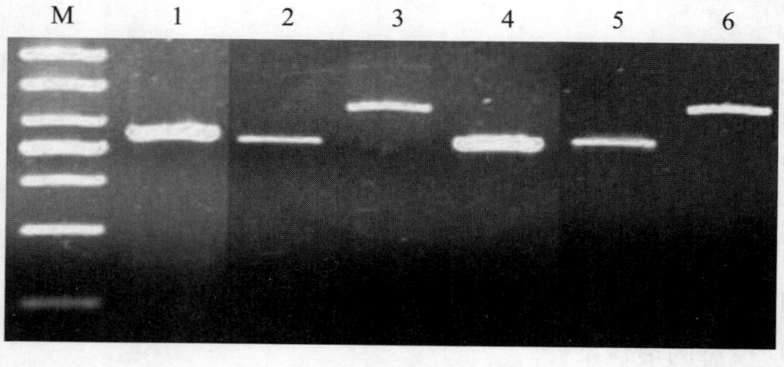


图 3. LPS 处理组和对照组 TLR4 和 TNF-α 的 mRNA 表达

Fig. 3　The mRNA level of TLR4 and TNF-α in control and LPS groups

M:DNA 分子质量标准;1:对照组 GAPDH;2:对照组 TLR4;3:对照组 TNF-α;4:LPS 处理组内参;5:LPS 处理组 TLR4;6:LPS 处理组 TNF-α

M:DNA marker; 1:The GAPDH of control group; 2:The TLR4 of control group; 3:The TNF-α of control group; 4:The GAPDH of LPS group; 5:The TLR4 of LPS group; 6:The TNF-α of LPS group

## 2.3　丹参酮ⅡA 对 LPS 诱导的牛肾细胞活性的影响

丹参酮ⅡA 对 LPS 诱导的牛肾细胞活性的影响结果见图 5。如图 5 所示,与对照组相比,LPS 处理组的细胞活性极显著下降($P<0.01$),说明 LPS 明显降低细胞活性。与 LPS 处理组相比,LPS + 丹参酮ⅡA 组的细胞活性极显著上升($P<0.01$),说明丹参酮ⅡA 对 LPS 诱导的牛肾细胞炎症有明显的抑制作用。

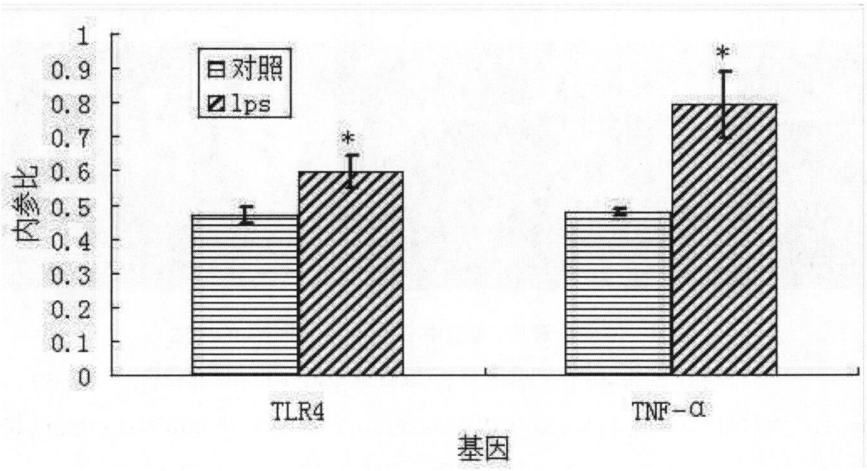


**图 4　LPS 处理组和正常组 TLR4、TNF-α 的内参比**

Fig. 4 The ratio of TLR4 and TNF-α in control and LPS groups

* 表示与对照组相比差异显著($P<0.05$) * $P<0.05$ vs. control group

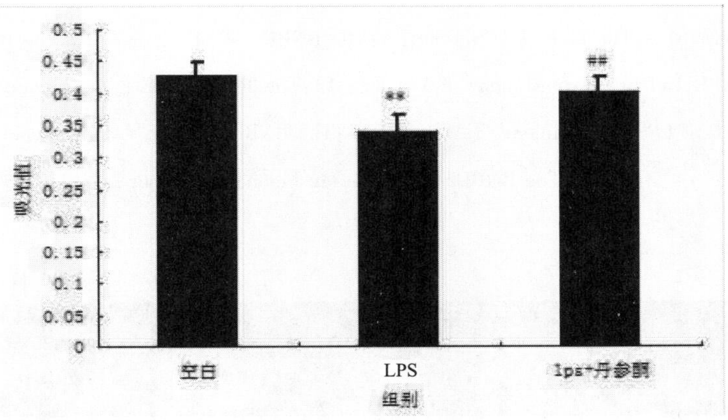


**图 5　丹参酮ⅡA 对 LPS 诱导的牛肾细胞活性的影响**

Fig. 5　Effect of　tanshinone ⅡA on cell activity in LPS-treated bovine kidney

* * 表示与对照组相比差异极显著($P<0.01$);##表示与 LPS 处理组相比差异极显著($P<0.01$)

* * $P<0.01$ vs. control group; ##$P<0.01$ vs. LPS group

## 2.4　丹参酮ⅡA 对 LPS 诱导牛肾细胞 β-防御素、TLR4、TNF-αmRNA 的表达情况

对照组、LPS 处理组和 LPS + 丹参酮ⅡA 组 β-防御素、TLR4、TNF-α mRNA 表达的电泳结果见图 6 和图 7,相对含量见图 8。从图 8 中可以发现,对照组、LPS 处理组和 LPS + 丹参酮ⅡA 处理组中均有 TAP、LAP、BNBD4、BNBD5、TLR4、TNF-α 的表达;LPS 处理组与对照组相比,TAP、LAP、BNBD4 的 mRNA 表达显著上升($P<0.05$),而 BNBD5、TLR4、TNF-α 的 mRNA 表达极显著上升($P<0.01$);LPS + 丹参酮ⅡA 组与 LPS 处理组相比,TAP、TLR4、TNF-α 的 mRNA 表达显著下降($P<0.05$)而 LAP、BNBD4 的 mRNA 表达极显著下降($P<0.01$),BNBD5 的 mRNA 表达没有显著变化($P>0.05$)。

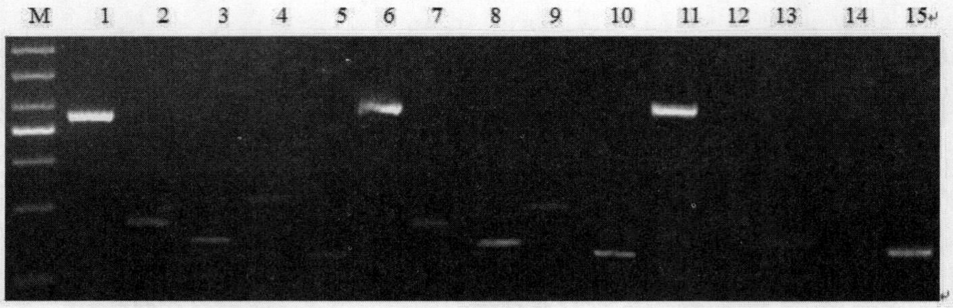


**图6　不同处理后,各组中β-防御素 mRNA 的表达**

**Fig. 6 The mRNA level of β-defensins in different groups**

M:Marker;1:对照组 GAPDH;2:对照组 TAP;3:对照组 LAP;4:对照组 BNBD4;5:对照组 BNBD5;
6:LPS 处理组 GAPDH;7:LPS 处理组 TAP;8:LPS 处理组 LAP;9:LPS 处理组 BNBD4;10:LPS 处理组 BNBD5;
11:LPS + 丹参酮ⅡA 处理组 GAPDH;12:LPS + 丹参酮ⅡA 处理组 TAP;13:LPS + 丹参酮ⅡA 处理组 LAP;
14:LPS + 丹参酮ⅡA 处理组 BNBD4;15:LPS + 丹参酮ⅡA 处理组 BNBD5

M:DNA marker; 1:The GAPDH of control; 2:The TAP of control; 3:The LAP of control;
4:The BNBD4 of control; 5:The BNBD5 of control; 6:The GAPDH of LPS group;
7:The TAP of LPS group; 8:The LAP of LPS group; 9:The BNBD4 of LPS group; 10:The BNBD5 of LPS group;
11:The GAPDH of LPS + tanshinone ⅡA group; 12:The TAP of LPS + tanshinone ⅡA group;
13:The LAP of LPS + tanshinone ⅡA group; 14:The BNBD4 of LPS + tanshinone ⅡA group;
15; The BNBD5 of LPS + tanshinone ⅡA group;

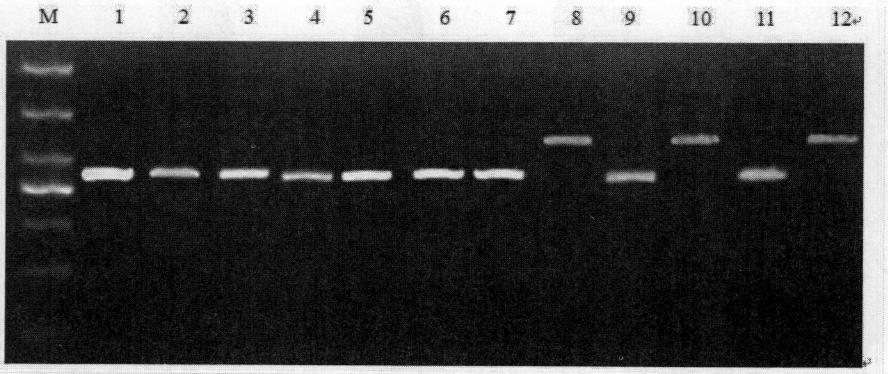


**图7　不同处理后,各组中 TLR4、TNF-α mRNA 的表达**

**Fig. 7 The mRNA level of TLR4 and TNF-α in different groups**

M:Marker;1:对照组 GAPDH;2:对照组 TLR4;3:LPS 处理组 GAPDH;4:LPS 处理组 TLR4;
5:LPS + 丹参酮ⅡA 处理组 GAPDH;6:LPS + 丹参酮ⅡA 处理组 TLR4;7:对照组 GAPDH;8:对照组 TNF-α;
9:LPS 处理组 GAPDH;10:LPS 处理组 TNF-α;11:LPS + 丹参酮ⅡA 处理组 GAPDH;12:LPS + 丹参酮ⅡA 处理组 TNF-α

M:DNA marker; 1:The GAPDH of control; 2:The TLR4 of control;
3:The GAPDH of LPS group; 4:The TLR4 of LPS group; 5:The GAPDH of LPS + tanshinone ⅡA group;
6:The TLR4 of LPS + tanshinone ⅡA group; 7:The GAPDH of control; 8:The TNF-α of control;
9:The GAPDH of LPS group; 10:The TNF-α of LPS group; 11:The GAPDH of LPS + tanshinone ⅡA group;
12:The TNF-α of LPS + tanshinone ⅡA group;

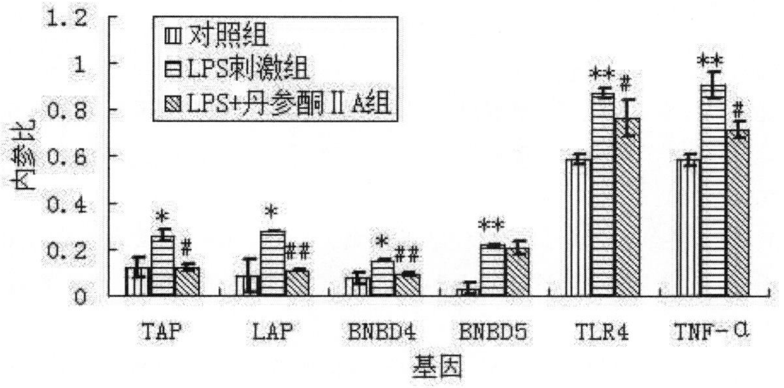


**图 8 不同处理后各组中 TLR4、TNF-α、β-防御素 mRNA 的内参比**

Fig. 8 The ratio of TLR4、TNF-α and β- defensins in different groups

* 表示与对照组相比差异显著($P<0.05$);* * 表示与对照组相比差异极显著($P<0.01$);

#表示与 LPS 处理组相比差异显著($P<0.05$);##表示与 LPS 处理组相比差异极显著($P<0.01$)

* $P<0.05$ vs. control; * * $P<0.01$ vs. control; # $P<0.05$ vs. LPS group; ## $P<0.01$ vs. LPS group

## 3 讨 论

LPS 作为革兰氏阴性菌细胞壁的主要成分,是阴性菌诱发炎症反应的主要毒力因子[9]。TLR4 是近年来发现的一类模式识别受体,属于Ⅰ型跨膜蛋白家族成员,可以特异性识别 LPS[10]。革兰氏阴性菌死亡后会释放 LPS,LPS 可以与细胞表面受体 TLR4 结合,刺激机体细胞产生大量促炎因子 TNF-α 等[11]。TNF-α 是机体对应激反应最早产生且在炎症反应中起核心作用的炎症介质,几乎介导所有内毒素的生物学效应[12-13]。本试验结果发现,牛肾细胞在 1μg/ml 的 LPS 刺激 24h 后,TLR4 和 TNF-αmRNA 表达水平显著上升,表明牛肾细胞已经发生炎症反应,炎症模型建立成功。与张桂林等[14]在牛子宫内膜上皮细胞上的实验结果相同,说明 1μg/ml 的 LPS 适合多种细胞炎症模型的建立。

丹参酮ⅡA 是从中药丹参中提取的脂溶性有效成分,作为活血、保心、抗炎的药物在临床上已使用多年,具有明确的抗菌、消炎、免疫调节等多种作用[15]。本试验结果发现 LPS 刺激牛肾细胞建立炎症模型后,添加 30μmol/L 丹参酮ⅡA 干预 24h,牛肾细胞 TLR4、TNF-α 的 mRNA 表达水平显著下降,与盛金良、李思扬和李卫萍[2-4]分别报道的结果一致,表明丹参酮ⅡA 确能通过下调 TLR4、TNF-α 的表达发挥抗炎作用。

β-防御素是 Diamond 等[16]首先在牛的气管黏膜上皮细胞中发现的,后来在牛组织中又相继发现了多种 β-防御素的表达,包括舌抗菌肽(LAP)、气管抗微生物肽(TAP)、牛嗜中性 β-防御素 4 和 5(BNBD4/BNBD5)[17]。β-防御素在机体的先天免疫和获得性免疫中起重要的调节作用[18]。β-防御素主要由上皮细胞产生,具有抗微生物活性和广泛的免疫调节作用[19]。张妮、廖伟等报道,LPS 在刺激细胞发生炎症反应的同时,可以诱导细胞 β-防御素的上调[7-8]。

赵俊丽等[20]也发现,LPS 确实能够通过 TLR4/NF-κB 途径诱导人肾小管上皮细胞 β-防御素的表达。而本试验结果发现,丹参酮ⅡA 在起抗炎作用的同时,可下调 LAP/TAP/BNBD4 的表达。分析原因,LPS 诱导机体炎症反应与 LPS 诱导机体 β-防御素表达的信号通路可能一致,而丹参酮在抑制细胞炎症反应的同时也抑制了 β-防御素的表达。

本实验室前期的试验发现,阻断 TLR4 以后,并不会影响 LPS 诱导的 BNBD5 mRNA 的表达变化[21]。而本试验发现,丹参酮ⅡA 并没有下调 BNBD5 的基因表达,与我们前期的试验结果一致。由此我们推测 BNBD5 表达的信号通路可能与其他防御素的不同。

综上所述,LPS 可以诱导体外培养的牛肾细胞 TLR4、TNF-α 和 β-防御素的 mRNA 高表达,而丹参酮ⅡA 发挥抗炎作用可能是通过下调 TLR4、TNF-α 和 β-防御素的 mRNA 表达来起作用,但 β-防御素中的 BNBD5

表现异常,其分子表达机制仍需进一步探讨。

## 参考文献

[1] 王华,徐德祥. 细菌脂多糖发育毒作用研究进展[J]. 国外医学卫生分册,2008,35(3):170 - 173.

[2] 盛金良,陈创夫,杨霞,等. 绵羊 Toll 样受体家族在肺泡巨噬细胞的分布及脂多糖(LPS)刺激对 TLR2、TLR4 表达的影响[J]. 华北农学报,2010,25(1):30 - 35.

[3] 李思扬. 脂多糖调节 TNF-α 在细胞滋养细胞的表达及临床意义[J]. 生殖免疫学,2008,24:730 - 732.

[4] 李卫萍,李元宏,郑绘霞,等. 脂多糖体外诱导 RAW264. 7 细胞产生 NF-κBp65、iNO 及 TNF-α[J]. 细胞与分子免疫学杂志,2013,29(8):823 - 825.

[5] 贾连群,冯峻屹,杨关林,等. 丹参酮ⅡA 对 LPS 诱导 EA. hy926 细胞 TLR4 和 TNF-α 的影响[J]. 细胞与分子免疫学杂志,2011,27(7):733 - 735.

[6] 李一龙,李蕾,王发龙,等. 丹参酮ⅡA 抑制脂多糖诱导后巨噬细胞 HMGB1 的释放[J]. 中华中医药杂志,2011,26(9):2059 - 2061.

[7] 张妮,房晓楠,崔南,等. LPS 诱导 L929 细胞 β 防御素 2 和 schlafen-2 基因的表达及其信号传导[J]. 中国免疫学杂志,2010,26:778 - 782.

[8] 廖伟,钱桂生,雷撼,等. 脂多糖诱导人气道上皮细胞 hBD-2 表达及核转录因子 JB 活性的变化[J]. 中国病理生理杂志,2007,23(1):71 - 75.

[9] Himanshu Kumar,Taro Kawai,and Shizuo Akira. Pathogen Recognition by the Innate Immune System[J]. International Reviews of Immunology,2011,30:16 - 34.

[10] Osamu Takeuchi,Shizuo Akira. Pattern Recognition Receptors and inflammation [J]. Cell,2010,140(6):805 - 820.

[11] John J. Bromfield,and I. Martin Sheldon. Lipopolysaccharide Initiates Inflammation in Bovine Granulosa Cells via the TLR4 Pathway and Perturbs Oocyte Meiotic Progression in Vitro[J]. Endocrinology,2011,152(12):5029 - 5040.

[12] Wen-Ming Chu. Tumor necrosis factor[J]. Cancer Letters,2013,328(2):222 - 225.

[13] Mariana Postal,Simone Appenzeller. The role of Tumor Necrosis Factor-alpha (TNF-α) in the pathogenesis of systemic lupus erythematosus[J]. Cytokine,2011,56(3):537 - 543.

[14] 张桂林,崔晓妮,曹荣峰. 脂多糖诱导奶牛子宫内膜上皮细胞核因子 κB 的表达[J]. 中国兽医科学,2010,40(10):1063 - 1066.

[15] 张萌涛,钱亦华,唐安琪. 丹参酮ⅡA 药理作用的研究进展[J]. 医学综述,2010,16(17):2661 - 2664.

[16] Diamond G,Zasloff M,Eck H,et al. Tracheal antimicrobial peptide,a cysteine rich peptide from mammalian tracheal mucosa: peptide isolation and cloning of a cDNA[J]. Proc Natl Acad Sci USA,1991,88(9):3952 - 3956.

[17] Davies D,Meade KG,Herath S,et al. Toll-like receptor and antimicrobial peptide expression in the bovine endometrium[J]. Reprod Biol Endocrinology,2008,18(6):53.

[18] 袁曦. 哺乳动物 β-防御素研究进展[J]. 中国畜牧兽医,2013,40(2):103 - 107.

[19] 高飞,徐来祥. 哺乳动物 β-防御素研究进展[J]. 生物学通报,2012,47(2):13 - 15.

[20] 赵俊丽,王俭勤,王晶宇. LPS 通过 TLR4 信号通路诱导人肾小管上皮细胞株(HK-2)表达 hBD-2[J]. 免疫学杂志,2013,29(3):222 - 225.

[21] 付开强,吕晓珮,李伟诗,等. 脂多糖诱导牛肾细胞 β-防御素-5 mRNA 表达的可能机制[J]. 中国免疫学杂志,2014,4:59 - 61.

# 代谢组学中核磁共振技术的数据分析方法总结

孙雨航,许楚楚,徐　闯,李昌盛,吴　凌,夏　成
(黑龙江八一农垦大学,动物科技学院,大庆 163319)

**摘　要**:本文有针对性地选择代谢组学中核磁共振检测技术,对其数据分析的主要过程,包括图谱数据预处理、数据转换和模式识别进行系统分析,重点讨论了主成分分析(Principal component analysis,PCA),偏最小二乘法判别分析(Partial least squares discriminant analysis,PLS-DA),正交的偏最小二乘法判别分析(Orthogonal partial least squares discriminant analysis,OPLS-DA)和正交信号校正偏最小二乘法判别分析(Orthogonal signal correction-partial least square discriminate analysis,OSC-PLS-DA)等4种模式识别方法,并进行对比分析,以期为后来的研究者提供有益参考。

**关键词**:代谢组学;核磁共振技术;数据预处理;多变量模式识别

# Summary data analysis of nuclear magnetic resonance in metabolomics

Yuhang SUN,Chuchu XU,Chuang XU,Changsheng LI,Cheng XIA*,Ling WU
(*College of Animal Science and Veterinary Medicine*,*Heilongjiang BaYi Agricultural University*,*Daqing* 163319,*China*)

**Abstract**: In this paper,nuclear magnetic resonance (NMR) in metabolomics was selected and the main process of data analysis,including the pre-processing of spectra data,transformation of spectra and pattern recognition were systematacially analysed. Four kinds of pattern recognition methods,principal component analysis (PCA),partial further squares discriminant analysis (PLS-DA),orthogonal partial further squares discriminant analysis (OPLS-DA) and orthogonal signal correction -partial further square discriminate analysis (OSC-PLS-DA) were respectively discussed and compared in order to provide beneficial reference for the later researchers.

**Keywords**: Metabolomics,nuclear magnetic resonance,pre-processing of data,multivariate pattern recognition

## 1　前　言

近年来,随着科学技术的不断创新,各种组学(Omics)技术应运而生[1]。1999年Nicholson首次给出了代谢组学(Metabolomics)的完整定义,是研究在内、外因素作用下,生物体所含内源性小分子代谢物(相对分子质量小于1 000)在种类和数量上的动态变化规律以及生理、病理变化的一门技术[2]。代谢组学主要包括3个检测平台,分别为气象色谱质谱联用,液相色谱质谱联用和核磁共振技术(nuclear magnetic resonance,NMR)[3]。其中,NMR已被广泛应用于生命科学的各个领域,包括植物种属分类[4,5],天然产物鉴定[6,7],疾病早期诊断[8],药物毒理和效果评价[9,10],以及寻找生物标记物[11]等。

尽管许多研究学者对NMR数据分析方法的使用和报道很多,尤其是在数学和化学领域,但是对于要应

**资助项目**:"十二五"农村领域国家科技课题(2012BAD12B05-2,2012BAD12B03-2),科技支撑计划(2013BAD21B01)

**第一作者**:孙雨航(1989－),女,满族,辽宁黑山人,在读硕士研究生,学士。临床兽医学专业,动物营养代谢病方向,E-mail:sunyuhang12345@sohu.com

**通讯作者**:夏　成(1964－),男,教授,博士生导师,主要从事动物营养代谢病研究,E-mail:xcwlxyf@sohu.com.

用这种方法于其他领域的非专业研究学者来说还是很难掌握。究其原因在于:(1)缺乏系统的数据分析报道,许多研究报告中对所使用的分析方法只是粗略提及,图形解析又不尽详细。(2)对于非专业研究人员来说,许多数据分析方法原理比较生涩难懂或与实际应用领域衔接不当。(3)数据分析方法种类繁多,分析软件又不尽相同,这就使呈现的图形(如图形横纵坐标标识)和数据结果不太一致,很难全面掌握。所以,本文根据笔者经验,并参考大量文献,尽量做到简单而全面地展现 NMR 数据分析的整个过程,为未来的研究者提供参考。

## 2 NMR 数据预处理

NMR 数据预处理是一个比较复杂的过程,是连接样品信息和数据分析的纽带。样品中以不同形式存在的 H 原子在磁场中因为能级跃迁而产生核磁共振吸收,形成核磁共振信号,最终以我们所熟悉的核磁共振图谱形式呈现。但是原始图谱不能直接用于化学计量分析,还需要转化成为数据矩阵,充分提取所获数据中的潜在信息,以消除或减小实验和分析过程带来的误差。

NMR 数据预处理主要分为归一化、中心化和尺度方差规模化[12]。从生物学角度来说,生物种类繁多,内源性代谢物浓度差异很大,浓度高的代谢物不一定比浓度低的代谢物更重要,而某些浓度非常低的代谢物在某些生理过程中恰恰不容忽视[13],所以为了保证所有的化合物在无偏性 NMR 检测后能够得到公平的分析,就需要对图谱数据进行标准化。简单来说,就是对于同种生物样品,限定样品含量[14],对于不同种样品,选用共有物质(如 DNA)[15],从而统一变量种类和尺度标准。

目前常规使用的数据预处理软件是 Topspin(Bruker GmBH, Karlsruhe, Germany)和 MestReNova(Mestrelab Research, Santiago de Compostela, Spain),它们都可以在网上免费下载。在 Topspin 软件中,所有的原始谱图进行傅里叶转换,根据样品数量自动进行分段积分,手动调零、校正基线和相位,保存成为包含所有峰值的完整图谱;再将完整谱图导入到 MestReNova 软件中,将其保存为“ASC Ⅱ”文件,最终得到包含所有代谢物化学位移的“txt”文本,用于后续统计分析。

## 3 多变量模式识别软件

为了从大量预处理后的 NMR 数据中得到潜在的、有价值的代谢物,需要运用降维的思想,借助多变量模式识别[16]。目前,用于 NMR 数据分析的软件通常是内部教学软件或者是昂贵的商用软件[17],主要包括 SMICA(Umetrics, Umeå, Sweden)[18] 和 R 语言(www. R-project. org)[12],前者最为常见,操作简便且智能化,但是分析模式较少;后者需要专业人员操作且指令复杂,但是模式识别和分析方法较多,还可以进行正交信号校正偏最小二乘法判别分析(Orthogonal signal correction-partial least squarediscriminate analysis, OSC-PLS-DA)和统计均匀聚类光谱分析[19]等。最近,研究学者又提出了一种新的核磁数据处理软件 MVAPACK,它可以完成数据预处理、模式识别和模型验证的整个过程,并已通过实践检验[17],相信不久的将来许多研究学者可以通过应用免费软件亲自进行数据分析。

## 4 主成分分析(Principal component analysis, PCA)

PCA 类似于聚类分析,是一种无师监督分析方法[20],在蛋白组学[21]和代谢组学[22]中都有应用。从数学角度来说,PCA 是一种高纬数据降纬的方法,就是将分散在一组变量的信息集中到某几个综合指标(主要成分)上[16,23],从而利用主要成分提取数据集的特征,形成一个二维或三维的得分图[24,25]。从生物学角度来说,PCA 就是在不分组的情况下,从整体上来描述样品的离散趋势[26]。

在 PCA 得分图中,通常以前两个主成分(PC1、PC2)为参数构建模型,主成分后的数值代表此成分对于模型分组的贡献率,数值越大表示其对分组的主导作用越大[26]。通常,所有的样品都将呈现在一个 95% 的置信区间中,区间以外的样品被视为异常值[27]。需要注意的是,对于异常值的处理,需要进行异常值检测[28],建立离群模型[29],综合考虑分析结果及其生物学意义,再决定是否剔除。

因此，当样品组间差异较大，组内差异较小时，PCA 可以很好地区分不同组样品。但是，当组间差异较小，组内差异较大（组内差异变量甚至大于样本含量）时，样本含量较大的一组将会主导模型，无法呈现组间差异，所以无师监督方法往往因无法忽略组内误差，过分关注与分组无关的指标而不利于分组，需要进行有师监督的模式识别来凸显组间差异。

此外，除了传统的以 PC1、PC2 作为横纵坐标建立模型，也有很多报道显示选择 t[1]、to[1]和 t[1]、t[2]、t[3]、t[4]等作为模型参数的，这给我们识图带来一定的困扰。其实，不论模型参数的表述形式如何，PCA 的分析原理及其表示的含义是不变的，那么也就是说参数代表的含义是相同的[30]，都是表示跟所有样品相关的主要指标，数字序号则代表是第几个主要指标（也称为成分）[31]。同理，对于其他几种模式识别方法亦然。

## 5 偏最小二乘法判别分析（Partial least squares discriminant analysis，PLS-DA）

PLS-DA 是一种有师监督分析方法（可以通过多次练习达到最佳效果），通过利用已知样品分组信息对多变量数据进行归类、识别和预测[32]。从模型构建上来说，PLS-DA 是对 PCA 模型的延伸，在 PCA 模型不分组的基础上利用虚拟变量分组，人为地将组别定义为 Y 变量，从而凸显组间差异[33]。在此过程中，给予 X 变量（代谢物的位移）及其响应变量（光谱强度）最大协方差，所有变量仍置于 95% 的置信区间内[34]，得到一个得分图。

此外，PLS-DA 模型还是一种线性分析法，其优势就在于其可以结合载荷图很容易比较出代谢物差异[35]，鉴别生物标记物[23]，而非线性方法则通常不能很好地解释模型的生物学意义，因为模型相关系数大小和方向的统计学意义可能与构成模型的相关变量不直接相关[36]。PLS-DA 模型也存在其自身的局限性，因为多变量和偏差系数的存在可能使其忽视真正的相关变量[37]。

## 6 正交的偏最小二乘法判别分析（Orthogonal partial least squares discriminant analysis，OPLS-DA）

OPLS-DA 的构建就是为了对 PLS-DA 进行修正[38]，通过移除与 Y 变量（分组）无关的 X 变量从而最大化组间差异[39]，或者通过分离平行变量（预测性变量和正交变量）来降低模型的复杂性，增加模型的可解释性[33]，也可以将其看做一个纯粹的数据预处理过程，或者是一个具有正交变量优势的简单 PLS-DA 模型[37]。

与 PLS-DA 模型相同，在 OPLS-DA 模型中，参数 $R^2$ 表示模型的解释率，$Q^2$ 表示模型的预测率，$R^2$ 和 $Q^2$ 的比值越接近于 1 表示模型越可靠[18]，通常 $Q^2$ 小于 $R^2$，$Q^2$ 为负值也比较常见，但是这意味着模型没有预测性[40]。随后对 OPLS-DA 模型进行内部验证，包括 7 倍交差验证和 1 000 次（200 次[18]）排列验证，以确定模型的准确性[41]；还可以应用受试者工作特征（Receiver Operating Characteristic，ROC）曲线进行 OPLS-DA 模型的外部验证[38]，以确定生物标记物的准确性，并计算其敏感性和特异性[42]。

对于筛选生物标记物，则需要 OPLS-DA 结合载荷图（Loading plot）、S 图（S plot）、变量投影重要性（variable importance projection，VIP）值、皮尔逊相关系数（Pearson's correlation coefficient，|r|），单因素方差分析和学生 $T$ 检验的 $P$ 值等其中之一或几个指标[41,43,44]，通过图形比对和重要变量上下调情况来确定差异代谢物，然后进行差异代谢物通路分析，最终完成生物标记物的鉴定。条件允许的情况下，还可以进行生物标记物的相关性分析，甚至是临界值计算等[43]。

## 7 OSC-PLS-DA

首先，我们要明确 OSC-PLS-DA 并不像个别报道中所述的那样等同于 OPLS-DA[31,45]，而 OPLS-DA 等同于 PLS-DA 结合 OSC[46]的说法也是不正确的，OPLS-DA 和 OSC-PLS-DA 模型的构建都是为了要优化 PLS-DA 模型，使组间差异更明显，但是它们各自发挥作用的途径却大不相同，前者是线性 PLS-DA 模型的修正，而后者是一种内部重复方法[47]。

OSC 被认为是一种基于 PLS 模型的数据过滤处理技术[48]，能够移除与 Y 响应矩阵不相关的 X 矩阵，从而使基于相关性 X 矩阵的 PLS 模型能够更加专一地分析有意义变量[49] 有报道指出,OSC-PLS-DA 就是在 PLS-DA[50] 或 OPLS-DA[51] 基础上结合正交信号修正过滤器所建立的分析模型,并且 OSC 还可以对 PCA 模型[52] 或 NMR 光谱[53] 进行优化。

在 OSC-PLS-DA 模型中,OSC 可以等于 0,1,2,3...,直到 PLS-DA 模型最佳为止[54]。通过这种加 OSC 的形式一次次去掉对模型分组无关的潜在变量,从而最大化组间差异[55]。

# 8 结 论

核磁数据预处理是保证所有数据能够被公平地进行多变量模式识别的基础;PCA 是首要的,是对所有变量的无偏性呈现,能够使人们从整体上把握样品水平;而其他 3 种方法是平行的,互相之间有比较也有优化,都能够直接应用于筛选生物标记物。当然,综合各模式分别讨论,不难发现样品分组趋势越来越明显,但这并不代表模型越来越好,我们需要明确每种识别模式各自的特点和适用范围,在应用时能够根据样品组成合理选择模式识别种类,使之既不过于复杂又能获得理想效果。

# References

[1] S. G. Oliver, M. K. Winson, D. B. Kell, et al. Systematic functional analysis of the yeast genome, *Trends in Biotechnology* 1998 (16) 373 – 378.

[2] J. K. Nicholson, J. C. Lindon, E. Holmes. Metabonomics: understanding the metabolic responses of living systems to pathophysiological stimuli via multivariate statistical analysis of biological NMR spectroscopic data, *Xenobiotica* 1999(29) 1181 – 1189.

[3] T. E. Wallaart, N. Pras, W. J. Quax, Isolation and identification of dihydroartemisinic acid hydroperoxide from Artemisia annua: a novel biosynthetic precursor of artemisinin, *Journal of Natural Products* 1999 (62) 1160 – 1162.

[4] Y. H. Choi, S. Sertic, H. K. Kim, et al . Classification of Ilex species based on metabolomic fingerprinting using nuclear magnetic resonance and multivariate data analysis, *Journal of Agricultural and Food Chemistry* 2005(53) 1237 – 1245.

[5] J. L. Ward, C. Harris, J. Lewis, et al. Assessment of H NMR spectroscopy and multivariate analysis as a technique for metabolite fingerprinting of Arabidopsis thaliana, *Phytochemistry* 2003 (62) 949 – 957.

[6] M. W. Lodewyk, M. R. Siebert, D. J. Tantillo. Computational prediction of 1H and 13C chemical shifts: A useful tool for natural product, mechanistic, and synthetic organic chemistry, *Chemical Reviews* 2011(112) 1839 – 1862.

[7] H. K. Kim, Y. H. Choi, R. Verpoorte. NMR-based metabolomic analysis of plants, *Nature Protocols* 2010 (5) 536 – 549.

[8] S. Tiziani, V. Lopes, U. L. Günther. Early stage diagnosis of oral cancer using 1H NMR-based metabolomics, Neoplasia (New York, NY) 2009 (11) 269.

[9] S. C. Connor, W. Wu, B. C. Sweatman, et al. Effects of feeding and body weight loss on the 1H-NMR-based urine metabolic profiles of male Wistar Han rats: implications for biomarker discovery, *Biomarkers* 2004 (9) 156 – 179.

[10] D. J. Crockford, E. Holmes, J. C. Lindon, et al . Statistical heterospectroscopy, an approach to the integrated analysis of NMR and UPLC-MS data sets: application in metabonomic toxicology studies, *Analytical Chemistry* 2006 (78) 363 – 371.

[11] H. Blasco, L. Nadal-Desbarats, P. -F. Pradat, et al. Untargeted 1H-NMR metabolomics in CSF Toward a diagnostic biomarker for motor neuron disease, *Neurology* 2014 (82) 1167 – 1174.

[12] A. M. Weljie, A. Bondareva, P. Zang, et al. 1H NMR metabolomics identification of markers of hypoxia-induced metabolic shifts in a breast cancer model system, *Journal of biomolecular* NMR 2011 (49) 185 – 193.

[13] 王敏,黄寅,张伟,等. 代谢组学信息获取与数据预处理瓶颈问题探讨,药学进展 2014 (38) 81 – 88.

[14] B. M. Warrack, S. Hnatyshyn, K. -H. Ott, et al. Normalization strategies for metabonomic analysis of urine samples, *Journal of Chromatography* B 2009 (87) 547 – 552.

[15] L. P. Silva, P. L. Lorenzi, P. Purwaha, et al Measurement of DNA Concentration as a Normalization Strategy for Metabolomic Data from Adherent Cell Lines, *Analytical chemistry* 2013 (85) 9536 – 9542.

[16] T. M. O'Connell. Recent advances in metabolomics in oncology, *Bioanalysis* 2012 (4) 431 – 451.

[17] B. Worley, R. Powers. MVAPACK: A Complete Data Handling Package for NMR Metabolomics, *ACS chemical biology*, 2014.

[18] Y. -S. Hong, Y. -T. Ahn, J. -C. Park, et al. 1H NMR-based metabonomic assessment of probiotic effects in a colitis mouse model, *Archives of Pharmacal Research* 2010 (33) 1091 – 1101.

[19] X. Zou, E. Holmes, J. K. Nicholson, et al. Statistical Homogeneous Cluster SpectroscopY (SHOCSY): an optimized statistical approach for clustering of 1H NMR spectral data to reduce interference and enhance robust biomarkers selection, *Analytical chemistry*, 2014.

[20] A. Patras, N. P. Brunton, G. Downey, A. Rawson, K. Warriner, G. Gernigon, Application of principal component and hierarchical cluster analysis to classify fruits and vegetables commonly consumed in Ireland based on *in vitro* antioxidant activity, Journal of *Food Composition and Analysis* 2011(24) 250 – 256.

[21] J. Rousu, D. D. Agranoff, O. Sodeinde, et al. Biomarker Discovery by Sparse Canonical Correlation Analysis of Complex Clinical Phenotypes of Tuberculosis and Malaria, *PLoS Computational Biology* 2013 (9) e1003018.

[22] G. Gürdeniz, L. Hansen, M. A. Rasmussen, et al. Patterns of time since last meal revealed by sparse PCA in an observational LC-MS based metabolomics study, *Metabolomics* 2013 (9) 1073 – 1081.

[23] R. Madsen, T. Lundstedt, J. Trygg, Chemometrics in metabolomics—a review in human disease diagnosis, Analytica Chimica Acta 2010 (659) 23 – 33.

[24] P. Banerjee, M. Dutta, S. Srivastava, et al. 1H NMR serum metabonomics for understanding metabolic dysregulation in women with idiopathic recurrent spontaneous miscarriage during implantation window, *Journal of Proteome Research* 2014.

[25] K. S. Solanky, N. J. Bailey, B. M. Beckwith-Hall, et al. Application of biofluid 1 H nuclear magnetic resonance-based metabonomic techniques for the analysis of the biochemical effects of dietary isoflavones on human plasma profile, *Analytical Biochemistry* 2003 (323) 197 – 204.

[26] Y. Sun, Z. Lian, C. Jiang, et al. Beneficial Metabolic Effects of 2′, 3′, 5′-tri-acetyl-N6-(3-Hydroxylaniline) Adenosine in the Liver and Plasma of Hyperlipidemic Hamsters, *PloS One* 2012 (7) e32115.

[27] J. Trygg, E. Holmes, T. Lundstedt, Chemometrics in metabonomics, Journal of proteome research 2007 (6) 469 – 479.

[28] P. J. Rousseeuw, M. Hubert, Robust statistics for outlier detection, Wiley Interdisciplinary Reviews: Data Mining and Knowledge Discovery 2011 (1) 73 – 79.

[29] M. Hubert, P. J. Rousseeuw, K. Vanden Branden, ROBPCA: a new approach to robust principal component analysis, *Technometrics* 2005 (47) 64 – 79.

[30] M. Mussap, R. Antonucci, A. Noto, et al. The role of metabolomics in neonatal and pediatric laboratory medicine, *Clinica Chimica Acta* 2013 (426) 127 – 138.

[31] M. Yang, X. Li, Z. Li, et al. Gene Features Selection for Three-Class Disease Classification via Multiple Orthogonal Partial Least Square, *PLoS One* 2013.

[32] B. Worley, S. Halouska, R. Powers, Utilities for quantifying separation in PCA/PLS-DA scores plots, *Analytical Biochemistry* 2013 (433) 102 – 104.

[33] E. Baldoni, M. Mattana, F. Locatelli, et alH Analysis of transcript and metabolite levels in Italian rice (*Oryza sativa* L.) cultivars subjected to osmotic stress or benzothiadiazole treatment, *Plant Physiology and Biochemistry* 2013(70) 492 – 503.

[34] H. -S. Son, G. -S. Hwang, K. M. Kim, et al. Metabolomic studies on geographical grapes and their wines using 1H NMR analysis coupled with multivariate statistics, *Journal of Agricultural and Food Chemistry* 2009 (57) 1481 – 1490.

[35] M. Li, J. Wang, Z. Lu, et al. NMR-based metabolomics approach to study the toxicity of lambda-cyhalothrin to goldfish (*Carassius auratus*), *Aquatic Toxicology* 2014 (146) 82 – 92.

[36] M. Rantalainen, M. Bylesjö, O. Cloarec, et al. Kernel-based orthogonal projections to latent structures (K-OPLS), *Journal of Chemometrics* 2007 (21) 376 – 385.

[37] H. Sadeghi-Bazargani, S. I. Bangdiwala, K. Mohammad, et al. Compared application of the new OPLS-DA statistical model versus partial least squares regression to manage large numbers of variables in an injury case-control study, *Sci Res Essays* 2011 (6) 4369 – 4377.

[38] F. Falahati Asrami, Alzheimer's Disease Classification using K-OPLS and MRI, *Academic Archive On-line* 2012.

[39] R. M. Salek, M. R. Pears, J. D. Cooper, et al. A metabolomic comparison of mouse models of the Neuronal Ceroid Lipofuscinoses, *Journal of Biomolecular NMR* 2011 (49) 175 - 184.

[40] A. M. Weljie, J. Newton, P. Mercier, et al. Targeted profiling: quantitative analysis of 1H NMR metabolomics data, *Analytical Chemistry* 2006 (78) 4430 - 4442.

[41] T. Chen, G. Xie, X. Wang, et al. Serum and urine metabolite profiling reveals potential biomarkers of human hepatocellular carcinoma, *Molecular & Cellular Proteomics* 2011 (10) M110.004945.

[42] K. M. Banday, K. K. Pasikanti, E. C. Y. Chan, et al. Use of urine volatile organic compounds to discriminate tuberculosis patients from healthy subjects, *Analytical chemistry* 2011 (83) 5526 - 5534.

[43] A. Zhang, H. Sun, Y. Han, et al. Exploratory urinary metabolic biomarkers and pathways using UPLC-Q-TOF-HDMS coupled with pattern recognition approach, *Analyst* 2012 (137) 4200 - 4208.

[44] F. Zhong, X. Liu, Q. Zhou, et al. 1H NMR spectroscopy analysis of metabolites in the kidneys provides new insight into pathophysiological mechanisms: applications for treatment with Cordyceps sinensis, *Nephrology Dialysis Transplantation* 2012 (27) 556 - 565.

[45] L. Wang, Y. Tang, S. Liu, et al. Metabonomic Profiling of Serum and Urine by 1H NMR-Based Spectroscopy Discriminates Patients with Chronic Obstructive Pulmonary Disease and Healthy Individuals, *PloS one* 2013 (8) e65675.

[46] W. Sui, L. Li, W. Che, et al. A proton nuclear magnetic resonance-based metabonomics study of metabolic profiling in immunoglobulin a nephropathy, *Clinics* 2012 (67) 363 - 373.

[47] H. Gu, Z. Pan, B. Xi, et al. Principal component directed partial least squares analysis for combining nuclear magnetic resonance and mass spectrometry data in metabolomics: Application to the detection of breast cancer, *Analytica Chimica Acta* 2011 (686) 57 - 63.

[48] H. Mao, M. Xu, B. Wang, et al. Evaluation of filtering effects of orthogonal signal correction on metabonomic analysis of healthy human serum^ 1H NMR spectra, *Acta Chimica Sinica-Chinese Edition-* 2007 (65) 152.

[49] L. Nadal-Desbarats, N. Aïdoud, P. Emond, et al. Combined 1 H-NMR and 1 H-13 C HSQC-NMR to improve urinary screening in autism spectrum disorders, *Analyst* 2014.

[50] A. Lodi, S. Tiziani, F. L. Khanim, et al. Proton NMR-based metabolite analyses of archived serial paired serum and urine samples from myeloma patients at different stages of disease activity identifies acetylcarnitine as a novel marker of active disease, *PloS One* 2013 (8) e56422.

[51] V. W. Davis, D. E. Schiller, D. Eurich, et al. Urinary metabolomic signature of esophageal cancer and Barrett's esophagus, *World J Surg Oncol* 2012 (10) 271 - 283.

[52] C. Gavaghan, I. Wilson, J. Nicholson, Physiological variation in metabolic phenotyping and functional genomic studies: use of orthogonal signal correction and PLS-DA, *FEBS Letters* 2002 (530) 191 - 196.

[53] M. I. Shariff, N. G. Ladep, I. J. Cox, et al. Characterization of urinary biomarkers of hepatocellular carcinoma using magnetic resonance spectroscopy in a Nigerian population, *Journal of Proteome Research* 2010 (9) 1096 - 1103.

[54] Y. Sun, C. Xu, C. Li, et al. Characterization of the serum metabolic profile of dairy cows with milk fever using 1H-NMR Spectroscopy, *Veterinary Quarterly* 2014 1 - 18.

[55] M. Rezaei-Tavirani, F. Fathi, F. Darvizeh, et al. Advantage of Applying OSC to 1H NMR-Based Metabonomic Data of Celiac Disease, *International Journal of Endocrinology and Metabolism* 2012 (10) 548.

# 复合抗菌肽“态康利保”对断奶仔猪红细胞免疫功能的影响研究

田春雷,任志华,邓俊良,袁　威,但启雄,金海涛,梁珍,高　爽
(四川农业大学动物医学院,动物疫病与人类健康重点实验室,环境公害与动物疾病四川省高校重点实验室,雅安 625014)

**摘　要**:为研究复合抗菌肽“态康利保”对断奶仔猪红细胞免疫功能的影响,试验采用日龄、体重(8.28 ±0.23)kg 的健康杜长大三元杂交仔猪 90 头,26 ~53 日龄,I-空白组饲喂基础日粮,II 组在基础日粮中添加 400mg/kg 黄芪多糖,III-V 组在基础日粮中分别添加 250mg/kg、500mg/kg 和 1 000mg/kg 的复合抗菌肽“态康利保”。分别于 32、39、46 和 53 日龄时每组随机选取 6 头仔猪前腔静脉采血,进行红细胞受体 $C_3b$ 花环试验、免疫复合物 IC 花环试验及血常规检验,观察复合抗菌肽“态康利保”对断奶仔猪红细胞免疫功能、红细胞数量及血红蛋白含量影响的动态变化。结果表明:复合抗菌肽“态康利保”可促使断奶仔猪红细胞 $C_3b$ 受体花环($C_3bRR$ 率)、红细胞总数和血红蛋白含量提高,IC 花环率(ICR 率)的明显变化($P<0.01$ 或 $P<0.05$)。说明复合抗菌肽“态康利保”可提高断奶仔猪红细胞免疫功能。

**关键词**:复合抗菌肽“态康利保”;断奶仔猪;红细胞免疫

# Study on the effect of the composite antibacterial peptide“Taikanglibao” on the immune function of erythrocytes in weaning piglets

TIAN Chunlei, REN Zhihua, DENG Junliang, YUAN Wei, DAN Qixiong, JIN Haitao, LIANG Zhen, GAO Shuang
*College of Veterinary Medicine, Sichuan Agricultural University, Ya'an, Sichuan 625014, China*

**Abstract**: To investigate the effect of the composite antibacterial peptide “Taikanglibao”on the immune function of erythrocytes in weaning piglets, Select the 21-day trial weaning weight similar (8.28 ±0.23kg) of Duroc crossbred piglets 90, based on similar principles of body weight, were randomly divided into five groups. During 26 to 53 days of age, the fed the basal diet, Ⅱ group added 400mg/kg astragalus polysaccharides on the basis of the diet, Ⅲ – Ⅴ group added 250mg/kg, 500mg/kg and 1 000mg/kg in the basal diet, respectively, composite antimicrobial peptides “Taikanglibao”, 32, 39, 46 and 53 respectively at the time of each group were randomly selected 6-day-old piglets before vena cava blood, peripheral blood lymphocyte subsets detection piglets, The immune adherence function of erythrocytes, the count of erythrocytes and the content of hemoglobin were examined in weaning piglets with erythrocyte $C_3b$ receptor rosette rate, erythrocyte immune complex rosette rate and blood routine examination. The results showed that the ratio of $C_3bR$ rosette, total number of red blood cell and content of hemoglobin were increased, and the ratio of IC rosette had obvious change by the composite antimicrobial peptides “Taikanglibao”. It indicates that the composite antimicrobial peptides “Taikanglibao”can cause increased immune function of erythrocytes in weaning piglets.

**Keywords**: Composite antibacterial peptide “Taikanglibao”, weaning piglets, erythrocyte immune

---

**基金项目**:“长江学者和创新团队发展计划”创新团队项目(IRT0848)

**作者简介**:田春雷(1990 – ),女(汉),内蒙古包头人,在读硕士,研究方向:动物环境公害性疾病。E-mail: 2364801669@qq.com

**通讯作者**:邓俊良(1966 – ),男(汉),四川仁寿人,教授,博导,主要从事动物环境公害性疾病的研究。E-mail: dengjl213@126.com

仔猪断奶期间,易受环境和饲料的转变而产生应激,容易引起较严重的腹泻,所以在断奶仔猪普通日两中都要添加抗生素等药物预防腹泻。大量研究表明,抗菌肽制剂抗腹泻作用的主要是其具有广谱抗菌作用,对大肠杆菌、沙门氏菌和产气荚膜梭菌等有害菌有抑制作用的优点,能够防止仔猪细菌性腹泻,改善断奶仔猪应激综合征[1-3]。抗菌肽是从动植物及人等多种生物体内组织和细胞中提取出的具有一定免疫作用的小分子类物质,又被称做肽抗生素(peptide antibiotics)或抗微生物肽(antimicrobial peptides)[4]。本试验所用的"态康利保"为四川华德生物工程有限公司研发生产的复合抗菌肽,由主要原料为猪防御素及海鞘抗菌肽。

早在1988年,郭峰等[5]人即发现红细胞膜粘附功能与机体免疫具有相关关系。目前,有关红细胞免疫功能与机体免疫功能的关系的研究报道日益增多[6-9],但鲜见有关抗菌肽对断奶仔猪红细胞免疫功能影响的研究,复合抗菌肽对断奶仔猪红细胞免疫功能影响的报道更是少见。为此,本试验通过复合抗菌肽"态康利保"对断奶仔猪红细胞免疫功能影响的研究,找到复合抗菌肽与断奶仔猪红细胞免疫功能的关系,最终为复合抗菌肽"态康利保"的实际应用提供理论依据及科学指导。

## 1 材料与方法

### 1.1 试验动物分组及饲养管理

经检查健康的21日龄断奶仔猪,网床(2m×2m)饲养,半漏缝式地板,舍温保持在22~25℃,不锈钢可调式料槽,乳头式饮水器。断奶后由乳猪全价料逐步更换为仔猪全价料,适应饲养至26日龄正式开始试验。粉料饲喂,自由采食、饮水。适应饲养至26日龄正式开始试验,试验周期28d。按猪场常规管理程序进行驱虫和免疫。基础日粮配方参照NRC(1998)配制。

将90头仔猪按照体重相近原则分成5组(Ⅰ-空白组,Ⅱ-400mg/kg黄芪多糖对照组,Ⅲ-250mg/kg抗菌肽组,Ⅳ-500mg/kg抗菌肽组,Ⅴ-1 000mg/kg抗菌肽组),每组18头。在26~53日龄,空白组饲喂基础日粮,第Ⅱ组添加黄芪多糖,第Ⅲ-Ⅴ组添加复合抗菌肽"态康利保";在21日龄时,所有仔猪注射猪瘟活疫苗(兔源),剂量为2头份/头。

动物分组、"态康利保"、药物、疫苗使用剂量、使用方法见表1。

表1 动物分组及处理

Table1 The groups and designs of animals

| 试验分组 | | 头数(头) | 添加药物 | | 药物用法 | 猪瘟疫苗(头份) |
|---|---|---|---|---|---|---|
| 空白组 | Ⅰ | 18 | — | — | — | 2头份/头 |
| 黄芪多糖对照组 | Ⅱ | 18 | 饲料添加黄芪多糖 | 400mg/kg | 连用28d | 2头份/头 |
| "态康利保"浓度梯度组 | Ⅲ | 18 | 饲料添加"态康利保" | 250mg/kg | 连用28d | 2头份/头 |
| | Ⅳ | 18 | | 500mg/kg | 连用28d | 2头份/头 |
| | Ⅴ | 18 | | 1000mg/kg | 连用28d | 2头份/头 |

### 1.2 试验药品与试剂

"态康利保" 由四川华德生物工程有限公司提供,规格:500g×袋;批号:20130620。

黄芪多糖 规格:100g×10袋;批号:20130330,纯度大于65%。

糖酵母多糖 规格:100mg×1袋。购于上海长江医院免疫实验室。

### 1.3 试验仪器

Abacus Junior VET动物血液分析仪(北京中西远大科技有限公司)、TDL-5-A型水平式离心机(上海安亭科学仪器厂)、CH30型显微镜、恒温水浴箱(天津市泰斯特仪器有限公司)等。

## 1.4 测定项目及方法

1.4.1 红细胞免疫功能 红细胞悬液的制备：将肝素钠抗凝血 2ml 离心分离红细胞，红细胞经生理盐水洗涤离心 3 次（每次 2 000r/min，5min），将红细胞配成 $1.25\times10^7$ 个/ml，制得红细胞悬液。

致敏酵母多糖的制备：冻干补体未致敏酵母多糖（购于上海长海医院免疫室）。取 2ml 酵母多糖悬液于 EP 管中，离心 2 000r/min，5 分钟，弃一半上清液，加入分离所得血清 1ml 混匀，37℃水浴 20min，即可制得致敏酵母多糖。

红细胞 $C_3b$ 受体花环（E-$C_3$bRR）试验：取致敏酵母多糖及红细胞悬液各 50μl，混匀，37℃水浴 30min；再加入 0.25% 的戊二醛 25μl，混匀，固定 5min；滴加 25μl 混合液于玻片上推片，立即烤干，滴加瑞氏染液 30s 至 1min，滴加缓冲液 5min，用清水清洗后，用滤纸吸干水，烤干，滴加香柏油后显微镜计数。一个红细胞结合 2 个或者 2 个以上酵母菌为一朵 $C_3b$ 受体花环（E-$C_3$bRR），计数 200 个红细胞，求出花环率。

$C_3b$ 受体花环百分率（%）=（红细胞花环数/200）×100%。

红细胞免疫复合花环（E-ICR）试验：除使用未致敏酵母多糖外，其余步骤均与红细胞 $C_3b$ 受体花环试验相同。

1.4.2 红细胞数量及血红蛋白含量 分别于 32d、39d、46d 及 53d 时对断奶仔猪进行颈静脉采血，抗凝，用 Abacus Junior VET 动物血液分析仪检测红细胞数量及血红蛋白含量。

## 1.5 数据统计

数据首先用 Excel 进行处理，然后用 SPSS19.0 统计软件进行统计分析，邓肯法（Duncan）进行多重比较。

# 2 结果与分析

## 2.1 断奶仔猪外周血红细胞 $C_3b$ 受体花环百分率及免疫复合物花环百分率

由表 2 可知，在整个试验期红细胞 $C_3b$ 受体花环百分率，各试验组极显著高于空白组（$P<0.01$）；除 32d、39d“态康利保”250mg/kg 组与黄芪多糖对照组差异不显著（$P>0.05$）外，其余时间点“态康利保”各剂量组与黄芪多糖对照组差异显著（$P<0.01$ 或 0.05）。红细胞 $C_3b$ 受体花环百分率随态康利保剂量增加而增加（$P<0.01$）。由此表明“态康利保”和黄芪多糖均有提高红细胞 $C_3b$ 受体花环百分率的作用，以态康利保组上升更明显，且有明显剂量依赖效应。

表 2 断奶仔猪外周血红细胞 $C_3b$ 受体花环及免疫复合物花环百分率（n=6）

Table 2 The percentage of E-$C_3$bRR of peripheral blood in weaned piglets（n=6）

| 项目 | 时间（日龄） | 空白组 | 黄芪多糖对照组 | 态康利保 | | |
|---|---|---|---|---|---|---|
| | | | | 250mg/kg 组 | 500mg/kg 组 | 1000mg/kg 组 |
| E-$C_3$bRR（%） | 32d | $7.29\pm0.84^D$ | $9.54\pm0.68^C$ | $9.21\pm0.83^C$ | $11.42\pm0.74^B$ | $13.72\pm0.48^A$ |
| | 39d | $9.16\pm0.43^D$ | $10.58\pm0.57^C$ | $11.29\pm0.44^C$ | $13.96\pm0.65^B$ | $16.63\pm0.29^A$ |
| | 46d | $10.54\pm0.16^D$ | $11.84\pm0.19^c$ | $12.33\pm0.27^C$ | $15.25\pm0.32^B$ | $18.38\pm0.29^A$ |
| | 53d | $11.13\pm0.50^E$ | $12.38\pm0.29^D$ | $14.46\pm0.32^C$ | $17.54\pm0.16^B$ | $18.50\pm0.20^A$ |
| E-ICR（%） | 32d | $12.38\pm0.29^A$ | $10.46\pm0.16^B$ | $10.50\pm0.14^B$ | $8.71\pm0.21^C$ | $7.71\pm0.44^D$ |
| | 39d | $12.46\pm0.16^A$ | $9.71\pm0.21^B$ | $9.75\pm0.39^B$ | $8.29\pm0.21^C$ | $7.17\pm0.19^D$ |
| | 46d | $11.88\pm0.60^A$ | $9.54\pm0.10^B$ | $8.92\pm0.17^b$ | $8.08\pm0.15^C$ | $7.01\pm0.62^D$ |
| | 53d | $11.75\pm0.29^A$ | $9.17\pm0.19^B$ | $8.29\pm0.21^C$ | $7.54\pm0.17^D$ | $6.74\pm0.19^E$ |

注：表中数据格式为 $\bar{x}\pm SD$。数据间上标有相同大写或小写字母者表示差异不显著（$P>0.05$），字母有相同而大小写不同表示差异显著（$P<0.05$），无相同字母表示差异极显著（$P<0.01$），下表同

在整个试验期红细胞免疫复合物花环百分率,各试验组极显著低于空白组($P<0.01$);除32d、39d"态康利保"250mg/kg组与黄芪多糖对照组差异不显著($P>0.05$)外,其余时间点"态康利保"各剂量组与黄芪多糖对照组差异显著($P<0.01$或0.05)。红细胞免疫复合物花环百分率随态康利保剂量增加而明显下降($P<0.01$)。由此表明"态康利保"和黄芪多糖均有降低红细胞免疫复合物花环百分率的作用,以态康利保组降低更明显,保组降低更明显,且具有明显剂量依赖效应。

## 2.2 断奶仔猪用药前后红细胞数量及血红蛋白含量

由表3可知,在32d,"态康利保"250、500mg/kg组RBC显著高于黄芪多糖组($P<0.05$);在39d,"态康利保"250mg、500mg/kg组RBC显著高于黄芪多糖组和态康利保1 000mg/kg组($P<0.05$),极显著高于对照组($P<0.01$);在53d,对照组显著高于"态康利保"1 000mg/kg组($P<0.05$),极显著高于黄芪多糖组($P<0.01$);其余各组在个日龄点均无显著性差异($P>0.05$)。

在32d,各组Hb均高于对照组,其中"态康利保"250mg/kg组显著高于对照组($P<0.05$);在39d,"态康利保"250mg、500mg/kg组极显著高于其他3组($P<0.01$),其余各组在个日龄点均无显著性差异($P>0.05$)。

表3 断奶仔猪红细胞数量及血红蛋白含量(n=6)

Table 3 RBC and Hb of weaned piglets(n=6)

| 项目 | 日龄 | 空白组 | 黄芪多糖对照组 | 态康利保 | | |
|---|---|---|---|---|---|---|
| | | | | 250mg/kg | 500mg/kg | 1000mg/kg |
| RBC$10^{12}$/L | 32d | $6.44 \pm 0.25^{AB}$ | $6.21 \pm 0.43^{Ba}$ | $7.08 \pm 0.14^{A}$ | $6.91 \pm 0.34^{A}$ | $6.83 \pm 0.18^{AB}$ |
| | 39d | $6.40 \pm 0.21^{B}$ | $6.63 \pm 0.27^{Ba}$ | $7.23 \pm 0.33^{A}$ | $7.17 \pm 0.21^{A}$ | $6.61 \pm 0.33^{Ba}$ |
| | 46d | $6.59 \pm 1.37^{A}$ | $6.43 \pm 0.39^{A}$ | $6.85 \pm 0.62^{A}_{A}$ | $6.53 \pm 0.32^{A}$ | $6.37 \pm 0.25^{A}$ |
| | 53d | $6.70 \pm 0.50^{A}$ | $5.98 \pm 0.24^{B}$ | $6.54 \pm 0.39^{AB}$ | $6.45 \pm 0.20^{AB}$ | $6.02 \pm 0.18^{Ba}$ |
| Hbg/L | 32d | $118.00 \pm 11.14^{Ba}$ | $124.00 \pm 12.00^{AB}$ | $141.00 \pm 3.61^{A}$ | $131.33 \pm 15.31^{AB}$ | $128.67 \pm 4.93^{AB}$ |
| | 39d | $120.00 \pm 8.72^{B}$ | $121.00 \pm 2.65^{B}$ | $154.00 \pm 2.00^{A}$ | $145.67 \pm 4.04^{A}$ | $123.67 \pm 5.13^{B}$ |
| | 46d | $127.33 \pm 27.68^{A}$ | $117.33 \pm 4.16^{A}$ | $134.67 \pm 13.57^{A}$ | $126.67 \pm 2.31^{A}$ | $116.67 \pm 2.89^{A}$ |
| | 53d | $123.00 \pm 7.81^{A}$ | $105.00 \pm 13.23^{A}$ | $118.67 \pm 7.51^{A}$ | $114.67 \pm 1.53^{A}$ | $107.00 \pm 4.36^{A}$ |

# 3 讨 论

红细胞是血液中最主要的细胞成分,现代科学研究发现,红细胞不仅可以运输氧和二氧化碳,还可以为机体供给能量,从而达到营养机体、代谢及调节酸碱平衡等作用。最新研究发现,红细胞可通过细胞膜上的$C_3b$受体发挥免疫作用。红细胞免疫功能是机体免疫功能的重要组成部分,红细胞膜上处于未结合状态的$C_3b$受体数量可反映红细胞的免疫功能状态,目前,已经明确红细胞膜上的补体受体有I型($CR_1$)和III型($CR_3$),主要发挥作用的是$CR_1$[10]。$CR_1$与血液循环中带有$C_3b$的免疫复合物结合,并运送至肝脏等内皮系统中清除,这就是红细胞免疫粘附(RCIA)机制。此外,红细胞能将抗原抗体复合物传递给单核巨噬细胞并使之激活,由此增强单核巨噬细胞对免疫复合物的作用使之与T细胞反应[11]。红细胞上$CR_1$表达降低或红细胞粘附功能下降会引起机体免疫功能下降,研究红细胞$CR_1$介导的免疫粘附功能对评价机体天然免疫功能状态乃至机体整体免疫都具有非常重要的作用,$CR_1$数量一般是由遗传特性决定的,而$CR_1$活性主要与其在红细胞膜上的分布状态有关$CR_1$在红细胞膜上呈散在分布和集簇状分布两种状态,而只有呈集簇状分布方式的$CR_1$才能与补体的结合点呈多价性,且连接更牢靠[12],基于此,郭峰于1982年所创立了红细胞膜$C_3b$受体花环($C_3b$ receptor rosette,$C_3b$RR)和免疫复合物花环(immune complex rosette,ICR)检测法[13,14],可以准确有效的检测红细胞表面未结合的$CR_1$数,这为现代红细胞研究奠定了基础。

赵翠燕等[15]研究发现黄芪多糖可以促进红细胞的生成，提高 $C_3b$ 受体花环率，降低 IC 花环率。钟妮娜等[6]也在试验中发现青刺果多糖可明显增强鸡红细胞膜上的 $CR_1$ 活性及吸附 IC 的能力，即对鸡红细胞免疫功能有增强作用。高学军[16]、李艳星等[17]在试验中也发现黄芪多糖可显著提高雏鸡红细胞免疫功能。

在本实验中，黄芪多糖对照组也显著提高了 $C_3b$ 受体花环率，降低了 IC 花环率，对红细胞免疫有明显的促进作用，这与赵翠燕等人的试验结论相一致，但在本次实验中，相同的使用时间下，中药组显著优于黄芪多糖组（$P<0.05$ 或 $P<0.01$），并且，与黄芪多糖组相比较，中药组极显著增加了断奶仔猪外周血血红蛋白含量及红细胞数（$P<0.01$），实验结果表明，复方中药“态康利保”与黄芪多糖相比，可以更加显著的提高断奶仔猪的红细胞免疫功能，且对断奶仔猪的造血功能有一定的促进作用。推测原因可能有以下两点，其一是“态康利保”改变了 $CR_1$ 在红细胞膜上的分布状态，使集簇状分布的 $CR_1$ 增多，从而增加了红细胞膜上 $CR_1$ 活性，促进了 $CR_1$ 表达，其二是“态康利保”可以提高断奶仔猪的整体免疫功能，降低了机体释放变性坏死的组织或细胞等抗原物质进入血液。

但试验中发现，“态康利保”仅在一定时间内可提高断奶仔猪的造血功能，随着饲喂时间的持续，断奶仔猪红细胞造血功能有所下降。针对长期使用会降低断奶仔猪造血功能的机制，还需要更加深入的试验研究。

## 4 结 论

经试验研究表明，复合抗菌肽“态康利保”在一定时间内可显著提高断奶仔猪的造血功能，提高断奶仔猪外周血红细胞表面 $C_3b$ 受体，降低红细胞免疫复合物，促进断奶仔猪红细胞免疫功能，对断奶仔猪的机体免疫功能有促进作用。且以 1000mg/kg 添加剂量最佳。

## 参考文献


[1] 都海明，戚广州，王建军，等. 抗菌脂肽对肉鸡生产性能和免疫机能的影响[J]. 江苏农业学报，2010，26(5)：1009－1014.

[2] 刘莉如. 天蚕素抗菌肽对蛋用仔公鸡生长，免疫及相关细胞因子 mRNA 表达水平影响的研究[D]. 乌鲁木齐：新疆农业大学，2012.

[3] 刘莉如，杨开伦，滑静，等. 抗菌肽对蛋用仔公鸡生长性能，免疫指标及空肠组织相关细胞因子基因 mRNA 表达的影响[J]. 动物营养学报，2012，24(7)：1345－1351.

[4] Brogden KA. Antimicrobial peptides：pore formers or metabolic inhibitors in bacteria[J]. Nature Reviews Microbiology，2005，3：238－250.

[5] 郭峰，曹漪明，杨虎天，等. SLE 患者红细胞免疫与 T 淋巴细胞免疫功能变化的初步观察[J]. 中国免疫学杂志，1988，4(1)：35－37.

[6] 钟妮娜，李超，殷中琼. 青刺果多糖对鸡红细胞免疫及外周血淋巴细胞免疫功能的影响[J]. 安徽农业科学，2007，35(31)：9937－9938.

[7] 张漾，周显青. 硫丹对小鼠红细胞免疫功能的影响[J]. 动物学杂志，2010，45(1)：50－58。

[8] 章亭，谭允育，潘彦舒. 四物汤对红细胞免疫及骨髓干细胞增殖能力的影响[J]. 北京中医药大学学报，2000，23(1)：36－38

[9] 芦珂. 高铜对雏鸭红细胞免疫黏附功能的影响[J]. 四川畜牧兽医，2011，38(12)：17－19

[10] McLare CA，Williom son JF，Stewart BJ，et al. Indels and imperfect duplication have drive the evolution of human complement receptor 1(CR1) and CR1-like from their precursor CR1 alpha：Importance of functional sets[J]. Human Immunology，2005，66(3)：258.

[11] Craig ML, Bankovieh AJ, Taylor RP. Visualization of the transfer reaction：tracking immune complex from erythrocyte complement recept 1 to macrophages[J]. Clinical Immunology，2002，105(1)：36－47.

[12] 黄盛东，郭峰. 红细胞免疫调控机理研究新进展[J]. 国外医学免疫学分册，1991，14(4)：191－195.

[13] 郭峰. 红细胞免疫功能的初步研究[J]. 中华医学杂志，1982，62(12)：715－716.

[14] 郭峰，张励力. 用 $C_3b$ 致敏酵母菌测定血清免疫复合物[J]. 上海免疫学杂志，1982，2(1)：32－37.

[15] 赵翠燕，徐嘉静，何建梅. 黄芪多糖对三黄鸡红细胞免疫功能的影响[J]. 广东农业科学，2012，39(23)：104－106.

[16] 高学军，李庆章. 黄芪多糖和香菇多糖对雏鸡红细胞免疫功能的影响[J]. 东北农业大学学报，2000，31(1)：69－71.

[17] 李星艳，罗燕，谷新利. 中药复方多糖对鸡红细胞免疫功能的影响[J]. 上海畜牧兽医通讯，2008，(4)：30－31.

# 复合营养舔砖对奶牛血清中微量元素水平的影响

王　慧,刘永明,王胜义,齐志明,刘世祥,李胜坤
(中国农业科学院兰州畜牧与兽药研究所农业部兽用药物创制重点实验室、甘肃省新兽药工程重点实验室、甘肃省中兽药工程技术研究中心,兰州 730050)

**摘　要**:为观察复合营养舔砖对奶牛血清中微量元素水平的影响,选择高、中、低产中国荷斯坦奶牛各 8 头,补饲复合营养舔砖,于第 1、2、3、4 月测定血清 Cu、Mn、Fe、Zn 的含量。结果表明,补饲后血清 Cu 的含量除第 1 月升高外,其余各月都有所降低;血清 Mn 的含量各组及各时间段均极显著高于补饲前($P<0.01$);血清 Fe 的含量,高、中产奶牛补饲前后无显著性差异,低产奶牛第 2、3 月明显高于补饲前($P<0.01$ 和 $P<0.05$);血清 Zn 的含量有所提高,其中高产奶牛较为明显;Cu/Zn 比值随着饲喂周期的增加而增大。表明复合营养舔砖对提高奶牛血清微量元素水平有促进作用。

**关键词**:复合营养舔砖,微量元素,奶牛,血清

# Effects of multinutrient blocks on the trace elements contents of serum in dairy cows

WANG Hui, LIU Yong-ming *, WANG Sheng-yi, QI Zhi-ming, LIU Shi-xiang, LI Sheng-kun

*Key Laboratory of Veterinary Pharmaceutical Development of Ministry of Agriculture, Key Laboratory of New Animal Drug Project of Gansu Province, Engineering & Technology Research Center of Traditional Chinese Veterinary Medicine of Gansu Province, Lanzhou Institute of Husbandry and Pharmaceutics Sciences of CAAS, Lanzhou 730050, China*

**Abstract**: The objective of this study was to evaluate the ability of China Holstein to regulate intake of multinutrient blocks with different trace mineral content, were available to dairy cows for ad libitum consumption. China Holsteins were chosen at random, and were divided into three groups, high producing dairy cow, middle producing dairy cow, and low producing dairy cow, respectively. After feeding for 1, 2, 3, and 4 months, the content of Cu, Mn, Fe and Zn in serum were detected. The results showed that the content serum Cu was reduced except the first month. The content of serum Mn in all groups was higher than before feeding period ($P<0.01$). The content of serum Fe in high and middle producing dairy cow had no significant difference compared with before feeding period, but higher than before feeding period in low producing dairy cow on second and third month ($P<0.01$ and $P<0.05$). The content of serum Zn was increased, especially the high producing dairy cow. The ratio of Cu/Zn was increased with the feeding cycle extend. In conclusion, the results of this experiment showed that multinutrient blocks have significant simulative effect on improving serum level of trace elements in dairy cows.

**Keywords**: multinutrient blocks, trace elements, dairy cows, serum

---

**基金项目**:农业科技成果转化资金项目(2010GB23260564),中央级科研院所基本科研业务费(1610322013003)

**作者简介**:王慧(1985-),男,陕西甘泉人,硕士,研究实习员,主要从事动物微量元素代谢病研究,E-mail:wanghui01@caas.cn

**通讯作者**:刘永明(1957-),男,研究员,硕士生导师,主要从事中兽药开发和微量元素代谢病研究,E-mail:myslym@sina.com

微量元素是动物维持生命和生产必不可少的营养素之一，主要以酶的必需组成部分（辅酶、辅基等）或激活剂形式直接或间接地参与机体几乎所有生理和生化过程，与动物的物质代谢、生长发育、繁殖、免疫功能等密切相关[1-4]。微量元素缺乏或过量都会导致动物生物化学的、结构的和功能性的病理学变化[1]。微量元素对于维持机体的正常生理功能十分关键。研究表明，Cu、Zn、Mn、Fe、Se 等复方微量元素可激活机体内多种酶，增强免疫力，强化代谢水平，和明显促进畜禽生长和发育的作用[5,6]。据熊桂林等报道，由于奶牛产奶量的不断提高，围产期疾病的发生与所遭受的氧化应激有关，Cu、Zn、Mn、Fe、Se 等微量元素参与机体的抗氧化作用，提高奶牛的抗氧化防御机制是预防围产期疾病发生的主要战略措施[7,8]。我国每年有长达 7 ~ 8 月的枯草期，在枯草季节，蛋白质和矿物质元素等营养物质摄取量明显不足，严重影响牛羊的生长发育，降低了经济效益。舔砖含有多种微量元素，保健促生长剂等成分，饲喂舔砖是给牛羊等草食家畜补充矿物质元素、非蛋白氮、可溶性糖等易缺乏养分的一种有效方式，弥补了天然牧草和秸秆饲料的营养不足，从而提高饲料利用率。因此，本试验通过补饲复合微量元素舔砖，研究其对奶牛血清微量元素水平的影响，为复合营养舔砖临床应用提供试验依据。

# 1 材料与方法

## 1.1 试验材料

试验选用的中国荷斯坦奶牛，由甘肃定西天辰奶牛场提供。试验用复合营养舔砖由中国农业科学院兰州畜牧与兽药研究所提供（批准文号：甘饲预字（2012）034002）。

## 1.2 试验处理

选择年龄，胎次，体重相近，体质健康，乳房整齐，无病，日产奶量分别为 20kg 以上、10 ~ 20kg、5 ~ 10kg 的中国荷斯坦高、中、低产奶牛 24 头，每组 8 头，同舍分栏饲养，分别按组补饲复合营养舔砖，试验期为 90 天，在试验前 1 天、试验第 30 天、第 60 天、第 90 天时颈静脉采血 10ml，分离血清，测定血清中微量元素的含量。

## 1.3 样品处理及元素总量的测定

取血清 1ml 加入消解罐中，依此加入 AR 级盐酸 6ml、硝酸 2ml、放入微波消解仪（MARS5，美国 CEM）中进行消化（优化的消解程序如表 2 所示）。在此优化条件下消解完毕后，加入高氯酸 1ml，在 110℃下排酸至 1ml，冷却后，完全转移消解液于 10.0ml 容量瓶中，用超纯水定容至刻度，同时做试剂空白，用原子吸收光谱仪（ZEEnit 700，德国 Analytik Jena）原子吸收分光光度仪测定血清中 Cu、Mn、Fe、Zn 元素含量。

表 1 微波消解仪工作条件

Table 1 Working conditions for microwave

| 工作步骤 Working step | 1 | 2 | 3 |
|---|---|---|---|
| 温度/℃ Temperature | 100 | 150 | 160 |
| 升温时间/min Heating up time | 4 | 5 | 5 |
| 保温时间/min Holding time | 3 | 5 | 15 |
| 功率/W Power | 400 | 400 | 400 |

## 1.4 数据处理

试验数据应用 SPSS17.0 统计软件中 One Way ANOVA 进行分析，LSD 法进行多重比较，显著水平为 $P < 0.05$，所有结果用“平均值 ± 标准差”表示。

# 2 结果与分析

## 2.1 复合营养舔砖对高产奶牛血清中微量元素含量的影响

表2显示:高产奶牛饲喂复合营养舔砖后,血清Cu的含量除第1月极显著升高外($P<0.01$),其余各月均有不同程度的下降,但与饲喂前无显著差异($P>0.05$);血清Mn的含量饲喂后均极显著提高($P<0.01$);血清Fe的含量在饲喂后有升高的趋势,但差异不显著($P>0.05$);血清Zn的含量除第2月外,其他各月均显著或极显著提高($P>0.05$ 或 $P>0.01$);随着饲喂周期的增加,Zn/Cu比值逐渐升高,其中第3、4月极显著高于饲喂前($P<0.01$)。

表2 复合营养舔砖对高产奶牛血清中微量元素含量的影响(μg/ml)

Table 2 Effects of multinutrient blocks on the trace elements contents of serum of high producing dairy cows

| 补饲期 Feedingperiod | Cu | Mn | Fe | Zn | Zn/Cu |
|---|---|---|---|---|---|
| 补饲前 Before feeding | 0.580 1±0.093 6 | 0.128 8±0.049 8 | 3.397 8±0.675 9 | 1.537 1±0.290 3 | 2.983 2±0.930 9 |
| 第1月 Feeding for one month | 0.698 4±0.231 6** | 0.475 5±0.116 1** | 4.698 4±0.592 6 | 2.216 1±0.437 2** | 3.472 6±1.326 1 |
| 第2月 Feeding for two months | 0.360 6±0.103 5 | 0.844 4±0.060 7** | 3.128 9±1.944 7 | 1.572 7±0.156 6 | 4.136 9±1.052 4 |
| 第3月 Feeding for three months | 0.534 4±0.065 2 | 1.084 5±0.022 7** | 2.193 2±0.731 5 | 1.981 1±0.677 2* | 5.153 4±1.920 8** |
| 第4月 Feeding for four months | 0.495 0±0.061 7 | 1.219 6±0.058 1** | 4.453 4±2.359 6 | 2.276 2±0.321 9** | 5.635 9±0.787 5** |

注:表中同一列数据右肩标*表示与补饲前相比差异显著($P<0.05$),**表示差异极显著($P<0.01$)。下表同

Note: for data in one column, "*" means $P<0.05$, and "**" means $P<0.01$ compare with before feeding. The same as below

## 2.2 复合营养舔砖对中产奶牛血清中微量元素含量的影响

表3显示:复合营养舔砖使中产奶牛血清Cu的含量有降低趋势,其中第2、3月极显著低于饲喂前($P<0.01$);血清Mn的含量极显著提高($P<0.01$);血清Fe的含量各试验期与饲喂前相比均无显著差异($P>0.05$);血清Zn的含量有升高趋势,但无显著差异($P>0.05$);Zn/Cu比值逐渐升高,其中第3、4月达到极显著水平($P<0.01$)。

表3 复合营养舔砖对中产奶牛血清中微量元素含量的影响(μg/ml)

Table 3 Effects of multinutrient blocks on the trace elements contents of serum of middle producing dairy cows

| 补饲期 Feeding period | Cu | Mn | Fe | Zn | Zn/Cu |
|---|---|---|---|---|---|
| 补饲前 Before feeding | 0.645 4±0.105 1 | 0.321 3±0.081 4 | 4.572 4±1.078 4 | 1.582 1±0.285 5 | 2.561 0±0.614 3 |
| 第1月 Feeding for one month | 0.809 5±0.243 8* | 0.846 8±0.083 1** | 4.117 3±0.753 4 | 1.624 0±0.985 0 | 2.900 5±0.586 3 |
| 第2月 Feeding for two months | 0.374 0±0.110 0** | 1.477 0±0.054 9** | 5.134 2±1.274 3 | 1.942 3±0.172 6 | 4.568 2±0.707 3** |
| 第3月 Feeding for three months | 0.405 1±0.065 7** | 0.999 9±0.029 5** | 4.767 2±2.491 8 | 1.981 8±0.461 4 | 5.487 0±1.066 9** |

（续表）

| 补饲期 Feeding period | Cu | Mn | Fe | Zn | Zn/Cu |
|---|---|---|---|---|---|
| 第4月 Feeding for four months | 0.527 5 ±0.109 1 | 1.324 0 ±0.025 1** | 3.574 1 ±0.651 5 | 1.926 4 ±0.410 0 | 5.866 9 ±1.136 4** |

## 2.3 复合营养舔砖对低产奶牛血清中微量元素含量的影响

由表4可见，复合营养舔砖对低产奶牛血清Cu含量的影响无显著差异；血清Mn的含量极显著提高（$P<0.01$）；血清Fe的含量除第2、3月分别极显著（$P<0.01$）和显著（$P<0.05$）提高外，其余试验期与饲喂前相比无显著差异（$P>0.05$）；血清Zn的含量有升高趋势，其中第2月达到显著水平（$P<0.05$）；Zn/Cu比值第2月极显著高于饲喂前（$P<0.01$），第3、4月显著高于饲喂前（$P<0.05$）。

表4　复合营养舔砖对低产奶牛血清中微量元素含量的影响（μg/ml）

Table 4　Effects of multinutrient blocks on the trace elements contents of serum of low producing dairy cows

| 补饲期 Feeding period | Cu | Mn | Fe | Zn | Zn/Cu |
|---|---|---|---|---|---|
| 补饲前 Before feeding | 0.633 7 ±0.095 4 | 0.339 7 ±0.067 6 | 4.572 4 ±1.078 4 | 1.582 1 ±0.285 5 | 2.480 6 ±0.574 0 |
| 第1月 Feeding for one month | 0.809 5 ±0.243 8 | 0.846 8 ±0.083 1** | 4.117 3 ±0.753 4 | 1.624 0 ±0.985 0 | 3.423 9 ±1.118 9 |
| 第2月 Feeding for two months | 0.499 2 ±0.056 5 | 0.650 3 ±0.087 1** | 5.134 2 ±1.274 3** | 1.942 2 ±0.172 6* | 5.268 1 ±2.126 3** |
| 第3月 Feeding for three months | 0.589 1 ±0.163 4 | 0.925 3 ±0.025 6** | 4.767 2 ±2.491 8* | 1.981 8 ±0.461 4 | 4.148 7 ±1.319 9* |
| 第4月 Feeding for four months | 0.515 0 ±0.115 3 | 1.383 4 ±0.018 7** | 3.574 1 ±0.651 5 | 1.926 4 ±0.410 0 | 4.395 7 ±1.276 7* |

# 3 讨　论

生物体对元素的代谢（吸收、转运、分布和排泄）都与疾病的发生发展情况密切相关，元素对许多基本的生命过程诸如信号传导，基因表达和代谢起着至关重要的作用。许多疾病的病理过程都伴随着微量元素的异常。微量元素在机体内发挥重要的生物学效应，许多酶的辅助因子和活性中心由金属离子组成，能够减少自由基的形成，增强杀菌能力，调节机体免疫力，提高抗病、抗应激能力，提高生产性能。微量元素缺乏会导致各种疾病[9]，日粮中平衡微量元素对反刍动物的营养优化非常重要[10]。

铜对奶牛的繁殖、生长和泌乳性能具有重要的作用[11]。铜是动物体内含铜金属酶的必需成分，大多以铜蛋白形式存在，铜也是CuZn-SOD的辅助因子和调节因子。锌对维持动物生长、发育和免疫功能起着重要的作用[12]。由于许多天然饲料缺锌，所以畜禽饲料中往往需要添加外源锌才能满足动物的生长和生理需要[13]。由肠道吸收的锌95%同血液中的白蛋白结合，因此血清锌能敏感地反映动物机体锌营养代谢状况[12]。大量研究表明铜、锌是两种拮抗性的元素，于体内吸收和转运过程中存在着互相竞争，互相抑制[14-18]。本试验结果显示，复合营养舔砖补饲后奶牛血清中Zn/Cu比值提高，表明复合营养舔砖对提高血清Cu、Zn的含量有促进作用。

锰是高等生命体所必需的微量元素，可催化金属活化酶，参于体内的造血过程，促进细胞内脂肪的氧化作用，参与机体碳水化合物、脂肪、蛋白质等多种代谢，可以促进动物生长，增强动物的繁殖性能[19]。同时，它对动物机体本身产生的副作用也较小，是一种理想的矿物质添加剂。本研究结果揭示复合营养舔砖对提高血清锰的水平有极显著的促进作用。

铁最主要的功能是作为血红蛋白和肌红蛋白中血红素的组成成分,与造血机能、氧的运输以及细胞内的生物氧化过程有着密切的关系。血浆中铁含量的稳定与否反映了机体内铁的营养状态。血浆中的铁和铁传递蛋白相结合,在铁代谢中起传递铁和增强机体抗感染能力的作用[20]。本研究发现复合营养舔砖有体高血清 Fe 含量的趋势。

## 4 结 论

复合营养舔砖对提高奶牛血清微量元素水平有促进作用,可以弥补饲草饲料中微量元素补充的不足,从而可以降低奶牛微量元素代谢相关疾病的发病率,改善奶牛健康。

## 参考文献

[1] Watanabe T,Kiron V,Satoh S. Trace minerals in fish nutrition[J]. Aquaculture,1997,151(1-4):185-207.

[2] 孙明茂,洪夏铁,李圭星,等. 水稻籽粒微量元素含量的遗传研究进展[J]. 中国农业科学,2006,39(10):1947-1955.

[3] 吕 林,罗绪刚,计 成. 矿物元素影响畜禽肉质的研究进展[J]. 动物营养学报,2004,16(1):12-19.

[4] 林仕梅,潘 瑜,罗 莉,等. 不同来源微量元素铁、锌、锰、铜对罗非鱼生长、代谢和非特异性免疫力的影响[J]. 动物营养学报,2011,23(5):763-770.

[5] 刘晋生,邓俊良,李友昌,等. 复方微量元素"生命元"对波尔山羊免疫功能的影响[J]. 中国兽医学报,2012,32(3):447-450.

[6] 胡延春,赵春蕊,刘 鹏,等. "一针肥"注射液中硒元素在山羊体内的药代动力学研究[J]. 中国兽医学报,2012,32(5):733-736.

[7] 熊桂林,王 林,顾建红,等. 奶牛不同生理阶段血清微量元素含量和氧化状态[J]. 中国兽医学报,2009,29(12):1613-1616.

[8] Castillo C,Hernandez J,Bravo A,*et al*. Oxidative status during late pregnancy and early lactation in dairy cows [J]. Vet J,2005,169(2):286-292.

[9] Braun J P,Trumel C,Bézille P. Clinical biochemistry in sheep:a selected review[J]. Small Ruminant Res,2010,92(1-3):10-18.

[10] Dove H. Balancing nutrient supply and nutrient requirements in grazing sheep[J]. Small Ruminant Res,2010,92(1-3):36-40.

[11] Mulligan FJ,O'Grady L,Rice DA, *et al*. Aherd health approach to dairy cow nutrition and production diseases of the transition cow[J]. Anim Reprod Sci,2006,96(3-4):331-353.

[12] 徐振华,李福昌. 不同锌源对断奶肉兔组织器官锌浓度及血清碱性磷酸酶活性的影响[J]. 中国畜牧杂志,2012,46(5):44-46.

[13] 曹家银,罗绪刚. 以组织锌、金属硫蛋白及其基因表达指标评价肉仔鸡对锌源的相对生物学利用率[J]. 畜牧兽医学报,2003,34(3):227-231.

[14] Oestreicher P,Cousins R J. Copper and zinc absorption in the rat:mechanism of mutual antagonism[J]. J Nutr,1985,115(2):159-166.

[15] Fischer P W F,Campbell J S,Giroux A. Effects of low copper and high zinc intakes and related changes in Cu,Zn-superoxide dismutase activity on DMBA-induced mammary tumor genesis[J]. Biol Trace Elem Res,1991,30(1):65-79.

[16] Barone A,Ebesh O,Harper R G,*et al*. Placental copper transport in rats:effects of elevated dietary zinc on fetal copper,iron and metallothionein[J]. J Nutr,1998,128(6):1037-1041.

[17] Willis M S,Monaghan S A,Miller M L,*et al*. Zinc-induced copper deficiency:a report of three cases initially recognized on bone marrow examination[J]. Am J Clin Pathol,2005,123(1):125-131.

[18] Horvath J,Beris P,Giostra E,*et al*. Zinc-induced copper deficiency in Wilson disease[J]. J Neurol Neurosur Ps,2010,81(12):1410-1411.

[19] 王夕国,李光玉,钟 伟,等. 锰的生物学功能及在畜禽中的研究进展[J]. 经济动物学报,2011,15(4):216-220.

[20] 焦小丽,叶占胜. 铁生物学效价的评定研究进展[J]. 天津农学院学报,2008,15(1):44-47.

# 国内流行犬瘟热病毒 DNA 重组疫苗的初步研究

吴 凌,孙雨航,许楚楚,李昌盛,夏 成

(黑龙江八一农垦大学,动物科技学院,大庆 163319)

**摘 要**:本研究构建了一株犬瘟热病毒(CDV) ZH-10 株 H 基因的真核表达载体 pCI-H,对表达产物的免疫原性进行了初步研究,并进行了小鼠的 DNA 免疫试验。结果显示:pCI-H 免疫组血清可与感染病毒的 MDCK 细胞发生特异性 IPMA 反应呈现特异的棕红色;ELISA 血清抗体滴度可达 1∶28～1∶79;病毒中和抗体可达 1∶11～1∶32。研究表明 H 蛋白具有免疫原性,但是诱导抗体效价不足,需进一步完善。

**关键词**:犬瘟热,ZH-10 株,H 基因,真核表达,DNA 重组疫苗

# A preliminary study of domestic popular canine distemper vaccine

Ling WU, Yuhang SUN, Chuchu XU, Changsheng LI, Cheng XIA*

(*College of Animal Science and Veterinary Medicine, Heilongjiang Bayi Agricultural University, Daqing* 163319)

**Abstract**: In this study, we constructed recombinant eukaryotic expression plasmid pCI-H and initially researched the immunogenicity of expression products, and subsequently the immune response was detected in mice induced by pCI-H DNA immunization. Results showed that the sera of immunized groups of pCI-H could react with MDCK cells infected by virus and had specifically bright red fluorescence. The tilters of antibody of ELISA and SN could relatively reach 1∶28 to 1∶79 and 1∶11 to 1∶32. The research indicated that H protein had the immunogenicity, which was insufficient to induce antibody, and should be further improved.

**Keywords**: canine distemper virus; ZH-10 strain; H gene; eukaryotic expression; recombinant DNA vaccines

犬瘟热(canine distemper, CD)是由犬瘟热病毒(Canine distemper virus, CDV)引起的一种急性、高度接触性传染病。CD 呈世界性分布,自然感染宿主范围不断扩大,且在不同物种之间中可以交叉传播,给养犬业、经济动物养殖业以及野生动物保护等都造成了严重的危害[1]。

CD 常规疫苗的广泛使用,使 CD 的流行得到有效的控制。但是在 CD 的防治过程中,越来越多的数据[2-4]显示 CDV 目前的主要流行株,尤其是来源于不同地区或国家的分离株和 Onderstepoort、Convac 等疫苗株在抗原性、细胞病变类型、毒力、遗传特性、临床症状等方面都存在一定的差异,使制备的疫苗不能完全保护动物免受异源毒株的感染,这无疑给 CD 的有效预防增加了困难。

因此,筛选出免疫原性良好的毒株,作为原始研究材料,进而构建出以野毒株抗原基因为骨架的新型疫苗更具有实际意义。

---

**第一作者**:吴凌(1966－),女,汉族,硕士研究生,硕士. 预防兽医学,E-mail:wuling8@163.com

**通讯作者**:夏成(1964－),男,教授,博士生导师; E-mail:xcwlxyf@sohu.com

# 1 材料与方法

## 1.1 试验材料

1.1.1 细胞、病毒、血清、抗体、载体和试验动物 SV40 转化的非洲绿猴肾细胞（MDCK 细胞）、犬肾传代细胞（MDCK 细胞），CDV 病毒液，重组质粒 pMD18-H，真核表达载体 pCI neo 质粒 DNA 由本室保存和提供；兔抗 CDV 高免血清，由本室制备；HRP 标记的羊抗鼠 IgG，购自北京中杉金桥生物技术有限公司；FITC 标记的兔抗鼠 IgG、羊抗兔 IgG 购自 Sigma 公司；真核表达载体 pCI 和克隆载体 pMD18-T 购自大连宝生物工程公司；受体菌 *E. coli* DH5α 由本实验室保存；5 ~6 周龄 BALB/c 小鼠由哈尔滨兽医研究所实验动物中心提供。

1.1.2 主要试剂 DMEM、培养基胎牛血清、Opti-MEM 购自 GIBCO 公司；TRIzol LS Reagent、First-Strand cDNA Synthesis Using SuperScript™ Ⅱ Reverse Transcriptase、RNaseOUT™ Reverse Transcriptase Inhibitor 、Lipofectamin™2000 购自 Invitrogen 公司；限制性内切酶、LA Taq™ DNA Polymerase、Random Primers：pd(N)$_6$、T4 DNA Ligase、dNTP Mixture、DNA Markers、青霉素、IPTG、X-gal 等均购自宝生物工程（大连）公司；小量胶回收试剂盒等购自上海华舜生物工程公司；去内毒素中量质粒提取纯化试剂盒购自 Promega 公司；细胞培养板购自 Nunc 公司；SeaPlaque GTG Agarose 购自 Lonza 公司；SDS-PAGE 及 Western Blot 试剂均购自 Sigma 公司；其他常规试剂均为国产分析纯。

## 1.2 试验方法

1.2.1 真核表达载体 pCI-H 的构建 分别用 *Kpn* I 和 *Not* I 双酶切已获得的插入外源基因的重组质粒 pMD18-H 和真核表达载体 pCI neo 质粒 DNA。经 1.2 % 琼脂糖凝胶电泳回收后，回收的 H 基因片段和 pCI neo 线性化质粒，经 T4 DNA Ligase 连接、转化 *E. coli* DH5α 感受态细胞，Amp 抗性筛选后，挑单菌落，提取质粒。拟获得的重组真核表达载体命名为 pCI-H。

1.2.2 pCI-H 的酶切及 PCR 鉴定 经 *Kpn* I 单酶切、*Kpn* I 和 *Not* I 双酶切鉴定 pCI-H。以酶切正确的质粒为模板，利用 HF/HR 引物针对 H 基因进行 PCR 鉴定。经 1 % 的琼脂糖凝胶电泳，筛选阳性克隆，送 Invitrogen 公司测序，验证其正确的 ORF。

1.2.3 pCI-H 大量制备和纯化 提取方法参照 Promega 公司 Wziard PureFection Plasmid DNA Purification System 试剂盒说明书进行。获得的质粒 DNA 经分光光度计测定 OD260/OD280，进而估计质粒的纯度。一般认为比值在 1.8 ~2.0，0.8 % 琼脂糖凝胶电泳中，超螺旋≥70 % 的质粒纯度较好。

1.2.4 pCI-H 转染 MDCK 细胞 转染前 8 h，消化 MDCK 细胞，铺 24 孔细胞培养板，500 μl/孔。待达到约 50% 细胞密度，尽弃培养液，用无血清和无抗生素的 DMEM 洗 3 遍后，每孔加入 500 μl Opi-MEM 作用2 h，然后按照 Lipofectamin™ 2000 的转染说明进行转染。

1.2.5 转录体 mRNA 的检测 转染 48 h 后，经反复冻融，收集 3 组细胞，用 TRIzol LS Reagent 提取细胞总 RNA，紫外分光光度计测定 RNA 的浓度，立即以提取的总 RNA 为模板，进行 RT-PCR 的操作，PCR 产物经 1 % 琼脂糖凝胶电泳检测目的基因是否转录。

1.2.6 间接免疫荧光(IIF)检测 pCI-H 的体外表达 将转染 pCI-H 的细胞以合适密度接种到 24 孔细胞培养板上，细胞长成单层后，弃去培养液，用 PBS 洗涤 2 次，自然干燥，加预冷的无水乙醇室温下固定 10 min 后晒干，利用 IIF 检测 pCI-H 在哺乳动物细胞中的表达情况。同时设置未转染正常细胞对照。

1.2.7 表达产物 SDS-PAGE 和 Western Blot 分析 刮取转染后 48 h 的 MDCK 细胞，连同培养液一起转入 EP 管中，250 × g 离心 10 min，小心吸去上清。用 1 ml PBS 重悬细胞，再次 250 × g 离心 10 min，吸去上清，向细胞沉淀中加入 20 μl PBS 重悬细胞，加入等体积的 2 倍 SDS 凝胶加样缓冲液，振荡混匀。样品置于沸水中煮 10 min 裂解细胞，然后进行 12 % SDS-PAGE 及 Western Blot 分析以验证 H 蛋白的表达情况。

1.2.8 重组真核表达载体 pCI-H 的动物免疫原性试验 取 10 只 5 ~6 周龄健康的 BALB/c 小鼠，随机分为

2 组，观察 3 天后，分别于后肢股四头肌肌肉注射 pCI-H 和空载体 pCI neo，免疫程序参照表 1；免疫前剪鼠尾采血，分离血清；4 免后 7 d 眼眶采血，分离血清备用。

**表 1　动物免疫程序**

Table 1　The procedures of animal immunization

| 免疫次数 | 免疫原 | 免疫量（μg／只） | 免疫周期 |
|---|---|---|---|
| 一免 | pCI-H：盐酸普鲁卡因 = 1：4 | 100 | 2 周 |
| 二免 | pCI-H：盐酸普鲁卡因 = 1：4 | 100 | 2 周 |
| 三免 | pCI-H：盐酸普鲁卡因 = 1：4 | 100 | 2 周 |
| 四免 | pCI-H：盐酸普鲁卡因 = 1：4 | 100 | 7 天 |

1.2.9　IPMA 检测抗血清与病毒的结合能力　为验证免疫血清是否能与病毒表达在感染细胞表面结合，以 ZH10 株以 200 TCID50/孔接种于 96 孔板培养的 MDCK 细胞，37 ℃感作 2 h，以 150 μl/孔加入 2 % 维持液。培养 48 h 后弃去培养液，用 PBS 洗涤 3 次；自然干燥，加入预冷的乙醇 200 μl/孔，室温固定 10 min 后晒干；弃去固定液，PBST 洗涤 3 次；加入四免后获得的抗血清及 CDV 标准阴性血清，37 ℃作用 45 min，PBST 洗涤 3 次；加入 1:500 倍稀释的 HRP-羊抗鼠 IgG，37 ℃作用 30 min，PBST 洗涤 3 次；用 AEC 作为底物进行显色，室温下放置 45 min。去掉 AEC，加入 0.05 mol/L NaAc，pH 5.0。光镜下观察，阳性细胞可被染成棕红色。非特异性染色所有细胞颜色相同，为无着色或棕黄色。

1.2.10　ELISA 检测血清中抗 H 抗体　用 MDCK 细胞大量增殖 ZH10 株病毒粒子，病毒培养上清液采用差速离心和蔗糖密度梯度离心浓缩纯化后作为诊断抗原。按常规的间接 ELISA 操作程序进行，最佳反应条件为：诊断抗原按 0.175 μg /孔稀释后，每孔 100 μl 加入酶标板，37 ℃包被 4 h，洗涤 4 次，每次 3 min，扣干，加封闭液 37 ℃作用 2.5 h，洗涤 4 次，每次 3 min，扣干；待检血清从 1:10 至 1:320 做对比稀释后加入酶标板中，37 ℃作用 1 h，洗涤，扣干；加入 1：10 000 稀释的酶标抗体，37 ℃作用 1 h，洗涤，扣干酶标板，加入显色液，37 ℃作用 15 min，加入 2 M $H_2SO_4$ 终止反应。在酶标仪上测定 $OD_{450nm}$值。按下列公式计算 P/N 值：P/N 值 = 阳性对照孔 $OD_{450nm}$均值 / 阴性对照孔 $OD_{450nm}$均值。间接 ELISA 判定标准，P/N 值≥2 者判为阳性，P/N 值<2 者判为阴性。

1.2.11　SN 检测血清中抗 H 抗体　采用固定病毒-稀释血清法，主要参照殷震《病毒学》（第二版）的介绍的方法进行：于试验的前一天傍晚铺设 MDCK 细胞的 96 孔板，次日早晨由-70 ℃冰箱取出冰冻保存的 CDV 病毒液，融化后按 $TCID_{50}$测定的效价进行稀释成 200 个 $TCID_{50}$，确保再与等量血清混合后，每个接种剂量中含有 100 个 $TCID_{50}$；然后将血清置于 56 ℃水浴灭活 30 min，以破坏补体和其他不耐热的非特异性杀病毒因子，随后做对比稀释；再取定量的病毒液和等体积不同稀释度的血清，置 1.5 mlEP 管中充分混合后于 37 ℃感作 1 h。为保证试验结果的准确，需要设置如下对照：（1）病毒对照：将病毒液配制成每个接种量含 0.2、2、20、200 $TCID_{50}$的浓度，再加上等量原倍待检血清，充分混合，此时病毒浓度分别降为 0.1、1、10、100 $TCID_{50}$；（2）血清毒性对照：即在细胞中加入原倍待检血清，确定被检血清本身对细胞没有任何毒性作用；（3）阳性和阴性血清对照：与待检血清进行平行试验，阳性血清应呈现预期中和效价，阴性血清应没有中和效价；（4）空白细胞对照：在整个试验过程中，此对照应一直保持良好的形态和特征。

待感作完毕后，迅速将病毒与血清的混合物接种 MDCK 细胞单层，病毒对照组最后接种。每孔 100 μl，每个稀释度做 4 孔，逐日观察并记录 CPE 的出现情况。被检血清孔出现 100 % CPE 判为阴性，50 % 以上细胞出现保护者为阳性，固定病毒-稀释血清法中和试验的结果是计算出能保护 50 % 细胞孔不产生细胞病变的血清稀释度，该稀释度即为该份血清中和抗体效价（滴度指 P/N≥2.1 时血清最大稀释倍数）。

# 2 结果

## 2.1 重组真核表达载体 pCI-H 的酶切及 PCR 鉴定

重组质粒 pCI-H 双酶切后，得到约 1 824 bp 和 5 428 bp 两个片段，单酶切线性化后得到大小约 7 185 bp 的 DNA 片段，与预期值大小一致（图 1）；pCI-H 的 H 片段 PCR 扩增得到约 1 824 bp 扩增片段，与 H 基因大小一致（图 2）。结果表明重组质粒 pCI-H 构建成功。

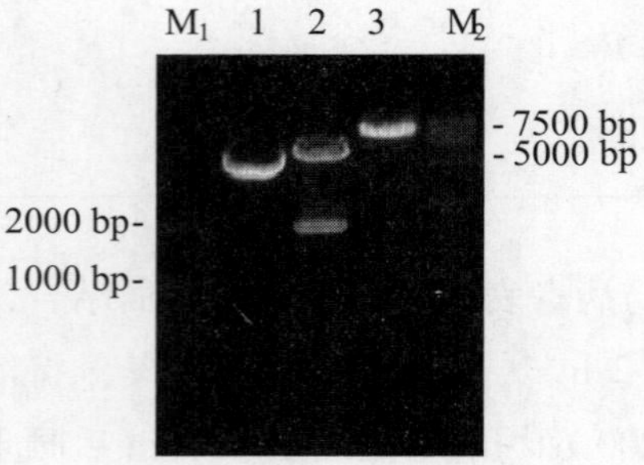


$M_1$: DNA分子量标准(DL2000); $M_2$: DNA分子量标准(DL15000); 1: pcDNA3.1-H质粒；2：经 KpnI 和 Not I 双酶切结果（约 1824bp 和 5361bp)；3：经 KpnI 单酶切结果(约 7185bp)

$M_1$: DNA molecular weight marker (DL2000); $M_2$: DNA molecular weight marker (DL15000); 1: recombinant plasmid pCI-H; 2: pCI-H digested by KpnI and Not I (1824bp and 5361bp); 3: pCI-H digested by KpnI ( 7185bp)

**图 1 重组质粒 pCI-H 的酶切鉴定结果**

Fig1 Restriction endonudeases digestion analysis of recombinant plasmid pCI-H

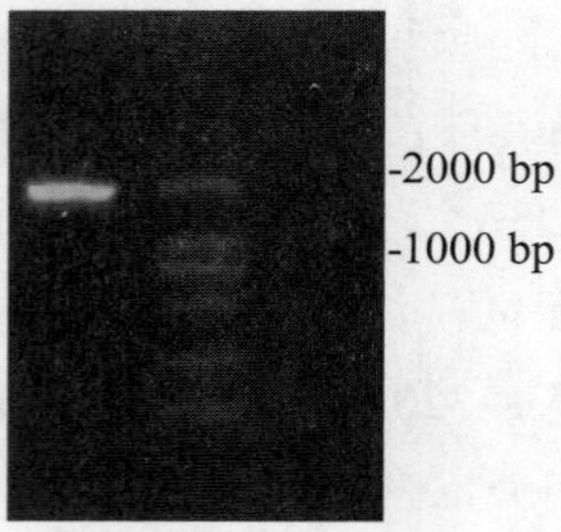


M：DNA分子量标准(DL2000)；1：pcDNA3.1-H 的 PCR扩增结果(1824bp)；2：阴性对照

M: DL 2000 Marker; 1: PCR amplified of pcDNA3.1-H(1824bp); 2: Negative control

**图 2 重组质粒 pcDNA3.1-H 的 PCR鉴定结果**

Fig 2 PCR amplified analysis of H gene of recombinant plasmid pcDNA3.1-H

## 2.2 重组真核表达载体 pCI-H DNA 的提取和纯化

经紫外分光光度计测定浓度为 3.88μg/μl，$OD_{260}/OD_{280}$ 为 1.96，无菌条件下取 0.5μl 溶于 49.5μl 灭菌水，从中取 5μl 与 5μl DNA 分子量标准（DL15000）进行 1% 琼脂糖凝胶电泳，进一步估计质粒超螺旋情况，可知琼脂糖凝胶电泳中超螺旋≥70%，说明所提质粒的浓度和纯度较好，可以满足转染和免疫动物的需要。

## 2.3 重组真核表达载体 pCI-H 在 MDCK 细胞中的表达及检测

2.3.1 表达产物 RT-PCR 检测 RT-PCR 扩增目的基因产物大小约为 1 824 bp，与理论值相符。

2.3.2 IIF 检测 pCI-H DNA 转染的 MDCK 细胞胞浆内呈现特异性亮绿色荧光，而原载体 DNA 转染后未见上述荧光信号，表明该真核表达重组质粒在细胞表面可以正确高水平表达 H 蛋白，并且表达产物具有较好的反应原性。

2.3.3 SDS-PAGE 电泳及 Western Blot 检测 结果显示：在 pCI-H 转染 MDCK 细胞样品中出现了大小约为 84 kD 的蛋白条带，且能与 CDV 多抗血清发生特异性反应，在蛋白印记中出现的特异性条带，而对照没有相应的条带（图 3）。这表明 H 蛋白获得了表达，并具有良好的反应原性。

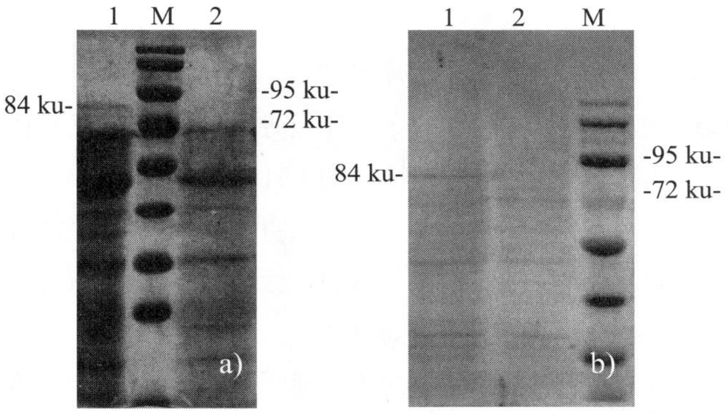


图 3 SDS－PAGE 电泳(左)和 Western blot (右)检测 H 蛋 白体外表达

Fig 3 Detection of the expression of pCI-H by SD- PAGE(Left) and Western Blot(Right)

M:预染蛋白 Marker;1:pCI-H 转染 MDCK 细胞;2:pCI neo 对照 M:prestained protein marker;

1:MDCK cells transfected by pCI－H;2:Control of cells transfected by empty vector

No steatosis was observed in the control group (A). Diffuse cytoplasmic vacuolization (arrow) indicative of steatosis was observed in mice treated with alcohol alone (B). Hepatocellular steatosis was alleviated in a dose-dependent manner by the administration of 0.25 mg/kg (C),0.5 mg/kg (D),and 1 mg/kg (E) BA for 14 days

## 2.4 重组真核表达载体 pCI-H 的动物免疫原性试验

2.4.1 IPMA 试验 如图 4 所示:病毒感染细胞能与制备抗血清结合,细胞呈棕红色,而未接种病毒的细胞对照无着色反应。表明该真核表达重组质粒在鼠体内获得正常表达,表达产物刺激机体产生的抗血清能够与犬瘟热病毒结合。

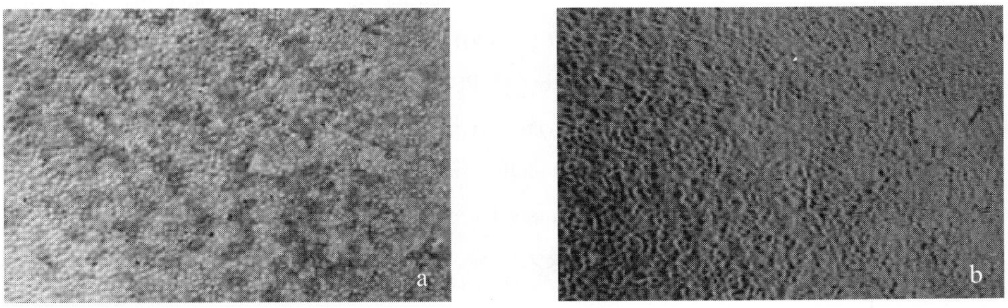


图 4 IPMA 试验结果 (x50)

Fig 4 The result of IPMA (x50)

a. 病毒感染的 MDCK-SLAM 细胞;b. 阴性对照

a. infected MDCK-SLAM cells;b. Negative control

2.4.2 ELISA 检测小鼠血清中的抗 H 的抗体 结果显示:pCI-H 免疫小鼠抗体滴度达 1 : 28 ~ 1 : 79,而原载体 pCI(+)质粒 DNA 为阴性,说明 pCI-H 能激发机体产生抗 H 的抗体。

2.4.3 免疫后诱导的血清中和抗体 试验组所有小鼠均可检测出 1 : 11 ~ 1 : 32 的病毒中和抗体,而对照组中和抗体检测均为阴性。证明 pCI-H 具有诱导产生血清中和抗体的能力。

## 3 讨 论

真核表达载体 pCI neo 是公认的安全性较高的表达载体[5-6],含有转录活性最高的人巨细胞病毒(CMV)早期启动子、Poly A 加尾信号、SV40 启动子、PUC 质粒骨架,Amp 和 Neo 抗性筛选标记[7]。Shoda 等发现当构建的质粒 DNA 中含有非甲基化脱氧核苷酸片段（CpG），又称 ISS 序列时，可增强免疫效果[8]，而 pCI neo 载体在其 Amp 抗性基因中就含有 2 个 ISS 序列。

一般认为 CDV 囊膜糖蛋白是产生中和抗体的主要靶抗原，可激发抗 CDV 体液免疫，其上具有的细胞毒性 T 淋巴细胞(CTL)表位，在小鼠和犬体内能诱发特异的 CTL 活性[9-10]，其在动物体内诱导的细胞免疫与高效价的中和抗体在抵抗 CDV 感染的过程中具有重要意义，并与病毒的清除呈正相关[11]。Nielsen 等[12]和 Cherpillod 等[13]分别以 pVIJ 和 PCI 作为表达载体，对 CDV 囊膜蛋白 H 和 F 的免疫原性进行比较，结果显示，前者的免疫原性好于后者，且不会引发攻毒后肥胖症。

本试验将 H 基因克隆到 pCI neo 表达载体上，成功构建了重组真核表达载体 pCI-H，从 mRNA、蛋白质水平水平上，将 H 基因在细胞中成功转录与转译，多重检测结果一致表明 pCI-H 表达产物具有良好的反应原性，使 pCI-H 在理论上成为可以产生良好免疫效果的核酸疫苗。

但在随后的动物试验中我们发现，免疫后的小鼠血清抗体滴度偏低，诱导的免疫反应较弱，分析原因可能是：1. 肌注的裸 DNA 只有少部分被肌细胞所吸收，绝大部分被核酸酶消化降解，未能诱导足够的免疫保护；2. 外源基因在宿主细胞内的表达水平低，或外源蛋白的天然结构在体外表达过程中发生了部分改变，虽然仍具有免疫活性，但产生的保护力降低；3. 核酸免疫多产生非分泌性蛋白，免疫后诱发的体液免疫具有滞后性。但由于未进行攻毒试验，对细胞免疫又未进行测定，所以目前无法很好解释。此外，免疫途径、免疫剂量、免疫次数、细胞因子及佐剂的类型和使用方法等多种因素都会影响疫苗的免疫效果，下步需在这些方面加以探讨，以期提高免疫诱导抗体水平。

综上所述，pCI-H DNA 重组疫苗能够诱导产生抗 CDV 的中和抗体，但抗体效价不足，有待进一步完善。

## 参考文献


[1] 张淼，刘清彪，刘宗架，等. 犬瘟热病原学研究进展[J]. 中国畜牧兽医，2009，36(11)：162－165.

[2] Martella V, Cirone F, Elia G, et al. Heterogeneity within the hemagglutinin genes of canine distemper virus (CDV) strains detected in Italy[J]. Veterinary microbiology, 2006, 116(4): 301－309.

[3] Calderon M G, Remorini P, Periolo O, et al. Detection by RT-PCR and genetic characterization of canine distemper virus from vaccinated and non-vaccinated dogs in Argentina[J]. Veterinary microbiology, 2007, 125(3): 341－349.

[4] Simon-Martinez J, Ulloa-Arvizu R, Soriano V E, et al. Identification of a genetic variant of canine distemper virus from clinical cases in two vaccinated dogs in Mexico[J]. The Veterinary Journal, 2008, 175(3): 423－426.

[5] 李凤琴. 犬瘟热病毒 N 蛋白基因核酸疫苗的构建及大熊猫 GM-CSF 基因对其免疫效果的影响研究[D]. 四川农业大学，2012.

[6] Witko S E, Kotash C S, Nowak R M, et al. An efficient helper-virus-free method for rescue of recombinant paramyxoviruses and rhadoviruses from a cell line suitable for vaccine development[J]. Journal of Virological methods, 2006, 135(1): 91－101.

[7] Liu M, Chen W, Ni Z, et al. Cross-inhibition to heterologous foot-and-mouth disease virus infection induced by RNA interference targeting the conserved regions of viral genome[J]. Virology, 2005, 336(1): 51－59.

[8] Shoda LK, Kegerreis KA, Suarez CE, et al. Immunostimulatory CpG-modified plasmid DNA enhances IL-12, TNF-alpha, and NO production by bovine macrophages[J]. Leukoc Biol, 2001, 70: 103－112.

[9] Sixt N, Cardoso A, Vallier A, et al. Canine distemper virus DNA vaccination induces humoral and cellular immunity and protects against a lethal intracerebral challenge[J]. Virol. 1998, 72: 8472－8476.

[10] Hirama K, Togashi K, Wakasa C, et al. Cytotoxic T-lymphocyte activity specific for hemagglutinin (H) protein of canine distemper virus in dogs[J]. J Vet Med Sci, 2003, 65: 109－112.

[11] Takahashi T, Tagami T, Yamazaki S, et al. Immunologic self-tolerance maintained by CD25＋ CD4＋ regulatory T cells constitutively expressing cytotoxic T lymphocyte-associated antigen 4[J]. The Journal of Experimental Medicine, 2000, 192(2): 303－310.

[12] Nielsen L, Søgaard M, Karlskov-Mortensen P, et al. Humoral and cell-mediated immune responses in DNA immunized mink

challenged with wild-type canine distemper virus[J]. Vaccine,2009,27(35):4791-4797.

[13] Cherpillod P, Tipold A, Griot-Wenk M, et al. DNA vaccine encoding nucleocapsid and surface proteins of wild type canine distemper virus protects its natural host against distemper[J]. Vaccine,2000,18(26):2927-2936.

# 牦牛口服凝胶剂的初步研究

汪小强，姜文腾，韩照清，李家奎

(华中农业大学，武汉 430070)

**摘　要**：为改进犊牦牛腹泻治疗的给药途径，以淀粉为主要基质制备口服凝胶剂。采用单因素方法考察各辅料的处方配比，以凝胶剂的稠度和涂展性为指标用正交试验法优化基质配方，确定淀粉、琼脂和明胶、吐温-80 和水的最佳配比。生产的凝胶性质稳定、无分层，常温下外观和涂展性无明显变化，在此基础上生产的口服庆大霉素和恩诺沙星凝胶制剂经动物实验安全有效，为该剂型药物在牦牛生产中的应用提供了基础。

**关键词**：牦牛，口服凝胶剂，正交试验，初步研究

## 1　目的及技术简介

不同的给药方式需要辅以一定的剂型，随着科技的进步，药物制剂的设计与发展进入了一个新阶段[1-2]。结合实际的需求，药物新剂型的研究开发对现在兽药和养殖的发展具有重要而特殊的作用。

凝胶剂是近年来兴起的一种药物新剂型，指药物与适量基质制成均一的、具有凝胶特性的半固体或稠厚液体[3]。水性凝胶剂是其中的一种，基质一般由水、甘油或丙二醇、海藻酸盐、明胶、纤维素衍生物、卡波姆、淀粉和琼脂等组成[4]。目前市场上药物的口服凝胶剂少且多数为人药，用于治疗畜禽的口服凝胶剂，未见相关报道[5-6]。

犊牦牛腹泻是一种发病率较高的常见临床疾病，严重危害畜牧业健康发展。其病因复杂，病原多样，细菌性、病毒、营养性、环境气候等均可引起；其中，致病性大肠杆菌、沙门氏菌等在发病犊牛中占据一定比例[7-8]。治疗时，为确保治疗效果，药物需要定期灌服或注射给药，但由于牦牛有一定的野性，保定给药困难，在临床实践中不宜操作。为此，本研究开展牦牛口服药物凝胶剂的研制，将药物混入可食用凝胶制剂中，使用时直接将凝胶剂挤入或涂抹于牦牛口腔，为牦牛疾病临床治疗提供方便。现报告如下。

## 2　技术形成经过简介

### 2.1　器材及试剂

2.1.1　器材　离心机、水浴锅、电炉、恒温培养箱、冰箱、显微镜、分析天平、烧杯、玻璃棒、pH 试纸等。

2.1.2 试剂　甘油、可溶性淀粉、吐温-80、琼脂、明胶、食用香精、庆大霉素及恩诺沙星粉剂。

### 2.2　口服凝胶药物制备方法

2.2.1　处方　可溶性淀粉，甘油，琼脂，明胶，吐温-80，山梨酸钾，蒸馏水适量，庆大霉素及恩诺沙星粉剂。

2.2.2　制备工艺　(1)称量：按配方取可溶性淀粉，山梨酸钾及吐温于烧杯 A，加入适量蒸馏水并搅拌均匀；另称取相应量的明胶，琼脂及甘油，加入剩余蒸馏水于烧杯(B)，并搅拌均匀。(2)加热：将 A 置于电炉，按单一方向搅拌加热，直至淀粉发生糊化并维持 3～5min；将 B 先 80℃水浴 30min，至琼脂和明胶完全溶解，形成透明均匀的液体。(3)混合制备：待 A 与 B 溶液温度相近约 70～80℃时，将 B 液缓缓倒入 A 中，并不断搅拌至完全冷却，即得凝胶剂。

### 2.3　处方优化

2.3.1　单因素试验　配方中各辅料在不同用量时将影响凝胶剂制备[9-11]，分别以淀粉、水、甘油、吐温、琼脂和明胶的不同水平进行考察，以其外观性状和稠度为判定。

结果显示，随着淀粉、琼脂和明胶的含量增加，其稠度相应的增大；同时，含水量少的凝胶剂(图 1)较含水量多(图 2)的更稠；吐温-80 的加入可改变凝胶剂的外观颜色，胶体更加细腻，同时对凝胶固化时间和稠度有一定的影响；甘油用量对凝胶剂

**基金项目**：国家肉牛牦牛产业技术体系(CARS-38)

无明显影响。

2.3.2 正交优化设计 根据单因素试验考察结果,以正交设计对口服凝胶剂配方进行优化[12];将其分为4个因素三个水平的考察(表1),采用$L_9(3^4)$正交表进行试验(表2),即淀粉含量为因素A,设11%、12.5%、14%三个水平,琼脂和明胶含量为因素B,设1.2%、1.4%、1.6%3个水平,吐温-80含量为因素C,设0.3%、0.4%、0.5%3个水平,水含量为因素D,设60%、70%、80%3个水平。以制得的凝胶的性质为评价标准,稠度分为合适(2)、较合适(1.5)、中等(1)、较差(0.5)、差(0)5个标准,涂展性分为好(1)、中等(0.5)、差(0)三个标准,常温下的稳定性分为好(1)、中等(0.5)、差(0)3个标准。

通过试验结果的均值、极差、直观分析显示,D因素(水)的影响最大,其次是B因素(琼脂+明胶),然后为A,C;即主要影响因素的次序依次为D>B>A>C。结合正交表相应的配比初步优选配方为$A_1B_1C_2D_3$(图3);并对此配方进行验证制备,效果良好。

**表1 试验因素水平表**

| 水平 | A | B | C | D |
|---|---|---|---|---|
| | 淀粉 | 琼脂+明胶 | 吐温-80 | 水 |
| 1 | 11% | 1.2% | 0.3% | 60% |
| 2 | 12.5% | 1.4% | 0.4% | 70% |
| 3 | 14% | 1.6% | 0.5% | 80% |

**表2 正交试验表**

| 试验号 | 因素 | | | | 评分 |
|---|---|---|---|---|---|
| | A | B | C | D | |
| 1 | 1 | 1 | 1 | 1 | 2.5 |
| 2 | 1 | 2 | 2 | 2 | 4 |
| 3 | 1 | 3 | 3 | 3 | 3 |
| 4 | 2 | 1 | 2 | 3 | 4 |
| 5 | 2 | 2 | 3 | 1 | 0 |
| 6 | 2 | 3 | 1 | 2 | 0 |
| 7 | 3 | 1 | 3 | 2 | 2.5 |
| 8 | 3 | 2 | 1 | 3 | 3 |
| 9 | 3 | 3 | 2 | 1 | 0 |
| $K1$ | 9.5 | 9 | 5.5 | 2.5 | |
| $K2$ | 4 | 7 | 8 | 6.5 | |
| $K3$ | 5.5 | 3 | 5.5 | 10 | |
| $\overline{K1}$ | 3.17 | 3 | 1.83 | 0.83 | |
| $\overline{K2}$ | 1.33 | 2.33 | 2.66 | 2.17 | |
| $\overline{K3}$ | 1.83 | 1 | 1.83 | 3.33 | |
| $R$ | 1.84 | 2 | 0.83 | 2.5 | |

图 1　60% 水含量凝胶剂

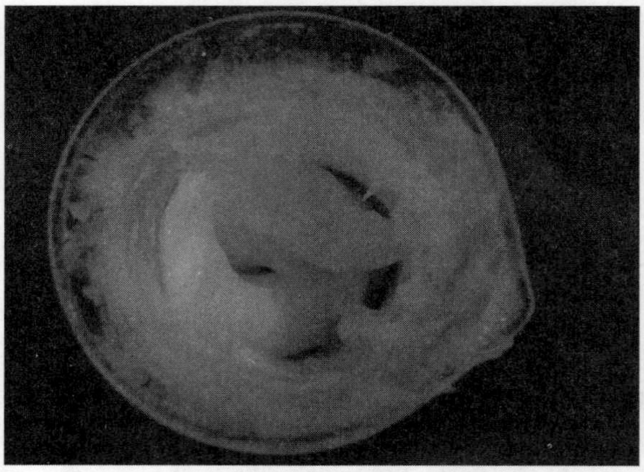

图 2　80% 水含量凝胶剂

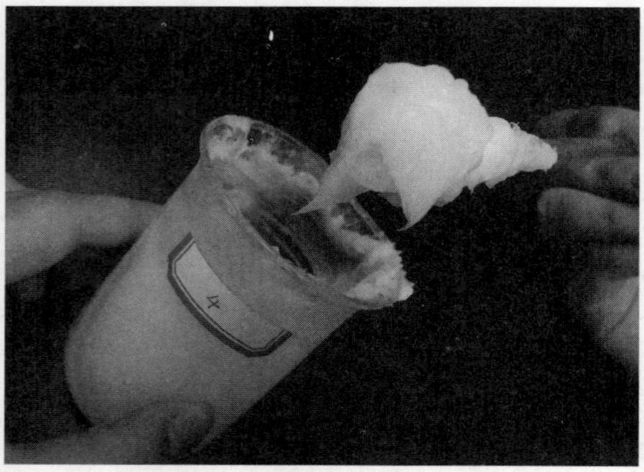

图 3　优选配方样品

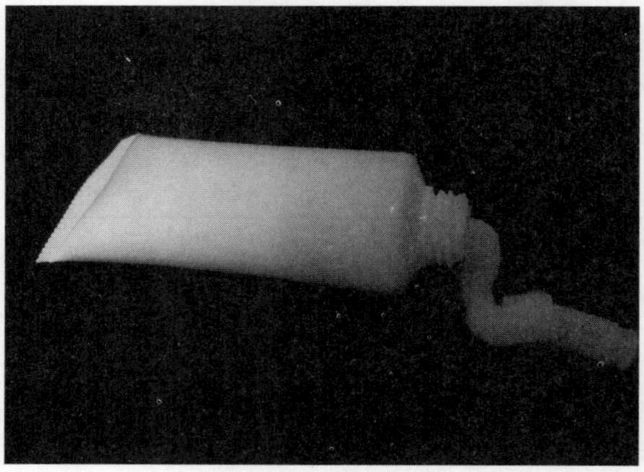

图 4　口服凝胶剂样品

### 2.4　初步质量检测

2.4.1　外观性状　观察制得的凝胶剂颜色、质地、黏稠度和延展性等。

2.4.2　离心试验　取该制备的 10g 样品置于离心管中,以 3 000r/min 转速离心 30min,观察凝胶剂有无分层。

2.4.3　pH 测定　取刚制备的凝胶样品,用精密试纸条直接测定。

2.4.4　温度试验　将样品分别于 -20℃、4℃冰箱、常温和 37℃放置;观察样品有无结晶物和回生[13],以及稠度、颜色变化(图 3)等。

### 2.5　动物试验

将制备的庆大霉素和恩诺沙星口服凝胶剂(图 4)进行小鼠和山羊的安全性试验,考察药物的安全性。

## 3　结果与应用潜力

### 3.1　口服凝胶剂产品质量检测结果

试制的口服凝胶剂基质呈乳白色、质地细腻,稠度适宜的半固体状,涂展性良好;凝胶制成品乳白色,加入药物后颜色会随药物的颜色有所变化;产品经离心后观察凝胶剂无分层现象,说明性质稳定;pH 在 5.4 ~ 6.5,达到产品要求;不同温度保存实验[14],发现在 -20℃的样品有结晶物,变硬,颜色变深,4℃样品则淀粉凝胶基质回生,变粗糙,易碎,不易搅拌,常温放置 3 个月未有明显变化,样品细腻,乳白色,而 37℃时样品较常温略稀,其他性状未有明显变化。

### 3.2　口服凝胶剂的动物试验结果

将制备的口服凝胶剂基质、庆大霉素和恩诺沙星口服凝胶剂进行小鼠和山羊的安全性试验,按照体重给予相应的药物凝胶制剂，发现对照组和实验组的生长没有明显差别，动物无不适反应，状态良好，说明药物安全可靠。

## 3.3 应用潜力

目前，兽药的剂型液体剂型、固体剂型、半固体剂型、气体剂型和特殊剂型，其中液体剂型、固体剂型最为常见，半固体剂型比较少见[15]。牦牛用药中，相较于液体剂型、固体剂型，半固体剂型的的口服凝胶剂克服了液体剂型、固体剂型通过拌料、饮水或注射给药的时药剂量不易掌握、注射给药费时费工、易造成应激的缺点，在牦牛疾病防治工作中具有广泛的应用价值。同时口服凝胶剂具有制作工艺简单，牧民易于掌握使用，原料价格低廉、简便易得的优点，大力开发牦牛口服凝胶药物是今后牦牛用药研制的一个方向。

## 参考文献


[1] 徐学剑. 设计药物新剂型的新思路与新方法[J]. 中国实用医药，2009，4(6)：211－222.

[2] 曾振灵，刘义明，黄显会. 兽药新剂型的研发现状与方向[J]. 中国家禽，2009，31(8)：5－11.

[3] 国家药典委员会. 中华人民共和国药典(二部)[Z]. 北京：化学工业出版社，2005 附录 12.

[4] 王淑华，林永强. 水性凝胶剂的制备及常用辅料[J]. 食品与药品，2006，8(2)：55－57.

[5] 李计萍，黄芳华，朱飞鹏，等. 新法规下关于中药改剂型为口服凝胶剂的一些思考[J]. 中国药事，2009，23(10)：962－964.

[6] 冯燕妮，李养学，张静. 幼泻宁口服凝胶的制备工艺研究[J]. 湖南中医杂志，2007，23(6)：66－67.

[7] 拉珍，索朗斯珠. 牦牛大肠杆菌病的研究进展[J]. 兽医研究，2013，292：8－10.

[8] 廖娟，王红宁，黄勇，等. 牦牛致病性大肠杆菌的分离鉴定及药敏试验[J]. 畜牧与兽医，2006，38(8)：49－50.

[9] 梁兴泉，郑安雄，封欣，等. 淀粉凝胶强度及其影响因素研究[J]. 化工时刊，2006，20(12)：10－12.

[10] Kcctels C. J. A. M，T. Van Vliet，P. Walstra. Gelation and Retrogardation of Concentrated Starch System：1. Gelation[J]. Food Hydrocollorids. 1996，10(3)：343－353.

[11] 初阳，陈涛，冯婉玉. 正交试验优化透明质酸钠如皋的处方工艺[J]. 中国医药导报，2011，8(6)：57－60.

[12] 张沂，任婷麟，王强. 正交设计优化复方黄芩凝胶的基质处方[J]. 海军总医院学报，2009，22(4)：193－195.

[13] Kcctels C. J. A. M，T. Van Vliet，P. Walstra. Gelation and Retrogardation of Concentrated Starch System：2. Retrogardation[J]. Food Hydrocollorids. 1996，10(3)：355－362.

[14] Lu T.，Jane J.，Keeling P. L. Temperature effect on retrogradation rate and crystalline structure of amylase[J]. Carbohydrate Polymers. 1997，33(1)：19－26

[15] 刘文利. 当前我国兽药剂型的现状与发展对策[J]. 兽医导刊，2013，6：41－43.

# 青藏高原地区牦牛弓形虫血清检测报告

李　坤[1],韩照清[1],高建峰[1],李家奎[1,2]

(1.华中农业大学动物医学院,武汉 430070;2.西藏大学农牧学院,林芝 860000)

**摘　要**:应用间接血凝试验(Indirect hemagglutination test,IHA)对在2012和2013年间采自青藏高原地区的905份和736份牦牛血清进行弓形虫血清学检测。结果为:2012年检出阳性血清196份,阳性率为21.66%;2013年检出阳性血清214份,阳性率为29.08%。统计数据表明青藏高原牦牛群中存在弓形虫的感染。该研究旨在摸清青藏高原牦牛群中弓形虫感染情况,为制定该病的防治提供理论依据。

**关键词**:青藏高原,牦牛,弓形虫,IHA

# Seroprevalence of *Toxoplasma gondii* infection in yaks on the qinghai-tibetan plateau, china

LI Kong[1], ZHAN Zhaoqing[1], GAO Jianfeng[1], LI Jiakui[1,2]

1. *College of Veterinary Medicine, Huazhong Agricultural University, Wuhan* 430070, *China*;

2. *Agriculture and Animal Hubandry College of Tibet University, Lizhi* 860000, *China*

**Abstract**: The seroprevalence of *Toxoplasma gondii* infection in yaks was surveyed on the Qinghai-Tibetan Plateau of China. A total of 905 and 706 serum samples were collected from the Qinghai-Tibetan and were tested by Indirect hemagglutination test (IHA). The results showed that 196(21.66%) and 214(29.08%) of samples were positive for *Toxoplasma gondii* in 2012 and 2013, respectively. The results of the present investigation indicate that *Toxoplasma gondii* is common in yaks on the Qinghai-Tibetan plateau. The purpose of this work was to investigate the seroprevalence of *Toxoplasma gondii* infection in yaks on the Qinghai-Tibetan plateau, and to provide the theoretical basis for disease control and prevention.

**Keywords**: Qinghai-Tibetan plateaut; China; yak; *Toxoplasma gondii*; IHA

弓形虫病(Toxoplasmosis)是一种由刚地弓形虫(*Toxoplasma gondii*)引起的感染人、畜、鸟类等多种动物的人畜共患病,该病呈世界性分布。弓形虫是一种有核细胞内机会性致病原虫[1],免疫功能正常的宿主感染后多呈无症状带虫状态,但免疫功能受损及先天感染者常导致严重的后果。弓形虫最早是在1908年于突尼斯的粪第梳趾鼠发现,1923年首次发现于人体。1955年,国内于恩庶在猫、兔体内检获虫体;1964年,谢天华在江西报道了人体内病例[2]。此外,1977年上海农科院报道了猪的群发病例[3]。弓形虫感染不仅给畜牧业带来重大的经济损失,更潜在威胁人的健康。动物感染弓形虫往往发生流产、不孕、死胎;而人同样对弓形虫易感,尤其是免疫缺陷或低下者、癌症患者、孕妇及幼儿等。孕妇感染弓形虫往往引发流产、早产、畸胎、死胎、弱智儿等一系列疾病;艾滋病患者感染常可导致严重的并发症,甚至死亡[4-6]。牛弓形虫病主要表现出共济失调和神经症状,常见呼吸困难、咳嗽、打喷嚏及发热;先天性感染弓形虫病的犊牛和羊羔的症状为

**基金项目**:国家肉牛牦牛产业技术体系(CARS-38)

发热、咳嗽、打喷嚏、鼻腔分泌物、痉挛性抽搐、惊厥、磨齿、头颈战颤及呼吸困难[6]。怀孕的牛感染弓形虫后常常出现流产、多产死胎,弱胎。目前,对青藏高原牦牛弓形虫感染情况进行系统调查研究的很少。故为摸清青藏高原牦牛群的弓形虫感染情况,笔者于 2014 年 1 月应用间接血凝法,对青藏高原地区牦牛进行了弓形虫血清学检测。检测结果表明,青藏高原牦牛存在一定程度的感染。

# 1 材料和方法

## 1.1 血清制备

2012 年和 2013 年在青藏高原牦牛群中随机采集血清 905 份和 736 份(具体见表 1 、表 2)。血清采回后放在-20℃保存,检测前在室温下自然解冻。

## 1.2 试剂

牛弓形虫间接血凝检测试剂盒及弓形虫抗原均购自中国农业科学院兰州兽医研究所。

## 1.3 试验方法

检测步骤采用弓形虫间接血凝诊断试剂盒说明书进行。

# 2 检测结果

表 1 2012 年青藏高原牦牛弓形虫血清检测结果

Table 1 Seroprevalence of *T. gondii* infection in yaks by IHA on the Qinghai-Tibetan plateau of China in 2012

| | 地区 | 检测血清数(份) | 阳性血清数(份) | 阳性率(%) |
|---|---|---|---|---|
| 西藏 | 日喀则 | 30 | 12 | 40.00 |
| | 拉萨 | 79 | 6 | 7.59 |
| | 林芝 | 91 | 7 | 7.69 |
| | 昌都 | 171 | 50 | 29.24 |
| | 那曲 | 33 | 4 | 12.12 |
| | 阿里 | 30 | 5 | 16.67 |
| | 总计 | 434 | 84 | 19.35 |
| 青海 | 杂多县 | 29 | 5 | 17.24 |
| | 囊谦县 | 41 | 8 | 19.51 |
| | 治多县 | 30 | 7 | 23.33 |
| | 称多县 | 36 | 14 | 38.89 |
| | 天峻县 | 45 | 14 | 31.11 |
| | 祁连县 | 30 | 8 | 26.67 |
| | 海晏县 | 48 | 2 | 4.17 |
| | 总计 | 259 | 58 | 22.39 |
| 四川 | 红原 | 212 | 54 | 25.47 |
| | 总计 | 905 | 196 | 21.66 |

由表 1、2 可知:2012 年,青藏高原地区牦牛弓形虫阳性率为 21.66%;西藏、青海、红原牦牛弓形虫阳性率分别为 19.35%、22.39%、21.66%;青海地区牦牛弓形虫血清阳性率略高于西藏和红原地区。2013 年青藏高原地区牦牛弓形虫血清阳性率为 29.08%;西藏、青海、红原牦牛弓形虫血清阳性率分别为 26.96%、26.38%、33.73%;红原地区牦牛弓形虫血清阳性率略高于西藏和青海地区。从中可以看出,2013 年三个地区牦牛弓形

虫阳性率较2012年牦牛弓形虫阳性率有一定的增长;青藏高原地区牦牛确实存在弓形虫感染。

表2 2013年青藏高原牦牛弓形虫血清检测结果

Table 2 Seroprevalence of *T. gondii* infection in yaks by IHA on the Qinghai-Tibetan plateau of China in 2013

| | 地区 | 检测血清数(份) | 阳性血清数(份) | 阳性率(%) |
|---|---|---|---|---|
| 西藏 | 日喀则 | 99 | 43 | 43.43 |
| | 拉萨 | 80 | 8 | 10.00 |
| | 林芝 | 51 | 11 | 21.57 |
| | 总计 | 230 | 62 | 26.96 |
| 青海 | 杂多县 | 36 | 8 | 22.22 |
| | 囊谦县 | 39 | 8 | 20.51 |
| | 治多县 | 52 | 16 | 30.77 |
| | 称多县 | 45 | 20 | 44.44 |
| | 祁连县 | 42 | 12 | 28.57 |
| | 海晏县 | 37 | 3 | 8.11 |
| | 总计 | 254 | 67 | 26.38 |
| 四川 | 红原 | 252 | 85 | 33.73 |
| | 总计 | 736 | 214 | 29.08 |

## 3 讨 论

弓形虫是一种对人畜危害较严重的机会性感染的原虫。1959年,我国首次检测出弓形虫病,20世纪70年代后期我国兽医工作者加大了对弓形虫病的重视。弓形虫病的调查研究逐渐深入,不同动物感染弓形虫不断被报道。其中,对于牛感染弓形虫的报道有:2007年,米晓云[7]等采用间接血凝的方法检测新疆地区牛血清72份,阳性23份,阳性率31.94%。2008年,陈才英[10]通过间接血凝的方法检测青海省大通县牦牛血清49份,阳性29份,阳性率为67.4%。2011年,赵全邦[11]等采用间接血凝的方法检测青海省德令哈地区牛血清97份,阳性13份,阳性率为13.4%。本次青藏高原牦牛弓形虫的血清检测结果低于陈才英报道的青海省大通县牦牛血清阳性率。本次检测表示,西藏地区黄牛弓形虫感染较严重,应引起足够重视。

弓形虫可经饮食(生牛奶、未煮熟的肉等)、污染的水源、接触感染禽畜、胎盘、输血等途径传播,并有家庭聚集现象[12]。弓形虫感染造成家畜生长缓慢,饲料利用率下降,生产性能低下,并且长期带虫,严重时死亡,造成繁殖动物生产异常,流产、死胎、非正常生产等。总之,弓形虫对畜牧业造成重大的经济损失,也威胁人的健康。防治弓形虫的关键是加强家畜的弓形虫检测,早发现、早处理。不仅可以减少畜牧业养殖业的经济损失,也是一种维护人们健康的重要公共卫生方法。

多种动物对弓形虫都有一定的免疫力,感染后往往不表现临床症状,弓形虫可在组织内形成包囊后而转为隐性感染[7]。包囊是弓形虫在中间宿主体内的最终形式,可存活数月甚至终生。因此某些地区家畜弓形虫感染率虽然比较高,但急性发病却不多。所以即使牛场等没有弓形虫的病史,仍然不可忽视对该病的防治工作。目前牛弓形虫的防治措施主要有:成牛与犊牛分开饲养,圈舍保持清洁,定期消毒;防止猫及其排泄物污染牛舍、饲料、饮水;做好流产胎儿及其排泄物的消毒处理[13]。

弓形虫病属于二类动物疫病,多种动物中都有弓形虫发病,但牛弓形虫病未受到足够的重视。青藏高原地区海拔高、气候环境复杂、草场广阔、牦牛数量众多。但由于地理环境和实验条件等因素的限制,对于青藏高原牦牛弓形虫的调查研究等很少。所以,为进一步完善青藏高原牦牛弓形虫血清学检测,我们将在今后加大样本采集数量,扩大样本采集范围,为全面掌握青藏高原牦牛弓形虫感染情况和综合防治提供参考依据。

## 参考文献

[1] 李祥瑞. 弓形虫病的流行的新趋势[J]. 动物医学进展,2010,31(8):234-236.
[2] 于思庶. 弓形虫病学[M]. 福州:福建科学技术出版刘,1992,3-6.
[3] 叶扣贯,范锋,崔洪平,等. 无锡地区商品屠宰弓形虫感染血清学调查[J]. 畜禽业,2004,5:43.
[4] 李淑梅,赵慧. 弓形虫简介[J]. 生物学教学,2012 37(11).
[5] 蔡广强,朱志森. 弓形虫病的研究进展[J]. 上海畜牧兽医通讯,2010(4).
[6] 罗才庆,袁匀,黄剑梅,等. 龙岩市部分地区牛、羊弓形虫病的血清学调查[J]. 中国动物保健,2013,15(7).
[7] 米晓云,巴音查汗,李文超. 新疆猪、牛、羊弓形虫病的血清学调查[J]. 中国兽医寄生虫病,2007,15 (2):22-24.
[8] 张居作,陈汉忠,徐君飞. 我国弓形虫的感染现状[J]. 动物医学进展. 2008,29(7):101-104.
[9] 王为升,张金生,陈伟,等. 石河子地区人畜弓形虫感染的血清学调查[J]. 动物医学进展,2011,32(2):120-122.
[10] 陈才英. 大通县家畜弓形虫病流行病学调查[J]. 青海畜牧兽医杂志,38(1).
[11] 赵全邦,胡广卫,李静,等. 青海省德令哈地区牛弓形虫病血清学调查[J]. 畜牧与兽医,2011,43(4).
[12] 王跃兵,杨向东,杨国荣,等. 弓形虫病研究概况[J]. 中国热带医学 2012,12(4).
[13] 刘文韬,路义鑫,鹿凌岩,等. 黑龙江省部分地区牛弓形虫病血清学调查[J]. 黑龙江畜牧兽医 2010(10).

# 犬传染性肝炎的病理变化研究

李思远[1],陶田谷晟[1],秦建辉[1],濮兴杰[2],左继鹏[3],肖 啸[1]

(1. 云南农业大学动物科学技术学院,昆明 650201;2. 德宏州盈江县畜牧兽医局,德宏 679300;
3. 云南农业大学动物医院,昆明 650031)

**摘 要**:犬传染性肝炎对幼犬的危害大,近些年对该病的研究也比较多但大多以病原、流行病学、治疗等方面进行研究。在病理变化方面研究的较少。随着动物机体、环境、病原的不断变化,有必要在前人的基础上对犬传染性肝炎的病理变化做进一步研究。通过解剖5只感染传染性肝炎、未及时治疗的病死犬,研究犬感染传染性肝炎后各器官的病理变化。分别采集剖检有明显病理变化的颌下淋巴结、肝脏、胆囊、脾脏、十二指肠、肾脏、脾脏和怀疑出现病变的心脏、胰腺。通过制作病理切片、镜下观察并分析其病理变化。结果显示从病理切片中看到,肝脏:肝细胞变性、坏死,坏死周围有淋巴细胞、单核细胞浸润,肝窦扩张与侯顺利报道的一致,淤血明显,肝索紊乱,肝细胞部分稍显浊肿,可见少许双核肝细胞,汇管区血管扩张,小胆管上皮淤胆。胆囊:胆囊黏膜出血、水肿,胆囊壁明显增厚;黏膜上皮细胞变性、坏死、脱落明显;各层组织中均有淋巴细胞、单核细胞浸润。十二指肠:小肠固有膜水肿并可见淋巴细胞灶性浸润;病死犬大部分黏膜上层细胞广泛性坏死,黏膜浅层自溶坏死,上皮缺失,黏膜下层、肌层无异常。肾脏:肾脏皮质淤血、出血,近曲小管部分浊肿,远曲小管水肿,肾小球体积增大,肾球囊腔变窄,间质中性粒细胞、巨噬细胞浸润,毛细血管袢充血。颌下淋巴结:淋巴结结构正常,血管增粗,血管纤维增生,大量染色质粗密、聚集成堆、常染成紫色、不均匀,在近核处一边常伸出半月状淡染的浆细胞和胞质内充满粗大、整齐、均匀、紧密排列的砖红色或鲜红色嗜酸性颗粒,嗜酸性粒细胞浸润,可见吞噬含铁血黄素组织细胞散在。都有明显的病理变化,其中肝脏、胆囊变化最为明显。脾脏:脾髓局部出血、淤血,部分红细胞溶解,散在中性粒细胞,髓质增生,组织细胞增生,部分吞噬含铁血黄素。以此得出犬传染性肝炎主要侵袭患病犬的肝胆、消化系统及免疫系统,为病理学研究和治疗的侧重点提供理论依据。

**关键词**:犬,传染性肝炎,病理切片,病理变化

犬传染性肝炎是由犬腺病毒1型引起的一种急性败血性传染病,主要发生于犬,也可见其他犬科动物。在犬主要表现为肝炎和眼睛疾患,犬不分年龄、性别、品种均可发病,但1岁以内的幼犬多发且死亡率高。犬传染性肝炎的传染源主要是病犬和康复犬。何英、叶俊华等报道康复犬尿液中排毒可达180~270d,是造成其他犬感染的重要疫源。传播途径主要是通过直接接触病犬的唾液、尿粪和接触被污染的用具而传播。最急性病例在出现呕吐、腹痛和腹泻等症状后几小时内死亡,急性型的患犬,怕冷,体温升高,精神沉郁,渴欲增加,呕吐、腹泻、粪便中带血。亚急性的患犬症状较轻微,咽炎和喉炎可致扁桃体肿大,淋巴结发炎肿大。特征性症状是角膜水肿,形成"蓝眼病"[1]。此病临床中比较常见,对幼犬的危害大,近些年对该病的研究也比较多但大多以病原、流行病学、治疗等方面进行研究。在病理变化方面研究的较少且研究水平停留在20世纪八九十年代。随着动物机体、环境、病原的不断变化,有必要在前人的基础上对犬传染性肝炎的病理变化做进一步研究。通过查阅文献大部分主要研究肝脏、肾脏、胃肠、脾脏的病理变化,文章在原有的基础上增加了对胆囊、心脏、淋巴结和胰腺的组织学研究,使得研究更系统。

## 1 材料方法

### 1.1 临床症状、诊断及病理解剖

1.1.1 临床症状和诊断 体温升高,精神沉郁,渐渐饮欲增加,呕吐白色黏液、腹泻。其中一例本地犬出现

肖 啸,邮箱:xiaoxiaokm55@163.com

“蓝眼病”的症状(图1)。用传染性肝炎(犬腺病毒1型)ICH试纸检测呈阳性。血常规检测发现白细胞、中性粒细胞升高,血小板明显下降。生化指标出现谷丙转氨酶、碱性磷酸酶严重升高,其他肝脏功能也有不同程度的恶化;肾脏功能未见明显异常。根据以上症状诊断为犬传染性肝炎。

图1 患犬出现单眼“蓝眼病”症状

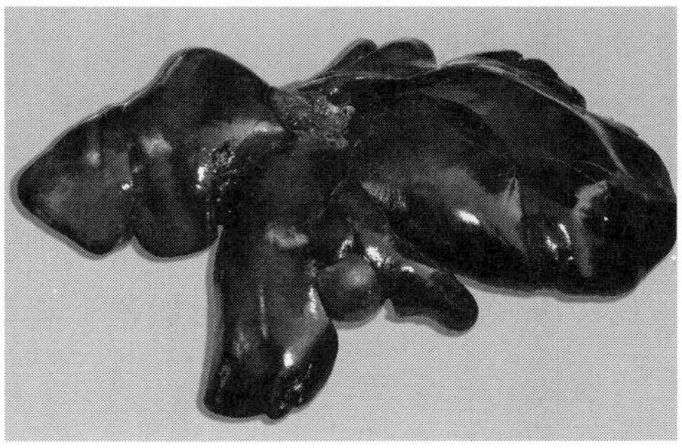

图2 患犬的肝脏

1.1.2 病理剖解 眼观病变:部分犬皮下水肿;颌下淋巴结肿大;肝脏肿大,边缘钝圆,质地脆,有大面积淤血和少量淡黄色坏死灶;胆囊充盈,胆囊壁增厚,浆膜有纤维素性渗出,黏膜出血(图2);心脏轻微肿大;胃、肠浆膜层出血,肠系膜淋巴结肿大;胰腺未见明显异常;脾脏肿大,有大面积淤血;肾脏肿大淤血明显;腹腔内积聚大量血样液体。

## 1.2 病料

搜集5例来自云南农业大学动物医院就诊患传染性肝炎(犬腺病毒1型)ICH试纸检测阳性的患犬,后未及时治疗死亡,通过尸体解剖,采集1.0$cm^3$的颌下淋巴结、肝脏、胆囊、脾脏、十二指肠、肾脏、心脏、脾脏、胰腺各一块,并且记录取样的位置,然后用生理盐水冲洗后分别放入10%甲醛溶液中冷藏保存。

表1 患病犬统计

| 编号 | 性别 | 年龄 | 品种 |
|---|---|---|---|
| 1 | 公 | 1岁 | 藏獒 |
| 2 | 公 | 11月龄 | 金毛 |
| 3 | 母 | 10月龄 | 泰迪 |
| 4 | 公 | 1.5岁 | 本地串 |
| 5 | 公 | 14月龄 | 藏獒 |

## 1.3 仪器及试剂

切片机;显微镜;恒温箱;10%的福尔马林;石腊;甲醛;固定液;蒸馏水;70%酒精;75%酒精;80%酒精;85%酒精;90%酒精;95%酒精;无水乙醇(Ⅰ);无水乙醇(Ⅱ);二甲苯(Ⅰ);二甲苯(Ⅱ);中性树胶等。

## 1.4 试验方法

常规石蜡纵横连续切片,切片厚度4~5μm,HE染色,Motic105M型显微镜下观察并拍照。

1.4.1 石蜡切片的制作

(1)水洗:取修好的组织块用流水冲洗24h;

(2)固定:病变组织块置于混合固定液中12h过夜;

(3)脱水:70%乙醇1h→75%乙醇1h→80%乙醇1h→85%乙醇1h→90%乙醇1h→95%乙醇1h→无水乙醇(Ⅰ)1h;

(4) 脱酒精:混合液(无水乙醇:二甲苯 =1∶1)30min;

(5) 透明:二甲苯(Ⅰ)30min 二甲苯(Ⅱ)30min;

(6) 浸蜡:52 ~54℃70min,54 ~56℃70min,56 ~58℃70min;

(7) 包埋:准备好包埋盒做好标记,往包埋盒中倒入液态蜡,然后用无齿镊取组织块放入石蜡中,置好方向,放置室温冷却。

(8) 切块,修块,粘块;

(9) 切片,摊片,烤片。

1.4.2 HE 染色 脱蜡、水化、染色、中性树胶封片。

1.4.3 观察切片 Motic105M 型显微镜下观察并拍照。

## 2 结 果

各器官组织切片结果。

### 2.1 肝脏

肝细胞变性、坏死,坏死周围有淋巴细胞、单核细胞浸润,肝窦扩张与侯顺利报道的一致,淤血明显,肝索紊乱,肝细胞部分稍显浊肿,可见少许双核肝细胞,汇管区血管扩张,小胆管上皮淤胆(图 3a)。

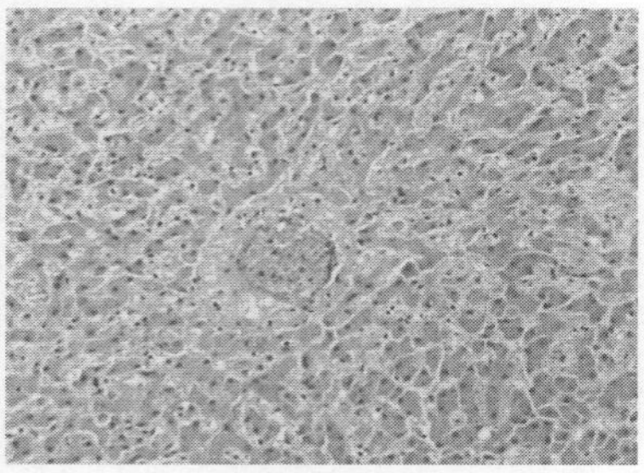

图 3a 肝脏 ×100H. E.

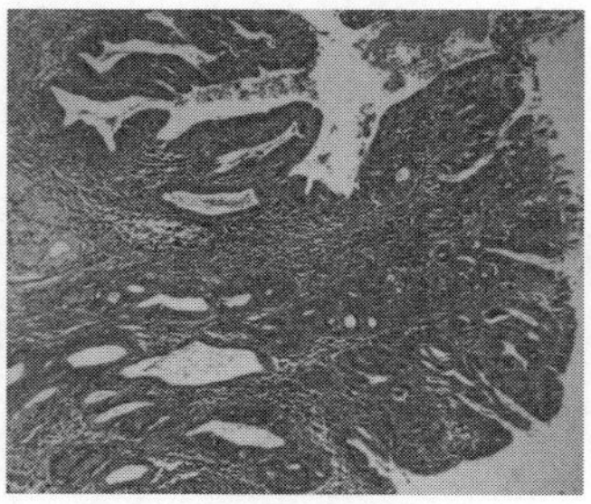

图 3b 胆囊 ×100 H. E.

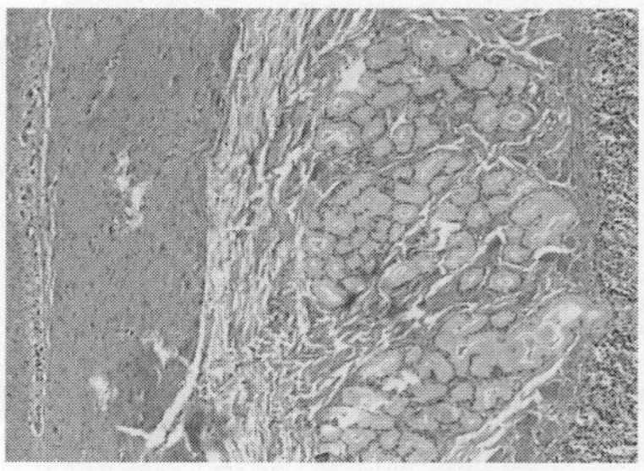

图 3c 十二指肠 ×100H. E.

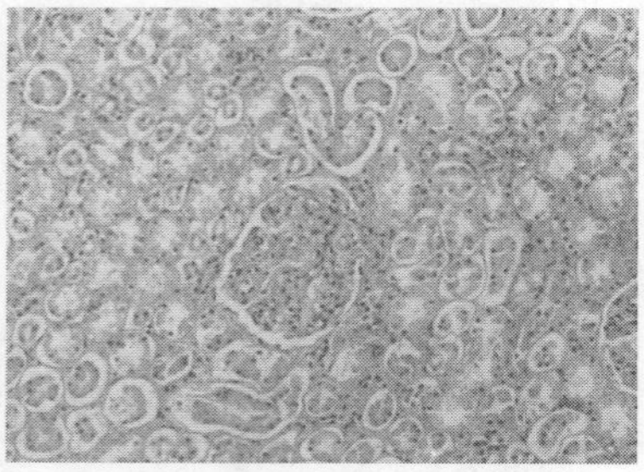

图 3d 肾脏 ×200H. E.

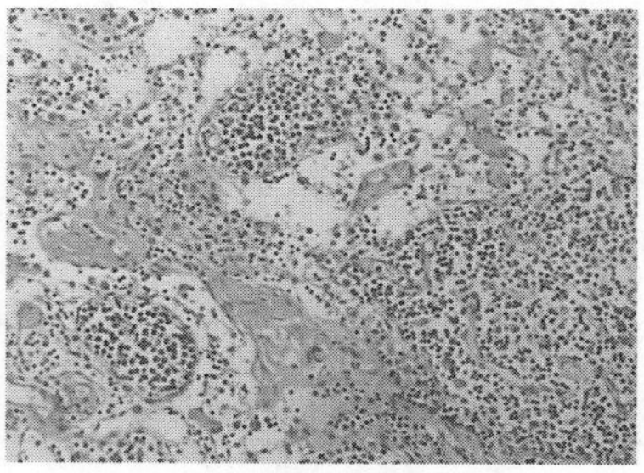

图 3e 颌下淋巴结 ×100H. E.

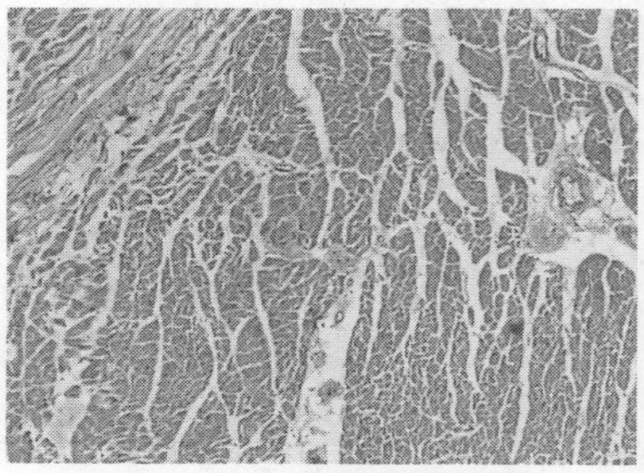

图 3f 心脏 ×100H. E.

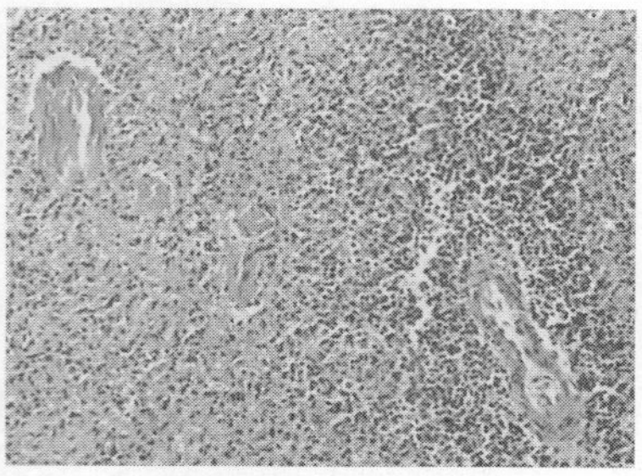

图 3g 脾脏 ×100H. E.

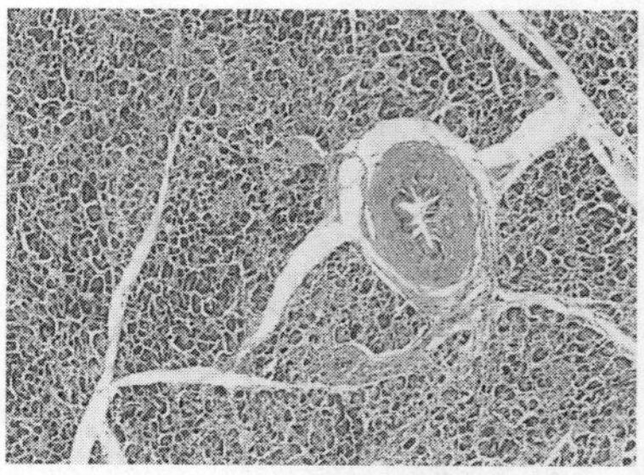

图 3h 胰腺 ×100H. E.

### 2.2 胆囊

胆囊黏膜出血、水肿，胆囊壁明显增厚；黏膜上皮细胞变性、坏死、脱落明显；各层组织中均有淋巴细胞、单核细胞浸润（图 3b）。

### 2.3 十二指肠

小肠固有膜水肿并可见淋巴细胞灶性浸润；病死犬大部分黏膜上层细胞广泛性坏死，黏膜浅层自溶坏死，与孙兆雯等报道一致，上皮缺失，黏膜下层、肌层无异常（图 3c）。

### 2.4 肾脏

肾脏皮质淤血、出血，近曲小管部分浊肿，远曲小管水肿，肾小球体积增大，肾球囊腔变窄，间质中性粒细胞、巨噬细胞浸润，毛细血管袢充血（图 3d）。

### 2.5 颌下淋巴结

淋巴结结构正常，血管增粗，血管纤维增生，大量染色质粗密、聚集成堆、常染成紫色、不均匀，在近核处一边常伸出半月状淡染的浆细胞和胞质内充满粗大、整齐、均匀、紧密排列的砖红色或鲜红色嗜酸性颗粒，嗜酸性粒细胞浸润，可见吞噬含铁血黄素组织细胞散在（图 3e）。

### 2.6 心脏

心肌充血，心内膜、心外膜未见异常（图 3f）。

### 2.7 脾脏

脾髓局部出血、淤血，部分红细胞溶解，散在中性粒细胞，髓质增生，组织细胞增生，部分吞噬含铁血黄素（图 3g）。

### 2.8 胰腺

胰腺腺泡及胰岛分布正常，未见异常（图 3h）。

## 3 讨 论

犬传染性肝炎感染主要表现全身性败血症变化，如病理剖检可见各实质器官出血、淤血，胃肠浆膜广泛性出血，腹腔内有血色液体等[1]。虽然犬传染性肝炎大多发生在 1 岁以内的幼犬，但是通过这次病例收集看到有成年犬感染。故通过这个病例应该对成年犬传染性肝炎的发病引起重视[2]。

肝细胞变性、坏死是犬传染性肝炎的基础性病变，同时受到病毒的侵袭时，机体的免疫细胞吞噬病原体从而可在组织学切片中看到大量淋巴细胞和单核细胞。由于在传染性肝炎中央静脉和小叶下静脉内膜显著肿胀、管腔狭窄或闭塞，血流受阻导致肝窦明显扩张、淤血，伴有不同程度的肝细胞混浊肿胀、变性和坏死，此与侯顺利等的报道一致。在切片中看到部分双核肝细胞是由于坏死的肝细胞通过细胞分裂的方式进行修复而出现的，从这也可看出病毒的侵袭部位。在病理剖检过程中看到胆囊充盈导致胆总管被阻塞以至于肝脏分泌的胆汁无法排入胆囊，从而在组织切片中看到汇管区血管扩张，小胆管上皮淤胆[3]。在此次观察中并未发现肝细胞核内出现嗜碱性包涵体，这与孙兆文等报道的不一致，可能是由于显微镜观察的倍数不同所致。

在病理解剖中看到胆囊充盈，胆囊壁增厚，胆汁浓稠等病理变化，与遇秀玲等的报道一致，并且符合犬传染性肝炎自然病例的发病特点。尤其是胆囊的病理改变具有诊断意义，剖检发现胆囊的这种变化，再结合临床与流行病学调查，即可做出初步诊断[4]。

小肠的病理变化一方面是由于病毒侵袭而出现淋巴细胞浸润，另一方面可能是由于胆汁过于浓稠，减少胆汁进入十二指肠，而不能有效的中和胃酸以致出现黏膜浅层自溶坏死，这可能与在剖检中看到的胃肠黏膜出血有关。

肾脏的病理变化非常明显,肾脏实质损伤严重,这与剖检中看到的肾脏肿大淤血相符,但从实验室的肾功能检测来看无明显异常不符。这种现象还需进一步研究。

颌下淋巴结的病理变化说明机体免疫系统在积极抵抗病毒的侵袭,并从切片中看到吞噬含铁血黄素组织细胞散在,这是由于淋巴结中的毛细血管出血,红细胞被巨噬细胞摄入并有其溶酶体降解的结果,这与血常规检测到血小板下降明显是对应的。

脾脏是体内重要的免疫器官和血液储存器官,所以在脾脏的病理变化中看到散在中性粒细胞和红细胞溶解。

心肌、胰腺并未见明显的病理变化,这表明 ICH 病毒不侵袭心血管系统和胰腺。

## 4 小 结

通过对患犬传染性肝炎病死犬进行病理剖检、采集病理变化明显的器官并做病理切片观察,发现犬腺病毒 1 型从病理组织学切片看到肝、胆、肾、肠上皮细胞都不同程度受损,说明该病毒对上皮细胞具有较高的亲和力[4];并对机体的消化系统和免疫器官有较大的损害。

## 参考文献

[1] 何英,叶俊华,赵玉军,等. 宠物医生手册[M]. 沈阳:辽宁科学技术出版社,2009,239 - 240.

[2] 陶永祥,梁军,胡碧丹,等. 1 例犬传染性肝炎的初步诊疗[J]. 畜牧与兽医,2006,38(10):40 - 41.

[3] 金履和,钟永明,袁瑞香,等. 162 例传染性肝炎病理及临床分析[J]. 山西医学杂志,1965,2:11 - 16.

[4] 遇秀玲,田克恭,郑振峰,等. 犬实验性感染 ICHV 的病理学研究[J]. 中国畜禽传染病,1994,3:50 - 52.

# 犬第三眼睑增生切除法与包埋法治疗效果的比较

陶田谷晟[1],李思远[1],秦建辉[1],濮兴杰[2],李晓鹏[3],肖　啸[1]*
(1. 云南农业大学动物科学学技术学院,昆明 650201;2 德宏州盈江县畜牧兽医局,德宏 679300;
3. 云南农业大学动物医院,昆明 650031)

**摘　要**:犬第三眼睑增生是由于腺体肥大等原因与瞬膜一起从眼内向外翻转而发生炎症的一种眼病,多采用切除法与包埋法进行治疗。为了研究犬第三眼睑增生切除法与包埋法的利弊,试验采用双眼患第三眼睑增生且不患其他疾病的犬只 20 只,分为切除组和包埋组进行手术并跟踪观察对比。结果表明,切除法具有手术简便、恢复时间短等优势,但是患犬治愈后会有干眼病等后遗症发生,对犬只的健康有极大的危害。而包埋法虽然手术复杂,恢复时间较长,但只要能够熟练的操作也能达到治愈效果,且对患犬不会产生其他不良影响。说明:虽然两种手法各有利弊,但是切除法的弊大于利,后遗症会带给犬只更多的不便,而包埋法的利大于弊,只要能够熟练掌握手术方法,做好术后的护理,就可以克服弊端。在今后的临床中,我们应尽量的劝前来就诊的畜主使用包埋法对第三眼睑增生进行治疗。

**关键词**:犬,第三眼睑增生,切除法,包埋法,对比

# Compare the treatment effect of excision method and embedding method for canine third eyelid hyperplasia

TAO Tian-gu-sheng[1] LI Si-yuan[1] QIN Jian-hui[1] PU Xing-jie[2] LI Xiao-peng[3] XIAO Xiao[1]*
(1. *Yunnan Agricultural University College of animal science and technology*, *Yunnan Kunming* 650201
2. *Yingjiang County in Dehong Prefecture Bureau of animal husbandry and veterinary medicine*, *Yunnan Dehong* 679300
3. *Veterinary Hospital of Yunnan Agricultural University*, *Yunnan Kunming* 650031)

**Abstract**: Canine third eyelid hyperplasia is an ophthalmopathy, because of gland hypertrophy and other reasons with the nictitating membrane together from the eye turned outwards and inflammation, many were treated by excision method and embedding method. In order to research on advantages and disadvantages of excision method and embedding method for canine third eyelid hyperplasia, test using both eyes have third eyelid hyperplasia and not suffering from other diseases 20 dogs, divided into excision group and embedding group to operation and tracking observation. Results show that the excision method has simple operation, recovery time is short and other advantages, but the dogs will have the sequelae, dry eye disease after cure, there is great harm to their health. While embedding method's operation is complex, recovery time is longer, but as long as skilled operation also can achieve the cure effect, and will not have other bad influence on the dog. Illustrate: Although the two methods have their advantages and disadvantages, but the disadvantages outweigh the advantages of excision method, sequelae will bring dogs more inconvenience, and embedding method has more advantages than disadvantages, as long as they can master the operation method, postoperative nursing well, can overcome the drawbacks. In future clinical practice, we

---

**基金项目**:云南省高端科技人才引进计划项目子项目(2009C1125-5)

**作者简介**:陶田谷晟(1991 - ),男,上海人,硕士研究生,主要从事动物病理学研究。E-mail:taotiann@163.com

* **通讯作者**:肖　啸(1966 - ),教授,男,硕士生导师,研究方向:小动物疾病学与经济性动物寄生虫学

should try to persuade the owner who comes to see a doctor using embedding method to treat the third eyelid hyperplasia.

**Keywords**: canine; third eyelid hyperplasia; excision method; embedding method; comparison

犬的第三眼睑增生是由于腺体肥大等原因与瞬膜一起从眼内向外翻转而发生炎症的一种眼病,使眼分泌物增多,呈现结膜炎症状。初期第三眼睑增生因为感染而肿大,结膜潮红,流泪等,逐渐在犬的下眼睑的内眼角处出现一个很小的肉牙状肿物,粉红色,也有的两只眼睛同时发生,粉红色的肿物在短时间内增大,通常一只眼睛发生不久,另一只眼睛也相继发生,也有的只有一只眼睛发生。肿物一般持续性存在,也有的滴眼药后消失一段时间然后再次出现,几乎没有脱出后自己彻底恢复的,严重者,脱出物呈暗红色、破溃,如果长时间不及时治疗者会继发引起角膜炎、结膜炎、角膜溃疡或穿孔,乃至失明(徐建国等,2013)。该病没有明显的季节性,多见于3~15月龄处于生长发育中的犬。经常发生于藏獒、斗牛、松狮、可卡、沙皮、比格、京巴以及小型杂交犬等(甄士伟等,2013)。

个别病例初期发病时使用氯霉素眼药水或羟苄唑眼药水与氢化可的松眼药水交替滴眼一日数次,几日后可见肿物消失(唐德琪,1998)。采用药物治疗无效或者发病时间拖延较长的病例,采取手术治疗。常规的手术治疗方法有两种,即切除法以及包埋法。

目前已经从切除法、包埋法的手术方法等方面有一定的研究。但还未比较过它们的治疗效果和利弊,本研究针对这两种方法进行试验,分别对比这两种方法的手术情况、手后情况以及各患犬情况,比较他们各自的利弊并得出结论,为本病之后的治疗提供参考。

## 1 材料与方法

### 1.1 实验材料

1.1.1 试验动物 本研究所用犬只均为双眼患第三眼睑增生且不患其他疾病的犬只20只,体重均控制在15kg以内,两组分别选取2~6岁患犬5只、6~9岁患犬5只。

1.1.2 试验分组 将20只犬根据手术方法不同分为切除组与包埋组两组。对所有20只犬只术前以及术后进行采血,进行血常规的测定并观察炎症。并对所有犬只的术前、术中、术后及恢复期做跟踪记录,观察情况。

1.1.3 手术设备及工具

1.1.3.1 切除法手术设备及工具 舒眠宁,苏醒灵3号,高频电刀,止血钳,开眼器,无菌棉布。

1.1.3.2 包埋法手术设备及工具 舒眠宁,苏醒灵3号,止血钳,无菌棉布,开眼器,手术尖刀,小镊子,3-0缝合线。

1.1.3.3 其他工具 干眼病检测:泪液检测试纸。

### 1.2 手术方法

1.2.1 切除法手术方法 首先,使用舒眠宁对患病犬只进行麻醉,卧侧保定,用无菌巾隔离患眼四周,插入电刀电极。用止血钳钳住腺体轻轻往外牵引,看到根部,使用电刀切除并在切口处轻压止血,之后慢慢放松止血钳。术后采用抗菌的滴眼液10日以上。

1.2.2 包埋法手术方法 使用舒眠宁对包埋组患病犬只进行麻醉,卧侧保定,用无菌巾隔离患眼四周,在患处两侧用手术线吊起,使腺体完全暴露,看到根部。在腺体后侧取出T形软骨并缝合开口。在突出腺体周围开口,并将腺体游离,切口的深度要穿透结膜和下面的组织,目的是将来缝合的时候有足够的张力挂住缝线。如果太浅,会造成结膜被缝线撕开。腺体本身不要切开,也不要切除腺体表面的结膜。采用3-0可吸收缝合线。闭合伤口要使用简单双层连续缝合,中间跨过腺体,这样就可以将腺体包埋到结膜的里面,最后一针要留一个开口,使泪液能够排出。打结最好打在远离角膜的睑结膜侧,避免对角膜的刺激和导致角膜溃疡的发生。对于单侧腺体脱出的病例,鉴于很多都会最终双侧脱出,建议将未脱出一侧做预防性包埋。手术后眼局部要使用三联抗生素眼膏(杆菌肽、多粘菌素 B、新霉素),全身要使用抗生素至少 7d(甄士伟等,2013)。

### 1.3 血常规指标采集

20 只患犬全部采血,采用前肢静脉采血,进行犬血常规检测,主要检测白细胞数目、淋巴细胞数目、单核细胞数目以及中性粒细胞数目。

### 1.4 干眼病检测

参照试剂盒说明书进行操作。

## 2 结果与分析

### 2.1 切除法与包埋法患犬情况比较

对本试验所用的 20 只患病犬只进行术前统计,两者在年龄、体重、性别、品种以及患病严重程度等情况都十分相近,且均未患干眼病,避免了个体情况不同对之后的手术结果产生影响,可以使研究按照预期进行,结果见表 1。

**表 1 切除法与包埋法患犬情况比较**

| 项 目 | 切除组 | 包埋组 |
|---|---|---|
| 年龄(岁) | 5.1 ±2.1 | 5.7 ±1.9 |
| 体重(kg) | 7.2 ±1.1 | 7.1 ±1.4 |
| 性别 | ♂6♀4 | ♂5♀5 |
| 品种 | 杂交小型犬 | 杂交小型犬 |
| 增生大小(cm) | 0.9 ±0.2 | 1.0 ±0.1 |
| 术前干眼病检测 | 无一例 | 无一例 |

### 2.2 切除法与包埋法手术情况比较

切除法的手术简单易操作,且创伤较小,所用的手术时间为包埋法的一半。所有的手术都使用相同的麻醉,心率和血压都在正常范围中,手术均成功。手术结束后可以观察到使用切除法的患犬结膜均有不同程度的充血;而使用包埋法的患犬结膜未见充血。结果见表 2。

**表 2 切除法与包埋法手术情况比较**

| 项 目 | 切除组 | 包埋组 |
|---|---|---|
| 手术时间(min) | 20.7 ±2.9 | 49.5 ±4.6 |
| 手术难易度 | 简单 | 复杂 |
| 术后恢复天数(d) | 7.1 ±0.9 | 13.9 ±1.0 |
| 出血情况 | 少 | 多 |
| 心率(次/分) | 98.5 ±13.2 | 119.7 ±15.4 |
| 血压(臂动脉,mmHg) | 收缩压:135.5 ±7.9;<br>舒张压:64.4 ±3.9 | 收缩压:131.5 ±5.2;<br>舒张压:58.2 ±6.9 |
| 创伤大小 | 小 | 大 |
| 麻醉方式 | 舒眠宁 | 舒眠宁 |
| 结膜充血情况 | 充血 | 无充血 |

## 2.3 切除法与包埋法手术后情况比较

跟踪两个对比组术后的情况,两组均未出现死亡。对患犬的精神情况均没有影响,饮食正常,感染情况比较轻微。切除组有三只患犬在术后患干眼症,包埋组无一例发生。患干眼病的犬只全部是6~9岁的老龄犬。结果见表3。

表3 切除法与包埋法手术后情况比较

| 项目 | 切除组 | 包埋组 |
|---|---|---|
| 手术后遗症(干眼病,例) | 3(6~9岁) | 0 |
| 手术死亡率(%) | 0 | 0 |
| 术后精神状态 | 良好 | 良好 |
| 术后饮食情况 | 良好 | 良好 |
| 术后感染情况 | 轻微 | 轻微 |

## 2.4 切除法与包埋法手术前后血常规对比

对本研究所用20只患犬进行术前以及跟踪术后3d、7d、14d血常规检测它们的炎症。术前切除组与包埋组的血常规检测均正常。术后3d两组白细胞数目均高于正常值,均有炎症发生,其中切除组淋巴细胞数目降低,单核细胞数目以及中性粒细胞数目上升,可以判断为急性炎症;包埋组淋巴细胞数目上升,单核细胞数目以及中性粒细胞数目上升,可以判断为慢性炎症。术后7d切除组血常规指标回归正常范围内,所有犬只基本恢复,包埋组的血常规指标继续上升,其淋巴细胞数目、单核细胞数目以及中性粒细胞数目均上升,炎症情况持续。术后14d切除组和包埋组的血常规指标都回归到正常范围内,此时包埋组的所有犬只也基本恢复。结果见表4。

表4 切除法与包埋法手术前后炎症(血常规)比较

| 方法 | 项目 | 手术前($\times10^9$/L) | 术后3d($\times10^9$/L) | 术后7d($\times10^9$/L) | 术后14d($\times10^9$/L) |
|---|---|---|---|---|---|
| 切除法 | 术后白细胞数目 | 15.4±0.8 | 17.8±0.9 | 15.6±0.9 | 15.4±0.8 |
| | 术后淋巴细胞数目 | 3.8±0.9 | 3.5±0.7 | 4.0±1.0 | 3.8±0.9 |
| | 术后单核细胞数目 | 0.6±0.3 | 0.9±0.3 | 0.6±0.3 | 0.6±0.3 |
| | 术后中性粒细胞数目 | 11.0±1.1 | 13.4±1.0 | 11.0±1.1 | 11.0±1.1 |
| 包埋法 | 术后白细胞数目 | 15.8±1.1 | 17.5±0.5 | 19.6±0.5 | 16.9±0.7 |
| | 术后淋巴细胞数目 | 3.17±0.9 | 3.5±0.7 | 4.2±0.8 | 3.4±0.7 |
| | 术后单核细胞数目 | 0.8±0.3 | 1.4±0.4 | 1.8±0.3 | 1.3±0.3 |
| | 术后中性粒细胞数目 | 11.9±1.2 | 12.6±1.1 | 13.6±1.0 | 12.3±0.5 |

# 3 讨论

犬的第三眼睑增生是由于腺体肥大等原因与瞬膜一起从眼内向外翻转而发生炎症的一种眼病,使眼分泌物增多,呈现结膜炎症状。采取手术治疗效果最为理想,目前已经有许多临床经验。

介绍切除法治疗的文献较多,如孙全煜的犬第三眼睑息肉的手术治疗(孙全煜,2010)、胡宇莉等的犬三种常见眼病的治疗(胡宇莉等,2003)、张谊的宠物犬第三眼睑增生症的诊治(张谊,2007)、潘应仙的犬第三眼睑腺增生的治疗(潘应仙,2012)、谭瀛的犬第三眼睑增生的手术疗法(谭瀛,2002)、潘懿的犬第三眼睑增生摘除(潘懿,2004)、李长信的犬第三眼睑增生症的手术疗法(李长信,2001)、张红申的犬第三眼睑腺摘除手术(张红申,2004)、范泉水的犬第三眼睑腺增生手术治疗的改进(范泉水,1999)等。而介绍包埋法治疗的文献较少,如薛岩等的犬第三眼睑脱出症的复位治疗(薛岩等,2007)、许建国等的包埋术治疗犬第三眼睑腺脱出(许建国等,2013)、高进东的犬第三眼睑腺脱出包埋术(高进东,2009)等。也有一些同时介绍了两种手术方法,如夏春峰的犬第三眼睑腺脱出的诊疗浅析(夏春峰,2009)、李德典的手术治疗犬第三眼睑突出(李德典,1997)等。

各文献中详细介绍了切除法以及包埋法的手术过程、方法等,但是很少有跟踪病例观察后遗症,也没有观察过术前术后的血常规情况。有的文献虽然同时介绍了两种方法,但并没有用数据详细对比两者优劣,存在薄弱环节。因此,详细研究两种手术方法并且做出正确的对比是十分必要的。

本研究中,切除法和包埋法对患病犬只的治疗各有利弊,两组的情况基本相同,具有可比性。切除法的手术比较简单易操作,且创伤较小,所需手术时间短,但术后观察到使用切除法的患犬结膜均有不同程度的充血。包埋法手术较复杂,需要更为丰富的临床经验。对比手术前后的两组血常规指标,可以得出恢复情况,术前切除组与包埋组的血常规检测均较正常。术后3d两组白细胞数目均高于正常值,均有炎症发生,切除组可判断为急性炎症,包埋组可判断为慢性炎症。术后7d切除组血常规指标回归正常范围内,包埋组的血常规指标继续上升,炎症情况持续。术后14d两组血常规指标都在正常范围内,此时包埋组的所有犬只也基本恢复。

但是,跟踪患犬治愈后的恢复情况时发现,切除组中有3只老龄患犬在手术后发生了干眼症,据国外统计,采用包埋术而发生干眼症的几率大约为14%,而采用部分腺体切割术发生干眼症的几率为48%(高进东,2009)。且国内曾有报道犬第三眼睑增生手术剪除后又快速生长的案例(栗新,1999)。

对比切除组和包埋组可以发现,切除法具有手术简便、恢复时间短等优势,但是患犬治愈后会有干眼病的情况发生,对犬只的健康有极大的危害。而包埋法虽然手术复杂,恢复时间较长,但只要能够熟练的操作也能达到治愈效果,且对患犬不会产生其他不良影响。同时,也有文章指出,为防止引起干眼病,在手术中采取不完全切除,适当保留部分腺体,也十分有效(夏春峰,2009)。

## 4 结 论

虽然两种手法各有利弊,但是切除法的弊大于利,后遗症会带给犬只更多的不便,而包埋法的利大于弊,只要能够熟练掌握手术方法,做好术后的护理,就可以克服弊端。在今后的临床中,我们应尽量的劝前来就诊的畜主使用包埋法对第三眼睑增生进行治疗。

## 参考文献


[1] 张谊. 宠物犬第三眼睑增生症的诊治[J]. 科技信息(科学教研),2007,30:350.

[2] 张红申,孙瑞刚,石金朴. 犬第三眼睑腺摘除手术[J]. 北方牧业,2004,19:27.

[3] 孙全煜. 犬第三眼睑息肉的手术治疗[J]. 农业科技与信息,2010,04:33-34.

[4] 李长信,曲静燕. 犬第三眼睑增生症的手术疗法[J]. 辽宁畜牧兽医,2001,03:34.

[5] 许建国,季珉珉,王传锋. 包埋术治疗犬第三眼睑腺脱出[J]. 畜牧与兽医,2013,11:115-116.

[6] 李德典. 手术治疗犬第三眼睑腺突出[J]. 中兽医学杂志,1997,01:30-31.

[7] 范泉水,齐桂凤,王度林,唐安建,路年生. 犬第三眼睑腺增生手术治疗的改进[J]. 云南畜牧兽医,1999,01:23.

[8] 胡宇莉,廖晓兵. 犬三种常见眼病的治疗[J]. 四川畜牧兽医,2003,05:50.

[9] 高进东. 犬第三眼睑腺脱出包埋术[A]. 中国畜牧兽医学会小动物医学分会(Association of Small Animal Medicine, CAAV)第四次学术研讨会、中国畜牧兽医学会兽医外科学分会第十六次学术研讨会论文集(1)[C],2009:5.

[10] 夏春峰. 犬第三眼睑腺脱出的诊疗浅析[J]. 中国畜牧兽医,2009,02:140-141.

[11] 栗新. 犬第三眼睑增生手术剪除后又快速生长的病例报告[J]. 黑龙江畜牧兽医,1999,03:47.

[12] 唐德琪. 犬第三眼睑增生症的防治[J]. 四川畜牧兽医,1998,01:36.

[13] 甄士伟,丁洪贵.第三眼睑腺脱垂的治疗[J].中国工作犬业,2013,04:19-20.1
[14] 谭瀛,孙凯,栾福权.犬第三眼睑增生的手术疗法[J].黑龙江畜牧兽医,2002,05:64.
[15] 潘应仙,郭云泽.犬第三眼睑腺增生的治疗[J].云南畜牧兽医,2012,02:45.
[16] 薛岩,倪俊兵,程素平.犬第三眼睑脱出症的复位治疗[J].畜牧兽医科技信息,2007,06:88.
[17] 潘懿,赵有福.犬第三眼睑增生摘除[J].广西畜牧兽医,2004,04:175.

# 犬真菌性皮肤病的诊断及各方法的疗效观察

陶田谷晟[1],秦建辉[1],李思远[1],濮兴杰[2],李晓鹏[3],肖　啸[1]
(1.云南农业大学动物科学技术学院,昆明 650201;2.德宏州盈江县畜牧兽医局,德宏 679300;
3.云南农业大学动物医院,昆明 650031)

**摘　要**:真菌性皮肤病是真菌所引起的犬的皮肤、黏膜、毛发、指甲的感染性疾病。发病率较高并具有传染性,是一种人畜共患病,容易复发或者再感染。本课题应用伍氏灯法以及刮片镜检来诊断真菌性皮肤病,并选用ZBPZ药物的方法对犬只进行真菌性的治疗,于此同时把48只犬分为A、B、C、D四组,A组12只犬内服Z(真维宁)、B(保肤康软胶囊),外用P(皮康舒)、Z(真菌克星)药物综合使用,B组12只使用内服ZB两种药,C组12只外用PZ两种,D组12只为对照组不用药正常饲喂,治疗3周,治疗前、治疗后和每个疗程结束时用伍氏灯检查。结果表明,A组ZBPZ药物综合用药治愈率90%,而B组60%,C组为45%。用ZBPZ法治疗犬真菌性皮肤病,疗效显著,犬的治愈率高。

**关键词**:ZBPZ方法,犬真菌性皮肤病,疗效观察

真菌病是由真菌引起的,外观一般有不同程度的脱皮屑、上皮碎屑、结痂和脱毛。本试验运用ZBPZ4种药物的方法对犬只进行真菌性皮肤病的治疗,观察并总结出最佳治疗方案。在临床诊断上可以利用显微镜观察毛发根部有无锯齿状断裂痕迹来辅助诊断。大多数的皮肤真菌病都会出现毛囊炎,接着毛囊口逐渐呈圆锥形扩大[1]。涉及毛囊的丘疹、脓包等都应考虑是否是真菌感染,发生毛囊炎的区域可是局限性或泛发性地出现[2]。真菌性皮肤病是一种人畜共患病,并且对妇女儿童极易感染[3]机体对真菌的免疫效应会使毛细血管的通透性增加,使未致敏的淋巴细胞致敏增生,并产生细胞毒作用。由于机体的保护和循环抗体间并无关联关系,因此细胞介导性免疫反应是机体的主要防御作用。支持这一点的主要依据是进行免疫抑制治疗的动物易发真菌病[4]。同样的增生也发生于毛囊内的细胞,毛根周围的细胞也角化,使毛发周围被角蛋质和菌丝包围,使毛囊呈喇叭状膨出,给人一种乳头状肿胀的印象[5]。

本课题主要通过伍氏灯法以及刮片镜检诊断病例,并使用ZBPZ药物治疗犬真菌性皮肤病,分组进行观察其治疗效果,这样可以证明,ZBPZ药物能提高动物机体的免疫力,增强皮肤的抵抗力,最后来评定ZBPZ药物对犬真菌性皮肤病的治疗价值和意义。

## 1　材料与方法

### 1.1　试验材料

1.1.1　试验动物　本课题所用的犬只是经过临床检查和伍氏灯的荧光检查确诊的48例患病犬只,并且体重均控制在15kg以下,将患部周围的毛剃除。

1.1.2　主要仪器　手电充电式伍氏灯,手套,电子天平,剃毛刀,玻片,显微镜等。

1.1.3　试验药物　Z真维宁片剂(复方盐酸特比萘芬片):中国绿亚生物科技有限公司,规格:每片0.8g,15片/瓶;B保肤康软胶囊:北京百林康源生物科技有限公司,规格:800mg×10粒×1板;P皮康舒洗剂:青岛康地恩药业有限公司,规格:100ml /瓶;Z真菌克星:美国摩登狗宠物用品制造有限公司,规格:110ml/瓶。

---

**基金项目**:云南省高端科技人才引进计划项目子项目(2009C1125-5)

**作者简介**:陶田谷晟(1991—),男,上海人,硕士研究生,主要从事动物病理学研究。E-mail:taootiann@163.com
*通讯作者:肖啸(1966—),教授,研究生生导师,研究方向:小动物疾病学与经济性动物寄生虫学

## 1.2 试验方法

1.2.1 试验分组 试验分为A、B、C、D四组,每一组12只犬。A组为4种药物综合治疗组,根据4种药物用量对12只犬只用药,7d为一疗程,3个疗程;B组只服用真维宁和保肤康软胶囊,根据药物用量用药,7d一疗程,3个疗程;C组只外用皮康舒洗剂和真菌克星,根据药物外用量用药,7d一疗程,3个疗程;D组为对照组,不用任何药物对患病犬只做处理,正常饲喂。

1.2.2 药物用法 真维宁:直接拌于食物中,小于7kg体重,0.5片/次,一日一次,7~15kg体重,1片/次,一日一次;保肤康软胶囊:将胶囊皮刺破混入食物中喂食,体重10kg以下每天一粒,10kg以上按每10kg1粒的比例增加喂量,最高可增至4粒;皮康舒洗剂:原药用水1:(100~200)稀释后,反复清洗患部,直接吹干即可,每日2~3次;真菌克星:避开动物眼睛,拨开毛发,逆向喷射患病部位,每天2~3次。

## 1.3 疗效判定指标

1.3.1 临床症状的表现

(1)治愈:表皮状况、精神状况、采食情况均回复正常,没有瘙痒的表现,疗程结束后,一周不复发。

(2)显效:药物疗程结束后,表皮的病变有所减轻或者暂时消失,皮损消退≥60%,当停药一周后又复发。

(3)好转:用药后患病表皮皮损消退20%~60%,没有观察到经常瘙痒。

(4)无效:用药后患病表皮皮损消退<20%或脱毛等临床症状和精神、食欲等均无变化,甚至更加严重。

有效率(%)=(痊愈病例数+显效病例数)/治疗总数病例×100%

1.3.2 荧光检查 伍氏灯检查是首选的检验方法。伍氏灯放出的紫外线通过含氧化镍的玻璃,在暗室内照射真菌致病的被毛、皮屑和甲屑,能发出荧光。把患病犬抱到暗室中用伍德式灯照射检查,将手提式伍德氏灯在暗室内开启,开灯5 min得到稳定波长以后再使用,光源与皮肤距离5~10cm左右,光线对准犬的皮肤患处,观察有无荧光出现。患处呈现各色荧光者为阳性,无荧光者为阴性,连续观察3~4个疗程。

真菌指数=照射部真菌产物数(发荧光皮肤面积)/照射面积×100%

当数值为10%时,此时,真菌指数记为1;以此类推,最高值可达7。

(1)无效:照射部真菌产物数占照射面积的比列为70%,记为7,表明达到高值。

(2)显效:照射部真菌产物占整个照射面积的比例为范围为70%~40%,记为7-4。

(3)好转:照射部真菌产物占整个照射面积的比例为范围为40%~20%,记为4-2。

(4)治愈:照射部真菌产物占整个照射面积的比例为范围为10%以下,记为1-0,达到治愈。

# 2 试验结果

## 2.1 真菌治疗前后的评价

A、B、C、D四组在3个疗程中真菌数的变化,见表1。

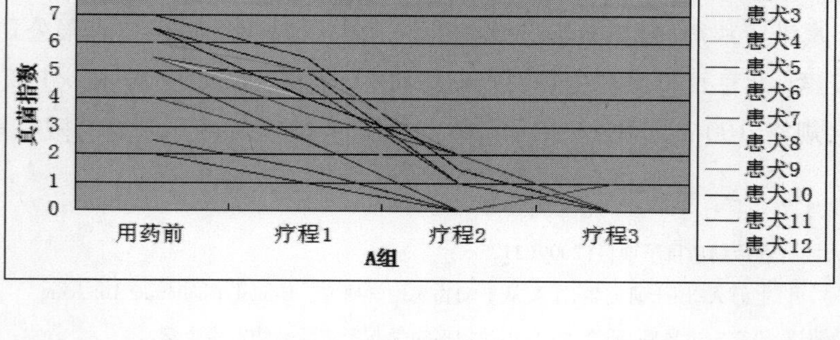

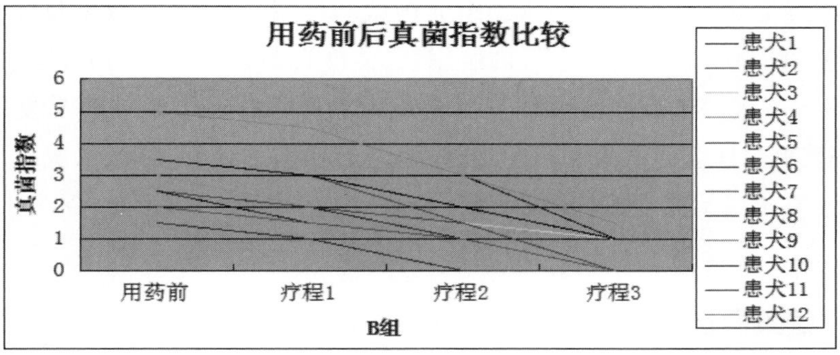


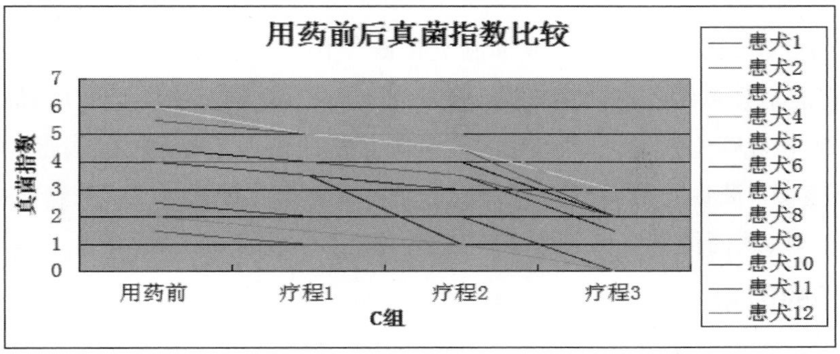


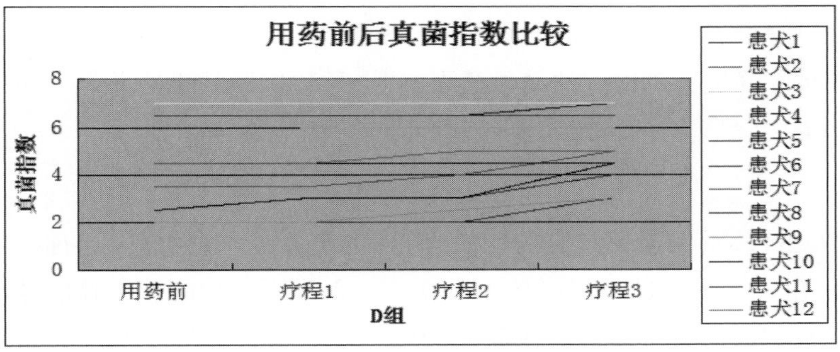


表 1　用药前后真菌指数比较

Table 1　Comparison of fungal index before and after treatment

从图表的显示我们可以看出，A 组的治愈效果比较好，而且治愈时间比较短，真菌降低的比较明显。B 组和 C 组治愈效果比较低，而且真菌下降的明显程度不大。D 组真菌没有下降，反而有些犬只真菌的数量和面积在增大。从 A、B、C、D 四组试验结果的对比下，说明 A 组治疗效果比较明显，基本上真菌的存活率为 0。

## 2.2　患犬临床疗效

A、B、C、D 四组进行平行比较，在疗程结束后的治疗结果见表 2。

表 2　患犬临床疗效的比较

Table 2　Comparison of clinical disease in dogs

| 组别 | 动物数量 | 痊愈 | 显效 | 好转 | 无效 | 有效 |
|---|---|---|---|---|---|---|
| A 组 | 12 | 11(90%) | 1(10%) | 0 | 0 | 12(100%) |
| B 组 | 12 | 8(60%) | 2(20%) | 2(20%) | 0 | 10(80%) |
| C 组 | 12 | 5(45%) | 4(35%) | 3(20%) | 0 | 9(80%) |
| D 组 | 12 | 0 | 0 | 0 | 12 | 0 |

注：治愈的标准为，背毛情况，皮肤恢复状况，红斑和红疹消失，脱皮掉毛消失，瘙痒消失，皮肤恢复健康色，背毛重新生长

通过对患犬的真菌治疗表明,A 组属于综合用药,内服和外用一起治疗,效果较好,治愈率较高,而且患犬恢复的比较快,治愈达到 90%。B 组只进行内服用药,体外真菌无法彻底消灭,导致犬的治愈率下降,容易复发,治愈达到 60%。C 组只进行外用涂擦,无法提高犬对真菌的抵抗力,也容易复发,治愈达到 45%,所以 B 组与 C 组治愈率较低。D 组为试验的对照组,没有进行任何的处理,临床状况跟治疗前一样。此结果说明 A 组的 ZBPZ 方法在临床治疗上比较可取。

## 3 分析讨论

### 3.1 ZBPZ 有治疗效果原因

药物中的真维宁片剂的主要成分是复方盐酸特比萘芬、亚麻酸、诱食剂。本品能使真菌细胞膜形成过程中麦角鲨烯环氧化反应受阻,从而达到杀灭或抑制真菌的作用[6]。保肤康软胶囊其主要成分是亚麻酸、亚油酸、维生素 E 等。它能增强皮细胞的再生能力,修复皮肤损伤,重建皮肤屏障效应,促进皮肤病的治疗。

皮康舒洗剂主要成分含有生物碱、酚类、毛茛黄素等。皮康舒具有抗菌、抗病毒、增强 T 细胞活力的作用,并能快速修复皮肤创伤、溃疡。真菌克星有效成分是酮康唑、氟虫腈、氟洛芬,它能高度选择性干扰真菌的细胞色素 P-450 的活性,从而抑制真菌细胞膜上麦角固醇的生物合成。

### 3.2 影响试验结果的因素

3.2.1 在用药的治疗期间,有犬只出现其他的疾病,必须给予较多的抗生素或其它药物,就不能确定 ZBPZ 药物是否真的作用于真菌病的治疗,因为抗生素也有治疗效果。

3.2.2 每个动物的个体差异也会对试验的结果又影响,会导致实验结果的差异。如果犬只得饲养环境较差,也会导致真菌病的复发,对犬只的护理因素也会影响实验的结果。

### 3.3 讨 论

本试验注重了内外结合,不仅能提高它的疗效,而且也增强了犬只对真菌的抵抗力,复发率低。其次 ZBPZ 治疗方法毒副作用小,对犬只减少了身体上的伤害。很多真菌病的感染伴随这犬的螨虫病,在 ZBPZ 方法的基础上加上伊维菌素,那么会促进犬皮肤病的治疗[7]。本方法与其他方法如复方蛇床子洗液治疗法[8]等效果相同或略高。

## 4 结 论

本试验运用伍氏灯法以及刮片镜检来确诊并通过真维宁、保肤康软胶囊、皮康舒、真菌克星的 ZBPZ 治疗方法,有效的治疗犬真菌性皮肤病,治愈率达到 90%。

## 参考文献

[1] 刘拂晓. 犬猫皮肤真菌病的综合防治 [J]. 特种经济动植物,2008(6):15－16.

[2] 李建鑫,李桂芹. 犬皮肤真菌病的发生与防治 [J]. 中国动物保健,2005,25(5):30.

[3] 王志. 人畜共患真菌病(上) [J]. 中国兽医杂志,1988,14(2):45－47.

[4] 祝俊杰. 犬猫疾病诊疗大全 [M]. 北京:中国农业出版社,2005.

[5] 安丽英. 兽医实验诊断 [M]. 北京:中国农业大学出版社,2000.

[6] 景涛,王世臣. 复方盐酸特比萘芬乳膏的制备与质量研究 [J]. 天津药学,2006,18(5):27.

[7] 李春波. 犬猫应用伊维菌素要适量 [J]. 黑龙江畜牧兽医,2000,5(3):46.

[8] 郭晓亮,赵秉权,王婷婷,等. 复方蛇床子洗液对猫真菌性皮肤病的治疗 [J]. 黑龙江畜牧兽医,2011,9(下):12.

# 如何取得执业兽医资格考试的好成绩

## ——对2009—2011年《兽医内科学》所涉试题的分析

徐庚全

（甘肃农业大学 动物医学院，兰州，730070）

执业兽医资格考试是由农业部组织，面向具有国家认可的兽医（畜牧兽医、中兽医、民族兽医等）专业大专以上学历人员，或不具有上述学历，而在2009年1月1日前，具有兽医师以上专业技术资格人员的全国统一命题考试。2009年在部分省市区试点，2010年开始在全国范围内实行。

本人曾参与编写《执业兽医资格考试单元强化自测与详解》（兽医临床），现结合2009、2010、2011三年的考试真题，就如何提高参试合格率谈点看法。

## 一、各年度大纲中试题覆盖面

2009年考试大纲8个单元中，试题覆盖面为100%；大纲要求的115种疾病中，涉及32种，平均覆盖率为27.83%（表1）。

表1 2009年试题（内科学）覆盖面

| 疾病单元（系统） | 大纲中疾病数量 | 采用疾病数量 | 百分率 |
|---|---|---|---|
| 消化系统 | 21 | 8 | 38.10% |
| 呼吸 | 11 | 4 | 36.36% |
| 血液循环 | 8 | 2 | 25.00% |
| 泌尿 | 8 | 2 | 25.00% |
| 神经 | 6 | 2 | 33.33% |
| 代谢性疾病 | 28 | 7 | 25.00% |
| 中毒性疾病 | 29 | 6 | 20.69% |
| 其他疾病 | 4 | 1 | 25.00% |
| 合计 | 115 | 32 | 27.83% |

2010年考试大纲16个单元中，试题覆盖率为100%；大纲要求的113种疾病中，涉及30种，平均覆盖率为26.55%（表2）。

表2 2010年试题（内科学）覆盖面

| 疾病单元（系统） | 大纲中疾病数量 | 采用疾病数量 | 百分率 |
|---|---|---|---|
| 口、咽、食管 | 5 | 1 | 20.00% |
| 反刍动物前胃、皱胃 | 8 | 2 | 25.00% |
| 其他胃肠 | 7 | 1 | 14.29% |
| 肝、腹膜、胰腺 | 3 | 2 | 66.66% |
| 呼吸 | 9 | 1 | 11.11% |
| 血液循环 | 6 | 2 | 33.33% |
| 泌尿 | 9 | 4 | 44.44% |
| 神经 | 5 | 1 | 20.00% |

(续表)

| 疾病单元(系统) | 大纲中疾病数量 | 采用疾病数量 | 百分率 |
|---|---|---|---|
| 糖、脂肪、蛋白质代谢障碍 | 8 | 4 | 50.00% |
| 矿物质代谢障碍 | 8 | 3 | 37.50% |
| 维生素、微量元素缺乏 | 10 | 3 | 30.00% |
| 中毒病概论及饲料毒物中毒 | 5 | 1 | 20.00% |
| 植物、霉菌毒素中毒 | 8 | 1 | 12.50% |
| 矿物质与微量元素中毒 | 9 | 1 | 11.11% |
| 其他中毒 | 7 | 2 | 28.58% |
| 其他内科病 | 6 | 1 | 16.67% |
| 合计 | 113 | 30 | 26.55% |

2011 年考试大纲 16 个单元中,覆盖 14 个单元,覆盖率为 87.5%;大纲要求的 113 种疾病中,涉及 32 种,覆盖率为 28.31%(表 3)。

表 3　2011 年试题(内科学)覆盖面

| 疾病单元(系统) | 大纲中疾病数量 | 采用疾病数量 | 百分率 |
|---|---|---|---|
| 口、咽、食管 | 5 | 1 | 20.00% |
| 反刍动物前胃、皱胃 | 8 | 1 | 12.50% |
| 其他胃肠 | 7 | 3 | 42.86% |
| 肝、腹膜、胰腺 | 3 | 0 | 0 |
| 呼吸 | 9 | 3 | 33.33% |
| 血液循环 | 6 | 2 | 33.33% |
| 泌尿 | 9 | 4 | 44.44% |
| 神经 | 5 | 0 | 0 |
| 糖、脂肪、蛋白质代谢障碍 | 8 | 2 | 25.00% |
| 矿物质代谢障碍 | 8 | 3 | 37.50% |
| 维生素、微量元素缺乏 | 10 | 2 | 20.00% |
| 中毒病概论及饲料毒物中毒 | 5 | 1 | 20.00% |
| 植物、霉菌毒素中毒 | 8 | 2 | 25.00% |
| 矿物质与微量元素中毒 | 9 | 1 | 11.11% |
| 其他中毒 | 7 | 3 | 42.86% |
| 其他内科病 | 6 | 2 | 33.33% |
| 合计 | 113 | 32 | 28.31% |

## 二、各年度试题所涉及疾病(细目)

2009 年(32)试题涉及疾病:口炎、瘤胃积食、皱胃变位、胃炎、肠炎、肠变位、肠便秘、肝炎;支气管炎、肺充血和肺水肿、支气管肺炎、胸膜炎;牛创伤性心包炎、贫血;尿道炎、尿石症;脑膜脑炎、日射病;营养代谢病(概论)、奶牛酮病、犬猫糖尿病、蛋鸡脂肪肝综合征、禽痛风、笼养蛋鸡疲劳综合症、铜缺乏症;中毒病(概论)、硝酸盐和亚硝酸盐中毒、黄曲霉毒素中毒、无机氟化物中毒、食盐中毒、有机磷农药中毒;肉鸡腹水综合症。

2010 年(30)试题涉及疾病:口炎;前胃弛缓、瘤胃积食;肠便秘;肝炎、胰腺炎;支气管肺炎;创伤性心包炎、心肌炎;肾炎、肾病、膀胱炎、尿石症;脊髓炎;奶牛酮病、猫脂肪肝综合征、糖尿病、禽痛风;佝偻病、血红蛋白尿病、笼养蛋鸡疲劳综合症;维生素 A 缺乏症、硒和维生素 E 缺乏症、维生素 $B_1$ 缺乏症;中毒病(概论);黄曲霉毒素中毒;食盐中毒;有机磷农药中毒、灭鼠药中毒;肉鸡腹水综合症。

2011 年(30)试题涉及疾病:口炎;瓣胃阻塞;胃炎、肠炎、肠便秘;支气管炎、支气管肺炎、大叶性肺炎;心力衰竭、贫血;肾炎、膀胱炎、膀胱麻痹、尿石症;奶牛酮病、糖尿病;笼养蛋鸡疲劳综合症、青草搐搦;维生素 A 缺乏症、硒和维生素 E 缺乏症、锰缺乏症;氢氰酸中毒;栎树叶中毒、黄曲霉毒素中毒;食盐中毒;洋葱中毒、瘤胃酸中毒、维生素 A 中毒;阿狄森氏病、库兴氏综合征。

## 三、各年度试题所涉要点及分值

2009 年(48 分):症状(17 分)、诊断(12 分)、治疗(11 分)、病因(7 分)、病名(1 分);

2010 年(40 分):诊断(18 分)、症状(8 分)、治疗(7 分)、病因(2 分)、发病机理(2 分)、病变(2 分)、预防(1 分);

2011 年(37 分):症状(18 分)、治疗(6 分)、病因(5 分)、诊断(5 分)、病名(2 分)、病变(1 分);

汇总 3 年试题(125 分):症状(43 分)、诊断(35 分)、治疗(24 分)、病因(14 分)、病名(3 分)、病变(3 分)、发病机理(2 分)、预防(1 分)。

## 四、各年度相关疾病出题频率

3 次出题的疾病(9 种):口炎、肠便秘、支气管肺炎、尿石症、奶牛酮病、糖尿病、笼养蛋鸡疲劳综合症、黄曲霉毒素中毒、食盐中毒;

2 次出题的疾病(16 种):瘤胃积食、肠炎、肝炎、支气管炎、牛创伤性心包炎、贫血、肾炎、膀胱炎、脊髓炎、脂肪肝综合症、禽痛风、维生素 A 缺乏症、硒和维生素 E 缺乏症、中毒病(概论)、有机磷农药中毒、肉鸡腹水综合症;

1 次出题的疾病(33 种):前胃弛缓、瓣胃阻塞、皱胃变位;胃炎、胃肠炎、肠变位;胰腺炎;肺充血和肺水肿、大叶性肺炎、胸膜炎;心力衰竭、心肌炎;肾病、尿道炎、膀胱麻痹;日射病;营养代谢病(总论);佝偻病、牛血红蛋白尿病、青草搐搦;维生素 $B_1$ 缺乏症、铜缺乏症、锰缺乏症;氢氰酸中毒、亚硝酸盐中毒;栎树叶中毒;无机氟中毒;灭鼠药中毒、洋葱中毒、瘤胃酸中毒、维生素 A 中毒;阿狄森氏病、库兴氏综合征。

## 五、各年度试题相关疾病所占分值

5 分:瘤胃积食、肠便秘、肝炎、支气管肺炎、奶牛酮病、笼养蛋鸡疲劳综合症;

4 分:贫血、尿石症、糖尿病;

3 分:口炎、皱胃变位、肾炎、脂肪肝综合征、禽痛风、硒和维生素 E 缺乏症、黄曲霉毒素中毒、食盐中毒、无机氟中毒、有机磷农药中毒;

2 分:瓣胃阻塞、胃炎、肠炎、支气管炎、肺充血和肺水肿、牛创伤性心包炎、膀胱炎、膀胱麻痹、牛产后血红蛋白尿病、维生素 A 缺乏症、中毒病(概论)、栎树叶中毒、肉鸡腹水综合征;

1 分:前胃弛缓、肠变位、胰腺炎、大叶性肺炎、胸膜炎、心力衰竭、心肌炎、肾病、尿道炎、脑膜脑炎、脊髓炎、日射病、营养代谢病(概论)、佝偻病、青草搐搦、维生素 $B_1$ 缺乏症、铜缺乏症、锰缺乏症、亚硝酸盐中毒、氢氰酸中毒、灭鼠药中毒、洋葱中毒、瘤胃酸中毒、维生素 A 中毒、阿狄森氏病、库兴氏综合征。

## 六、出题次数与分值的相关性分析

涉题 3 次、分值 5 分的疾病:肠便秘、支气管肺炎、奶牛酮病、笼养蛋鸡疲劳综合症;涉题 3 次、分值 4 分的疾病:尿石症、糖尿病;涉题 3 次、分值 3 分的疾病:口炎、黄曲霉毒素中毒、食盐中毒。

涉题2次、分值5分的疾病:瘤胃积食、肝炎;涉题2次、分值4分的疾病:贫血;涉题2次、分值3分的疾病:肾炎、脂肪肝综合征、禽痛风、硒和维生素E缺乏症、有机磷农药中毒;涉题2次、分值2分的疾病:肠炎、支气管炎、创伤性心包炎、膀胱炎、维生素A缺乏症、中毒病(概论)、肉鸡腹水综合征;

涉题1次、分值3分的疾病:皱胃变位、无机氟中毒;涉题1次、分值2分的疾病:瓣胃阻塞、胃炎、肺充血和肺水肿、膀胱麻痹、牛产后血红蛋白尿病、栎树叶中毒;涉题1次、分值1分的疾病:前胃弛缓、肠变位、胃肠炎、胰腺炎、大叶性肺炎、胸膜炎、心力衰竭、心肌炎、肾病、尿道炎、脑膜脑炎、脊髓炎、日射病、营养代谢病(概论)、佝偻病、青草搐搦、维生素B1缺乏症、铜缺乏症、锰缺乏症、亚硝酸盐中毒、氢氰酸中毒、灭鼠药中毒、洋葱中毒、瘤胃酸中毒、维生素A中毒、阿狄森氏病、库兴氏综合征。

## 七、组题原则(各类比例)

### (一)考试科目及所占比例

1. 临床科目(100分),其中:兽医内科学(20分)。

2. 综合应用(100分):其中:兽医内科学(20分)。

### (二)各题型及各动物所占比例的总数

1. 题型比例总数(400分):$A_1=230$;$B_1=60$;$A_2=55$;$A_3/A_4=55$。

2. 动物总比例(100):猪=20;牛羊=25;鸡=15;犬猫=30;马=7;其他=3。

### (三)各科目题型及所占比例

1. 临床科目(100分):各题型比例:$A_1=60$,$B_1=13$,$A_2=12$,$A_3/A_4=15$。

兽医内科学(20分):$A_1=11$,$B_1=2$,$A_2=1$,$A_3/A_4=6$。

2. 综合应用(100分):各题型比例:$A_1=28$,$B_1=9$,$A_2=28$,$A_3/A_4=35$;其中,猪=20,牛羊=25,鸡=15,犬猫=30,马=7,其他3。兽医内科学(20分):$A_1=3$,$B_1=2$,$A_2=7$,$A_3/A_4=8$;猪=4,牛=3,鸡=3,犬猫=10。

3年所涉125分中,犬猫32分,牛羊29分,鸡鸭15分,猪14分,马8分,涉及多种动物的27分,符合组题要求。

## 八、题例解析

### (一)A1型题

1. 可用于治疗奶牛酮病的激素是(　　)(2011)

A. 泌乳素　　B. 甲状腺素　　C. 甲状旁腺素

D. 促甲状腺素　　E. 促肾上腺皮质激素

[答案]E

[考点]奶牛酮病的治疗

[解题分析]泌乳素促进乳腺发育与泌乳;甲状腺素促进一般组织代谢,提高神经兴奋性和身体发育;甲状旁腺素主要作用在骨骼、肾脏,增加血液中的钙离子浓度;促甲状腺素调节甲状腺的内分泌功能;在治疗奶牛酮病抗酮疗法中用促肾上腺皮质激素,故选E。

### (二)A2型题

2. 某奶牛场部分奶牛产犊1周后,只采食少量粗饲料,病初粪干,后腹泻,迅速消瘦,乳汁呈浅黄色,易起泡沫;奶、尿液和呼出气有烂苹果味。病牛血液生化检测可能出现(　　)(2009)

A. 血糖含量升高　　B. 血酮含量升高　　C. 血酮含量降低

D. 血清尿酸含量升高　　E. 血清非蛋白氮含量升高

[答案]B

［考点］奶牛酮病的诊断。

［解题分析］血糖升高多示糖尿病；血酮降低无示病意义；血清尿酸升高多见于痛风症、肾小球肾炎等；血清非蛋白氮升高多示尿毒症；依发病时机与典型症状，可认为该病为奶牛酮病，血检结果应是血酮升高、血糖下降，故选B。

3. 奶牛，5岁，产犊后第3周发病，仅采食少量粗饲料，先便秘后腹泻，迅速消瘦，乳汁、尿液和呼出的气体呈烂苹果味，需要进一步检查的项目是（　　）（2010）

A. 尿蛋白　　B. 血清钙　　C. 血清酮体

D. 尿胆素原　　E. 血清无机磷

［答案］C

［考点］奶牛酮病的症状。

［解题分析］肾炎、肾病以及一些感染和非感染性疾病等可出现蛋白尿；血清钙、血清磷检测多用于甲状旁腺功能、维生素D缺乏、肾脏疾病等诊断；尿胆素原在酸性尿时减少，碱性尿时增加；依发病奶牛年龄、时机及典型症状可判定为奶牛酮病，故选C。

4. 5岁奶牛，产犊10天后表现食欲减退，迅速消瘦，精神沉郁，凝视。轻度腹痛。有时出现转圈运动，血液指标变化最可能是（　　）（2011）

A. 血钠升高　　B. 血钠降低　　C. 血糖升高

D. 血糖降低　　E. 血钾升高

［答案］D

［考点］奶牛酮病的诊断

［解题分析］血钠升高见于组织水肿，肺水肿及心力衰竭；血钠降低见于肾病、心力衰竭等；血糖升高可见于糖尿病；血钾升高多见于肾衰竭；依病牛年龄、发病时期和症状可怀疑为奶牛酮病，该病表现为血糖降低，金属离子中血钙水平稍降低，故选D。

**（三）B1型题**

A. 酮体　　B. 糖原　　C. 蛋白质

D. 胆固醇　　E. 葡萄糖

5. 6岁奶牛，分娩5天后精神沉郁，凝视，食欲减退，迅速消瘦。呼出气体有烂苹果味。诊断该病，需首先检查的血液生化指标是（　　）（2011）

［答案］A

［考点］奶牛酮病的症状

［解题分析］血液中胆固醇高多示高脂血症、肾病综合征、糖尿病等；依病牛年龄、发病时期和症状可怀疑为奶牛酮病，诊断时常快速简易定性法检测血、尿、乳中的酮体，故选A。

## 九、结　论

从3年真题看，动物的分布大致符合动物总比例的要求，分值以犬猫最高，牛羊次之，这与《兽医内科学》发展趋势中“重视群发性内科病的研究；提高小动物内科病的诊疗技术”完全吻合。随着畜牧业生产规模的不断扩大及专业分工的细化，特别是经济动物饲养向工厂化形式的转变，像奶牛酮病、笼养蛋鸡疲劳综合症等群发性营养代谢病的发生会逐步增多。另外，随着社会的变化，宠物也逐步变为伴侣动物，犬猫的尿石症、糖尿病等也愈来愈多。所以本科、专科阶段《兽医内科学》的课堂教学就势必要顺应这个变化的趋势。对于大多兽医专业本、专科毕业生而言，择业的最广泛途径是农业技术推广单位或服务养殖企业，这些职业都要求具备执业兽医资格，持证上岗。上述所列疾病，不仅是兽医职业资格考试的重点，同样也是本科阶段《兽医内科学》教学的重点。

3 年试题中,涉及症状 43 分、诊断 35 分、治疗 24 分、病因 14 分、病名 3 分等。在有限的时间内要全面、系统、扎实的掌握这些内容,对好多同学是有难度的,因此,需要掌握的重点就应该是各种疾病的病名、症状(特别是典型症状或示病症状)、诊断(实验诊断的项目及部分项目的重要指标)、治疗(原则、投药方法、重要药品)和病因。

因为在组题时要严格遵守组题原则所要求的动物比例、题型比例,往往在组一套题时,需要准备两套题的素材,所以涉及的内容非常广泛、零碎,有时不得不出一些较偏、生僻部分的内容。要想在考试中获得高分,在复习时必须做到全面系统。

## 参考文献

[1] 全国执业兽医资格考试委员会. 2009 年全国执业兽医资格考试大纲[M]. 北京:中国农业出版社,2009. 5:56 – 68.

[2] 全国执业兽医资格考试委员会. 2010 年全国执业兽医资格考试大纲[M]. 北京:中国农业出版社,2010. 5:61 – 75.

[3] 全国执业兽医资格考试委员会. 2011 年全国执业兽医资格考试大纲[M]. 北京:中国农业出版社,2011. 5:61 – 75.

[4] 全国执业兽医资格考试委员会. 2012 年全国执业兽医资格考试大纲[M]. 北京:中国农业出版社,2011. 5:61 – 75.

[5] 中国兽医协会. 2009 年执业兽医资格考试应试指南 [M]. 北京:中国农业出版社,2012. 5:5 – 6.

[6] 中国兽医协会. 2010 年执业兽医资格考试应试指南(下) [M]. 北京:中国农业出版社,2010. 5:5 – 6.

[7] 中国兽医协会. 2011 年执业兽医资格考试应试指南 [M]. 北京:中国农业出版社,2012. 5:5 – 6.

[8] 中国兽医协会. 2012 年执业兽医资格考试应试指南 [M]. 北京:中国农业出版社,2012. 5:5 – 6.

[9] 徐庚全,岳海宁. 2012 年执业兽医资格考试单元强化自测与详解—临床兽医[M]. 北京:中国农业出版社,2012. 5:98 – 188.

[10] 徐庚全,岳海宁. 2013 年执业兽医资格考试单元强化自测与详解—临床兽医[M]. 北京:中国农业出版社,2012. 5:98 – 188.

[11] 徐庚全,岳海宁. 2014 年执业兽医资格考试单元强化自测与详解—临床兽医[M]. 北京:中国农业出版社,2012. 5:98 – 188.

# 生育三烯酚对高脂模型小鼠胆固醇代谢影响

文　琼,Froilan Bernard Matias,何莎莎[1],李荣芳[1],黄海滨[1],文利新[1,2]
(1.湖南农业大学动物医学院,长沙 410128;2.湖南烟村生态农牧科技股份有限公司,长沙 410128)

**摘　要**:探讨生育三烯酚对小鼠高脂模型胆固醇代谢的影响。选取体质健康的3周龄昆明系雄性小鼠60只,随机分6组。其中,正常组包括生育三烯酚组和空白对照组,分别给予0、50、100、200mg/kg生育三烯酚和等剂量的生理盐水,饲喂基础日粮;高脂组分为4组,通过灌胃分别给予200mg/kg的生育三烯酚,饲喂高脂饲料60d。每5d称一次体重。(1)高脂对照组中,饲喂生育三烯酚组小鼠体重显著升高($P<0.05$)。(2)血清生化指标结果显示,生育三烯酚对ALT和AST活性有一定的抑制作用;正常组小鼠添加生育三烯酚后,TC、TG、LDL和FFA活性下降,HDL活性上升,组间差异不显著($P>0.05$);高脂组内,小鼠添加生育三烯酚后TC活性升高,TG、LDL和FFA活性下降,当生育三烯酚剂量为200mg/kg时小鼠HDL的活性反而降低,但组间差异都不显著($P>0.05$)。(3)高脂组内肝脏中TC活性随生育三烯酚剂量升高而降低,组间差异不显著($P>0.05$);TG活性随生育三烯酚剂量升高而升高,低、中剂量差异极显著($P<0.01$)结论:生育三烯酚可以改善小鼠肝组织损害及肝功能损伤,促进肝细胞胆固醇代谢能力。

**关键词**:生育三烯酚,胆固醇代谢,高脂血症模型

# Effects of tocotrienols on cholesterol metabolism in hyperlipidemia-induced mice

WEN Qiong[1],Froilan Bernard Matias[1,3],HE Shasha[1],LI Rongfang[1],HUANG Haibin[1],WEN Lixin[1,2]*
1. *College of Veterinary Medicine,Hunan Agricultural University,Changsha City* 410128;
2. *Hunan Sevoc Ecological Agriculture and Husbandry Technology Co. ,LTD,Changsha City* 410128;
3. *College of Veterinary Science and Medicine,Central Luzon State University,Science City of Muñoz* 3120
*Nueva Ecijia,Philippines*

**Abstract**: This research studied the effects of tocotrienols on cholesterol metabolism in hyperlipidemia-induced mice. Three-week-old Kunming male mice were randomly divided into six groups. Two groups were fed with basal diet and given with 200mg/(kg · day) tocotrienol and same volume of saline,respectively,while 4 groups were fed with high-fat diet and were given with 0,50,100 and 200 mg/(kg · day) tocotrienol,respectively. The experiment was conducted for 60 days and the mice were weighted every 5 days. (1)High-fat diet group given with tocotrienol showed significantly higher body weight ($P<0.05$). (2) Serum biochemical parameters showed that tocotrienol can inhibit the activity of ALT and AST; basal diet fed groups given with tocotrienol showed decreased levels of TC, TG,LDL and FFA,while increased HDL level,however,no significant difference between groups ($P>0.05$); high-fat diet fed groups given with tocotrienol showed increased level of TC, TG, and LDL, while FFA was decreased. Tocotrienols dose of 200mg/kg showed reduced HDL activity however no significant differences ($P>$

---

**项目基金**:长沙绿叶生物有限公司,烟村农牧生态公司资助,项目编号 08LY01 13038

**作者简介**:文　琼,女,湖南衡阳人,在读硕士研究生,主要从事动物营养与保健相关方面的研究,E-mail:516407301@qq.com
何莎莎,女,湖南邵阳,在读硕士研究生,主要从事动物营养与保健相关方面的研究,E-mail:357979775.qq.com
* **通讯作者**:文利新,E-mail:sfwlx8015@sina.com

0.05) were found between groups. (3) High-fat diet fed groups showed reduction of TC level as the dose of tocotrienol increases, no significant difference ($P > 0.05$) was found between groups; on the other hand, TG activity level increased when the dose of tocotrienol was increased, low and medium tocotrienol dose groups showed highly significant difference ($P < 0.01$). Tocotrienols may reduce liver tissue damage, prevent liver function impairment and improve the ability of liver cells to metabolize cholesterol.

**Keywords**: Tocotrienols, cholesterol metabolism, hyperlipidemia model

胆固醇又名胆甾醇是以环戊烷多氢菲为骨架的衍生物,这类分子物质不仅广泛存在于动物机体的组织和细胞中,也是维生素 $D_3$、前列腺素的前体物,还是血液中脂类转移系统的必需因子[1-3]。但若体内胆固醇堆积过多,使血液中胆固醇浓度过高,那会给人们带来种种健康问题,高胆固醇血症是冠心病和动脉粥样硬化重要致病因素[4]。所以,人类就面临一个矛盾问题:胆固醇既是动物机体生长与代谢所必需的,同时又要防止胆固醇过多或过低,过量和不足都会对机体产生一定不良影响[5](陈丽筠,1988 年)。农业在中国占有举足轻重的地位,谷类及油料资源极为丰富,而生育三烯酚主要存在于谷物或植物油中,资源丰富,功能繁多。VE 是生育酚和生育三烯酚的总称,2001 年,Yamamoto 等在大西洋鲑的鱼卵中分离出一种新的 VE[6],这种新 VE 被命名为生育三烯酚。生育三烯酚主要存在于米糠油、棕榈油和大麦油中。生育三烯酚在结构上不同于生育酚,其碳链末端多含有一个双键。研究发现,当处于 0℃时,生育三烯酚的抗氧化能力比生育酚高,这可能与生育三烯酚含有双键有关。生育酚和生育三烯酚每种都有 α、β、γ、δ4 种衍生物[7]。两者的基本结构相同都是色原醇环和一个含 16 碳的侧链,不同点在于生育三烯酚的侧链多含 3 个不饱和双键,而生育酚的侧链则没有此结构。近年来生育三烯酚已成为国内外研究的热点,大量研究表明生育三烯酚在抗氧化性能、降低胆固醇、抑制癌症等方面具有一定的生理作用,并在某些方面还要优于生育酚。生育三烯酚是一种安全的食品添加剂,在保健食品应用上具有重要价值。因此,本研究通过探讨生育三烯酚对小鼠体重、胆固醇、血液生化指标,揭示生育三烯酚对高脂模型小鼠胆固醇代谢,为了解和开发生育三烯酚提供理论和实用价值,为更好开发生育三烯酚这一资源,促进动物和人类健康提供理论依据。

## 1 材料与试剂

### 1.1 主要试剂与仪器

生育三烯酚:由湖南烟村生态农牧科技股份有限公司提供,纯度为 50%。生育三烯酚溶解到米糠油中,终浓度为 200mg/ml。

深圳迈瑞生物医疗电子股份有限公司的谷丙转氨酶检测试剂盒、谷草转氨酶检测试剂盒、胆固醇检测试剂盒、甘油三酯检测试剂盒、低密度脂蛋白检测试剂盒和高密度脂蛋白检测试剂盒,普利莱基因技术有限公司的甘油三脂(组织细胞)酶法测试盒、总胆固醇(组织细胞)酶法测试盒以及游离脂肪酸超敏测试盒,BS-190 全自动生化分析仪,酶标仪(infinite M200PRO)、WH-2 微型漩涡混合仪,低温高速离心机,匀浆器等。

### 1.2 饲料以及配方

基础饲料:购于湖南斯莱克景达实验动物有限公司。

高脂饲料:由 1% 胆固醇、10% 猪油(自制)、0.2% 胆酸钠、10% 蛋黄粉、78.8% 基础饲料组成,由湖南斯莱克景达实验动物有限公司制成圆条状颗粒;-4℃保存。

## 2 方法与步骤

### 2.1 动物饲养及处理

试验动物为昆明系小鼠(3 ~4 周龄),雄性,SPF 级,体重 20 ~ 30 g,购于湖南省斯莱克景达实验动物有限公司。试验小鼠常规预饲养 5d 后,将 60 只小鼠随机分成 6 组,每组 10 只,分为正常组(A/C)和高脂组

(B/D/E/F)。具体分组见表1-1。每天早上9:00准时给小鼠灌胃,室温保持在20~25℃,自由采食和饮水,饲喂60d。

表1 小鼠灌胃剂量表

Table 1 Doses of Tocotrenol on mice

| 分组 | 生育三烯酚剂量(mg/(kg/d)bw) | 生理盐水剂量(ml) |
|---|---|---|
| 空白组A | 0 | 0.6 |
| 高脂空白组B | 0 | 0.6 |
| 生育三烯酚组C | 200 | 0 |
| 高剂量+高脂组D | 200 | 0 |
| 中剂量+高脂组E | 100 | 0 |
| 低剂量+高脂组F | 50 | 0 |

每天观察各组小鼠的生长情况至第60天,采用眼球摘除法收集小鼠血液,3 000rpm离心5min,取血清置于-20℃冰箱中冻存,用全自动生化分析仪检测相关血液生化指标。颈椎脱臼处死小鼠,解剖暴露内脏,观察小鼠肝脏外观变化,取出肝脏用生理盐水洗净淤血后称重。取肝脏制备10%的组织匀浆,检测相关指标测定方法按照试剂盒说明进行。

### 2.2 数据处理

数据结果以$x \pm SD$表示,统计分析采用SPSS 17.0软件包建立数据库,多组均数进行方差齐性检验,组间比较采用单因素方差分析。

## 3 结果和分析

### 3.1 小鼠体重结果

由表2可知,空白对照组与高脂对照组的小鼠相比,两组小鼠体重差异不显著($P>0.05$),高脂对照组小鼠体重降低;正常组内,生育三烯酚组小鼠体重高于空白组小鼠体重,但组间差异不显著($P>0.05$)。高脂对照组与高脂高剂量组、高脂中剂量组及高脂低剂量组相比,发现对高脂小鼠饲喂生育三烯酚后,小鼠的体重升高,且灌胃200mg/kg、100mg/kg生育三烯酚小鼠体重显著升高($P<0.05$)。结果表明,一定剂量的生育三烯酚可促进小鼠体重增加。

表2 生育三烯酚对小鼠体重的影响

Table 2 Effect of tocotrienol on mice weight

| 分组 | 小鼠增重(g) |
|---|---|
| A | 19.037 0±3.207 6 |
| B | 17.038 0±3.958 3 |
| C | 19.175 0±3.154 9 |
| D | 19.408 0±3.725 9* |
| E | 21.537 0±4.657 1* |
| F | 18.650 0±3.570 5 |

注:数据表示的是平均值±标准差(n=8),与正常空白组比较,▲表示差异显著($P<0.05$),▲▲表示差异极显著($P<0.01$);与高脂空白组比较,*表示差异显著($P<0.05$),**表示差异极显著($P<0.01$),下同

## 3.2 小鼠血液生化分析结果

表3 生育三烯酚对小鼠血清 ALT、AST 活性的影响

Table 3 Effect of tocotrienol on ALT、AST activity of mice

| 组别 | FFA(umol/l) | AST(V/L) | ALT(V/L) |
|---|---|---|---|
| A | 0.949 4 ±0.335 3 | 37.088 0 ±4.243 1 | 99.288 0 ±6.589 7 |
| B | 1.226 1 ±0.126 9▲ | 65.725 0 ±30.061▲▲ | 167.275 0 ±17.521 6▲ |
| C | 0.905 2 ±0.244 2 | 36.300 0 ±10.316 3 | 91.562 0 ±17.546 3 |
| D | 1.128 2 ±0.233 3 | 54.912 0 ±10.722 3 | 124.800 0 ±22.866 4 |
| E | 0.887 2 ±0.159 0** | 41.262 0 ±5.799 5* | 105.288 0 ±17.170 9 |
| F | 1.013 8 ±0.124 3 | 41.262 0 ±5.799 5* | 115.362 0 ±22.530 5 |

由表3可知,在正常组内,饲喂生育三烯酚的小鼠 FFA 值要低于空白对照组,但组间差异不显著($P>0.05$),表明生育三烯酚对小鼠血液中 FFA 活性有一定的抑制作用。高脂对照组与空白对照组相比,高脂对照组的 FFA 值要显著高于对照组($P<0.05$)。在高脂组内,饲喂生育三烯酚的小鼠体内 FFA 值都要小于高脂对照组,且中剂量组显著低于对照组($P<0.01$);高剂量、中剂量和低剂量三组相比,中剂量组小鼠的 FFA 值最低,表明中剂量的生育三烯酚对血液中 FFA 活性抑制效果最好。

空白组和高剂量组相比较,高剂量的生育三烯酚可降低小鼠 ALT、AST 的量,但影响不显著($P>0.05$);空白组与高脂对照组相比较发现,高脂组的 ALT、AST 显著要高于空白组($P<0.05$)。高剂量高脂组、中剂量高脂组和低剂量高脂组与高脂对照组相比较,这3组小鼠的 ALT、AST 呈下降趋势,中剂量高脂组的 ALT、AST 值最低,ALT 活性最高,组间差异显著($P<0.05$)但 AST 结果不显著($P>0.05$),说明饲喂生育三烯酚后对高脂小鼠的 ALT、AST 活性有一定影响,对小鼠肝脏有一定保护作用。

表4 生育三烯酚对小鼠血清 TC、TG、HDL 和 LDL 活性的影响

Table 4 Effect of tocotrienol on TC、TG、HDL and LDL activity of mice

| 分组 | TC(mmol/l) | TG(mmol/l) | HDL(mmol/l) | LDL(mmol/l) |
|---|---|---|---|---|
| A | 2.568 8 ±0.398 9 | 1.298 8 ±0.271 3 | 2.350 0 ±0.106 0 | 0.211 2 ±0.050 0 |
| B | 3.623 7 ±1.152 7▲ | 1.587 5 ±0.335 9 | 3.413 8 ±1.137 6▲ | 0.470 0 ±0.550 0▲ |
| C | 2.530 0 ±0.535 4 | 0.950 0 ±0.183 3 | 2.496 0 ±0.375 8 | 0.203 8 ±0.453 4 |
| D | 3.872 5 ±0.442 1 | 1.436 2 ±0.286 2 | 3.248 8 ±0.360 8 | 0.370 0 ±0.046 3* |
| E | 4.968 8 ±0.566 8** | 1.352 5 ±0.282 4 | 3.862 5 ±0.586 7 | 0.392 5 ±0.177 2 |
| F | 4.991 3 ±0.407 9** | 1.317 5 ±0.374 3 | 3.738 7 ±0.352 7 | 0.461 3 ±0.685 4 |

由表4可知,在正常组间内,生育三烯酚组 TC、TG 和 LDL 的活性低于空白组,生育三烯酚组 HDL 的活性高于空白组,但组间结果都不显著($P>0.05$)。表明生育三烯酚对小鼠的胆固醇代谢具有一定的降低作用。在高脂组间内,生育三烯酚组 TG、LDL 活性低于高脂对照组的,TC、HDL 活性高于高脂对照组的;且随着生育三烯酚剂量的增加,小鼠体内 TC、LDL 活性降低,TG、HDL 活性升高,但组间差异不显著。表明生育三烯酚对高脂小鼠血液中胆固醇有一定的降低作用,有利于机体清除多余的脂类。

### 3.3 对小鼠肝脏总胆固醇以及甘油三酯的影响

表 5 生育三烯酚对小鼠肝脏总胆固醇、甘油三脂活性的影响

Table 5 Effect of tocotrienol on TC、TG activity in liver of mice

| 分组 | TC(umol/l) | TG(umol/l) |
|---|---|---|
| A | 0.4480 ±0.0140 | 0.2650 ±0.0344 |
| B | 0.5403 ±0.0584▲▲ | 0.2792 ±0.0493 |
| C | 0.4386 ±0.0108 | 0.2427 ±0.0148 |
| D | 0.5320 ±0.0265 | 0.2759 ±0.0124 |
| E | 0.5413 ±0.0372 | 0.1946 ±0.0146 * * |
| F | 0.5653 ±0.0389 | 0.1601 ±0.0997 * * |

由上表可以看出,高脂对照组中小鼠肝脏总胆固醇要极显著高于空白对照组的($P<0.01$);在正常组内,饲喂生育三烯酚的小鼠肝脏总胆固醇要低于空白组,但结果不显著($P>0.05$)。在高脂组内,饲喂高剂量的生育三烯酚组的肝脏在正常组间,生育三烯酚组的小鼠肝脏甘油三酯活性低于空白组,但组间差异不显著($P>0.05$),说明生育三烯酚对小鼠肝脏中的甘油三酯有一定的抑制作用。在高脂组间,饲喂生育三烯酚的小鼠肝脏甘油三酯活性低于高脂对照组,且中剂量组和低剂量组与高脂对照组相比,组间差异极显著($P<0.01$);高剂量组、中剂量组和低剂量组三组间相比,随着生育三烯酚的剂量增加,小鼠肝脏中甘油三酯值越大,表明低剂量的生育三烯酚对高脂小鼠肝脏中甘油三酯抑制最强。

## 4 讨 论

高血脂症是脂肪代谢或运转异常使血浆一种或多种脂质高于正常,即血中 TC 和/或 TG 过高或 HDL-C 过低。人类高脂血症动物模型表现为不同种类的动物有不同水平的血脂异常。如高莹等人[8]用高脂饲料造成的 SD 大鼠高脂血症模型以类似人类 TC 和 TG 均升高的混合型高脂血症为主要表现。

本试验在给正常小鼠和高血脂小鼠灌胃不同剂量的生育三烯酚后,小鼠体重增加,表明生育三烯酚对小鼠有一定的增肥作用。在给健康小鼠灌胃 200mg/kg 的生育三烯酚后,小鼠血清中 TC、TG、LDL 及 FFA 和肝脏中 TC、TG 都有一定的降低,血清中 HDL 升高,说明生育三烯酚能够促进健康小鼠的胆固醇代谢。试验表明,生育三烯酚的剂量与调节胆固醇代谢的能力并不成正向关系,最佳剂量为 100mg/kg。动物机体内血清中 TC、LDL 的升高可导致发生动脉粥样硬化的几率升高,而 HDL 的降低可使机体发生动脉粥样硬化的机率增加[9,10]。Steinberg D 等研究发现,生育三烯酚可以抑制 OX-LDL 的形成,阻止形成粥样斑快,从而预防动脉粥样硬化的形成[11]。Theriault A 和 Nogchi N 等发现生育三烯酚有较强的抗炎症作用,炎症反应可引起动脉粥样硬化的形成[12,13]。血液中胆固醇含量过高可引起心血管疾病。1986 年 Qureshi 等研究发现,大麦可以降低血液中胆固醇的含量,而大麦中富含生育三烯酚[14]。羟甲基戊二酸单酰 CoA(HMG-CoA)还原酶是机体合成胆固醇的关键酶。研究表明,生育三烯酚能抑制甲羟戊酸的合成,降低了胆固醇合成的中间产物的含量,抑制肝脏中 HMG-CoA 还原酶活性,从而减少胆固醇的生成[15,16]。但 α-生育酚具有相反作用,能诱导此酶的活化。

生育三烯酚可以降低高脂模型小鼠血液中的 ALT 和 AST。其中,中剂量的生育三烯酚对降低高脂模型小鼠血液 ALT 和 AST 最明显,血清中 ALT 和 AST 活性的检测一直被认为是衡量肝实质损伤的“金标准”[17,18]。ALT 和 AST 是肝细胞产生氨基酸代谢的重要酶[19-22]。当肝脏受损,肝细胞膜被破坏,通透性增加,ALT 和 AST 细胞中渗出进入血液循环,使血液中 ALT 和 AST 升高从而表明生育三烯酚对肝脏具有一定的保护功能。

## 5 结 论

综上所述生育三烯酚对高血脂小鼠的胆固醇代谢具有一定的调节作用,减少高脂饲料对小鼠机体的影响,并在一定程度上保护肝脏。

## 参考文献


[1] 郑建仙.低能量食品;[M].北京:中国轻工业出版社,2001.

[2] 于守洋,崔洪斌.中国保健食品的进展;[M].北京,人民卫生出版社,2001:408-435.

[3] 郑建仙.功能性食品;[M].第三卷.北京,中国轻工业出版社,2001:275-289.

[4] LRCP. The lipid research clinic coronary primary trial results. The relation of reduction in incidence of coronary heart disease to cholesterol lowing [J]. J American Medical Association, 1984, 251: 365-374.

[5] 陈丽君.代谢(三)脂质生物化学;[M].北京:科学出版社,1988.169-170.

[6] Yamamoto Y, Fujisawa A, Hara A, etal. An unusual vitamin E constituent (α-tocopherol) provides enhanced antioxidant protection in marine organisms adapted to cold water environments[J]. Ecology: pnas. 2001, 98: 13144-13148.

[7] IUPAC-IUB. Joint comission on biochemical nomenclature (1982), nomenclature of tocopherols and related compounds. recommendations[J], Eur Biochem. 1981, 123: 473-475.

[8] 高莹,李可基,唐世英,等.几种高脂血症动物模型的比较 [J].卫生研究,2002,3 1(2) 97-99.

[9] 刘玮玮.植物裕醇醋对高脂血症大鼠降血脂功能的研究[D].杭州:浙江大学,2007.

[10] 王志兵,肥胖的诊治进展[J].首都医药,2005,12(20):18-19.

[11] Steinberg D. 1997. Low density lipoprotein oxidation and its pathobiological significance [J]. J Biol Chem. 272(34): 20963-20966.

[12] Teriault A, Chao JT, and Gapor A. 2002. Tocotrienol is the most effective vitamin E for reducing endothelial expression of adhesion molecules and adhesion to monocytes [J]. Atherosclerosis. 160(1): 21-30.

[13] Noguchi N, Hanyu R, Nonaka A, et al. 2003. Inhibition of THP-1 cell adhesion to endothelial cells by alpha-tocopherol and alpha-tocotrienol is dependent on intracellular concentration of the antioxidants [J]. Free Radic Biol Med. 34(12): 1614-1620.

[14] Qureshi AA, Burger W C, Peteson D M, et al. The structure of an inhibitor of cholesterol biosynthesis isolated from barley [J]. J Biol Chem, 1986, 261 (23): 10544-10550.

[15] Qureshi AA, Sami S A, Salser W A, et al. Dose-dependent suppression of serum cholesterol by tocotrienol-rich fraction (TRF25) of rice bran in hypercholesterolemic humans [J]. Atherosclerosis, 2002, 161 (1): 199-207.

[16] Chen CW, Cheng H H. Rice bran oil diet increases LDL-receptor and HMG-CoA reductase mRNA expressions and insulin sensitivity in rats with streptozotocin/ nicotinamide-induced type 2 diabetes [J]. J Nutr, 2006, 136 (6): 1472-1476.

[17] 麻晓林,陈伟,金榕兵.肝损伤后检测肝功能指标的实验研究[J].创伤外科杂志,2003,5(1):20-22.

[18] 高利宏,敖林,胡冉,等.4种药物致小鼠肝脏毒性的基因表达谱聚类分析[J].解放军预防医学杂志,2008,28(1):19-25.

[19] 宋国培.肝功能—血清酶学检测的临床意义[J].临床肝胆病杂志,2003,19(4):195-197.

[20] Daniel S. Pratt, M. d. Marshall M. Kaplan M. D. Evaluation of abnormal liver-enzyme results in asymptomatic patients [J]. N Engl J Med, 2000, 324(17): 1266-1271.

[21] 彭双清,伍一军.毒理学代替法[M].北京:军事医学出版社,2008:118-121.

[22] 刘衍忠,刘东霞,高曙光,等.草铵膦的大鼠肝脏毒性[J].环境与健康杂志,2006,23(5):416-417.

# 糖适康配合胰岛素治疗犬糖尿病的疗效与临床观察分析

秦建辉[1]，李思远[1]，陶田谷晟[1]，段靖峰[2]，肖　啸[1]
(1. 云南农业大学动物科学技术学院，昆明 650201；2. 云南农业大学动物医院，昆明 650031)

**摘　要**：目前犬的糖尿病发病率逐年升高，主要发生于中、老龄犬，并且患糖尿病及其并发症的发病率和死亡率呈逐渐增加的趋势。30 年前，国内患病率约为 1.9%，而 1999 年，患病率上升了 3 倍，约为 5.8% 。临床反应单纯用胰岛素治疗会对犬机体带来很大的副作用，且对临床症状的治疗效果不够理想。糖适康具有控制糖进入血液速度和调节血糖代谢的作用，本试验就针对糖适康配合胰岛素治疗犬糖尿病与单纯使用胰岛素治疗糖尿病的疗效效果作出观察分析。挑选 20 只典型糖尿病小型犬，其中 10 只用于糖适康配合胰岛素治疗（试验 1）；另外 10 只单纯用胰岛素治疗做对照试验（试验 2）。其结果为糖适康配合胰岛素治疗犬糖尿病可有效抑制患犬消瘦，且可使其体重回升，而单使用胰岛素则不能；用糖适康配合胰岛素治疗糖尿病，在确保血糖在正常范围内的同时，其使用胰岛素的频率比单纯使用胰岛素治疗所使用胰岛素的频率少（糖适康配合胰岛素比单用胰岛素治疗的平均有效抑制血糖时间长 6h）；试验 1 的治疗消除了“三多一少”的临床症状。

**关键词**：糖尿病，胰岛素，糖适康，治疗，犬

# The clinical efficacy of Tangshi Kang with insulin-treated diabetes small dogsand observation and analysis

QIN Jianhui[1] LI Siyuan[1] TAO Tiangusheng[1] DUAN Jingfeng[2] XIAO Xiao[1]
1. *Animal Science and Technology*, *Yunnan Agricultural University*, *Kunming* 650201;
2. *Animal Hospital of Yunnan Agricultural University*, *Kunming* 650031

**Abstract**: Currently the incidence of diabetes in dogs increased year by year, Occurs mainly in aging dogs, and suffering from diabetes and its complications, morbidity and mortality showed a growing trend. 30 years ago, the domestic prevalence rate of about 1.9% in 1999, the prevalence rate increased by three times, about 5.8%. Clinical response with a great deal of side effects of insulin therapy alone would dog the body, and for the treatment of clinical effect is not well. Tangshi Kang has control of sugar into the blood velocity and the regulation of glucose metabolism. The fitness test on Sport for sugar diabetes with insulin therapy dogs and simple to use insulin treatment to observe and analyze the effects of diabetes. A typical selection of 20 small dogs with diabetes, including 10 for Tangshi Kang with insulin therapy (test 1); Another 10 treated with insulin alone do controlled trials (trial 2). The result is Tangshi Kang canine diabetes with insulin therapy can effectively suppress dogs suffering from weight loss, and can make weight rebound, the single-use insulin you cannot; the Tangshi Kang for treating diabetes with insulin, the blood glucose within the normal range to ensure at the same time, the frequency of their use of insulin therapy with insulin than simply using a frequency less insulin (insulin with Tangshi Kang than the average insulin therapy alone effectively inhibited glucose long time 6 hours); the test 1 Eliminate the clinical symptoms of “Little more than three”.

**Keywords**: diabetes; insulin; Tangshi Kang; treatment; dogs

犬糖尿病的分类没有公认的国际标准[1]，宠物临床上将糖尿病大体普遍分为 I 型糖尿病（胰岛素依赖型糖尿病 IDDM），Ⅱ型糖尿病（非胰岛素依赖型糖尿病 NIDDM）和继发性糖尿病。犬糖尿病 95% 以上都属于 I

型糖尿病(胰岛素依赖型糖尿病)[2]。

胰岛素依赖型糖尿病是由于主动免疫过程破坏了胰岛素 β-细胞,使胰岛素分泌减少而致,临床症状可能在数月或数年后出现。当胰岛素分泌储备能力降低到正常水平的20%以下时,就会出现明显的葡萄糖不耐量现象。外周组织利用葡萄糖受阻,肝葡萄糖又产生过剩,导致禁食高血糖及糖耐量异常。如果紊乱的代谢通过注射胰岛素得以纠正,则表明有些 β-细胞尚有功能,这个周期称为密月期,时间维持数月,长的可达1年[3]。

非胰岛素依赖型糖尿病如果糖尿病在早期得以诊断,口服降糖药可发挥更大作用[4],可暂时治愈,但在一段时间后可能复发。继发性糖尿病常常伴有胰腺疾病、某些内分泌紊乱、胰岛素受体异常以及特定的遗传综合征[5]。

一旦得了糖尿病,就不可能真正治愈,即便临床症状得以暂时消除,也会随时复发[6]。接下来本文就针对使用糖适康配合胰岛素治疗犬糖尿病与单纯使用胰岛素治疗糖尿病的疗效效果作出观察分析,并为犬糖尿病的治疗提供有效依据。

## 1 糖适康的作用机理

成分:

| | | | |
|---|---|---|---|
| Fibersol-2 25 g | 南瓜多糖 7.5 g | 葡萄糖酸锌 120 mg | 半胱氨酸 200mg |
| Phase-2 200 mg | 烟酸铬 700 μg | 甘氨酸 300 mg | 牛磺酸 240mg |
| 菊粉 5 g | 氨基酸螯合钒 300 μg | 谷氨酸 150 mg | 热量≥120 kcal |

糖适康能通过延缓肠道内碳水化合物的消化,控制餐后小肠对糖的吸收;又能通过促进 GTF(耐糖因子)合成,提高细胞胰岛素受体的敏感度,增强细胞对葡萄糖的摄取。既具有控制糖进入血液速度的作用,又具有促进血糖代谢的作用,能有效抑制餐后血糖的快速升高。

## 2 用糖适康配合胰岛素试验治疗糖尿病的材料与方法(并单用胰岛素治疗做对照试验)

半年内在动物医院确诊并治疗的患犬,挑选临床症状(多饮、多尿、多食、消瘦等)显现的小型患犬,消除患犬的并发症。再根据实验室进一步确诊(到目前为止,在犬糖尿病潜在病因的鉴定上没有快速的实验室诊断方法,而对于该病的诊断则通常是在疾病的后期才被做出[1])。

试验中将临床选出的20只小型患犬分两组,其中选出10只患犬分别按年龄从小到大编号1~10号,用糖适康配合胰岛素治疗(试验1)。剩余10只患犬分别按年龄从小到大编号1~10号,只用胰岛素治疗(试验2)。治疗步骤相同,药物用量根据患犬的体重调整,试验期间管理水平一致。

### 2.1 试验材料

2.1.1 试验仪器及试剂 80-2 离心沉淀器,上海手术器械厂;BS-180 全自动生化仪,Mindray;BA600 尿液分析仪,深圳市锦瑞电子有限公司;电脑。

2.1.2 试验药物 糖适康与中效胰岛素 NPH 配合治疗,单纯使用中效胰岛素 NPH 治疗。

### 2.2 试验方法过程

2.2.1 采样 采集病犬的血液,尿液。

2.2.2 样本检测

血糖检测:将采集的血液立即用离心沉淀器以3 000r/min 离心 2min,离心后将采血管平稳放入 BS-180 全自动生化仪,用电脑控制进行项目 GLU 检测。

尿糖检测:将尿检试卡浸入新鲜尿液中至全部指示条,或用一次性注射器吸取后淋在尿检试卡上,擦干试卡背面和侧面尿液后，把卡放在 BA600 尿液分析仪检口上检测。

2.2.3　确诊糖尿病　根据临床典型“三多一少”症状初步确诊后，并利用实验室诊断血糖（空腹血糖超过 11.10mmol/L），尿糖呈强阳性，葡萄糖耐量试验异常，胰岛素测定及血清果糖胺测定等进一步确诊[4]。

2.2.4　试验治疗方法　每只患犬输液治疗解除临床症状。根据每只犬的情况画出其血糖曲线，以方便控制糖尿病给药作出参考。试验 1 使用中效胰岛素 NPH0.5U/kg，饲喂前给药（如超过 1.5U/kg 仍无作用，表明有胰岛素抵抗）[2]，同时使用糖适康来配合胰岛素治疗，喂食前饲喂，按 5g/10kg 的标准每天一次。试验 2 单用胰岛素治疗做对照试验。观察分析患犬临床症状及血糖的变化。

# 3　试验评定标准与试验记录结果分析

## 3.1　评定标准

通过治疗记录对比每只患犬 3 周的临床病例差别，判断糖适康配合胰岛素治疗进展及特点，其对临床症状多尿、多食、消瘦的治疗效果的好坏，以及之后与单用胰岛素治疗相比胰岛素使用频率是否明显减少等作出评定。

## 3.2　试验记录分析与结果分析

3.2.1　试验 1 与对照试验 2 对抑制病犬体重减少结果记录及分析

表 1　试验 1

| 编号 | 患犬 | 发病时体重(kg) | 治疗 3 周后体重(kg) | 体重变动(kg) |
|---|---|---|---|---|
| 1 | ♀4 岁泰迪 | 4.1 | 4.1 | 0 |
| 2 | ♀5 岁泰迪 | 4.9 | 4.9 | 0 |
| 3 | ♀5 岁西施 | 5.2 | 5.3 | +0.1 |
| 4 | ♀6 岁京巴 | 7.1 | 7.2 | +0.1 |
| 5 | ♀7 岁西施 | 3.0 | 3.0 | 0 |
| 6 | ♂7 岁泰迪 | 2.5 | 2.5 | 0 |
| 7 | ♂8 岁小鹿 | 3.6 | 3.6 | 0 |
| 8 | ♀9 岁泰迪 | 4.6 | 4.6 | 0 |
| 9 | ♂10 岁京巴 | 7.7 | 7.9 | +0.2 |
| 10 | ♀10 岁京巴 | 8.5 | 8.7 | +0.2 |
| | 平均 | 5.12 | 5.18 | +0.06 |

表 2　试验 2

| 编号 | 患犬 | 发病时体重(kg) | 治疗 3 周后体重(kg) | 体重变动(kg) |
|---|---|---|---|---|
| 1 | ♀5 岁泰迪 | 4.3 | 4.0 | −0.3 |
| 2 | ♀5 岁蝴蝶 | 4.0 | 3.8 | −0.2 |
| 3 | ♀6 岁京巴 | 7.8 | 7.4 | −0.4 |
| 4 | ♀6 岁京巴 | 7.1 | 6.7 | −0.4 |
| 5 | ♀6 岁蝴蝶 | 3.6 | 3.3 | −0.3 |
| 6 | ♂7 岁京巴 | 8.3 | 8.0 | −0.3 |

(续表)

| 编号 | 患犬 | 发病时体重(kg) | 治疗3周后体重(kg) | 体重变动(kg) |
|---|---|---|---|---|
| 7 | ♂7岁蝴蝶 | 3.6 | 3.4 | -0.2 |
| 8 | ♀8岁泰迪 | 4.6 | 4.3 | -0.3 |
| 9 | ♀9岁小鹿 | 4.7 | 4.5 | -0.2 |
| 10 | ♂10岁西施 | 4.5 | 4.4 | -0.1 |
| | 平均 | 5.25 | 4.98 | -0.27 |

(试验犬均为小型犬,其体重方差值小,可用其平均值来表示)

**表3　平均体重对照**

| 平均值 kg | 试验1 | 试验2 |
|---|---|---|
| 发病时 | 5.12 | 5.25 |
| 1周治疗后 | 5.13 | 5.15 |
| 2周治疗后 | 5.15 | 5.07 |
| 3周治疗后 | 5.18 | 4.98 |
| 每周平均每只增加体重 | +0.02 | -0.09 |

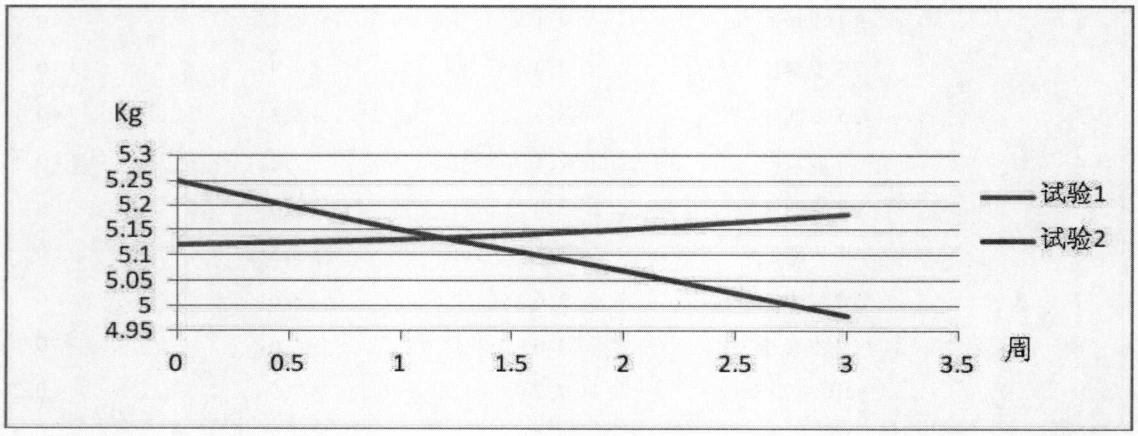


**图1　糖尿病犬平均体重曲线**

可以明显看出试验1比试验2对抑制由于糖尿病导致体重下降有较好的疗效。可以看出试验1中每只患犬平均体重每周增加0.02kg,试验2中每只患犬平均体重每周减少0.09kg。试验1使患犬消瘦得到抑制,体重得到恢复,而试验2消瘦未能得到较好的抑制。

3.2.2　试验1与试验2对所用胰岛素频率做比较　通过3周的治疗,将患糖尿病的小型犬的血糖降至正常范围后,用糖适康配合中效胰岛素来维持患犬血糖的正常范围与单用中效胰岛素来维持患犬血糖正常范围作比较。比较试验1与试验2胰岛素的使用的频率,及其维持血糖的稳定性。

将试验1与试验2的患犬每隔6h监测其血糖,且计算其血糖平均值(因其血糖变化方差值很小,所以可用血糖平均值表示)画出血糖曲线作比较。

表4　血糖平均值

| 时间(h) | 试验1平均血糖(mmol/L) | 试验2平均血糖(mmol/L) |
|---|---|---|
| 0 | 7.0 | 6.7 |
| 6 | 4.5 | 4.2 |
| 12 | 5.1 | 7.2 |
| 18 | 6.5 | 3.9 |
| 24 | 5.6 | 7.0 |
| 30 | 4.5 | 4.5 |
| 36 | 5.5 | 7.4 |
| 42 | 6.9 | 4.3 |
| 48 | 4.7 | 7.3 |
| 54 | 5.2 | 4.5 |
| 60 | 6.7 | 7.2 |
| 66 | 5.7 | 4.6 |
| 72 | 4.3 | 7.4 |
| 平均血糖方差 | 0.845562 | 2.056213 |

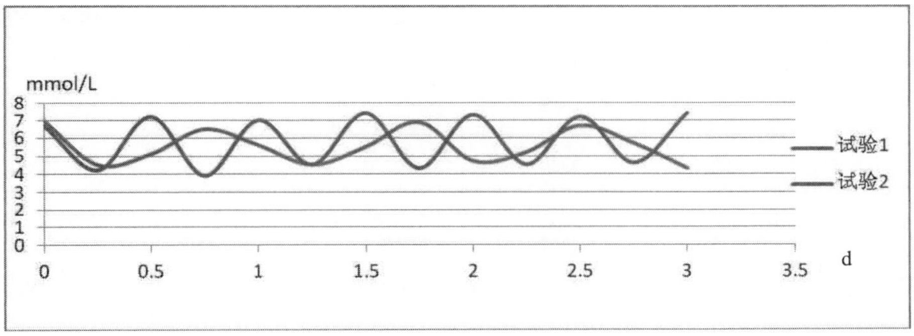


图2　糖尿病犬平均血糖曲线

通过曲线对比说明糖适康配合胰岛素治疗比单纯胰岛素治疗周期曲线长(单用胰岛素平均有效抑制血糖时间为12h,糖适康配合胰岛素治疗平均有效抑制血糖时间为18h),所以在保证血糖在正常范围内的同时,试验1比试验2所使用胰岛素的频率少,可以减少长期使用胰岛素带来的副作用,同时在经济上也得到了节省。

3.2.3　试验1与试验2对糖尿病临床症状“三多一少”的治疗结果作比较　通过临床观察试验1的治疗消除了“三多一少”的临床症状;试验2的临床症状只是得到改善,而未完全消除。

因为病犬表现多饮症状是由于血糖升高之后,吸收周围组织中的水分,机体处于失水状态,所以会变现为多饮;血糖升高,大量的葡萄糖不能被机体利用,排入尿中,引起尿渗透压升高,而肾小管吸收水减少,故表现为多尿。多饮多尿都与血糖有关,只有血糖下降到正常,其症状才能消失。由于血液中的糖利用不足,导致能量的供应来自脂肪和蛋白质。时间长了,脂肪就会消耗过多,导致偏瘦。为了补偿损失的糖分,维持机体活动,病犬常易饥、多食。所以试验1与试验2的临床症状对比可以用平均血糖值,平均体重值说明。平均血糖值(表4,图2)可见试验1的平均血糖在7.0~4.3mmol/L范围内上下波动,其波长约为30(h),其平均血糖方差值为0.845 562;试验2的平均血糖在7.4~3.9mmol/L范围内上下波动,其波长约为18(h),其平均血糖方差值为2.056 213。试验1小于试验2的平均血糖方差值,且试验1波长周期大于试验2,说明试验1比试验2控制血糖稳定。又根据平均体重值(表3,图1),试验1中每只患犬平均体重每周增加0.02kg,试验2中每只患犬平均体重每周减少0.09kg,在对抑制患犬消瘦中,试验1明显优于试验2。血糖即维持在正常范围,消瘦得到抑制,且体重又得到回升,所以试验1的治疗消除了“三多一少”的临床症状,而试验2无法快速有效抑制消瘦,且血糖波动幅度较大,所以临床症状只是得到改善,而未完全消除。

## 4 总 结

糖适康配合胰岛素治疗犬糖尿病可有效抑制患犬消瘦,且可使其体重回升,而单使用胰岛素则不能;用糖适康配合胰岛素治疗糖尿病,在确保血糖在正常范围内的同时,其使用胰岛素的频率比单纯使用胰岛素治疗所使用胰岛素的频率少(糖适康配合胰岛素比单用胰岛素治疗的平均有效抑制血糖时间长6h);试验1的治疗消除了“三多一少”的临床症状;试验2的临床症状只是得到改善,而未完全消除。所以得出使用糖适康配合胰岛素治疗小型犬糖尿病比单使用胰岛素治疗的效果要好。

## 参考文献

[1] 景秋红,犬糖尿病的分类及相关研究[J].现代农业,2009,(10):33-34.

[2] 周桂兰,高德仪.犬猫疾病实验室检验与诊断手册[M].北京:中国农业出版社,2010:266-280.

[3] 詹达忠.犬糖尿病的类型与并发症[J].养殖技术顾问,2012,(7):137.

[4] 梁永春.小型犬糖尿病的病因学与临床诊疗[J].中国畜牧兽医,2010,37(2):197-199.

[5] 施振声.小动物临床手册(第四版)[M].北京:中国农业出版社,2004:457-459.

[6] 张大宁.单纯降糖不等于糖尿病治疗[J].糖尿病天地:文摘刊,2013,(1):43-43.

# 一例犬右侧前后肢同时骨折内固定手术治疗

张斌恺[1]，李思远[1]，陶田谷晟[1]，王江豪[1]，秦建辉[1]，肖　啸[1,2]

（1. 云南农业大学动物科学技术学院，昆明 650201；2. 云南农业大学动物医院，昆明 650031）

**摘　要**：内固定手术是指用金属螺钉、钢板、髓内针、钢丝或骨板等物直接在断骨内或外面将断骨连接固定起来的手术，内固定术可以较好地保持骨折的解剖复位，比单纯外固定直接而有效，近些年来随着我国小动物行业医疗水平取得了巨大的进步，单纯采用简单外固定的治疗方法已经渐渐不适用，大量新的改良的内固定方法逐渐取代了外固定方法大量应用于犬的骨折手术中。本试验就是结合一例犬肱骨远端骨折、胫骨中端骨折内固定手术的过程从各个方面介绍了这个病例的具体情况（从术前到术中到术后）。结果表明，该犬采用塑形钢板加柯氏针对肱骨远端骨折进行固定和采用 DCP 接骨板对胫骨中端骨折内固定整复后，效果良好，术后第 4 天该犬就能正常下地行走，2 周后基本恢复正常。通过这个病例的研究，证明了采用内固定的手术疗法可使骨折很快愈合、治疗效果也能达到预期目的。

**关键词**：犬，骨折，内固定

## 1　引　言

骨折是骨组织的完整性或连续性因外力作用或病理因素而遭受破坏的状态，骨折的同时伴有周围软组织不同程度的损伤。治疗骨折的方法一般有整复和固定。固定方法又可分为外固定和内固定两种，外固定操作相对简单，组织损伤小，愈合时间较短，但由于某些特殊部位的骨折以及部分动物及主人配合性较差，外固定会受到很多因素的限制。内固定手术技术要求较高，组织损伤大，愈合时间相对较长，临床应用不普遍，但对于某些特殊部位的骨折，必须采用内固定矫形治疗[1]。

对于内固定手术治疗，随着我国小动物医学事业的不断发展，以前简单的使用柯氏针无塑形钢板做内固定甚至只是做简单外固定的治疗方法已经远远不能满足如今的临床要求，这些进步得益于大批致力于兽医事业的研究者们呕心沥血的探索研究和医学技术的进步，大量新的改良的内固定方法层出不穷[2]。现就对一例由于车祸导致右侧肱骨和胫骨同时骨折的 4 月龄杜宾犬内固定手术治疗的情况及操作步骤总结。

## 2　材料与方法

### 2.1 手术材料

2.1.1　试验动物　杜宾犬，4 月龄，雄犬，体重 7 kg。该犬主将其带出，在马路上因未系绳子，而狗非常活泼，在过马路时，被车撞，右侧前后肢不愿着地。主人一周后带来检查。经检查，体温 39.5℃，呼吸约 27 次/min，脉搏约 200 次/min。患肢重度跛行，患部皮肤完整，X 光摄片显示犬右侧肱骨（图 1）和胫骨骨折（图 2）。

2.1.2　手术器械　骨科器械：骨钳、电动骨钻及钻头、螺钉导向器、骨科用手术剪、骨锤、骨挫、骨撬、骨膜分离器、钢丝剪等。

常用外科手术器械：手术刀柄及刀片、手术剪、止血钳、持针钳、医用缝线等。

手术所需要的材料：不同直径的髓内针三根、不同宽度的接骨板（试验动物使用的是塑形钢板和 DCP 接骨板）、接骨板配套的螺丝 6 颗，医用钢丝等。

2.1.3　麻醉药　丙泊酚、异氟烷。

肖　啸，E-mail：xiaoxiaokm@163.com

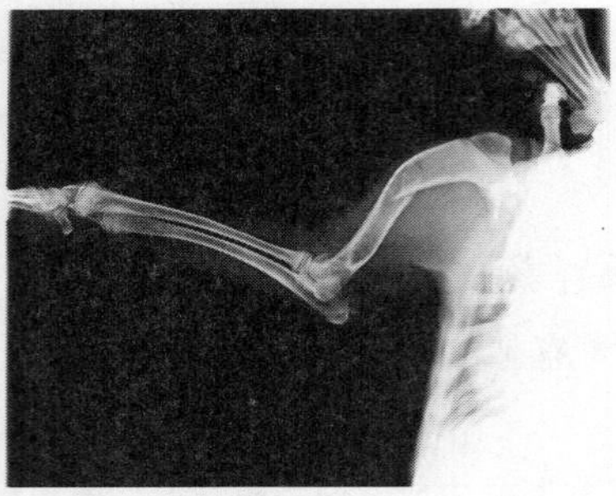

图1　右前肢正位

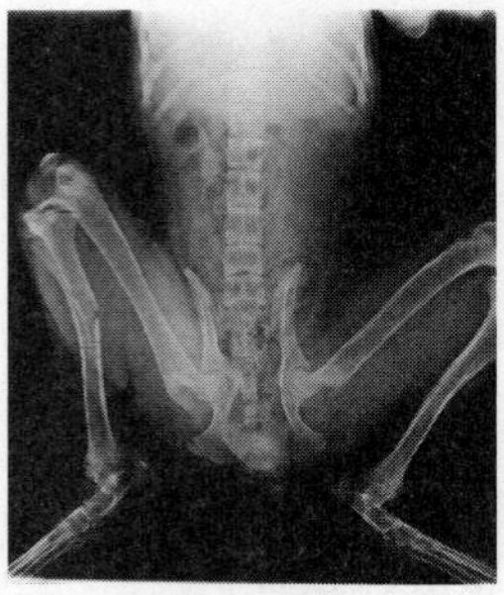

图2　右后肢正位

2.1.4　其他设备　X 光机(MIKASAHF200A)[3]、血常规分析仪、生化结果分析仪、呼吸机。

术前评估结果见表1和表2。

表1　血常规结果

| 项目名称 | 结果 | 参考范围 |
|---|---|---|
| WBC 白细胞计数 | H 35.0 × $10^9$/L | 6.0 ~ 17.0 |
| Lymph#淋巴细胞数目 | 4.0 × $10^9$/L | 0.8 ~ 5.1 |
| Mon#单核细胞数目 | 0.8 × $10^9$/L | 0.0 ~ 1.8 |
| Gran#中性粒细胞数目 | H 30.2 × $10^9$/L | 4.0 ~ 12.6 |
| RBC 红细胞数目 | 7.06 × $10^{12}$/L | 5.50 ~ 8.50 |
| HGB 血红蛋白 | 160g/L | 110 ~ 190 |
| HCT 红细胞压积 | 49.5% | 39.0 ~ 56.0 |
| PLT 血小板数目 | 412 × $10^9$/L | 117 ~ 460 |

表2　生化结果

| 项目名称 | 项目全称 | 结果浓度 | 项目单位 | 结果描述 | 参考范围 |
|---|---|---|---|---|---|
| ALT | 谷丙转氨酶 | 110.2 | U/L | 高 | 10.0 ~ 100.0 |
| ALP | 碱性磷酸酶 | 123.9 | U/L | 正常 | 23.0 ~ 212.0 |
| TP | 总蛋白 | 60.4 | g/L | 正常 | 52.0 ~ 82.0 |
| UREA | 尿素氮 | 6.55 | mmol/L | 正常 | 2.50 ~ 9.60 |
| CREA | 肌酐 | 49.8 | μmol/L | 正常 | 44.0 ~ 159.0 |
| GLU | 血糖 | 7.13 | mmol/L | 正常 | 4.10 ~ 7.94 |

该病例的术前结果满足做手术的条件，于是决定尽快进行内固定修复手术。

## 2.2　手术方法

2.2.1　术前准备　包括需要的各种手术器械、骨科器械、术者准备、患犬准备及术前评估。

术者准备：明确手术方案，具备熟练的操作技能。

患犬准备：手术前需要做生化、凝血试验、血常规、血气等化验项目并对体温、呼吸、脉搏做相关检查，以确保患犬能够经得起这个麻醉。术前禁食 24 h，禁水 12h。术前静脉输注 5% 糖盐水、维生素 C、能量合剂、止血敏等。

注射前 15min 皮下注射阿托品 0.5 mg，然后，先用丙泊酚静脉推注，诱导麻醉，当动物反应迟钝后，用喉头镜将导气管插入气管,用异氟烷加上呼吸机进行麻醉,在将患肢在上的侧卧保定,对右侧肱骨及胫骨上部外侧常规剃毛消毒,准备手术。

2.2.2　手术通路　肱骨干和胫骨干手术通路。

肱骨干手术通路,进刀位置选择在右前肢内侧,避开大血管、神经干。

胫骨干手术通路,手术切口选择在胫部外侧,纵向切开皮肤至小腿远端,分离皮下组织,结扎血管(走向从前下方到后上方),暴露骨折断端,分离骨骼周围组织及骨膜。

2.2.3　手术过程　分为肱骨骨折手术过程和胫骨骨折手术过程,首先是肱骨骨折手术过程,患犬左侧卧并固定在手术台上,术部常规剃毛、消毒 2 次,术部皮肤切口 3cm,钝性分离三角肌后缘和臂三头肌前缘,用止血纱布吸净切口视野,暴露骨折段,清除骨折区的血凝块、挫伤组织和骨碎片,翘起肱骨两断端,对准并压迫到正常解剖位置。注意避开动静脉和神经。用骨尺量肱骨长度,剪下相应长度的柯氏针,并把一端修成尖三角形,用持骨钳钳住固定两断端,用骨钻沿两断端口周边钻 4 个 0.8mm 的孔,对正骨折端后用 0.5mm 不锈钢丝扎紧断端,整平接头。用剪好的柯氏针夹在骨钻中,从肩关节端旋入髓内,并轻敲柯氏针至肩关节端留下 3mm,以备日后取针。然后用生理盐水清洗创口和校正肱骨,撒上氨苄青霉素粉,复位肌肉,常规缝合皮肤切口,在切口皮肤处用 3% 碘酊涂布 3min 后,涂上创口冻胶。

胫骨骨折手术过程,根据胫骨骨折的手术通路切开皮肤,对骨折断端进行整复,整复前先在胫骨下方预置 2 根钢丝,然后采用牵拉和撬动的方法将骨折断端对合,对合后迅速用钢丝进行简单的加固,待 2 根钢丝加固后,2 个骨折断端已基本对合,最后拧紧钢丝,用 4 孔接骨板在骨折的上下两端进行固定,上紧螺丝后用生理盐水对创腔进行反复冲洗,取出创腔内的血凝块和其他异物,涂上油剂阿莫西林。常规闭合手术创口,碘酊消毒后作加压绷带[4]。

## 3　结果与分析

术后的 X 光摄片见图 3、图 4:

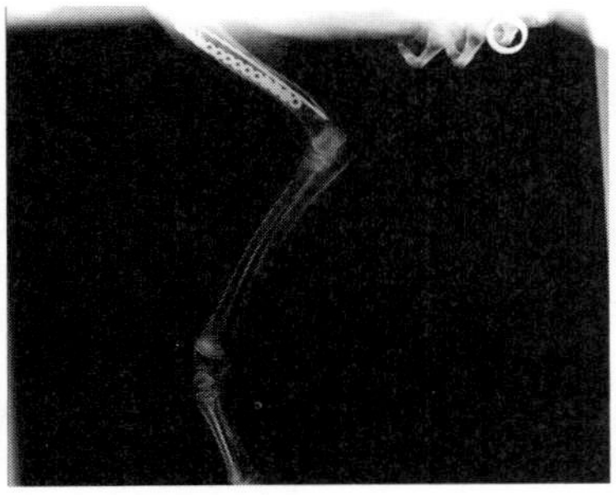

图 3　术后前肢 CR 成像

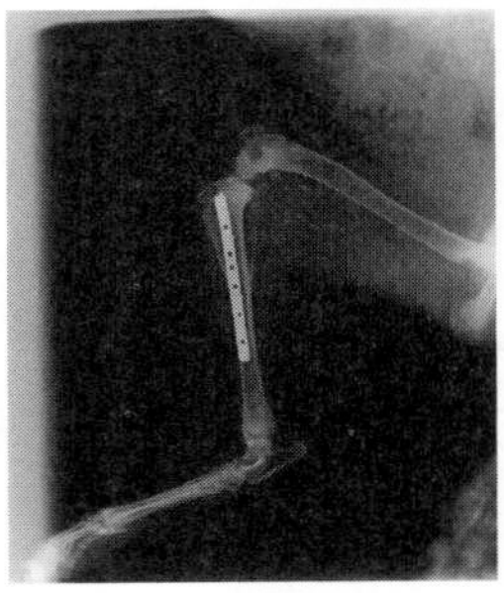

图 4　术后后肢 CR 成像

两周后能正常站立并行走,说明患犬恢复良好。

## 4　讨　论

把髓内针埋入皮下,这样不易造成感染,有利于伤口的愈合,也有利于髓内针的固定,髓内针要以不穿过骨密质为宜,过长会损伤软组织及肌肉,过短则达不到固定目的。采用髓内针内固定术,切口小,对软组织及骨折端骨膜损伤小,不影响骨折端供血,固定物在髓腔内,不增加皮肤切口的张力,取出方便,而接骨板和指骨板取出时损伤较大,对小型犬髓内针也要优于接骨板和指骨板,接骨板的尺寸取决于动物体格大小及接骨板的功能。对于成年犬术后只要不影响运动,可以不拆除接骨板。

此次手术采用塑形钢板加柯氏针和 DCP 接骨板,对骨折断端进行固定,固定效果比较理想。且由于该试验动物的骨折程度较为严重,若选择外固定作为手术的通路,则骨折整复会受到较大的影响,而使用接骨板内固定可使整复效果大幅提升,且采用内固定的手术通路可使骨折恢复速度和效果更加良好,此次手术,术后第 4 天,患犬就能正常下地行走也说明了此次手术是成功有效的。

## 参考文献

[1] 曹杰,王春璈.实验性犬肱骨骨折内固定术[J].中国兽医杂志,2004,10:43.

[2] 何英,叶俊华.宠物医生手册[M].第一版.沈阳:辽宁科学技术出版社,2003.240.

[3] J. Kevin Kenly Hester 编著,谢富强主译.犬猫 X 射线与 B 超诊断技术[M].第四版.沈阳,辽宁科学技术出版社,2006.286 - 287.

[4] 迈克尔沙尔编著,林德贵主译.犬猫临床疾病图谱[M].沈阳:辽宁科学技术出版社,2004.257.

# 牛源胸腺嘧啶依赖型金黄色葡萄球菌小菌落突变株的分离及鉴定

朱立力，王奇惠，曲伟杰
（云南农业大学动物科学技术学院，昆明 650201）

**摘　要**：对某牛场患奶牛乳房炎的奶样中分离出一株疑似金黄色葡萄球菌小菌落突变株（SCVs）进行分离鉴定、生理生化特性研究、药敏试验和补偿实验，为金黄色葡萄球菌 SCVs 引起的奶牛慢性乳房炎的预防和控制及其致病机制的研究奠定基础。通过形态观察、生理生化特性研究、金黄色葡萄球菌相关保守基因片段（*nuc*、*nucA*、16 SrDNA）多重 PCR 扩增鉴定，药敏试验和补偿实验。证实分离出的金黄色葡萄球菌为一株黄色葡萄球菌 SCVs，该菌含有金黄色葡萄球菌菌种特异性基因 nuc 和 nucA。菌落形态主要表现为菌落细小、生长缓慢、溶血能力下降；凝固酶活性下降；耐盐能力降低；革兰染色为革兰阳性球菌，呈葡萄状排列；补偿试验鉴定该金黄色葡萄球菌 SCVs 为胸腺嘧啶依赖型。结果表明，本研究成功分离鉴定出一株胸腺嘧啶依赖型金黄色葡萄球菌 SCVs。

**关键词**：金黄色葡萄球菌，小菌落突变株，分离鉴定，胸腺嘧啶依赖型

# Isolation and identification of a thymine-dependent *Staphylococcus aureus* small colony variants from bovine milk with mastitis

ZHU Lili, WANG Qihui, QU Weijie
(*College of Animal Science and Technology*, *Yunnan Agricultural University*, *Kunming* 650051, *China*)

**Abstract**: The purpose of this study is to identify and isolate *Staphylococcus aureus* small colony variant (*S. aureus* SCVs) from milk samples of dairy cow with mastitis, and analysis the physiological and biochemical characteristics, as well as investigate its antibiotic resistance and auxotroph type, so as to provide a basis for the prevention and control bovine mastitis caused by *S. aureus* SCVs. Possible *S. aureus* milk samples were isolated from milk samples of dairy cow with mastitis for bacterial colony morphology, biochemical characteristics and susceptibility test. Identification of the gene (*nuc*, *nucA*, 16SrDNA) of *S. aureus* was done by multiple PCR amplification and a series of auxotroph type tests. One of the isolates from milk samples of dairy cow with mastitis was identified as *S. aureus* SCVs with the gene (*nuc*, *nucA*, 16SrDNA) of *Staphylococcus aureus* confirmed by PCR multiple amplification. Its major phenotypes included small colony, slow growth, decreased hemolysis, decreased coagulase, decreased salt tolerance ability. This clinical isolate of *Gram positive cocci* in grape like clusters, and complementation test showed that the growth of this clone was thymine-dependent. One thymine-dependent *Staphylococcus aureus* small colony variant strain was isolated and identified, a series of biochemical tests were conducted, which laid the groundwork for the control and prevention of the cow mastitis disease caused by *S. aureus* SCVs and research of its pathogenic mechanism.

---

**基金项目**：国家自然科学基金资助（No. 31260629）

**作者简介**：朱立力（1989－），男，云南红河人，硕士研究生，研究方向：动物营养代谢病

**通迅作者**：曲伟杰（1978－），男，博士，副教授，E-mail：qwj1204@sohu.com

**Keywords**: *Staphylococcus aureus*; small colony variants; isolation and identification; $CO_2$-dependent; Thymine-dependent

奶牛乳房炎是一种由微生物因素、物理因素、化学因素、环境因素等刺激而引起的一类多因素炎症反应,其每年给我国的奶牛业造成巨大的经济损失,严重影响着我国动物性食品安全的发展。该病可由多种非特定病原微生物引起,其中以金黄色葡萄球菌、链球菌、大肠杆菌等最为常见,另外部分真菌和支原体也能引发奶牛乳房炎(尹柏双等,2010)。

金黄色葡萄球菌在患乳房炎的奶牛乳汁中检出率最高,该属细菌在自然条件下可以发生变异,抵抗力较强,容易对抗生素产生耐药性。Jensen(1957)首次报道了金黄色葡萄球菌小菌落突变株(SASCVs)。SCVs是一种由正常菌群变异后形成的生长缓慢、表形独特、致病性变化的细菌亚群。相关研究表明(Balwit JM等,1994),一些非特异性吞噬细胞(如上皮细胞)对金黄色葡萄球菌具有内摄作用,被吸收的正常金黄色葡萄球菌可迅速破坏细胞,使细胞裂解,而SASCVs却能在哺乳动物细胞中持续存在,从而可能躲避抗生素和宿主免疫机制的攻击(Proctor RA 等,1995)。临床上的多种类型 SASCV 相较野生株在真核细胞中存活更长时间(Kahl B 等,1998)。由于 SASCVs 有别于正常野生株的生存方式,导致引起隐性或复发性感染概率增大。

目前国内对奶牛乳房炎进行了大量的研究,但大多数都集中于宏观的防治措施和营养管理等方面,对慢性和隐形奶牛乳房炎的发病原因和机制报道较少。本试验成功由患慢性乳房炎的奶牛奶样中,分离并鉴定出一株 SASCVs,并进行了生理生化特性、补偿试验等,为慢性和隐形奶牛乳房炎的研究提供了新的思路,奠定一定的基础,报道如下。

## 1 材料与方法

### 1.1 材料

1.1.1 菌株来源 由某牛场患慢性奶牛乳房炎的奶牛乳样中分离培养出一株疑似金黄色葡萄球菌 SCVs,质控菌株 ATCC25923 购自杭州天和微生物试剂有限公司。

1.1.2 主要试剂和仪器 葡萄球菌属生化编码鉴定管购自杭州天和微生物试剂有限公司;胸腺嘧啶、硫胺素、甲萘醌和血红素均购自 Sigma 公司; PCR Mastermix、细菌基因组 DNA 提取试剂盒、琼脂糖凝胶 DNA 回收试剂盒、DNA Marker 均购自天根生化科技北京有限公司;抗生素药敏纸片购自杭州天和微生物试剂有限公司;无菌脱纤绵羊血购自广州蕊特; TSB 、TSA、MHA 培养基均购自北京奥博星生物技术有限责任公司;其他试剂均为进口分装或国产分析纯。凝胶成像系统购自 Syngene;高速离心机购自 Eppendorf 公司;高压灭菌锅购自上海博迅实业有限公司医疗设备厂。

### 1.2 方法

1.2.1 细菌形态观察 用接种棒挑取已分离疑似 SASCVs 在羊血平板上培养生长 24 h 后,观察菌落大小、颜色和溶血状况,以金黄色葡萄球菌质控菌株(ATCC 25923)为阳性对照。

1.2.2 多重 PCR 扩增 利用 nuc、nucA 和 16S rDNA 基因的引物设计的多重 PCR 鉴定方法鉴定该疑似 SASCVs。引物设计:根据文献(吕国平等,2012)设计引物并委托上海生工生物工程技术服务有限公司进行合成,具体引物信息:见表 1。模版制备:使用细菌基因组 DNA 提取试剂盒制备。配制 PCR 扩增体系:25 μl PCR Mastermix 套装反应液,上、下游引物(20 μmol/L)各 0.5 μl,PCR 模版 4 μl,加超纯水至 50 μl。PCR 扩增反应条件:94 ℃预变性 2 min;94 ℃变性 30 s,54 ℃退火 45 s,72 ℃延伸 50 s,34 个循环;72 ℃延伸 3 min。以金黄色葡萄球菌质控菌株(ATCC25923) 为阳性对照。对 PCR 扩增产物进行琼脂糖凝胶电泳,利用凝胶成像系统对目的片段进行检测;割取含目的 DNA 条带的琼脂糖凝胶,利用胶回收试剂盒进行回收,纯化后的产物送上海生工生物工程公司测序。

表1 引物信息

| 目的基因 | 序列 | 长度(bp) | 多重 PCR 引物 Tm |
|---|---|---|---|
| 16SrDNA | 上游:5'-GGCGTTGCTCCGTCAGGCTT-3'<br>下游:5'-CGCTGGCGGCGTGCCTAAT-3' | 375 | 54℃ |
| *nucA* | 上游:5'-CGCTTGCTATGATTGTGGTAGCC-3'<br>下游:5'-TTCGGTTTCACCGTTTCTGGCG-3' | 239 | |
| *nuc* | 上游:5'-TCGTCAAGGCTTGGCTAAAGTTGC-3'<br>下游:5'-TCAGCGTTGTCTTCGCTCCAAA-3' | 126 | |

1.2.3 药敏试验 根据文献(Besier,2007),主要通过药敏片法进行了以下几种药敏实验,磷霉素(50 μg/片)、甲氧苄啶/磺胺甲基异噁唑(25 μg/片)、庆大霉素(10 μg/片)、苯唑西林(1 μg/片)、氨苄西林(10 μg/片)、万古霉素(30 μg/片)、链霉素(10 μg/片)、卡那霉素(30 μg/片)、青霉素(10 μg/片)、头孢西丁(30 μg/片)。在 TSA 平板上挑取新鲜培养的单菌落,接种于 TSB 液体培养基,37 ℃水浴摇菌培养箱培养 18~24h 后,用棉签均匀涂布于 MH 平板,贴上药敏纸片,置于35℃培养,同时用质控金黄色葡萄球菌 ATCC25923 为阳性对照。因 SASCVs 生长较慢,在培养 24 和 48 h。2 个时间点各测一次抑菌圈的直径,以保证试验结果的准确性。对照纸片扩散法解释标准,判断细菌的抗药性。综合分析两次测量的结果,得出结论。

1.2.4 生理生化实验 将疑似 SASCVs 接种于质量分数为 0.85%、5%、10%、15%及 25% NaCl 培养基中(陈燕飞,2007),37 ℃培养;于培养 24、48h 时观察其生长情况,同时用标准金黄色葡萄球菌 ATCC25923 为阳性对照。将 PCR 鉴定含目的基因的小菌落在 TSA 平板上划线培养 36 h 后接种至葡萄球菌属生化编码鉴定管,于 37 ℃恒温箱培养,分别于接种 24、48、72 h 记录结果,同时用标准金黄色葡萄球菌 ATCC25923 为阳性对照。将 PCR 鉴定含目的基因的小菌落在 TSA 平板上划线培养 36 h 后进行革兰染色试验、触酶试验和血浆凝固酶试验(杨烨建等,2013)。

1.2.5 补偿实验 将 PCR 鉴定含目的基因的小菌落依据文献(何娟梅等,2009),分别进行了胸腺嘧啶、硫胺素、甲萘醌和血红素的补偿实验。操作如下:配置含如下浓度的 MH 平板,胸腺嘧啶(1 μg/ml)、硫胺素(100 μg/ml)、甲萘醌(1 μg/ml)、血红素(1 ug/ml),其中由于胸腺嘧啶只能微溶于冷水,故需直接加入 MH 中,121 ℃高压 30 min 倒入平板。其余药物先配置成原液后,按一定比例稀释成以上浓度,0.122 μm 过滤除菌后涂布于 MH 平板上。细菌接种于 TSB,培养 5 h,倍比稀释成 $10^{-8}$ ~ $10^{-1}$共 8 个稀释度,取 $10^{-8}$ ~ $10^{-5}$稀释度的菌液,滴加于以上各平板。设菌液滴在 MH 平板上作为空白组,设标准金黄色葡萄球菌 ATCC25923 在 MH 平板为阳性组,37 ℃培养 24 h,比较三组细菌菌落大小。

# 2 结果与分析

## 2.1 细菌形态观察

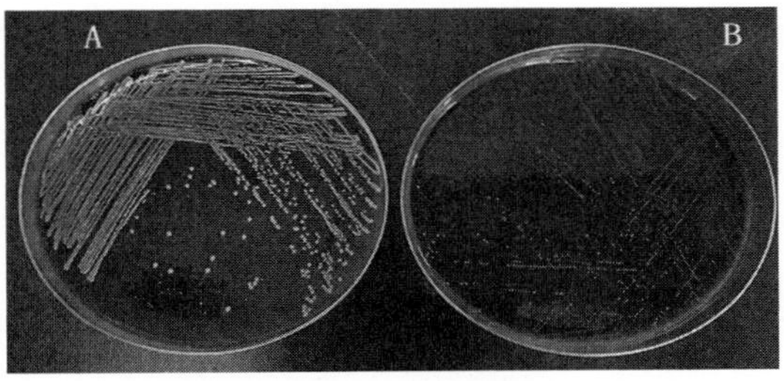


图1 金黄色葡萄球菌标准株与疑似金黄色葡萄球菌 SCVs 在普通羊血平板上培养 24h 生长情况对比

注:A 为金黄色葡萄球菌质控菌株(ATCC 25923),B 为疑似金黄色葡萄球菌 SCVs

在绵羊血平板上培养 24 h 后,疑似菌落较标准株生长缓慢,溶血能力下降,颜色变浅,见图 1。

## 2.2 多重 PCR 鉴定方法鉴定该疑似 SASCVs

利用多重 PCR 扩增后,疑似 SASCVs 及阳性对照株均能通过琼脂糖凝胶成像系统检测得到 126、239、375 bp 3 个目的片段,与预期扩增位置一致,见图 2。胶回收产物测序后得到 3 段序列通过美国 NCBI 网站 BLAST 软件比对,证明为 *nuc*、*nucA*、16SrDNA 基因片段。表明分离到的疑似 SCVs 菌株为金黄色葡萄球菌 SCVs。

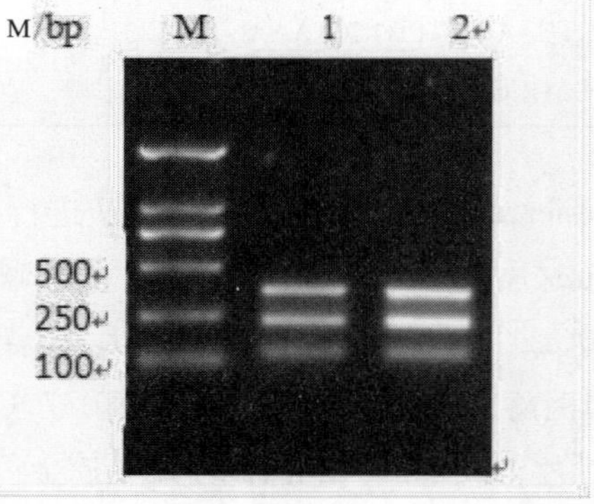


**图 2 金黄色葡萄球菌标准株及金黄色葡萄球菌 SCVs 多重 PCR 产物琼脂糖凝胶电泳图**

注:M 为 Marker,1 为疑似金黄色葡萄球菌 SCVs,2 为标准金黄色葡萄球菌株

## 2.3 药敏试验

在培养 24 h 后,金黄色葡萄球菌 SCVs 抑菌圈并不明显,标准株抑菌圈明显可见,金黄色葡萄球菌 SCVs 培养 48 h 后,抑菌圈清晰可见,金黄色葡萄球菌 SCVs 培养 72 h 后,抑菌圈直径与 48 h 相同。金黄色葡萄球菌 SCVs 及标准株药敏试验结果如表 2。进行重复试验后,结果一致。从表 2 中可以看出,金黄色葡萄球菌 SCVs 对甲氧苄啶/磺胺甲基异噁唑(SXT)并不敏感,而很多报道(Kahl BC 等,2003)都指出,胸腺嘧啶依赖性金黄色葡萄球菌 SCVs 对 SXT 具有抗性,从这个角度可以初步证明此金黄色葡萄球菌 SCVs 为胸腺嘧啶依赖型。

**表 2 疑似金黄色葡萄球菌 SCVs 与标准金黄色葡萄球菌药敏试验结果**

| 药物名称 | 金黄色葡萄球菌标准株 | 疑似金黄色葡萄球菌 SCVs |
|---|---|---|
| 磷霉素(50μg/片) | S | R |
| 甲氧苄啶/磺胺甲基异噁唑(25μg/片) | S | S |
| 庆大霉素(10μg/片) | S | R |
| 苯唑西林(1 μg/片) | S | S |
| 氨苄西林(10μg/片) | S | R |
| 万古霉素(30μg/片) | S | S |
| 链霉素(10μg/片) | S | R |
| 卡那霉素(30μg/片) | S | R |
| 青霉素(10μg/片) | S | R |
| 头孢西丁(30μg/片) | S | S |

注:S 代表敏感,R 代表不敏感

## 2.4 生理生化试验

2.4.1 耐盐能力实验 较阳性株相比，疑似 SASCVs 耐盐能力下降，阳性株与疑似 SASCVs 在含 25% TSA 平板上均不能生长，但阳性株在 15% 的 NaClTSA 平板中能正常生长，疑似 SASCVs 生长受到抑制。

2.4.2 生化编码鉴定管鉴定 在培养 24 h 后，标准株甘露醇由原来的紫色完全变成黄色，结果为阳性，而疑似 SASCVs 较标准株利用甘露醇、甘露糖、果糖、乳糖的的能力下降，在培养 24 h 后，SCVs 发酵管中的甘露糖及果糖开始由紫色慢慢变淡，48 h 后甘露糖和乳糖完全变黄，SCVs 不利用甘露醇及乳糖。

2.4.3 革兰染色试验、触酶试验和血浆凝固酶试验 革兰染色为革兰阳性球菌，呈葡萄状排列；触酶试验结果阳性；血浆凝固酶试验 2 ~ 16h 均为阴性，18 h 呈阳性。

## 2.5 补偿试验

补偿试验发现分离到的 SASCV 在含硫胺素、甲萘醌及血红素的培养基中依然生长不良，在含胸腺嘧啶培养基中生长良好，较原株大，比标准株稍小，证明该 SASCVs 为胸腺嘧啶依赖型 SASCVs。

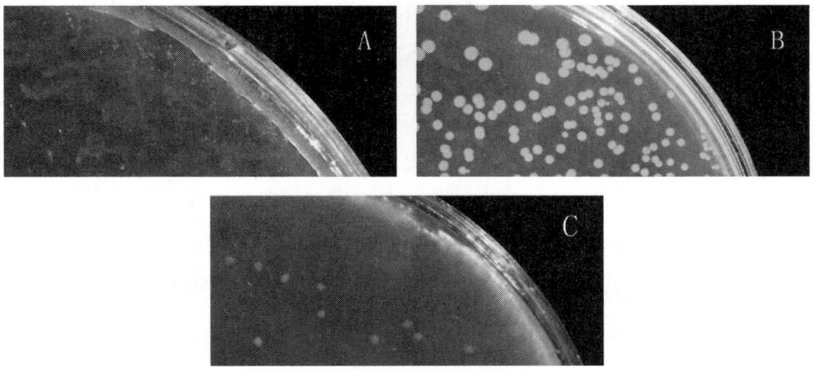


**图 3 胸腺嘧啶补偿试验对比图**

注：A 为空白组，B 为阳性组，C 为试验组

# 3 讨 论

从 Jacobsen 等（1910）小菌落突变株首次被报道起，经过近 1 个世纪的研究，很多菌种都发现了小菌落变异株，并有了深入的研究，这其中以 SASCVs 的研究最为全面。SASCVs 在人医上已得到广泛报道（刘修权等，2011），其作为一种利于持续复发性感染的细菌致病形式，容易引起隐形感染（Melter 等，2012），其与慢性疾病的关联性已得到证明。在与奶牛乳房炎有关的报道中，Sompolinsky 等（1974）的一篇报道中，首次发现 SASCVs 与慢性乳房炎有关。Atalla 等（2008）报道关于 SASCVs 基因转录谱方面的深入研究表明 SASCVs 可能是导致慢性乳房炎的一个重要原因。

SASCVs 的典型特征为在固体琼脂培养基上菌落细小，相较野生株色素生成和溶血能力降低，凝固酶活性降低，本试验中分离到的一株 SASCVs 的表型符合以上的特征。SASCVs 出现不利用甘露醇等非典型的生化反应，故在运用生化实验鉴定菌株时，容易出现假阴性结果，故本实验通过一个多重 PCR 检测法对 *nuc*、*nucA* 两个特异性基因及 16S rDNA 测试以确定其为金黄色葡萄球菌。相关研究表明，SASCVs 实质上是一种与呼吸链相关的营养缺陷型突变株。

金黄色葡萄球菌主要靠需氧呼吸进行产能，呼吸链系统的完整性是完成需氧呼吸的基础，呼吸链（respiratory chain）系统通过多步氧化还原反应释放 ATP 供能。SASCVs 的出现正是由于呼吸链中某一环节的功能单位合成下降，导致 ATP 生成受阻，供给菌体能量不足，从而出现 SASCVs 表型。根据呼吸链功能异常影响 ATP 生成受阻方式的不同，目前研究领域基本上把 SASCVs 分为两个大类型，即电子传递缺陷型 SCVs 和胸腺嘧啶合成缺陷型 SCVs。

电子传递缺陷型 SASCVs 是最常见的营养缺陷类型,大量研究表明,甲萘醌或血红素合成受阻时,呼吸链将会终止,ATP 合成受限,导致细菌获能不足而出现 SASCVs 表型。在生成甲萘醌的反应中,需要硫胺素的参与,故硫胺素一旦合成受阻也会导致野生株生成该种 SASCVs,在本试验中为分离出的该株 SASCVs 补充甲萘醌、血红素以及硫胺素并不能使其恢复至野生株表型。而在补充胸腺嘧啶的补偿实验中,该 SASCVs 比空白菌株有了显著增大,比标准株稍小,基本恢复野生表型,从而从另一个角度证明了其为 SASCVs,而且类型属于胸腺嘧啶依赖型 SASCVs。

胸腺嘧啶依赖型 SASCVs 与电子传递缺陷型 SASCVs 有基本相似的表型,根据最新报道,胸腺嘧啶缺陷型 SASCVs 的产生是由于其 thyA 基因的缺失突变引起,此基因编码胸苷酸合成酶(Besier 等,2007),胸苷酸合成酶催化 5,10-二甲基四氢叶酸将尿嘧啶脱氧核苷酸(dUMP)甲基化而合成胸腺嘧啶脱氧核苷酸(dTMP)的反应(Stroud 等,1987),而 dTMP 是合成 DNA 的所必需的,所以通过相关可干扰胸苷酸合成酶合成的药物的治疗(如磺胺甲基异噁唑可干扰四氢叶酸合成,而四氢叶酸是胸苷酸合成酶的重要辅酶),可抑制金黄色葡萄球菌的生长,产生胸腺嘧啶依赖性 SASCVs。在长期的使用此类抗生素的过程中,胸腺嘧啶依赖性 SASCVs 通过吸收胞外胸腺嘧啶(由金黄色葡萄球菌 DNA 酶降解坏死细胞释放的 DNA 产生)的方式(Smith 等,2005),获取 DNA 合成的原料,继续存活,由于此途径获取胸腺嘧啶有限,故菌体生长缓慢,呈现在固体琼脂上菌落细小等特殊表型。当通过人为进行补偿试验,使胞外有大量胸腺嘧啶存在时,胸腺嘧啶依赖性 SASCVs 可恢复成与野生株相关的表型。

SASCVs 与慢性疾病的关系依然有待进一步研究,其产生机制也有待进一步验证,其在菌株分离鉴定、耐药机制,感染治疗等多个方面还需要深入研究,特别是由 SASCVs 引起的奶牛乳房炎方面的研究目前十分匮乏。本试验成功从患慢性乳房炎奶牛奶样中分离出一株胸腺嘧啶依赖型 SASCVs,无疑为开展 SASCVs 感染的预防和治疗奠定了基础,也为奶牛慢性乳房炎的研究开创了新的方向。

## 参考文献

[1] 尹柏双,李国江.奶牛乳房炎的研究新进展[J].中国畜牧医,2010,37(2):182-184.

[2] 吕国平,王苋,秦丽云.金黄色葡萄球菌的多重 PCR 快速鉴定方法[J].环境与健康杂志,2012,29(10):939-941.

[3] 刘修权,曲伟杰,高健等.金黄色葡萄球菌小菌落突变株研究进展[J].中国兽医杂,2011,47(3):54-57.

[4] 何娟梅,朱军,巢国祥等. 一株金黄色葡萄球菌小菌落突变株的分离鉴定[J]. 微生物学报,2009,49(10):1397-1402.

[5] 陈燕飞.渗透压对细菌的影响[J].太原师范学院学报(自然科学版),2012,11(3):136-139

[6] 杨烨建,雷建女.金黄色葡萄球菌试管法血浆凝固酶试验影响因素的探讨[J].实用预防医学,2013,20(7):882-893.

[7] Atalla H, Gyles C, Jacob C L, et al . Characterization of a *Staphylococcus aureus* small colony variant(SCV) associated with persistent bovin emastitis[J] . Foodborne Pathog Dis,2008,5(6):785-799.

[8] Balwit JM, van Langevelde P, Vann JM, et al. Gentamicinresistant menadione and hemin auxotrophic *Staphylococcus aureus* persist within cultured endothelial cells[J]. The Journal of Infectious Diseases,1994,170 (4):1033-1037.

[9] Besier S, Ludwig A, Ohlsen K, et al. Molecular analysis of the thymidine-auxotrophic small colony variant phenotype of *Staphylococcus aureus*[J]. Int J Med Microbiol,2007,297(4):217-225.

[10] Besier S, Smaczny C, Mallinckrodt CV, et al. Prevalenceand clinical significance of *staphylococcus aureus* small colony variants in cystic fibrosis lung disease. Journal of Clinical Microbiology,2007,45 (1):168-172.

[11] Jacobsen K A. Mitteilungen uiber einen variablen Typhusstamm (Bak t. typhi mutabile) sowie Uiber eigentiim lichen hemmende Wirkung des gewihnlichen Agar, verursacht durch Autok- lavierung[J] . Centralbl f Bakt,1910,56:208.

[12] Jensen J. Biosynthesisofhematin com pounds in a hem in requiring strain of Micrococcus pyogenes var. aureus. I. The significance of coenzyme A for the terminal synthesis of catalase[J] . J Bacteriol,1957,73(3):324-333.

[13] Melter O, Radojevic B. Small colony variants of *Staphylococcus aureus*-review[J]. Folia Microbiol(Praha),2010,55(6):548-558.

[14] Proctor RA, van Langevelde P, Kristjansson M, et al. Persistent and relapsing infections associated with small colony variants of *Staphylococcus aureus*[J]. Clinical Infectious Diseases, 1995, 20 (1): 95 - 102.

[15] Smith K M, Slugoski M D, Loewen S K, et al. The broadly selective human Na +/ nucleoside cotransporter (hCNT3) exhibits novel cation - coupld nucleoside transport characteristics[J]. J Biol Chem, 2005, 280: 25436 - 25449.

[16] Sompolinsky D, Cohen M, Ziv G. Epidemiological and biochemical studies on thiam ine-less dwarf-colony variants of *Staphylococcus aureus* as etiological agent s of bovine mastitis[ J] . Infect Im mun, 1974, 9(2): 217 - 228.

[17] Stroud RM, Santi DV, Hardy LW, et al. Atomic structure of thymidylate synthase: target for rational drug design[J]. Science, 1987, 235 (4787): 448 - 455.

[18] Kahl B, Herrmann M, Everding A S, et al. Persistent infection with small colony variant strains of *Staphylococcus aureus* in patients with cystic fibrosis[J]. J Infect Dis, 1998, 177: 1023 - 1029.

[19] Kahl BC, Duebbers A, Lubritz G, et al. Population dynamics of persistent *Staphylococcus aureus* isolated from the airways of cystic fibrosis patients during a 6-year prospective study[J]. Journal of Clinical Microbiology, 2003, 41 (9): 4424 - 4427.

# 应用蛋白质组学技术对奶牛乳热血浆生物标志物的筛选及其生物信息学分析

舒　适,夏　成,王朋贤,孙宇航,许楚楚,李昌盛

(黑龙江八一农垦大学动物科技学院,大庆 163319)

**摘　要**:奶牛乳热分为临床型低血钙症,即奶牛乳热(Milk Fever),以及亚临床型低血钙症(Subclinical Hypocalcemia)。乳热是以分娩或产后母牛血浆钙浓度降低,产后瘫痪甚至昏迷为主要临床特征的一种营养代谢性疾病。本研究通过应用荧光差异双向凝胶电泳(fluorescence two-dimensional differential gel electrophoresis,2D-DIGE)技术和质谱分析技术分离和鉴定乳热奶牛、亚临床低血钙奶牛和健康奶牛之间的血浆差异表达蛋白,探究这些蛋白在乳热发生过程中所起的作用。结果显示:本试验获得了奶牛乳热和亚临床低血钙症血浆差异表达蛋白的2D-DIGE电泳图谱,运用质谱技术和生物信息学技术鉴定出13种奶牛乳热血浆差异表达蛋白,分别为AOCC、IgM、ALB、ACT、FG、C、FGA、FGB、FGG、A2M、FX、SERPIN和未知蛋白。通过网络生物信息学的Networks分析、GO分析和Pathway分析和搜索,结果提示A2M、FGA、FGB、C(C4A)、ALB、IgM、FX和SERPIN,共8种蛋白,探究了这些差异蛋白与疾病可能存在的关系,及证实这些蛋白在疾病发生中所起的作用,尚需进一步验证。本试验为阐明奶牛乳热发生机制和判定病情与防治效果提供了新的方向。

**关键词**:奶牛乳热,亚临床低血钙症,荧光差异双向凝胶电泳,质谱分析,差异蛋白

# Select the biomarker of milk fever using proteomic technology and bioinformatics

SHU Shi, XIA Cheng, WANG Pengxian, SUN Yuhang, XU Chuchu, LI Changsheng

*College of Animal Science and Veterinary Medicine, Heilongjiang Bayi Agricultural University, Daqing* 163319, *China*

**Abstract**: Hypocalcemia in dairy cows is divided into clinical type, named as milk fever, and subclinical type. Milk fever is an important nutritional and metabolic disease characterized by the plasma Ca concentration under the normal level, paralysis and even coma around calving. To find the differential expressed proteins among the milk fever cows (MF), subclinical hypocalcemia (SH) cows and healthy (C) cows by fluorescence two-dimensional differential gel electrophoresis (2D-DIGE) combined with matrix-assisted laser desorption /ionization time-of-flight mass spectrometry (MALDI-TOF-MS), and to explore effect of these proteins on milk fever in dairy cows. Electrophoretogram of plasma samples from cows with MF, SH and C groups were obtained by 2D-DIGE, 13 differential expressed proteins were separated and identified by MALDI-TOF-MS and the search of NCBI database, and results showed that AOCC, IgM, ALB, ACT, FG, C, FGA, FGB, FGG, A2M, FX, SERPIN and unnamed protein. Eight proteins searched and analyzed in the online analysis of Networks, GO, and Pathway, were A2M, FGA, FGB, C(C4A), ALB, IgM, FX and SERPIN. It suggest that these proteins were maybe related to the milk fever and hypocalcaemia. However, it need to verify these proteins and their relationship with milk fever in the future, and offer a new strategy to interpret mechanism of milk fever and to assess state of an illness and effect of prevention and cure in future.

**Keyworld**: milk fever; subclinical hypocalcemia; 2D-DIGE; MALDI-TOF-MS; differ proteins

---

**资助项目**:本研究由"国家自然科学基金面上项目"(项目编号 30972235)资助

**作者简介**:舒　适(1986-),女,汉族,浙江省,硕士研究生,硕士。临床兽医学专业,动物营养代谢病,E-mail:519296311@qq.com

**通讯作者**:夏　成,E-mail:xcwlxyf@sohu.com

奶牛乳热(Milk fever)又称为产后瘫痪或生产瘫痪,一般发生在分娩时或分娩前后,是一种常见但是发病机理复杂的营养代谢紊乱性疾病,常伴有严重的低血钙症、食欲不振、抽搐等临床症状[1]。奶牛体内正常的血钙浓度一般为2.24~3.05mmol/L。临床上根据患病奶牛血钙水平及临床症状可分为临床型低血钙症,即乳热和亚临床型低血钙症。血钙浓度降低可减弱免疫细胞对病原刺激的应答能力,这有利于感染性疾病的发生,比如乳腺炎,因为它可以减弱平滑肌收缩而使乳头不能正常收缩而导致乳房炎。它也可以降低平滑肌的收缩,使瘤胃和皱胃弛缓,从而导致真胃变位以及采食量降低[2]。对于奶牛乳热发生机制的研究,已经从最初的病理、生理、生化水平上发展到如今的分子水平。但其精确的发生机制仍未完全清楚。

为了更加深入和清楚的探索奶牛乳热及亚临床低血钙症的发生机制,需要借助更多先进的研究方法和技术来深入研究以往传统方法无法解决的乳热发生的病因学问题。比较蛋白质组学的新理论和新技术已成为探讨动物疾病发生动态变化的一个全景式研究手段,它能够全面的发现处在不同患病程度的动物体内蛋白质的整体变化。荧光双向差异凝胶电泳(fluorescence two-dimensional differential gel electrophoresis,2D-DIGE)是比较蛋白质组学中经典的分离和鉴定技术。利用荧光染色技术,2D-DIGE 可在一块胶上同时展现3个样本的蛋白质组情况。通过引入内标,2D-DIGE 能很好的降低样品间的生物学差异以及实验过程中的技术误差,全面直接的展示样品之间的差异程度。

在奶牛发生乳热过程中,应用 DIGE 技术监测奶牛机体蛋白质组的变化,获得患病动物血浆差异蛋白,从比较蛋白质组学的角度探索奶牛乳热、亚临床低血钙的发生机制,将对丰富和发展奶牛乳热的新理论和知识具有重要的科学价值。

# 1 材料和方法

## 1.1 试验动物分组和样品处理

在黑龙江某集约化和规范化的奶牛场,根据其血浆钙浓度及有无临床症状,选取乳热病牛(MF,血钙水平低于1.80mmol/L 且有明显临床症状)、亚临床低血钙组(SH,血钙水平在1.80~2.00mmol/L 且无任何临床表现)和健康对照组奶牛(C,血钙水平在2.00~3.00mmol/L 且无任何临床表现)各9头作为试验动物,应用生物学重复试验方法,分为3个乳热组(MF1、MF2、MF3)、3个亚临床低血钙组(SH1、SH2、SH3)和3个对照组(C1、C2、C3)。采取其血液,并加入 EDTA 抗凝剂分离血浆,-80℃保存。试验奶牛年龄、胎次、体重和血钙浓度见表1。试验奶牛饲喂日粮组成与营养水平:精料8~9 kg,青贮17~20 kg,干草3.5~4 kg,脂肪300~400 g,DM 55.60%,粗蛋白16%,产奶净能7.35 MJ/DM,脂肪5.60%,NDF 39.10%,ADF 20.30%,钙180 g,磷116 g。

**表1 试验奶牛的年龄、胎次和血钙浓度**

Table 1 The age,parity and plasma Ca concentration of tested cows

| 组别 | | 年龄 | 胎次 | 血钙水平(mmol/L) | 血钙平均值(mmol/L) |
|---|---|---|---|---|---|
| MF | MF1 | 6 | 4 | 1.05 | 1.52±0.40 |
| | | 4 | 2 | 1.74 | |
| | | 4 | 2 | 1.76 | |
| | MF2 | 5 | 3 | 1.55 | 1.66±0.09 |
| | | 6 | 4 | 1.70 | |
| | | 5 | 3 | 1.72 | |
| | MF3 | 6 | 4 | 1.56 | 1.61±0.06 |
| | | 6 | 4 | 1.59 | |
| | | 4 | 2 | 1.68 | |

(续表)

| 组别 | | 年龄 | 胎次 | 血钙水平(mmol/L) | 血钙平均值(mmol/L) |
|---|---|---|---|---|---|
| | | 5 | 3 | 1.77 | |
| | SH1 | 4 | 2 | 1.91 | 1.89 ±0.12 |
| | | 4 | 2 | 2.00 | |
| | | 5 | 3 | 1.81 | |
| SH | SH2 | 5 | 3 | 1.83 | 1.87 ±0.09 |
| | | 4 | 2 | 1.97 | |
| | | 5 | 3 | 1.84 | |
| | SH3 | 4 | 2 | 1.87 | 1.88 ±0.04 |
| | | 4 | 2 | 1.92 | |
| | | 4 | 2 | 2.19 | |
| | C1 | 5 | 3 | 2.20 | 2.26 ±0.11 |
| | | 5 | 3 | 2.38 | |
| | | 6 | 4 | 2.20 | |
| C | C2 | 4 | 2 | 2.21 | 2.28 ±0.12 |
| | | 4 | 2 | 2.42 | |
| | | 6 | 4 | 2.22 | |
| | C3 | 4 | 2 | 2.22 | 2.23 ±0.02 |
| | | 6 | 4 | 2.25 | |

## 1.2 血钙检测

血浆钙浓度:应用全自动生化分析仪(modullarDPP,德国罗氏),血浆钙试剂盒(651564-01,德国罗氏)检测。

## 1.3 荧光双向差异凝胶电泳(2D-DIGE)试验

1.3.1 样品处理 本研究试验动物分为3组,即MF组、SH组和C组。每组9头试验奶牛,即9个样品。根据生物学重复试验,每个实验组内3个样品混合成为一个样本(见表1),即MF组:MF1、MF2和MF3;SH组:SH1、SH2和SH3;C组:C1、C2和C3。9个样本等蛋白量混合作为内标。3组样本被随机标记为Cy3和Cy5,内标荧光标记为Cy2。具体试验设计见表2。

表2 DIGE试验设计

Table 2 The design of DIGE study

| 胶号 | Cy2 | Cy3 | Cy5 |
|---|---|---|---|
| Gel1 | Pool | MF1 | SH1 |
| Gel2 | Pool | SH2 | C1 |
| Gel3 | Pool | C2 | MF2 |
| Gel4 | Pool | MF3 | C3 |
| Gel5 | Pool | SH3 | MF1 |

按照Bradford法测定样品蛋白浓度以确定样品凝胶上样量,之后对样品进行高丰度蛋白质的去除,应用

GE 去血清/血浆白蛋白/IgG 等高丰度蛋白亲和柱去除高丰度蛋白质，使凝胶的分辨率提高，以获取更多有价值的地风度蛋白点。再应用密里博 3kd 超滤管对去高丰度蛋白后的样品进行除盐，10 000 × g 离心 30min。

1.3.2 第一向固相梯度等电聚焦电泳 应用专业的胶条槽放置胶条，将处理过的待测血浆样品与水化液混合后，放入胶条，室温放置 30 ~ 60min。用 Amersham Biosciences 的 IPGPhor 等电聚焦仪进行第一向等电聚焦电泳。

1.3.3 第二向 SDS 聚丙烯酰胺凝胶电泳（SDS-PAGE） 将通过第一向等点聚焦后的胶条取出，将其放入平衡缓冲液中。配制分离 14-100ku 范围蛋白的浓度为 12% 的均一的 SDS 聚丙烯酰胺分离凝胶，将已经平衡的 IPG 胶条小心的放入 SDS 凝胶顶部，用 EttanDALT Ⅱ 垂直电泳仪，电泳直至溴酚蓝前沿泳至凝胶的底部时，终止电泳。

1.3.4 图像扫描和分析 应用 Typhoon 9410 扫描仪在 488/520nm，532/580nm，633/670nm 波长分别对 Cy2，Cy3，Cy5 荧光染料标记的图像进行扫描。应用 DeCyder v. 6. 5 图像分析软件对 DIGE 图像进行分析和差异点寻找。快速生物变化分析模块用来捕捉蛋白点，并且与 3 块胶上的所有蛋白进行匹配。这些结果应用 $t$ 检验来分析其显著性。差异点满足条件：$P \leqslant 0.05$，差异倍数 1.5 倍以上。

1.3.5 质谱分析 根据样品第二次定量结果，对样品进行双向凝胶电泳以得到制备胶。将分析胶上差异蛋白点与制备胶相对比，找到制备胶上的相应差异蛋白点，用解剖刀小心切下进行胶内酶解，也可将其冻干备用。然后对质谱结果进行蛋白质鉴定，鉴定搜库的数据库为 NCBI 全库。

# 2 结 果

## 2.1 血浆样品双向凝胶电泳图谱

每块对比荧光胶上的蛋白点分布模式基本一致，选取 3 块差异蛋白点较多的荧光凝胶进行扫描，可获得 3 组试验奶牛血浆蛋白凝胶图像（图 1 ~ 3），通过定量分析，筛选显著性差异表达蛋白位点（$P < 0.05$）。试验结果共找到 110 个差异蛋白点。

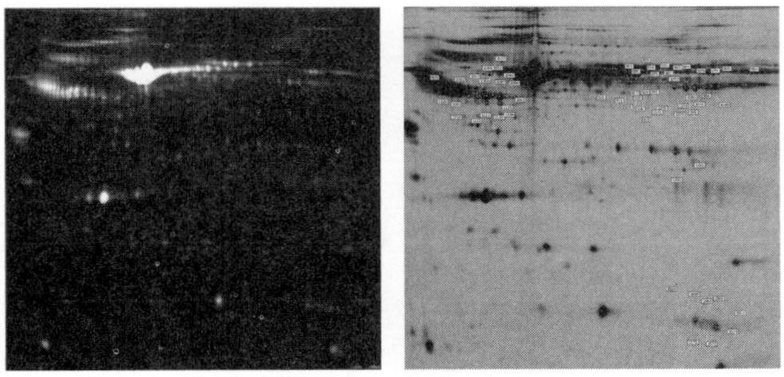

图 1 乳热组与亚临床低血钙组奶牛血浆样品荧光标记差异凝胶电泳图

Fig. 1 The plasma samples fluorescent differential gel electrophoresis of MF and SH

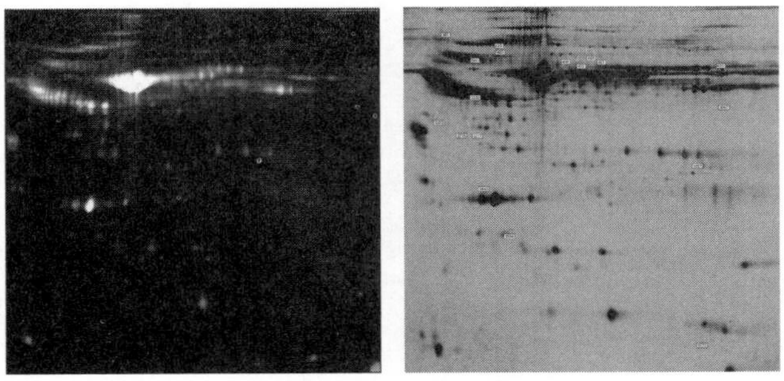

图 2 亚临床低血钙组与对照组奶牛血浆样品荧光标记差异凝胶电泳图

Fig. 2 The plasma samples fluorescent differential gel electrophoresis of SH and C

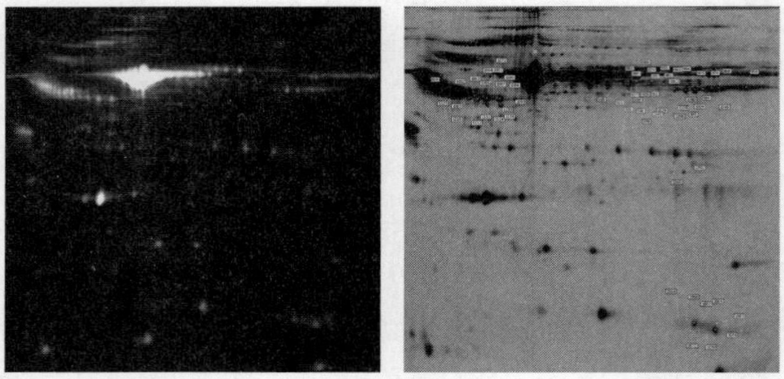

图3 乳热组与对照组奶牛血浆样品荧光标记差异凝胶电泳图

Fig. 3 The plasma samples fluorescent differential gel electrophoresis of MF and C

## 2.2 蛋白质肽质量指纹图谱鉴定

根据双向电泳图像软件的分析结果,选取两两对比中蛋白点较多的三块胶,从胶中切取80个差异蛋白点(fold chang >2, $P<0.05$),胶内酶解进行4800 MALDI-TOF-MS 分析,得到66个质谱鉴定结果以及66套肽质量指纹图谱。对所得到的66套肽质量指纹图谱,在NCBI数据库进行搜索,匹配有意义的蛋白质(蛋白得分 >65, $P<0.05$),并至少有一个肽段匹配,进行综合分析,成功鉴定13种不同的差异表达蛋白,表3显示:

表3 质谱鉴定结果

Table 3 Results of MS

| NCBI 编号 | 蛋白名称 | 蛋白点数量 | 蛋白质得分(mean ± *SD*) | 分子量(*ku*) | 等电点 |
|---|---|---|---|---|---|
| gi\|229552 | albumin | 11 | 262.73 ± 193.01 | 66 087.6 | 5.76 |
| gi\|1245695 | alpha 1-antichymotrypsin; ACT [Bos taurus] | 1 | 176 | 28 552.9 | 5.54 |
| gi\|177872 | alpha-2-macroglobulin | 1 | 82 | 70 750.8 | 5.47 |
| gi\|195539525 | amine oxidase, copper containing 3 [Bos taurus] | 1 | 262 | 84 702.8 | 5.57 |
| gi\|6980814 | Chain A, The Crystal Structure Of Modified Bovine Fibrinogen (At ~4 Angstrom Resolution) | 7 | 278.86 ± 136.48 | 42 688.1 | 8.21 |
| gi\|119904982 | PREDICTED: coagulation factor X [Bos taurus] | 3 | 137.67 ± 21.50 | 54 475.2 | 5.35 |
| gi\|38649251 | Complement component 4A (Rodgers blood group) | 2 | 75 | 192 710.4 | 6.54 |
| gi\|50402019 | complement component C3d [Bos taurus] | | 215 | 34 421.7 | 6.68 |
| gi\|3789962 | fibrinogen A-alpha chain [Bos taurus] | 6 | 334.17 ± 110.11 | 41 905.4 | 6.3 |
| gi\|357 | fibrinogen beta chain [Bos taurus] | 19 | 371.89 ± 134.15 | 48 468.6 | 8.47 |
| gi\|27806893 | fibrinogen, gamma chain [Bos taurus] | 7 | 392.43 ± 64.25 | 50 212.2 | 5.54 |
| gi\|2232299 | IgM heavy chain constant region [Bos taurus] | 4 | 558.75 ± 267.19 | 47 884.7 | 5.68 |
| gi\|31340900 | serpin peptidase inhibitor, clade A, member 3 [Bos taurus] | 2 | 393.00 ± 90.51 | 46 175.2 | 5.57 |
| gi\|12846939 | unnamed protein product [Mus musculus] | 1 | 73 | 15 192.8 | 8.96 |

结果显示的 13 种差异蛋白分别为 albumin(白蛋白,ALB)、antichymotrypsin(抗胰凝乳蛋白酶,ACT)、alpha-2-macroglobulin(α-2 巨球蛋白,A2M)、amine oxidase copper containing(胺氧化酶,AOCC)、Fibrinogen(纤维蛋白原 FG)、coagulation factor X(凝血因子 X,FX)、Complement component(补体成分,C)、fibrinogen A-alpha chain(纤维蛋白原 A 链,FGA)、fibrinogen beta chain(纤维蛋白原 β 链,FGB)、fibrinogen gamma chain(纤维蛋白原 γ 链,FGG)、IgM(免疫球蛋白 M)、serpin peptidase inhibitor(丝氨酸蛋白酶抑制剂,SERPIN)和 unnamed protein(未知蛋白)。

差异表达蛋白在三组比较中的情况见表 4。

**表 4　三组奶牛之间差异表达蛋白的比较**

**Table 4　Comparison of different proteins between three groups**

| 差异蛋白 | | 乳热组(M) | 乳热组(S) | 对照组(C) |
|---|---|---|---|---|
| AOCC | | C↑ | — | M↓ |
| IgM | | C↑ | C | M↓S↑ |
| Unnamed protein | | C↑ | — | M↓ |
| ALB | | C↓S↓ | M↑ | M↑ |
| ACT | | C↓S↓ | M↑ | M↑ |
| FG | | C↓S↓ | M↑C↓ | M↑S↑ |
| C | C3d | C↓ | — | M↑ |
| | C4A | — | C↓ | S↑ |
| FGA | | C↓S↓ | M↑ | M↑ |
| FGB | | C↓S↓ | M↑C↓ | M↑S↑ |
| FGG | | C↓S↓ | M↑ | M↑ |
| SERPIN | | S↑ | M↓C↓ | S↑ |
| A2M | | — | C↓ | S↑ |
| FX | | S↓ | M↑ | — |

注:↑:表示上调蛋白;↓:表示下调蛋白;—:表示无变化。表中 M 代表乳热组,出现时表示出现行的蛋白与乳热组比较蛋白表达情况,S 表示亚临床低血钙组,C 表示对照组。如第一行 AOCC 后,乳热组下单元格为“C↑”,表示:与对照组相比,AOCC 在乳热组中表达上调

## 2.3　生物信息学分析

2.3.1　Networks 分析　应用 *Cytoscape* 软件对试验结果的 13 种蛋白进行牛类基因网络搜索,并得到 Networks 分析结果图。由于 *BOND* 数据库中列出的相互作用蛋白质都是已经得到研究结果的蛋白质,根据搜索结果,本试验结果中得 14 种蛋白质,只有 3 种蛋白质可在数据库中搜索得到结果,分别是 FGB,FGA 和 A2M。并得到其相关的 Networks 分析结果,见图 4。

根据结果可知:A2M 与钙离子相关,通过软件中自带的网络数据连接搜索得到,A2M 参与凝血过程,而钙离子,是凝血过程中一个必要因子,所以,A2M 的变化可能与钙离子的变化相关从而引起疾病的发生。而 α-氨基葡萄糖苷酶与 A2M 之间的相互作用并未搜索到,但 α-氨基葡萄糖苷酶显示与神经系统的开发相关。而 A2M 本身有研究证明它可能在儿童时期的对凝血酶有控制作用。差异倍数较高的 FGA 和 FGB,本身组合成为纤维蛋白原,自然存在于机体凝血过程中,并担任重要作用。

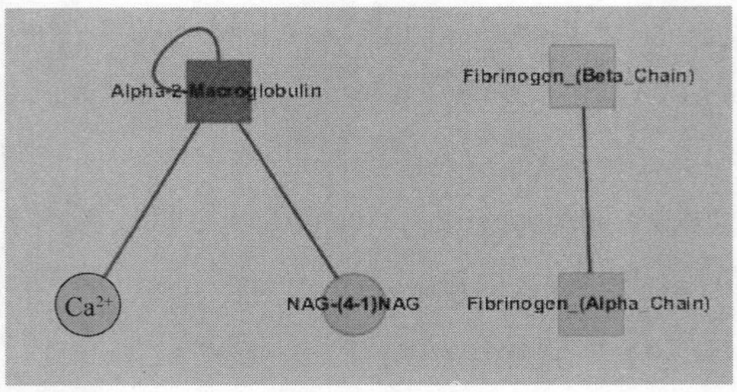


图 4　Networks 分析结果

Fig. 4　Results of Networks Analysis

注:图中节点正方形图标表示差异显著蛋白($P<0.05$),三角形为差异不显著蛋白($P>0.05$),圆形为结果蛋白在基因网络中的相关蛋白;红色为差异倍数在 2 以下,绿色为差异倍数在 2 以上,粉色为结果蛋白在基因网络中的相关蛋白。

Note: the square node means the difference proteins($P<0.05$), thetriangle means non-difference proteins($P>0.05$), the rotundity means the related difference proteins; the red node is under the two fold change, the green is above the two fold change, and the pink is the related difference proteins.

2.3.2　GO 分析结果　应用 Cytoscape 软件中 BinGO 插件对 13 种蛋白进行 GO 分析,并得到 GO 分析结果。在结果中得到的 13 种蛋白质,经输入软件搜库得知,已经注释的蛋白质有 5 种,分别为 A2M, FGB, FGA, ALB 和 C(C4a)。分析结果显示,5 种差异蛋白在 cellular component(细胞组件)、molecular function(分子功能)和 biological process(生物过程)3 个层面上的分别涉及 6 个、14 个和 100 个注释结果,结果见表 5 ~7。

表 5　GO 分析中 cellular component 结果

Table5　Results of cellular component analysis from GO

| GO-ID | Description | Genes |
|---|---|---|
| 5615 | extracellular space | A2M FGA C4A FGB ALB |
| 44421 | extracellular region part | A2M FGA C4A FGB ALB |
| 5576 | extracellular region | A2M FGA C4A FGB ALB |
| 5577 | fibrinogen complex | FGA FGB |
| 5938 | cell cortex | FGA FGB |
| 31091 | platelet alpha granule | FGB |

根据结果可知:分析结果涉及 5 个蛋白质,并得到 6 个注释结果,所涉及的蛋白质均存在于细胞外,而且 FGA 和 FGB 存在于纤维蛋白原复合体中,而纤维蛋白原参与凝血过程,提示可能以该方式与钙离子相关,这与 networks 分析的结果相一致。

表 6　GO 分析中 molecular function 结果

Table 6　Results of the molecular function analysis from GO

| GO-ID | Description | Genes |
|---|---|---|
| 30674 | protein binding, bridging | FGA FGB |
| 5102 | receptor binding | FGA FGB |
| 15643 | toxin binding | ALB |
| 43120 | tumor necrosis factor binding | A2M |

（续表）

| GO-ID | Description | Genes |
|---|---|---|
| 19959 | interleukin-8 binding | A2M |
| 19958 | C-X-Cchemkine binding | A2M |
| 19966 | interleukin-1 binding | A2M |
| 43499 | eukaryotic cell surface binding | FGB |
| 8144 | drug binding | ALB |
| 51087 | chaperone binding | FGB |
| 5504 | fatty acid binding | ALB |
| 43498 | cell surface binding | FGB |
| 19956 | chemokine binding | A2M |
| 33293 | monocarboxylic acid binding | ALB |

分子功能主要表示涉及蛋白质各自的生物功能，分析涉及 4 种蛋白质，并得到 14 个注释结果，而结果整体看来所涉及的蛋白质基本都具有结合作用，并且白蛋白有利于脂肪酸的结合，这提示可能会与导致其他营养代谢病相关。

表 7　GO 分析中 biological process 结果

Table 7　Results of the biological process analysis from GO

| GO-ID | Description | Genes |
|---|---|---|
| 50789 | regulation of biological process | A2M FGA C4A FGB ALB |
| 65007 | biological regulation | A2M FGA C4A FGB ALB |
| 6950 | response to stress | FGA C4A FGB ALB |
| 50896 | reponse to stimulus | FGA C4A FGB ALB |
| 9611 | response to wounding | FGA C4A FGB |
| 65008 | regulation of biological quality | FGA FGB ALB |
| 50776 | regulation of immune response | A2M C4A |
| 2682 | regulation of immune system process | A2M C4A |
| 48583 | regulation of response to stimulus | A2M C4A |
| 7596 | blood coagulation | FGA FGB |
| 50817 | coagulation | FGA FGB |
| 7599 | hemostasis | FGA FGB |
| 30168 | platelet activation | FGA FGB |
| 51258 | protein polymerization | FGA FGB |
| 50878 | regulation of body fluid levels | FGA FGB |
| 42060 | wond healing | FGA FGB |
| 43623 | cellular protein complex assembly | FGA FGB |
| 1775 | cell activation | FGA FGB |
| 6461 | protein complex assembly | FGA FGB |
| 70271 | protein complex biogenesis | FGA FGB |

(续表)

| GO-ID | Description | Genes |
|---|---|---|
| 34622 | cellular macromolecular complex assembly | FGA FGB |
| 34621 | cellular macromolecular complex subunit organization | FGA FGB |
| 65003 | macromolecular complex assembly | FGA FGB |
| 43933 | macromolecular complex subunit orgnization | FGA FGB |
| 22607 | cellular component assembly | FGA FGB |
| 44085 | cellular component biogenesis | FGA FGB |
| 45916 | negative regulation of complement activation | A2M |
| 1869 | negative regulation of complement activation, lectin pathway | A2M |
| 2921 | negative regulation of humoral immune response | A2M |
| 1868 | regulation of complement activation, lectin pathway | A2M |
| 45824 | negtive regulation of innate immune response | A2M |
| 2698 | negative regulation of immune effector process | A2M |
| 10955 | negative regulation of protein maturation by peptide bond cleavage | A2M |
| 30449 | regulation of complement activation | A2M |
| 2673 | regulation of acute inflammatory response | A2M |
| 10953 | regulation of protein maturation by peptide bond cleavage | A2M |
| 70613 | regulation of protein processing | A2M |
| 50777 | nagative regulation of immune response | A2M |
| 2920 | regulation ofhumoral immune response | A2M |
| 31348 | negative regulation of defense response | A2M |
| 2683 | negative regulation of immune system process | A2M |
| 45088 | regulation of innate immune response | A2M |
| 50727 | regulation of inflammatory response | A2M |
| 48585 | negative regulation of response to stimulus | A2M |
| 2697 | regulation of immune effectorproess | A2M |
| 1897 | cytolysis bysymbint of host cells | ALB |
| 51801 | cytolysis of cells in other organism involved in symbiotic interaction | ALB |
| 51715 | cytolysis of cells of anither organism | ALB |
| 44004 | disruption by symbiont of host cells | ALB |
| 19836 | hemolysis by symbiont of host erythrocytes | ALB |
| 44179 | hemolysis of cells in other organism | ALB |
| 52331 | hemolysis of cells in other organism involved in symbiotic interaction | ALB |
| 1907 | killing by symbiont of host cells | ALB |
| 51659 | maintenance mitochondrion location | ALB |
| 52332 | modification by organism of cell membrane in other organism in symbiotic interaction | ALB |
| 52025 | modification by sybiont of host cell membrane | ALB |

（续表）

| GO-ID | Description | Genes |
|---|---|---|
| 52043 | modification by sybiont of host cellular componet | ALB |
| 52111 | modification by sybiont of host structure | ALB |
| 52188 | modification of cellular component in other organism involved in symbiotic interaction | ALB |
| 52185 | modification of structure of other organism involved in symbiotic intercation | ALB |
| 44003 | modification by sybiont of host morphology or physiology | ALB |
| 51657 | maintenance of organelle location | ALB |
| 51818 | disruption of cells of other organism involved in symbiotic interaction | ALB |
| 51883 | killing of cells in other organism involved symbiotic interaction | ALB |
| 51646 | mitochondrion localization | ALB |
| 51817 | modification of morphology or phsiology of other organism involved in symbiotic interaction | ALB |
| 31640 | killing of cells of another organism | ALB |
| 51701 | interaction with host | ALB |
| 1906 | cell killing | ALB |
| 9267 | cellular response to starvation | ALB |
| 42594 | response to starvation | ALB |
| 44403 | sybiosis, encompassing mutualism through parasitism | ALB |
| 31669 | cellular response to nutrient levels | ALB |
| 51651 | maintenance of location in cell | ALB |
| 44419 | interspecies interaction between organisms | ALB |
| 71496 | cellular response to external stimulus | ALB |
| 31668 | cellular response to extracellular stimulus | ALB |
| 51640 | organelle localization | ALB |
| 51235 | maintenance of location | ALB |
| 31667 | response to nutrient levels | ALB |
| 9991 | response to extracellular stimulus | ALB |
| 6958 | complement activation, classical pathway | C4A |
| 2455 | humoral immune response mediated by circulating immunoglobulin | C4A |
| 2541 | activation of plasma proteins involved in acute inflammatory response | C4A |
| 6956 | complement activation | C4A |
| 6959 | humoral immune response | C4A |
| 19724 | B cell mediated immunity | C4A |
| 16064 | immunoglobulin mediated immune response | C4A |
| 2250 | adaptive immune response | C4A |
| 2460 | adaptive immune response based on somatic recombination of immune receptors built from immunoglobulin suerfamily domains | C4A |

(续表)

| GO-ID | Description | Genes |
|---|---|---|
| 2449 | lymphocyte mediated immunity | C4A |
| 51605 | protein maturation by peptide bond cleavage | C4A |
| 2526 | acute inflammatory response | C4A |
| 2443 | leukocyte mediatred immunity | C4A |
| 2253 | activation of immune response | C4A |
| 16485 | protein processing | C4A |
| 51604 | protein maturation | C4A |
| 51592 | response to calcium ion | FGB |
| 10038 | response to metal ion | FGB |
| 10035 | response to inorganic substance | FGB |

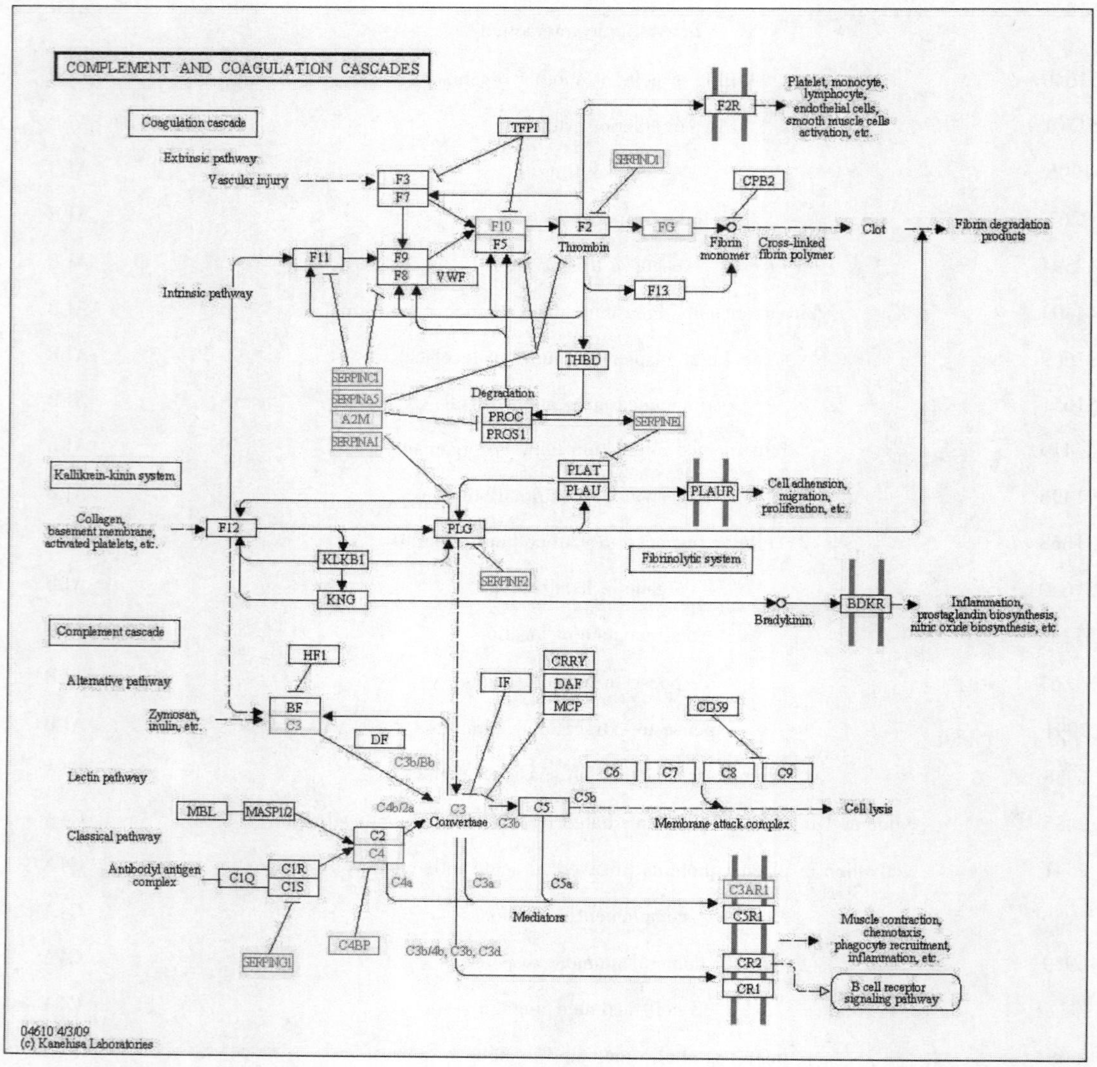


图 5　KEGG pathway 凝血与补体连锁通路，编号 map04610

Fig. 5　The Complement and Coagulation Cascades pathway from KEGG PATHWAY Database, ID: map04610

注：图中标红色的为差异蛋白

Note: the red means the difference proteins

生物过程主要表示涉及蛋白质参与哪些生理过程，分析得到 100 个注释结果，结果显示差异蛋白基本参与应答类的生物过程，其中 FGA 和 FGB 存在于凝血过程中，再次印证之前的分析结果。

2.3.3 Pathway 分析结果　应用 KEGG(Kyoto Encyclopedia of Genes and Genomes) pathway 数据库搜索结果中的 13 种蛋白。由于 KEGG 数据库中列出的相互作用蛋白质都是已经得到研究结果的蛋白质，根据搜索结果，本试验结果中得到 13 种蛋白质，只有 7 种蛋白质可在数据库中搜索得到结果，分别是 A2M、FGB、IgM、FGA、FX、SERPIN、C(C4A 和 C3d)，并得到其相关的 Pathway 分析结果，见图 5 ~6。

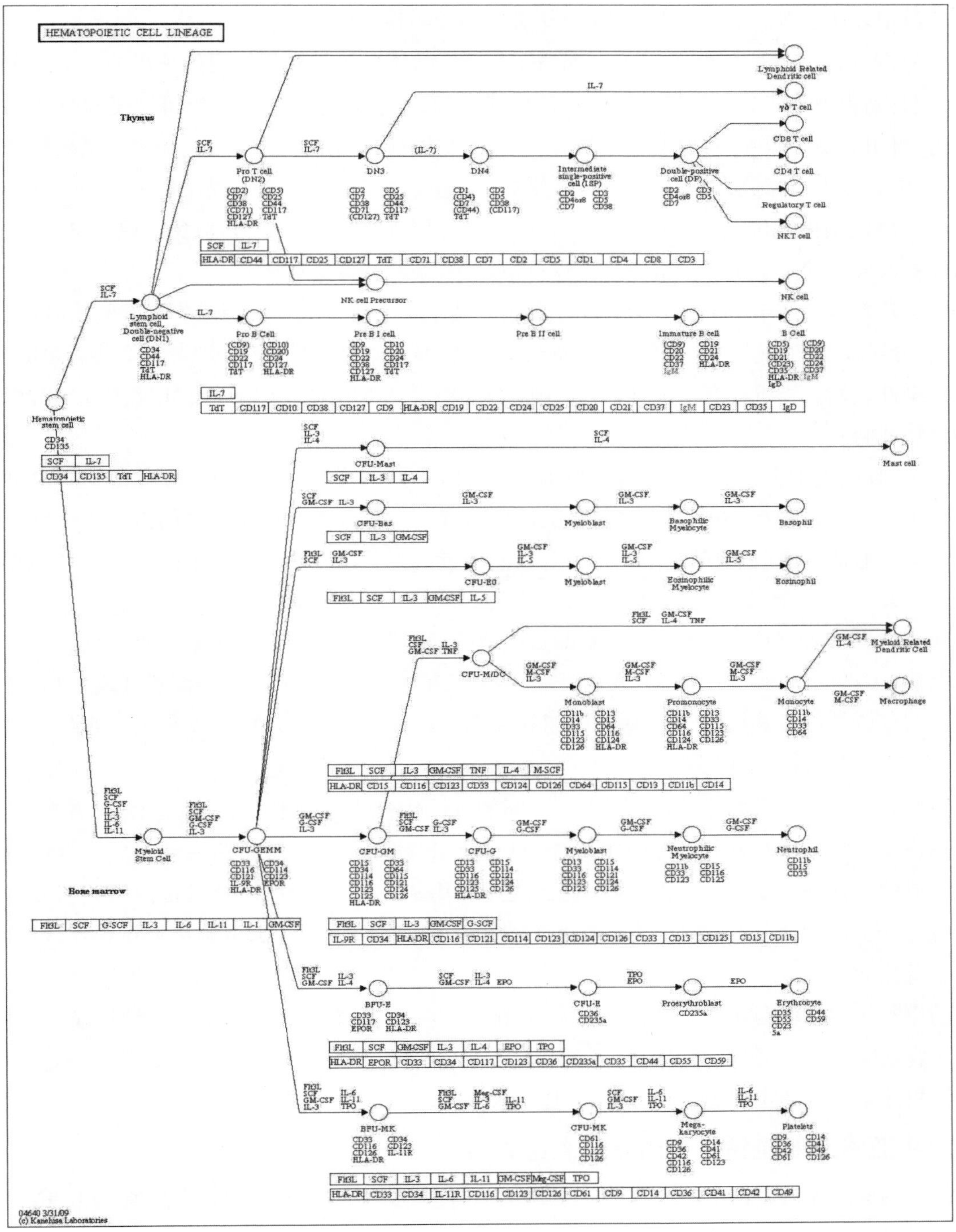


图 6　KEGG pathway 造血细胞系通路，编号 map04640

Fig. 6　The Hematopoietic cell lineage pathway from KEGG PATHWAY Database, ID: map04640

注：图中标红色的为差异蛋白

Note: the red means the difference proteins

根据结果可知,有 6 种蛋白涉及,提示这些蛋白可能通过该过程对钙离子产生影响,导致疾病的发生,而此再次印证之前的讨论结果。

造血细胞系通路中涉及 IgM 这一种蛋白,如图可见,免疫球蛋白 M 出现在免疫过程中,这可能提示当奶牛发生疾病时所引起的炎症反应而引起的免疫球蛋白 M 表达的变化。

# 3 讨 论

## 3.1 与奶牛乳热发生相关的差异表达蛋白

在乳热组中表达上调的差异蛋白有 3 种,分别为 AOCC、IgM 和未知蛋白。

胺氧化酶(AOCC)是可以对一些主要一元胺,二元胺以及组胺起氧化作用的一组酶类。此类酶除能分解组胺外,在肠黏膜中还能分解由氨基酸脱羧所生成的胺,起着解毒作用[3]。当机体受到理化刺激或发生过敏反应时,可引起细胞脱颗粒,导致组胺释放,与组胺受体结合而产生生物效应。所以当机体发生刺激或过敏反应时,组胺会大量释放,之后机体会产生氨氧化酶进行分解。免疫球蛋白 M(Immunoglobulin M,IgM)主要由脾脏和淋巴结中浆细胞分泌合成,为 190ku。它主要分布在血液中,是对免疫原最早出现的抗体,所以它是机体初次应答的重要抗体。抗体出现越早,对机体保护也就越有效[4]。

在奶牛乳热发生过程中 AOCC 和 IgM 两个蛋白表达上调。这可能提示当奶牛分娩和泌乳时,随着体内钙大量流失,骨钙动员、肠钙吸收等不足以补充流失的钙,而发生低血钙症时,奶牛免疫功能也随之减退,可能会引发感染性疾病。当机体有炎症存在时,组胺释放促使 AOCC 释放增多。IgM 增多以应对机体可能存在的各种疾病威胁。

本试验中,乳热组中表达下调的差异蛋白有 7 种,分别为 ALB、ACT、FG、C(C3d)、FGA、FGB 和 FGG。

血清白蛋白是一条含有 580 ~585 个氨基酸的单肽链。它是血液中存在最多的蛋白质,合成于肝脏[5]。本试验虽去除了血清白蛋白,但结果中白蛋白却以差异蛋白的形式出现,这也可能提示血浆中的白蛋白与奶牛乳热的发生有关,也为此蛋白与奶牛乳热疾病的关系奠定了研究的基础。

α-抗胰凝乳蛋白酶(ACT)糖蛋白是丝氨酸蛋白酶的抑制剂,基因位于 14q3211,近年研究证实它参与了阿尔茨海默病(Alzheimerdisease,AD)的发病,该病又称老年性痴呆,如 ACT 可与淀粉样蛋白相互作用,影响淀粉样蛋白的分解并加速其形成[6]。ACT 可能导致老年痴呆症,乳热奶牛时常表现精神沉郁和瘫痪等症状,是否与它有关需要进一步证实。

纤维蛋白原是一个血浆糖蛋白,在炎症反应和凝血过程中有重要的作用,并且它有一个潜在的功能是调节和粘附在许多呼吸道病原体上[7],同时它包括 FGA、FGB 和 FGG。纤维蛋白原转变成纤维蛋白是凝血过程中一个重要的过程,而 $Ca^{2+}$ 在这个过程中起到促进作用。在凝血过程中,纤维蛋白原大部分转化成纤维蛋白,同时消耗大量的 $Ca^{2+}$,可能是血浆钙浓度降低的一个因素。从而引起亚临床低血钙症,也可能严重至临床型奶牛乳热症。

补体的生理作用为辅助和补充特异性抗体,介导免疫溶菌、溶血作用[8]。本研究结果显示,该补体成分在奶牛发生乳热时表达下调,可能提示奶牛发生乳热时,机体防御能力增强,使得补体系统中各种成分激活,大量的辅助机体抵御各种疾病的侵袭,从而使其表达降低。

## 3.2 与奶牛亚临床低血钙相关的差异表达蛋白

在本研究的亚临床低血钙组中没有表达上调的差异蛋白,而表达下调的蛋白质有 SERPIN、A2M、IgM、FG、C(C4A)和 FGB。

Serine peptidase inhibitor 是一类以其独特的功能、相同结构和功能性质而被证实的抑制蛋白酶类[9]。本研究结果显示,丝氨酸蛋白酶抑制剂在奶牛发生亚临床低血钙时表达下调,可能提示由于其与钙离子同时存在于凝血与补体连锁通路中,所以当奶牛机体血浆钙浓度降低时,联系两者的凝血与补体连锁通路发挥了作用,使其双方表达下降。

巨球蛋白是一类分子质量大于400ku的球蛋白。alpha-2-macroglobulin是其中之一,分子量约720ku,是一种蛋白酶抑制因子。它主要是在肝脏上皮试纸细胞合成[10]。研究表明,它在儿童期对凝血酶起着重要的作用[11]。根据生物信息学分析中Networks分析结果显示A2M与$Ca^{2+}$相关,提示A2M参与了$Ca^{2+}$的调控机制。

### 3.3 奶牛乳热与亚临床低血钙相关的差异表达蛋白

本研究结果显示,相对于亚临床低血钙组,乳热组中SERPIN表达上调。但是SERPIN相对于对照组中,在亚临床低血钙组中表达下调。故SERPIN表达水平为:乳热组 > 亚临床低血钙组 < 对照组,但乳热组与对照组之间无差异。因此,SERPIN表达的变化与亚临床低血钙症、乳热之间关系并不一致,造成这一问题的机理还需要深入探讨。

然而,相对于亚临床低血钙组,乳热组表达下调的蛋白为ALB、ACT、FX、FG、FGA、FGB和FGG。

根据GO分析中生物过程(biological process)的结果显示,FGA、C4A、FGB、ALB参与机体的应激应答(response to stress)和刺激应答(response to stimulus)。奶牛发生乳热或亚临床低血钙时都会引起炎症反应,从而导致各种围产期疾病。在发生疾病过程中机体产生应激应答或刺激应答,而为了保护机体,相应的蛋白就会消耗,故FGA、C4A、FGB、ALB等蛋白含量就会降低。

## 4 结 论

本实验应用荧光差异凝胶电泳和质谱分析获得了奶牛乳热、亚临床低血钙奶牛和健康奶牛血浆差异表达蛋白点用图谱和66套血浆差异蛋白点的一、二级质谱图,并应用NCBI数据库鉴定了66个差异蛋白点为10种蛋白质,分别为: ALB、ACT、A2M、AOCC、FG、FX、C、FGA、FGB、FGG、IgM、SERPIN、unnamed protein。而且对结果进行讨论得到,这些差异蛋白提示它们可能与奶牛乳热发生有关,但仍需进一步验证。为丰富和发展奶牛血浆蛋白组学研究奠定了基础。

## 参考文献


[1] 李英,李增和,陈云鹏. 奶牛产期瘫痪的原因与预防措施[J]. 中国牛业科学,2011,37(1):94-95.

[2] Timothy A. Reinhardt, John D. Lippolis, Brian J. McCluskey, Jesse P. Goff, Ronald L, Horst. Prevalence of subclinical hypocalcemia in dairy herds[J]. The Veterinary Journal,2011,188:122-124.

[3] 王红珊,王红梅,吴琳,简美贞. 二胺氧化酶在肝癌组织中的基因表达[J]. 临床医药实践杂志,2007,16(9).

[4] Marianne D.,Lene J.,E. B. An in vivo characterization of colostrum protein uptake in porcine gut during early lactation[J]. J. Proteomics. 2011,74:101-109.

[5] Peters,T. 1975. Putman FW. ed. The plasma proteins. Academic Press. 133-181.

[6] Eriksson S,Janciauskiene S,Lannfelt L. A1-antichymotripsin regulates Alzheimer beta-amyloid peptide fibril formation[J]. Proc Natl Acad Sci USA,1995,92:2313-2317.

[7] Gary O'Donovan,Edward Kearney,Roy Sherwood,et al. 2012. Fatness,fitness,and cardiometabolic risk factors in middle-aged white men. Metabolism-Clinical and Experimental 61(2):213-220.

[8] Goralski KB,Sinal CJ. 2007. Type 2 diabetes and cardiovascular disease:getting to the fat of the matter. Can J Physiol Pharmacol 85:13-32.

[9] Whasun Lim,Ji-Hye Kim. Avian SERPINB11 Gene:Characteristics,Tissue-Specific Expression,and Regulation of Expression by Estrogen[J],BOR Papers in Press. 2011,(8):1-2.

[10] Andus T,Gross V,Tran-Thi T-A,et al. The biosynthesis ofacute-phase proteins in primary cultures of rat hepatocytes[J]. Eur J Biochem,1983,133(5):61-71.

[11] 侯秀荣,包承鑫 1A-2 巨球蛋白在儿童期的抗凝血作用[J]. 中国危重病急救医学,1994,6(6):380-382.

# 云南家畜皮肤弹性正常值的测定

黄绍义*,张莹[1],雷晓琴[1],郭成裕[1**],李子龙[2**]

(1. 云南农业大学动物科学技术学院,昆明 650201;2. 云南呈贡县职业中学,昆明 650500)

**摘　要**:2012 年 10 月至 2014 年 1 月,对云南马、驴、骡、水牛、黄牛、乳牛、绵羊、山羊、猪等 9 种健康动物的皮肤弹性正常值进行了测定,结果表明马、驴、骡颈侧部皮肤皱褶恢复原状所需的时间分别为(0.92 ±0.10)s、(0.87 ±0.11)s、(0.82 ±0.13)s,肩胛后胸侧部皮肤皱褶恢复原状所需的时间分别为(0.52 ±0.12)s、(0.47 ±0.10)s、(0.44 ±0.09)s;水牛、黄牛、乳牛颈侧部皮肤皱褶恢复原状所需的时间分别为(1.04 ±0.12)s、(1.06 ±0.10)s、(1.08 ±0.11)s,肋弓后腹侧部皮肤皱褶恢复原状所需的时间分别为(0.53 ±0.12)s、(0.60 ±0.11)s、(0.55 ±0.10)s;绵羊、山羊腰背部皮肤皱褶恢复原状所需的时间分别为(0.48 ±0.12)s、(0.37 ±0.10)s;猪肋弓后腹侧部皮肤皱褶恢复原状所需的时间为(0.54 ±0.12)s。

**关键词**:家畜,皮肤弹性,正常值

# Measurements on the normal values of livestock's skin elasticity in Yunnan Province

HUANG Shaoyi[1*],ZHANG Ying[1],LEI Xiaoqin[1],GUO Chengyu[1**],LI Zilong[2**]

1. *Faculty of Animal Science and Technology*,*Y A U*,*Kunming* 650202

2. *Chenggong Vocational Middle School of Agriculture*,*Kunming* 650500

**Abstract**: From October 2012 to January 2014, the normal value of skin elasticity of 9 kinds of healthy animals (horse, donkey, mule, buffalo, cattle, cow, sheep, goat, pig) in Yunnan were measured. The results indicated as follow: Horses, donkeys, mules sides of neck skin folds restitution time were (0.92 ±0.10)s、(0.87 ±0.11)s、(0.82 ±0.13)s, after the shoulder side of the chest skin folds restitution time were (0.52 ±0.12)s、(0.47 ±0.10)s、(0.44 ±0.09)s; Buffalos, cattles, cows sides of neck skin folds restitution time were (1.04 ±0.12)s、(1.06 ±0.10)s、(1.08 ±0.11)s, the costal arch ventral skin folds after the reinstatement of the time were (0.53 ±0.12)s、(0.60 ±0.11)s、(0.55 ±0.10)s; Sheep, goat skin folds back the time required for restitution were (0.48 ±0.12)s、(0.37 ±0.10)s; After the ventral part of the pigs' costal arch skin folds restitution time were (0.54 ±0.12)s.

**Keywords**: Livestock, skin elasticity, normal value

在兽医临床上,动物皮肤弹性常用于检查引起动物机体脱水的疾病(如胃肠炎、便秘疝、瘤胃积食、真胃阻塞等)和失血性疾病(如大出血等),它是判断动物脱水或失血程度的重要参考指标[1]。同时也用于慢性皮肤病(如湿疹、螨病等),是判断皮肤受损程度的参考指标。然而,到目前为止,在兽医临床上尚未见到一个有关反映家畜和家禽皮肤弹性客观指标的正常值,以供临床兽医工作者参考。因此,笔者于 2012 年 10 月

* **作者简介**:黄绍义,(1990 - ),男,山东省菏泽市,硕士研究生,临床兽医方向。E-mail:13199090726@126.com

** **通讯作者**:郭成裕,教授,E-mail:guolidoc@126.com;李子龙,讲师,E-mail:augustaaa@126.com

至2014年1月，对云南主要家畜的皮肤弹性正常值进行了测定，现报告如下，供参考。

# 1 材料与方法

## 1.1 供测动物及测定期间的外界温度

1.1.1 马、驴、骡 2012年10月，在宾川县牛井镇对64匹年龄为4～13岁（公30匹、母34匹）的大理马进行了测定；对62匹年龄为3～12岁（公30匹、母32匹）的云南驴进行了测定；对66匹年龄为3～13岁（公32匹、母30匹）的骡进行了测定。测定期间的外界平均温度为18.4℃。

1.1.2 水牛、黄牛 2014年1月，在寻甸县河口乡对62头3～11岁（公30头、母32头）云南水牛进行了测定；对63头3～12岁（公30头、母33头）云南滇中黄牛进行了测定。测定期间的外界平均温度为11.6℃。

1.1.3 乳牛 2012年10月，在大理市喜洲镇对33头3～12岁（均为母牛）黑白花奶牛进行了测定。测定期间外界平均温度为16.4℃。

1.1.4 绵羊、山羊 2014年1月，在寻甸县河口乡对68只2～8岁（公30只、母38只）云南细毛羊进行了测定；对70只2～8岁（公30只、母40只）云南山羊进行了测定。测定期间外界的平均温度为11.6℃。

1.1.5 猪 2014年1月，在寻甸县河口乡对62头3～4月龄（公10头、去势40头、母12头）杜（杜洛克猪）长（长白猪）乌（云南乌金猪）三元杂交猪进行了测定。测定期间外界的平均温度为11.6℃。

## 1.2 供测动物的其他情况

1.2.1 供测的家禽家畜都是发育良好，营养中等或中上等，饮食欲正常，无任何异常的健康动物。

1.2.2 供测的家畜都经过体温、脉搏频率（心率）、呼吸频率三项检查，确认属于正常范围之后才进行皮肤正常值的测定。每天都在北京时间8:00～11:00进行测定。

## 1.3 计时工具

瑞士产HEUER牌秒表。

## 1.4 测定方法

1.4.1 测定部位 根据动物种类的不同，我们选取了不同的部位测定动物皮肤的弹性[2]。马、驴、骡的测定部位为颈侧部和肩胛后胸侧部两个部位；水牛、黄牛、和牛的测定部位为颈侧部和肋弓后腹侧部两个部位；绵羊、山羊的测定部位为腰背部；猪的测定部位为肋弓后腹侧部。用手将皮肤捏成皱褶的方向。

颈侧部 在颈部1/2处的颈侧部，用手将该部皮肤捏成皱褶的方向与颈部的纵轴垂直。

肩胛后胸侧部 在距离肩胛骨的肩胛后角3cm处的胸侧部，用手将该部皮肤捏成皱褶的方向与胸部的纵轴垂直。

肋弓后腹侧部 在距离肋弓后3cm的腹侧部，用手将该处的皮肤捏成皱褶的方向与腹部的纵轴垂直。

腰背部 在腰背部，用手将该部皮肤捏成皱褶的方向与腰背部的纵轴垂直。

1.4.2 测定的方法（以动物左侧为例） 畜主或助手保定好动物，让动物保持自然站立姿势。检查者站立于被检查动物的左侧，面向动物的被检查部位；检查者左手持秒表（检查颈侧部时，右手持秒表），用右手检查颈侧部时，用左手拇指、食指、中指和无名指将动物检查部位的皮肤捏成皱褶，皮肤皱褶高为3cm，然后放开皮肤皱褶，在放开皮肤皱褶的同时按动秒表，并观察皮肤皱褶变化情况，当皮肤皱褶恢复原状时，则按停秒表，记录秒表所示时间，即为皮肤皱褶恢复原状所需的时间[3]。

# 2 结 果

测定结果见附表

附表　云南家畜皮肤弹性正常值($\bar{x} \pm s$)

Table.　The normal values of livestock's skin elasticity in Yunnan

| 动物 | n | 皮肤褶皱恢复原状所需时间(s) | | | |
|---|---|---|---|---|---|
| | | 颈侧部 | 肩胛后胸侧部 | 肋弓后腹侧部 | 腰背部 |
| 马 | 64 | 0.92 ±0.10 | 0.52 ±0.12 | | |
| 驴 | 62 | 0.87 ±0.11 | 0.47 ±0.10 | | |
| 骡 | 66 | 0.82 ±0.13 | 0.44 ±0.09 | | |
| 水牛 | 62 | 1.04 ±0.12 | | 0.53 ±0.12 | |
| 黄牛 | 63 | 1.06 ±0.10 | | 0.60 ±0.11 | |
| 乳牛 | 33 | 1.08 ±0.11 | | 0.55 ±0.10 | |
| 绵羊 | 68 | | | | 0.48 ±0.12 |
| 山羊 | 70 | | | | 0.37 ±0.10 |
| 猪 | 62 | | | 0.54 ±0.12 | |

* 注:由于所测定的每一种动物左侧和右侧相同部位的皮肤皱褶恢复原状所需要的时间平均数间的差异在统计学上无显著差异,故表中未列出右侧各部位的正常值

## 3 讨　论

3.1　对于马、驴、骡、水牛、黄牛、乳牛的颈侧部和马、驴、骡的肩胛后胸侧部皮肤皱褶恢复原状所需时间的平均数间差异显著性检验以及水牛、黄牛、奶牛和猪的肋弓后腹侧部皮肤皱褶恢复原状所需时间的平均数间差异显著性检验,笔者采用 $F$ 检验,因 $F$ 检验不显著,笔者就未再继续进行平均数间的多重比较。而同种动物不同部位皮肤皱褶恢复原状所需时间的平均数间差异显著性检验,笔者采用 $T$ 检验。结果表明:马、驴、骡颈侧部皮肤皱褶恢复原状所需时间比水牛、黄牛、乳牛的快,但差异不显著($P > 0.05$)。马、驴和骡肩胛后胸侧部皮肤皱褶恢复原状所需时间,水牛、黄牛、乳牛和猪肋弓后腹侧部皮肤皱褶恢复原状所需时间,绵羊和山羊腰背皮肤皱褶恢复原状的时间差异均不显著($P > 0.05$)。而马、驴、骡肩胛后胸侧部皮肤皱褶恢复原状所需时间比颈侧部的皮肤快($P < 0.05$);水牛、黄牛、乳牛肋弓后腹侧部皮肤皱褶恢复原状所需时间比颈侧部的皮肤快($P < 0.05$),这是否与两个部位皮肤结构差异和皮下脂肪厚度的不同等因素有关,还待今后作进一步的研究。

3.2　由于我们对每种动物测定头数较少,故未进行不同性别,年龄之间的比较和分析,这有待今后作进一步研究。

## 参考文献

[1]　东北农学院.兽医临床诊断学[M].第二版.北京:农业出版社,1995.

[2]　东北农业大学.兽医临床诊断学[M].第三版.北京:中国农业出版社,2009.46 - 47.

[3]　王俊东,刘宗平.兽医临床诊断学[M].第二版.北京:中国农业出版社,2008.

# 重金属镉离子人工抗原的合成与鉴定

韩盈盈,李小兵,刘国文,孔 涛,郎广平,王 哲
(吉林大学畜牧兽医学院,长春 130062)

**摘 要**:选用螯合剂 Isothiocyanobenzy-EDTA(ITCBE)络合重金属镉离子,再分别与载体蛋白 BSA 和 KLH 相连接,初步制备免疫抗原 Cd-ITCBE-KLH 和检测抗原 Cd-ITCBE-BSA。通过 BCA 法测定完全抗原浓度,并利用 SDS-PAGE 电泳、紫外分光光度法及石墨炉原子吸收法对所合成抗原进行初步鉴定;BCA 法测出 Cd-ITCBE-BSA、Cd-ITCBE-KLH 及 ITCBE-BSA 的实际蛋白浓度依次为:1.426、1.504、0.890mg/ml。SDS-PAGE 电泳和紫外分光光度法定性的说明抗原合成成功,石墨炉原子吸收法测得完全抗原中镉离子含量最高可达 44.56μg/ml,其他无金属抗原、载体蛋白和 HEPES 中镉离子含量几乎为零,定量的说明了抗原合成成功。

**关键词**:镉,人工抗原,制备,鉴定

# Synthesis and identification of artificial antigen for heavy metal cadmium ions

HAN Yingying, LI Xiaobing, LIU Guowen, KONG Tao, LANG Guangping, WANG Zhe
*Collgeg of Veterinary Medicine, Jinlin University, Changchun* 130062, *China*

**Abstract**: Cadmium ions was coupled to Keyhole Limpet Hemocyanin (KLH) and Bovine Serum Albumin (BSA) via Isothiocyanobenzy-EDTA to gain complete antigen (Cd-ITCBE-KLH and Cd-ITCBE-BSA). The antigen concn. was measured by BCA method; ultraviolet scan of spectrophotometer and SDS-PAGE electrophoresis qualitative identification was done and the content of cadmium ions in antigen was determined by GFAAS method. The actual protein concn. of Cd-ITCBE-BSA, Cd-ITCBE-KLH and ITCBE-BSA were in the sequence of 1.426, 1.504 and 0.890mg/ml, ultraviolet scan and electrophoresis identification proved that the antigen synthesis was successful. Concentration of cadmium ion in complete antigen was measured as high as 44.56μg/ml, and that of cadmium ion in ITCBE-BSA, BSA, KLH and HEPES were almost zero, which indicated the hapten was surely coupled into protein, and further testified the antigen synthesis was successful.

**Keywords**: Cadmium; Artifical antigen; Sythesis; Identification

重金属是指相对密度较高的金属化学元素。重金属污染给生态环境和人类带来极大威胁,已成为世界性问题。因此,重金属污染是食品、环境、卫生监测的重要内容[1]。重金属一旦残留于环境中将会持续很长一段时间,当其与土壤或沉淀物结合时,包括:气候、环境湿度、水和土壤中的 PH 的改变以及向环境中释放有机质等因素,都会造成重金属的动员并且大大的增加其毒性作用。镉是大气环境中毒性较强的重金属污

**基金项目**:吉林省世行贷款农产品质量安全项目(2011-242)

**作者简介**:韩盈盈,女,在读硕士,研究方向:兽医内科学
**通讯作者**:王 哲,E-mail:wangzhe500518@sohu.com

染物,土壤中的镉化学活性很强,极易被作物吸收而进入食物链,从而在人和动物体内富集,尤其是富集于肝脏和肾脏。镉的毒性作用及其机制在国内外已有许多报道[2,3]。实验室研究结果表明,镉对机体免疫系统有强大的抑制作用[4],可使子代小鼠的T细胞计数等免疫指标表现出迟发、刺激和抑制作用[5]。当其在人和动物体内的量累积到一定水平就会导致镉中毒,镉中毒损害呼吸器官功能、影响钙磷代谢,使骨骼更多地吸收镉而不是钙,导致骨质疏松易骨折,同时也大大增加癌症、肾功能紊乱和高血压的风险,影响人类健康[6,7]。重金属免疫学检测是重金属检测的一种新方法,与传统的检测方法相比,具有检测速度快、易于操作、处理量少、高灵敏度和高特异性等优点,可用于环境及食品中重金属残留的现场检测和常规检测,自1985年Reardan等为重金属检测提供了一种新的免疫学方法以来[8],重金属快速免疫学检测技术成为国内外实验室研究的热点,Khosraviani等利用竞争ELISA法测定了环境水样中镉离子的含量[9]。Darwish等采用一步竞争性免疫检测法对人血清中镉离子进行了分析[10],但是这些方法特异性方面存在不足。本试验利用双功能螯合剂初步制备了镉螯合剂人工抗原,并通过SDS-PAGE电泳、紫外分光光度法等几种方法对所合成抗原进行了初步鉴定。目的是制备出高特异性的镉的单克隆抗体,为建立胶体金快速免疫检测法奠定基础。

# 1 材料与方法

## 1.1 材料与试剂

硝酸镉(99.999%)、牛血清白蛋白(bovine serum albumin, BSA)、钥孔血蓝蛋白(keyhole limpet hemocyanin, KLH)购自美国Sigma-Aldrich公司;Isothiocyanobenzyl-EDTA购自日本同仁化学研究所;N-2-Hydroxyethylpiperazine-N'-2'-ethanesulfonic Acid(HEPES)购自美国Promega公司;盐酸(优级纯)、硝酸(优级纯)购自鼎国生物技术有限公司;Centricon-30超滤离心管购于Millipore公司,其他试剂均为分析纯。

## 1.2 仪器与设备

紫外分光光度计(UV-2501PC,日本岛津);紫外可见光分光光度计(Varian Cary 500 UV-Vis-NIR, Varian, Palo Alto, CA);石墨炉原子吸收分光光度计(Varian Spectr AA 220Z, Varian, Palo Alto, CA)。磁力搅拌器(GL-3250B, QILINBEIER);电子分析天平(BS-124S SARTORIUS);低速离心机(TGL-16B, Anke);低温超速离心机(3K06123B, KUBOTA);酶标仪(ELX800,美国Bio-TEK);超纯水仪为Millipore产品;TS-1脱色摇床为江苏海门市麒麟医用仪器厂产品;Power Pac200电泳仪为美国Bio-Rad公司产品;Gel Doc2000™/Chemi Doc™凝胶成像系统为美国Bio-Rad公司产品。

## 1.3 抗原的合成

取5.6mg ITCBE溶于5.6ml新鲜DMSO中,制备ITCBE溶液。取14.8mg的$Cd(NO_3)_2 \cdot 4H_2O$溶于14.8ml的去离子水中,制备$Cd(NO_3)_2 \cdot 4H_2O$溶液。取KLH与BSA各10mg分别溶于5ml pH 7.5的Hepes缓冲液中,制备KLH和BSA溶液,并用1mol/L的NaOH或HCl调其pH至7.5。

1.3.1 Cd-ITCBE-KLH的制备　取0.36ml ITCBE溶液于青瓶中,搅拌状态下逐滴加入1.234ml的$Cd(NO_3)_2 \cdot 4H_2O$溶液,调pH至8.5,4℃搅拌过夜形成半抗原溶液。取2ml KLH于青瓶中,将已配制好的半抗原溶液逐滴加入其中,4℃搅拌24h制备Cd-ITCBE-KLH完全抗原溶液。

1.3.2 Cd-ITCBE-BSA和ITCBE-BSA的制备　取1.125ml ITCBE溶液于青瓶中,搅拌状态下逐滴加入3.856ml $Cd(NO_3)_2 \cdot 4H_2O$溶液,调其pH至7.5,4℃搅拌过夜。取2.5ml BSA于另一青瓶中,逐滴加入上述螯合剂溶液,4℃搅拌24h,得到Cd-ITCBE-BSA完全抗原溶液。同样取2.5ml BSA于另一青瓶中,搅拌情况下逐滴加入1.125ml ITCBE溶液,4℃搅拌24h,形成ITCBE-BSA无金属抗原溶液。

上述偶联反应结束后,用Centeicon-30超滤离心管对蛋白质复合物进行分离纯化。

## 1.4 抗原的鉴定

1.4.1 抗原浓度测定　参照BCA试剂盒的操作步骤,以BSA作为标准蛋白,以570nm处的OD值作为$y$

值,以浓度为 $x$ 值。构建标准曲线,

1.4.2 抗原 SDS-PAGE 电泳 用 5% 的浓缩胶,15% 的分离胶,对 BSA、ITCBE-BSA、Cd-ITCBE-BSA 进行垂直 SDS-PAGE 电泳,上样量为 10μl/孔。

1.4.3 完全抗原、无金属离子抗原、载体蛋白的紫外分光光度法检测 以 0.1mol/L,pH 为 7.5 的 HEPES 作为空白对照,并且利用该 HEPES 缓冲液对 Cd-ITCBE-BSA、ITCBE-BSA、BSA、Cd-ITCBE-KLH 及 KLH 做稀释处理,制备 0.2mg/ml 的待检溶液。在 260 ~ 500nm 波长范围内进行紫外扫描。

1.4.4 抗原中镉离子含量的检测 以 0.1 mol/L,pH 7.5 的 HEPES 作为空白对照,参照国际 GB/T 17141—1997,用石墨炉原子吸收光谱法检测抗原中镉离子的浓度。[9]

# 2 结 果

## 2.1 BCA 法检测蛋白浓度

利用 BCA 法测定分离纯化好的完全抗原及无金属抗原 Cd-ITCBE-BSA、Cd-ITCBE-KLH 及 ITCBE-BSA 的蛋白浓度依次为:1.426、1.504、0.890mg/ml。

## 2.2 抗原 SDS-PAGE 电泳

由图 1 可知泳道 2 的 Cd-ITCBE 比泳道 1 的载体蛋白 BSA 迁移速度慢,泳道 3 的 Cd-ITCBE-BSA 比泳道 2 的 Cd-ITCBE-BSA 迁移速度慢,这说明泳道 2 中 Cd-ITCBE 的分子量大于泳道 1 中 BSA 的分子量,而泳道 3 中 Cd-ITCBE-BSA 的分子量大于泳道 2 中 Cd-ITCBE 的分子量,所以条带依次滞后,且每条泳道中的蛋白条带均无杂带,表明合成抗原成分较均一。初步说明抗原合成成功。

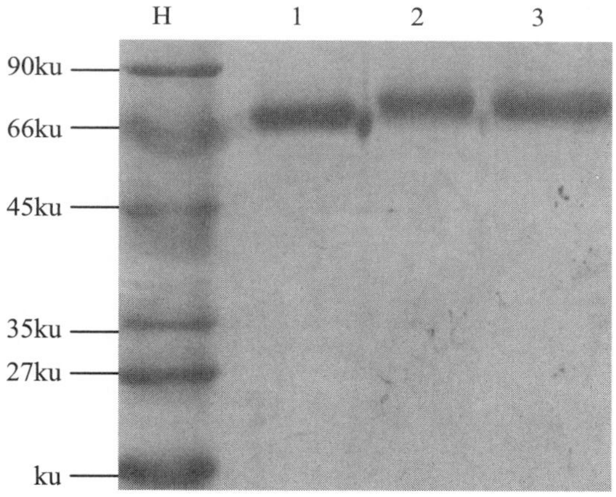


图 1 BSA 及其相应抗原的 SDS-PAGE 电泳

Fig. 1 The SDS-PAGE electrophoretogram of BSA and corresponding antigen

注:泳道 H 为 marker;泳道 1 为 BSA;泳道 2 为 ITCBE-BSA;泳道 3 为 Cd-ITCBE-BSA

Note:H:Marker;1:BSA;2:ITCBE-BSA;3:Cd-ITCBE-BSA

## 2.3 紫外分光光度法检测图谱

由图谱可知:BSA、KLH 的最大吸收峰在 278nm、279nm,免疫抗原 Cd-ITCBE-KLH 最大吸收峰在 270nm,检测抗原 Cd-ITCBE-BSA、ITCBE-BSA 最大吸收峰分别为 268nm 与 271nm 。与载体蛋白 KLH、BSA 相比,免疫抗原和检测抗原最大吸收峰发生了改变,且吸收峰值单一,可进一步说明抗原合成成功。

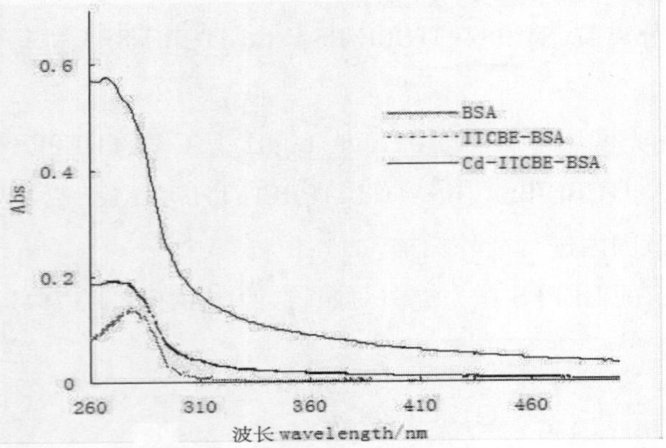


图 2　检测抗原的紫外图谱

Fig. 2　The UV spectra of BSA, ITBBE-BSA and Cd-ITCBE-BSA

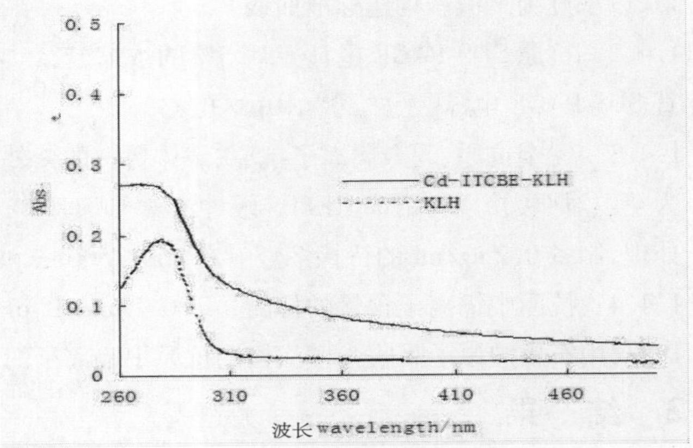


图 3　免疫抗原的紫外图谱

Fig. 3　The UV spectra of KLH and Cd-ITCBE-KLH

## 2.4　石墨炉原子吸收光谱法测定抗原中镉离子含量

如表 1 所示:通过该方法检测镉完全抗原中镉离子含量达 44.56μg/ml,制备的无金属抗原以及 HEPES 中镉离子含量几乎为零,表明抗原合成成功。

表 1　人工抗原中镉离子含量

Table 1　The concentration of $Cd^{2+}$ in artificial antigen

| 人工抗原 | 镉离子含量(μg/ml) |
|---|---|
| Cd-ITCBE-BSA | 28.55 ±1.23 |
| Cd-ITCBE-KLH | 44.56 ±0.47 |
| ITCBE-BSA | 0.013 ±0.00 |
| BSA | $1.15 \times 10^{-3}$ ±0.22 |
| KLH | $0.29 \times 10^{-3}$ ±0.11 |
| HEPES | $1.11 \times 10^{-3}$ ±0.15 |

## 3　结论与讨论

镉特异性单克隆抗体的制备是重金属镉免疫学检测方法建立的基础,而镉特异性单克隆抗体制备的关键在于镉人工抗原的成功合成。由于镉离子属于小分子物质,不能形成抗原表位,不能被机体免疫系统识别产生免疫应答,所以必须通过某种方式将其制备成完全抗原才能实现此目的。本试验就是通过双功能螯合剂 ITCBE 螯合镉离子形成金属螯合物后,再与载体蛋白 KLH 与 BSA 相偶联,得到相应的免疫抗原和检测抗原。SDS-PAGE 电泳图谱结果显示出完全抗原、无金属抗原及载体蛋白的条带依次滞后,初步说明抗原偶联成功;紫外分光光度法检测结果表明完全抗原 Cd-ITCBE-KLH 与 Cd-ITCBE-BSA 的最大吸收峰值都与载体蛋白有明显差异,同样说明抗原偶联成功;并通过石墨炉原子吸收分光光度法对其进行了定量分析,通过这一系列鉴定,证实抗原合成成功。为镉单克隆抗体的成功制备以及重金属免疫学检测方法的建立奠定了基础。

## 参考文献

[1]　Blake DA, Jones RM, Blake RC, et al H. Antibody-based sensors for heavy metal ions. Biosensors and Bioelectronics 2001; 16: 799 - 809.

[2] Buchko GW, Hess NJ, Kennedy MA. Cadmium mutagenicity and human nucleotide excision repair protein XPA: CD, EXAFS and 1H/15N-NMR spectroscopic studies on the zinc (II)-and cadmium (II)-associated minimal DNA-binding domain (M98 – F219). Carcinogenesis 2000;21:1051 – 1057.

[3] Bialkowski K, Bialkowska A, Kasprzak KS. Cadmium (II), unlike nickel (II), inhibits 8-oxo-dGTPase activity and increases 8-oxo-dG level in DNA of the rat testis, a target organ for cadmium (II) carcinogenesis. Carcinogenesis 1999;20:1621 – 1624.

[4] Yücesoy B, Turhan A, Üre M, et al A. Effects of occupational lead and cadmium exposure on some immunoregulatory cytokine levels in man. Toxicology 1997;123:143 – 147.

[5] 黄旭,钟赛贤. 小鼠孕期染镉对子代免疫功能的影响[J]. 卫生毒理学杂志 1998;12:86 – 88.
HUANG Xu, ZHONG Sai-qin. Effects of exposure of cadmium during pregnancy on immune function of offspring in mice. [J]. Health Toxicology. 1998;12:86 – 88. (in Chinese)

[6] Satarug S, Baker JR, Urbenjapol S, et al. A global perspective on cadmium pollution and toxicity in non-occupationally exposed population. Toxicology Letters 2003;137:65 – 83.

[7] 贾广宁. 重金属污染的危害与防治[J]. 有色矿冶 2004;20:39 – 42.
JIA Guang-ning. On the damage and control of heavy metal pollution. [J] Nonferrous Metals. 2004;20:39 – 42.

[8] Bontidean I, Lloyd JR, Hobman JL, et al. Bacterial metal-resistance proteins and their use in biosensors for the detection of bioavailable heavy metals. Journal of inorganic biochemistry 2000;79:225 – 229.

[9] Khosraviani M, Pavlov AR, Flowers GC, et al. Detection of heavy metals by immunoassay: optimization and validation of a rapid, portable assay for ionic cadmium. Environmental science & technology 1998;32:137 – 142.

[10] Darwish IA, Blake DA. Development and validation of a one-step immunoassay for determination of cadmium in human serum. Analytical Chemistry 2002;74:52 – 58.

# 猪链球菌宿主多样性调查

臧莹安[1],郭海翔[1],吴斯宇[1],张创峰[1],李　淼[2,3],宋　帅[2,3],李春玲[2,3]
(1.仲恺农业工程学院动物科学系,广州510225;2.广东省农业科学院兽医研究所,广州510640;
3.广东省兽医公共卫生公共实验室,广州510640)

**摘　要**:利用PCR方法对表观健康的猪肉、禽、鱼和狗随机抽取317份样品进行猪链球菌检测。结果显示猪肉样品中4份呈阳性,阳性率为5.4%;禽口腔拭样中15份呈阳性,阳性率为10.6%;鱼和犬样品皆为阴性。进一步对阳性株进行分型鉴定,19株中9型7株,22型4株,1型、2型、16型及24型各1株,还有4株未知型号。对分离出来的19株菌株进行毒力岛GI4、GI8、GI12、毒力基因lin和rmp检测,结果为猪肉中分离出来的一株2型猪链球菌对GI8、GI12、lin阳性。分别用从猪肉中分离出的2型、鹧鸪中分离出的9型、鸽子中分离的9型、鸡中分离的1型、研究室保存的2型菌株9801对斑马鱼进行攻毒试验,结果各半数致死量($LD_{50}$)依次为$4.27\times10^4$CFU、$2.72\times10^4$CFU、$5.18\times10^3$CFU、$2.25\times10^4$CFU、$6.38\times10^5$CFU。

**关键词**:猪链球菌,分型鉴定,半数致死量

# Survey of host diversity of *Streptococcus suis*

ZANG Ying'an,GUO Haixiang[1],WU siyu[1],ZHANG ChuangFeng[1],LI Miao[2,3],
SONG Shuai[2,3],LI ChunLing[2,3]
1. *Zhongkai University of Agriculture and Technology*,*Guangzhou* 510225;
2. *Institute of Veterinary*,*Guangdong Academy of Agricutral Sciences*,*Guangzhou* 510640;
3. *Guangdong Key Laboratory of Veterinary Public Health*,*Guangzhou* 510640

**Abstract**:PCR technology was used to detect streptococcus suis from 317 samples of apparent healthy pork、poultry、fish and dog. The result was 4 samples positive of pork,the positive rate was 5.4%,15 samples positive of poultry,the positive rate was 10.6%; Negative on dog and fish. Further to serotype positive strains,there were 7 strains for 9 type,4 strains for 22 type,and 1 type、2 type、16 type、24 type were 1 strain,4 unknow. One type 2 strain isolated from pork contains the pathogenicity island of GI8、GI12 and virulence genes of lin. Finally,the zebrafish challenged with the 2 type isolated from pork,9 type isolated from partridge,9 type isolated from pigeon、1 type from chicken and type 2 HA9801 keeped by the laboratory. The LD50 were ordinal $4.27\times10^4$CFU、$2.72\times10^4$CFU、$5.18\times10^3$CFU、$2.25\times10^4$CFU、$6.38\times10^5$CFU。

**Keywords**: *Streptococcus suis*; isolation and identification;$LD_{50}$

猪链球菌广泛存在于自然界,已有不少关于猫、犬、牛、羊、马、鹿等啮齿类动物,甚至鸟类和鱼类感染本病的报道[1,2]。苍蝇也能在猪场内或不同场间传播本病[3],这些动物的排泄物和分泌物又成为新的感染源,进一步加大了更大范围感染的机会[4]。可见猪链球菌具有非常广泛的宿主范围,但各种动物对该菌的易感

**作者简介**:臧莹安(1971-),女,博士,教授,主要从事兽医临床及兽药研发教学与科研工作。E-mail:quietmail@126.com
**通讯作者**:李春玲,E-mail:lclclare@163.com。本研究受到广东省科技计划项目(项目编号:2012A020602052)、国家大学生创新计划项目(1134712046)和广东省社会发展项目(项目编号:2011B031500004)的资助

风险和水平等缺乏必要的研究。所以,本研究拟对广东地区的与人类密切相关的动物(猪、禽、鱼、狗)进行猪链球菌携带情况本底调查,进一步掌握广东地区链球菌在不同宿主间的分布情况及分子流行病学特点,为广东地区猪链球菌的防控提供科学数据。

# 1 材料和方法

## 1.1 材料

1.1.1 主要仪器 PCR 仪,购自上海宝生物有限公司;DYY-6D 型电泳仪电源,购自北京市六一仪器;全自动数码凝胶图像分析系统—Tanon—1600,购自上海天能科技有限公司; HZQ-F16 振荡培养箱,购自中国·哈尔滨市东联电子技术开发有限公司。

1.1.2 培养基 新生牛血清、血平板均购自广东环凯微生物科技有限公司;托—休二氏液体培养基(Todd-Hewitt broth,简称 THB)和托—休二氏固体培养基(Todd-Hewitt Agar,简称 THA)均为自行配制。

1.1.3 PCR 引物序列 针对 SS 、SS2、SS1、SS7、SS9,根据 S. suis 荚膜多糖基因(Capsular polysaccharide, CPS) cps1I、cps2J、cps7H 和 cps9H 序列分别设计 5 对特异性引物(5’-3’)分别为:CAGTATTTACCGCATGG-TAGATAT GTAAGATACCGTCAAGTGAGAA;GTTGAGTCCTTATACACCTGTT CAGAAAATTCATATTGTCCACC;GGCGGTCTAGCAGATGCTCG GCGAACTGTTAGCAATGAC;GAATCAATCCAGTCAGTGTTGG CTAATTCGAT-ACGAAGCTAAAC;GGCTACATATAATGGAAGCCC CCGAAGTATCTGGGCTACTG。片段大小分别为:294,461,441,541,388bp。

1.1.4 猪链球菌分型试剂盒 猪链球菌分型试剂盒,购自杭州微生物试剂有限公司。

1.1.5 斑马鱼 80 日龄健康斑马鱼 800 尾,雌、雄各半,购自西朗斑马鱼养殖场。

## 1.2 方法

1.2.1 样品的采集 从广州一大型农贸市场采取表观健康猪肉、活禽、活淡水鱼、部分冰冻咸水鱼以及佛山一狗养殖场的表观健康狗随机采样;猪肉在采样时用消毒后的手术刀切开猪肉暴露切口,再采取棉签拭样,其他的皆为口腔拭样。将棉签拭样放入装有 3mlTHB 液体培养基的试管中,于 37℃恒温摇床 180r/min 中震荡培养 18h。

1.2.2 猪链球菌的分离 THB 液体培养基中的菌液划线接种于牛血清琼脂平板,于 37 度的温箱中恒温培养 18 ~24h。符合猪链球菌菌落特征的可以初步判断为猪链球菌。

1.2.3 猪链球菌的 16S rRNA 的 PCR 鉴定 PCR 扩增采用 25μl 反应体系。在反应管中依次加入 10 × PCR Buffer2.5μl、25 mmol/L $MgCl_2$ 1.0μl、2.5mmol/L dNTPs2.0μl、10 pmol/μl 引物 1.0μl、模板 1.0μl、5 U/μl *Taq* 酶 0.2μl,然后加 $ddH_2O$ 调整终体积至 25μl。94℃变性 5 min 后进入循环,94℃45 s,54℃ 45s,72 ℃ 1min,35 个循环后 72 ℃保温 10 min。将得到的 PCR 产物用 1.5% 的琼脂糖凝胶进行电泳。取 7μl PCR 产物和 1μl 6 × buffer 上样缓冲液充分混匀置 2.0% 琼脂糖凝胶(含溴化乙锭替代物 Goldview)电泳。电压 5 V/cm,电流 120mA,用凝胶成像仪观察结果并拍照。

1.2.4 猪链球菌的血清分型 分别采用 PCR 方法(参照猪链球菌的 16S rRNA 的 PCR 鉴定)和凝集方法进行。

1.2.5 猪链球菌基因岛的测定 参照猪链球菌 16SrRNA 的 PCR 鉴定方法,但退火温度设为 55℃,mrp 在 72℃延伸 1min,GI4、GI8、GI12 在 72℃延伸 1.2min,lin 在 72℃延伸 3min。

1.2.6 猪链球菌对斑马鱼的攻毒试验 分别选取从猪肉、鸽子、鸡、鹧鸪中分离到的 2 型、9 型、1 型、9 型猪链球菌以及由广东省农科院兽医研究所猪病室保存的 H 本次试验腔注射接种 10μl 悬液。同时设对照组,注射等量 PBS。接种后,各组分开饲养于不同的水族箱中,定时观察。接种后 96h,统计死亡数量,计算半数致死量。

## 2 试验结果

### 2.1 猪链球菌的 PCR 检测

74 份表观健康的猪肉,经过 PCR 扩增确定 4 份为猪链球菌阳性样品,总阳性率为 5.4%。141 份家禽口腔拭样,经过 PCR 扩增确定 15 份为猪链球菌阳性样品,总阳性率为 10.6%。68 份鱼口腔拭样,和 34 份犬口腔拭样,经过 PCR 扩增,未检出阳性样品

### 2.2 猪链球菌的血清型分型结果

19 株猪链球菌分型结果为:1、2、16 和 24 型各 1 株,19 型 7 株,22 型 4 株其他未知型号 4 株。

### 2.3 基因岛检测结果

从猪肉中分离出来的一株 2 型猪链球菌对 GI8、GI12、lin 阳性。

### 2.4 斑马鱼攻毒试验结果

注射高剂量菌液的斑马鱼在接种后 6h 游动开始变缓慢,12h 后陆续出现病症,表现为腹部鼓胀,肛门和腮部周围出血,部分死亡。剖检死鱼可观察到鱼体内脏器出血,腹腔积水多,为典型的细菌性败血症病理变化。注射低剂量菌液的斑马鱼出现病变较迟。或病变不明显。对照组斑马鱼与注射 PBS 前无异,没有死亡(表 1)。

表 1 斑马鱼攻毒试验结果

| 菌株及来源 | 组别 | 感染菌量(CFU) | 菌量对数 | 死亡数(只) | 死亡率(死亡数/感染数) |
|---|---|---|---|---|---|
| 猪肉中分离的 2 型 | 1 | $6\times10^7$ | 7.778 | 13 | 13/20 |
| | 2 | $6\times10^6$ | 6.778 | 12 | 2/20 |
| | 3 | $6\times10^5$ | 5.778 | 11 | 0/20 |
| | 4 | $6\times10^4$ | 4.778 | 10 | 0/20 |
| | 5 | $6\times10^3$ | 3.778 | 8 | 0/20 |
| | 6 | $6\times10^2$ | 2.778 | 8 | 0/20 |
| | 对照 | 注射灭菌 TSB | | 0 | 0/20 |
| 实验室保存菌株 9801 | 1 | $2.8\times10^7$ | 7.447 | 17 | 17/20 |
| | 2 | $2.8\times10^6$ | 6.447 | 15 | 2/20 |
| | 3 | $2.8\times10^5$ | 5.447 | 13 | 0/20 |
| | 4 | $2.8\times10^4$ | 4.447 | 11 | 0/20 |
| | 5 | $2.8\times10^3$ | 3.447 | 9 | 0/20 |
| | 6 | $2.8\times10^2$ | 2.447 | 7 | 0/20 |
| | 对照 | 注射灭菌 TSB | | 0 | 0/20 |
| 鸽子中分离的 10 型 | 1 | $4.25\times10^7$ | 7.628 | 16 | 13/20 |
| | 2 | $4.25\times10^6$ | 6.628 | 13 | 2/20 |
| | 3 | $4.25\times10^5$ | 5.628 | 10 | 1/20 |
| | 4 | $4.25\times10^4$ | 4.628 | 8 | 3/20 |
| | 对照 | 注射灭菌 TSB | | 0 | 0/20 |

（续表）

| 菌株及来源 | 组别 | 感染菌量（CFU） | 菌量对数 | 死亡数（只） | 死亡率（死亡数/感染数） |
|---|---|---|---|---|---|
| 鸡中分离的 1 型 | 1 | $5\times10^7$ | 7.699 | 19 | 19/20 |
| | 2 | $1\times10^7$ | 7.000 | 15 | 15/20 |
| | 3 | $5\times10^6$ | 6.699 | 11 | 11/20 |
| | 4 | $1\times10^6$ | 6.000 | 9 | 9/20 |
| | 5 | 5×105 | 5.699 | 6 | 6/20 |
| | 对照 | 注射灭菌 TSB | | 0 | 0/20 |
| 鹧鸪中分离的 9 型 | 1 | $5\times10^7$ | 7.699 | 20 | 20/20 |
| | 2 | $1\times10^7$ | 7.000 | 13 | 11/20 |
| | 3 | $5\times10^6$ | 6.699 | 10 | 8/20 |
| | 4 | $1\times10^6$ | 6.000 | 7 | 4/20 |
| | 5 | $5\times10^5$ | 5.699 | 5 | 8/20 |
| | 对照 | 注射灭菌 TSB | | 0 | 0/20 |

记录斑马鱼死亡数，根据 $LD_{50}$ 计算公式 $\log LD_{50}=\Sigma 1/2(X_i+X_{i+1})(P_{i+1}-P_i)$ 计算出 Hps 对斑马鱼半数致死量（$LD_{50}$），其中 $X_i$，$X_{i+1}$ 表示相邻两组的剂量对数，$P_{i+1}$，$P_i$ 表示动物死亡百分率。计算结果为猪肉中分离出的 2 型、研究室保存的 9 801 菌株、鸽子中分离的 9 型、鸡中分离的 1 型、鹧鸪中分离出的 9 型的猪链球菌的半数致死量分别为 $4.27\times10^4$CFU、$6.38\times10^5$CFU、$5.18\times10^3$CFU、$2.25\times10^4$CFU、$2.72\times10^4$CFU。

## 3 讨论与分析

本次调查中，74 份猪肉样品有 4 份检出猪链球菌，阳性率为 5.4%；141 份禽口腔拭样有 15 份检出，阳性率为 10.6%，其中鸡、鸭、鸽、鹌鹑和鹧鸪口腔拭样的阳性率分别为 2.27%、10%、20.83%、14.71% 和 10.53%，鸽子感染率最高，且多为 7 型，鸭、鹌鹑和鹧鸪的感染率都比猪高。鱼和犬样品中未发现猪链球菌。需完善市场禽类准入的检测方法或检测标准，尽量减少人群接触猪链球菌[5,6]。养殖场应结合猪场实际情况，对飞鸟采取相关措施，防止其携带病菌进入猪舍。另外，禽类养殖场工作人员也要注意防范，身体有伤口时尽量避免接触禽类。

猪链球菌共有 35 种血清型，其中 1、2、7、9 型毒力较强，是检测的重点[7]。对阳性菌株中采用 PCR 方法鉴定此 4 种血清型，其他型采用玻片凝集鉴定，确定其中 1 型、2 型、16 型、24 型各 1 株，9 型有 7 株，22 型有 4 株，检出的强毒力菌株占了大多数，可见防范猪链球菌在日常生活中不可忽视。猪链球菌主要通过伤口或经口感染人，人感染猪链球菌与其所从事的职业有密切关系，与猪或猪肉密切接触的人员，如屠宰场的工人及猪场饲养员较其他人群感染的几率高得多，国外学者将人感染猪链球菌称为人类的动物源性职业病[8]。

基因组岛（genomic island）又称基因岛，是毒力岛（Pathogenicity island）和一类在组成结构、进化来源上与毒力岛类似，功能却不仅限于编码细菌毒力的基因组结构的统称[9]。本次试验检测了分离到的 19 株猪链球菌，仅分离出 1 株 2 型猪链球菌对 GI8、GI12、lin 阳性。没有分离到我国强毒株所独有的 GI4 毒力岛。目前的研究未能证明毒力岛与菌株的致病性有直接挂钩，但是毒力岛能经基因水平转移获得，可使细菌基因组进化在短期内发生“量的飞跃”，直接或间接增强细菌的生态适应性，与病原菌的致病性密切相关[10,11]。

濮俊毅[12]等人此前已经证实斑马鱼能够作为研究猪链球菌感染的动物模型。本试验用猪肉中分离出的 2 型、鹧鸪中分离出的 9 型、鸽子中分离的 9 型、研究室保存的 9 801 菌株对斑马鱼进行攻毒，结果各半数致死量（$LD_{50}$）依次为 $4.27\times10^4$CFU、$2.72\times10^4$CFU、$5.18\times10^3$CFU、$6.38\times10^5$CFU。斑马鱼在接种 12h 后陆续出现典型的细菌性败血症症状,肛门周围和腹部出血,腹腔内积水等。猪链球菌对斑马鱼的 $LD_{50}$ 试验,国内外鲜有报道,用斑马鱼代替猪、兔子、豚鼠等试验动物进行测定,大大减少试验的成本。本次试验通过猪链球菌感染斑马鱼半数致死量的测定获得了猪链球菌对斑马鱼的感染剂量范围,为进一步研究猪链球菌毒力提供了参考依据。

## 参考文献


[1] 汪华,胡晓抒,朱凤才,等.人猪链球菌感染性综合征的流行病学调查[J].现预防学,2000,27 (3):312 -314.

[2] Yu H,Jing H,Chen Z,et al. Human Streptococcus suis outbreak,Sichuan,China[J]. Emerg Infect Dis,2006,1(6):914 - 920.

[3] 刘纪成,张敏,李建柱,等.猪链球菌毒力因子的研究进展[J].黑龙江畜牧兽医(科技版),2012:23 - 25.

[4] Rigden DJ,Botzki A,Nukui M,et al. Design of new benzoxa - zole-2-thione-derived inhibitors of Streptococcus pneumoniae hyaluronan lyase:structure of a complex with a 2-phenylindole[J]. Glycobiology,2006,16(8):757 - 765.

[5] 王海丽,赵德明,葛长城,等.猪链球菌的分离鉴定及其生物学特性试验[J].中国兽医杂志,2012,48(7):36 - 39.

[6] 毕祥乐,张春梅,朱晓飞.猪链球菌病综合防制方略[J].湖南农机,2011,38(5):213 - 218.

[7] Straw B E,Zimmerman J J,Allaire S D,et al. Diseases of Swine[M]. 9th ed. Iowa :Blackwell Publishing Professional,2006 : 769 - 784.

[8] Leelarasamee A,Nilakul C,Tien G S,et al. Streptococcus suis toxic-shock syndrome and meningitis[J]. Med Assoc Thai,1997, 80(1):63 - 68.

[9] Vela A I,Moreno M A,Cebolla J A,et al. Antimicrobial Susceptibility of Clinical Strain s of *Streptococcus suis* Isolated from Pigs in Spain[J] . Vet Microbiol,2005,105(2):143.

[10] Peters E D J,Hall M A L,Box A T,et al. Novel gene cassettes and integrons[J]. Antimicrob Agents Chemother,2001,45(6): 2961 - 2963.

[11] Courvalin P. Transfer of antibiotic resistance genes between gram positive and gram negative bacteria [J]. Antimicrob Agents Chemother,1994,38(6):1447 - 1451.

[12] 濮俊毅,黄新新,陆承平.用斑马鱼检测猪链球菌2 型的致病力[J].中国农业科学,2007,40,(11):2655 - 2658.

# 自制中药复方片剂对小鼠急性毒性试验研究

王　静，何生虎，葛　松，郭树强

（宁夏大学农学院，银川 750021）

**摘　要**：为了验证中药复方片剂的安全性，确定中草药配伍提取后制成的片剂的毒性，试验采用经典的小鼠急性毒性试验方法测定小鼠半数致死量（$LD_{50}$），为临床安全合理用药提供可靠的依据。结果表明：复方中药片剂对小鼠腹腔的 $LD_{50}$ = 42 884 mg/kg，其 $D_{min}$ = 10 000mg/kg，$D_{max}$ = 80 000mg/kg，MLD = 16 822mg/kg，$LD_{50}$的95%可信限为35 222.77～50 545.23 mg/kg。剖解注射剂量为47 604mg/kg的小鼠，肝脏肿大其边缘呈暗红色，易碎，肾脏肿大不明显，而剖解61 742mg/kg与80 079mg/kg的小鼠肝脏和肾脏均肿大，肝脏表面呈现紫红色，其切面呈暗红色，质地软易碎，肠系膜略有充血现象，其他脏器暂无眼观变化，根据毒理学评价标准，$LD_{50}$ >10 000 mg/kg属于无毒性物质，本试验中复方中药片对小鼠安全无毒。

**关键字**：急性毒性；腹腔注射；半数致死量（$LD_{50}$）

# Study on the acute toxicity of the self-made traditional chinese herbal compound medicine in mice

WANG Jing，HE Shenghu，Ge Song，GUO Shuqiang

（*College of Agriculture*，*Ningxia University*，*Yinchuan*，750021，*China*）

**Abstract**：In order to evaluate the applied reliability and ascertain the toxicity of the self-made traditional Chinese herbal compound medicine，the experiment of acute toxicity and the determination of $LD_{50}$of the self-made traditional Chinese herbal compound medicine was conducted in mice with different doses. The result showed that the $LD_{50}$ of traditional Chinese herbal compound medicine was 42 884mg per kg weight，the minority of the doses was 10 000mg per kg weigh，the maximum of the doses was 80 000mg per kg weight，the MLD was 16 822mg per kg weight and the 95% confidence interval was from 352 22. 77 to 5 0545. 23 mg per kg weight. Dissected symptom of died mice showed that the liver of 47 604 mg per kg weight was enlarged and fragiler with dark red of edge，but the kindney's pathological change wasn't obvious. However，the liver and kindney from 61 742 to 80 079 mg per kg weight were enlarged and its surface was purple，the mesentery bleeding slightly，but the other organs didn't show obvious pathological changes. According to the toxicological evaluation standard，when $LD_{50}$ is greater than 10 000 mg per kg weight，the medicine belongs to the non-toxic matter，the results confirmed that the self-made traditional Chinese herbal compound medicine was safe and non-toxic for mice.

**Keywords**：Acute toxicity；Intraperitioneal injection；Median lethal dose（$LD_{50}$）

随着我国奶牛养殖业的不断发展，奶牛饲养数量的不断增加，母牛在产犊后，气血亏损，生殖器官受到损

---

**基金项目**：宁夏科技攻关项目：奶牛疫病防控技术研究“（C）kjnx2009-3”

**作者简介**：王　静（1989－），女，硕士研究生，主要从事临床兽医诊断技术研究

**通信作者**：何生虎（1959－），男，教授，硕士，博士生导师，主要从事临床兽医诊断技术和动物营养代谢病研究，E-mail：heshenghu308@163.com

伤,机体抵抗力减弱,消化机能降低,而乳腺机能却在逐渐恢复,泌乳量逐日上升,体质的恢复和产乳之间的矛盾较为突出[1],极易引起各种疾病,特别是奶牛产后疾病在临床上较为常见,其发病率高、治愈率低。奶牛产后疾病包括:奶牛产后瘫痪、胎衣不下、子宫内翻及脱出子宫内膜炎、乳房炎等。若不能及时对其进行合理的诊治,不但会直接影响了奶牛的生产性能、降低受胎率、延续发情、奶牛胎间距延长甚至会导致奶牛死亡[2],因此,奶牛产后保健对奶牛业的发展具有极其重要的意义。

目前,有关奶牛产后疾病的治疗,国内外研究学者总结出了很多的治疗方法,但存在诸多问题,其奶牛产后疾病的治疗主要是以抗生素疗法为主,而长期使用抗生素不仅会使奶牛机体产生许多副作用,损伤机体内脏器官,同时也会导致耐药菌株的产生和药物残留等问题,从而降低了治疗效果,且对食品安全以及人和动物健康都产生着不同程度的影响[3],并且抗生素还会使奶牛的免疫功能受到抑制,容易造成重复感染。现在越来越多的学者注重中草药疗法,对奶牛产后疾病的疗效进行各方面的研究,常用中药制剂主要有散剂、汤剂、针剂、子宫灌注剂等剂型。本试验从中草药黄芪、当归、枸杞子、益母草、炙甘草等提取有效成分,研制出一种具有含药量高,服用剂量小,使用方便,吸收快,能提高奶牛机体免疫力等特点的中药复方口服片剂,从而达到对奶牛产后疾病的保健作用。为了验证中药复方片剂的安全性,确定中草药配方提取后制成的片剂的安全性,试验采用小鼠急性毒性试验方法测定小鼠半数致死量($LD_{50}$),对复方中药制剂进行药物安全性评价,为中草药复方片剂在奶牛产后疾病中的广泛应用提供可靠的依据。

## 1 试验材料

### 1.1 试验动物

ICR 系小白鼠 160 只,约 7 周龄,雌雄各半,体重 18~22g,由宁夏医科大学动物实验中心购入。在通风良好的情况下,饲养在室温 20~25℃的鼠笼中,观察 5d,预实验随机分成 5 组,8 只/组,其中 1 组小白鼠为空白对照组;正式试验随机分成 10 组,10 只/组,其中 1 组为小白鼠空白对照组。自由饮水及采食。实验前禁食 12~16h,给药后 4h 内禁食,但不禁水。

### 1.2 试验药品

复方中药制剂由宁夏大学农学院临床兽医实验室研制;原药采用黄芪、当归、枸杞子、灵芝、党参、益母草等有效成分提取物,按一定浓度配制而成,1ml 相当于原药物 2g,经除杂过滤后为红棕色透明澄清液体。

### 1.3 试验器材

D228-1 电子计数秤(上海高致精密仪器有限公司)、1ml 注射器、小鼠笼、小烧杯、外科器械,无菌蒸馏水等。

## 2 试验方法

### 2.1 小鼠口服(灌胃)急性毒性预试验

选取 30 只 ICR 系小鼠,随机分成 5 组,雌雄各半,6 只/组,其中一组小白鼠为空白对照组,用 12 号灌胃针进行灌服,灌服剂量分别为 20 000、40 000、80 000、160 000mg/kg 体重,给药后采用笼边观察法,先观察 30min,在给药后的 4h 再观察一次,以后每天早晚各观察一次,连续观察 7d,对动物的全身精神状态做以观察,主要内容有被毛组织光滑度、眼角膜是否有血丝、红肿现象、小鼠活动是否有异常行为,以及小鼠的神经系统、呼吸系统、消化系统、泌尿生殖系统等,并记录各组小鼠死亡数和存活数,试验过程中对中毒死亡后的小鼠应进行解剖,观察小鼠各组织器官的变化。

### 2.2 小鼠腹腔注射急性毒性预试验

选取 30 只 ICR 系小鼠,随机分成 5 组,雌雄各半,6 只/组,其中一组小白鼠为空白对照组,以 2 倍稀释的药物浓度,即各组剂量为 10 000、20 000、40 000、80 000mg/kg 体重腹腔注射小鼠,观察时间及内容同小鼠灌胃方法。根据各组小鼠死亡数和存活数，确定出大致死亡浓度，即估计致死量（$D_{min}$和$D_{max}$）。当出现$LD_{100}$时，若前一组剂量不为$LD_{100}$时，则该剂量为$D_{max}$；当出现$LD_0$时，若后一组剂量不为$LD_0$时，则该剂量为$D_{min}$。腹腔注射后观察小鼠中毒症状和死亡数量，根据公式$i=(\log D_{max}-\log D_{min})/(n-1)$，计算出来组间距i值（i值即组间距，剂量的对数值等差分布，剂量成等比分布），根据D min和D max设计出正式试验各组的剂量分布，即$D_1=D\ \min$，$D_2=D_1(1+i)$、$D_3=D_2(1+i)$、$D_4=D_3(1+i)\cdots D_9>D\ \max$，以此类推即可得到正式试验组各组的剂量值。

### 2.3 小鼠腹腔注射急性毒性正式试验

随机抽取100只小白鼠（18-22g），雌雄各半，随机分为10组，10只/组，其中一组小鼠为空白对照组。各组分别按照上述预试验计算出来结果，将本试验研制的复方中药液以不同浓度梯度剂量为准对小鼠进行腹腔注射给药，按每只小鼠的体重计算给药量（mg/kg体重），给药后采用笼边观察法，先观察30min，在给药前18h和给药后4h内禁食，自由饮水，在给药后的4h再观察一次，以后每天早晚各观察一次，连续观察7d，并记录7d内小鼠的各体征动态、给药前后体重变化、行为活动以及小鼠的死亡数。对中毒后的小鼠应剖解观察各组织器官的变化并取毒性靶器官做组织病理切片。按改良寇氏法计算出半数致死量[4]（$LD_{50}$）。

## 3 试验结果

### 3.1 小鼠口服（灌胃）急性毒性预试验结果

用灌胃针灌服复方中药提取液之后，在10min后观察小鼠，大部分小鼠表现很兴奋，个别小鼠蜷缩在角落，眼睛微闭，呼吸频率加快，活动量减少；用药4h后观察小鼠，注射20 000mg/kg剂量组的小鼠小鼠精神状态基本恢复正常，小鼠的采食量、饮水量、活动均正常，并且小鼠被毛平整，反应灵敏，与对照组相比较粪便形态无区别，但是尿液呈现黄色。18h后观察40 000mg/kg剂量组的小鼠小鼠精神状态基本恢复正常，小鼠的采食量良好、饮水量、活动均正常，小鼠被毛平整，反应灵敏，但80 000mg/kg剂量组的各别小鼠被毛松弛，活动懒散，有各别呈嗜睡状态。42h后观察，40 000mg/kg剂量组的小鼠死亡一只，80 000mg/kg剂量组的小鼠死亡3只，而160 000mg/kg剂量组的小鼠全部死亡，剖解死亡后小鼠发现小鼠肝脏肿大表面呈现紫红色，切面呈暗红色，质地软易碎，肾脏无明显眼观变化，预实验结果见表1。

表1 小鼠口服注射急性毒性预实验结果

Table 1 The pre-experimental results of the injection oral acute toxicity in mice

| 剂量（mg/kg） | 死亡数/动物数（只） |
|---|---|
| 空白对照组 | 0/6 |
| 20 000 | 0/6 |
| 40 000 | 0/6 |
| 80 000 | 3/6 |
| 160 000 | 6/6 |

由上表可知，预实验中估计致死量$D_{min}$为40 000mg/kg，$D_{max}$为80 000mg/kg，按照毒理学评价标准，$LD_{50}>10\ 000$mg/kg时药物属于无毒性物质，而本试验的最小致死量已经远远超于10 000mg/kg，说明复方中药制剂对小鼠安全无毒。

### 3.2 小鼠腹腔注射急性毒性试验结果

3.2.1 小鼠腹腔注射急性毒性预试验结果　由表2可知，预实验中估计致死量Dmin为10 000mg/kg，Dmax为80 000mg/kg。按寇氏法设计根据公式$i=(\log D_{max}-\log D_{min})/(n-1)$，计算出来组间距$i=0.113$，组间剂量比为 1 比 0.77,所以正式试验各组剂量,分别为 10 000、12 970、16 822、21 818、28 297、36 703、47 604、61 742、80 079、10 3862mg/kg 体重,在给药前 18h 和给药后 4h 内禁食,自由饮水。一次给药后连续观察 7d 内动物的行为、状态,并统计死亡数量。按改良寇氏法计算半数致死量($LD_{50}$)。

**表 2　小鼠腹腔注射急性毒性预实验结果**

**Table 2　The pre-experimental results of the intraperitoneal injection acute toxicity in mice**

| 剂量(mg/kg) | 死亡数/动物数(只) |
|---|---|
| 空白对照组 | 0/6 |
| 10 000 | 0/6 |
| 20 000 | 1/6 |
| 40 000 | 2/6 |
| 80 000 | 6/6 |

3.2.2　小鼠腹腔注射急性毒性正式试验结果　用一次性注射器腹腔注射复方中药提取液制剂 6min 后,大部分小鼠表现出兴奋不安,痉挛,缩成一团,眼睛微闭,嗜睡,被毛粗乱,平滑肌松弛,呼吸急促,对外界刺激略微迟钝,部分小鼠饮食饮水量少。用药 4h 后观察小鼠以上症状仍持续,注射 10 000 ~ 21 818mg/kg 组个别小鼠有恢复迹象,有采食饮水行为;注射 28 297 ~ 47 604mg/kg 组个别小鼠蜷缩在角落,被毛松散,呼吸频率加快,活动量减少;用药 8h 后观察小鼠,注射 61 742 ~ 80 079mg/kg 组个别小鼠有濒死迹象,呈极度嗜睡状态,呼吸呈深度腹式呼吸,翻正反射迟钝甚至消失,对外界刺激无反应。18h 后观察注射 10 000 ~ 21 818mg/kg 组的小鼠逐渐恢复,活跃程度不如空白对照组,采食量减少,尿液呈黄色,粪便形状良好,但相比空白对照组颜色略黄且湿润。注射 36 703 ~ 61 742mg/kg 组的小鼠被毛无光泽、松弛,活动量少,平滑肌松弛,采食饮水量相对于对照组明显减少,并有死亡迹象。80 079mg/kg 组的小鼠死亡只数为 3 只,解剖后小鼠肝脏肿大,颜色呈紫红色,嘴角并有血迹现象。46h 后观察注射 80 079mg/kg 的小鼠全部死亡,除此之外,其他各组剂量的小鼠均有死亡;随着剂量组剂量的增大,ICR 小鼠的死亡率也是逐渐增加,脏器病变也随之显著,直到用药后的 108h 后,存活的小鼠采食、饮水均正常,活动自如,各项体征和各系统情况与对照组相比无差别。针对死亡小鼠剖解发现,小鼠肝脾脏肿大、被膜紧张、边缘钝圆,尤其以肝脏肿大最为明显,颜色呈暗紫红色,纵切面呈暗红色,肝脏质地变软易碎,肠系膜略有充血现象,其他脏器未见有异常情况。各剂量组死亡统计见表 3。

**表 3　小鼠腹腔注射急性毒性正式试验结果**

**Table 3　The experimental results of the intraperitoneal injection acute toxicity in mice**

| 组别 | 剂量 d(mg/kg 体重) | lgd(x) | 死亡数/动物数(只) | 死亡率 P |
|---|---|---|---|---|
| 1 | 10 000 | 4.000 0 | 0/10 | 0.0 |
| 2 | 12 970 | 4.112 9 | 0/10 | 0.0 |
| 3 | 16 822 | 4.225 9 | 1/10 | 0.1 |
| 4 | 21 818 | 4.338 8 | 2/10 | 0.2 |
| 5 | 28 297 | 4.451 7 | 2/10 | 0.2 |
| 6 | 36 703 | 4.564 7 | 3/10 | 0.3 |
| 7 | 47 604 | 4.677 6 | 5/10 | 0.5 |
| 8 | 61 742 | 4.790 6 | 6/10 | 0.6 |
| 9 | 80 079 | 4.903 5 | 10/10 | 1.0 |

结果表明，小鼠的最大耐受量 $LD_0$ = 12 970mg/kg，最小致死量 MLD = 16 822mg/kg，根据改良寇氏法计算公式：$LD_{50} = lg - 1[X_m - i(\sum P - 0.5)]$，式中 $Xm$ 为最大剂量的对数，P 为各组动物的死亡率，$\sum P$ 为各组动物死亡率总和，i 为组间距(相邻两组对数剂量的差值)，根据表 4 计算出 $LD_{50}$ = 42884mg/kg。根据公式 $Slog\ LD_{50} = d\ \sqrt{\sum p(1-p)/(n-1)}$，式中 d 为对数组距，p 为各组死亡率，n 为每组动物数，得出标准误 Slog $LD_{50}$ = 0.039 7，$LD_{50}$ 的 95% 的可信限范围为 35 222.77 ~ 50 545.23 mg/kg。根据毒理学评价标准，$LD_{50}$ > 10 000 mg/kg 属于无毒性物质，本试验中复方中药片对小鼠安全无毒。各组小鼠用药前后体重变化情况见表 4(备注：1 为空白对照组，2 ~ 10 组为试验组)。

**表 4　小鼠腹腔注射急性毒性试验给药前后体重变化情况($\bar{x} \pm s, n = 10$)**

Table 4　Weight changes before and after dosing of the intraperitoneal injection acute toxicity in mice

| 组别 | 1 | 2 | 3 | 4 | 5 | 6 | 7 | 8 | 9 | 10 |
|---|---|---|---|---|---|---|---|---|---|---|
| 用药前 | 19.32 ± 1.22 | 19.35 ± 1.20 | 19.78 ± 1.24 | 20.21 ± 1.21 | 20.75 ± 1.20 | 20.43 ± 1.26 | 20.89 ± 1.24 | 21.32 ± 1.35 | 20.81 ± 1.23 | 21.09 ± 1.23 |
| 用药后 | 22.41 ± 1.20 | 21.51 ± 1.22 | 21.06 ± 1.25 | 20.01 ± 1.22 | 20.13 ± 1.23 | 19.86 ± 1.23 | 19.98 ± 1.31 | 20.11 ± 1.29 | 19.53 ± 1.19 | 19.38 ± 1.21 |

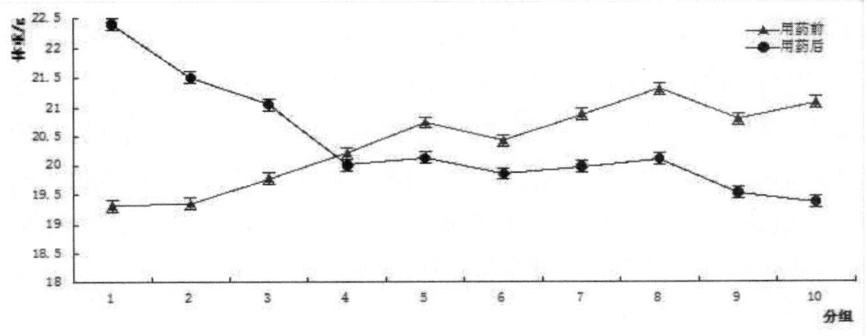


**图 1　小鼠体重变化表**

Fig. 1　Weight changes of mice

由图 1 可知，1 组为对照组，小鼠体重明显增加，2、3 组在给药前后小鼠的体重变化明显，在小鼠一切行为活动恢复正常的同时，且小鼠的体重较给药前有所增加；4 ~ 10 组在给药前后小鼠的体重明显减轻，并且随着剂量的增加，小鼠给药前后的体重减轻趋势也愈加明显。

## 4　讨论与分析

现在越来越多的学者注重采用中草药疗法对奶牛产后疾病的效果研究，从中医学的角度来看[5]，奶牛产后疾病主要在由于气血不畅或者气血不足而引起的，经产奶牛大多气虚血虚，中气不足，或有中气下陷，导致外感风寒风热，会继发各种产科疾病，使得脾胃虚弱，导致产奶量降低。中草药具有清热解毒、抑菌消炎、活血化瘀、收敛止血、改善血液循环、增强子宫收缩[6] 以及增强机体免疫力的功效[7]，并且药物残留低，无毒、刺激性低，并兼有药物与营养剂的双重功效。此复方中药制剂，原药主要是黄芪、当归、枸杞子、灵芝、党参、益母草等有效成分提取物，按一定浓度配制而成，具有抑菌消炎、改善微循环、增加机体免疫力，提高机体非特异和特异性免疫的功能和兴奋子宫等的作用，中草药在临床中的应用，首先要保证安全可靠，而急性毒性试验是药物安全性评价的重要环节，是任何一类新药都不可缺少的试验项目之一，其试验结果将为药物后续的毒理学试验和药效学评价提供可靠的依据。

### 4.1 关于药物毒性的讨论

根据中草药制剂的有关规定[8],每种中药在两部以上古籍中记载无毒,便视为无毒性作用,该配方中的单味中药在文献《本草》、《别目》、《本草纲目》均未见有毒性记载,说明选取的药物安全可靠。

因药物与毒物之间并不存在绝对的界限,而只能根据动物中毒剂量的大小做相对比较。若剂量过大,会引起动物中毒,用量适当对疾病起到一定的良好疗效。本试验对复方中药制剂进行了急性毒性试验,口服给药时,剂量达到40 000mg/kg体重与对照组差异不显著,且无不良临床症状和死亡。按照药物学急性毒性试验标准[9],在小鼠急性毒性口服试验中,口服剂量>10 000 mg/kg体重,在观察108h后,小鼠健活,则认为该药物安全无毒,由此可证实本试验 $D_{min}$ =40 000mg/kg复方中药制剂对小鼠安全无毒。而腹腔注射后,由于小鼠腹腔面积相对较大,药物直接以原型的形式进入血液循环[8],其吸收几乎不受吸收部位的影响,并且药物的毒性也不会被机体的消化酶等降解,因此小鼠会因注射剂量的增大,其死亡率也会随之增大,小鼠中毒症状也愈加明显,肝、肾脏的病理变化也随着剂量的增大而增强,这与牛建荣等报道的一致[10]。

### 4.2 关于急性毒性试验小鼠给药体积及前后体重变化的讨论

在急性毒性实验中给药体积的也是非常重要的,啮齿类动物口服给药和腹腔注射的最大体积对水溶性或混悬液一般不超过20ml/kg,小鼠腹腔的最大容积应不超过50ml/kg[11],因此对给药体积较大的不便一次给完的应将给药剂量等量分配,每隔6~8h给药一次,在24h内给完。根据药物毒理学急性毒性实验相关规定,此种给药方法安全可行,符合药物毒理学研究规定,基本不影响试验效果。

在急性毒性试验前、中、后(包括试验中死亡动物)应各称取体重一次,以观察给药后的变化,一般认为在保持动物正常饮食及生长环境的条件下,出现体重下降应看做是药物毒效应的一种表现。而引起动物体重减轻的原因比较复杂,譬如药物引起腹泻而影响动物食欲,从而进一步影响了食物的吸收和被利用,或对机体消化机能造成障碍而引起动物厌食或绝食、绝饮,造成肾功能及其他组织脏器的急性损伤,都可能使动物体重下降。本试验小鼠的体重随药物剂量的增加而出现减轻的趋势,在剖检后发现药物毒性作用于肝脏、肾脏变化较为明显,可认为小鼠的体重减轻可能与肝脏、肾脏的损伤有关。

## 参考文献

[1] 付尚杰.奶牛产后保健[J].黑龙江畜牧兽医,2013,5:44-47.

[2] 王秀芳.2种常见奶牛产后疾病的症状与防治[J].畜牧与饲料科学,2013,34(1):123-124.

[3] 莫伟伟.新型奶牛子宫内膜炎中药复方的研制[D].呼和浩特:内蒙古农业大学,2006.

[4] 徐叔云,卞如濂,陈修.药理实验方法学;[M].第二版.北京:人民卫生出版社,1994.

[5] 温赞.补中益气汤和当归四物散对经产奶牛的保健效果观察[J].中国奶牛,2013,9:54-55.

[6] 汤春华,礼中,建清.草药制剂防治奶牛繁殖障碍疾病的效果[J].江苏农业科学,2001,(1):61-62.

[7] 陈品球,蒋继琰.中药子宫灌注液治疗奶牛子宫内膜炎的效果观察[J].中兽医学杂志,2006,(4):21-22.

[8] 陈忠伟,刘伟,赵武.复方中药制剂对小鼠的急性毒性试验[J].中国畜牧兽医,2007,34(6):112-113.

[9] 徐叔云,卞如濂,陈修.药理实验方法学[M].第2版.北京:人民卫生出版社,1994.

[10] 牛建荣,李剑勇,等.复方葛黄消炎液刺激性及急性毒性试验[J].动物医学进展,2006,27(7):111-113.

[11] 刘昌孝,孙瑞元.药物评价实验设计与统计学基础[M].北京:军事医学科学出版社,1998.

# 自体结扎与传统结扎在公猫去势手术中的效果比较

张斌恺[1]，李思远[1]，陶田谷晟[1]，王江豪[1]，秦建辉[1]，肖　啸[1,2]
（1. 云南农业大学动物科学技术学院，昆明 650201；2. 云南农业大学动物医院，昆明 650031）

**摘　要**：公猫去势术是临床上常见的手术之一，为了探索自体结扎与传统结扎在临床手术中的应用效果，方便动物医生在临床手术时更有针对性地选择对公猫进行手术的方法。现开展两个手术对比研究，随机对 10 个年龄、体型相近的实验公猫分组：自体结扎组 5 只；传统结扎组 5 只，在手术前、手术中、手术后，分别对其手术时间、手术难易度、术后食欲恢复天数、术后炎症病例数和血常规等方面进行比较。结果表明：自体结扎较传统结扎手术时间较短，手术难度较低，术后食欲恢复较快，无炎症病例，术前自体结扎组与传统结扎组的血常规检测均正常。术后 3d 传统结扎组白细胞数目高于正常值，淋巴细胞数目、单核细胞数目以及中性粒细胞数目均在正常范围内，可以判断为急性炎症。术后 7d 传统结扎组的血常规指标都回归到正常范围内。自体结扎组的血常规均在正常范围内。结论：用自体结扎法对公猫去势，手术操作简单、手术时间短、对动物伤害较小、术后恢复较快，在临床上建议推广。

**关键词**：公猫去势术，自体结扎，传统结扎

## 1　引　言

去势术（male animal castration）：摘除雄性动物的睾丸或破坏其生殖机能，使动物失去性欲或繁殖能力的一种方法叫去势术[1]。公猫去势的必要性包括：（1）控制群体数量。（2）降低发生前列腺疾病发生的机率。（3）减少会阴疝的发生率。（4）预防肛周腺瘤的发生。（5）降低由激素引起的皮肤病的发生率。（6）猫去势后可减少猫本身特有的臭气。（7）减少猫发情时的性行为，如猫在夜间的叫声会对周围环境造成污染。（8）体重增加。（9）延长寿命等。

传统结扎是在切除动物睾丸后用手术线结扎精索[2]，临床上偶有少数去势公猫术后发炎，随着临床实践发展，动物医生们发明了一种自体结扎的方法，即切除睾丸后，用动物本身的精索打成一个死结[3]，这种方法显著降低了炎症发生的概率，现已广泛用于临床实践。现就自体结扎与传统结扎的操作方法、恢复时间、愈后健康状况等方面进行比较研究。

## 2　材料与方法

### 2.1　病例收集及分组

对 10 只公猫随机分组：传统结扎组 5 只；自体结扎组 5 只。

### 2.2　术前准备

诊断器械：听诊器、温度计、秒表各 1 支。

维持器械：输液器械 1 套（一次性输液管、一次性输液针）、一次性注射器（1ml 和 5ml 各 6 支，带注射针头）。

手术器械：电动剃毛器 1 把、手术刀（刀柄 、刀片）1 把、手术剪 1 把、止血钳 2 把、持针钳 1 把、创巾钳 4 把、手术镊 1 把，缝针、缝线、纱布 、创巾、棉签、棉球、医用胶带等。

药物准备：新洁尔灭消毒液、碘伏、酒精、葡萄糖氯化钠 2 瓶、硫酸阿托品、舒眠宁、苏醒灵 3 号、肾上腺素。纱布做高压灭菌，金属器械做新洁尔灭溶液浸泡灭菌。

---

肖　啸，E-mail：xiaoxiaokm@163.com

### 2.3 麻醉、保定与消毒

俯卧保定,全身麻醉,局部剃毛消毒。

### 2.4 术式

将两侧睾丸同时用手推挤到阴囊底部,用食指、中指和拇指固定一侧睾丸,并使阴囊皮肤绷紧。在距阴囊缝际一侧0.5~0.7cm处平行阴囊缝际作一2~3cm皮肤切口,切开肉膜和总鞘膜,显露睾丸。术者左手抓住睾丸,右手用剪刀剪断阴囊韧带,向上撕开睾丸系膜,然后将睾丸引出阴囊切口外,充分显露精索。

### 2.5 结扎精索、切断精索、去掉睾丸

2.5.1 自体结扎 左手中指和无名指夹持精索,向后压住阴囊,拇指食指夹持睾丸尾精索向外轻拉,中指和食指自然撑开,中间露出精索约4~5cm。

右手用18cm的持针钳挑起精索再扭转一定度数,同时拇指食指把睾端精索送入持针钳,而后扣死,持刀割去睾丸,左手拇指食指把缠绕在持针钳的精索向前推,右手使持针钳轻微转动,使精索结套滑出持针钳[4]。

用左手的拇指和食指将绕在持针钳上的精索环往前推,使其自体打结。观察断端有无出血,确保没有问题后将其还纳回鞘膜腔内。

同理将另一侧的睾丸摘除,检查伤口,对合皮肤,不缝合,涂碘伏消毒。

2.5.2 传统结扎 在精索的近心端钳夹第一把止血钳,在第一把止血钳的近睾丸侧的精索上,紧靠第一把止血钳钳夹第二、三把止血钳。用4-0号丝线,紧靠第一把止血钳钳夹精索处进行结扎,当结扎线第一个结扣接近打紧时,松去第一把止血钳,并使线结恰位于第一把止血钳的精索压痕处,然后打紧第一个结扣和第二个结扣,完成对精索的结扎,剪去线尾。

在第二把与第三把钳夹精索的止血钳之间,切断精索。用镊子夹持少许精索断端组织,松开第二把钳夹精索的止血钳,观察精索断端有无出血,在确认精索断端无出血时,方可松去镊子,将精索断端还纳回鞘膜管内。

在同一皮肤切口内,按上述同样的操作,切除另一侧睾丸。在显露另一侧睾丸时,切忌切透阴囊中隔。

2.5.3 术后护理及营养控制 注射苏醒灵3号。

## 3 结果与分析

### 3.1 自体结扎法和传统结扎法术后情况比较

表1 自体结扎法和传统结扎法手术情况比较

| 项 目 | 自体结扎组 | 传统结扎组 |
|---|---|---|
| 手术时间(min) | 3±1.03 | 7±2.10 |
| 手术难易度 | 简单 | 复杂 |
| 术后食欲恢复天数(d) | 1.5±0.5 | 3.5±0.5 |
| 术后炎症病例数 | 0 | 1 |

表1是对去势猫进行手术并记录的相关数据,可以发现,自体结扎法去势和传统结扎法去势手术有十分大的差异。由于自体结扎法的手术比较简单易操作,所以自体结扎法所用的手术时间不到传统结扎法的一半。所有的手术都使用相同的麻醉,心率和血压都在正常范围中,手术均成功。

### 3.2 自体结扎法和传统结扎法手术前后炎症(血常规)对比

表2 自体结扎法手术前后血常规比较

| 项目 | 手术前 | 术后3d | 术后7d | 参考范围 |
|---|---|---|---|---|
| 白细胞数目($\times10^9$/L) | 16.4±0.8 | 16.6±0.9 | 16.4±0.8 | 5.5~19.5 |
| 淋巴细胞数目($\times10^9$/L) | 5.8±0.9 | 6.0±0.7 | 5.8±0.9 | 0.8~7.0 |
| 单核细胞数目($\times10^9$/L) | 0.6±0.7 | 0.6±0.6 | 0.6±0.7 | 0~1.9 |
| 中性粒细胞数目($\times10^9$/L) | 11.0±1.1 | 11.0±1.1 | 11.0±1.1 | 2.1~15 |

表3 传统结扎法手术前后血常规比较

| 项目 | 手术前 | 术后3d | 术后7d | 参考范围 |
|---|---|---|---|---|
| 白细胞数目($\times10^9$/L) | 16.8±1.1 | 19.6±0.5 ↑ | 16.9±0.7 | 5.5~19.5 |
| 淋巴细胞数目($\times10^9$/L) | 5.17±0.9 | 5.2±0.8 | 6.4±0.6 | 0.8~7.0 |
| 单核细胞数目($\times10^9$/L) | 1.5±0.3 | 1.8±0.5 | 1.7±0.3 | 0~1.9 |
| 中性粒细胞数目($\times10^9$/L) | 11.9±1.2 | 13.6±1.0 | 12.3±0.5 | 2.1~15 |

对研究所用10只公猫进行术前以及检测术后3d、术后7d血常规检测它们的炎症,结果见表2、表3。术前自体结扎组与传统结扎组的血常规检测均正常。术后3d传统结扎组白细胞数目高于正常值,有炎症发生,淋巴细胞数目、单核细胞数目以及中性粒细胞数目均在正常范围,可以判断为急性炎症。术后7d传统结扎组的血常规指标都回归到正常范围内。自体结扎组的血常规均在正常范围内,判断为无炎症发生。

## 4 讨 论

自体结扎法手术操作简单,因其没有手术线等异物刺激,所以术后恢复时间较短,自体结扎法的手术比较简单、易操作,术后基本不出血[5],且手术不需要用到手术缝针和缝线,操作很简单,所以手术时间短。

传统结扎法手术操作较复杂,且时间较长;传统结扎法能将精索结打紧,术后扎结不易脱落,但结扎线属于异物,机体会产生免疫排斥反应,因此,传统结扎术后恢复时间长、可能引起炎症。

## 5 结 论

用自体结扎法对公猫去势,手术操作简单、手术时间短、对动物伤害较小、术后恢复较快,在临床上建议推广。

## 参考文献

[1] 谷风柱.国内公畜去势术研究进展[J].山东农业大学学报.1993,24(1):118-122.
[2] 韩文彩,朱余军,陈龙.犬消声术和公犬去势术的过程与体会[J].中国畜牧兽医文摘.2013.29(3):175.
[3] 许道庆.公猫去势术[C].全国兽医外辩学第13次学术研讨会小动物医学第1次学术研讨会暨奶牛疾病第3次学术讨论会论文集.2006,10:571-572.
[4] 唐如勋.公畜去势的精索打结法[J].中国兽医杂志.1995,21(5):35.
[5] 刘忠诚,于连富,李凤山.宠物几种不出血手术的术式[J].黑龙江畜牧兽医.1998,3:33.

# 美国 CORNELL 大学访学杂记

陈进军

(广东海洋大学农学院,湛江 524088)

经一系列选拔程序,我作为广东省地方公派高级访问学者,经签证申请、面签、艰难漫长达 49 天的行政审查之后拿到了赴美 J-1 签证,于 2013 年 10 月 17 日顺利到达位于美国纽约州伊萨卡市的 Cornell 大学,开始了为期 6 个月的访学生活。现将访学感受以杂记的形式与各位同仁分享。

## 一、安顿生活、及时报到

因我有国家公派加拿大访学(Guelph 大学,2001 ~ 2002)的经验,故这次我提前托网友帮忙租好了住房,加之房东接机,还有二手自行车让我骑,且次日就是周末,所以住、吃基本生活很快就安定下来了。虽然有些许“没车、没房、没多钱,回到了新中国成立前”的感觉,但想到为了留学深造,那就以“流血”的精神适应、坚持吧!

到达 Cornell 大学的第一个星期一上午,我即到该校微生物系 Hay 实验室报道,和 Anthony Hay 博士就可能的研究计划从斑马鱼繁育养殖体系建立,到我所带的样品检测等方面进行了交流,最后确定利用宏基因组学原理和高通量测序技术,研究有机污染物(PCBs)与斑马鱼肠道微生物群落的互作关系。随后到系办,并通过学校人力资源部登记中心(Boarding Centre)办理了 Cornell 大学身份证、内部邮箱等。

## 二、Cornell 大学科学研究运作概况

1. 信息发达使科研效率大大提高　这里的信息太方便了,这是我的第一感受;美国人太幸运了,Anthony 这样说。的确,在 Cornell 大学,随便在数平方公里大的校园任何角落或图书馆打开 Google,找到的文献都可以全文下载,有的操作视频也能快捷得到。

2. 学术氛围浓厚是优秀校风的核心　学术气氛浓厚是在骨子里的,何谓校风? 这就是! 以首席科学家(principal investigator,PI)名字命名的实验室每 2 周有 1 次学术例会,系里每周有 2 次 Seminar,院里有、群众学术团体也有各个层面的学术活动。这些学术报告和活动一般提前安排(公布相关简况和日程表)在每个学期(春季学期 1 月下旬至 5 月下旬、秋季学期 8 月下旬至 12 月份圣诞节)的学期中期之前,以便于师生参与。

3. 科研腐败成为不可能　在 Cornell 大学,包括 PI 在内,科研人员不接触任何形式的研究经费,经费按 PI 户头,由系(院)专人统一管理。只要是某实验室的项目经费,该实验室负责人即该 PI 所带领团队的研究人员(包括研究生、博士后、访问学者)均可以通过实验员,直接去系里实验用品仓库记账领用;库里没有的,小额的即刻填单购货,而大宗的如 1 千美元及以上的采购,则由项目负责人签字后交由管理人员直接购买,到货很快很及时。当然,PI 会定期得到反馈:你的户头上还剩多少钱了! 根本没有借款、报账、冲账之说,谁个能科研腐败! 全社会信用体系完善联网,社会主流是诚信的,管理人员敬业、负责,哪来的以权谋私、回扣之说! 科教人员以兴趣相聚,一股正气,喜欢科研,喜欢探索,非眼前利益所驱动,人格为何? 这就是!

## 三、教育先进取决于教师人格高尚

1. 办学条件优良　大学教学经费来源多样化、非常充足,设施设备先进、配套、耐用:图书馆、大一些的建筑门廊、厅堂等部位,布置了足够多的雅座,学生或教工读书、上网、交谈、简单会客、简单快餐与饮酌,均可随意安排。决不会出现不够座或抢座的情况。任何厕所(restroom)都有足够的厕纸、洗手液、擦手纸巾及烘手机。公共场所如教室、会议室、博物馆、艺术馆、体育场及投影、电脑、仪器等等设施设备,用者爱护,加之质量

保证、经久耐用,毫无问题。

2. 教职工实行合约管理　即工资按合约岗位约定,保证完成教学任务,按约定年薪分 9 个月发给工资,再无额外课时酬金(PI 可以从自己负责的科研项目经费中支取 1 个月薪水给自己);科研无硬性要求,也无激励措施,校、院、系对职员发表的各类各层次的论文无奖励,获得了项目、成果、专利无奖励(院长可决定给予拉来特大捐赠款项如 1 000 万美元者特别奖励),但科研是兴趣所致,是"百花齐放"的宽松环境所致,教工们为科学精神和荣誉而搞科研,比如化学系有 2 位诺贝尔奖获得者,所得好处仅仅是可免费在校园各区停车;教师每上一节课,课前、课中、课后辅导哪怕占用一天时间,教工们依旧津津而为之,其耐性与敬业育人的精神令我费解! 就这样一个无利益驱动的宽松环境,人才辈出,科研照样搞得红红火火,该校综合排名照样世界前茅。想通这个问题,对我们而言,绝不可能! 没有利益驱动,没有严格管理,没有思想工作,还那么卖命,不可能的事! 那不是傻子吗!?

3. 学生可任意选课　当然有选课系统、有时间节点。并不因跨专业、跨系、跨学院选课而让学生自己另外再交选课费,而是由该生一开始注册缴费的那个学院给学生交纳课时费,因为学生已经给他的学院交纳了足够多的学费了(本科生每生 4.2 万美元/每年,硕士生每生约 6 万美元/每年;博士生不但不收费,反而会有每年约 4 万美元的奖学金或薪水)。课程讲授采用一门课多师制,特别是学位课和必修课必须由 2 名以上教师分别主讲各自精通的部分。

4. 实践教学足够重视　本科生课程实验在公共实验室即教学实验室,由实验师或经申请合格并认可的博士后、博士研究生以助教身份担任。创新或相关的学分可申请到某(几)个科研实验室完成,学生可获得助研补贴(每小时 6 ~ 10 美元)或学分,其时间由学生本人在网上登记,自觉诚信遵守,无人监督。本科生学业很重,不是课多,而是课外阅读与作业、课程报告多,又杜绝抄袭,所以几年下来,本科生很厉害,毕业后很容易就业,甚至可以直接申请相关专业的博士学位,生命科学类的本科毕业生可申请医学大夫(MD)或兽医大夫(DVM)文凭。学生很看重校外实习、见习,因为这些货真价实的经历是将来得到工作推荐和深造推荐的必要条件;在教授特别是名教授实验室的助研经历,也是深造或到科技公司工作得到推荐的重要条件。

5. 教学质量的评估与教育环境　众所周知,美国的本科教育是各层次教育中最厉害的,Cornell 大学的本科生教育质量更是深受社会认可。本科生的教学质量评估坚持学生评教,但这个评教不是为了给教师好看或难堪,而是在课程教学周期中间,进行一次学生填表(就像机读答题卡)评估,征集学生建设性建议和意见,以便后续内容上的更好。不管是教,还是学,还是管理,各方都把自己当成是高尚的人、完全可信的人,而不是相互提防!

整个校园、草坪、教室、图书馆、实验室、标示、马路和人行横道……都是十分人性化的,都是培育人才、崇尚科学和传承历史文化的舒适、宽松环境,教职工们努力敬业,不是单一为竞争、攀比,不违背人性本质和教育规律:工作日午间不休息,安排好、做好各项科教工作,教师全身心搞好教学活动与任务,周末和节假日绝不加课、加班,也不鼓励加班开课和搞实验研究。

## 四、考察美国兽医(动物医学)专业教育

总体认识:美国共有 28 所大学有兽医学院。学生须在取得生命科学类学士学位或完成 3 年的兽医预科后,才有资格报考进入为期 4 年的兽医大夫(Doctor of Veterinary Medicine,DVM)专业学习。由于竞争特别激烈,常常 10 个以上的考生,只能录取 1 个,且除了笔试外,还要进行严格的面试,所以一般考生都会在获得学士学位后,自己找动物诊疗机构锻炼 1 年左右,才敢去参加 DVM 入学考试。所以,美国各大学的兽医学院不属于本科层次教育,称之为本科毕业后教育。在美国,兽医是备受崇敬的职业和行业,是学费最贵的专业之一。

考入兽医学院的学生在第四学年,完全是实践之前所学过的所有知识,学生到教学动物医院进行兽医临床实践,与动物及其主人直接交流,这一学年为 12 个月,每 2 ~ 3 周,学生还要去不同的兽医诊疗机构和服务

点实践。

1. Cornell 大学兽医学院 Cornell 大学兽医学院(College of Veterinary Medicine, Cornell University)在全美排名第一。每年招收 DVM 学生不多于 100 名。该兽医学院包括担负教学科研任务的 5 个系和 1 个动物健康研究所,教职员工逾 300 名;另有 1 个纽约州动物疫病诊断中心、1 个教学动物医院及兽医科学类图书馆。

单就其教学动物医院而言,由 3 个部分构成:伴侣动物医院、马医院和农畜医院,各类医务人员逾 200 名,面向社会服务,面向学生实训。其场地、设施设备和功能堪称胜过我国的三级甲等医院。令我吃惊的是,动物肿瘤、牙科、宠物主人心理抚慰等这些我国动物医院基本不涉及的功能以及核磁共振、CT 这些极其昂贵尖端的仪器在这里也很常见。在这里,我和教学动物医院院长兼兽医学院副院长 Lorin Warnick 教授预约进行了超时交谈,涉及中美兽医地位与历史、学生专业实践教学运行和联合人才培养可能性等内容。

2. Tennessee 大学兽医学院 Tennessee 大学兽医学院的 Hwa-Chain R. Wang 终身教授曾于 2012 年应邀到广东海洋大学访问过,所以我专门抽出 4 天时间,远赴距 Cornell 大学近 20 小时高速大巴车程的 Tennessee 大学兽医学院(College of Veterinary Medicine, University of Tennessee)及其教学动物医院考察学习。

Tennessee 大学兽医学院每年招收 DVM 学生 80 名。该兽医学院包括担负教学科研任务的 4 个系和 1 个兽医中心,教职员工近 150 名;另有 1 个教学动物医院及兽医与农业科学类图书馆。

其教学动物医院包括宠物医院和大动物医院,设施设备水平、医务人员规模及诊疗功能与 Cornell 相近,场地面积要比 Cornell 大学教学动物医院的还要大一些。Wang 教授陪同我单在教学动物医院就参观了 1 整天,其场地之大、设施设备之先进、功能科室之丰富令我惊叹,不时见到学生、特别是女生在处置大、小患畜甚至野生动物。期间我和 Wang 教授及兽医学院院长 James Philip Thompson 教授和 Misty Renee Bailey 女士在兽医专业建设、双语教学和人才培养与学术研究等方面进行了交流,建立了直接联系,加深了友谊。

## 五、人文感受及其他

1. Cornell 大学的人员招募机制 在进入前,先形成人才需求及其岗位描述(由 Search Committee 讨论形成,一岗一描述。该委员会由系主任指定人员组成)→提交学院院长,同意→广告招募→系主任审查资料、筛选,确定被面试人员→候选人员作报告(面向全系教工和研究人员包括研究生和访问学者)→系里教授(tenure faculty)投票→系主任审定→报学院院长批准。全部环节学校不介入。人员流动无"档案"概念。

2. 校友捐款大有成就感 各种大额捐款来自各类基金会、公司及成功校友。系里有专人负责这项工作。捐赠者的名字、照片、画像甚至塑像会被特别突出地、永久地放置于显眼位置,材料特别讲究,可能是铜材或石材,或以他们的名字命名楼宇、花坛及奖学金、奖教金,甚至人才引进基金、院长或系主任基金,等。

3. 学风自由而活跃 学生除了自由选课、攻读主修专业(Major),还可以选修其他专业(Minor)。此外,各种文体竞赛活动、咖啡沙龙等很丰富。比如学生自己形成的 Dragon Day 校园游行在每年 3 月份的最后一个星期五举行,按既定路线,由警察维护秩序,用时约 1 小时,主题涉及环保、和平和社会福利与收入等方面,自由、轻松、好玩。又比如每年 5 月下旬期末考试结束后,学生自己利用校园北区和西区之间的大坡草坪,举办 Slope Day 晚间 Party,有限度地放纵一番,不同肤色、不同专业的学生在一起聚餐、演唱、跳舞、喝点啤酒,热闹、轻松而不失雅趣。

4. 总体感悟 中美之差别,已不再是物质的了。除了汽车、飞机,很难在美国找到不是"Made in China"的商品,特别是电子和轻工产品。

美国的主要优势在于:一是交通安全,美国最重要的一个交通规则是在任何"STOP"标识和交叉路口白实线前,必须将车停稳,然后谨慎通过,没警察、没交通灯、没监控探头,但人人自觉遵守。二是食品安全,价廉物美的动、植物食品原料,确实令人放心。三是环保意识深入骨髓。最后也是最根本的是人们内心实实在在的公义、人格和博爱,这可能与其人口相对较少、资源丰富、环境优越、福利保障和法制执行力强等有关。

# 昆明地区犬戊型肝炎流行的血清学调查

汪登如[1] 严玉霖[2] 陈 玲[2] 高 洪[2]

（1. 昆明爱美森宠物医院，昆明 650224；2. 云南农业大学，昆明 650201）

**摘　要**：为了解昆明地区犬戊型肝炎（HE）流行的情况，共采集血清样本 268 份，其中包括昆明主城区家庭散养犬血清样本 163 份，养犬场犬血清样本 65 份，城市流浪犬血清样本 40 份。采用双抗原夹心 ELISA 方法对上述血清样本进行 HEV 抗体的检测。结果显示，268 份犬血清中 HEV 抗体平均阳性率为 48.51%，城市流浪犬 HEV 抗体阳性率较高，不同品种不同年龄犬 HEV 抗体阳性率有差异显著。结果表明昆明地区犬戊型肝炎已经普遍流行，尤其以流浪犬 HEV 的感染较为严重。

**关键词**：犬戊型肝炎，血清学调查

# The seroprevalence of canine hepatitis E prevalence in Kunming district

WANG Dengru[1], YAN Yulin[2], CHEN Ling[2], GAO Hong*

(1. *Kunming Amth pet hospital*, *Kunming*, 650224; 2. *Yunnan Agricultural University*, *Kunming*, 650201)

**Abstract**: The aim of this study was to investigate the prevalence of canine hepatitis E virus (HEV) infection in Kunming region. The 268 canine serum from August 2008 to August 2009 in Kunming region were collected and detected the antibody to HEV (anti-HEV) by ELISA. The result showed that the total positive rate of anti-HEV was 48.51 percentage, which the stray dog of city was highest, and the differences of varying in different species and growth phases were significant. It indicated that the canine HEV was widespread in Kunming region, especially in the city stray dog.

**Keywords**: canine hepatitis E, serosurvey

戊型肝炎（Hepatitis E，HE）是由戊型肝炎病毒（Hepatitis E virus，HEV）引起的急性病毒性肝炎，是一种人畜共患病[1]。该病主要经粪—口途径传播，也有报道可以通过口－鼻和血液传播，有流行和散发两种形式，HE 的病死率较甲型、乙型、丙型和丁型肝炎高，人群感染后死亡率在 0.5%～3%，孕妇感染病情较重，死亡率高达 15%～25%[2,3]。在流行病学中，戊型肝炎的暴发是不常见的，常见的是零星的急性病例，并且常常发生在环境卫生条件比较差，饮水常被粪便污染的发展中国家。我国自 1980 年以来，新疆、辽宁、吉林、内蒙古和山东等地均有 HE 的流行[4]。近年来，越来越多的研究证实猪、牛、羊、犬、鸡、鼠等多种与人类密切接触的动物可感染 HEV[5-9]，并且动物中分离的 HEV 与人体内分离的 HEV 的基因序列高度一致[3,4]，提示上述动物在戊肝的传播中可能发挥作用。

犬是人类最亲密的朋友，随着人民生活水平的提高，越来越多的人们开始养犬。近年来，昆明市宠物犬数量迅速增加，据不完全统计约 30 多万只，主要集中在主城区家庭散养和城郊养犬场饲养。人与犬的关系越来越密切，与犬接触的人群数量越来越多，同时也给对于像戊型肝炎这样的人畜共患病的防控带来了难度。为了解昆明地区犬 HEV 感染的情况，本研究对昆明地区犬 HEV 的感染进行了血清学调查，以期为犬戊型肝炎的防控提供理论依据，引导人们科学健康地养犬。

---

**作者简介**：汪登如（1975－），兽医硕士，执业兽医师，研究方向：小动物临床医学，E-mail：amthpet@sohu.com

# 1 材料与方法

## 1.1 血清样本

2008 年 8 月 ~2009 年 8 月,共获得血清样本 268 份,其中昆明主城区家庭散养犬血清样本 163 份,养犬场犬血清样本 65 份,城市流浪犬血清样本 40 份, -20℃冰柜中保存备用。

## 1.2 检测试剂

双抗原夹心法 HEV IgG 抗体检测试剂盒(ELISA),购自北京万泰生物药业股份有限公司。

## 1.3 检测方法

双抗原夹心 ELISA 方法,严格按照检测试剂盒的说明书进行操作,通过配制洗涤液,稀释样品,加样,温育,洗板,加酶标二抗,显色,终止等操作步骤,最后于 450nm 处测定孔的 OD 值。

## 1.4 结果判定

以空白孔调零,当阳性对照孔平均值 $OD_{450} \geq 0.8$,阴性对照孔平均值 $OD_{450} \leq 0.1$ 时试验成立。被检血清样品 $OD_{450} \geq 0.24$ 时判为阳性,$OD_{450} < 0.24$ 时判为阴性。

## 1.5 统计分析

不同时间或不同分组间抗体阳性率的比较采用 Pearson $\chi^2$检验,以 $P<0.05$ 时差异显著,$P<0.01$ 差异极显著,统计分析使用 SPSS10.0。

# 2 结果

## 2.1 不同来源的犬血清中 HEV 抗体检测结果

对 2008 年 8 月 ~2009 年 8 月期间的 268 份血清样本进行检测,结果显示,HEV 抗体阳性血清有 130 份,平均阳性率为 48.51%,其中主城区家庭散养犬 HEV 感染率为 48.47%,养殖场犬 HEV 感染率为 33.85%,流浪犬 HEV 感染率为 72.5%(表 1)。检测结果表明,昆明地区犬 HEV 的感染较为普遍,尤其城市流浪犬最为严重。

表 1 血清来源及 HEV 抗体阳性率

Table 1 The source of serum and positive rates of HEV specific antibody

| 项 目 | 家庭散养犬 | 养殖场犬 | 流浪犬 | 总 计 |
|---|---|---|---|---|
| 样本数(份) | 163 | 65 | 40 | 268 |
| 阳性数(份) | 79 | 22 | 29 | 130 |
| 阳性率(%) | 48.47 | 33.85 | 72.50 | 48.51 |

## 2.2 不同品种的犬血清中 HEV 抗体检测结果

268 份待检血清样本中,涉及的犬品种主要有昆明犬、金毛猎犬、贵宾犬、萨摩耶、雪橇犬、京巴犬,通过对不同品种的犬只血清进行检测,统计结果如下表所示,不同品种间 HEV 抗体阳性率差异显著($P<0.05$,$P<0.01$),其中昆明犬感染率最高,为 80%,其次为贵宾犬,为 66.66%,雪橇犬感染率最低,为 21.45%。

## 2.3 不同年龄的犬血清中 HEV 抗体检测结果

取年龄明确不同品种的犬血清 138 份,根据 HEV 抗体检测结果,分析犬 HEV 抗体阳性率与不同年龄阶段犬的关系。结果表明,不同年龄阶段的犬感染 HEV 的阳性率差异显著($P<0.01$),其中 2 ~8 岁的成年犬感染率最高为 68.75%,其次是 8 岁以后的老龄犬 78.79%,2 岁以前的幼龄犬感染率较低为 42.11%(表 3,图 1)。

表 2　不同品种犬 HEV 抗体阳性率

Table 2　The HEV antibody positive rates of different species dogs

| | 样本数(份) | 阳性数(份) | 阳性率(%) |
|---|---|---|---|
| 昆明犬 | 80 | 64 | 80.00 |
| 贵宾犬 | 12 | 8 | 66.66 |
| 金毛猎犬 | 46 | 23 | 50.00 |
| 萨摩耶 | 29 | 10 | 34.48 |
| 京巴犬 | 33 | 10 | 30.30 |
| 其他品种 | 40 | 9 | 22.50 |
| 雪橇犬 | 28 | 6 | 21.43 |
| 总计 | 268 | 130 | 48.51 |

表 3　不同年龄犬 HEV 抗体阳性率

Table 3　The seroprevalence of HEV antibody indifferent growth phases dogs

| | 样本数(份) | 阳性数(份) | 阳性率(%) |
|---|---|---|---|
| 2 岁以前(幼龄犬) | 57 | 24 | 42.11 |
| 2 ~ 8 岁(成年犬) | 48 | 33 | 68.75 |
| 8 岁以后(老龄犬) | 33 | 26 | 78.79 |
| 总计 | 138 | 83 | 60.15 |

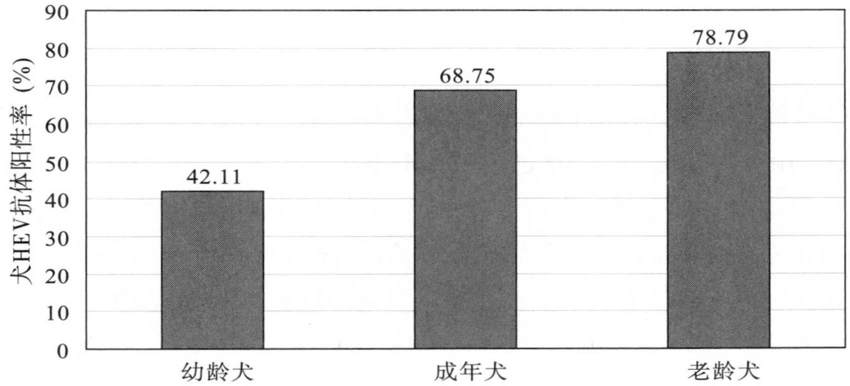


图 1　不同年龄犬 HEV 抗体阳性率

Figure 1　The seroprevalence of HEV antibody in different growth phases dogs

## 3 讨　论

目前尚未见到关于昆明地区犬感染 HEV 的情况报道,本研究通过采集 2008 年 8 月 ~2009 年 8 月昆明地区 268 只犬的血清,对不同来源的犬血清 HEV 抗体进行了血清学调查,结果显示昆明地区犬 HEV 感染的平均阳性率为 48.51%,其中城市流浪犬 HEV 的感染率最高为 72.5%,其次是主城区家庭散养犬,HEV 感染率为 48.47%,养殖场犬 HEV 感染率较低为 33.85%。以上数据表明,昆明地区犬只的 HEV 感染较为普遍,尤其以流浪犬的感染较为严重,根据戊型肝炎的流行病学特点,HEV 的感染与犬只平时的生活环境密切相关[3-6],卫生条件好、生物安全措施得当的养犬场,HEV 的感染率较低,反之则较高。因此建议政府部门加大对城市散养犬和流浪犬的监管力度,对于家庭散养犬,主人应在所属辖区办理养犬证,定期进行犬疫病的诊断监测,定期进行相关疫苗的免疫接种,以保障爱犬的健康和自身健康。对于与日俱增的城市流浪犬,提倡广大市民发扬爱心,不要遗弃自己的爱犬,尽是将流浪犬送到犬只收容所,以减少戊型肝炎等人畜共患病传染给人类的途径。

为了进一步明确不同品种犬 HEV 的流行情况,我们对所采集到的 268 份犬血清按品种进行了分类统计,涉及的犬品种主要有昆明犬、金毛猎犬、贵宾犬、萨摩耶、雪橇犬、京巴犬等,统计结果表明不同品种的犬对 HEV 均易感,且不同品种间 HEV 抗体阳性率差异显著($P<0.05$,$P<0.01$),其中昆明犬感染率最高,其次为贵宾犬,雪橇犬感染率最低。本研究结果可以为探讨不同品种犬对 HEV 的易感性提供参考,但 HEV 易感性是否与犬品种相关,还有待于深入研究。为了方便流行病学调查统计,根据犬的年龄特点,我们还将犬分为三个生长阶段,幼龄犬(2 岁以前),成年犬(2~8 岁),老龄犬(8 岁以后)。本次调查收集了年龄明确的犬血清 138 份,从不同年龄阶段的犬 HEV 感染的分布来看,从幼龄犬,成年犬,老龄犬三个年龄阶段的犬血清中均有 HEV 阳性抗体的检出,说明不同年龄的犬对 HEV 均易感,不同年龄阶段的犬感染 HEV 的阳性率差异显著。成年犬和老龄犬的感染率较高,而幼龄犬感染率相对较低,这可能与母源抗体存在的因素有关。此调查结果与人类戊型肝炎的发病特点相一致,即人类的 HE 与年龄有关,幼年时感染 HEV 多为亚临床型,青壮年时期感染 HEV 多为临床型,且较易感[9,10]。本研究未做犬性别与 HEV 感染率的关系分析,但刘俊峰等认为雌性宠物犬 HEV 的感染率明显高于雄性犬[13]。

迄今为止,国内已有其他地区犬 HEV 流行的报道,不同地区不同时间的犬 HEV 的感染情况也不尽一致。如上海地区犬 HEV 感染率在 2006 年为 17.82%[11],2007 年上升至 35.56%[12],2008 年又回落为 21.4%[13];而在 2003 年南京地区犬 HEV 的感染率仅为 9.79%[15];2005 年福建省犬 HEV 的感染率为 38.78%[16];2006 年广西地区犬 HEV 的感染率为 25.71%[17];在本研究中,昆明地区犬 HEV 感染率相对较高,为 48.51%,这可能与血清样本的差异有关,也与 HEV 在不同的地理气候环境中的数量有关。值得注意的是,从临床采集的 HEV 抗体呈阳性的血清中,大部分犬本身并未表现出明显的肝炎临床症状(如黄疸、肝脏肿大、肝硬化等),提示犬 HEV 的感染大多为隐性感染或亚临床感染,这一结论与与以前的报道一致[11,13]。

目前在没有安全有效地预防犬戊型肝炎疫苗的情况下[14],犬以及与犬密切接触的人群应注意保持良好的清洁卫生,定期进行 HEV 的诊断和肝功能的检测。HEV 主要经粪—口途径传播,特别是经被 HEV 污染的水和食物传播[4,7,8,10],因此应特别注意犬粪便的消毒,人类饮用水和食物的卫生。此次的血清学调查结果应当引起有关部门和兽医工作者的高度重视,云南省养殖规模不断扩大,野生动物的种类和数量位居全国之首,被誉为“动物王国”,因此非常有必要进一步深入开展不同种属动物间 HEV 的分子流行病学研究及 HEV 致病机理等方面的研究,以便尽快制定并采取行之有效的措施进行预防和控制,阻止 HEV 的蔓延。

## 参考文献


[1] 余锐萍,李文贵,王英华,等.病毒性肝炎——值得警惕的重要人兽互传病[J].科技导报(北京),2007,25(4):44-52.

[2] Balayan MS, Usmanov RK, Zamyatina NA, et al. Brief report: experimental hepatitis E infection in domestic pigs [J]. J Med Virol, 1990, 32: 58-62.

[3] Meng XJ. Novel strains of hepatitis E virus identified form humans and other animal species: is hepatitis E a zoonosis [J]. J Hepatol, 2000, 33: 842-849.

[4] Yu Y, Sun J, Liu M, et al. Seroepidemiology and genetic characterization of hepatitis E virus in the northeast of China [J]. Infect Genet Evol, 2009, 9(4): 554-561.

[5] Zhuang H, Cao XY, Liu CB, et al. Epidemiology of hepatitis E in China [J]. Gastroenterol Jpn, 1991, 26 Suppl 3: 135-138.

[6] Meng XJ. Hepatitis E virus: Animal reservoirs and zoonotic risk [J]. Vet Microbiol, 2009, Mar 20. [Epub ahead of print].

[7] Liu J, Zhang W, Shen Q, et al. Prevalence of antibody to hepatitis E virus among pet dogs in the Jiang-Zhe area of China [J].

Scand J Infect Dis,2009,41(4):291 - 295.

[8] Shao ZJ,Li JH,Zheng YJ,et al. Epidemiological screening for hepatitis E virus in bile specimens from livestock in northwest China [J]. J Clin Microbiol,2009,47(3):814 - 816.

[9] 多海刚,田克恭. 人与动物感染戊型肝炎病毒的研究进展[J]. 实验动物的科学与管理,2002,19(2):34 - 38.

[10] 罗铭,张立伐. 戊型肝炎病毒学与流行病学研究进展[J]. 实用肝脏病杂志,2008,11(5):352 - 355.

[11] 周锦萍,孙泉云,刘佩红,等. 上海地区多种动物戊型肝炎血清学调查[J]. 动物医学进展,2006,27(12):85 - 88.

[12] 瞿浩生,施建标,倪惠军,等. 上海浦东地区犬戊型肝炎病毒流行情况调查[J]. 中国动物检疫,2007,24(6):30.

[13] 刘俊峰,华修国,张文,等. 上海地区宠物犬中戊型肝炎病毒流行病学调查[J]. 安徽农业科学,2008,36(14):5895 - 5896,5941.

[14] 鞠龚讷,孙泉云,葛杰,等. 动物戊型肝炎研究进展[J]. 上海畜牧兽医通讯,2007,5:13 - 14.

[15] 丁福,孟继鸿,张兰芳,等. 与人类关系密切的 7 种动物对戊型肝炎病毒易感性的初步研究[J]. 中国人兽共患病杂志,2004,20(1):52 - 55.

[16] 王灵岚,陈高,何似,等. 福建省六种哺乳动物戊型肝炎病毒感染调查[J]. 中国人兽共患病杂志,2006,22(2):188 - 188,131.

[17] 韦献飞,梁靖瑞,唐荣兰,等. 广西地区猪、鼠、狗戊型肝炎病毒感染血清学分析[J]. 中国兽医公共卫生,2007,23(2):228 - 229.

# 乳酸菌代谢产物乳酸对 LPS 诱导的 MIMVEC 细胞信号通路 NF-κB 的调控作用

刘 静,薛九州,朱志宁,任晓明

(北京农学院动物科学技术学院,北京 102206)

**摘 要**:旨在探讨乳酸抑制脂多糖(LPS)致大鼠肠黏膜微血管内皮细胞(MIMVEC)中核因子 κB(NF-κB)P65 表达的作用。原代培养 MIMVEC,分为空白对照组、LPS 阳性对照组、乳酸高浓度组、中浓度组和低浓度组,高、中、低浓度的乳酸预处理细胞 3h 后,LPS 处理细胞 4h,用 ELISA 方法检测 TNF-α 和 IL-6 的表达量。用荧光定量 PCR 方法对 NF-κB mRNA 进行定量检测;用 Western Blotting 的方法检测 NF-κB p65 蛋白表达的变化。荧光定量 PCR 结果显示,LPS 处理细胞 4h 的乳酸高、中、低浓度组中 NF-κB mRNA 的表达量是空白组对照的 1.51 倍、2.62 倍、3.00 倍,而 LPS 组中 NF-κB mRNA 的表达量是空白对照组的 7.36 倍;Western Blotting 的结果显示,各组蛋白提取物中均存在分子量约为 65ku 的、与兔抗 NF-κB P65 多克隆抗体发生免疫阳性反应的蛋白条带,且乳酸高浓度组细胞中 NF-κB P65 的表达与空白对照组相比差异不显著($P>0.05$);中浓度组和低浓度组细胞中 NF-κB P65 的表达与空白对照组相比差异显著($P<0.05$)。LPS 组 NF-κB P65 的表达与其他各组相比均差异极显著($p<0.01$)。结果提示,乳酸对 LPS 所致的 MIMVEC 细胞 NF-κB 激活具有明显的抑制作用,下调了转录因子 NF-κB 的转录活性,进而抑制了炎性因子 TNF-α 和 IL-6 的表达。

**关键词**:乳酸,肠黏膜微血管内皮细胞,LPS,NF-κB

脂多糖(LPS)作为内毒素的主要成分,诱导血管内皮细胞产生白细胞介素(IL)-6 和肿瘤坏死因子(TNF)-α,这些因子是参与内毒素所致休克、全身炎症反应综合征等的重要炎性介质[1-3]。抑制这些因子的合成,便可抑制 NF-κB 细胞信号转导通路的激活。为深入探讨益生菌代谢产物乳酸抑制 LPS 激活 NF-κB 细胞信号转导通路的作用,本试验测定了不同条件下 NF-κB p65 亚基 mRNA 和蛋白的表达情况。

## 1 材料与方法

### 1.1 试验动物与试验设计

将原代培养的大鼠肠黏膜微血管内皮细胞(MIMVEC)分为空白对照组(A 组)、LPS 对照组(B 组)及 2.5(C 组)、5.0(D 组)、7.5(E 组)μl/ml 乳酸处理组 5 组。A 组:加维持培养液;B 组:加入 1μg/ml 的 LPS;C 组:终浓度 2.5μl/ml 的乳酸预处理 3h 后,加入 1 μg/ml 的 LPS;D 组:终浓度 5.0μg/ml 的乳酸预处理 3h 后,加入 1μg/ml 的 LPS;E 组:终浓度 7.5μl/ml 乳酸预处理 3h 后,加入 1μg/ml 的 LPS;于培养开始后第 4 小时裂解细胞。

### 1.2 主要试剂和仪器

DMEM 培养基:Gibco 公司产品;Hank's 干粉:Sigma 公司产品;克隆专用胎牛血清、Ⅱ型胶原酶:PAA Laboratories GmbH 公司产品;TNF-α、IL-6ELISA 试剂盒:上海森雄科技实业有限公司产品;兔 NF-κB p65 亚基多克隆抗体:英国 abcam 公司产品;HRP 标记的羊抗兔 IgG:武汉博士德公司产品;TRIZOL 通用型 RNA 快速提取试剂盒:inventrogen 公司产品;20×EvaGreen:美国 Biotium 公司产品;荧光 PCR 仪、SDS-PAGE 凝胶电泳仪、电转仪均为美国 Bio-Rad 公司产品。

### 1.3 试验方法

1.3.1 MIMVEC 原代培养 取出生 24h 之内的 SD 大鼠乳鼠空肠,Hank's 液冲洗,纵向切开,冲洗干净内容物,剪成小段,用 1g/L Ⅱ型胶原酶消化,暴露出微血管网,离心,取沉淀剪成碎块,接于 6 孔细胞培养板中,每孔加 2ml 完全培养基，置于 5% $CO_2$、37℃培养箱中培养，2～3d 换液 1 次。细胞生长至 80% 融合时传代。完全培养基：在基础培养基中按比例加入以下辅助成分：15% 优质胎牛血清（FBS，Hyclone）、0.584 g/L 谷氨酰胺、$1 \times 10^5$ IU/L 注射用青霉素（华北制药股份有限公司）、0.1 g/L 注射用链霉素（华北制药股份有限公司），100 mg/L 内皮细胞生长添加物（ECGS，Upstate）。消化液：预热 37℃，0.25% 胰蛋白酶（Gibco）

1.3.2　试验细胞的准备及处理　试验使用第 3 代细胞，将生长至融合期的微血管内皮细胞以 2.5g/L 胰蛋白酶消化，细胞悬液接种于 6 孔细胞培养板，置于 5% $CO_2$、37℃培养箱中培养。待细胞融合成单层细胞开始试验。

1.3.3　ELISA 法测定 TNF-α、IL-6 的表达量　细胞培养及处理方法同前，按照说明书进行操作。

1.3.4　用荧光定量 PCR 法测定细胞中 NF-κB p65 亚基 mRNA 的表达　用 TRIZOL 通用型 RNA 快速提取试剂盒提取样品 RNA，最后加入 30μl 无 RNase 的水，溶解 RNA。引物由北京擎科生物技术有限公司合成，分装并保存于 -20℃，用时溶解于 DEPC 处理的灭菌超纯水中。引物序列见表 1。DNase I 消化样品 RNA 中的 DNA：模板（RNA），10μg；RNase Inhibitor，4μl；DNase，I buffer，10μl；DNase 1，10μl；加 DEPC 处理水至 100μl；混匀，37℃，90min。RNA 样品经 1% 的琼脂糖凝胶电泳验证可见 28S 和 18S 两条明亮条带，无 DNA 条带污染。RNA 反转录为 cDNA，按照以下反应体系进行：模板（RNA）3μg；引物（50μM）T18，4.0μl；DEPC 处理水至 25μl，混匀，70℃5min，立即冰浴；5 × buffer，8.0μl；dNTP（10mmol/L），4.0μl；RNase Inhibitor，1.0μl，混匀，37℃5min；M-MuLV，2.0μl，42℃60min，70℃10min。定量 PCR 检测按照以下反应体系进行：模板（cDNA）/ $ddH_2O$，1.0μl；10μM 引物 F/R，0.5μl；2 × PCR Mix，12.5μl；Eva green；1μl；$ddH_2O$，10μl，混匀，置于荧光定量 PCR 仪中。

**表 1　试验所用基因及内参基因引物序列**

| Primer 名称 | 序列（5' to 3'） | 碱基数 | 产物长度 |
|---|---|---|---|
| β-actin 190/193 | FW：GAAGTGTGACGTTGACATCCG | 21 | 282bp |
| | RV：GCCTAGAAGCATTTGCGGTG | 20 | |
| NF-κB p65 1043/1044 | FW：TGG AGA ACT TTG AGC CTC T | 19 | 208bp |
| | RV：GCTGACAGAAGACACGAG | 18 | |

1.3.5　用 Western Blotting 检测 NF-κBp65 的蛋白表达　细胞培养及处理方法同前。提取细胞总蛋白，BCA 法测定蛋白含量。将 10μl（每个上样孔含 40μg 抽提的蛋白）加样于 10% SDS-聚丙烯酰胺凝胶中电泳 2h，然后转移至 PVDF 膜（Millipore IP USA）。5% 脱脂奶粉封闭 1h，加入兔抗 NF-κB p65 一抗（1∶1 000 稀释，abcam 公司）、β-actin 一抗（1∶1 000 稀释，北京博奥森公司），37℃孵育 1h，洗涤后加入辣根过氧化酶标记的羊抗兔 IgG（1∶5 000 稀释，北京博奥森公司），室温孵育 1h。洗涤后按照 DAB 染色试剂盒（北京康为世纪，cw0125）的说明书染色，显色充分后脱水封固，用凝胶图像扫描仪进行半定量分析。

## 1.4　数据处理

实时荧光定量 PCR：目的基因的相对表达量用 $\Delta\Delta C_T$ 法计算，$\Delta C_T$（目的基因）$= C_T$（目的基因）$- C_T$（内参基因），$\Delta\Delta C_T = \Delta C_T$（处理组）$- \Delta C_T$（空白组）。目的基因的相对表达水平 $= 2^{-\Delta\Delta CT}$，数据用 Excel 进行统计分析，荧光实时定量 PCR 结果均用均值 ± 标准误表示，其中各基因的表达量所示结果均经过内参 β-actin 表达量进行校正。

Western Blotting：检查 PVDF 膜上的预染 Marker 在膜上的对应位置及电泳起始点，确定目标条带所在的位置。应用 Gel-Pro Analyzer 6.0 软件对大鼠 NF-κB p65 和 β-actin 的 Western Blotting 结果进行半定量分析，蛋白半定量值 = NF-κB p65 蛋白含量/β-actin 蛋白含量。数据均采用均值 ± 标准误表示，统计分析应用

SPSS11.5 统计软件,多组间比较采用单因素方差分析。显著性差异水平为 $P<0.05$,极显著性差异水平为 $P<0.01$。

## 2 结 果

### 2.1 MIMVEC 细胞的鉴定

倒置显微镜下观察,细胞呈典型的铺路石状,无重叠生长现象(图 1)。经免疫化学法检测,内皮细胞Ⅷ因子相关抗原为阳性,因此确定该细胞为 MIMVEC,图 2。

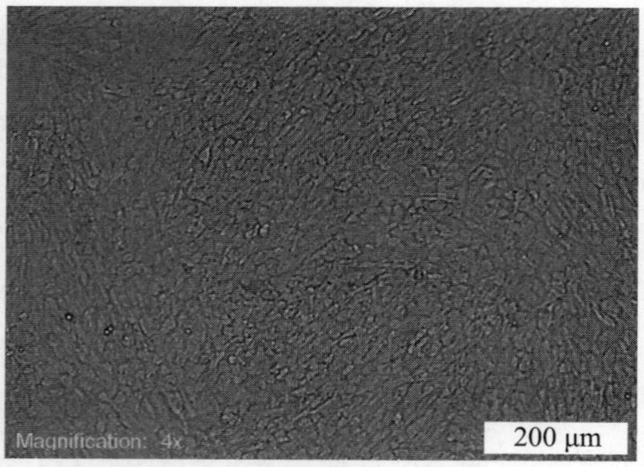


图 1 生长至融合的 MIMVEC 细胞

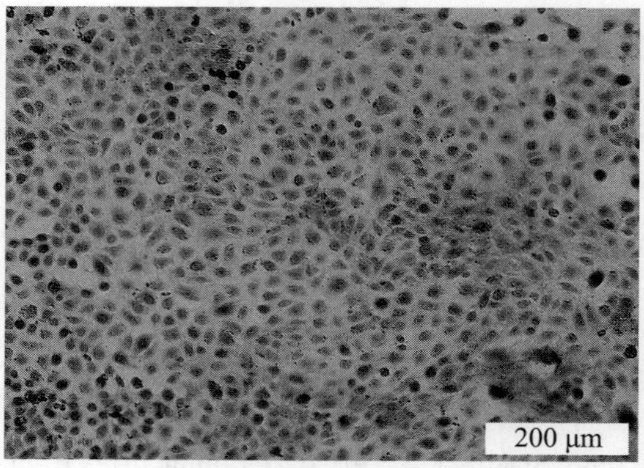


图 2 免疫化学法鉴定内皮细胞Ⅷ因子阳性

### 2.2 LPS 对乳酸预处理的 MIMVEC 细胞 TNF-α 表达量的影响

ELISA 的结果表明,TNF-α 表达量在第 9h 附近达到峰值,高浓度乳酸预处理组的细胞 TNF-α 各时间段表达量与空白对照组相比差异不显著,与 LPS 对照组相比差异极显著;中、低浓度乳酸预处理组细胞的 TNF-α 各时间段表达量与空白对照组相和 LPS 组相比均差异显著。

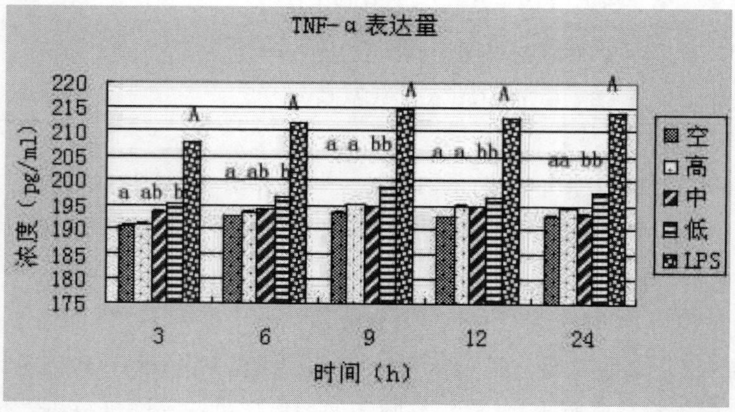


标有不同的小写字母表示差异显著($P<0.05$),标有不同的大小写字母表示差异极显著($P<0.01$)

### 2.3 LPS 对乳酸预处理的 MIMVEC 细胞 IL-6 表达量的影响

ELISA 的结果表明,乳酸预处理组的细胞 IL-6 各时间段表达量与空白对照组相比差异显著,与 LPS 组相比差异极显著。

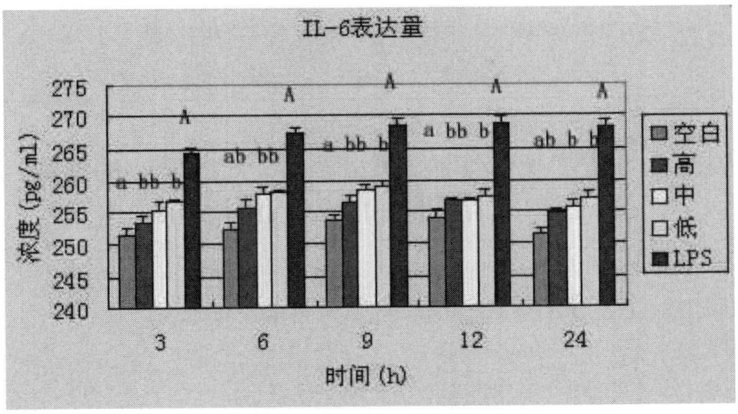


标有不同的小写字母表示差异显著($P<0.05$),标有不同的大小写字母表示差异极显著($P<0.01$)

## 2.4 LPS 对乳酸预处理的 MIMVEC 细胞 NF-κB p65mRNA 表达的影响

荧光定量 PCR 的扩增曲线符合标准的“S”型荧光增长曲线,目的基因(图 3)及内参基因(图 4)扩增的动力学曲线整体平行性较好,基线平而无明显上扬趋势现象且拐点清楚。熔解曲线图示梯度模板熔解曲线集中。目的基因(图 5)与内参基因(图 6)的扩增产物 $T_m$ 值较为均一,熔解曲线上呈明显单峰,表明在实时荧光定量 PCR 过程中,荧光强度均来自于特异性的扩增产物,各个基因及内参基因没有非特异性扩增及引物二聚体产生。

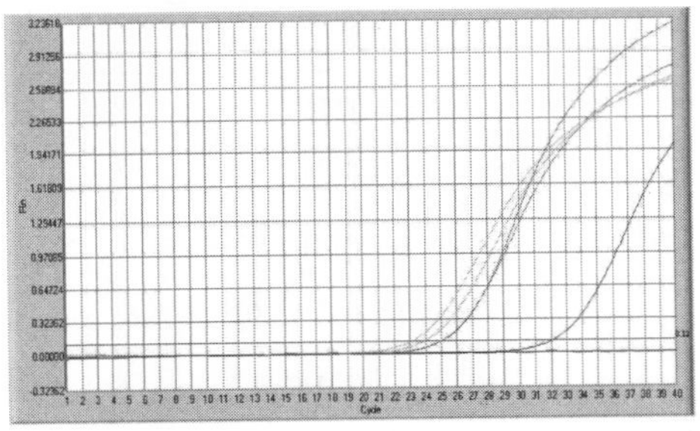

图 3　NF-κB p65 亚基基因的扩增曲线

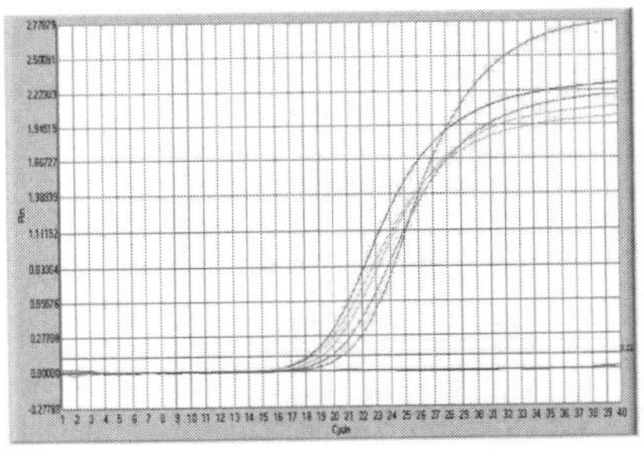

图 4　内参 β-actin 的扩增曲线

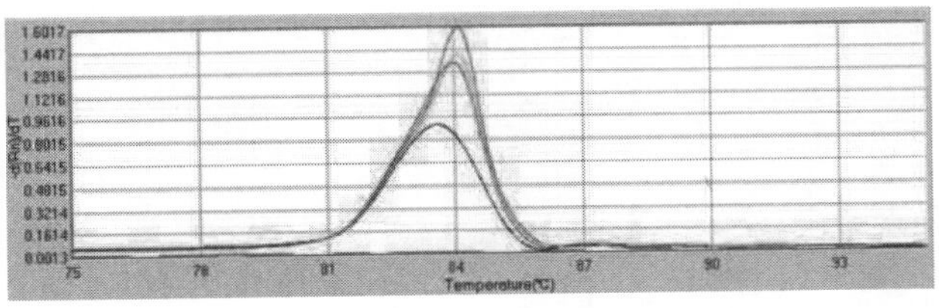

图 5　NF-κBp65 亚基的熔解曲线

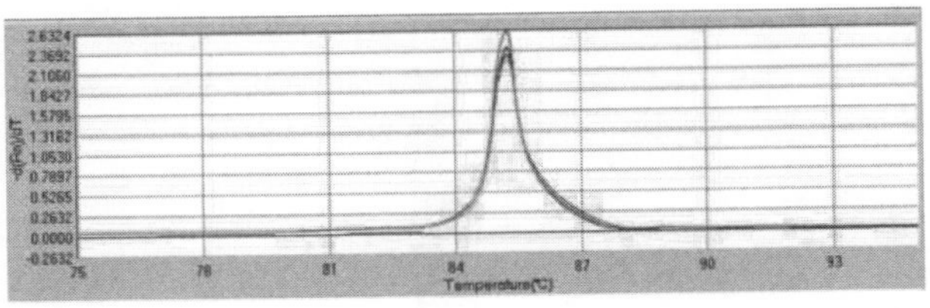

图 6　β-actin 的熔解曲线

相对定量的结果表明,高、中、低浓度乳酸处理组的 NF-κB p65 mRNA 的表达量分别为空白对照组表达量的 1.51 倍、2.62 倍、3.00 倍,LPS 对照组的 NF-κB p65mRNA 的表达量是空白对照组表达量的 7.36 倍。

表 2　NF-κB 基因在不同处理组中的实时荧光定量 PCR 结果

| 组　别 | $\Delta C_t$ | $\Delta\Delta C_t$ | $2^{-\Delta\Delta Ct}$ |
|---|---|---|---|
| 空白对照组 | 8.066 7 ±0.043 33 | 0.000 0 ±0.043 3 | 1.00 |
| LPS 对照组 | 5.183 3 ±0.085 70 | −2.883 3 ±0.085 7 | 7.36 |
| 乳酸高浓度组 | 7.476 7 ±0.069 36 | −0.590 0 ±0.069 3 | 1.51 |

| | | | |
|---|---|---|---|
| 乳酸中浓度组 | 6.676 7 ±0.018 56 | −1.390 0 ±0.018 5 | 2.62 |
| 乳酸低浓度组 | 6.433 3 ±0.035 28 | −1.633 3 ±0.035 2 | 3 |

### 2.5 LPS 对乳酸预处理的 MIMVEC 细胞 NF-κB p65 蛋白表达的影响

Western Blotting 结果显示各组中均有分子量约 65kDa 且与多克隆兔抗 NF-κB p65 抗体发生免疫阳性反应的蛋白条带(图 7);NF-κB p65 在 LPS 对照组的表达显著高于乳酸预处理组和空白对照组的表达,乳酸预处理组的表达与空白对照组相比差异不显著。(表 3)

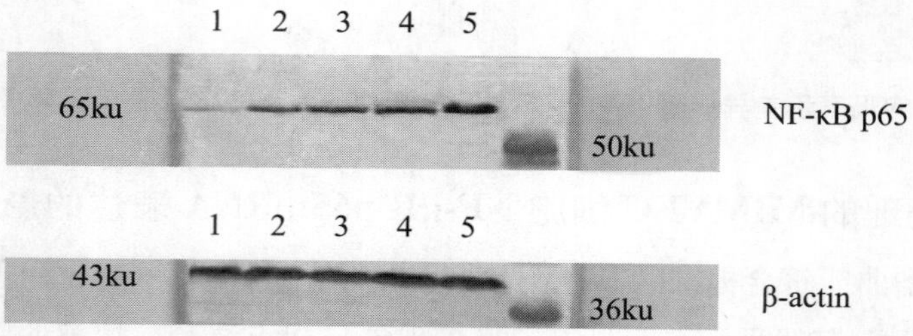


图7 NF-κB p65 的蛋白 Western Blotting

1 为空白对照组;2、3、4 为乳酸高、中、低浓度组;5 为 LPS 对照组

表3 NF-κB p65 的蛋白 Western Blotting 半定量平均值(n = 3)

| | NF-κB 相对定量值 |
|---|---|
| 空白对照组 | 1.00 1 ±0.101 6[a] |
| 乳酸高浓度组 | 1.014 6 ±0.202 5[a] |
| 乳酸中浓度组 | 1.026 3 ±0.101 3[b] |
| 乳酸低浓度组 | 1.028 0 ±0.101 9[b] |
| LPS 对照组 | 2.828 0 ±0.176 7[A] |

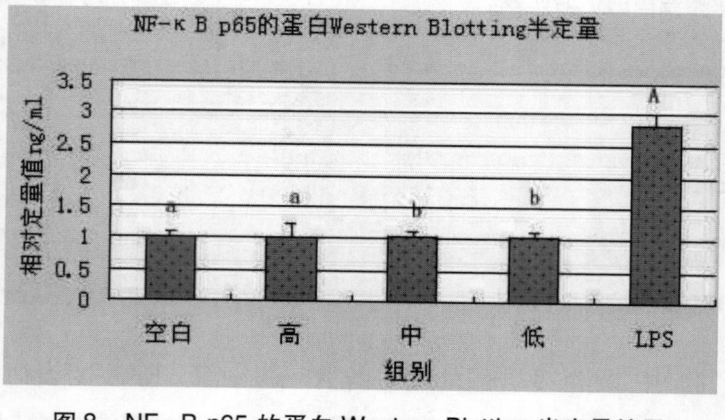


图8 NF-κB p65 的蛋白 Western Blotting 半定量结果

标有不同的小写字母表示差异显著($P<0.05$),标有不同的大小写字母表示差异极显著($P<0.01$)

## 3 讨 论

NF-κB 是细胞内重要的核转录因子,通过调控多种基因的表达而参与免疫反应、细胞凋亡、细胞增殖和肿瘤发生等过程[7-10]。已知在炎症介质表达的调控中,NF-κB 被激活和核移位是关键的一步[11-12]。内皮细胞作为主要炎症反应细胞,在 LPS 引起的内毒素休克等病理过程中具有重要的作用[13]. 炎症反应与内皮细胞功能障碍密切相关[14]。

笔者所在实验室的以往研究结果表明，乳酸菌的代谢产物乳酸对预防和治疗由大肠杆菌内毒素所致小鼠肠道疾病有可靠的效果，并且掌握了乳酸使用的有效安全浓度以及 TNF-α、IL-6 等相关细胞因子的表达情况[15-16]。为深入了解乳酸抑制 LPS 激活 NF-κB 通路的作用，本试验测定了 NF-κB p65 亚基 mRNA 和蛋白的表达情况。

关于 LPS 激活 NF-κB 表达过程中的时相，一些研究已经证明大约在 4h 前后，NF-κB 的 mRNA 和蛋白表达达到高峰[17]。用 LPS 处理后，NF-κB 的基因和蛋白表达均出现了升高现象，但是未经乳酸预处理的组 NF-κBp65 亚基的 mRNA 及蛋白表达升高的更为显著，这说明乳酸对抑制 LPS 激活 NF-κB 转录和表达产生了影响。而经过乳酸预处理组 NF-κBp65 亚基的 mRNA 表达与空白对照组相比，并无显著差异，这说明乳酸对 LPS 攻击下的细胞所起的保护作用明显，其中，中、低浓度预处理组的 NF-κB 的蛋白表达显著高于空白组和高浓度组，这也从一定程度上证明乳酸的保护作用在一定范围内呈现剂量依赖性。

NF-κB 家族及其介导的细胞信号转导通路在细胞凋亡中的作用是国内外研究的热点。近些年的研究证明 NF-κB 具有促细胞凋亡的作用，并发现 NF-κB 亚单位的种类及数量在细胞凋亡中起着决定性的作用，通过抑制 NF-κB 信号转导途径的激活可以减缓细胞凋亡。对 NF-κB 及相关信号转导通路具体机制的研究，将有助于加深对不同疾病发生的分子机制的认识，也可为治疗这些疾病开拓新的途径[18-20]。

## 参考文献


[1] 谢艳萍，王建春. Toll 样受体与信号转导. 国外医学呼吸系统分册，2003，23(4)：198 – 200.

[2] 梁自文，杨宗城，罗向东. 人脐静脉内皮细胞脂多糖刺激后上调基因表达谱的分析. 第三军医大学学报，2003，25(4)：279 – 282.

[3] 陈主初. 病理生理学. 北京：人民卫生出版社，2001，46 – 48；202 – 225.

[4] 张卫华，高煜. 国内外益生菌产品发展状况. 口岸卫生控制. 2005. 9(6)：44 – 46.

[5] 程林春. 微生态制剂的应用. China Animal Husbandry & Veterinary Medicine，2003，6(30)：22 – 24.

[6] 杨汝德，陈琼，陈惠音. 乳酸菌发酵制品研究的现状与发展. 2003，77(19)：79 – 83.

[7] Tak P，Firestein G S. NF-kappa B：a key role in inflammatory disseases. J Clin Invest. 2001，107(1)：7 – 11.

[8] Wulczyn F G，Kramg/kgann D，Scheidereit C. The NF-κB/Rel and IκB gene families：mediators of immune response and inflammation. J Mol Med，1996，74(12)：749 – 769.

[9] Pahl H L. Activators and target genes of Rel/NF-keppaB transcription Factors. Oncogene，1999，18(49)：6853 – 6866.

[10] Karin M. Nuclear factor-κB in cancer development and progression. Nature，2006，441(7092)：431 – 436.

[11] Senftleben U，Karin M. The IKK/NF-κB pathway. Crit Care Med，2002，30：S105 – 111.

[12] Zingarelli B，Sheehan M，Wong HR. Nuclear factor-κB as a therapeutic expression by PPM-18，a novel anti0inflammatory agent，*in vitro* and *in vivo*. Biochem J，1997，328：363 – 369.

[13] 姜勇. 内毒素激活内皮细胞的信号机制的研究进展. 中华医学杂志，1999，79(1)：76 – 78.

[14] Kerekes G，Szekanecz Z，Der H，etal. Endothelial dysfunction and atherosclerosis in rheumatoid arthritis：a multiparametric analysis using imaging techniques and laboratory markers of inflammation and autoimmunity. J Rheumatol，2008，35：398 – 406.

[15] 朱珊珊，崔雯，任晓明. 乳酸菌产酸动态测定及小鼠乳酸耐量初探. 中国农学通报. 2009，25(06)：33 – 36.

[16] 朱珊珊，崔雯，任晓明. 乳酸杆菌黏附作用的研究进展. 中国畜牧兽医. 2009，36(7)：219 – 220.

[17] 徐静，张鑫，孙艺平等. LPS 对 PC1 2 细胞 NF-κB 的激活时程. 大连医科大学学报. 2010，32(2)：156 – 159.

[18] CalzadoM A，Bacher S，Schmitz M L. NF-kappaB inhibitors for the treatment of inflammatory diseases and cancer. Curr Med Chem，2007，14(3)：367 – 376.

[19] Camandola S，Mattson M P. NF-kappa B as a therapeutic target in neurodegenerative diseases. Expert Opin Ther Targets，2007，11(2)：123 – 132.

[20] O'sullivan B，Thompson A，Thomas R. NF-kappa B as a therapeutic target in autoimmune disease. Expert Opin Ther Targets，2007，11(2)：111 – 122.

# 两种活性羰基类物质对ECV304的细胞毒性及其机制初探

穆颍颍[#],张天宇[#],李盼盼,武召珍,董　强*
(西北农林科技大学动物医学院,杨凌712100)

**摘　要**:探讨GO和MGO对上皮细胞的细胞毒性及其机制。方法:MTT法检测两种RCS对ECV304细胞活力的影响,评价其细胞毒性;DCFH-DA探针法检测氧化应激;使用罗丹明123检测线粒体膜电位及线粒体形态变化。GO和MGO对ECV304细胞的IC50分别为3.6mmol/L±0.1mmol/L和1.3mmol/L±0.1mmol/L;氧化应激水平小幅度增高,线粒体膜电位随发生改变;线粒体形态发生改变。GO和MGO的细胞毒性与氧化应激水平增加相关性较低而与线粒体损伤相关。

**关键词**:RCS,细胞毒性,线粒体膜电位,氧化应激

奶牛的亚临床子宫内膜炎和胚胎死亡在临床上有较高发病率,分别为42%和50%[1],其主要原因为急性或持久性的慢性炎症导致子宫感染[3]。晚期蛋白质氧化产物(Advanced oxidation protein products,AOPP)和糖基化晚期终末产物(Advanced glycation end products,AGEs)分别促进了急性和慢性炎症的发生[4]。当机体发生氧化应激及羰基应激时,蛋白质氧化损伤增强,AOPP和AGEs生成增加激活其受体,导致活性氧(Reactive oxygen species,ROS)水平上升,诱发子宫内膜炎[6]。AGEs是由活性羰基化合物(reactive carbonyl species,RCS)攻击氨基酸、多肽或蛋白质,经Schiff base和Amadori等一系列复杂的化学重排反应后形成的不可逆的交联产物[7]。甲基乙二醛(Methylglyoxal,MGO)和乙二醛(Glyoxal,GO)是体内毒性较大的RCS。MGO和GO来源于糖酵解的中间产物磷酸丙糖和糖异生中酮体的分解,另外反刍动物瘤胃内的微生物菌群在高糖环境下会大量合成MGO[8,9]。在形成AGEs的过程中,活性羰基化合物不但造成羰基应激,而且引起氧化应激,导致上皮细胞代谢障碍和细胞功能障碍[10]。本文以内皮样细胞ECV304细胞为研究对象,初步探讨MGO和GO对内皮细胞的细胞毒性及其机制,为探讨AGEs与子宫内膜炎症的关系提供依据。

## 1　材料与方法

### 1.1　材料

1.1.1　主要试剂　MTT购自Amresco公司,DMEM培养基购自Gibco公司,胎牛血清购自Hyclon公司,MGO、GO、氨基胍、罗丹明123、DCFH-DA等购自Sigma公司,其他常用试剂购自国药。

1.1.2　主要仪器　EVOS f1型数码倒置显微镜(AMG公司)、VICTORTM X5型全波长荧光酶标仪(PerkinElmer公司)、Epoch型微孔板分光光度计(BioTek公司)、漩涡混合器(江苏海门其林医用仪器厂)。

### 1.2 方法

1.2.1　细胞培养　ECV304细胞(一种永生化的人源内皮细胞),购自武汉大学细胞保藏中心。以$10^4$个细胞每孔接种于96孔板,以含10%胎牛血清的DMEM培养基培养。培养于37 ℃的5% $CO_2$细胞培养箱内。

1.2.2　MTT法检测细胞活力　活细胞可通过线粒体能量代谢过程中的琥珀酸脱氢酶将MTT(四甲基偶氮唑盐)还原形成蓝紫色甲臜沉积于细胞内,通过对甲臜产物的比色分析来可用于测定细胞的活力。接种于96孔板的细胞达到平台期后,与不同浓度GO或MGO(溶于含10%胎牛血清的DMEM培养基)孵育24 h后,弃上清,加入200 μl含MTT(0.5 mg/ml)的培养基。在37℃孵育4h后,弃上清,形成的晶体用200 μl二甲亚砜溶解20 min,振荡混匀。在490 nm处用Epoch型微孔板分光光度计读取吸光度值[11]。细胞活力百

**基金项目**:国家自然科学基金项目(31001090);陕西省博士后基金

**作者简介**:#共同第一作者。穆颍颍(1989-),女,河南许昌人,硕士研究生主要从事氧化损伤和生化分子毒理学研究;张天宇(1989-),男,内蒙古赤土峰人,硕士研究生,主要从事氧化损伤和生化分子毒理学研究。

分率 =(试样组 OD 值/阴性对照组 OD 值)×100%,半致死浓度(IC50)以 Reed-Muench 法得出。

1.2.3 DCFH-DA 法测定细胞内 ROS 细胞内生成的 ROS 用非特异性 ROS 探针(DCFH-DA)检测[12]以不同浓度 GO 或 MGO 孵育,对 2 mmol/L 和 4 mmol/L 另设置同浓度氨基胍干预组孵育细胞,37℃,5% $CO_2$ 条件下孵育 3 h。弃上清,每孔加入 150 μl 含 10 μmol/L DCFH-DA 的培养基,孵育 20min 后洗去多余探针,荧光强度由荧光酶标仪读取,ex/em 为 485 nm/535 nm。

1.2.4 FOX 法测定培养基中 $H_2O_2$ 的积累 使用 FOX 法检测培养基中 $H_2O_2$ 积累量[13],可以灵敏检测培养基中微摩级的 $H_2O_2$ 水平变化。96 孔板中加入 20 μl 培养基样品后加入 180 μl FOX 试剂振荡混匀,室温下孵育 20 min。在 560nm 处以微孔板分光光度计读取吸光度值,代入以 $H_2O_2$ 制作的标准曲线,结果表示为 $H_2O_2$ 浓度。

1.2.5 线粒体膜电位检测与线粒体形态观察 线粒体膜电位(mitochondrial membrane potential,MMP)采用荧光染料罗丹明 123(Rhodamine 123)检测[14]。培养于 96 孔板的细胞在含有 GO 或 MGO 的培养基中孵育 3 h 后,弃上清,每孔加入 150 μl 含 10 μmol/L 罗丹明 123 的培养基,培养箱中孵育 30 min 后洗去多余探针,荧光强度由荧光酶标仪读取,ex/em 为 490 nm/520 nm。以荧光倒置显微镜观察线粒体形态并拍照[13]。

# 2 结 果

## 2.1 GO 与 MGO 对 ECV304 细胞活力的影响

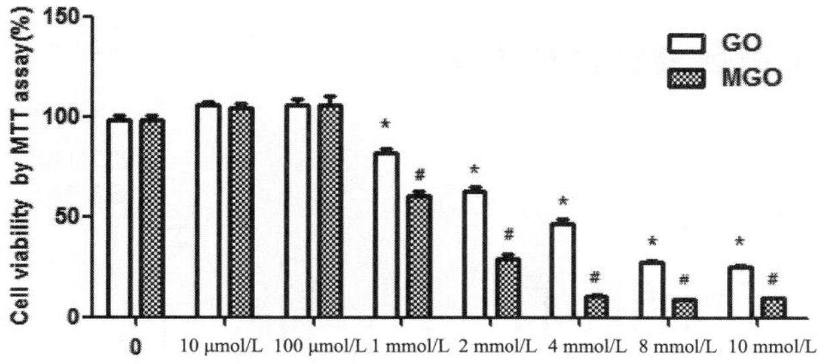


图 1 MTT 法检测 GO 及 MGO 对 ECV304 细胞活力的影响

GO 孵育组与无 GO 对照组比较* $P<0.05$;MGO 孵育组与无 MGO 对照组比较* $P<0.05$

结果显示,到达 1 m 时细胞活力出现下降,并且细胞活力随药物浓度增加而下降,100 μmol/L 的 GO 或 MGO 不影响细胞活力,(图 1)。MGO 的 IC50 约为 1.3 mmol/L ±0.1 mmol/L,而 GO 的 IC50 约为 3.6 mmol/L ±0.1 mmol/L。

## 2.2 GO 与 MGO 诱导的细胞内 ROS 水平

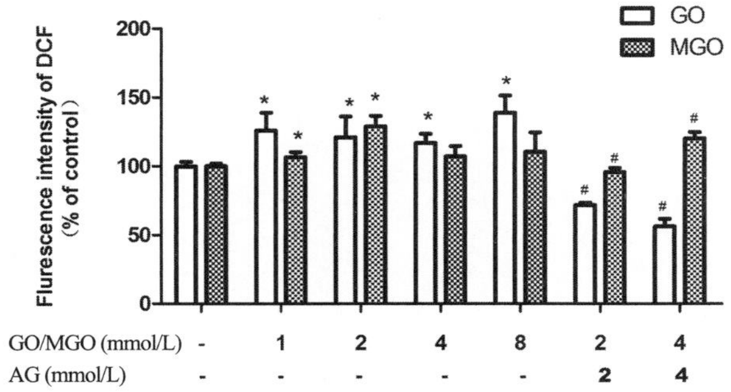


图 2 DCF 荧光探针法检测细胞内 ROS 水平

GO 或 MGO 孵育组与未处理组对比 * $P<0.05$,氨基胍干预组与同浓度 GO 或 MGO 组对比# $P<0.05$

自 1mmol/L 起,GO 均可使 ROS 水平显著升高,而低浓度 MGO(1 ~2mmol/L)可使 ROS 水平增加,而高浓度 MGO 并未增加 ROS 水平。2 m 氨基胍显著降低 2 mmol/L GO 或 MGO 引起的 ROS 水平上升。4 m 氨基胍也降低 4 mmol/L GO 处理组的 ROS 水平。

## 2.3 培养基中 $H_2O_2$ 的积累

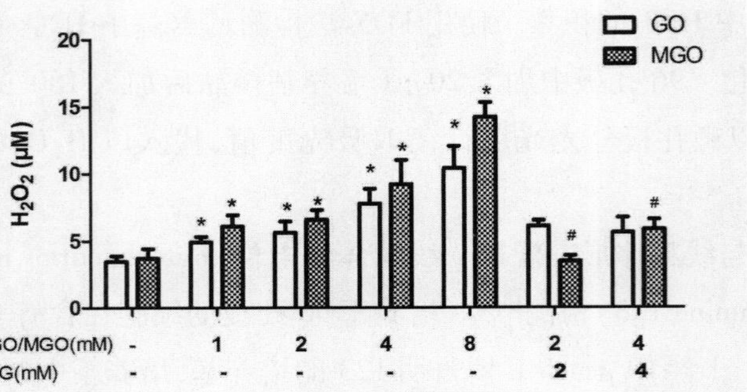


图3 GO 或 MGO 孵育细胞 3 h 后培养基中过氧化氢的积累

GO 或 MGO 孵育组与未处理组对比 * $P<0.05$,氨基胍干预组与同浓度 GO 或 MGO 组对比 # $P<0.05$

GO 或 MGO 均使培养基中的 $H_2O_2$ 浓度显著上升,其上升程度均与 MGO 或 GO 呈浓度梯度效应。氨基胍可降低 MGO 诱导的 $H_2O_2$ 水平,但未能降低 GO 诱导的 $H_2O_2$ 水平(图3)。

## 2.4 GO 或 MGO 引起的线粒体跨膜电位变化

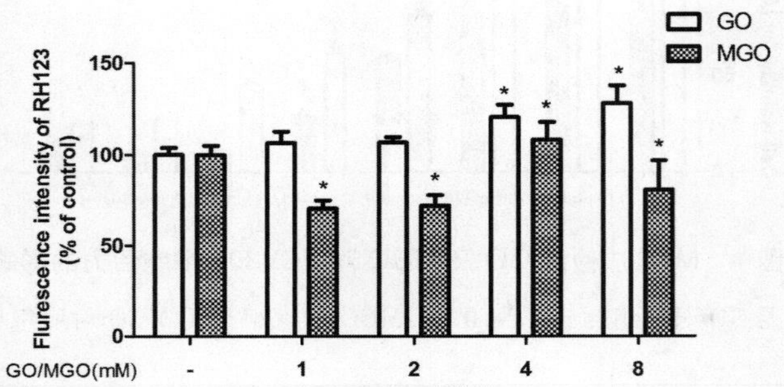


图4 线粒体膜电位检测

GO 或 MGO 孵育组与未处理组对比 * $P<0.05$

除 4 mmol/L MGO 可使 MMP 升高外,1 、2、8 mmol/L MGO 均可使 MMP 降低。4 mmol/L 和 8 mmol/L GO 可使 MMP 膜电位升高。

## 2.5 线粒体形态变化

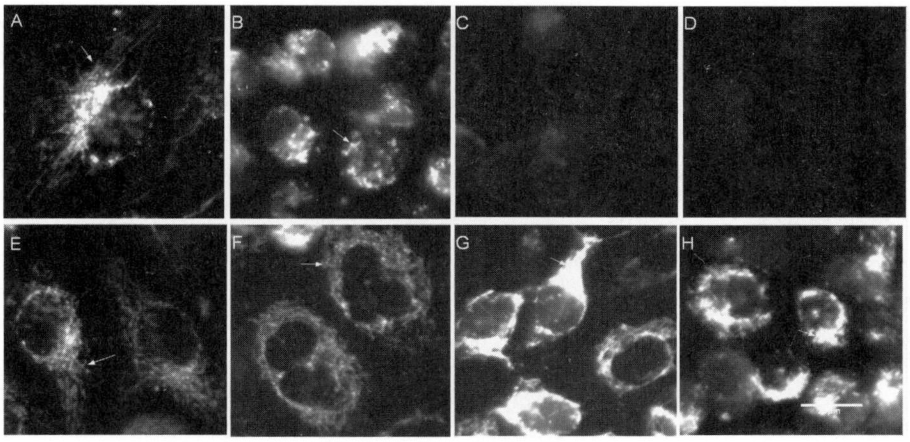


**图5 不同浓度 MGO 及 GO 孵育引起的线粒体形态变化**

A. E. 为未处理组;B. C. D. 分别为 2 mmol/L、4 mmol/L、8 mmol/L MGO 孵育组; F. G. H 分别为 2、4、8 m GO 孵育组

正常细胞线粒体较细长,呈发散状分布,轮廓清晰(图 5A,E)。MGO 或 GO 孵育 24h 后,MGO(图 5B)或 GO(图 5G,H)可使细胞的线粒体由正常的线状改变为膨大的球形。

细胞死亡后膜电位消失,无罗丹明聚集(图 5C,D)。

## 3 讨 论

GO 和 MGO 对 ECV304 细胞的 IC50 分别为 3.6 mmol/L ± 0.1 mmol/L 和 1.3 mmol/L ± 0.1 mmol/L, MGO 对 ECV304 细胞的细胞毒性稍大于 GO。

ECV304 细胞在 MGO 诱导的条件下 ROS 水平无显著变化,前人研究也得到了相同的结果[15],可能是因为其细胞毒性与 ROS 没有紧密的相关性。GO 引起 ROS 水平的上升,但是增加水平较低(增加 20% ~ 40%),与 GO 浓度没有呈浓度梯度效应,这可能与细胞的抗氧化系统发挥作用有相关性[16]。然而在人造骨细胞上,MGO 首先诱导 ROS 的增加并进一步激活下游凋亡信号因子,表明 ROS 的诱导产生与细胞种类有关[17]。最近报道指出,MGO 能够直接损伤小鼠肌细胞线粒体造成线粒体功能障碍[18],同时伴随 ROS 增加,这表明 MGO 诱导的细胞凋亡和线粒体损伤与 ROS 水平有一定相关性。

由于 $H_2O_2$ 可以自由通过细胞膜并扩散到胞外,因此环境中的 $H_2O_2$ 浓度可以显示出细胞内的 $H_2O_2$ 水平,可以看出细胞内过氧化氢水平是随着 GO 或 MGO 诱导浓度增加而增加的(从 3.7μmol/L 到 14.4μmol/L)。$H_2O_2$ 可能是线粒体基质中的超氧阴离子($O_2^-$)经过 $Mn^{2+}$ 依赖的超氧化物歧化酶的歧化作用或通过非酶歧化作用产生的[19,20],因此尽管 ROS 总体水平变化不大,但 $H_2O_2$ 的结果暗示 $O_2^-$ 的生成是增加的。有报道,200 μmol/L $H_2O_2$ 处理 ECV304 细胞 4h 可激活 NF-κB 通路诱导的细胞凋亡[21],而本试验的 $H_2O_2$ 积累量较低,表明 ROS 水平和 $H_2O_2$ 积累量可能不是细胞凋亡的主要原因。

氨基胍为 RCS 的捕获剂,能有效抑制 AGEs 的形成。本试验中,氨基胍可有效抑制 GO 与 MGO 引起的氧化应激水平,表明 RCS 诱导的氧化应激可能与 AGEs 的形成有关[22]。

健康细胞线粒体膜电位 $\triangle\psi m$ 维持在正常水平,线粒体膜电位去极化为线粒体损伤诱导凋亡的主要标志,实验中 MGO 导致细胞的 MMP 下降提示 MGO 可能造成线粒体损伤[23]。由于 GO 反而引起的膜电位超极化,此现象可能是由于影响了线粒体外膜阴离子电压通道(Voltage dependent anion channels, VDAC)的功能造成线粒体超极化型损伤,此模型为 VDAC 关闭后阻碍 ADP 进入线粒体,中断 ATP 合成使呼吸链氧化还原电势增高[24,25],引起线粒体膨胀及线粒体通透性改变。当细胞色素 C 释放到胞质内就会激活凋亡信号通路启动细胞凋亡[26]。

本研究确定了 GO 和 MGO 对 ECV304 细胞的 IC50,MGO 的细胞毒性大于 GO,并且发现 GO 与 MGO 对

ECV304 细胞的细胞毒性与氧化应激水平增加相关性较低而与线粒体损伤途径相关,但其对线粒体膜通道蛋白的作用需进一步研究证实。

## References


[1] Mann, G. and G. Lamming, The influence of progesterone during early pregnancy in cattle. *Reproduction in Domestic Animals*, 1999. 34(3-4): p. 269-274.

[2] Kaufmann, T., et al., Prevalence of bovine subclinical endometritis 4h after insemination and its effects on first service conception rate. *Theriogenology*, 2009. 71(2): p. 385-391.

[3] Singh, J., et al., The immune status of the bovine uterus during the peripartum period. *The Veterinary Journal*, 2008. 175(3): p. 301-309.

[4] Kalousová, M., et al., Advanced glycoxidation end products in chronic diseases—clinical chemistry and genetic background. Mutation Research/Fundamental and Molecular Mechanisms of Mutagenesis, 2005. 579(1): p. 37-46.

[5] Celi, P., et al., Relationship between late embryonic mortality and the increase in plasma advanced oxidised protein products (AOPP) in dairy cows. *Reproduction, Fertility and Development*, 2011. 23(4): p. 527-533.

[6] Merlo, M., et al. Embryonic mortality and plasma advanced oxidation protein products (AOPP) increase in dairy cows. in DAIRY RESEARCH FOUNDATION' Current Topics in Dairy Production, Symposium. 2008.

[7] Wu, L. and B. H. Juurlink, Increased methylglyoxal and oxidative stress in hypertensive rat vascular smooth muscle cells. *Hypertension*, 2002. 39(3): p. 809-814.

[8] PHILLIPS, S. A. and P. J. Thornalley, The formation of methylglyoxal from triose phosphates. *European Journal of Biochemistry*, 1993. 212(1): p. 101-105.

[9] Russell, J., Glucose toxicity in Prevotella ruminicola: methylglyoxal accumulation and its effect on membrane physiology. *Applied and Environmental Microbiology*, 1993. 59(9): p. 2844-2850.

[10] Wautier, J.-L. and A. M. Schmidt, Protein glycation a firm link to endothelial cell dysfunction. *Circulation Research*, 2004. 95(3): p. 233-238.

[11] Bai, L., et al., Modulation of Sirt1 by resveratrol and nicotinamide alters proliferation and differentiation of pig preadipocytes. *Molecular and Cellular Biochemistry*, 2008. 307(1-2): p. 129-140.

[12] Wang, H. and J. A. Joseph, Quantifying cellular oxidative stress by dichlorofluorescein assay using microplate reader. *Free Radical Biology and Medicine*, 1999. 27(5): p. 612-616.

[13] Long, L. H., M. V. Clement, and B. Halliwell, Artifacts in cell culture: rapid generation of hydrogen peroxide on addition of (-)-epigallocatechin, (-)-epigallocatechin gallate, (+)-catechin, and quercetin to commonly used cell culture media. *Biochemical and Biophysical Research Communications*, 2000. 273(1): p. 50-53.

[14] Hail, N. and R. Lotan, Mitochondrial permeability transition is a central coordinating event in N-(4-hydroxyphenyl) retinamide-induced apoptosis. *Cancer Epidemiology Biomarkers & Prevention*, 2000. 9(12): p. 1293-1301.

[15] Tu, C.-Y., et al., Methylglyoxal induces DNA crosslinks in ECV304 cells via a reactive oxygen species-independent protein carbonylation pathway. *Toxicology in Vitro*, 2013. 27(4): p. 1211-1219.

[16] Yu, B. P., Cellular defenses against damage from reactive oxygen species. *Physiological Reviews*, 1994. 74(1): p. 139.

[17] Chan, W. H., H. J. Wu, and N. H. Shiao, Apoptotic signaling in methylglyoxal-treated human osteoblasts involves oxidative stress, c-Jun N-terminal kinase, caspase-3, and p21-activated kinase 2. *Journal of Cellular Biochemistry*, 2007. 100(4): p. 1056-1069.

[18] Wang, H., J. Liu, and L. Wu, Methylglyoxal-induced mitochondrial dysfunction in vascular smooth muscle cells. *Biochemical Pharmacology*, 2009. 77(11): p. 1709-1716.

[19] Andreyev, A. Y., Y. E. Kushnareva, and A. Starkov, Mitochondrial metabolism of reactive oxygen species. *Biochemistry (Moscow)*, 2005. 70(2): p. 200-214.

[20] Bielski, B. and D. Cabelli, Highlights of current research involving superoxide and perhydroxyl radicals in aqueous solutions.

*International Journal of Radiation Biology*, 1991. 59(2): p. 291 – 319.

[21] Bowie, A. G., P. N. Moynagh, and L. A. O' Neill. Lipid Peroxidation Is Involved in the Activation of NF-κB by Tumor Necrosis Factor but Not Interleukin-1 in the Human Endothelial Cell Line ECV304 lack of involvement of $H_2O_2$ in NF-κB activation by either cytokine in both primary and transfrommed endothelial cells. *Journal of Biological Chemistry*, 1997. 272(41): p. 25941 – 25950.

[22] Chappey, O., et al., Advanced glycation end products, oxidant stress and vascular lesions. *European Journal of Clinical Investigation*, 1997. 27(2): p. 97 – 108.

[23] Ly, J., D. Grubb, and A. Lawen, The mitochondrial membrane potential ($\Delta\Psi m$) in apoptosis; an update. *Apoptosis*, 2003. 8(2): p. 115 – 128.

[24] Scarlett, J. L., et al., Changes in mitochondrial membrane potential during staurosporine-induced apoptosis in Jurkat cells. *FEBS Letters*, 2000. 475(3): p. 267 – 272.

[25] Heiden, M. G. V., et al., Bcl-x < sub > L </sub > Prevents Cell Death following Growth Factor Withdrawal by Facilitating Mitochondrial ATP/ADP Exchange. *Molecular Cell*, 1999. 3(2): p. 159 – 167.

[26] Lam, M., N. L. Oleinick, and A. -L. Nieminen, Photodynamic therapy-induced apoptosis in epidermoid carcinoma cells reactive oxygen species and mitochondrial inner membrane permeabilization. *Journal of Biological Chemistry*, 2001. 276(50): p. 47379 – 47386.

# 儿茶素对甘油醛白蛋白非酶糖基化的抑制作用

武召珍#,杨立军#,冯翠霞,董　强*

(西北农林科技大学 动物医学院,杨陵 712100)

**摘　要**:探讨天然产物儿茶素能否抑制甘油醛/Fenton 反应诱导 BSA 的非酶糖基化。Girard's reagent T 实验检测儿茶素抑制甘油醛代谢产物的能力;构建甘油醛/Fenton 反应诱导的 BSA 非酶糖基化模型,探讨儿茶素、NAC、AG 和吡哆胺对非酶糖基化的抑制作用。结果显示,儿茶素和 NAC 可抑制甘油醛在 Fenton 反应条件下形成的代谢产物;甘油醛在 Fenton 反应条件下形成的 Schiff base 可被 NAC 和 AG 抑制;儿茶素、NAC 和 AG 均可抑制甘油醛在 Fenton 反应下诱导 BSA 产生的 CML。

**关键词**:儿茶素,甘油醛,Fenton 反应,Schiff base,羧甲基赖氨酸

蛋白非酶糖基化是蛋白质的氨基与还原糖的羰基,经亲电加成反应形成可逆的 Schiff base,进一步经由 Amadori 重排后最终形成不可逆的晚期糖基化终末产物(Advanced glycation end products,AGEs)的一系列非酶促反应[1]。AGEs 的前体为活性羰基化合物(Reactive dicarbonyl species,RCS),具有活泼的羰基,如乙二醛(Gloxal,GO)。GO 是细胞代谢中普遍存在的副产物,主要为体内果糖自氧化和脂质过氧化的代谢产物[2]。生理条件下奶牛血液及肝脏中的葡萄糖经糖酵解转变为甘油醛,再经进一步的自氧化和代谢转化便可为 GO,Fenton 反应由于产生大量羟基自由基而加剧代谢产物 GO 的形成[3],这些体内的 RCS 与蛋白进一步发生非酶糖基化反应,引起机体氧化应激和羰基反应,产生大量的活性氧(Reactive oxygen species,ROS),引发机体的炎症反应和组织损伤[4]。因此,抑制活性羰基化合物诱导的蛋白非酶糖基化,对预防动物体内慢性持久性炎症显得尤为重要。传统的活性羰基捕获剂氨基胍(Aminoguanidine),因此无法应用于临床。天然产物儿茶素为绿茶中提取的黄酮类化合物,具有很强的抗氧化性,且对机体没有毒副作用。最近研究表明黄酮类化合物特有的环状结构有捕获羰基的能力[6,7]。本文将构建 BSA 的非酶糖基化模型,探讨儿茶素对非酶糖基化的抑制作用,为天然产物抑制蛋白非酶糖基化在分子水平上提供理论依据。

## 1　材料与方法

### 1.1　材料

1.1.1　主要试剂　BSA,甘油醛,儿茶素,AG,吡哆胺,N-乙酰-L-半胱氨酸(N-acetyl-L-cysteine,NAC),Girard's reagent T,2,4-二硝基苯肼(2,4-dinitrophenylhydrazine,DNPH)均购自 Sigma 公司;羧甲基赖氨酸($N^{\varepsilon}$-carboxymethyllysine,CML)鼠单克隆抗体购自 R&D 公司。

1.1.2　主要仪器　电泳仪,冷冻离心机,UV1102 紫外分光光度计,全波长酶标仪等。

### 1.2　方法

1.2.1　羰基检测试验(Girard's reagent T 试验)　采用 Girard's reagent T 实验[8]检测反应体系中的羰基化合物。将 2 mmol/L GO 或 Fenton 反应条件下 20 mmol/L 的甘油醛在 200 mmol/L PBS(pH 7.5)中 37℃条件下孵育 2 h,取 50 μl 反应液加入到 950 μl 新鲜配置的 120 mmol/L pH 9.3 的硼酸钠的硼酸钠缓冲液中,取 200

---

**基金项目**:国家自然科学基金项目(31001090);陕西省博士后基金

**作者简介**:#为并列第一作者。武召珍(1989 - ),女,硕士研究生,主要从事氧化损伤和生化分子毒理学的研究,E-mail:wuzhaozhen9810@163.com;
杨立军(1988 - ),男,硕士研究生,主要从事氧化损伤和生化分子毒理学的研究,E-mail:ylj172712@163.com

**通讯作者**:董　强,E-mail:ardour@126.com

μl 混合液加入 800 μl Girard's reagent T。测定 326 nm 的吸光度,最后通过标准曲线计算反应体系中 GO 的浓度。

1.2.2 构建甘油醛/Fenton-BSA 模型 以 pH 7.4 0.1 M PBS 2 mg/ml 的 BSA 溶液,500 μmol/L 甘油醛和 Fenton 试剂(200 μmol/L $Fe^{2+}$/EDTA + 1 m mol/L $H_2O_2$)孵育,以及 0.03%(*W/V*)叠氮化钠,37℃暗环境孵育 24 h。

1.2.3 SDS-PAGE 电泳 反应混合液中加入适量上样缓冲液,沸水中煮 10 min,配制 10% 分离胶,5% 浓缩胶,上样 10 μg,浓缩胶恒压 80 V,30 min;分离胶恒压 120 V,90 min。

1.2.4 DNPH 法检测 Schiff base 取 0.5 ml 甘油醛/Fenton-BSA 样品孵育液与等量 0.1% DNPH 室温孵育 1 h,加入 1 ml 20% TCA,5 000 rpm 离心 3 min,弃上清,以乙酸乙酯和乙醇(1∶1)混合液洗涤 3～4 次,自然晾干,1 ml 6 mol/L 盐酸胍溶解,374 nm 测定吸光度,以 22 000 L/mol 消光系数计算 Schiff base 生成量[9]。

1.2.5 Western blot 检测 运用 Western blot 对孵育产生的 CML 进行定性检测,取 20 μg 孵育的糖基化反应混合物,采用 10%(*W/V*)胶进行 SDS-PAGE 蛋白质电泳,之后将蛋白转到 PVDF 膜上,再用 5% 的脱脂奶粉室温封闭 2 h。一抗选用抗 CML 小鼠单克隆抗体 1∶1 000 倍稀释后 4℃孵育过夜,TBST 洗膜 3 次,每次 10 min。二抗选用山羊抗小鼠交联辣根过氧化物酶 1∶1 500 倍稀释后室温孵育 2 h,TBST 洗膜 3 次,每次 10 min,ECL 发光检测 CML。

## 2 结 果

### 2.1 儿茶素对甘油醛代谢产物的抑制作用

全波段扫面结果显示 Girard's reagent T 在 326 nm 不出现吸收峰(图 1),Girard's reagent T 与 20 m mol/L 甘油醛反应后在 326 nm 几乎不出现吸收峰。GO 和 Girard's reagent T 反应后在 326 nm 处出现吸收峰;甘油醛在 Fenton 反应条件下孵育 3 d 后,Girard's reagent T 与形成的代谢产物反应后在 326 nm 处出现与 GO 反应后相似的吸收峰。

Girard's reagent T 实验检测羰基清除能力的结果显示(表 1),儿茶素、NAC 以及 AG 均可显著降低 Fenton 反应中甘油醛诱导 BSA 产生的代谢产物。儿茶素对甘油醛代谢产物的清除率可达 84.77%(儿茶素与甘油醛摩尔浓度比为 4∶1 时),显著高于阳性对照 AG(75.51%)和相同摩尔浓度比的 NAC(61.42%)。

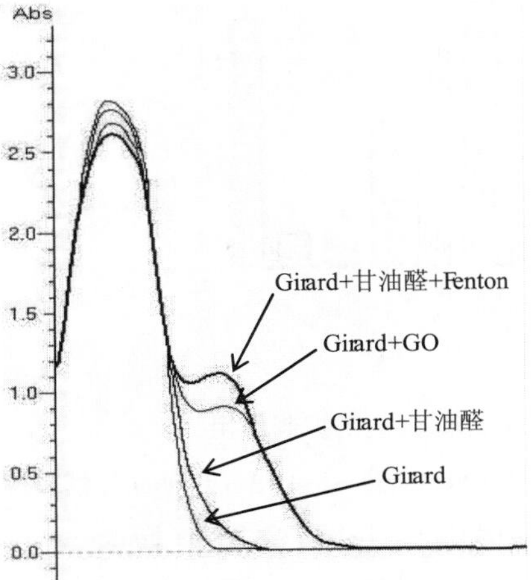


图 1 甘油醛全波段扫描结果

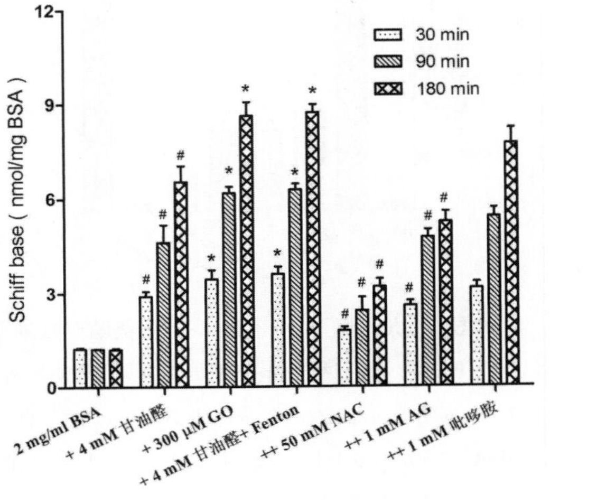


图 2 NAC 对甘油醛/Fenton 反应诱导的 Schiff base 的影响

注:* 与 20 mg/mL BSA 对比,$P < 0.05$;# 与 20 mg/mL BSA + 4 mM 甘油醛 + Fenton 对比,$P < 0.05$。

表 1　儿茶素对甘油醛代谢产物的抑制作用($\overline{X} \pm SD$)

| 孵育时间(d) | 3 d | 5 d | 7 d |
|---|---|---|---|
| 3 mmol/L GO | 4.93 ±0.13 | 4.77 ±0.24 | 4.89 ±0.17 |
| 20 mmol/L 甘油醛 + Fenton | 4.86 ±0.18 | 4.91 ±0.09 | 4.73 ±0.18 |
| +5 mmol/L 儿茶素 | 0.74 ±0.11 * | 0.82 ±0.20 * | 0.68 ±0.24 * |
| +2.5 mmol/L 儿茶素 | 2.13 ±0.09 * | 2.38 ±0.15 * | 2.25 ±0.11 * |
| +5 mmol/L NAC | 1.87 ±0.11 *# | 1.91 ±0.27 *# | 1.78 ±0.19 *# |
| +2.5 mmol/L NAC | 2.79 ±0.06 *# | 2.86 ±0.13 *# | 2.74 ±0.10 *# |
| +2.5 mmol/L AG | 1.19 ±0.17 *# | 1.22 ±0.09 *# | 0.99 ±0.23 *# |

注:* 与 20 mmol/L 甘油醛 + Fenton 对比,$P<0.05$;# 与 20 mmol/L 甘油醛 + Fenton + 5 mmol/L 儿茶素对比,$P<0.05$

## 2.2　NAC 对甘油醛 Fenton 反应诱导的 Schiff base 形成量的影响

图 2 结果表明,2 mg/ml BSA 在 37℃条件下孵育 1 h 即可产生少量 Schiff base,甘油醛可诱导 BSA 产生一定量的 Schiff base,300 μmol/L GO 和 4 mmol/L 甘油醛/Fenton 均可使 Schiff base 生成量增多,且形成量随时间延长而增加,3 h 时分别增长约 7.7 倍和 7.1 倍。50 mmol/L NAC(最大抑制率 2 h,67.61%)和 1 mmol/L AG(最大抑制率 3 h,40.23%)均可显著抑制产生的 Schiff base,1 mmol/L 吡哆胺对产生的 Schiff base 则没有显著的抑制作用。

## 2.3　儿茶素和 NAC 对 CML 形成的抑制作用

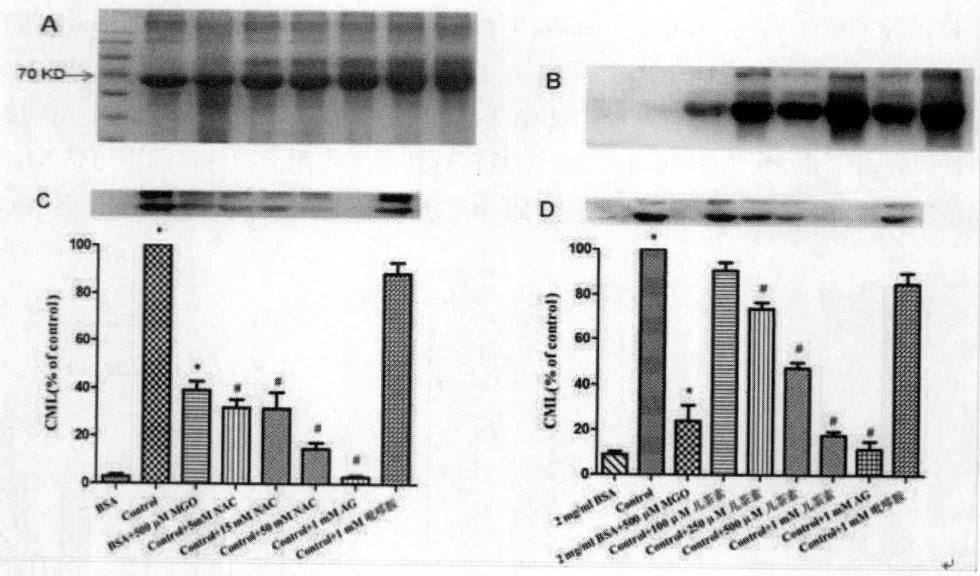


图 3　儿茶素和 N-乙酰-L-半胱氨酸对甘油醛 Fenton 反应诱导的 CML 形成的抑制作用

注:A. 甘油醛/Fenton 反应诱导的 BSA 蛋白质电泳结果 1:marker;2:BSA;3:BSA + Fenton;4 - 8:BSA + Fenton 分别加 0.5 mmol/L、1 mmol/L、2 mmol/L、4 mmol/L、8 mmol/L 甘油醛。B. Western blot 检测 CML。1:BSA;2:BSA + Fenton;3、5、7 分别为 BSA 加 500 μmol/L、1 mmol/L、2 mmol/L 甘油醛;4、6、8 分别为 BSA + Fenton 加 0.5 mmol/L、1 mmol/L、2 mmol/L 甘油醛。C. NAC 对 CML 形成的抑制。Control:BSA + 甘油醛/Fenton;D. 儿茶素对 CML 形成的抑制。Control:BSA + 甘油醛/Fenton。* 与 2 mg/mL BSA 组对比,$P<0.05$;# 与 Control 对比,$P<0.05$

SDS-PAGE 电泳结果(图 3A)显示,甘油醛使 BSA 的蛋白条带发生改变,出现特异性条带,Fenton 反应加剧这一变化,蛋白条带逐渐变宽弥散,随甘油醛浓度的增大呈梯度效应。Western blot 检测结果(图 3B)表明,BSA 不产生 CML,甘油醛孵育 BSA 会产生一定量的 CML,Fenton 反应体系会使甘油醛孵育 BSA 产生的 CML 显著增加。NAC 和儿茶素都会显著降低 Fenton 反应条件下甘油醛诱导 BSA 形成的 CML(图 3C,图 3D)。当儿茶素与甘油醛/Fenton 浓度比为 1∶2 时,可显著抑制 CML 的形成,浓度比 2∶1 时达到最大抑制率 82.76%,抑制率随儿茶素浓度的增加逐渐增强。NAC 和 AG 同样显著抑制 Fenton 反应条件下甘油醛诱导 BSA 形成的 CML,吡哆胺对 CML 的形成则表现出较弱的抑制作用。

## 3 讨 论

葡萄糖的中间代谢产物甘油醛可诱导蛋白发生非酶糖基化反应,引起机体氧化应激[11],导致 ROS 的产生,从而诱发慢性持久性炎症,如奶牛子宫内膜炎、子宫内膜异位以及乳房炎等疾病[10]。Fenton 反应诱导的氧化应激环境可使甘油醛代谢转化为活性更高的 GO,导致细胞毒性,并形成 AGEs,促使炎性因子释放(文献)[11]。本试验结果表明,甘油醛可能在 Fenton 反应条件下快速生成代谢产物 GO,经非酶糖基化反应产生毒性更大的 AGEs。甘油醛直接衍生形成的 AGE-2 为毒性晚期糖基化终末产物(toxic AGEs,TAGEs)的主要结构,AGE-2 可与晚期糖基化终末产物受体(Receptor of AGEs,RAGE)结合,产生大量 ROS 并释放促炎因子,诱导细胞凋亡和组织炎症[12]。鉴于 Fenton 反应下甘油醛对组织和细胞的毒性作用,本研究试图验证儿茶素和 ROS 清除剂 NAC 可抑制 AGEs 的形成,从而阻断 ROS 和促炎因子的释放。Fenton 反应条件下甘油醛快速自氧化形成 GO,儿茶素和 NAC 可以显著抑制上述反应形成的 GO,这与儿茶素很强的羰基捕获能力[7]以及儿茶素与 NAC 的抗氧化性有着直接的关系。甘油醛/Fenton 自氧化形成的 GO 诱导 BSA 产生 Schiff base 和 CML,NAC 显著抑制 Schiff base 的产生,表明 NAC 可通过抑制 GO 来降低 Schiff base 的量,也可能因为 NAC 有效清除 ROS,减少了氧化应激[13];儿茶素和 NAC 显著抑制 CML 的产生,一方面因为它们对 GO 的抑制作用,另一方面推测其在 CML 的形成过程中有效清除了羟基自由基和过氧阴离子自由基[14]。

总之,儿茶素和 NAC 通过捕获 GO,以及它们清除 ROS 和羟基自由基的作用,抑制 Fenton 反应条件下甘油醛诱导 BSA 的非酶糖基化反应,降低 AGEs 的产生,对抑制机体氧化应激和炎症因子做出了积极的贡献。

## 参考文献

[1] Baynes J W, Thorpe S R. Glycoxidation and lipoxidation in atherogenesis[J]. Free Radical Biology and Medicine, 2000, 28(12): 1708 – 1716.

[2] Odani H, Shinzato T, Matsumoto Y, *et al*. Increase in Three α, β-Dicarbonyl Compound Levels in Human Uremic Plasma: Specific <i> in Vivo </i> Determination of Intermediates in Advanced Maillard Reaction[J]. Biochemical and biophysical research communications, 1999, 256(1): 89 – 93.

[3] Feng C, Wong S, Dong Q, *et al*. Hepatocyte inflammation model for cytotoxicity research: fructose or glycolaldehyde as a source of endogenous toxins[J]. Archives of physiology and biochemistry, 2009, 115(2): 105 – 111.

[4] Murase T, Haramizu S, Ota N, *et al*. Tea catechin ingestion combined with habitual exercise suppresses the aging-associated decline in physical performance in senescence-accelerated mice[J]. American Journal of Physiology-Regulatory, Integrative and Comparative Physiology, 2008, 295(1): R281 – R289.

[5] Edelstein D, Brownlee M. Aminoguanidine ameliorates albuminuria in diabetic hypertensive rats[J]. Diabetologia, 1992, 35(1): 96 – 97.

[6] Lo C Y, Li S, Tan D, *et al*. Trapping reactions of reactive carbonyl species with tea polyphenols in simulated physiological conditions[J]. Molecular nutrition & food research, 2006, 50(12): 1118 – 1128.

[7] Sang S, Shao X, Bai N, *et al*. Tea polyphenol (–)-epigallocatechin-3-gallate: a new trapping agent of reactive dicarbonyl species[J]. Chemical research in toxicology, 2007, 20(12): 1862 – 1870.

[8] Mitchel R, Birnboim H. The use of Girard-T reagent in a rapid and sensitive method for measuring glyoxal and certain other α-dicarbonyl compounds[J]. Analytical biochemistry, 1977, 81(1): 47 – 56.

[9] Levine R L, Garland D, Oliver C N, *et al*. Determination of carbonyl content in oxidatively modified proteins[J]. Methods in

Enzymology,1990,186:464.

[10] Celi P,Merlo M,Da Dalt L,*et al.* Relationship between late embryonic mortality and the increase in plasma advanced oxidised protein products (AOPP) in dairy cows[J]. Reproduction,Fertility and Development,2011,23(4):527－533.

[11] 董强. 杏仁皮提取物对果糖及其两种代谢产物大鼠肝细胞毒性的保护作用[D]. 杨凌:西北农林科技大学,2010.

[12] Takeuchi M,Yamagishi S. Alternative routes for the formation of glyceraldehyde-derived AGEs (TAGE) in vivo[J]. Medical Hypotheses,2004,63(3):453－455.

[13] Lu C C,Yang J S,Huang A C, *et al.* Chrysophanol induces necrosis through the production of ROS and alteration of ATP levels in J5 human liver cancer cells[J]. Molecular Nutrition & Food Research,2010,54(7):967－976.

[14] He X,Yi Z,Tian Y,*et al.* Ability of catechin to eliminate O2-* and * OH[J]. Zhong nan da xue xue bao Yi xue ban = Journal of Central South University Medical Sciences,2006,31(1):138－140.

# 一例犬子宫蓄脓合并子宫角扭转的诊治

张　倩，董　强

（西北农林科技大学动物医学院，杨凌 712100）

子宫蓄脓也称子宫积脓，是未绝育中老年母犬常见的疾病，患犬子宫壁增厚，子宫腔增大，子宫内充满脓性分泌物[1]。犬子宫角扭转在临床较少见，以单侧扭转为主，一般怀孕母犬妊娠末期、分娩期和使用催产剂，以及子宫积液和子宫息肉等时可发生[2]。

犬子宫蓄脓合并子宫角扭转的情况在临床并不多见，报道的也并不多[3]。笔者现将临床所见一例犬子宫蓄脓合并子宫角扭转病例的诊治作一介绍。

## 1　病例基本情况

2013 年 4 月，西北农林科技大学西安教学宠物医院收治了一例杂种犬，雌性，9 岁，体温 38.4 ℃，体重 6.1 kg，未绝育。主述患犬精神沉郁，最近一周烦渴，厌食，腹围明显增大，两岁多生育过一次，阴道未见分泌物排出，疑有子宫疾患前来就诊。

## 2　诊　断

视诊患犬腹部胀满，腹壁紧张。X 光片显示腹腔有三处管状浓密阴影。触诊可摸到膨大的子宫角。血常规检查，白细胞 $22.33\times10^9$/L（参考值（6～17）$\times10^9$/L），肝肾功能正常。经过 5d 抗菌消炎的保守治疗后，症状未见好转，白细胞升高到 $71.12\times10^9$/L，碱性磷酸酶也由 237U/L 升高到 510U/L（参考值 10～254U/L），总蛋白 94g/L（参考值 54～82g/L）。B 超显示子宫腔内有大量积液，子宫壁增厚且不规则，诊断为子宫蓄脓[1,4]（图 1）。

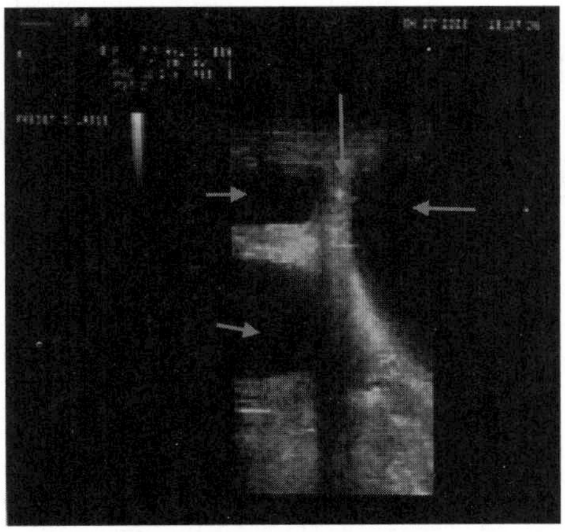

**图 1　子宫 B 超图**

子宫腔增大，充满低回声内容物（水平指向的三个箭头），子宫角不对称增大，子宫壁增厚且不规则（竖直指向的箭头）

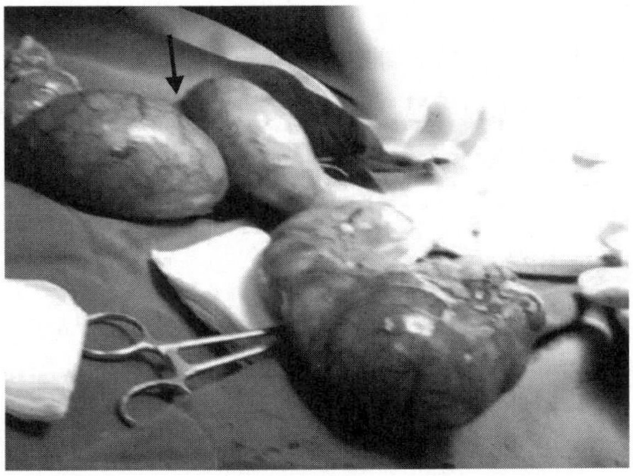

**图 2　子宫扭转（箭头所指部位为扭转处）**

**作者简介**：张　倩，女，硕士研究生，主要从事小动物老年病的研究，E-mail：zhangqian92122@163.com

**通讯作者**：董　强，E-mail：ardour@126.com

## 3 治 疗

### 3.1 保守疗法

肌内注射拜有利 0.1ml/kg 和布维林 1 ml,阴唇注射氯前列醇 1 ml。但保守治疗后患犬精神沉郁,体温升高至 39.8 ℃,呕吐,给以抗菌消炎、补液、补充营养等治疗。

### 3.2 手术疗法

采取手术方法摘除子宫卵巢。

3.2.1 术前准备 前臂头静脉埋置静脉留置针,输液。肌内注射阿托品 0.04mg/Kg,丙泊酚静推以诱导麻醉,待患犬角膜反射消失,立即插入气管插管,连接上麻醉机和心电监护仪,用异氟烷维持麻醉。

3.2.2 手术 术部常规剃毛消毒,腹白线切口打开腹腔。分别将两侧蓄脓的子宫逐步引出切口,发现一侧子宫角在腹腔中发生了扭转(图 2),用“三钳法”[5]将患犬两侧卵巢子宫切除,常规缝合腹腔切口,关闭腹腔。

3.2.3 术后护理 术后输液治疗,抗菌消炎 4 日,带伊丽莎白项圈防止患犬舔咬切口,切口处贴上切口贴,套上弹性绷带。10 日后拆线,伤口愈合良好。15 日后回访,患犬恢复正常,饮食、精神状态俱佳。

## 4 分析与讨论

开放型子宫蓄脓可采用保守疗法[6],但治疗要尽早而且要谨慎,避免子宫收缩药引起子宫破裂,甚至犬只死亡,造成治疗失败[1];闭锁型子宫蓄脓采取保守疗法,需先开张子宫颈,促进子宫内容物排出,通过手术的方法摘除卵巢子宫是治疗该病的行之有效的方法[1,4,5]。有文献报道可以采用腹腔镜辅助卵巢子宫切除术作为子宫蓄脓犬的有效治疗方法,手术成功率较高,创口小,但是选择的子宫蓄脓犬的子宫直径都是中等大小的[7]。

此病例中,子宫颈是闭合的,且有一侧子宫角发生扭转,单纯采用保守疗法,促进子宫内脓汁的排出,抗菌消炎,补充营养等,不但没有达到理想效果,反而使炎症加重,且扭转一侧子宫角的脓汁无法排出,病情恶化,增加了手术的风险。不及时治疗的子宫蓄脓的病犬,肝肾功能受损加重,甚至发生败血症,后果更严重[8]。

本病例中,犬的子宫角发生扭转的原因可能是子宫蓄脓的并发症,也可能是治疗过程中发生。犬子宫角扭转很难进行早期诊断,在评估手术风险后应及时进行手术摘除子宫卵巢。

子宫蓄脓 B 超诊断一般用凸探头,因线性探头频率一般比较小,主要用于位置比较浅的部位和小器官。而且凸探头检查子宫比较直观,方便数据的测量,比如测子宫壁的直径等。本病例中,犬个体较小,最终用线性探头也不影响最终诊断为子宫蓄脓。

## 参考文献

[1] kumari Baithalu R, Maharana B R, Mishra C, et al. Canine Pyometra[J]. Veterinary World, 2010, 3(7): 340-342.

[2] Chambers B A, Laksito M A, Long F, et al. Unilateral uterine torsion secondary to an inflammatory endometrial polyp in the bitch [J]. Australian veterinary journal, 2011, 89(10): 380-384.

[3] Schlafer D H. Diseases of the canine uterus[J]. Reproduction in Domestic Animals, 2012, 47(s6): 318-322.

[4] 张波. 利用 X 光及 B 超诊断和治疗犬子宫蓄脓[J]. 吉林畜牧兽医, 2011, 32(8): 42-43.

[5] 杨艳杰, 董文秀, 李志超. 犬的卵巢子宫切除术[J]. 黑龙江畜牧兽医: 下半月, 2011 (1): 107.

[6] Jena B, Rao K S, Reddy K C S, et al. Comparative efficacy of various therapeutic protocols in the treatment of pyometra in bitches[J]. Veterinarni Medicina, 2013, 58(5): 271-276.

[7] Adamovich-Rippe K N, Mayhew P D, Runge J J, et al. Evaluation of Laparoscopic-Assisted Ovariohysterectomy for Treatment of Canine Pyometra[J]. Veterinary Surgery, 2013, 42(5): 572-578.

[8] 钱存忠, 侯加法. 犬子宫蓄脓的病因学及病理变化特征[J]. 畜牧与兽医, 2006, 37(9): 49-52.

# 犬痛风的临床症状及实验室检查

代飞燕[1],段超华[2],陶田谷晟[1],张斌恺[1],肖 啸[1]
(1.云南农业大学动物科学学院,昆明 650201;2.云南省公安消防总队搜救犬基地,昆明 650031)

**摘 要**:为了研究犬痛风的临床症状及实验室诊断方法,试验对12只痛风犬进行临床症状统计和血样采集,并进行血常规、血液生化和血气检测,并与参考指标做对比,得出犬痛风临床特征和实验室检测特征;试验结果表明:犬痛风平均发病年龄为6.6岁,发病年龄呈两极化,且多发于春季。临床特征为游走性疼痛,关节肿大;血液检测特征为血液pH值低于7.458,尿酸盐浓度升高至90mmol/L以上。犬痛风严重时会导致犬肝肾损伤,肾功能衰竭,肝肾机能障碍以及关节肿大、变形,运动障碍和机能障碍。

**关键词**:犬,痛风,尿酸

# Clinical symptom and its diagnosis of canine gout

DAI Feiyan[1],DUAN Chaohua[2],TAO Tiangusheng[1], ZHANG Binkai[1],XIAO Xiao[1]
1. *College of Animal Science and Technology*,*Yunnan Agricultural University*,*Kunming* 650201;
2. *Yunnansheng Gongan Xiaofang Zongdui Soujiuquan Jidi*,*Kunming* 650031

**Abstract**:Currently,there are much more pet dogs diagnosed as gout and it is very difficult to cure. The objective of this paper is to record the clinical symptom and its laboratory diagnosis of canine gout. 12 gout dogs were collected from clinical hospital and blood samples were collected for CBC,blood biochemistry and blood gas detection. The results showed that the dog gout onset age of 6.6 years on average,onset age polarized,and happens in the spring. Clinical features of the ill dogs were wandering pain,joint swelling. Characteristics of blood tests for blood pH below 7.458,uric acid salt concentration increased to more than 90 Umol/L. severe gout caused kidney damage,kidney failure,liver and kidney dysfunction and joint swelling,deformation,movement disorders,and functional disorder.

**Keywords**:Dog gout; Uric acid; pH; Clinical features

## 引 言

痛风是指血液中尿酸盐大量蓄积,不能被迅速排出体外,形成尿酸血症,进而尿酸盐沉积在关节囊、关节软骨、软骨周围及胸腹腔、各种脏器表面和其他间质组织上的一种代谢病[1]。痛风可分为内脏型和关节型[2]。内脏型痛风是指尿酸盐沉积在内脏器官表面;关节型痛风是指尿酸盐沉积在关节腔及其周围[3,4]。能使尿酸生成过多或排泄障碍的因素均可导致痛风的发生;饲料中蛋白质尤其核蛋白和嘌呤碱含量太多[5]。如用动物的内脏(胸腺、肝、肾、脑、胰)、肉屑、鱼粉及熟鱼或大豆粉、豌豆等作为蛋白质来源,由于核

代飞燕,女,汉族(1971-),云南省大理人,兽医学硕士,云南农业大学动物科学技术学院教师,从事临床兽医工作15年,现致力于犬猫实验室诊断及小动物皮肤病和小动物影像学研究。邮箱:daifeiyan113@163.com
肖 啸,男,汉族(1966-),云南昆明人,云南农业大学动物科学技术学院教授,研究生导师,云南农业大学动物医院院长,研究方向为动物病理学和经济性动物寄生虫研究、小动物疾病学及重大动物疫病防控技术,邮箱:xiaoxiaokm55@163.com

酸和嘌呤的代谢终产物尿酸生成太多,而引起尿酸症[1,6,7]。肾脏损伤。凡具嗜肾性、能引起肾机能损伤的病原微生物,如传染性支气管炎病毒等均可引起肾脏损伤而造成尿酸盐排泄受阻,并致痛风发生[7]。维生素A缺乏能引起肾小管和输尿管上皮角化、脱落,造成堵塞,使尿酸排泄受阻[3]。磺胺类药物中毒,可引起结晶尿和肾损伤[7,8]。

# 1 试验材料与试验方法

## 1.1 试验动物及材料

本次试验所检测的发病犬是来自云南农业大学附属动物医院具有典型临床症状的12例痛风病例,一次性使用真空贮血管、一次性使用采血针、离心沉淀器、血气分析仪、自动化血液分析仪、全自动生化分析仪。

## 1.2 试验方法

1.2.1 试验分组 跟踪12只患痛风犬,并根据临床症状进行统计分析。

1.2.2 血样的采集和处理 采用前臂内静脉采血法,每只犬采血3ml两支,一支用全血做血气分析和血常规,另一支则放入负压储血管保存,静置1小时后放入离心机中离心,以4 000r/min离心10min,分离血清,取上层血清移入各试剂杯中,用半自动生化分析仪检测。

1.2.3 检测指标 待测血液做血气分析、血常规和生化指标检测,主要检测犬血液pH,白细胞数目,红细胞数目,红细胞压积,谷丙转氨酶,谷草转氨酶,碱性磷酸酶,尿酸,尿素氮,肌酐。

# 2 结果与分析

表1 试验犬临床综合报告

| 编号 | 品种 | 年龄 | 性别 | 发病时间 | 临床症状 |
|---|---|---|---|---|---|
| 1号犬 | 沙皮犬 | 9岁 | ♂ | 1月 | 肥胖,游走性疼痛,喘,精神萎靡,低烧,缩头夹尾,舌头发紫 |
| 2号犬 | 西施犬 | 6岁 | ♂ | 3月 | 肥胖,游走性疼痛,喘,精神萎靡,缩头夹尾,舌头发紫 |
| 3号犬 | 斗牛犬 | 11岁 | ♀ | 1月 | 肥胖,游走性疼痛,精神萎靡,低烧,缩头夹尾 |
| 4号犬 | 北京犬 | 8岁 | ♂ | 2月 | 游走性疼痛,喘,精神萎靡,低烧,缩头夹尾,舌头发紫 |
| 5号犬 | 北京犬 | 7岁 | ♀ | 12月 | 肥胖,游走性疼痛,,精神萎靡,缩头夹尾 |
| 6号犬 | 西施犬 | 7岁 | ♂ | 11月 | 游走性疼痛,喘,精神萎靡,低烧,缩头夹尾 |
| 7号犬 | 德国牧羊犬 | 2月 | ♀ | 7月 | 关节肿大,低烧,精神萎靡,食欲不振 |
| 8号犬 | 斗牛犬 | 8岁 | ♂ | 2月 | 游走性疼痛,精神萎靡,缩头夹尾,食欲不振 |
| 9号犬 | 巴哥犬 | 7岁 | ♀ | 1月 | 肥胖,游走性疼痛,喘,精神萎靡,低烧,缩头夹尾,舌头发紫 |
| 10号犬 | 北京犬 | 9岁 | ♂ | 12月 | 肥胖,游走性疼痛,喘,精神萎靡,低烧,缩头夹尾,舌头发紫 |
| 11号犬 | 松狮犬 | 7岁 | ♀ | 5月 | 游走性疼痛,喘,精神萎靡,低烧,缩头夹尾,喜卧 |
| 12号犬 | 阿拉斯加犬 | 3月 | ♂ | 1月 | 关节肿大,低烧,精神萎靡,食欲不振 |

由表1可知犬痛风平均发病年龄为6.6岁,发病年龄呈两极化,多发于幼龄犬和老龄犬。春季发病较多,无性别差异。综合临床症状可知,犬痛风有肥胖,精神萎靡,食欲不振,缩头夹尾,关节肿大,喜卧,喘,舌头发紫,游走性疼痛,低烧等症状。

表2　犬血液 pH 值和部分血常规检测报告

| 编号 | 血液 pH | 白细胞数目 | 红细胞数目 | 红细胞压积 |
|---|---|---|---|---|
| 1 号犬 | 7.28 | $26.4\times10^9$/L | $8.62\times10^{12}$/L | 60.1% |
| 2 号犬 | 7.15 | $34.5\times10^9$/L | $8.42\times10^{12}$/L | 55.8% |
| 3 号犬 | 7.30 | $36.8\times10^9$/L | $8.67\times10^{12}$/L | 62.2% |
| 4 号犬 | 7.35 | $42.7\times10^9$/L | $8.74\times10^{12}$/L | 63.4% |
| 5 号犬 | 7.26 | $46.4\times10^9$/L | $8.37\times10^{12}$/L | 55.3% |
| 6 号犬 | 7.33 | $38.6\times10^9$/L | $8.52\times10^{12}$/L | 58.9% |
| 7 号犬 | 7.18 | $49.6\times10^9$/L | $8.60\times10^{12}$/L | 59.3% |
| 8 号犬 | 7.06 | $33.4\times10^9$/L | $8.28\times10^{12}$/L | 54.7% |
| 9 号犬 | 6.94 | $32.0\times10^9$/L | $8.24\times10^{12}$/L | 54.4% |
| 10 号犬 | 7.02 | $27.2\times10^9$/L | $8.38\times10^{12}$/L | 55.8% |
| 11 号犬 | 7.40 | $32.6\times10^9$/L | $7.99\times10^{12}$/L | 50.4% |
| 12 号犬 | 7.08 | $31.4\times10^9$/L | $8.57\times10^{12}$/L | 57.3% |

表3　犬肝肾生化检测报告

| 犬血清 | 肝 | | | 肾 | | |
|---|---|---|---|---|---|---|
| | 谷丙转氨酶 U/L | 谷草转氨酶 U/L | 碱性磷酸酶 U/L | 尿酸 Umol/L | 尿素氮 Mmol/L | 肌酐 Umol/L |
| 1 号犬 | 87 | 39 | 94 | 94 | 8.60 | 101 |
| 2 号犬 | 76 | 34 | 97 | 99 | 6.42 | 94 |
| 3 号犬 | 88 | 37 | 78 | 108 | 7.91 | 86 |
| 4 号犬 | 140 | 58 | 158 | 119 | 10.10 | 118 |
| 5 号犬 | 104 | 51 | 72 | 97 | 9.32 | 89 |
| 6 号犬 | 94 | 48 | 78 | 102 | 9.57 | 97 |
| 7 号犬 | 68 | 29 | 69 | 97 | 8.42 | 67 |
| 8 号犬 | 82 | 43 | 81 | 99 | 8.68 | 78 |
| 9 号犬 | 96 | 47 | 92 | 110 | 7.34 | 92 |
| 10 号犬 | 109 | 51 | 89 | 93 | 6.58 | 96 |
| 11 号犬 | 24.7 | 45.1 | 17.8 | 49.9 | 6.90 | 128.1 |
| 12 号犬 | 89 | 40 | 97 | 95 | 7.67 | 99 |

表4　各组犬血液 pH、部分血常规和部分血液生化平均数

| 血液 pH | 白细胞数目 | 红细胞数目 | 红细胞压积 | 谷丙转氨酶 | 谷草转氨酶 | 碱性磷酸酶 | 尿酸 | 尿素氮 | 肌酐 |
|---|---|---|---|---|---|---|---|---|---|
| 7.196 | 35.97 | 8.45 | 57.3 | 88.14 | 43.51 | 85.23 | 96.91 | 8.13 | 95.4 |

由表2、3、4、5、6 可知犬患痛风后血液 pH 降低，白细胞升高，红细胞上限，血液浓稠，尿酸含量非常高，谷丙转氨酶，谷草转氨酶，碱性磷酸酶，尿素氮，肌酐均有不同程度的升高 。说明，犬痛风对犬肝肾功能及其他机能产生了影响。

## 3 讨 论

3.1 试验结果表明犬痛风平均发病年龄为6.6岁,发病年龄呈两极化,且多发于春季,无性别差异。综合临床症状为:肥胖、精神萎靡、食欲不振、缩头夹尾、关节肿大、喜卧、喘、舌头发紫、游走性疼痛、低烧。

3.2 犬痛风高发于:肝肾功能不全,大量饲喂蛋白质的幼犬,肝肾功能退化、肥胖、运动量小的老龄犬。此类犬血检显示,血液浓稠,红细胞上限,血液 pH 降低,红细胞压积升高,血沉减慢。个别病例白细胞升高。尿检结果是血液中尿酸盐浓度升高至90mmol/L以上(正常值< =70 mmol/L),血液非蛋白氮含量升高。肝功检测结果显示,轻度痛风,肝功指标变化不太明显;重度痛风,肝功指标变化明显。肾功能检测结果是轻度痛风,尿素氮略有升高,肌酐正常。重度痛风,尿素氮、肌酐升高幅度较大。

3.3 犬痛风应以预防为主,根据宠物生长生育和需要,合理调配饲料,不宜过多饲喂动物性蛋白饲料,禁止用动物腺体组织饲喂,调整日粮中钙磷比例,宠物添加维生素 A,均有一定的预防作用。掺入沙丁鱼饲喂,可防止痛风发生。增加运动可降低本病的发病率。

3.4 本试验病例有限,无法得出年龄、品种、运动量、气温、湿度等因素对犬痛风的影响。

## 4 结 论

犬痛风平均发病年龄为6.6岁,发病年龄呈两极化,且多发于春季。临床特征为游走性疼痛,关节肿大;血液检测特征为血液 pH 低于7.458,尿酸盐浓度升高至90Umol/L以上。犬痛风严重时会导致犬肝肾损伤,肾功能衰竭,肝肾机能障碍以及关节肿大、变形,运动障碍和机能障碍。

## 参考文献

[1] 孙维平,刘小宝.何海健.宠物疾病诊治[M].北京:化学工业出版社,2011,172 - 173.

[2] 中国兽医协会.执业兽医资格考试应试指南(兽医全科类)[M].北京:中国农业出版社,2012,996 - 1002.

[3] 赵兰利.幼犬痛风病的诊治报告[J].中国养犬杂志,1996(7):14.

[4] 蒋小刚,沃妮娜,朱丽娜,等.犬痛风病的诊疗一例[J].上海畜牧兽医通讯,2010(6):83.

[5] 郑旭 王杰.痛风的发病机制及其诊断治疗[J].2008(16):195 - 196.

[6] 凌明亮.禽痛风的发病机制及其防治方法[J].湖北畜牧兽医,2005(2):26 - 29.

[7] 罗学宏.哪些药物引起高尿酸血症[J].求医问药,2009(12).

[8] 庄桂玉,秦刚,阎作诗,等.幼犬痛风的诊断与防治[J].山东畜牧兽医,2008(29:):62 - 63.

# 奶牛围产期疾病的控制

吴心华

（宁夏大学农学院，银川 750021）

**摘　要**：奶牛围产期疾病主要有低血钙症和营养负平衡。它们是规模化奶牛场分娩母牛和新生犊牛致死的主要原因，围产期疾病的发生与奶牛干奶期、围产前期、分娩期、围产后期的营养、管理和护理技术有着紧密的联系，本文旨在对奶牛围产期疾病的控制进行综述，从而提高奶牛的饲养管理和健康发展。

**关键词**：奶牛，低血钙症，营养负平衡，控制

## 1　围产期疾病

奶牛围产期疾病主要有低血钙症和营养负平衡。其中低血钙是多数疾病发生的始动因素，分娩应激是导火索，能量负平衡是一个必然的过程。

低血钙症可以直接诱发奶牛胎衣不下、产后瘫痪、子宫全脱、消化道运动迟缓，子宫收缩力降低，子宫复旧迟缓、真胃变位、乳房水肿和产前瘫痪等疾病。营养负平衡的主要原因是奶牛进入围产前期启动分娩，雌激素分泌量增加、抑制采食量，导致干物质摄入量大幅度下降，致使营养物质的吸收普遍下降，特别是维生素和微量元素的吸收量减少，造成胎盘营养不良，免疫力降低，在分娩过程中和产后恢复期，经常表现出产后无名高热、乳房炎、阴道炎和子宫炎。

由于初乳的形成和产后大量泌乳，从消化道吸收的营养物质不能满足泌乳的需要，母牛便出现能量负平衡，开始分解体脂肪来供能，从而出现酮血症，特别是产前肥胖奶牛，产后酮血症发病明显增加，酮血症严重阻抑了产后各系统生理机能的恢复，加重疾病的发生。进入分娩期，母牛受到分娩应激的刺激，促使低血钙症和能量负平衡等发生。干奶期、围产前期和分娩期的饲养管理因素除了导致分娩母牛发病，还造成新生犊牛发病甚至死亡。新生犊牛的疾病发生可追溯到母牛干奶期、围产前期和分娩期的营养、管理和免疫注射及产后护理等几个主要环节。犊牛经常发生的高度致死性疾病多数属于传染病，如大肠杆菌病、巴氏杆菌病、沙门氏杆菌病、支原体病、黏膜病毒病、轮状病毒病、冠状病毒病、隐孢子虫病、球虫病、钩虫病、锥虫病等，这些疾病主要发生在哺乳初期的 20 日龄以内。犊牛断奶后 2 周内，经常发生断奶应激综合症。断奶应激容易诱发犊牛精神沉郁、瘤胃鼓气、腹泻、肺炎、水中毒、外伤、营养不良、被毛紊乱和毛色发红等现象。

## 2　围产期疾病的控制

奶牛围产期疾病有低血钙症、能量负平衡和分娩应激综合征。

### 2.1　钙的代谢与调控

奶牛每生产 1L 牛奶需要 1g 多的钙。因此，高产奶牛在开始泌乳时会突然发生钙紧缺。身体内储存的钙主要在骨骼中，而且不能被快速动员使用。其实，奶牛可以缓慢动员骨钙以调节身体对钙的需求。然而，直到泌乳前 10d，奶牛的钙沉积和钙动员的水平是相同的。此外，获得钙还依赖于肠道吸收的有效性。因此奶牛在泌乳的前 10d 发生产后瘫痪的可能性最大。身体对钙缺乏的最先反应症状是肌肉功能障碍，因为肌肉需要钙离子传导神经信号。因此，发生产乳热时奶牛无法站立并显示出全身肌肉软弱，如颈部肌肉无力。奶牛还有嗜睡或昏睡的现象。虽然这一症状称为产乳热，但事实上奶牛在这种情况下体温通常下降。如不及时处理，产后瘫痪很可能致命。静脉注射钙盐是最常用的恢复血钙的方法。一些提高奶牛肠道钙吸收从而防止分娩时产乳热发生的措施已逐渐被人们采用。这些措施主要是在产前一段时期限制钙的摄入并给予

维生素 D 刺激肠道对钙的吸收和骨钙的动员。

奶牛围产期易发生低血钙,分娩当天发生率最高。奶牛围产期低血钙的发生不仅与围产期外源性钙的摄入不足有关,而且与钙代谢调节相关激素未能动员骨钙有关,因为钙与维持平滑肌的功能关系密切,即使亚临床低钙血症的牛也有平滑肌收缩性减弱的体征,当产后的子宫平滑肌收缩正常,子宫就能迅速复位,而低钙血症常导致子宫脱垂、胎衣不下及子宫感染。血钙浓度太低,使得子宫肌肉麻痹、分娩时胎儿活动异常,分娩启动失调,造成胎位异常,不能努责,分娩困难甚至难产。致使骨骼肌兴奋性下降,奶牛瘫痪。同时,胃肠平滑肌蠕动下降,使瘤胃蠕动变弱导致奶牛采食量下降。DMI 降低的另外一个原因是瘤胃液渗透压的升高。瘤胃内短链脂肪酸(SCFA)、葡萄糖和乳酸含量增加,灌服大量盐类泻剂使瘤胃渗透压升高,瘤胃内流增加,进而导致奶牛采食量下降。

奶牛产前 2 周开始饲喂低钙高磷饲料,激活甲状旁腺机能,提高机体吸收钙的能力。分娩后提高日粮中的钙含量,并至少口服 2 次钙制剂,同时喂一些健胃药,促进胃肠对钙的吸收,产后勿急于挤空牛奶,第 1 次挤奶量为正常的 1/3,以后逐渐增加,4d 后达到正常挤奶量。一方面降低日粮中钾、钠水平,围产期少用或不用含钾高的苜蓿,饲喂含钾低的玉米和玉米青贮料;添加阴离子 $Cl^-$、$S^{2-}$ 等诱导奶牛轻度的酸中毒(酸化日粮),从而增加了机体对 $Ca^{++}$ 的吸收,可以减少奶牛生产瘫痪的发生。干奶期保持日粮钙含量不变,只是添加阴离子盐至 10 ~ 15mg 当量/每千克干物质,即可促进骨钙溶解,补充血钙,使血钙升高。

## 2.2 低血钙症

2.2.1 分娩奶牛低血钙的形成　分娩奶牛急性低血钙的形成可分为干奶期钙供给不足性低血钙和围产前期钙调节失灵性低血钙,调节性低血钙是在干奶期钙供给满足的条件下进行的调节技术。

(1)奶牛干奶期钙供给不足性低血钙　怀孕奶牛干奶期钙及矿物质供给不足往往发生分娩期急性低血钙、低血磷。干奶期主要目的是恢复瘤胃微生物群系,保证胎儿增长和母体骨钙沉积,控制奶牛体况,修复乳腺等,如果在干奶期钙等矿物质及维生素,优质蛋白供给不足,进入围产前期,母牛必然出现低血钙,这种类型低血钙的形成原因主要是干奶期预混料供给不足或者质量差,造成钙磷供给不足,没有满足胎儿生长需要和母体骨骼贮备,控制办法就是严格执行干奶期管理,阻止无乳牛进入干奶圈舍,供给优质平衡矿物质维生素及优质蛋白质,限制能量摄入,禁止肥胖。

奶牛的血钙变化是随着生产循环规律的变化而变化。奶牛钙磷等矿物质及维生素需求量是随着怀孕以后胎儿的逐渐生长和母体骨骼贮备而发生的变化,随着分娩后泌乳量的变化,血钙也会有规律的发生变化。胎儿 1 ~ 3 月龄需要较少的钙。1 ~ 3 月龄,胎儿体重增加甚微,主要是脑及心肝脾肺肾等实质器官的发育,只需要一定量的优质蛋白质和丰富的矿物质维生素供给。这一时间,母牛都处在泌乳高峰期,需要大量的能量、蛋白质和丰富的矿物质维生素供给。胎儿 4 ~ 6 月龄,胎儿生长很慢,只需要适量的矿物质、维生素、蛋白质和能量。此时,母牛处于泌乳中后期,产奶量逐渐下降,母牛瘤胃机能变得强大,能够很好地利用日粮中的营养物质合成能量,用于自身复膘增重,恢复体况,此阶段是怀孕母牛恢复体膘的最佳时段,适当的分群和营养调配可以较好的控制母牛体况。胎儿 7 ~ 8.5 月龄,母牛停止泌乳进入干奶期,胎儿迅速增长,需要大量的矿物质维生素和优质蛋白质,由于母牛停止泌乳,不需要很多的能量,所以,此阶段,母牛需要更多的矿物质维生素用于骨骼内的贮备来满足下一个泌乳期的泌乳需要。

生产中,如果在母牛怀孕以后 7 ~ 8.5 个月(干奶期)矿物质维生素给母牛供给不足,就会影响胎儿的生产需要和母牛骨骼贮备,导致胎儿和母体钙磷供给不足,进入围产期就会直接发生低血钙,低血磷,我们把这种钙不足引起的低血钙称为供给不足性低血钙,这种低血钙往往引起严重的产后瘫痪和胎衣不下等疾病。头胎牛由于自身的骨骼还没有发育完善,需要在围产期不断供给钙磷等,经常见到供给不足性低血钙。

(2)围产前期调节失灵性低血钙　围产前期调节血钙是干奶期管理的重要技术。母牛在干奶期供给优质平衡的矿物质维生素,已经满足胎儿生长需要和母体骨骼贮备,进入围产前期需要采取低钙日粮或者添加阴离子盐促使骨钙溶解在分娩前启动快速溶解来补充血钙，以满足母牛分娩后大量泌乳的钙量需求，这项技术在成母牛上已经被广泛使用。围产前期调节性低血钙是在干奶期钙量供给充足，在围产前期使用的低钙阴离子盐日粮，加速骨钙溶解，使血钙升高的技术失败而造成的低血钙，称为调节性低血钙。

奶牛的正常血钙范围为8.8～10.4 mg/dl，当血钙浓度水平<6 mg/dl时，一般可做出临床低血钙症的诊断。如果血钙测定范围为6.2～7.5 mg/dl，可认定亚临床低钙血症。母牛血钙升高的主要途径包括：消化道吸收的钙，骨中溶解的钙，减少消化道和泌尿系统的排泄量。目前围产前期调节血钙的方法有低钙日粮法和日粮阴离子盐法。低钙日粮法是进入围产前期，采取降低日粮钙含量，饲喂10～13d，母牛血液中钙离子浓度忽然下降到一定浓度，这个信号就会引起母牛甲状旁腺激素分泌，在低血钙信号的刺激下，甲状旁腺素反馈性的兴奋，产生大量的甲状旁腺激素，甲状旁腺激素促进骨钙溶解，钙离子进入血液，使血液浓度升高，从而缓解了分娩过程中母牛的低血钙症。机体还有降钙素调解血钙的升高，当血钙浓度升高到一定程度时，降钙素就会大量分泌，抑制甲状旁腺素的分泌，制止骨钙继续溶解。降钙素的大量分泌就可促进血液中钙离子浓度的下降。体内地维生素D，它可以促进血钙的升高和降低，起着调节血钙稳定的作用。机体就是靠着甲状旁腺素、降钙素和维生素D有效地进行着血液浓度的调节，使之始终处于正常浓度的范围之内，保持了动物的健康。在奶牛分娩阶段，往往只见到甲状旁腺素和维生素D有明显的变化，而降钙素没有明显的改变。因此，母牛围产期血钙浓度主要受甲状旁腺素的影响，甲状旁腺素浓度升高，母牛骨钙溶解加快，血钙升高。奶牛分娩期经常发生急性低血钙或亚临床型低血钙。急性低血钙会引起母牛产后瘫痪，子宫脱垂造成奶牛死亡。亚临床型低血钙直接造成母牛分娩无力，产程延长，子宫破裂，产道损伤，胎衣不下，真胃变位，阴道脱出，酮血病，子宫复旧不全，产后发热，乳房炎等严重疾病。

目前大型牧场发生的低血钙，主要是调节失灵性低血钙，较少发生供给不足性低血钙。规模化牧场在奶牛干奶期钙及矿物质维生素供给是充足的，胎儿和母牛骨骼贮备基本完成，进入围产前期采用血钙调节技术有饲喂低钙日粮或日粮中添加阴离子盐，含量为10～15mg/kg干物质，饲喂10～12d，使尿液pH达到6.0～6.5时，能够促进血钙溶解加快，补充血钙，预防母牛低血钙症的发生。

奶牛进入围产期前期，要通过改变日粮和饲养管理实现瘤胃黏膜的快速增生和黏膜面积的增大，同时要通过调节机制实现骨钙溶解加快，补充血钙的目的。进入围产前期，就进入分娩准备阶段，此时，胎儿已经完全发育成熟，胎儿的目的是尽快娩出母体子宫，不需要继续供给生长的营养物质，由于雌激素水平的不断升高和黄体酮水平逐渐下降，母牛采食中枢不断受到抑制，采食量明显减低，降低额度达到干奶期额度的20%～30%，同时由于初乳形成需要营养，造成母牛能量负平衡。

由于初乳形成需要大量的钙离子及分娩过程肌肉活动需要大量钙离子，分娩后急剧泌乳更需要大量钙离子等原因，奶牛在进入分娩期往往进过一个急剧的低血钙过程。急性低血钙过程往往是围产期疾病发生的始要因素，所以，母牛进入围产期，通过消化道摄入的钙明显减少，如果血中的钙不能迅速溶解进行补充，就会发生急性低血钙而造成严重疾病，甚至导致母子死亡，所以，母牛进入围产前期的另一个重要任务就是调节骨钙快速溶解来补充血钙浓度。

奶牛产后瘫痪与其体内钙的代谢密切相关，产后泌乳的启动，大量钙质从乳中排出，此时血浆钙含量必须通过增加胃肠吸收和骨钙析出得到补充，但母牛分娩前后胃肠机能紊乱，使钙的吸收率降低，同时母牛雌激素水平升高，阻止骨钙及时转送到血液，造成钙的负平衡，最终导致低血钙性瘫痪的发生。由于第一天大量血钙进入牛奶所致。这一症状部分是因为围产期饲料中钙含量过量（围产期料中以低于0.7%为宜），或钙磷比例不当所造成的，若不及时治疗可引起瘫痪甚至死亡。

（3）缺失调节性低血钙　成年怀孕母牛从干奶期开始饲喂高钙，丰富的矿物质，维生素，优质蛋白日粮，进入围产前期仍然饲喂干奶期日粮，进入分娩期，由于初乳形成及产后大量泌乳需要大量的钙不能从消化道来满足，此时，骨钙溶解速度慢，不能及时溶解补充血钙，造成急性低血钙，这种围产前期缺乏对钙进行调控的低血钙症称缺失调节性低血钙。生产中由于头胎牛自身骨骼还需要进一步生长,往往在产前一直使用高钙高矿物质,维生素,低能量日粮,不进行低钙日粮调节。为了预防妊娠水肿,可以添加阴离子盐或减少日粮苏打,食盐,苜蓿的喂量。

2.2.2　低血钙的控制　头胎怀孕母牛怀孕200d进入干奶圈舍,应该饲喂干奶期日粮,进入围产前期饲喂围产前期日粮,进入分娩期饲喂产后日粮。成年怀孕母牛怀孕220d进入干奶圈舍,应该饲喂干奶期日粮,进入围产前期饲喂围产前期日粮,进入分娩期饲喂产后日粮。同时要严格分群:建立青年牛干奶圈舍,围产前期牛舍和分娩牛舍;建立成年母牛干奶圈舍,围产前期牛舍和分娩牛舍;各阶段饲喂停留时间要严格管理:干奶期45d,围产前期13d,分娩期7d(其中分娩启动2d,分娩期1d,产后护理4d)。分娩期采用产后日粮饲喂。干奶期日粮特点:丰富的矿物质、维生素,优质蛋白,低能量,优质秸秆饲料,控制精料喂量。围产前期日粮特点:成年母牛喂给丰富的矿物质、维生素,优质蛋白,高能量,优质粗饲料,但须是低钙,低钠、低钾的阴离子盐日粮,增加精料喂量。青年牛喂给丰富的矿物质、维生素,优质蛋白,高能量,优质粗饲料,但须是低钠、低钾的阴离子盐日粮,增加精料喂量。

## 2.3　能量负平衡

2.3.1　能量负平衡形成原因　母牛产后能量负平衡的主要原因:母牛分娩前过于肥胖;日粮质量差,特别是青贮质量差;进入围产期干物质采食量下降和瘤胃内膜吸收面积不足。母牛分娩前过于肥胖,主要原因是低产牛群中出现的无乳牛不建立独立群,长期饲养在干奶牛群中,其次是低产期分群不细致,干奶期过长。日粮质量差,精料配比不能满足需要或供给超标,青贮丁酸含量高,日粮缺乏钴等微量元素容易诱发酮血病。围产前期干物质采食量下降及瘤胃内膜吸收面积不足,围产前期干物质采食量下降是个必然的过程,通过在围产期增加精料量和精料中淀粉,限制干物质采食量来提高产后瘤胃黏膜的吸收面积,提高产后干物质采食量及营养吸收的有效措施。

2.3.2　能量负平衡的控制　在怀孕4~6月龄严格分群,控制体况为3.25~3.5分。

(1)建立无乳牛群。

(2)严格怀孕母牛220d进入干奶期饲喂。

(3)干奶期严格分群。

(4)围产前期促进瘤胃黏膜增长,促进瘤胃自我吸收。

## 2.4　分娩应激综合征

奶牛分娩应激综合征是由分娩应激所引发的分娩过程中损伤性疾病和分娩后疾病的总称。分娩应激是发病的直接诱因,营养不平衡造成的低钙血症是引起围产期相关疾病的主要原因,特别是热应激分娩,产后疾病高发,治疗效果最差,死亡率最高,对奶牛损害最严重。

分娩应激是指临产母牛在分娩过程中,对分娩环境变化、胎儿及母体神经内分泌变化、软硬产道变化、助产刺激等特殊生理过程中所发生的生理和行为上的特异性或非特异性反应。

2.4.1　分娩应激综合征的临床病理变化

(1)微循环缺血　如果分娩异常,皮质类固醇长期增多,则又可引起微循环缺血导致休克的发生和重要器官的损害,并能使机体的免疫中枢发生损伤,减缓抗体的产生,使母牛容易发生潜在性的疾病,如难产、低血钙、感染和严重脱水和电解质流失。

(2)胃肠道缺血运动迟缓　分娩应激可加重分娩时胃肠道贫血之后发生的淤血、水肿、缺血、出血、微循环缺血和痉挛等,致使胃肠黏膜上皮细胞变性和坏死,降低胃肠道的屏障功能,并可使肠道内的毒性物质透过黏膜入血,引起毒血症和顽固性前胃迟缓等。

(3)免疫力下降　围产期容易出现乳房炎、产后瘫痪、胎衣不下、真胃变位、酮血病和奶牛趴窝综合征等。

（4）代谢性酸中毒　如果分娩障碍，由于微循环灌流量减少，导致组织细胞缺氧，无氧酵解加强，使乳酸等酸性代谢产物蓄积，同时又由于尿少不能充分排出，而产生代谢性酸中毒。

（5）疼痛与脱水　强烈的分娩应激会导致母牛严重脱水、能量大量消耗、产道损伤、感染和剧烈疼痛等。

（6）低血钙　分娩应激促使催乳素大量分泌和动员机体贮备，骨胳代谢活跃，表现为母畜的骨盆软弱，骨钙动员加强。如果甲状旁腺代谢异常，不能有效期启动骨钙代谢，往往出现急性低血钙症。如急性低血钙、生产瘫痪、子宫收缩力降低，胎衣不下，子宫感染，产后乏情等疾病发生均与内分泌代谢紊乱有关。

（7）产道感染　糖皮质激素也能抑制组织再生，影响产道创伤愈合，导致子宫复旧不全，子宫内膜炎等感染性疾病。

2.4.2　分娩应激的控制

（1）调整营养和饲喂技术，提高奶牛抗应激能力。

（2）加强围产前期环境控制，降低应激原刺激。

（3）促进分娩。

## 3　小　结

干奶期严格控制在45d以内，怀孕不足220d的母牛留在干奶圈舍，只有怀孕达到220d的母牛才能进入围产前期。围产前期为13d，采用血钙调控技术和瘤胃调控技术饲养。在不发生瘫痪的牛群中不采用围产前期的低钙日粮饲喂，尤其是初次产犊的母牛。但是，为了减少初产母牛水肿，可以添加阴离子盐。经产母牛可以采用低钙日粮，配合阴离子盐，饲喂时间为10～13d。母牛进入分娩期，就应该饲喂产后泌乳期日粮。严格分娩期的分群管理。分娩期为分娩启动期2d，分娩期1d，分娩后护理期4d。奶牛出现分娩症状就进入分娩牛舍，异常牛采用诱导分娩技术，进入产拦，自由分娩，及时初产，胎衣排出后进入分娩后护理牛舍，需要4d的程序化护理。

# 二、中国畜牧兽医学会兽医内科与临床诊疗学分会论文摘要

## Biofilm formation and biofilm-associated genes assay of *Staphylococcus aureus* isolated from bovine subclinical mastitis in China

HE Jianzhong[1,2], WANG Anqi[1], LIU Gang[1], GAO Jian[1], Tariq Ali[1] and HAN Bo[1*]

[1]*College of Veterinary Medicine, China Agricultural University, 100193 Beijing, China*; [2]*Key Laboratory of Tarim Animal Husbandry Science & Technology of Xinjiang Production & Construction Corps*, 843300 *Xinjiang Alar, China*

**Abstract**: *Staphylococcus aureus* is one of the most common pathogens responsible for contagious bovine mastitis. Genes involved in biofilm formation is a special defensive mechanism of this pathogen to combat the host immune response and remain stable in hostile environment. The present study was designed with objectives to investigate strategies involving biofilm formation and biofilm associated genes (BAGs) of *S. aureus* strains, and to assess the consistency of two phenotype test methods. One hundred and two *S. aureus* strains were isolated from bovine subclinical mastitis cases from 32 commercial dairy farms in nine provinces of China. These isolates were screened for biofilm-producing capacity by Congo Red Agar (CRA) and Semi Quantitative Adherence Assay (SQAA) methods. Thirteen BAGs including *rbf*, *SigB*, *SasG*, *icaA*, *sarA*, , *icaR*, *icaD*, *clfA*, *clfB*, *fib*, *fnbpB*, *bap* and *fnbpA* were amplified by PCR assay. The results of current study revealed that *rbf* (95.1%) and *SigB* (94.1%) were the most prevalent BAGs, followed by *SasG* (89.2%), *icaA* (88.2%), *sarA* (87.3%), *icaR* (84.3%), *icaD* (82.5%), *clfA* (64.7%), *clfB* (45.1%), *fib* (43.1%) and *fnbpB* (19.6%). However, *bap* and *fnbpA* genes were not detected in any strain. By CRA method, 78.4% strains of *S. aureus* produced biofilm and 48.0% of strains were biofilm-positive by SQAA. Therefore, the data concluded that majority of *S. aureus* strains were capable to produce biofilm, controlled by eleven associated genes, and CRA detection rate was higher than SQAA for biofilm producing capacity of *S. aureus*.

* Corresponding author: hanbo@cau.edu.cn, 本文发表于 Pak Vet J 2014(on line)

## 5-Hydroxytryptamine levels in the pulmonary arterioles of broilers with induced pulmonary hypertension and its relationship to pulmonary vascular remodeling

LI Ying, ZENG Jianying, TANG Zhaoxin, PAN Jiaqiang*

*College of Veterinary Medicine, South China Agricultural University, Guangzhou, China*

**Abstract**: This experiment was performed to explore the relationship between 5-hydroxytryptamine (5-HT) levels in pulmonary arterioles and in pulmonary vascular remodelling in broilers. Pulmonary arterial hypertension was induced by injecting cellulose microparticles intravenously. Pulmonary hypertension syndrome (PHS) morbidity, right ventricle/total ventricle weight ratio (RV/TV), packed cell volume (PCV), haemoglobin concentration (HB), vessel wall area to vessel total area ratio (WA/TA) and mean tunica media thickness in pulmonary arterioles (mmol/LTPA) were measured. Proliferating cell nuclear antigen (PCNA), argyrophilic nucleolar organizer region proteins (Ag-NORs) and 5-HT content in pulmonary arterioles were determined. The results showed that injecting cellulose microparticles intravenously in broilers could successfully increase the PHS morbidity, significantly elevate RV/TV, PCV and HB, significantly increase mmol/LTPA and WA/TA, and significantly increase the argyrophilic particles in smooth muscle cell nucleoli, PCNA-positive cells in the medial layer, and the 5-HT content in pulmonary arterioles. Correlation analysis showed that the level of 5-HT was strongly positively correlated with PCNA and Ag-NORs. The results indicated that the increase of 5-HT in the tunica media could possibly promote the proliferation of smooth muscle cells in pulmonary arterioles and thus the occurrence of pulmonary vascular remodelling.

# β-Hydroxybutyrate activates the NF-κB signaling pathway to promote the expression of pro-inflammatory factors in calf hepatocytes

SHI Xiaoxia[a,c], LI Xinwei[a,c], LI Dangdang[a], LI Yu[a], SONG Yuxiang[a], DENG Qinghua[a], WANG Zhe[a], LI Xiaobing[a], LIU Guowen[a*]

[a] *College of Veterinary Medicine, Jilin University, Changchun* 130062, *China*

**Abstract**: *β*-hydroxybutyrate (BHBA) is the major component of ketone bodies in ketosis. Dairy cows with ketosis often undergo oxidative stress. BHBA is related to the inflammation involved in other diseases of dairy cattle. However, whether BHBA can induce inflammatory injury in dairy cow hepatocytes and the potential mechanism of this induction are not clear. The *NF-κB* pathway plays a vital role in the inflammatory response. Therefore, this study evaluated the oxidative stress, pro-inflammatory factors and *NF-κB* pathway in cultured calf hepatocytes treated with different concentrations of BHBA, pyrrolidine dithiocarbamate (PDTC, an *NF-κB* pathway inhibitor) and N-acetylcysteine (NAC, antioxidant). The results showed that BHBA could significantly increase the levels of oxidation indicators (MDA, NO and *iNOS*), whereas the levels of antioxidation indicators (*GSH-Px*, *CAT* and *SOD*) were markedly decreased in hepatocytes. The *IKKβ* activity and phospho-*IκBα* (*p-IκBα*) contents were increased in BHBA-treated hepatocytes. This increase was accompanied by the increased expression level and transcription activity of *p*65. The expression levels of *NF-κB*-regulated inflammatory cytokines, namely *TNF-α*, *IL*-6 and *IL*-1*β*, were markedly increased after BHBA treatment, while significantly decreased after NAC treatment. However, the *p-IκBα* level and the expression and activity of *p*65 and its target genes were markedly decreased in the PDTC + BHBA group compared with the BHBA (1.8 mmol/L) group. Moreover, immunocy to fluorescence of *p*65 showed a similar trend. The present data indicate that higher concentrations of BHBA can induce cattle hepatocyte inflammatory injury through the *NF-κB* signaling pathway, which may be activated by oxidative stress.

**Keywords**: oxidative stress, NF-κB pathway, inflammatory injury, bovine hepatocytes

* Corresponding authors: Guowen Liu, E-mail: liuguowen2008@163.com

## A survey on cestodes and protozoa of dogs in China

LI ZHAN[a] ZHNAG Limei[a], JING Zhihong[b], LIU Jianzhu [a*]

*a. College of Animal Science & Veterinary Medicine, Shandong Agricultural University, Tai'an, P. R. China*

*b. College of Foreign Languages, Shandong Agricultural University, Tai'an, P. R. China*

**Abstract**: The present study conducts a survey on cestode and Protozoan of dogs in Harbin, Urumqi, Qinhuangdao, Ji'nan, Tai'an, Zhengzhou, Shanghai, Guangzhou, Shenzhen and Haikou in China using brine flotation and centifugal sedimentation methods. A total of 358 faecal samples of household dogs (165 males and 193 females) were examined by faecal flotation for the presence of helminth eggs. Faecal samples (simple random sampling) were taken by local veterinary clinics between April 2011 and October 2012 in the framework of routine clinical procedures from 358 dogs (165 males and 193 females). Positive results with the presence of at least one parasite corresponded to 5.59% *Diplogonoporus grandis*, 6.15% *Echinococcus*, 8.38% *Taenia* canine, 6.98% *Isospora canine*, 2.23% *Cryptosporidium* and 7.82% *Balantidium coli*. There was no significant difference in overall prevalence between genders and sexes, except for *Balantidium coli*, which was more common in male dogs than female dogs ($P < 0.05$). Shanghai showes significant difference in *Echinococcus* compared with other cities ($P < 0.05$). In the meanwhile, statistically significant differences were observed among different regional groups in *Balantidium coli* in Shanghai, Tai'an and Guangzhou ($P < 0.05$). The data were analyzed using the statistical software package SPSS version 17.0 (SPSS Inc., Chicago, IL, USA). Identification of a risk factor required at a 95% confidence level ($P < 0.05$), as well as a biologically plausible association between the factor and seroreactivity to Parasites. The results of this survey indicate that the control and treatment of Parasites have been effective in some cities of China. However, further implementation of integrated strategies is necessary to prevent and control Parasites in dogs.

**Keywords**: china, cestode, Protozoan, dogs

* Corresponding author, E-mail: liujz@sdau.edu.cn

## Berberine protects against lipopolysaccharide-induced endometritis in mice

FU Kaiqiang, LV Xiaopei, LI Weishi, CAO Rongfeng *

*Department of Clinical Veterinary Medicine, College of Animal Science And Technology,*

*Qingdao Agricultural University, Qingdao 266109, People's Republic of China*

**Abstract**: Endometritis is a common disease in animal production, and influencing the breeding all over the world. Berberine is one of the main alkaloids isolated from *Rhizoma coptidis*. Previous reports showed that berberine has the potential of anti-inflammatory. However, the anti-inflammatory effects of berberine in a mouse model of lipopolysaccharide (LPS)-induced endometritis and its underlying molecular mechanisms remain to be elucidated. The purpose of the present study was to investigate the effects of berberine on LPS-induced mouse endometritis. Berberine was administered intraperitoneally at 1h before and 12h after induction of LPS. The results show that berberine significantly attenuates the infiltration of neutrophils, suppresses myeloperoxidase activity and decreases the production of NO, TNF-α and IL-1β. Furthermore, berberine inhibited the phosphorylation of NF-κB p65 subunit and the degradation of its inhibitor IκBα. These findings suggest that berberine exerts potent anti-inflammatory effects on LPS-induced mouse endometritis and might be a potential therapeutic agent for endometritis.

**Keywords**: berberine; endometritis; LPS; NF-κB

* Corresponding author, E-mail: rfcao@126.com

## Cadmium induction of reactive oxygen species activates the mitochondrial and AKT/mTOR pathways, leading to neuronal cell apoptosis

YUAN Yan[1], HU Feifei[1], JIANG Chenyang, ZHANG Kangbao, BIAN Jianchun, LIU Xuezhong, GU Jianhong, and LIU Zongping*

*College of Veterinary Medicine, Yang Zhou University, Yangzhou, China*

**Abstract**: Cadmium (Cd), a widely toxic metal, induces apoptosis in neuronal cells. In our previous study, we found that Cd induced the generation of reactive oxygen species (ROS) and mitochondrial pathway played an important role in Cd-induced apoptosis in both rat primary cerebral cortical neurons and PC12 cells. Studies have demonstrated that Cd induces apoptosis of neuronal cells (PC12 and SH-SY5Y) in part by activation of protein kinase B and mammalian target of rapamycin (Akt/mTOR) pathways. However, whether Cd activates mitochondrial and AKT/ mTOR pathways via induction of ROS in neuronal cells remain elusive. In this study, primary rat cerebral cortical neurons and PC12 cells were exposed to Cd, which significantly decreased the B-cell lymphoma 2/Bcl-2 associated X protein ((Bcl-2/Bax) ratio and increased the percentage of apoptotic cells, release of cytochrome c, cleavages of caspase-3 and poly (ADP-ribose) polymerase (PARP), and nuclear translocation of apoptosis-inducing factor (AIF). In addition, exposure to Cd decreased protein expression of PTEN (phosphatase and tensin homologue on chromosome 10) and induced phosphorylation of Akt, mTOR and its downstream effector molecules, p70 S6 kinase 1 (S6K1), eukaryotic initiation factor 4E-binding protein 1 (4E-BP1). Rapamycin, an inhibitor of mTOR, partially attenuated Cd-induced phosphorylation of Akt, mTOR, S6K1, and 4E-BP1, as well as apoptosis of the neuronal cells. Pretreatment with N-acetyl-L-cysteine (NAC), a ROS scavenger, effectively reversed decrease of Bcl-2/Bax ratio, release of cytochrome c, cleavages of caspase-3 and PARP, and nuclear translocation of AIF, phosphorylation of Akt, mTOR, S6K1, and 4E-BP1, as well as apoptosis of the neuronal cells. All together, these results demonstrated that the ROS-mediated AKT/mTOR pathway and mitochondrial apoptotic pathway (involving caspase-dependent and caspase-independent pathways) plays an important role in Cd-induced neuronal apoptosis.

**Keywords**: cadmium, apoptosis, neuronal cell, mitochondrial pathway, mammalian target of rapamycin

* Corresponding author, E-mail: liuzongping@yzu.eolu.cn

## Cepharanthine attenuates lipopolysaccharide-induced mice mastitis by suppressing the NF-κB signaling pathway

ZHOU Ershun, FU Yunhe, WEI Zhengkai, CAO Yongguo, ZHANG Naisheng and YANG Zhengtao*

*Department of Clinical Veterinary Medicine, College of Veterinary Medicine, Jilin University, Changchun* 130062, *P. R. China*

**Abstract**: Cepharanthine (CEP), a biscoclaurine alkaloid isolated from *Stephania cepharantha Hayata*, has been reported to have potent anti-inflammatory properties. However, the anti-inflammatory effects of CEP on a mouse model of lipopolysaccharide (LPS)-induced mastitis and its underlying molecular mechanisms remain to be elucidated. The purpose of the present study was to investigate the effects of CEP on LPS-induced mouse mastitis. The mouse model of mastitis was induced by inoculation of LPS through the canals of the mammary gland. CEP was administered intraperitoneally at 1 h before and 12 h after induction of LPS. The results show that CEP significantly attenuates the infiltration of neutrophils, suppresses myeloperoxidase activity, and reduces the levels of TNF-α, IL-1β, and IL-6 in LPS-induced mouse mastitis. Furthermore, CEP inhibited the phosphorylation of NF-κB p65 subunit and the degradation of its inhibitor IκBα. All the results suggest that CEP exerts potent anti-inflammatory effects on LPS-induced mouse mastitis. Accordingly, CEP might be a potential therapeutic agent for mastitis.

**Keywords**: cepharanthine, mastitis, LPS, NF-κB

* YANG Zhengtao, E-mail: yangzhengtao01@sina.com

## Changes of serum biochemical parameters in periparturient dairy cows and cows with subclinical hypocalcemia

WANG Jianguo[1], LI Xiaobing[2], ZHAO Baoyu[1], LIU Guowen[2]*, WANG Zhe[2]

1. *College of Veterinary Medicine, Northwest A&F University, Yangling* 712100, *China*;

2. *College of Veterinary Medicine, Jilin University, Changchun* 130062, *China*

**Abstract**: Subclinical hypocalcemia is a common metabolic disorder of dairy cows that occurs during early lactation. Many studies have focused on the involvement of calcitropic hormones in hypocalcemia; however, little is known about calcitonin, PTH and 1,25(OH)$_2$D$_3$ concentrations in dairy cows with subclinical hypocalcemia. Other mineral elements may also play prominent roles in regulating Ca homeostasis and may be involved in the pathogenesis of subclinical hypocalcemia in dairy cows; these hypotheses have not been verified so far. The aim of the study was to measure the serum concentrations of macrominerals elements, major calcitropic hormones and signal molecules related to energy metabolism levels of healthy periparturient cows and to compare the state of energy metabolism and the concentrations of macromineral elements and major calcium-regulating hormones in the serum of cows with subclinical hypocalcemia and healthy cows. Blood samples were taken from 219 multiparous Holstein cows near parturition. Fifty-one cows were identified as having subclinical hypocalcemia given their parity, body condition score (BCS) and total blood Ca levels (total blood Ca 1.4 – 2.0 mmol/L); 51 cows with normal Ca levels (total blood Ca 2.0 – 2.5 mmol/L within 24 h postpartum) and no obvious signs of other diseases were selected as healthy cows (the control group). The blood samples were collected from the coccygeal vein in the morning before feeding. Nine blood biochemical parameters were evaluated by means of individual analysis method.

The results showed that the serum concentrations of Ca and K were lowest at parturition. Calcium concentrations remained low during the first 2 wk postpartum then rose gradually; K levels increased gradually to reach their baseline levels within 2 wk postpartum. Mg serum concentrations were lowest at 1 wk postpartum and subsequently returned to their baseline levels. Serum Na levels peaked at parturition (day 0) then decreased gradually to their baseline levels. Serum P concentrations peaked during the first week prepartum and returned to baseline levels at parturition. Serum Cl concentrations peaked during the first week postpartum then decreased gradually and remained low during lactation; Serum PTH concentrations peaked at calving and returned to baseline levels during the first 4 wk postpartum. Serum 1,25 $(OH)_2D_3$ concentrations were highest during the first week postpartum and remained high for the first 4 wk postpartum. Calcitonin concentrations did not significantly fluctuate during the peripartum period; The levels of macromineral elements and major calcium-regulating hormones (with the exception of calcitonin) change dramatically in dairy cows during the periparturient period, particularly at or near calving. In contrast to the healthy cows, subclinically hypocalcemic dairy cows had significantly lower serum levels of calcium, potassium, sodium, magnesium, chlorine, 1,25 $(OH)_2D_3$ and glucose ($P < 0.05$) and significantly higher levels of serum phosphorus, non-esterified fatty acids (NEFAs) and β-hydroxybutyrate (BHBA) ($P < 0.05$). There were no significant differences in calcitonin and PTH concentrations in the two groups of cows.

The present data indicate that the subclinically hypocalcemic cows experienced a blunted PTH response and a more severe negative energy balance, which can aggravate the subclinical hypocalcemia and may increase the animals' risk of acquiring other metabolic diseases.

**Keywords**: Dairy cow, Peripartum period, Subclinical hypocalcemia

* Correspondence to Professor LIU Guowen, E-mail: liuguowen2008@163.com

# Characterization of the serum metabolic profile of dairy cows with milk fever using [1] H-NMR spectroscopy

SUN Yuhang[1], XU Chuchu[1], LI Changsheng[1],
XIA Cheng[1,2]*, XU Chuang[1], WU Ling[1] and ZHANG Hongyou[1]

[1] *College of Animal Science and Veterinary Medicine, Heilongjiang BaYi Agriculture University, Daqing* 163319, *China*

[2] *Synergetic Innovation Center of Food Safety and Nutrition, Northeast Agricultural University, Harbin* 150030, *China*

**Abstract**: Milk fever (MF), also known as clinical hypocalcemia, is a common calcium metabolism disorder in perinatal cows that is characterized by hypocalcemia, muscle weakness, paralysis of the limbs, and depression, with coma occurring in severe cases. Currently, information regarding the overall effects of such treatments on metabolism in cows with MF is scant. The purpose of the research was to study the metabolic profiling of serum samples from cows with MF, and explore other underlying pathological mechanisms of this disease. In our current study, Thirty-two Holstein cows (day 0 postpartum) with parities of three to six, body condition scores of approximately 3.25, and similar milk yields were selected for our study. The cows were maintained in free-stall housing in Heilongjiang, China. Production data were generated using the Dairy Comp 305 dairy management software system (Valley Agricultural Software, Tulare, CA, USA). Based on their clinical presentation and serum calcium levels, 24 healthy cows were assigned to the control group (No MF symptoms and calcium concentration > 2.5 mmol/L), and 8 cows were assigned to the MF group (MF symptoms and calcium concentration < 1.4 mmol/L). All serum samples were placed under the signal acquisition probe of an Avance III 500-MHz Digital NMR Spectrometer (Bruker, Ettlingen, Germany), and the NMR spectra were recorded using a Call-Purcell-Meiboom-Gill (CPMG) sequence (90 [τ-180-τ] n-acquisition) with 64 timed scans and a 10-ms total relaxation period (transverse relaxation and longitudinal relaxation) to inhibit the protein signal. The free induction decay (FID) signal was measured following transverse relaxation, and the FID signals were Fourier transformed to obtain the $^{1}H$ NMR spectra. After pre-processing of data, the multivariate pattern recognition, including the principal component analysis (PCA), partial least squares-discriminant analysis (PLS-DA) and orthogonal signal correction-partial least square-discriminate analysis (OSC-PLS-DA) were carried out. Then a one-way analysis of variance was performed using the SPSS statistical software (IBM, Armonk, NY, USA). The level of statistical significance was set at $P < 0.05$ for the identification of differentially expressed metabolites. In the OSCs plot, data points to the left (Group MF) of the vertical axis were negative, and those to the right side (Control group) were positive. In the loading plot, the data points above the horizontal axis were positive, and those under the horizontal axis is negative. Therefore, if a positive or negative shift occurred in the loading plot for a metabolite in Group MF, the metabolite was considered to be decreased or increased, respectively, compared to the level of the metabolite in control group. Nine metabolites in Group MF were present at significantly different levels, compared with those of control group. Glucose, alanine, glycerol, phosphocreatine, and gamma-aminobutyrate (GABA) were decreased, and β- hydroxybutyrate (BHBA), acetone, pyruvate, and lysine were increased in cows with MF. Most of these were sugars and amino acids involved in various energy metabolism pathways. The different metabolites in cows with MF reflected the pathological features of negative energy balance and fat mobilization, suggesting that MF is associated with altered energy metabolism. The $^{1}H$-NMR spectroscopy can be used to understand the pathogenesis of MF and identify biomarkers of the disease.

**Keywords**: dairy cow, $^{1}H$-NMR spectroscopy, hypocalcemia, milk fever, multivariate statistical analysis, pathogenesis

* Corresponding author, E-mail: xiacheng2001@gmail.com

# Cyanidin-3-O-β-glucoside ameliorates lipopolysaccharide-induced acute lung injury by reducing TLR4 recruitment into lipid rafts

FU Yunhe, ZHOU Ershun, WEI Zhengkai,
WANG Wei, WANG Tiancheng, ZHANG Naisheng, YANG Zhengtao *

*Department of Clinical Veterinary Medicine, College of Veterinary Medicine, Jilin University, Changchun* 130062, *P. R. China*

**Abstract**: Cyanidin-3-O-β-glucoside (C3G), a typical anthocyanin pigment that exists in the human diet, has been reported to have anti-inflammatory properties. The aim of this study was to detect the effect of C3G on LPS-induced acute lung injury and to investigate the molecular mechanisms. Acute lung injury was induced by intratracheal administration of LPS in mice. Alveolar macrophages from mice were stimulated with LPS and were treated with C3G. Our results showed that C3G attenuated lung histopathologic changes, myeloperoxidase (MPO) activity, TNF-α, IL-1β and IL-6 production in LPS-induced acute lung injury model. *In vitro*, C3G dose-dependently inhibited TNF-α, IL-1β, IL-6, IL-10 and IFN-β production, as well as NF-κB and IRF3 activation in LPS-stimulated alveolar macrophages. Furthermore, C3G disrupted the formation of lipid rafts by depleting cholesterol and inhibited TLR4 translocation into lipid rafts. Moreover, C3G activated LXRα-ABCG1-dependent cholesterol efflux. Knockout of LXRα abrogated the anti-inflammatory effects of C3G. In conclusion, C3G has a protective effect on LPS-induced acute lung injury. The promising anti-inflammatory mechanisms of C3G is associated with up-regulation of the LXRα-ABCG1 pathway which result in disrupting lipid rafts by depleting cholesterol and reducing translocation of TLR4 to lipid rafts, thereby suppressing TLR4 mediated inflammatory response.

**Keywords**: Cyanidin-3-O-β-glucoside, TLR4, lipid raft, LXR, ABCG1

Correspond authors, Zhengtao Yang, E-mail: yangzhengtao01@ sina. com; Naisheng Zhang, E-mail: zhangns61@ sohu. com

发表杂志: Biochemical Pharmacology

## Cyanidin-3-O-β-glucoside inhibits lipopolysaccharide-induced inflammatory response in mouse mastitis model

FU Yunhe, WEI Zhengkai, ZHOU Ershun, ZHANG Naisheng, YANG Zhengtao

*Department of Clinical Veterinary Medicine, College of Veterinary Medicine, Jilin University, Changchun* 130062, *P. R. China.*

**Abstract**: Cyanidin-3-O-β-glucoside (CAS No. 7084-24-4), a typical anthocyanin pigment that exists in the human diet, has been reported to have anti-inflammatory properties. However, the effect of C3G on LPS-induced mastitis and the molecular mechanisms has not been investigated. In this study, we detected the protective effects of C3G on LPS-induced mouse mastitis model and investigated the molecular mechanisms in LPS-stimulated mouse mammary epithelial cells. Our results showed that C3G could attenuate mammary histopathologic changes, and myeloperoxidase (MPO) activity; inhibit TNF-α, IL-1βand IL-6 production; caused by LPS. Meanwhile, C3G dose-dependently inhibited TNF-α and IL-6 in LPS-stimulated mouse mammary epithelial cells. C3G suppressed LPS-induced NF-κB and IRF3 activation. Furthermore, C3G disrupted the formation of lipid rafts by depleting cholesterol. Moreover, C3G activated LXR-ABCG1-dependent cholesterol efflux. Knockdown of LXRα abrogated the anti-inflammatory effects of C3G. In conclusion, C3G has a protective effect on LPS-induced mastitis. The promising anti-inflammatory mechanisms of C3G is associated with up-regulation of the LXRα-ABCG1 pathway which result in disrupting lipid rafts by depleting cholesterol, thereby suppressing TLR4 mediated NF-κB and IRF3 signaling pathways induced by LPS.

**Keywords**: Cyanidin-3-O-β-glucoside, NF-κB, IRF3, TLR4, lipid raft, LXR, ABCG1

Correspond Authors. Zhengtao Yang, E-mail: yangzhengtao01@ sina. com,发表杂志: Journal of Lipid Research

# Dietary selenium alters the transcriptome of selenoprotein in chicken liver

WANG Lili, CAO Changyu, ZHANG Cong, LI Jinlong

*Northeast Agricultural University*, *Harbin* 150030

**Abstract**: Selenium (Se) can promote growth and appetite, however the mechanism behind its action remains unclear. Selenoprotein are the *biological* effector molecules of Se, and selenoproteins are known to directly regulate appetite through unknown pathways. The liver, the largest digestive gland in the body is known as the "Se storehouse". Characterize the relationship between the dietary Selenium (Se) deficiency or excess and the transcriptome of selenoprotein in chicken liver at $6^{th}$ week. 1-day-old chickens received deficiency Se (0.028 mg Se/kg), normal Se (0.15mg Se/kg) or excess Se (3.0 or 5.0 mg Se/kg) in their diets for 6 weeks. Detect the activities of antioxidants including glutathione peroxidase (GPX), total antioxidant capacity (T-AOC), Methyl Di Aldehyde (MDA), thioredoxin reductase (TrxR) with antioxidant kits. Establish optimized quantitative RT-PCR array to alternative gene chips, choose six reference genes to detect the high-throughput expression of chicken transcriptome of selenoprotein, the data analysis on mRNA expression were analyzed with GeneCopoeia-FulenGen qPCR Array data analysis system online (http://www.genecopoeia.com/product/qpcr/analyse2/index.php). The antioxidant including GPX, T-AOC, MDA, TrxR activities in chicken liver were increased by dietary Se increased except the excess Se (5.0 mg Se/kg) group decreased compared with the normal Se group. In the 7 selenoenzymes and 21 selenoproteins, 7 selenoprotein-synthesis factors (*selenocysteine-tRNA-specific* (*EEFSEC*), *Sec-tRNA synthase* (*SEPSECS*), *selenocysteine* 1 *associated protein* 1 (*SECP*43-1), *O-phosphoseryl-tRNASec kinase* (*PSTK*), *selenophosphate synthetase* 1 (*SEPHS*1), *selenophosphate synthetase* 2 (*SECISPS*2), *seryl-tRNA synthetase* (*SARS*)) and 4 GPX genes (*GPX*1, *GPX*2, *GPX*3, *GPX*4) decreased in the deficiency Se and excess Se (5.0 mg Se/kg) groups, increased in the excess Se (3.0 mg Se/kg) group. 3 TrxRs (*TXNRD*1, *TXNRD*2, *TXNRD*3), 5 endoplasmic reticulum genes (*small* 15-*kDa selenoprotein* (*SEP*15), *selenoprotein S* (*SELS*), *selenoprotein K* (*SELK*), *selenoprotein N*1 (*SEPN*1), *selenoprotein T* (*SELT*) and other selenoproteins (*selenoprotein I* (*SelI*), *selenoprotein X*1 (*SEPX*1), *selenoprotein U* (*SelU*) and *selenoprotein O* (*SELO*) increased in the excess Se group (3.0 mg Se/kg). The change of 3 *deiodinases* (*DIO*1, *DIO*2, *DIO*3) and 2 Cytoplasm Sels (*selenoprotein W* (*SELW*), *plasma selenoprotein P*1 (*SEPP*1)) genes were not very obvious. 7 selenoprotein-synthesis factors and 4 GPX genes were the most sensitive genes to dietary Se level; 3 TrxRs, 5 endoplasmic reticulum genes and other selenoproteins were sensitive genes to dietary Se level; 3 deiodinases and Cytoplasm selenoproteins were not very sensitive to dietary Se level.

**Keywords**: selenium deficiency or excess, selenoenzyme, selenoprotein, transcriptome, chicken liver

Corresponding author, E-mail: jinlongli@neau.edu.cn.

# Differential characteristics and *in vitro* angiogenesis of bone marrow-and peripheral blood-derived endothelial progenitor cells: evidence from avian species

TAN Xun*, BI Shicheng, LIU Xi

*Department of Veterinary Medicine, College of Animal Sciences, Zhejiang University, Hangzhou* 310058, *P. R. China*

**Abstract**: Endothelial progenitor cells (EPCs) represent a cell population in the peripheral circulation being capable of differentiation into endothelial cells. Accumulating evidence suggests that EPCs play a critical role in neoangiogenesis and endothelial maintenance and repair, and EPC-based cell therapy is emerging as a promising novel therapeutic approach for various cardiovascular disorders. It is most commonly accepted that EPCs in circulation are originated from bone marrow (BM), and that vascular injury triggers the mobilization of EPCs from the BM to peripheral blood (PB). Accordingly, both BM and PB are widely used for the isolation and culture of EPCs. Although both BM- and PB-derived EPCs (BM-EPCs and PB-EPCs, respectively) have been proved to incorporate into the vasculature of experimentally induced wounds or ischemic areas, the therapeutic potential of BM-EPCs remains conflicting. The present study was conducted to compare the phenotypes and *in vitro* angiogenic capacity of EPCs derived from BM and PB from an avian species, fast-growing meat-type chicken (broiler). Mononuclear cells were isolated from broiler BM and PB (BMMNCs and PBMNCs) and cultured in EGM-2 medium. Cells at days 7 to 14 were used for experiments. The expression of progenitor and endothelial markers, the number of DilacLDL/lectin dual-positive cells, and adipogenic and osteogenic differentiation were determined. The migration and *in vitro* angiogenic ability between BMMNC- and PBMNC-derived cells were compared. PBMNCs developed typical EPC appearances, with initial spindle shape followed by a cobblestone appearance, whereas BMMNC-derived cells retained their constitutive spindle-like morphology throughout the experiment. Cells derived from both sources expressed CD133, CD31 and VEGFR-2, while PBMNCs-derived cells had decreased CD133 expression. Nevertheless, the number of DilacLDL/lectin dual-positive cells was not different between groups. Adipogenic and osteogenic lineages were verified in BMMNC- but not in PBMNC-derived cells. PBMNC-derived cells formed tubular networks on Matrigel. However, BMMNC-derived cells formed few tube-like structures which were not morphologically comparable to those developed by their counterparts. Our results suggest that the so called "EPCs" derived from BMMNCs are not "true" EPCs, supporting previous findings on mammals that BM may not serve as an optimal isolation source of EPCs.

**Keywords**: Broiler, Endothelial progenitor cells, Bone Marrow, Peripheral blood

Corresponding author, E-mail: tanxun@zju.edu.cn

## Effects of different selenium levels on gene expression of a subset of selenoproteins and antioxidative capacity in mice

ZHANG Qin, CHEN Long, GUO Kai, ZHENG Liangyan, LIU Bitao, GUO Cuili, TANG Zhaoxin*

**Abstract**: This study aimed to evaluate how excess selenium induces oxidative stress by determining antioxidant enzyme activity and changes in expression of selected selenoproteins in mice. BALB/c mice ( n = 20 per group) were fed a diet containing 0. 045 ( Se-marginal), 0. 1 ( Se-adequate), 0. 4 ( Se-supernutrition), or 0. 8 ( Se-excess) mg Se/kg. Gene expression was quantified in RNA samples extracted from the liver, kidney, and testis by real-time quantitative reverse transcription-polymerase chain reaction. We found that glutathione peroxidase ( GPx) and catalase activities decreased in livers of mice fed the marginal or excess dose of Se as compared to those in the Se-adequate group. Additionally, superoxide dismutase and glutathione reductase activities were significantly reduced only in mice fed the excess Se diet, compared to animals on the adequate Se diet. Se-supernutrition had no effect on hepatic mRNA levels of GPx isoforms 1 and 4 ( GPx1 and GPx4), down-regulated GPx isoform 3 ( GPx3), and upregulated selenoprotein W ( SelW) mRNA expression. The excess Se diet led to decreased hepatic mRNA levels of GPx1, GPx3 and GPx4 but no change in testicular mRNA levels of GPx1, GPx3 or SelW. Dietary Se had no effect on testicular mRNA levels of GPx4. Thus, our results suggest that Se exposure can reduce hepatic antioxidant capacity and cause liver dysfunction. Dietary Se was found to differentially regulate mRNA levels of the GPx family or SelW, depending on exposure. Therefore, these genes may play a role in the toxicity associated with Se.

* Corresponding author, E-mail: tangzx@ scau. edu. cn

## Effects of niacin on *Staphylococcus aureus* internalization into bovine mammary epithelial cells by modulating NF-κB activation

WEI Zhengkai, FU Yunhe, ZHOU Ershun, TIAN Yuan, YAO Minjun, LI Yimeng, YANG Zhengtao, CAO Yongguo*

*College of Veterinary Medicine, Jilin University, Changchun* 130062

**Abstract**: Niacin is a precursor of coenzymes NAD and NADP and plays a critical role in electron transfer during the metabolic process. In addition to its nutrimental function, niacin has long been used for the treatment of lipid disorders and cardiovascular disease. However, the effect of niacin on *Staphylococcus aureus* ( *S. aureus* ) internalization into bovine mammary epithelial cells ( bMEC ) remains unclear. Here we sought to examine the effect of niacin on *S. aureus* internalization into bovine mammary epithelial cells ( bMEC) and to investigate the potential mechanism. In this study, the growth of *S. aureus* supplemented with niacin ( 0. 5 – 2 mmol/L) was monitored turbidimetrically at 600 nm for 24 h and cell viability was measured by MTT assay. Gentamicin protection assay was carried out to determine the effect of niacin on *S. aureus* internalization into bMEC. To determine the potential mechanism, tracheal antimicrobial peptide (TAP) and β-defensin (BNBD5) expressions were detected by quantitative Real-time polymerase chain reaction (qRT-PCR). The activation of nuclear factor-kappa B (NF-κB) was determined by Western blotting. The results showed that niacin (0.5-2 mmol/L) did not affect *S. aureus* growth and bMEC viability, whereas it inhibits *S. aureus* internalization ranging from 13% to 42% and down-regulated the mRNA expression of TAP and BNBD5 compared to the control group. No exactly relationship was discovered between *S. aureus* internalization into bMEC and antimicrobial peptide expression, while niacin inhibited *S. aureus*-induced NF-κB activation in a dose manner. These dates suggest that inhibiting NF-κB activation may be the potential mechanism of niacin on modulating *S. aureus* internalization into bMEC.

**Keywords**: *Staphylococcus aureus*, niacin, bovine mammary epithelial cell, bovine mastitis, antimicrobial peptides, NF-κB

Correspond authors. Dr. Yongguo Cao, E-mail: ygcao82@ jlu. edu. cn, 发表于 Microbial Pathogenesis

# Glycyrrhizin inhibits lipopolysaccharide-induced inflammatory response by reducing TLR4 recruitment into lipid rafts in RAW264.7 cells

FU Yunhe, ZHOU Ershun, WEI Zhengkai, SONG Xiaojing, LIU Zhicheng, WANG Tiancheng, WANG Wei, ZHANG Naisheng, LIU Guowen, YANG Zhengtao *

*Department of Clinical Veterinary Medicine, College of Veterinary Medicine, Jilin University, Changchun 130062, P. R. China*

**Abstract**: The aim of this study was to investigate the effect of glycyrrhizin on LPS-induced endotoxemia in mice and clarify the possible mechanism. An LPS-induced endotoxemia mouse model was used to confirm the anti-inflammatory activity of glycyrrhizin *in vivo*. *In vitro*, RAW264.7 cells were stimulated with LPS in the presence or absence of glycyrrhizin. The expression of cytokines was determined by ELISA. Toll-like receptor 4 (TLR4) was determined by Western blot analysis. Nuclear factor-kB (NF-κB) and Interferon regulatory factor 3 (IRF3) activation were detected by Western blotting and luciferase assay. Lipid raft staining was detected by immunocy to chemistry. *In vivo*, the results showed that glycyrrhizin can improve survival during lethal endotoxemia. *In vitro*, glycyrrhizin dose-dependently inhibited the expression of TNF-α, IL-6, IL-1β and RANTES in LPS-stimulated RAW264.7 cells. Western blot analysis showed that glycyrrhizin suppressed LPS-induced NF-κB and IRF3 activation. However, glycyrrhizin did not inhibit NF-κB and IRF3 activation induced by MyD88-dependent (MyD88, IKKβ) or TRIF-dependent (TRIF, TBK1) downstream signaling components. Moreover, glycyrrhizin did not affect the expression of TLR4 and CD14 induced by LPS. Significantly, we found that glycyrrhizin decreased the levels of cholesterol of lipid rafts and inhibited translocation of TLR4 to lipid rafts. Moreover, glycyrrhizin activated ABCA1, which could induce cholesterol efflux from lipid rafts. Glycyrrhizin exerts an anti-inflammatory property by disrupting lipid rafts and inhibiting translocation of TLR4 to lipid rafts, thereby attenuating LPS-mediated inflammatory response.

Learning the anti-inflammatory mechanism of glycyrrhizin is crucial for the anti-inflammatory drug development.

**Keywords**: cytokine, glycyrrhizin, NF-κB, IRF3, TLR4, lipid raft

* Correspond author, Zhengtao Yang, E-mail: yangzhengtao01@sina.com
发表杂志:Biochimica et Biophysica Acta (BBA) - General Subjects

# Glycyrrhizin inhibits the inflammatory response in mouse mammary epithelial cells and mouse mastitis model

FU Yunhe, ZHOU Ershun, WEI Zhengkai, LIANG Dejie, WANG Wei, WANG Tiancheng, GUO Mengyao, ZHANG Naisheng, YANG Zhengtao *

*Department of Clinical Veterinary Medicine, College of Veterinary Medicine, Jilin University, Changchun* 130062, *P. R. China*

**Abstract**: Glycyrrhizin, a triterpene glycoside isolated from licorice root, has been known to exhibit anti-inflammatory activities. However, the effect of glycyrrhizin on mastitis has not been reported. The purpose of this study was to investigate the anti-inflammatory effect and mechanism of glycyrrhizin on LPS-induced mastitis in mouse. An LPS-induced mouse mastitis model was used to confirm the anti-inflammatory activity of glycyrrhizin *in vivo*. The primary mouse mammary epithelial cells were used to investigate the molecular mechanism and targets of glycyrrhizin. *In vivo*, glycyrrhizin significantly attenuated mammary gland histopathologic changes, myeloperoxidase (MPO) activity, and the infiltration of neutrophilic granulocytes; down-regulated the expression of TNF-α, IL-1β and IL-6; caused by LPS. *In vitro*, glycyrrhizin dose-dependently inhibited LPS-induced the expression of TNF-α, IL-6, and RANTES. Western blot analysis showed that glycyrrhizin suppressed LPS-induced NF-κB and IRF3 activation. However, glycyrrhizin did not inhibit NF-κB and IRF3 activation induced by MyD88-dependent (MyD88, IKKβ) or TRIF-dependent (TRIF, TBK1) downstream signaling components. Moreover, glycyrrhizin did not act though affecting the function of CD14 or expression of TLR4. Finally, we showed that glycyrrhizin decreased the levels of cholesterol of lipid rafts and inhibited translocation of TLR4 to lipid rafts. Moreover, glycyrrhizin activated ABCA1, which could induce cholesterol efflux from lipid rafts. In conclusion, we find that the anti-inflammatory effects of glycyrrhizin may be due to its ability to activate ABCA1. Glycyrrhizin might be a useful therapeutic reagent for mastitis and other inflammatory diseases treatment.

**Keywords**: glycyrrhizin, NF-κB, IRF3, TLR4, lipid raft, ABCA1

* Correspond author, Zhengtao Yang, E-mail: yangzhengtao01@sina.com, 发表杂志:FEBS Journal

# $^{1}$H NMR and GC/MS based plasma metabolic profiling of dairy cows with ketosis

SUN Lingwei, ZHANG Hongyou*, XIA Cheng, XU Chuang, BAO Kai

*College of Animal Science and Veterinary Medicine, Heilongjiang Bayi Agricultural University, Daqing* 163319, *China*

**Abstract**: In some past researches, the exploration for pathogenesis of ketosis mainly focuses on physiology, biochemistry and pathology. However, there is little report about the metabolomics of ketosis. Therefore, this report has an important significance in this study to reveal the global change of life functions during the development of ketosis, enrich and develop the its pathogenesis. Ketosis is a metabolic disorder in dairy cattle. Primary or secondary ketosis can result during early lactation. Generally, high milk production during lactation and/or inadequate energy intakes result in a negative energy balance, which induces ketosis. Because ketosis can lead to substantial economic losses in the dairy industry, it is of outmost importance to prevent ketosis in dairy cattle.

Through the application of metabolomics technology combines principal component analysis (PCA) and orthogonal partial least-squares discriminant analysis (OPLS-DA), changes of endogenous metabolites with clinical and subclinical ketosis cows were analyzed, which to study the affects of ketosis on cow metabolic process, and find the differences metabolites.

In this study, 80 lactation Holstein cows were selected as experimental animals 7 to 21 days postpartum, at the same time experimental cow plasma samples were collected according to the plasma β-hydroxybutyrate (BHBA) concentration and clinical symptoms. Clinical ketosis groups of cows (CK, $n = 24$) showed obvious clinical symptoms postpartum and plasma BHBA concentration is more than 1.60 mmol/L, subclinical ketosis groups of cows (SK, $n = 33$) didn't show clinical symptoms postpartum and plasma BHBA concentration is more than 1.20 mmol/L; healthy groups of cows (C, $n = 23$) had no clinical symptoms and plasma BHBA concentration is less than 1.00 mmol/L. Experimental animals selected had no other concurrency and secondary diseases in addition to ketosis. The plasma metabolites were detected by $^{1}$H nuclear magnetic resonance ($^{1}$H NMR) and gas chromatography/mass spectrometry (GC/MS) technique. Changes of metabolite contents were analyzed among 3 groups using principal component analysis and partial lease squares. Finally, metabolites were analyzed by using bioinformatics technology biology.

The plasma metabolic profiles of the three groups were obtained by $^{1}$H NMR. Compared with the healthy control group, 23 different metabolites were obtained in subclinical and clinical ketosis groups, in which Acetylacetonate (ACAC), β-hydroxybutyric acid (BHBA), acetone and acetic acid, etc are increased, while histidine, lysine, glutamic acid, glutamine, lactic acid and glucose, etc are decreased. By comparison, 28 different metabolites were obtained between clinical ketosis group and subclinical ketosis group, in which BHBA, ACAC, and acetone, etc are increased, while citric acid, formic acid, histidine, alanine, proline, tyrosine, low density lipoprotein, and very low density lipoprotein, etc are decreased. The plasma metabolome was measured by GC/MS, which led to the detection of 267 variables. There were 40 types of metabolites without difference among 3 groups. Compared with the healthy control group, 32 different metabolites were obtained in subclinical and clinical ketosis groups, in which BHBA, α-aminobutyric acid, sitoesterol, isoleucine, leucine, glycine, myristic acid and palmitic acid, etc are increased, while glucose, lactose, glyoxylic acid, alanine, glutamic acid, lactic acid, etc are decreased. By comparison, 13 different metabolites were obtained between clinical ketosis group and subclinical ketosis group, in which heptadecanoic acid, stearic acid and 3-hydroxy valeric acid, etc are increased, while proline, serine, proline, α-aminobutyric acid and 3,4-docosahexaenoic acid, etc are decreased. Through the KEGG database analysis, these metabolites primarily were related with amino acid metabolism, fat metabolism and carbohydrate metabolism.

The results showed $^{1}$H NMR and GC/MS combined with pattern recognition technique can effectively get the different metabolites of diagnosing clinical and subclinical ketosis. The substances of different content and contributing to classification may be the potential metabolic marker and objective indicators for diagnosing as ketosis, and panoramic reveal that widely metabolic disorder on cows in the process of ketosis, which lays the foundation for the exploration of ketosis mechanism and obtaining new biomarkers in the future.

**Keywords**: metabolomics, multivariate statistical analysis, ketosis, $^{1}$H NMR, GC/MS

* Correspondence author, ZHANG Hongyou, E-mail: zhy478@163.com

## $^{1}$H NMR-based plasma metabolic profiling of dairy cows with type I and type II ketosis

LI Ying[1], XU Chuang[1], XIA Cheng[1,2], ZHANG HongYou*, SUN LingWei[1], GAO Yang[1], XU ChuChu[1]

Chuang Xu[1], Ying Li[1], Cheng Xia[1,2], HongYou Zhang, LingWei Su[1], ChuChu Xu[1]

1. *College of Animal Science and Veterinary Medicine, Heilongjiang Bayi Agricultural University, Daqing*, 163319, *China*;

2. *Synergetic Innovation Center of Food Safety and Nutrition, Northeast Agricultural University, Harbin* 150030, *China*

**Abstract**: This study identified differences in plasma metabolites among three groups of dairy cows: type I ketotic (K1), type II ketotic (K2), and healthy control cows (C). Plasma metabolomic profiles were analyzed by $^{1}$H-nuclear magnetic resonance technology ($^{1}$H NMR). The data were processed by principal component analysis (PCA) and orthogonal partial least-squares discriminant analysis (OPLS-DA). The results revealed that OPLS-DA was more effective at distinguishing amongst the three groups. Additionally, there were seven different metabolites between K2 and C, 19 different metabolites between K1 and C, and 24 different metabolites between K1 and K2. Therefore, the combination of $^{1}$H-NMR and multivariate statistical analyses can effectively distinguish the differential metabolites among the K1, K2, and C groups, thereby providing important information on the pathogenesis, early diagnosis, and prevention of type I and type II ketosis in dairy cows.

**Keywords**: $^{1}$H NMR; metabolomics; type I ketosis; type II ketosis; multiple analysis

* Corresponding author: ZHANG HongYou, E-mail: zhy478@163.com

# Immune responses and allergic reactions in piglets by injecting glycinin

WANG Xichun, LI Bao, WU Jinjie*, KOU Yanan, XU Shuliang, SUN Zhikuo, FENG Shibin, MA Liangyou

*College of Animal Science and Technology, Anhui Agricultural University, Hefei 230036, P. R. China*

**Abstract**: Glycinin, one of the main soybean antigen proteins, is researched by many nutritionists and it can cause allergic reactions in young animals. But its effect is considerably more severe in weaned piglets, showing symptoms of diarrhea. At present, in order to eliminate the allergic effects, various food processing methods were used to process soybean. However, there are few reports about the allergic reactions induced by glycinin which are alleviated with a prior immunized method in piglets. So we exploring a new way to prevent the allergic reaction induced by soybean antigen protein. The aim of this study was to investigate the effects of a prior immunization with glycinin on growth performance, serum immunoglobulin contents, small intestinal histamine release and mucosal histology in piglets. Forty piglets (7 d of age) were randomly divided into four groups of ten piglets each. Piglets of Groups Im (immunized) and Im + S (immunized and sensitized) were immunized twice by hypodermic injection with glycinin at 500 μg/kg body weight (BW) at day 7 and 21 of age. At day 23, Groups Im + S and S (sensitized) were intramuscularly injected with 2 500 μg glycinin per kg BW. The piglets of Group C (control) received a physiological saline solution by hypodermic injection. All piglets were weaned at the age of 23 d and blood samples were taken on days 7, 21 and 35. At the end of the trial, five piglets per group were slaughtered and the intestine was collected for evaluating histamine levels and mucosal histology. Compared with Group C, in Group S the average daily gain (ADG), feed intake and gain: feed ratio were decreased ($P < 0.01$), and serum levels of IgG and IgE piglets were increased ($P < 0.05$). Furthermore, in this Group the histamine levels in the duodenum and ileum were significantly decreased ($P < 0.05$), and the structure of duodenal and ileal mucosa was severely damaged. But in Groups Im and Im + S the ADG was increased ($P < 0.05$), serum IgE levels were decreased ($P < 0.05$), intestinal histamine levels were increased ($P < 0.05$), and the intestinal mucosa was not damaged. The results suggest that prior immunization with glycinin can increase ADG and serum IgG levels, decrease serum IgE levels and histamine release. Therefore, this method is also potentially able to protect the structural integrality of the intestinal mucosal epithelia and alleviate allergic reactions in piglets.

**Keywords**: glycinin; immunoglobulins; histamine; mucosa histology; piglets

* Corresponding author, E-mail address: wjj@ahau.edu.cn.

# Influence of different factors on swainsonine production in fungal endophyte from locoweed

ZHANG Leilei[1], HE Shenghu[1], YU Yongtao[1*], ZHAO Qingmei[2,3], GE Song[1]

1. *College of Agronomy Ningxia University*, *Yinchuan* 750021, *China*; 2. *College of Biological Science and Engineering*, *Beifang University of Nationalities*, *Yinchuan* 750021, *China*; 3. *Key Laboratory of Fermentation Brewing Engineering and Biotechnology State Nationlities Affairs Commission*, *Yinchuan* 750021, *China*

**Abstract**: Locoweeds are toxic plants of the genera *Astragalus* and *Oxytropis* containing swainsonine. Swainsonine is the principal agent responsible for locoism in animals. Recently, the fungal endophytes are isolated frequently from locoweeds, *Undifilum* spp, are believed that responsible for the production of swanisonine in locoweeds. If the swainsonine production in *Undifilum* spp is inhibited, the locoweed's toxicity would be reduced or eliminate. Currently, the mechanism was few known about the biosynthesis of swainsonine in Locoweed's endophytic fungi. The research about influence of different factors on swainsonine production in *Undifilum oxytropis* will be benefit for building the important basis for further to elucidate the swainsonine biosynthetic pathways and molecular regulation mechanism. *U. oxytropis* were inoclated in liquid medium of different pH or containing different concentration of PEG, L-pipecolic acid, L-lysine and α-ketoglutaric acid, respectively. To estimate the influence of different factors on swainsonine production in *U. oxytropis*, the swainsonine of cultured fungal mycelia and zymotic fluid were extracted and detected. after shaking culture. The results showed that is no significant difference about growth of different pH treatment of *U. oxytropis* groups. However, low pH (pH4.5) will inhibit the synthesis of swainsonine. PEG promoted the growth of *U. oxytropis*, but inhibited the synthesis of swainsonine. L-pipecolic acid could increase significantly the mycelial dry mass. When the initial concentration of L-pipecolic acid was $10^{-3}$ mol/L and $10^{-2}$ mol/L, the production of swainsonine in *U. oxytropis* was inhibited. However, when the initial concentration of L-pipecolic acid was $10^{-4}$ mol/L, the production of swainsonine in *U. oxytropis* was increased significantly. When the initial concentration of L-lysine was $10^{-3}$ mol/L, the production of swainsonine in *U. oxytropis* was increased significantly. However, when the initial concentration of L-pipecolic acid was $10^{-1}$ mol/L, $10^{-2}$ mol/L and $10^{-4}$ mol/L, the production of swainsonine in *U. oxytropis* was inhibited significantly. When the initial concentration of α-ketoglutaric acid was $10^{-1}$ mol/L, $10^{-2}$ mol/L and $10^{-3}$ mol/L, the growth of production of swainsonine in *U. oxytropis* was inhibited significantly. Low pH or PEG added to medium could significantly inhibit the production of swainsonine in *U. oxytropis*. It was significant that the influence of L-pipecolic acid, L-lysine and α-ketoglutaric acid on the swainsonine producing of fungal endophyte. However, the extent of influence on fungal swainsonine producing was highly correlated with the concentration of L-pipecolic acid, L-lysine and α-ketoglutaric in liquid media.

**Keywords**: locoweed, fungal endophyte, *undifilum oxytropis*, swainsonine, biosynthesis

**基金项目**:国家自然科学基金(31201962);宁夏自然科学基金(NZ1118)

**作者简介**:张蕾蕾(1989-),女,陕西人,硕士在读,研究方向:兽医临床诊断技术。E-mail:andybeilei@163.com

**通讯作者**:余永涛(1980-),男,宁夏平罗人,副教授,博士,硕士研究生导师,研究方向:动物中毒病与营养代谢病。E-mail:yyt1211@163.com

# Insulin suppresses the AMPK signaling pathway to regulate lipid metabolism in primary cultured bovine hepatocytes

DING Hongyan, LI Yu, LI Xinwei, WANG Zhe, LIU Guowen * and LI Xiaobing

*Key Laboratory of Zoonosis, Ministry of Education, College of Veterinary Medicine, Changchun* 130062, *China*

**Abstract**: Insulin plays a pivotal role in the regulation of lipid metabolism. However, the mechanism of insulin action on lipid metabolism is still completely unclear in dairy cows. Hepatocytes were isolated from newborn female cows and co-cultured *in vitro* with various concentrations of insulin in the presence or absence of AICAR (an AMPKα activator). The results showed that insulin increased the phosphorylation of insulin receptor and decreased phosphorylation of AMP-activated protein kinase alpha (AMPKα). Moreover, insulin increased the expression and transcriptional activity of the carbohydrate responsive element-binding protein (ChREBP) and sterol regulatory element-binding protein (SREBP), resulted in the upregulation of lipogenic genes and decreased peroxisome proliferators-activated receptor-α (PPARα), resulted in the downregulation of lipid oxidation genes. While contrast results were observed after hepatocytes were treated with AICAR. In addition, the triglyceride (TG) content was significantly increased in insulin-treated groups. Collectively, these results indicated that insulin suppressed the AMPKα signaling pathway through insulin receptor to mediate a decreased lipid oxidation and an increased lipid synthesis in bovine hepatocytes, which elucidates the pathological mechanism of lipid metabolism disorder induced by hyperinsulinemia which occurs in insulin resistance of dairy cows with negative energy balance (NEB).

**Keywords**: Insulin; AMPKα; lipid oxidation; lipid synthesis; bovine hepatocytes

E-mail: xbli@jlu.edu.cn (X. B. Li) and guowenliu2008@163.com (G. W. Liu)

# Isolation and characterization of peripheral blood-derived endothelial progenitor cells from broiler chicken

BI Shicheng, TAN Xun *, ALI Shah Qurban

*Department of Veterinarymedicine, Zhejiang University, Hangzhou* 310058, *China*

**Abstract**: Endothelial progenitor cells (EPCs) are bone marrow-derived immature cells being capable of differentiating into endothelial cells and are involved in endothelial homeostasis as well as physiological and pathological angiogenesis (Asahara et al., 1999). So far, two morphologically distinct EPC subpopulations, named as early EPCs (eEPCs) and endothelial outgrowth cells (EOCs, also termed late-outgrowth EPCs or endothelial colony-forming cells), have been identified in mammals. However, the characteristics of EPCs in avian species remain unknown. The resent study was conducted to isolate EPCs from peripheral blood of broilers. PBMNCs from clinically healthy 4- to 6-week-old Arbor Acres broilers were prepared by density-gradient centrifuging and resuspended in EBM-2 medium supplemented with growth factors as provided in the EGM-2 Bullet Kit, and 10%

fetal bovine serum (FBS), 100 IU/ml penicillin and 100 μg/ml streptomycin. Cells were seeded onto 6-well tissue plates coated with rat tail collagen (5 $\mu g/cm^2$, Sigma-Aldrich) at $1\times10^7$ cells per well, and incubated at 39℃ in 5% $CO_2$. Non-adherent cells were removed after 48 h of plating and the medium was refreshed every 2 days thereafter. Cell morphology was monitored daily using a phase contrast microscope. Quantative PCR was performed to detecte the expression of progenitor and endothelial markers. As an alternative approach, cell phenotype was characterized by Dil-ac-LDL uptake and FITC-UEA-l binding. Cell proliferation was determined by a 3- (4,5-dimethylthiazol-2-yl)-2, 5-diphenyl-2H-tetrazolium bromide (MTT) assay. On day 14 of plating, cells were submitted to an in vitro angiogenesis assay on BD Matrigel™ Basement Membrane Matrix. During a 2-week culture, one heterogeneous cell population dominated by spindle-shaped cells (eEPCs) and one homogenous cell population exhibiting cobblestone-like morphology (EOCs) appeared sequentially. Quantitative PCR revealed the expression of progenitor and endothelial cell markers such as CD133, VEGFR-2 and CD31 in both populations. However, another progenitor marker, CD34 gene, was undetectable in either freshly isolated PBMNCs or cultured cells. The endothelial phenotype of the EOCs was further identified by acetylated low-density lipoprotein/lectin double staining, and in vitro tube formation. Collectively, this work demonstrates that chicken EPCs can be isolated and cultured from PBMNCs, and suggests that EPCs obtained from peripheral blood may originate mainly from CD34- subpopulation.

**Keywords**: endothelial progenitor cells; peripheral blood; broiler chicken

Conresponding author, E-mail: tanxun@ zju. edu. cn

## Lipopolysaccharide increases toll-like receptor 4 and downstream toll-like receptor signaling molecules expression in bovine endometrial epithelial cells

FU Yunhe, LIU Bo, FENG Xiaosheng, LIU Zhicheng, LIANG Dejie, LI Depeng, ZHANG Naisheng, YANG Zhengtao

*Department of Clinical Veterinary Medicine, College of Veterinary Medicine, Jilin University, Changchun* 130062, *P. R. China*

**Abstract**: The endometrium is easily contaminated with bacteria and the endometrial epithelial cells (EECs) play an important role in defence against invading pathogens which recognized pathogen-associated molecular patterns (PAMPs) *via* pattern recognition receptors (PRRs). Toll-like receptor 4 (TLR4) can recognize lipopolysaccharide (LPS) from Gram-negative bacteria and initiates innate immune responses. In this study, we stimulated bovine EECs with LPS from *Escherichia coli* (*E. coli*). The expression of TLR4 was detected by quantitative real-time polymerase chain reaction (qRT-PCR) and western blot. The expression of downstream TLR4 signaling molecules was detected by qRT-PCR. The results showed that TLR4 and downstream adaptor molecules, transcription factors and cytokines were up-regulated when bovine EECs were stimulated with LPS. Furthermore, the expression of TOLLIP and β-defensin 5 were up-regulated when cells were stimulated with LPS. The results demonstrated that both MyD88 dependent and independent pathways in TLR4 were activated by LPS in bovine EECs. Bovine EECs have the immune repertoires required in defending against *E. coli* and play an important role in innate immune defence of the bovine endometrium.

**Keywords**: TLR; bovine; endometrium; epithelial cell; LPS

Correspond authors. Zhengtao Yang, E-mail: yangzhengtao01@ sina. com,发表杂志: Veterinary Immunology and Immunopathology

## NEFA induce dairy cows hepatocytes apoptosis through mitochondrial-mediated ROS-JNK/ERK signaling pathway

LI Yu, DING Hongyan, LI Xinwei, LI Xiaobing, WANG Zhe*, and LIU Guowen

*Key Laboratory of Zoonosis, Ministry of Education, College of Veterinary Medicine, Jilin University, Changchun* 130062, *China*

**Abstract**: Elevated plasma nonesterified fatty acids (NEFA) and hepatocytes damage are features of ketosis in dairy cows. Oxidative stress is involved in the pathogenesis of liver damage induced by NEFA. However, the exact mechanism between reactive oxygen species (ROS) and the liver apoptosis injury induced by NEFA remain poorly understood. It is well accepted that apoptosis induced cellular damage by oxidative stress. In the present study, we identified the signaling transduction pathway in bovine hepatocytes responsible for apoptosis induced by NEFA. The results demonstrated that NEFA depleted intracellular glutathione (GSH), increased the content of malondialdehyde (MDA), contributed to ROS generation, transcriptional activation of p53, transcriptional inhibition of Nrf2, loss of mitochondrial membrane potential (MMP), release of apoptosis-inducing factor (AIF) and cytochrome (cyt *c*) to cytosol, leading to hepatocytes apoptosis. In addition, NEFA triggered apoptosis in bovine hepatocytes via the regulation of c-Jun N-terminal kinase (JNK) and extracellular signal-regulated protein kinases (ERK1/2), Bcl-2-associated X protein (Bax), B-cell lymphonma gene 2 (Bcl-2), caspase 9 and poly ADP-ribose polymerase (PARP). Pre-treatment with inhibitors SP600125, PD98059, and antioxidants *N*-acetylcystein (NAC) indicated that NEFA-ROS-JNK/ERK-mitochondrial signaling pathway plays a crucial role in hepatocytes apoptosis induced by NEFA. Meanwhile, the results suggested that transcription factors p53 and Nrf2 is the downstream of NEFA-ROS-JNK/ERK and involved in hepatocytes apoptosis induced by NEFA. In conclusion, these findings supported that NEFA-ROS-JNK/ERK-mitochondrial pathway plays an important role in NEFA-induced bovine hepatocytes apoptosis and strongly imply a new clue that the inhibitors of SP600125, PD98059, and NAC maybe developed as medicine for prevention hyperlipidemia-induced apoptosis damage in ketotic dairy cows.

**Keywords**: NEFA, apoptosis, ROS-JNK/ERK-Mitochondrial, transcription factors, dairy cows, hepatocytes

* E-mail: guowenliu2008@ 163. com (G. W. Liu) and wangzhe500518@ sohu. com (Z. Wang)

# Non-esterified fatty acids activate the ROS-p38-p53/Nrf2 signaling pathway to induce bovine hepatocyte apoptosis *in vitro*

SONG Yuxiang, LI Xinwei, LI Na, SHI Xiaoxia, DING Hongyan, ZHANG Yuhang, LI Xiaobing, LIU Guowen*, WANG Zhe

*Key Laboratory of Nutrition Metabolism, Ministry of Education, College of Veterinary Medicine, Jilin University, Changchun 130062, China*

**Abstract**: A high plasma concentration of non-esterified fatty acids (NEFAs) is an important pathogenic factor that leads to ketosis and fatty liver in dairy cows. NEFAs may be associated with oxidative stress in dairy cows with ketosis or fatty liver and the subsequent induction of hepatocyte damage. However, the molecular mechanism of NEFAs-induced oxidative stress and whether NEFAs cause apoptosis of hepatocytes are unclear. Therefore, the aim of this study was to investigate the molecular mechanism of NEFAs-induced oxidative liver damage in bovine hepatocytes. The results showed that NEFAs increased oxidative stress, resulting in p38 phosphorylation. High activated p38 increased the expression, nuclear localization and transcriptional activity of p53 and decreased the nuclear localization and transcriptional activity of Nrf2 in bovine hepatocytes treated with high concentrations of NEFAs. High concentrations of NEFAs also promoted the apoptosis of bovine hepatocytes. Both N-Acety-L-Cysteine (NAC) and glucose (GLU) could attenuate the NEFA-induced apoptotic damage. These results indicate that NEFAs activate the ROS-p38-p53/Nrf2 signaling pathway to induce apoptotic damage in bovine hepatocytes.

**Keywords**: Non-esterified fatty acids; oxidative stress; signal pathway; apoptosis

* E-mail: liuguowen2008@163.com. xbli@jlu.edu.cn.

# Oral immunization of chicken with recombinant *Lactobacillus casei* vaccine against early ALV-J

CAI Dongjie[1], ZHANG Limei[a,1], DONG Hong[b], JING Zhihong[c], LI Zhan[a] & LIU Jianzhu[a*]

*a. College of Animal Science & Veterinary Medicine, Shandong Agricultural University, Taian, P. R. China*

*b. Beijing Key Laboratory of Traditional Chinese Veterinary Medicine, Beijing University of Agriculture, Beijing, P. R. China*

*c. College of Foreign Languages, Shandong Agricultural University, Taian, P. R. China*

**Abstract**: In this study, to exploit a safe and effective vaccine for the prevention against avian leukemia subgroup J, we have developed the investigation of the protective antibodies induced by prokaryotic expression of avian leukemia virus subgroup J gp85 envelope protein and its anti-ALV-J infection role in chicks. At first, the gp85 gene of the subgroup J avian leukosis virus was cloned into the expressing vector pMG36e with Emr labellings, which can shuttle and express in *E. coli* and *Lactobacillus casei*. The recombinant vector pMG36e-gp85 was subsequently transformed into the competent cells *Lactobacillus casei* for expressing proteins we need. Through the technique of western blotting analysis and sequencing, the recombinant protein was determined. Oral immunization of chicks was carried out with the recombinant strain *Lactobacillus casei* harboring pMG36e-gp85 vaccine. Specific anti-ALV IgG in the serum of chicks was detected by indirect ELISA every week after the fist immunization. Seven weeks after immunization with the recombinant *Lactobacillus casei*, the chicks were challenged with standard-type ALV-J (NX0101). The results show that antibody levels can rise significantly after immunization with the recombinant *Lactobacillus casei* (pMG36e-gp85/L. casei) in chicks($P<0.05$). The average viremia of control group was 86%, while the oral group was 40%. Thus, recombinant *Lactobacillus casei* vaccine has some resistance on ALV-J. Our study was the first utilizing of the *Lactobacillus casei* vector for expressing avian leukemia virus envelope protein gp85 gene, providing reliable scientific basis for making recombinant *Lactobacillus casei* oral vaccine of subgroup J avian leukemia, but also providing new methods and ideas of prevention for other enveloped virus in avian disease.

**Keywords**: ALV-J; GP85; *Lactobacillus casei*; pMG36e; oral immunization

* Corresponding author. E-mail: liujz@ sdau. edu. cn

These authors contributed equally to this work.

## Phylogenetic group, virulence factors and antimicrobial resistance of *Escherichia coli* associated with bovine mastitis

LIU Yongxia△, LIU Gang△, LIU Wenjun, LIU Yong, Tariq Ali, CHEN Wei, YIN Jinhua, HAN Bo

*College of Veterinary Medicine, China Agricultural University, Beijing* 100193, *P. R. China*

**Abstract**: *Escherichia coli* is one of the important pathogens involved in the etiology of bovine mastitis. A total of 70 *E. coli* isolates recovered from clinical and subclinical mastitis samples were characterized with respect to their phylogenic group, virulence factors, and antimicrobial susceptibility. Based on the presence of specific genes *chuA*, *yjaA*, and *TspE*4. *C*2, these isolates were found to belong to three different groups: group A (25), group B1 (41) and group D (4). Twenty five (35.7%) isolates harbored at least one virulence gene, and the most prevalent virulence genes were *f17A*, *irp*2, *astA*, *iucD*, and *colV*. The *irp*2-coding gene was more often detected in group A compared to group B1 isolates, in contrast *colV* was identified more often in group B1 isolates. The majority of isolates (87.1%) were resistant to at least one antimicrobial compound. Forty-seven isolates (67.1%) were resistant to streptomycin, and those from. group B1 were more resistant to streptomycin than isolates from group A. This latter feature was supported by the distribution of streptomycin resistance genes observed in group B1 compared to group A.

**Keywords**: *Escherichia coli*; phylogenetic group; virulence factors; antimicrobial resistance; bovine mastitis

Corresponding author: Bo Han, E-mail: hanbo@ cau. edu. cn

△These authors contributed equally to this work.

本文发表于 Res Microbiol. 2014 May; 165(4): 273-277.

## Preparation and characterization of eudragit L100 microcapsules containing zinc gluconate

LUO Lijuan[a], HU Yanchun[a*], FU Hualin[a], LUO Biao[a], QIAO Yan[a], LIU Xi[b], LAN Lan[b], DENG Shijin[b], ZUO Zhicai[a], DENG Junliang[a], JIANG Zhongrong[b*]

[a] *College of Veterinary Medicine, Sichuan Agricultural University, Ya'an* 625014 *China*;

[b] *Livestock Research Institute of Ganzi Prefecture, Sichuan Province, Kangding* 626000, *China*

**Abstract**: Eudragit L100 microcapsule was prepared to encapsulate zinc gluconate trihydrate (zinc gluconate) using emulsion solvent evaporation method and the optimum condition was investigated by response surface methodology (RSM) analysis in order to improve the bioavailability of oral zinc preparations. The drug-loading rate (DL) of the microcapsules was detected by flame atomic absorption spectrophotometer, and the drug release property in vitro and some other characteristics were studied. The results showed that the optimum preparation conditions for Eudragit L100 microcapsules based on encapsulation efficiency (EE) were: 0.64 g zinc gluconate, 125 ml liquid paraffin, 1.26 g talcum powder and 2.57 g span-80. Prepared microcapsules had round and smooth surface, uniform diameter distribution of 250 μm, DL of 17.97% and EE of 93%, and the release rate of zinc gluconate from microcapsules in vitro conformed to the standard of enteric preparation. In conclusion, the preparation method is reliable and simple, and comprehensive evaluation of the microcapsules is good, which provide a reliable basis for clinical applications research.

**Keywords**: eudragit L100, microcapsules, zinc gluconate, emulsion solvent evaporation method, RSM optimization, flame atomic absorption spectrophotometer

## Preparation and identification of monoclonal antibodies against bovine haptoglobin

WANG Caihong, GU Cheng, GAO Jing, LI Chunqiu, WANG Jianfa, WU Rui, SUN Dongbo*

*College of Animal Science and Veterinary Medicine, Heilongjiang Bayi Agricultural University, Daqing* 163319, *P. R. China*

**Abstract**: Haptoglobins are members of the alpha 2-globulin family of proteins, mainly expressing in the liver, skin, lung, kidney and adipose tissue. Haptoglobin functions to bind free plasma hemoglobin, preventing oxidative damage. Additionally, haptoglobin is an acute-phase protein, whose expression level increases in the inflammatory process. The accumulating reports indicated that bovine haptoglobin (BoHp) was a potential biomarker in many inflammatory diseases of dairy cows caused by infectious microorganisms, involving in footrot, mastitis, enteritis, peritonitis,

* Corresponding author. Yanchun Hu, E-mail address: yanchunhu@126.com

Zhongrong. Jiang, E-mail address: 34986829@qq.com

endocarditis, abscesses, endometritis, interdigital dermatitis, and so on. Thus, BoHp has the potential use as an early diagnostic marker of inflammatory diseases in dairy cattle. In order to provide some basis for development of rapid diagnostic reagents of BoHp, female 8-week-old BALB/c mice were immunized with purified recombinant proteins of the predicted immunodominant region of bovine haptoglobin (pirBoHp). A total of two monoclonal antibodies (McAbs), named 1B3 and 6D6, were prepared by conventional B lymphocyte hybridoma technique. Tites of ascites fluid and cell-culture supernatant of the McAb 1B3 were $1:9.6\times10^{8}$ and $1:8.2\times10^{4}$, respectively, and that of the the McAb 6D6 were $1:4.4\times10^{5}$ and $1:1.0\times10^{4}$, respectively. The subtype of the McAbs 1B3 and 6D6 was IgG1κ. In western blot, the McAbs 1B3 and 6D6 could recognize α-chain of the native BoHp from plasma of dairy cows. These data indicated that the McAbs 1B3 and 6D6 will have a potential use for developing diagnostic reagents of Bo Hp.

**Keywords**: haptoglobin; hybridoma, monoclonal antibody

* Corresponding author, E-mail: dongbosun@ 126. com

# Preparation and regeneration of protoplasts of the swainsonine-producting endophytic fungi, *Undifilum oxytropis* from locoweed

ZHANG Leilei[1], HE Shenghu[1], YU Yongtao[1*], ZHAO Qingmei[2,3], GE Song[1]

1. *College of Agronomy Ningxia University, Yinchuan* 750021, *China*;
2. *College of Biological Science and Engineering, Beifang University of Nationalities, Yinchuan* 750021, *China*;
3. *Key Laboratory of Fermentation Brewing Engineering and Biotechnology State Nationlities Affairs Commission, Yinchuan* 750021, *China*

**Abstract**: Locoweeds are toxic plants of the genera *Astragalus* and *Oxytropis* containing swainsonine. Swainsonine is the principal agent responsible for locoism in animals. Recently, the fungal endophytes are isolated frequently from locoweeds, *Undifilum* spp, are believed that responsible for the production of swanisonine in locoweeds. Currently, the mechanism was few known about the biosynthesis of swainsonine in Locoweed's endophytic fungi. The swainsonine-pruducing endophytic fungi, *Undifilum oxytropis*, was used to develop protoplasts in this research. This study will provide a important basis for further to elucidate the swainsonine biosynthetic pathways and build a non-toxic fungus. The effects of some factors on formation and regeneration of protoplasts were investigated, including culture method, mycelium age, digesting time and temperature. The results showed that the highest preparation rate was achieved when prepared through mixture solution of 1 % cellulase + 1 % snailase + 0.1 % lysing enzyme, at 35 ℃, 80 r/min for 3 h. The regeneration colonies were achieved when cultured in YCM medium for 9 d at 22 ℃ and the highest regeneration rate (8.5 %) was achieved when cultured in YCM medium for 15 d at 22 ℃. The color of regeneration colonies was different from wild strain and the form of regeneration mycelium was the same as wild strain when cultured in YCM medium. The regeneration mycelium could still produce swainsonine. The preparation and regeneration of protoplasts of the swainsonine-producting endophytic fungi, *Undifilum oxytropis*, which would providing important tool for the research of genetic transformation and genetic manipulation and science basis for further to elucidate the synthesis of swainsonine in molecular level.

**Keywords**: swainsonine, *Undifilum oxytropis*, protoplasts, preparation, regeneration

**基金项目**:国家自然科学基金(31201962);宁夏自然科学基金(NZ1118)。

**作者简介**:张蕾蕾(1989 - ),女,陕西人,硕士在读,研究方向:兽医临床诊断技术。E-mail:andybeilei@ 163. com

**通讯作者**:余永涛(1980 - ),男,宁夏平罗人,副教授,博士,硕士研究生导师,研究方向:动物中毒病与营养代谢病。E-mail:yyt1211@ 163. com。

# Proteomic analysis on protein expression profiling induced by fluoride and sodium sulfite in mice testis

ZHANG Jianhai, LIANG Chen, QIE Mingli, LI Zhihui, LUO Guangying, SUN Zilong, WANG Jundong *

*Shanxi Key Laboratory of Ecological Animal Science and Environmental Medicine, Shanxi Agricultural University, Taigu, 030801*

**Abstract**: Fluoride and sodium sulfite are two well known environmental contaminants, which pose a serious threat to male reproduction. The present study aims to determine the effects of exposure to fluoride and sodium sulfite on the mouse testis proteome as a first step toward the investigation of the potential damage mechanism in terms of MAPK signaling pathway. 72 sexually male Kunming mice were randomly assigned to four groups (n = 18): Control group (given distilled water), NaF group (150mg NaF/L), $Na_2SO_3$ group (500mg $Na_2SO_3$/L) and Combination group (150mg NaF/L and 500mg $Na_2SO_3$/L). After 98 consecutive days, a two-dimensional gel electrophoresis approach was employed to investigate the changes in protein expression in the testes. Proteomic analysis was conducted using PDQuest 8.0 software and MALDI-TOF-TOF, which led to the identification of 19 differentially abundant proteins including 2 down-regulated proteins: A. 14-3-3, plays a role in MAPK signaling pathway by recognition of a phosphoserine or phosphothreonine motif. B. peroxiredoxin-6, involved in redox regulation of the cell, may play a role in the regulation of phospholipid turnover as well as in protection against oxidative injury. And 17 up-regulated proteins included in inositol monophosphatase 1, ubiquitin C-terminal hydrolase L3, Chain n, sperm surface protein Sp17, cytochrome b-5, uncharacterized protein LOC691496, Prdx2 protein, Rho GDP dissociation inhibitor (GDI) alpha, ubiquitin carboxyl-terminal hydrolase isozyme L1, alpha-globin, Chain M, adenylate kinase isoenzyme 1 isoform 2, L-lactate dehydrogenase B chain, 14 ku phosphohistidine phosphatase, cytochrome c oxidase subunit 5A, and cytochrome b-c1 complex subunit 1. Among these proteins, 14-3-3 exerted critical roles in the regulation of cellular phenomena in MAPK signaling pathway through phosphor-dependent binding to a lot of intracellular proteins that were targeted by various classes of protein kinases. Hence, the change of 14-3-3 expression possibly has an influence on the male reproduction, which must be vigorously further investigated.

**Keywords**: proteomic analysis, fluoride and sodium sulfite, testis

* Corresponding author, E-mail: wangjd@ sxau. edu. cn

## Prevalence and risk factors of *Giardia doudenalis* in dogs from China

YANG Dubao[a,1], ZHANG Qingfeng[a,1], ZHANG Limei[a,1], DONG Hong[b], JING Zhihong[c], LI Zhan[a] & LIU Jianzhu[a*]

*a. College of Animal Science & Veterinary Medicine, Shandong Agricultural University, Taian, P. R. China*

*b. Beijing Key Laboratory of Traditional Chinese Veterinary Medicine, Beijing University of Agriculture, Beijing, P. R. China*

*c. College of Foreign Languages, Shandong Agricultural University, Taian, P. R. China*

**Abstract**: The aim of this study was to carry out a survey for the presence of *Giardia duodenalis* infection in canine using ELISA and PCR and to identify risk factors for infection. Samples from 318 dogs' faeces living in nine cities in China were used in present study. Each sample was tested for the presence of *G. duodenalis* specific antigens using ELISA and 197 out of 318 samples were further examined for the presence of *G. duodenalis* using PCR. The overall rate of canines infected with giardiasis in present study was 16. 04% and 15. 22 % using ELISA and PCR, respectively. No significant difference was found between sex and *Giardia* positivity. Young dogs (up to 1 year) and living in communities were identified as risk factors for infection by multivariate logistic regression analysis. In conclusion, giardiasis in dogs was present in nine cities in China; as risk factors, young dogs (up to 1 year) and living in communities were of great significance. *Giardia*-infected canine should be treated for hygienic management to prevent transmission in dog-to-human.

**Keywords**: dog, *Giardia*, ELISA, PC*R*, risk factors, *China*

* Corresponding author. E-mail: liujz@ sdau. edu. cn

These authors contributed equally to this work.

## Regulatory effect of NOD1 receptor on neutrophil function in dairy cows

WEI liangjun, TAN Xun*, LIU Xi

*Department of Veterinary Medicine, College of Animal Sciences, Zhejiang University, Hangzhou* 310058, *China*

**Abstract**: Polymorphonuclear neutrophils (PMN) represent the most abundant circulating immune cells and act as the first leukocytes which rapidly migrate to the site of infection for phagocytosis and killing of invading microorganisms. Dysfunction of neutrophils has been linked to diverse inflammatory diseases in humans and in mastitis in dairy cows. Recently, both NOD1 and NOD2 have been identified in bovine neutrophils by our group, and we found a significant down-regulation of NOD1, but not NOD2 protein, in the neutrophils of periparturient dairy cows that are known to be susceptible to increased incidence and severity of infectious diseases such as mastitis and metritis, However, it remains to be elucidated whether NOD1 is required for effective neutrophil immune responses to bacterial infection. The present study was conducted to investigate the effect of intracellular receptor NOD1 on the functionality of PMN in dairy cows. PMN was isolated from peripheral blood of clinically healthy heifers. The cells were pre-incubated with NOD1 selective inhibitor ML130 for 2 h, and then exposed to 100 ng/ml lipopolysaccharide (LPS) for further 4 h. Expression of NOD1 protein was determined by Western blot and the mRNA of proinflammatory cytokines (TNF-α, IL-1β and IL-8), chemotaxis cytokine CXCL2 and adherence molecular (CD11b/CD18 and CD62L) evaluated by qPCR. Annexin V/PI staining was employed to detect cell apoptosis. The capacity of phagocytosis and oxidative burst was evaluated by flow cytometry and cell migration evaluated using Transwell chamber. Upon LPS stimulation, inhibition of NOD1 by ML130 significantly inhibited the expression of NOD1 protein and down-regulated the mRNA of TNF-α, IL-1β, IL-8, CXCL2 and CD62L. However, the CD11b/CD18 mRNA expression was not affected by NOD1 inhibition. Inhibition of NOD1 also diminished the phagocytosis and oxidative burst as well as migratory capacity of PMN and increased the number of apoptotic cells in response to LPS stimulation. These data suggest that the immune function of PMN in dairy cows is NOD1-dependent.

**Keywords**: Dairy cow; Neutrophil; NOD1 receptor

Corresponding author, E-mail: tanxun@ zju. edu. cn

## Reproductive and developmental toxicities caused by swainsonine from locoweed

WU Chenchen, LIU Xiaoxue, MA Feng, GENG Pengshuai, CAO Dandan, YAN Dujian, ZHAO Baoyu

*College of Animal Veterinary Medicine, Northwest A & F University, Yangling* 712100, *P. R. China*

**Abstract**: Swainsonine relevance: Swainsonine is the primary toxin in locoweeds and causes intention tremors, generalized depression, nervousness, proprioceptive deficits, aberrant behavior, reproductive dysfunction, emaciation and death.

Aim of the Study: The objective of the present study was to evaluate the potential reproductive and developmental toxicities caused by swainsonine in mice.

Materials and Methods: The F0 and F1 mice were administered swainsonine by intraperitoneal injection of 1/10LD50, 1/20LD50 and 1/30LD50 value of swainsonine. These mice were subsequently evaluated for reproductive and developmental toxicity.

Results Experimoent 1: The LD50 value was determined as 5. 25 mg/kg body weight for swainsonine. Experiment 2: The T-IV group mice exhibited significantly fewer estrous cycles and an increased number of estrous of mice compared to the T-I, T-II and T-III groups ($P < 0.05$). Experiment 3: For the F0, F1 and F2 generation mice, the mice in groups I, II and III had significantly higher spleen, liver and kidney indices and had significantly lower body weight compared to group IV ($P < 0.05$). For the F0 and F1 generations, the number of live pups on PND 0, 4 and 15 and the copulation indices of the group I, II and III mice were significantly decreased compared to the group IV mice ($P < 0.05$). The fertility and gestation indices of the F0-I, F0-II and F0-III group mice were significantly increased compared to the F1-I, F1-II and F1-III group mice ($P < 0.05$).

Conclusions: The reproductive/developmental toxicity effects of swainsonine were prolonged estrous cycle, decreased number of live pups on PND 15, decreased copulation index and fertility index and the accumulation of swainsonine in the body. Additionally, the weight of spleen was significantly increased and the body weight was significantly decreased for the treatment group. Swainsonine may cause reproductive and development toxicities in both the parent and offspring mice.

**Keywords**: swainsonine; locoweed; mouse; reproductive and development

Corresponding author, E-mail: zhaobaoyu12005@ 163. com (Zhao Baoyu), wucen95888@ 163. com (Wu Chenchen)
Wu Chenchen and Liu xiaoxue contributed equally to this work.

# Role of cholecystokinin in anorexia induction following oral exposure to the 8-Ketotrichothecenes deoxynivalenol, 15-Acetyldeoxynivalenol, 3-Acetyldeoxynivalenol, fusarenon X, and nivalenol

Wenda WU, Haibin ZHANG

*College of Veterinary Medicine, Nanjing Agricultural University, Nanjing* 210095, *P. R. China*

**Abstract**: Cereal grain contamination by trichothecene mycotoxins is known to negatively impact human and animal health with adverse effects on food intake and growth being of particular concern. The head blight fungus *Fusarium graminearum* elaborates five closely related 8-ketotrichothecene congeners: (1) deoxynivalenol (DON), (2) 3-acetyldeoxynivalenol (3-ADON), (3) 15-acetyldeoxynivalenol (15-ADON), (4) fusarenon X (FX), and (5) nivalenol (NIV). While anorexia induction in mice exposed intraperitoneally to DON has been linked to plasma elevation of the satiety hormones cholecystokinin (CCK) and peptide $YY_{3-36}$ ($PYY_{3-36}$), the effects of oral gavage of DON or of other 8-keotrichothecenes on release of these gut peptides have not been established. The purpose of this study was to (1) compare the anorectic responses to the aforementioned 8-ketotrichothecenes following oral gavage at a common dose (2. 5mg/kg bw) and (2) relate these effects to changes plasma CCK and $PYY_{3-36}$ concentrations. Elevation of plasma CCK markedly corresponded to anorexia induction by DON and all other 8-ketotrichothecenes tested. Furthermore, the CCK1 receptor antagonist SR 27897 and the CCK2 receptor antagonist L-365, 260 dose-dependently attenuated both CCK- and DON-induced anorexia, which was consistent with this gut satiety hormone being an important mediator of 8-ketotrichothecene-induced food refusal. In contrast to CCK, $PYY_{3-36}$ was moderately elevated by oral gavage with DON and NIV but not by 3-ADON, 15-ADON, or FX. Taken together, the results suggest that CCK plays a major role in anorexia induction following oral exposure to 8-ketotrichothecenes, whereas $PYY_{3-36}$ might play a lesser, congener-dependent role in this response.

**Keywords**: mycotoxin; trichothecene; anorexia; deoxynivalenol; 3-acetyldeoxynivalenol; 15-acetyldeoxynivalenol; fusarenon X; nivalenol

To whom correspondence should be contacted with E-mail: haibinzh@ njau. edu. cn

# Selenium nlocks PCV2 replication promotion induced by oxidative stress by improving GPx1 expression

CHEN Xingxiang[a], REN Fei[a], HESKETH John[b], SHI Xiuli[a], LI Junxian[a], GAN Fang[a], HUANG Kehe[a]*

[a]*Institute of Nutritional and Metabolic Disorders in Domestic Animals and Fowls, Nanjing Agricultural University, Nanjing* 210095, *China.* [b]*Institute for Cell and Molecular Biosciences, University of Newcastle, The Medical School, Framlington Place, Newcastle Upon Tyne NE*2 4*HH*, *UK*

**Abstract**: Porcine circovirus type 2 (PCV2) is recognized as a key infectious agent in post-weaning multisystemic wasting syndrome (PMWS), but not all pigs infected with PCV2 will develop PMWS. The aim of this work was to explore the relationships between PCV2 infection, oxidative stress and selenium in a PK-15 cell culture model of PCV2 infection. The results showed that oxidative stress induced by H2O2 treatment increased PCV2 replication as measured by PCV2 DNA copies and the number of infected cells. Furthermore, PCV2 replication was inhibited by selenomethionine (SeMet) at a high concentration (6μM) and the increase in PCV2 replication by oxidative stress was blocked by SeMet at physiological concentrations (2 or 4μM). PCV2 infection caused a decrease of glutathione peroxidase 1 (GPx1) activity but an increase of GPx1 mRNA levels, suggesting that GPx1 may represent an important defense mechanism during PCV2 infection. SeMet did not significantly block the promotion of PCV2 replication in GPx1-knockdown cells. This observation correlates with the observed influence of SeMet on GPx1 mRNA and activity in GPx1-knockdown cells, indicating that GPx1 plays a key role in blocking the promotion of PCV2 replication. We conclude that differences in morbidity and severity of PMWS caused on different pig farms may be related to variations in oxidative stress and that selenium has a potential role in control of PCV2 infection.

**Keywords**: selenium; selenoproteins; porcine circovirus type 2; GPx1 knockdown, oxidative stress

* Corresponding author: HUANG Kehe, E-mail: khhuang@ njau. edu. cn

# Selenoprotein W may influence the mRNA level of dome selenoprotein by reactive oxygen species in chicken myoblasts

YAO Haidong, ZHAO Wenchao, ZHANG Ziwei, LI Shu, XU Shiwen*

*Department of Veterinary Medicine, Northeast Agricultural University, Harbin* 150030, *P. R. China*

**Abstract**: Selenoprotein W (*Selw*) plays crucial roles in skeletal muscle. In the present study, we examined the mRNA expression levels of selenoproteins following the silencing or overexpression of *Selw* in chicken myoblast. The results showed that silencing of *Selw* increased *Gpx*3, *Gpx*4, *Txnrd*1, *Selt*, *Selh* and *Sepp*1 ($P<0.05$), but reduced *Sels* and *Sep*15 ($P<0.05$). And Reactive oxygen species (ROS) scavenger, N-acetyl-L-cysteine (NAC), influence the effect of Selw on the expression of some selenoproteins (*Gpx*3, *Gpx* 4, *Txnrd*1, *Selt*, *Selh*, $P>0.05$). However, the overexpression of *Selw* increased the expression of *Sepn*1, *Selt*, *Selh*, *Selm*, *Selpb* and *Sepx*1 ($P<0.05$), but *decreased Gpx*1 ( $P<0.05$ ). In addition, $H_2O_2$ increased the expressions of selenoproteins ( $P<0.05$ ), except the decreased Gpx1 and *Selw* ( $P<0.05$ ). And this response was influenced by the overexpression of *Selw*. The results showed that as one highly expressed selenoprotein in myoblast, variation of *Selw* influence the expression of other selenoproteins. It indicated that under oxidative condition, myoblast consumed *Gpx*1 and *Selw* and increased other selenoproteins to improve the oxidative imbalance. And *Selw* may regulate some selenoproteins ( *Gpx*3, *Gpx* 4, *Txnrd*1, *Selt*, *Selh*) by the pathway of ROS. It indicated that Selw reserves irreplaceable roles in myoblast.

**Keywords**: selenoprotein W; myoblasts; chicken; reactive oxygen species

* E-mail: shiwenxu@ neau. edu. cn

# Shotgun proteomic analysis of plasma from dairy cattle suffering from footrot: characterization of potential disease-associated factors

ZHANG Hong, WANG Caihong, GU Cheng, GAO Jing, WANG Xinyu, WANG Jianfa, WU Rui, SUN Dongbo *

*College of Animal Science and Veterinary Medicine, Heilongjiang Bayi Agricultural University, Daqing* 163319, *P. R. China*

**Abstract**: Footrot is an acute and highly infectious disease of cattle that develops between the claws of the hoof and is caused by the Gram-negative anaerobic bacterium *Fusobacterium necrophorum*, which is present in the rumen and feces of normal cattle and their environment. Plasma is an amorphous and important component of blood and changes in the quantity and quality of plasma proteins are associated with physiological or pathological states in humans and other animals. However, the plasma protein profiles of cattle with footrot are not fully understood, and there are still a great many unknown potential disease-associated proteins. In this study, the plasma proteome of healthy dairy cattle and those with footrot was investigated using a shotgun LC-MS/MS approach. In total, 648 proteins were identified in healthy plasma samples, of which 234 were non-redundant proteins and 123 were high-confidence proteins; 712 proteins were identified from footrot plasma samples, of which 272 were non-redundant proteins and 138 were high-confidence proteins. The high-confidence proteins showed significant differences between healthy and footrot plasma samples in molecular weight, isoelectric points and the Gene Ontology categories. 22 proteins were found that may differentiate between the two sets of plasma proteins, of which 16 potential differential expression (PDE) proteins from footrot plasma involved in immunoglobulins, innate immune recognition molecules, acute phase proteins, regulatory proteins, and cell adhesion and cytoskeletal proteins 6 PDE proteins from healthy plasma involved in regulatory proteins, cytoskeletal proteins and coagulation factors. Of these PDE proteins, haptoglobin, SERPINA10 protein, afamin precursor, haptoglobin precursor, apolipoprotein D, predicted peptidoglycan recognition protein L (PGRP-L) and keratan sulfate proteoglycan (KS-PG) were suggested to be potential footrot-associated factors. The PDE proteins haptoglobin, PGRP-L and KS-PG were highlighted as potential biomarkers of footrot in cattle. The haptoglobin has been shown to be a useful biomarker for monitoring the occurrence and severity of inflammatory responses in cattle with mastitis, pneumonia, enteritis, peritonitis, endocarditis, abscesses, endometritis and hoof disease . Here, haptoglobin was verified as plasma inflammatory biomarkers of footrot in dairy cattle. The predicted PGRP-L may be necessary for recognition of the innate immune activators of the Gram-negative anaerobic bacterium *F. necrophorum*. The keratan sulfate in blood has been shown to be a marker of cartilage catabolism. The KS-PG, which was found in the plasma from footrot-affected dairy cattle, may reflect catabolism of hoof cartilage, and it has been suggested to be a potential marker for evaluation of foot damage in dairy cattle. The resulting protein lists and potential differentially expressed proteins may provide valuable information to increase understanding of plasma protein profiles in cattle and to assist studies of footrot-associated factors.

**Keywords**: dairy cattle, footrot, plasma proteomics, disease-associated factors

* Corresponding author, E-mail: dongbosun@126.com

# Shotgun proteomic analysis of serum from dairy cattle affected by hoof deformation

GU Cheng, WANG Caihong, GAO Jing, LI Chunqiu, WANG Xinyu, WANG Jianfa, WU Rui, SUN Dongbo *

*College of Animal Science and Veterinary Medicine, Heilongjiang Bayi Agricultural University, Daqing* 163319, *P. R. China*

**Abstract**: Hoof deformation (HD) is caused by a variety of pathogenic factors. It is characterized by abnormal growth of the horny sheath and changes in the shape of the hooves, which are obviously different from those of normal dairy cows. With the rapid increases in intensive rearing and demand for milk production of dairy cows, the incidence of HD has been increasing, and is reportedly as high as 29.5% in the Beijing and Tianjin areas of China. The presence of HD in dairy cattle can have a deleterious effect on the weight-bearing surface and internal structure of the hooves, resulting in the occurrence of a variety of hoof diseases. Although some studies have been carried out to investigate HD-associated factors, global changes in the physiological indicators of HD-affected dairy cows remain unclear. In this study, The serum proteome of two pooled samples from unaffected dairy cattle (UA) ($n = 26$) and those with hoof deformation (HD) ($n = 40$) was investigated using a shotgun liquid chromatography and tandem mass spectrometry (LC-MS/MS) approach. In total, 393 proteins were identified in the UA serum sample, of which 224 were high-confidence proteins; 409 proteins were identified from the HD serum sample, of which 225 were high-confidence proteins. The high-confidence proteins showed clear differences between the UA and HD samples in functional analysis based on Gene Ontology annotations. Twenty-four proteins were found that showed potential differential expression (PDE) between the two sets of high-confidence serum proteins. Among the PDE proteins, 18 from the HD sample were involved in binding, catalytic activity, structural molecule activity, protein binding transcription factor activity, and enzyme regulatory activity; of these, desmoplakin, junction plakoglobin, and annexin A2 were highlighted as potential biomarkers of hoof deformation in dairy cattle. Furthermore, selected proteins were validated by enzyme linked immunosorbent assay or western blot . The proteins identified provide valuable information to increase understanding of serum protein profiles in cattle and to assist studies of the factors associated with hoof deformation.

**Keywords**: dairy cows; hoof deformation; serum proteomics

* Corresponding author, E-mail: dongbosun@126.com

## *Staphylococcus aureus* and *Escherichia coli* elicit different innate immune responses from bovine mammary epithelial cells

FU Yunhe, ZHOU Ershun, LIU Zhicheng, LI Fengyang, LIU Bo, LIANG Dejie, WANG Wei, WANG Tiancheng, CAO Yongguo, ZHANG Naisheng, YANG Zhengtao

*Department of Clinical Veterinary Medicine, College of Veterinary Medicine, Jilin University, Changchun* 130062, *P. R. China*

**Abstract**: *Escherichia coli* and *Staphylococcus aureus* are the most important pathogenic bacteria causing bovine clinical mastitis and subclinical mastitis, respectively. However, little is known about the molecular mechanisms underlying the different host response patterns caused by these bacteria. The aim of this study was to characterize the different innate immune responses of bovine mammary epithelium cells (MECs) to heat-inactivated *E. coli* and *S. aureus*. Gene expression of Toll-like receptor 2 (TLR2) and TLR4 was compared. The activation of nuclear factor kappa B (NF-κB) and the kinetics and levels of cytokine production were analyzed. The results show that the mRNA for TLR2 and TLR4 was up-regulated when the bovine MECs were stimulated with heat-inactivated *E. coli*, while only TLR2 mRNA was up-regulated when the bovine MECs were stimulated with heat-inactivated *S. aureus*. The expression of tumor necrosis factor-α (TNF-α), interleukin (IL)-1β, IL-6 and IL-8 increased more rapidly and higher when the bovine MECs were stimulated with heat-inactivated *E. coli* than when they were stimulated with heat-inactivated *S. aureus*. *E. coli* strongly activated NF-κB in the bovine MECs, while *S. aureus* failed to activate NF-κB. Heat-inactivated *S. aureus* could induce NF-κB activation when bovine MECs cultured in medium without fetal calf serum. These results were confirmed using TLR2- and TLR4/MD2-transfected HEK293 cells and suggested that differential TLR recognition and the lack of NF-κB activation account for the impaired immune response elicited by heat-inactivated *S. aureus*.

**Keywords**: *Escherichia coli*; *Staphylococcus aureus*; Bovine mammary epithelial cells; TLR2; TLR4; NF-κB

Correspond authors. Zhengtao Yang, E-mail: yangzhengtao01@sina.com

发表杂志：Veterinary Immunology and Immunopathology

## T-2 toxin regulates steroid hormone secretion of rat ovarian granulosa cells through cAMP-PKA pathway

WU Jing*, YUAN Liyun, YI Jin'e, TIAN Ya'nan

*College of Veterinary Medicine, Hunan Agricultural University, Changsha* 410128, *P. R. China*

**Abstract**: T-2 toxin is a secondary metabolite produced by *Fusarium* genus and has been found to present as common contaminant in food and feedstuffs of cereal origin. Very little information regarding the ovarian effects of T-2 toxin is available. In porcine model T-2 toxin was found to inhibit the steroidogenesis. However, the mechanism has not been well understood. In this study we used rodent model to investigate the effects of T-2 toxin on cAMP-

PKA pathway and explored potential mechanism for T-2 toxin induced reproductive toxicity in rats. We first analyzed the the effects of T-2 toxin on progesterone and estrogen production in ratgranulosa cells (GC). For this purpose the GC were cultured for 48 h in 10% fetal bovine serum-containing medium followed by 24 h in serum-free medium containing FSH (10ng/ml), and rostenedione (3ng/ml) and T-2 toxin (at various doses/combinations). T-2 toxin dose-dependently inhibited the growth of cells and the steroid hormone production. cAMP-PKA pathway is critical of steroidogenesis and we found that cAMP levels was dose-dependently inhibited by T-2 toxin (1-100nmol/L). Furthermore, we found agonists of cAMP-PKA pathway 8-Br-cAMP and 22R-HC induced progesterone production in GC was abolished by T-2 toxin treatment at low dose (1 nmol/L) suggesting the granulosa cell is highly sensitive to the T-2 toxin. cAMP-stimulated steroidogenic acute regulatory protein (StAR) is rate limiting de novo protein in progesterone synthesis. Exposure to T-2 toxin caused significant suppression of StAR expression as determined by Western and semiquanttive RT-PCR suggesting StAR is a sensitive target for T-2 toxin. Taken together, our results strongly suggest T-2 toxin inhibits steroidogenesis by targeting at StAR protein through suppression of cAMP-PKA pathway. The antisteroidogenesis effects were observable at low T-2 dose (1 ng/ml) suggesting T-2 toxin has an endocrine disruptive effect.

**Keywords**: T-2 Toxin, granulosa cells, steroid production, cAMP-PKA pathway

* Corresponding author, E-mail: wu23jing@ yahoo. com. cn

# The effect of selenium deficiency on the DNA methylation in the tissues of chicks

ZHANG Ziwei, YAO Haidong, LI Shu, Xu Shiwen *

*Department of Veterinary Medicine, Northeast Agricultural University, Harbin* 150030, *P. R. China*

**Abstract**: Selenium (Se), an essential trace element, was indicated to play important role in the regulation of DNA methylation. Se deficiency induced hypomethylation in tissue of mice. However, the effect of Se on DNA methylation in poultry was less reported. In the present study, two groups of day-old layer chicks (n = 60/group) were fed a corn-soy basal diet (33 mg Se/kg; produced in the Se-deficient area of Heilongjiang, China) or the diet supplemented with Se (as sodium selenite) at 0. 15 mg/kg for 55 d. Then the DNA methylation level, DNA methyltransferases (DNMTs), Demethylase (MBD2) were examined. The results indicated that DNA methylation level were lower in muscle tissues, brain tissues, immune tissues and livers of chicks in Se deficiency groups than that of control group ($P < 0.05$). Se deficiency decreased the mRNA levels of DNMT1、DNMT3A、DNMT3B, but increased MBD2 in muscle tissues, brain tissues, immune tissues and livers of chicks ($P < 0.05$). It indicated that Se deficiency influenced the level of DNA methylation in chicks.

**Keywords**: chicks, selenium deficiency, DNA methylation, DNMTs, demethylase MBD2

* Correspond authors. E-mail address: shiwenxu@ neau. edu. cn

# Thymol inhibits *Staphylococcus aureus* internalization into bovine mammary epithelial cells by inhibiting NF-κB activation

WEI Zhengkai, ZHOU Ershun, GUO Changming, FU Yunhe, YU Yuqiang, LI Yimeng, YAO Minjun, ZHANG Naisheng, YANG Zhengtao*
*College of Veterinary Medicine, Jilin University, Changchun* 130062, *P. R. China*

**Abstract**: Bovine mastitis is one of the most costly and prevalent diseases in the dairy industry and is characterised by inflammatory and infectious processes. *Staphylococcus aureus* (*S. aureus*), a Gram-positive organism, is a frequent cause of subclinical, chronic mastitis. Thymol, a monocyclic monoterpene compound isolated from *Thymus vulgaris*, has been reported to have antibacterial properties. However, the effect of thymol on *S. aureus* internalization into bovine mammary epithelial cells (bMEC) has not been investigated. In this study, the growth of *S. aureus* supplemented with thymol (16-64 μg/ml) was monitored turbidimetrically at 600 nm for 24 h and cell viability was measured by the Cell Counting Kit-8 (CCK-8). Gentamicin protection assay was carried out to determine the effect of thymol on *S. aureus* internalization into bMEC. To determine the potential mechanism, tracheal antimicrobial peptide (TAP) and β-defensin (BNBD5) expressions were detected by quantitative real-time polymerase chain reaction (qRT-PCR). The activation of nuclear factor-kappa B (NF-κB) was determined by Western blotting. Our results showed that thymol (16-64 μg/ml) could reduce the internalization of *S. aureus* into bMEC and down-regulate the mRNA expression of TAP and BNBD5 in bMEC infected with *S. aureus*. In addition, thymol was found to inhibit *S. aureus*-induced nitric oxide (NO) production in bMEC and suppress *S. aureus*-induced NF-κB activation in a dose-dependent manner. In conclusion, these results indicated that thymol inhibits *S. aureus* internalization into bMEC by inhibiting NF-κB activation.

**Keywords**: Staphylococcus aureus; thymol; antimicrobial peptides; bovine mammary epithelial cell; bovine mastitis; NF-κB

* Corresponding authors. E-mail address: yangzhengtao01@sina.com

# 1α,25-(OH)$_2$D$_3$调控破骨细胞分化过程中MMP-9蛋白的表达

顾建红,仝锡帅,王东,陈　阳,卞建春,刘学忠,袁　燕,刘宗平
(扬州大学兽医学院,扬州 225009)

**摘　要**:破骨细胞(osteoclast,OC)是体内唯一负责骨吸收的细胞,在骨重建过程中发挥着至关重要的作用。研究表明,在OC降解细胞外基质过程中,基质金属蛋白酶(matrix metalloproteinase,MMP)发挥了重要作用。然而,维生素D作为骨代谢的重要调节因子,是否能够直接调控OC形成和活化过程中MMP-9蛋白的表达仍不十分清楚。为了探讨1α,25-(OH)$_2$D$_3$调控OC形成及分化过程中MMP-9蛋白表达的变化,本研究采用30 μg/L RANKL+25 μg/L M-CSF联合诱导RAW264.7分化为OC。培养过程中添加不同浓度1α,25-(OH)$_2$D$_3$($10^{-10}$、$10^{-9}$、$10^{-8}$mol/L),通过MTT法测定细胞增殖情况,TRAP染色及骨吸收陷窝鉴定OC的形成,Western Blot检测MMP-9蛋白表达水平。结果显示,培养48 h添加RANKL、M-CSF各组细胞增殖率均极显著高于对照组(A组)($P<0.01$);在此基础上添加1α,25-(OH)$_2$D$_3$剂量依赖性抑制细胞增殖。诱导培养5 d,A组(对照组)无

OC 形成，其余各组均有 TRAP 阳性多核 OC 形成。添加 1α,25-(OH)$_2$D$_3$能显著增加 OC 数量。诱导培养 9 d，除 A 组(对照组)骨片无明显吸收陷窝外，其余各组均出现吸收陷窝，且吸收陷窝面积随着 1α,25-(OH)$_2$D$_3$浓度的升高而增加。诱导培养 48 h，添加 RANKL、M-CSF 组 MMP-9 蛋白表达水平均极显著高于对照组(A 组)($P<0.01$)；添加 $10^{-9}$、$10^{-8}$ mol/L 1α,25-(OH)$_2$D$_3$组 MMP-9 蛋白表达水平显著或极显著高于 B 组($P<0.05$ 或 $P<0.01$)。结果表明，1α,25-(OH)$_2$D$_3$在生理剂量范围内能剂量依赖性地促进 OC 形成及骨吸收活性，上调其 MMP-9 蛋白表达水平，调节骨代谢。

**通讯作者**：刘宗平，E-mail：liuzongping@yzu.edu.cn

## H9 亚型禽流感病毒感染引起 MDCK 细胞内抗氧化功能的变化

郑良焰[1]，陈　龙[1]，徐家华[2]，潘家强[1]，唐兆新[1]

(1. 华南农业大学兽医学院，广州 510642；2. 肇庆大华农生物药品有限公司，广州 526238)

**摘　要**：将传代 MDCK 细胞以 $8\times10^4$/cm$^2$的密度接种于 6 孔细胞培养板，H9 亚型禽流感病毒以不同剂量 0(对照组)、$10^{-3}$(高剂量组)、$10^{-4}$(中剂量组)、$10^{-5}\times$EID50(低剂量组)分别接种，检测 H9 亚型禽流感病毒对 MDCK 细胞内抗氧化功能的影响。结果表明，与对照组相比，高剂量 H9 亚型禽流感病毒接种 MDCK 细胞使细胞内过氧化氢($H_2O_2$)和羟基自由基(·OH)含量均显著增加($P>0.05$)，细胞内超氧化物歧化酶(SOD)和谷胱甘肽过氧化物酶(GSH-Px)活力均显著下降($P>0.05$)，细胞内的丙二醛(MDA)含量显著增加($P>0.05$)。说明 H9 亚型禽流感病毒可显著提高 MDCK 细胞内活性氧自由基(ROS)含量，并降低抗氧化酶活力，阻碍 ROS 在细胞内的清除机制，引起细胞脂质过氧化作用，从而损伤细胞。

**关键词**：H9 亚型禽流感病毒，MDCK 细胞，抗氧化功能

**通讯作者**：唐兆新，E-mail：tangzx@scau.edu.cn

## HPLC 法测定贯叶连翘提取物中金丝桃素的含量

王建舫，穆　祥

(北京农学院兽医学(中医药)北京市重点实验室，北京 102206)

**摘　要**：本试验旨在建立一种高效液相色谱—紫外检测法(HPLC-UV)测定贯叶连翘提取物中金丝桃素含量的方法。采用 YMC-Pack ODS-A 色谱柱(5 μm，150 mm×4.6 mm)，以乙腈-0.02M 磷酸二氢钠(85∶15)为流动相，流速 1.0 ml/min，检测波长 588 nm，柱温 25 ℃，结果表明金丝桃素在 6～36 μg/ml 范围内线性关系良好($r=0.9996$)，平均加样回收率为 98.86%($n=6$)，峰面积的相对标准偏差(RSD)为 3.15%。此方法专属性强、重复性好，可用于贯叶连翘提取物中金丝桃素的含量测定。

**关键词**：贯叶连翘，金丝桃素，高效液相色谱，含量测定

## Determination of hypericin in *Hypericum perforatum* L. extract by HPLC

WANG Jianfang，MU Xiang

*Beijing Key Laboratory of Traditional Chinese Veterinary Medicine，Beijing University of Agriculture，Beijing* 102206，*China*

**Abstract**：This experiment was conducted to establish a method for determining the content of hypericin in

Hypericum perforatum L. extracts by HPLC-UV. The chromatographic separation was carried out on a YMC-Pack ODS-A column (5 μm, 150 mm × 4.6 mm). The mobile phase composed of Methyl Cyanide-0.02M sodium dihydrogen phosphate (85:15) was used at the flow rate of 1 mL/min, and the column temperature was at 25℃. Hypericin showed good linearity ($r = 0.9996$) in the range of 6 - 36 μg/mL, and the average recovery was 98.86% with RSD 3.15%. The HPLC system presented high specificity and good repeatability and could be used for determination of *Hypericum perforatum* L. extract.

**Keywords**: *Hypericum perforatum* L., Hypericin, HPLC, Assay

## IFN-τ调节奶牛子宫上皮细胞 BoLA-Ⅰ的表达

朱 喆,吴 岳,刘宏靖,肖 雅,祝 晶,邓干臻*

(华中农业大学,武汉 430070)

**摘 要**:不孕是制约畜牧业生产的主要问题,围植入期着床不稳定或失败是造成动物不孕的主要原因。而着床失败主要与妊娠识别、子宫容受态形成、合子基因突变等有关。IFN-τ是由反刍动物胚胎滋养层细胞分泌的,具有Ⅰ型干扰素的一般功能,其受体在子宫上皮细胞上。IFN-τ是反刍动物妊娠识别的信号分子,但是其具体调节机制尚不明确。BoLA-Ⅰ的表达情况,决定了在妊娠过程中胎儿作为一种同种异体抗原是否会被母体接受而不出现免疫排斥。本研究旨在通过研究IFN-τ,BoLA-Ⅰ及奶牛子宫上皮细胞的免疫调节通路和功能研究进一步认识奶牛子宫容受态形成和着床机制,提升解决动物不孕的科学思想和方法。本试验运用细胞培养方法培养奶牛子宫上皮细胞,建立奶牛子宫内膜上皮细胞体外表达模型,用外源性的IFN-τ的干预,设计不同时间浓度梯度。采用实时荧光定量PCR的方法检测IFN-τ对奶牛子宫上皮细胞*BoLA*-Ⅰ中典型的*BoLA-A*,*Heavy chain*,*BoLA-N*∗03101,*BoLA-N*∗03701,*BoLA-N*∗01201,*A*11以及非典型的*MIC*1,*BoLA-NC*1∗,*BoLA-NC*3∗基因表达的影响。进一步筛选出干预结果显著的BoLA-A,Heavy chain,MIC1,制备多克隆抗体,用Western blot的方法检测其蛋白表达在干预下的调节情况。结果显示,*MIC*1,*BoLA-A*,*Heavy chain*,*BoLA-N*∗01201,IFN-τ对其表达上调作用很明显,*BoLA-NC*1∗,*BoLA-NC*3∗,*A*11,有上调,作用不明显,时间浓度关联性不大,*BoLA-N*∗03101,*BoLA-N*∗03701几乎没有影响。运用Western blot的方法检测其蛋白的表达水平,与定量结果一致。结果提示IFN-τ是通过调控子宫上皮细胞表达BoLA-Ⅰ来调节子宫着床内环境的。

**关键词**:IFN-τ,BoLA-Ⅰ,奶牛子宫上皮细胞,妊娠免疫

* E-mail:151225445@qq.com

## Label free 方法筛选降解苦马豆素蛋白

王 妍,翟阿官,李勤凡,王建华*

(西北农林科技大学动物医学院,杨凌 712100)

**摘 要**:动物疯草中毒病严重制约着我国西部畜牧业的生产。动物采食疯草后可引起慢性神经机能障碍。苦马豆素(swainsonine,SW)是疯草中的主要毒性成分。但疯草具有返青早,枯萎迟,蛋白含量高,对动物适口性好等优点,更适合草资源相对匮乏,生态环境脆弱的西部草场利用。为去除SW,本实验室前期分离出了能够降解SW的*Arthrobacter* sp. HW08菌,并对其特性进行了研究,证明该菌的胞内酶具有降解SW的能力。本研究旨在探查*Arthrobacter* sp. HW08降解SW的相关差异表达蛋白谱,筛选降解SW的功能蛋白。本试验利用label free定量蛋白质组学技术分析比较SW诱导与未经SW诱导的HW08菌,根据蛋白质的基因本体注释对所鉴定蛋白进行分类分析;并通过实时荧光定量PCR法验证相关差异蛋白。结果显示,共

鉴定到 2043 个蛋白，经 iBAQ 方法差异性比较，选择 2 倍以上（$t$ 检验 $P<0.05$）的结果获得 129 个与降解 SW 相关的差异表达蛋白。SW 诱导后 45 个蛋白表达量上调，84 个蛋白表达量下调。其中，10 个为诱导前 HW08 菌中特有，8 个为诱导后 HW08 菌特有。对 129 个差异蛋白按其 GO 注释分别进行细胞组件、分子功能及生物学过程分类后发现，差异表达蛋白主要来自于细胞、细胞膜及细胞器，以催化和结合功能为主，参与代谢、细胞生物过程居多，其次参与生物调节、应对刺激等生物学过程。根据这些差异蛋白参与的生物学过程和功能，从中筛选了 12 对蛋白。通过实时荧光定量 PCR 验证得到了与降解 SW 相关的 4 组蛋白，分别是异构酶（A1R5X7）、氧化还原酶（A1R5X8）、脱氢酶（A1R6C3）、乙酰基转移酶（A0JZ95）。它们在 KEGG 中分别参与有机物分解过程、细胞代谢过程，异构酶活性；羧酸代谢过程，醛糖和酮糖互换，氧化还原酶活性；糖异生/降解脱氢酶、己内酰胺降解脱氢酶；苯甲酸降解、乙苯降解等乙酰转移酶作用。结果提示这些蛋白为进一步分析 *Arthrobacter* sp. HW08 降解 SW 的生物学机制，构建转基因工程菌提供科学数据。

**关键词**：Arthrobacter sp. HW 08，苦马豆素，Label free 法，降解蛋白，疯草

* E-mail：jhwang1948@ sina. com

## N-乙酰-L-半胱氨酸对牛源无乳链球菌诱导的小鼠肝脏氧化损伤的保护作用

杨　峰，王旭荣，李新圃，罗金印，王　玲，李宏胜*

（中国农业科学院兰州畜牧与兽药研究所，兰州 730050）

**摘　要**：无乳链球菌（*S. agalactiae*）是引发奶牛乳房炎的主要病原菌之一。本试验旨在研究不同血清型的牛源无乳链球菌感染小鼠后诱导的肝脏组织氧化损伤及 N-乙酰-L-半胱氨酸（NAC）在氧化损伤过程中所起的作用。60 只小鼠随机分为对照组、NAC 处理组、Ⅰa 型无乳链球菌感染组、Ⅱ型无乳链球菌感染组、（NAC + Ⅰa 型无乳链球菌）处理组和（NAC + Ⅱ型无乳链球菌）处理组，试验期为 10d。采用多重 PCR 方法鉴定试验菌株的血清型，比色法测定各组试验小鼠肝脏组织中的总抗氧化能力（TAC）和丙二醛（MDA）含量，并进行统计分析。结果显示：与对照组相比，Ⅰa 型和Ⅱ型无乳链球菌感染组肝脏 TAC 显著降低（$P<0.01$），MDA 含量显著升高（$P<0.01$），NAC 处理组肝脏 TAC 显著升高（$P<0.01$），MDA 含量显著降低（$P<0.01$）；（NAC + Ⅰa 型无乳链球菌）处理组和（NAC + Ⅱ型无乳链球菌）处理组分别与与Ⅰa 型和Ⅱ型无乳链球菌感染组相比，TAC 显著升高（$P<0.01$），MDA 含量显著降低（$P<0.01$），而与对照组相比，TAC 显著降低（$P<0.01$），MDA 含量显著增高（$P<0.01$）。结果表明，Ⅰa 型和Ⅱ型牛源无乳链球菌感染小鼠后，均可够诱导小鼠肝脏组织氧化损伤；100 mg/kg 体重剂量的 NAC 预处理能够显著增强小鼠肝脏中 TAC，显著减小 MDA 含量，可以减轻该菌诱导的氧化损伤，但不足以使感染小鼠恢复至正常水平。

**关键词**：N-乙酰-L-半胱氨酸，Ⅰa 型，Ⅱ型，无乳链球菌，氧化损伤

* E-mail：lihsheng@ sina. com

## SAA 与 HP 在奶牛子宫内膜上皮细胞炎性反应中的表达及意义

张世栋，严作廷*，王东升，董书伟，邝晓娇，魏立琴

（中国农业科学院兰州畜牧与兽药研究所，兰州 730050）

**摘　要**：血清淀粉样蛋白（serum amyloid A，SAA）和触珠蛋白（haptoglobin，HP）是主要的急性相蛋白，在机体受到感染和出现炎症时会表达增强。SAA 对机体感染革兰氏阴性和阳性细菌后都具有明显的免疫调节作用。HP 在机体中不仅能调节白细胞先天免疫反应，还具有抗炎、直接抑菌作用和分子伴侣活性。之前大量研究表明，SAA 和 HP 都是由肝脏产生，随着体液循

环至机体各个部位。但近几年的研究表明,在反刍动物中除肝脏以外的许多组织中也都有 SAA 和 HP 的表达,其基因表达变化可直接反应机体身体部位的病理状态。本研究中,将体外培养的奶牛子宫内膜上皮细胞(endometrial epithelial cell,EnEpC)用不同剂量的细菌内毒素 LPS 诱导炎性反应后,用 real-time qPCR 检测了细胞中 SAA(gb | AF540564.1)和 HP(gb | NM_001040470.2)基因的表达变化。结果显示,1 μg/ml,5 μg/ml 和 10 μg/ml 的 LPS 处理 EnEpC 后,SAA 基因的相对表达量显著升高,分别是对照的(4.18 ±0.58)倍,(4.57 ±0.32)倍和(5.95 ±0.38)倍;HP 基因的相对表达量也显著升高,分别是对照的(7.766 ±0.979)倍,(5.35 ±0.722)倍和(3.916 ±0.399)倍。结果表明,奶牛 EnEpC 中 SAA 和 HP 基因的表达变化与细胞炎性反应直接相关。该结果对奶牛子宫内膜炎的机制研究具有重要的理论意义,同时 SAA 和 HP 基因亦可作为奶牛子宫内膜炎诊断的潜在分子生物学标志物。

**通讯作者**:E-mail:yanzuoting@ yahoo. com. cn

## 不同硒浓度日粮对小鼠肝脏和睾丸组织中部分硒蛋白 mRNA 水平的影响

郑良焰,张　琴,刘碧涛,郭　凯,余文兰,刘正伟,陈　叶,郭翠丽,潘家强,唐兆新*
(华南农业大学兽医学院,广州 510642)

**摘　要**:选取 80 只 7 周龄雄性 BALB/c 小鼠随机分为Ⅰ、Ⅱ、Ⅲ、Ⅳ组,每组 20 只,Ⅰ、Ⅱ、Ⅲ、Ⅳ组分别在基础日粮中添加 0.045、0.1、0.4 和 0.8 mg/kg 硒。结果表明:1)Ⅰ组日粮能显著降低小鼠肝脏中的硒沉积量($P<0.05$)。2)谷胱甘肽过氧化物酶 1(glutathione peroxidase 1,GPx1)在Ⅰ组中的 mRNA 水平明显低于Ⅱ、Ⅲ组($P<0.05$),而Ⅱ、Ⅲ组在肝脏中差异显著($P<0.05$),在睾丸中差异不显著($P>0.05$)。3)Ⅰ、Ⅲ组谷胱甘肽过氧化物酶 4(glutathione peroxidase 4,GPx4)的 mRNA 水平在肝脏中相对差异不显著($P>0.05$),均显著低于常规对照组($P<0.05$),而在睾丸中三组间无显著差异($P>0.05$)。4)硫氧还蛋白还原酶 2(thioredoxin reductase 2,TrxR2)的 mRNA 相对表达量在肝脏和睾丸中随着日粮中硒添加量的增加而显著提高($P<0.05$),呈线性关系。综上所述,在肝脏和睾丸组织中,饲喂低硒饲料能显著降低硒沉积量,GPx1、GPx4 和 TrxR2 的 mRNA 相对表达量;而饲喂高硒饲料能显著降低 GPx1 的 mRNA 水平($P<0.05$),同时显著增加 TrxR2 的 mRNA 相对表达量($P<0.05$),表明硒对 GPx1、GPx4 和 TrxR2 mRNA 相对表达量的调节存在组织差异性。
**关键词**:硒,谷胱甘肽过氧化物酶 1(GPx1),谷胱甘肽过氧化物酶 4(GPx4),硫氧还蛋白还原酶 2(TrxR2),mRNA,肝脏,睾丸

* **通讯作者**: E-mail:tangzx@ scau. edu. cn

## 抵抗素对肝脏糖脂代谢影响的研究概况

任　毅,袁贵强,苟丽萍,王　婵,万涛梅,左之才*
(四川农业大学动物疫病与人类健康四川省重点实验室,雅安 625014)

**摘　要**:抵抗素(Resistin),是 RSTN 基因编码的产物,是一种肽激素,富含半胱氨酸的分泌蛋白,属于 RELM 家族,也称之为 ADSF(脂肪组织特异性的分泌因子)。抵抗素在调节机体能量代谢中起着重要作用,可影响机体组织对胰岛素的敏感性,损害机体糖耐量。肝脏是机体物质代谢的重要器官,在很大程度上影响着机体能量平衡。本文就抵抗素在肝脏糖脂代谢中的作用及其可能机制作相关的介绍。
**关键词**:抵抗素,胰岛素抵抗,糖代谢,脂代谢

* **通讯作者**:E-mail:zzcjl@ 126. com

# 低水平激光辐照可通过抑制 PMN 黏附减轻 LPS 诱导的大鼠乳腺炎病变

王建发，王跃强，贺显晶，孙东波，武　瑞*
（黑龙江八一农垦大学动物科技学院，大庆 163319）

**摘　要**：乳腺炎是乳用动物乳腺受到物理、化学和微生物等刺激所引起的一种炎症反应。它是全球广泛流行和造成乳品行业经济损失最严重的疾病之一。多形核嗜中性粒细胞（PMN）募集在乳腺炎发病过程中具有重要的作用，其募集作用主要受到炎性细胞因子（TNF-α、IL-1β、IL-6 和 IL-8）、黏附分子（CD62L 和 CD11b）以及细胞间粘附分子-1（ICAM-1）和血管内皮细胞间粘附分子-1（PECAM-1）等黏附受体水平的影响。低水平激光辐照疗法（LLLT）治疗炎症反应的临床应用效果已得到广泛证实，但其缓解乳腺炎病变的机理尚不明确。为探讨 LLLT 治疗乳腺炎的机理，本研究应用乳头内注射脂多糖（LPS）的方法建立产后大鼠乳腺炎模型，并用 LLLT（650nm，2.5mW，30mW/$cm^2$）对大鼠乳腺进行辐照，每天 2 次，每次辐照 30min。分别采用 ELISA 法检测血液中 TNF-α、IL-1β、IL-6 和 IL-8 含量，流式细胞术检测外周血 PMN 表面 CD62L 和 CD11b 的阳性率，RT-PCR 法检测乳腺组织中 ICAM-1 和 PECAM-1 的 mRNA 转录水平，HE 染色法观察乳腺组织病理变化和腺泡内 PMN 数量，ELISA 法检测乳腺组织髓过氧化物酶（MPO）活性的变化。结果表明，LLLT 辐照可以显著降低 LPS 诱导的 IL-1β 和 IL-8 的分泌（$P<0.05$），降低 PMN 细胞表面 CD62L 的表达量，提高 PMN 细胞表面 CD11b 的表达量，抑制 LPS 诱导的 ICAM-1 基因表达，进而降低乳腺腺泡内 PMN 数量和乳腺组织 MPO 活性，缓解 LPS 引起的乳腺腺泡形成空泡和腺泡壁增厚等病理变化。由此可见，LLLT 可通过抑制 PMN 黏附减轻 LPS 诱导的大鼠乳腺炎病变，可作为一种治疗奶牛乳腺炎的候选物理疗法。

**通讯作者**：E-mail：fuhewu@126.com

# 恩诺沙星注射液对肉牛细菌性呼吸道感染的疗效试验

苟丽萍，吴　鹏[1]，王正义[1]，王　婵[1]，左之才[1,2]
（1. 四川农业大学动物医学院，雅安 625014；
2. 四川农业大学动物疫病与人类健康四川省重点实验室，雅安 625014）

**摘　要**：为观察恩诺沙星注射液对肉牛细菌性呼吸道感染的临床疗效，本试验选取 60 头患细菌性呼吸道感染肉牛随机分成 3 组，每组 20 头，即恩诺沙星注射液高剂量组（Ⅰ组，剂量 5.0 mg/kg. bw）和低剂量组（Ⅱ组，剂量 2.5 mg/kg. bw）、泰拉菌素注射液对照组（Ⅲ组，剂量 2.5 mg/kg. bw）。用药第 0d、7d 采集患牛鼻腔拭子进行细菌分离鉴定。观察患牛临床症状并对第 0d，3d，5d，7d，14d 进行临床症状评分。结果显示，Ⅰ、Ⅱ、Ⅲ组肉牛细菌性呼吸道感染的治愈率分别为 85%、75% 和 80%，有效率分别为 95%、90% 和 90%。结论，使用恩诺沙星注射液能有效清除 70% 以上肉牛细菌性呼吸道感染病原菌，对肉牛细菌性呼吸道感染疾病有良好疗效，临床推荐使用剂量为 2.5 mg/kg. bw。

**关键词**：恩诺沙星，肉牛，细菌性呼吸道感染，疗效

**通讯作者**：左之才，E-mail：zzcjl@126.com

## 氟中毒对雄性小鼠下丘脑—垂体—性腺轴显微和超微结构的影响

韩海军,王　冲,孙子龙,王俊东
(山西农业大学,太谷 030801)

**摘　要**:多年来对氟中毒的研究表明:氟可对多种组织器官造成损伤,其中包括对生殖系统的损伤。生殖系统的脏器发育、生殖功能以及第二性征的维持都依赖性激素的分泌,而其分泌主要受下丘脑—垂体—性腺轴(Hypothalamic-pituitary-gonad axis,HPGA)的调节,通过一系列正、负反馈调节而维持其功能。已知氟可引起动物 HPGA 相关激素分泌的紊乱,然而,氟对 HPGA 结构基础的影响尚未有所报道。探讨氟中毒对雄性小鼠 HPGA 显微及超微结构的影响,为今后分子机制研究奠定基础。将 12 只性成熟雄性昆明小鼠随机分为 4 组:对照组、低氟组(含 25mg/L NaF)、中氟组(含 50mg/L NaF)、高氟组(含 100mg/L NaF),通过饮水摄氟 8 周后,取出下丘脑、垂体和睾丸组织,利用 HE 染色和透射电子显微镜观察组织显微和超微结构的变化。HE 染色结果显示,氟中毒可导致下丘脑神经元细胞核固缩,浓染。垂体中细胞核变形,浓染,胞浆减少,嗜酸性、嗜碱性和嫌色细胞呈不均匀分布。睾丸曲细精管各级细胞排列紊乱,层数减少,管壁变薄,管腔内精子数目减少,高氟组间质细胞出现明显的空泡,并且随着染氟剂量的增加,病变程度加重。透射电镜结果显示,氟中毒可导致下丘脑中的神经元细胞核染色质聚集,髓鞘呈葱皮样变,严重的可见髓鞘球形成;垂体中促性腺激素细胞核染色质聚集,线粒体肿胀,呈空泡样,内质网扩张。高氟可导致睾丸中精原细胞和精母细胞核染色质浓缩,胞质内线粒体肿胀,形成空泡;精子细胞胞质内线粒体形成空泡,核膜破裂、不完整,部分细胞缺少顶体颗粒;支持细胞胞浆密度降低,线粒体空泡化;间质细胞局部核膜不完整,线粒体嵴断裂,形成很多空腔,胞浆减少。氟中毒可造成雄性小鼠下丘脑、垂体、睾丸组织显微和超微结构不同程度的改变,提示氟中毒可能通过生殖内分泌系统 HPGA 影响雄性生殖系统。

**关键词**:氟中毒,HPGA,显微,超微结构

**通讯作者**:王俊东,E-mail:wangjd53@ outlook. com

## 广东、湖南两地鸡饲料中矿物元素含量分析

朱余军,韩文彩,张春红,张建峰,潘家强
(华南农业大学兽医学院,广州 510642)

**摘　要**:矿物元素作为饲料添加剂在世界各国普遍应用,但也存在潜在的不安全因素。市场上的饲料品质良莠不齐,饲料中矿物元素的添加量往往难以达到最适的标准。饲料中矿物元素供应不足会影响机体正常发育;而过量同样会影响机体健康,甚至造成中毒。在家畜中,猪体内的矿物元素含量约为其体重的 3%,容易出现矿物元素缺乏。试验从广东、湖南两地养猪场中采集饲料样本,对饲料中的常见矿物元素含量进行了测定,采集的饲料按所适用猪的年龄和生长阶段划分为乳猪(初生至 2 月龄)料、小猪(3 ~ 4 月龄)料、中猪(5 ~ 12 月龄)料、哺乳母猪(1 ~ 6 岁)料各 35 份。结果如下:两地猪饲料中的铜含量均达到中国饲养标准,平均含量为中国饲养标准的 10 ~ 20 倍,均未超过美国营养需要(NRC)的最高限量(200 ~ 250 mg/kg);铁含量均达到或稍超过中国饲养标准,且未超过 NRC 规定的最高限量(3 000 mg/kg);锰含量均高于中国饲养标准,平均含量为中国饲养标准的 20 ~ 30 倍,但未超过 NRC 规定的最高限量(400 mg/kg);锌含量达到中国饲养标准量,平均含量为中国饲养标准的 2 ~ 3 倍,且远低于 NRC 规定的最高限量(3 000 mg/kg);多个样本的乳猪、小猪、中猪和哺乳母猪料中的钙含量达不到中国饲养标准的要求,提示要注意补充钙;两地猪饲料中的磷含量均比中国饲养标准低,提示要注意磷的补充。

**通讯作者**:E-mail:panjq@ scau. edu. cn

# 厚朴酚通过下调 TLR4 介导的 NF-κB 和 MAPK 信号通路抑制 LPS 刺激的小鼠乳腺上皮细胞炎性因子的产生

王　巍,梁德洁,宋晓静,周二顺,曹永国,杨正涛,张乃生
(吉林大学,长春 130062)

**摘　要**:家畜乳腺炎是由乳腺组织感染导致的炎症,主要由细菌感染引起。对乳腺炎的治疗通常采用抗生素,但存在耐药性产生,药物残留等问题。而中药因其副作用小,不产生耐药性及其广泛的抗炎及提高免疫力等方面的作用,逐渐在家畜乳腺炎的治疗中占据重要地位。厚朴酚是从中药厚朴中分离的一种木脂素类成分,为和厚朴酚的异构体,已被证实存在抗炎活性。然而,厚朴酚对患有乳腺炎的家畜的抗炎效果及抗炎作用的分子机制尚不清楚。本研究的目的是通过 LPS 刺激小鼠乳腺上皮细胞构建乳腺炎模型,通过不同浓度的厚朴酚药物处理,观察厚朴酚的抗炎效果并探讨其抗炎机制。具体方法是将小鼠乳腺上皮细胞分为对照组,LPS 刺激组和给药组(在 LPS 刺激前加入 12.5,25 或 50 μg/ml 的厚朴酚)。通过 ELISA 方法测定小鼠乳腺上皮细胞促炎性细胞因子的分泌情况,并通过 Westernblot 方法对 NF-κB 和 MAPK 两条信号通路的蛋白以及 TLR4 进行蛋白含量的测定。结果显示,与 LPS 刺激组相比,厚朴酚给药组能显著抑制 LPS 刺激的乳腺上皮细胞分泌促炎性细胞因子 TNF-α 和 IL-6,且这种抑制作用随给药浓度的增加而增加,呈浓度依赖性抑制。同时,厚朴酚给药组显著抑制 NF-κB (IkB-α 和 NF-κB p65)和 MAPKs (ERK、JNK 和 p38MAPKs)这两条信号通路的磷酸化及 TLR4 的表达,同样呈浓度依赖性。结果表明,厚朴酚能缓解 LPS 刺激的乳腺上皮细胞的炎症反应,而其作用机制是通过抑制 TLR4/NF-κB/MAPKs 信号通路系统来抑制炎性因子的表达,从而抑制炎症反应。因此,厚朴酚可以作为治疗家畜乳腺炎的药物。

**关键词**:厚朴酚,脂多糖(LPS),乳腺炎,炎性因子,NF-κB,MAPKs

**通讯作者**:张乃生,E-mail:zhangns@ jlu. edu. cn

# 黄白双花口服液药效学研究

王胜义,王　慧,刘永明,齐志明,荔　霞,董书伟
(中国农业科学院兰州畜牧与兽药研究所,农业部兽用药物创制重点实验室,兰州 730050)

**摘　要**:通过黄白双花口服液药物的抑制小肠运动试验、抗炎试验、镇痛试验、抑制番泻叶致小鼠腹泻试验、脾虚小鼠胸腔巨噬细胞吞噬试验和脾虚小鼠免疫脏器指数以及免疫因子测定,来阐明黄白双花口服液的临床治疗湿热型犊牛腹泻的作用机制。试验结果表明黄白双花口服液能显著抑制碳末在小肠内的推动($P<0.05$)和明显减少番泻叶引起的小鼠腹泻次数($P<0.05$),说明其具有涩肠止泻作用;能显著抑制二甲苯引起的小鼠耳廓肿胀($P<0.01$),说明具有较强的抗炎作用;能显著减少醋酸致小鼠疼痛的扭体次数($P<0.01$),说明具有较强的镇痛作用;能显著提高脾虚小鼠的腹腔巨噬细胞吞噬率($P<0.01$)并能不同程度的提高 IFN-γ 和 IL-2 在血清中的含量,说明具有提高脾虚小鼠的免疫功能。试验结果表明,黄白双花口服液具有明显的止泻作用,能抑制小肠推动是其抑制腹泻的重要机制之一。另外腹泻与肠道黏膜的损伤和各种炎性介质的释放有关,所以临床可见动物的努责弓腰,因而药物的抗炎镇痛效果也是其发挥作用的重要机制之一。动物的机体免疫力是疾病恢复的重要因素,因此药物对机体免疫力的调节也是其发挥作用的重要途径。以上这些功效可全面调节动物的胃肠道功能,对犊牛的腹泻起到标本兼治的作用。

**关键词**:黄白双花口服液,肠运动,抗炎,镇痛,止泻,免疫功能

**通讯作者**:刘永明(1957.6 – ),E-mail:myslym@ sina. com

# 黄曲霉毒素对猪生长性能和免疫指标的影响

张麦收，杨亮宇[2]，袁小松[2]，刘知奇[2]，周　勇[2]*

(1. 山东省安丘市慧康饲料有限公司，安丘；2. 云南农业大学，昆明 650201)

**摘　要**：本试验旨在探讨猪饲喂含不同浓度黄曲霉毒素(AF)饲料对其生长性能和免疫指标的影响，试验选用 40 头体重(20 ± 0.5) kg 接近的三元杂交猪(DLY)，随机均分为 4 组：1 号组为对照组，饲喂无 AF 的基础日粮；2 号组饲喂添加 50 μg /kg AF 的基础日粮；3 号组饲喂添加 100 μg /kg AF 的基础日粮；4 号组饲喂添加 150 μg /kg AF 的基础日粮。试验时长为 60d，期间猪自由采食和饮水，记录饲料摄取量、体重，采血进行血细胞计数和肝功能测定试验，并测定免疫球蛋白 IgG 和 IgM 变化水平，60d 时对试验猪剖检以观察内脏病变与称重。

结果发现与 1 号组相比 3、4 号组试验猪的平均日增重明显减少(分别为 0.63、0.52 和 0.50 kg/d，$P=0.05$)，平均日采食量有所减少(分别为 1.68、1.49 和 1.43，$P=0.06$)。与 4 号组相比 1、2、3 号组试验猪的白细胞计数显著增高(分别为 $23.5\times10^3/\mu l$、$16.3\times10^3/\mu l$、$17.5\times10^3/\mu l$ 和 $18.3\times10^3/\mu l$，$P<0.05$)，IgG 含量显著降低(分别为 $(57.89\pm1.85)$ mg/ml、$(83.51\pm2.71)$ mg/ml、$(79.37\pm2.52)$ mg/ml、$(75.31\pm2.08)$ mg/ml，$P<0.05$)，IgM 也显著降低(分别为 $(8.51\pm0.11)$ mg/ml、$(11.79\pm0.21)$ mg/ml、$(10.17\pm0.12)$ mg/ml、$(9.89\pm0.18)$ mg/ml，$P<0.05$)。剖检发现与 1 号组相比 3、4 号组试验猪肝的纤维化极度明显($P\leqslant0.05$)。研究表明，当饲料 AF 含量大于 100 μg /kg 时猪的采食量和增重量有所减少，肝脏损伤严重；当饲料 AF 含量大于 150 μg /kg 时会导致试验猪免疫指标发生变化，出现全身性炎症，危害猪生长程度进一步加大。

**关键词**：黄曲霉毒素，猪，生长性能，免疫指标

**通讯作者**：E-mail：panjq@ scau. edu. cn

# 鸡传染性贫血病病毒 TaqMan 荧光定量 PCR 检测方法的建立

郭翠丽，张春红，郭鹏举，张建峰，沈海燕，潘家强*

(华南农业大学兽医学院，广州 510642)

**摘　要**：鸡传染性贫血病(chicken infectious anemia，CIA)又称蓝翅病、出血综合征或贫血性皮炎综合征，是由鸡传染性贫血病毒(chicken infectious anemia virus，CIAV)引起的以雏鸡再生障碍性贫血和全身性淋巴组织萎缩为主要特征的免疫抑制病。目前，CIA 的诊断方法主要包括：病毒分离与鉴定、PCR 诊断方法、巢式 PCR、间接免疫荧光法、酶联免疫吸附试验、免疫过氧化物酶法、病毒中和试验，核酸探针技术等。但这些方法虽各有优点，但在敏感性、特异性和时效性等方面均存在自身不足。荧光定量 PCR 技术具有快速、灵敏、特异性强、可定量等优点，非常适用于病毒感染的检测，尤其是早期感染检测。本研究根据 CIAV 基因组保守区域设计一对特异性引物和探针，通过优化反应条件，建立了一种快速检测 CIAV 的 Taqman 荧光定量 PCR 方法，同时验证其特异性、灵敏性和重复性。结果显示，本试验建立的 Taqman 荧光定量 PCR 方法灵敏度可达 $1.8\times10^1$ 拷贝/μl，远高于常规 PCR 方法；并与禽类其他病毒性疾病无交叉反应，具有高特异性。本研究建立的 Taqman 荧光定量 PCR 方法特异性强、灵敏度高、重复性好，可同时检测大量临床样品，适用于 CAV 的诊断与流行病学分析。

# 江西不同地方品种鸡 apoA-Ⅰ、apoB 基因多态性与其脂肪代谢的关联性分析

夏安琪，郭小权，曹华斌，张彩英，黄爱民，刘　平，胡国良*
（江西农业大学动物科学技术学院，南昌 330045）

**摘　要**：本试验以崇仁麻鸡、余干黑鸡、海蓝褐蛋鸡、绿壳蛋鸡、宁都三黄鸡、安义瓦灰鸡和泰和乌骨鸡各 50 羽作为试验材料，根据已公布鸡的 apoA-Ⅰ、apoB 基因序列设计引物，利用 PCR-DNA 直接测序技术检测 apoA-Ⅰ、apoB 基因 SNPs 和基因型分析，探讨 apoA-Ⅰ、apoB 基因的多态性及其与脂肪代谢的关系。结果如下：

1. 在 apoA-Ⅰ基因序列起始密码子 ATG 上游 163bp 处存在一个 A/T 突变，该突变产生 AA、AB、BB 基因型。基因型分析结果表明，余干黑鸡以 AB 基因型为主导基因型，绿壳蛋鸡、安义瓦灰鸡和宁都三黄鸡以 BB 基因型为主导基因型；余干黑鸡和泰和乌骨鸡以 A 基因为主要等位基因，绿壳蛋鸡、安义瓦灰鸡和宁都三黄鸡以 B 基因为主要等位基因；不同品种鸡 apoA-Ⅰ基因的突变位点为中度多态位点。

2. 在 apoB 基因序列的第 26 个外显子 123bp 处产生 T/G 突变，并产生 TT、TG、GG 基因型。其中崇仁麻鸡和泰和乌骨鸡以 TT 基因型为主导基因型，余干黑鸡、绿壳蛋鸡、安义瓦灰鸡和宁都三黄鸡的 TG 基因型为主导基因型；不同品种鸡均以 T 基因为主要等位基因；不同品种鸡 apoA-B 基因的突变位点为中度多态位点。

3. 关联性研究表明含有 B 等位基因的崇仁麻鸡和余干黑鸡的肝脂率、腹脂重、腹脂比率显著高于海蓝褐蛋鸡，且差异显著（$P<0.05$），初步推断 B 等位基因与崇仁麻鸡和余干黑鸡的肝脂率和腹脂重有一定相关性；含有 G 等位基因的余干黑鸡的腹脂重、腹脂比率显著低于崇仁麻鸡的腹脂重、腹脂比率，且差异显著（$P<0.05$），说明 G 等位基因与崇仁麻鸡和余干黑鸡的腹脂重和腹脂比率有一定相关性。

综合上述研究结果，初步推断 apoA-Ⅰ基因和 apoB 基因与崇仁麻鸡和余干黑鸡的脂肪代谢具有显著相关性；并可应用于鸡脂肪性状的分子标记辅助选择育种方案中。

**关键词**：鸡，apoA-Ⅰ，apoB，基因多态性，脂肪代谢

* **通讯作者**：E-mail：hgljx@163.com

# 基于石墨烯与蒁堇纳米复合物的电化学适配体传感器检测伏马菌素 $B_1$

施志玉，郑亚婷，吴文达，张海彬*
（南京农业大学动物医学院动物中毒病实验室，南京 210095）

**摘　要**：电化学适配体传感器具有特异性强，测试费用低，不受样品颜色、浊度的影响，所需仪器设备相对简单、方便易行等优点。本文基于石墨烯与蒁堇纳米复合物的信号增强放大作用和 DNA 适配体的特异性构建了一种灵敏、简便的检测玉米中伏马菌素 $B_1$ 的无标签电化学适配体传感器。通过金纳米颗粒修饰玻碳电极，捕获 DNA 通过金硫键与金纳米颗粒连接，适配体 DNA 与捕获 DNA 特异反应后连接石墨烯和蒁堇纳米复合物，样品中的伏马菌素 $B_1$ 与蒁堇竞争结合适配体 DNA，用循环伏安法检测蒁堇电化学信号的变化对伏马菌素 $B_1$ 进行定量分析。利用透射电子显微镜表征合成的金纳米颗粒，紫外可见光光谱表征合成的石墨烯和蒁堇纳米复合物。在优化捕获 DNA 浓度、石墨烯片和蒁堇纳米复合物与适配体 DNA 反应时间等条件后，该适配体传感器检测伏马菌素 $B_1$ 的线性范围为 1 pg/ml 到 100 ng/ml，线性方程为 $y = 1.175 + 0.067X$，相关系数为 0.998 2，检测限可达 1 pg/ml。在添加毒素的玉米样品检测中回收率高于 93.77%，变异系数 <5%，其重复性、重现性、稳定性与特

异性试验的变异系数均 <5%。该电化学适配体传感器简单易行,特异性强,可以用于伏马菌素 $B_1$ 的检测。

**关键词**:伏马菌素 $B_1$,电化学适配体传感器,石墨烯硫堇纳米复合物

* **通讯作者**:E-mail:haibinzh@ njau. edu. cn

## 利巴韦林对猫细小病毒体外抑制作用的研究

郑良焰[1],刘碧涛[1],任常宝[2],唐兆新[1*]

(1. 华南农业大学 兽医学院,广州 510642;2. 肇庆大华农生物药品有限公司,肇庆 526238)

**摘 要**:通过研究抗病毒药物利巴韦林对猫细小病毒(FPV)的体外抑制作用,为利巴韦林治疗猫细小病毒的临床应用提供实验依据。试验采用细胞病变效应(CPE)法和细胞增殖分析,评价利巴韦林对猫细小病毒(FPV)的抑制作用。结果表明利巴韦林对猫肾(F81)细胞的半数毒性($TC_{50}$)浓度为 179. 60μg/ml,10μg/ml 浓度的利巴韦林对 FPV 的体外抑制率最高,为 54. 35%。安全剂量内,利巴韦林抗 FPV 作用与其剂量呈正相关,抑制 FPV 的半数有效浓度($EC_{50}$)为 15. 38 μg/ml,治疗指数(TI)为 11. 68。体外实验证实利巴韦林对猫细小病毒有抑制作用。

**关键词**:利巴韦林,猫细小病毒,抗病毒作用,细胞病变(CPE),MTT 分析法

* **通信作者**:E-mail:tangzx@ scau. edu. cn

## 硫氧还蛋白对 $H_2O_2$ 诱导的 BRL-3A 细胞损伤的保护作用

余文兰[1],王朵朵[2],郭剑英[1],胡莲美[1],潘家强,唐兆新[1*]

(1. 华南农业大学兽医学院,广州 510642;2. 中国科学院华南植物园,广州 510642)

**摘 要**:硫氧还蛋白(thioredoxin,Trx)是一类广泛存在于生物体内的低分子量的蛋白质,对维持体内稳定的氧化还原状态具有重要的作用,目前,Trx 对活性氧自由基诱导肝细胞损害的保护机理的研究尚未报道。本研究利用不同浓度过氧化氢($H_2O_2$)作用于 BRL-3A 大鼠肝细胞,用四甲基偶氮唑蓝比色法检测 BRL-3A 细胞存活数量确定 $H_2O_2$ 最适损伤浓度;试验随机分为正常对照组、$H_2O_2$ 处理组和硫氧还蛋白(2.5 和 5μmol/L)预处理组,用过氧化物丙二醛(MDA)、超氧化物歧化酶(SOD)、过氧化氢酶(CAT)试剂盒测定细胞的氧化应激变化。结果表明:1 000μmol/L $H_2O_2$ 对 BRL-3A 细胞增殖具有明显的抑制作用;2. 5μmol/L 硫氧还蛋白对 $H_2O_2$ 损伤的 BRL-3A 具有保护作用,它能明显抑制 $H_2O_2$ 诱导 BRL-3A 细胞损伤产生丙二醛(MDA)($P<0.05$),并明显增加细胞超氧化物化物歧化酶(SOD)及过氧化氢酶(CAT)的活性($P<0.05$),从而保护肝细胞。综上指标表明,硫氧还蛋白对 $H_2O_2$ 诱导 BRL-3A 细胞的氧化应激损坏具有保护作用。

**通讯作者**:E-mail:tangzx@ scau. edu. cn

## 慢性氟中毒对小鼠精子 ATP 生成途径的影响

张 雯,韩海军,孙子龙*

(山西农业大学,太谷 030801)

**摘 要**:氟中毒是我国危害最严重的地方性疾病。氟中毒可造成氟斑牙、氟骨症的同时,还严重地影响着人类及动物的的生殖健康。精子是影响生殖最主要的因子,而精子获得足够的能量是成功受精的基础。ATP 是精子运动的能量来源,一般情况

下,在哺乳动物的精子细胞里有两条产生能量的代谢途径,无氧酵解和有氧呼吸。迄今为止,有关慢性氟中毒对精子呼吸状态影响的相关研究还鲜有报道。本文旨在探讨精子在不同氟浓度下,有氧呼吸与无氧酵解对精子 ATP 产量的贡献,进而确定氟致精子活力降低的动力依据。将性成熟昆明小鼠随机分为 4 组:对照组(蒸馏水),低氟组(含 25mg/L NaF)、中氟组(含 50mg/L NaF)、高氟组(含 100mg/L NaF),通过饮水摄氟 180d 后,取出精子悬液,测定其密度,活力等基本指标,通过建立无氧酵解与有氧呼吸两种呼吸模型,利用化学发光法测定不同模型下 ATP 的产量。试验结果表明长期摄氟可导致精子的密度和活力下降,在中氟与高氟影响下有氧呼吸产生的 ATP 显著下降;而无氧酵解条件下,低氟组与中氟组 ATP 产量下降,高氟组 ATP 的产量有所回升,但都差异不显著。结果提示,慢性氟中毒可导致精子密度与活力下降,降低有氧呼吸 ATP 的产量,说明氟化物可能是通过影响精子有氧呼吸途径,进而影响精子的动力来源的。

**关键词**:氟,精子,ATP,有氧呼吸,无氧酵解

* **通讯作者**:E-mail:sunzilong2000@163.com

# 猫细小病毒的分离与鉴定

刘碧涛[1],任常宝[2],张晓战[1],许冬蕾[1],郑良焰[1],潘家强,唐兆新[1,2]*

(1. 华南农业大学 兽医学院,广州 510642;2. 肇庆大华农生物药品有限公司,肇庆 526238)

**摘 要**:猫细小病毒病是由猫细小病毒(Feline panleukopenia virus,FPV)引起的一种高度接触性急性传染病,具有很高的发病率和死亡率。患病动物表现为高热、呕吐、腹泻、出血性肠炎和白细胞数目严重减少等症状。该病每年对猫科等动物养殖业带来重大经济损失。目前,市场上还没有应对 FPV 的特效药和有效的治疗办法,临床上以免疫预防为主。但是国内还没有真正批准上市的 FPV 疫苗。

为了丰富猫细小病毒(FPV)流行病学资料和制备灭活疫苗,通过猫肾传代细胞(F81)培养方法从疑似感染 FPV 病猫粪便中分离病毒。对分离到的病毒进行形态学观察、理化特性试验、细胞敏感性试验、PCR 检测及测序分析和动物感染试验等进行鉴定。结果显示,磷钨酸负染后,电镜下可以观察到呈圆形,直径为 20 nm 左右,表面无囊膜,实心和少量空心的病毒粒子;病毒悬液接种于 F81、MDCK、Vero、BHK 细胞后,只有 F81 细胞发生明显细胞病变;病毒经 5-碘脱氧尿核苷(5-IUDR)处理后,TCID50 下降超过 2 个数量级;VP2 基因核苷酸序列与 GenBank 已发表的 FPV 标准株 VP2 基因序列同源性达到 99.0%;动物感染试验显示攻毒猫出现典型的 FPV 感染症状,血清中抗体含量显著升高,并且排毒。证明该分离毒株为猫细小病毒。本试验成功分离到猫细小病毒,进一步充实了我国 FPV 病毒库,也为灭活疫苗的制备奠定了基础。

**关键词**:猫细小病毒,分离,鉴定

* **通讯作者**:E-mail:tangzhaoxin@hotmail.com

# 钼、镉联合诱导对鸭血常规的影响

陈 花,曹华斌,张彩英*,胡国良,徐雄卫,宗益波,熊 金,龚婷,李山威

(江西农业大学动物科学技术学院,南昌 330045)

**摘 要**:“三废”的污染常造成镉、钼等重金属联合污染环境,国内外学者对镉、钼的毒性进行了大量的研究,但这些研究主要集中在它们的单独毒性效应上,而对于镉、钼联合毒性研究则相对较少,特别是镉、钼联合诱导对鸭血常规的影响尚未见报道。本试验旨在探讨钼、镉联合诱导对鸭血常规的影响。挑选 360 羽(公母比例 1∶1)11 日龄“江南 2 号”麻鸭,随机分成 6 组,分别为对照组、低钼组、高钼组、镉组、低钼镉组和高钼镉组,对照组饲喂基础日粮,其余各组分别在每 kg 基础日粮中添加 Mo 15mg、Mo 100mg、Cd 4mg、Mo 15mg + Cd 4mg 以及 Mo 100mg + Cd 4mg。钼源为七钼酸铵[$(NH_4)_6Mo_7O_{24}\cdot 4H_2O$],镉源为

硫酸镉($3CdSO_4 \cdot 8H_2O$)。试验周期为120d,在试验的第30、60、90、120天分别随机取10羽鸭翅静脉采血,测量血常规。结果显示与对照组相比,低钼组和镉组有降低RBC、HGB、HCT、MCV、MCH和MCHC的趋势,但差异不显著($P>0.05$);高钼组、低钼镉组和高钼镉组,均导致RBC、HGB、HCT、MCV、MCH和MCHC降低($P<0.05$),具有明显的剂量—效应和时间—效应关系,镉钼呈现协同作用;PLT各组间差异不显著($P>0.05$)。结果提示钼、镉联合引发小细胞低色素贫血症,有剂量—效应和时间—效应关系,镉钼呈现协同作用。

**关键词**:鸭,钼,镉,血常规

* **通信作者**:E-mail:zhangcaiying0916@163.com

# 奶牛产后灌服丙二醇与钙磷镁合剂对比研究

侯引绪[1*],严宝英[2],魏朝利[3],曾光祥[4]

(1.北京农业职业学院,北京 102442;2.陕西省动物疾病预防控制中心,西安 710016;

3.华秦源(北京)动物药业有限公司,北京 102206;4.北京三元绿荷奶牛养殖中心圣兴达奶牛场,北京 102442;)

**摘　要**:分娩是奶牛生产过程中的一个最大生理性应激,所导致的奶牛产后生理功能紊乱、产后爬卧不起、产后猝死、代谢负平衡(低血钙、低血糖 、高血酮等)给奶牛养殖造成了巨大的经济损失。研究、创新奶牛产后(围产后期)保健技术,不仅对提升奶牛健康、控制代谢病发生、降低死淘率和提升奶牛单产有双重重要意义,也是奶牛场非常需要的一种科技需求。笔者承继前人产后灌服丙二醇的研究基础,设计了"奶牛产后灌服丙二醇与钙磷镁合剂的对比研究试验,对二种灌服措施对奶牛泌乳性能、SCC、血钙、血磷、血糖、血酮等指标的影响作了临床对比研究。结果表明:

1.试验1组(奶牛分娩后第1~3天灌服丙二醇400g/d),与试验2组(分娩当天一次性灌服钙磷镁合剂+复方口服补液盐400g)相比,二种灌服对分娩后第4日的血钙、血糖、血磷、血酮的影响不存在显著差异。

但试验2组(灌服钙磷镁+口服补液盐组)只有1头亚临床型低血钙症患牛,发病率为6.67%;而试验1组(灌服丙二醇组)有4头亚临床型低血钙症患牛,亚临床型低血钙症发病率为26.67%。

2.分娩后第18天的测定结果表明,两组的血酮(β-羟丁酸)含差异常显著($P<0.05$),试验2组血酮含量显著低于试验1组;血清钙、血清磷、血糖含量均无显著差异($P>0.05$)。

由此可见,说明分娩后第1d、第5 d分别给奶牛灌服钙磷镁合剂+复方口服补液盐,比分娩后第1~3d连续3d灌服丙二醇在缓解产后能量负平衡,防控产后奶牛隐性酮病的效果显著优于丙二醇灌服组。

3.虽然两组在分娩后3~42d内的最高产奶日时间、最高产奶日产奶量、日平均产奶量、总产奶量、SCC方面差异不显著。但实验2组奶牛日平均产奶量比实验1组提高4.54 kg/d。试验2组奶牛比实验1组奶牛多产奶2 372 kg(182.46 kg/头),呈显一定程度的增产趋势。

综合分析表明,产后灌服丙二醇和产后灌服钙磷镁合剂+复方口服补液盐对促进奶牛产后奶牛生理代谢恢复和防控产后代谢负平衡均有较好作用,但灌服复合制剂的效果优于单一制剂灌服。建议奶牛产后灌服保健采用:丙二醇+钙磷镁合剂+复方口服补液。

* **通讯方式**:E-mail:hyx003@163.com

# 奶牛血清钙检测试剂盒研制与初步应用

宋国希，贺显晶，孙东波，王建发，武　瑞*

（黑龙江八一农垦大学动物科技学院，大庆 163319）

**摘　要**：奶牛低钙血症是危害养牛业的常见疾病。针对目前血清钙检测方法操作复杂、成本较高、不适合基层现场应用等问题，本试验旨在研制一种适合临床使用的血清钙快速检测试剂盒。本试验在 EDTA 络合滴定原理的基础上，加入一种可增加变色范围的惰性染料；并通过正试验，筛选出最适剂量的指示剂、掩蔽剂和缓冲体系，组装成试剂盒；通过分光光度法确定试剂滴加顺序；通过回收率、重复性、特异性、干扰性、稳定性、开口稳定性实验对试剂及试剂盒进行综合评价。本试剂盒主要由 3 种试剂组成，即 A、B、C，其中 A 是缓冲体系，B 是掩蔽体系，C 是显色体系。回收率 100%，重复率 100%。在钙标准溶液中加入干扰物质 3mmol/L $Mg^{2+}$、3mmol/L $Fe^{2+}$、3mmol/L $Cu^{2+}$ 对钙离子的检测没有影响。不适合检测患有溶血和黄疸性疾病奶牛的血清样品。试剂盒在避光 4℃ 条件下可稳定保存 30d，开口试剂可保存 7d。选择大庆市 4 家奶牛场，运用奶牛血清钙邻甲酚酞络合铜标准检测法与研制的试剂盒进行对比试验，结果表明：试剂盒的符合率为 99.06%，应用血清钙试剂盒检测的低钙血症发病率为 5.25%，运用邻甲酚酞络合铜标准法检测的低钙血症发病率为 5.25%，假阳性率为 0。综上所述，该奶牛血清钙检测试剂盒可以代替邻甲酚酞络合铜标准方法，适合在兽医临床上推广应用。

**通讯作者**：E-mail：fuhewu@ 126. com

# 宁夏某地区牛支原体病的流行病学调查及分子进化分析

郭澍强，罗海峰，葛　松，王　静，何生虎*

（宁夏大学农学院，银川 750021）

**摘　要**：牛支原体（*Mycoplasma bovis*）是一种能够导致牛发生多种疾病，且感染率极高的病原，2013 年 11 月宁夏某牛场成年奶牛出现以肺炎、乳房炎及犊牛出现重呼吸困难、关节炎等为主要特征的病例，发病的犊牛和成年奶牛使用青霉素、氨苄西林、四环素、头孢类抗生素、柴胡、安乃近等药物治疗，无明显效果。结合病史、病理剖检及实验室初步诊断，怀疑为牛支原体感染。分离培养的菌落在 40 倍显微镜下呈“油煎蛋”样菌落，将分离到的 3 株可疑菌株纯培养后，使用牛支原体特异性引物和支原体 16S rRNA 通用引物扩纯培养物并测序，测序结果显示 3 个分离株以 16 SrRNA 通用引物扩增后测序结果与 GenBank 中登陆的牛支原体 HuBei-1 同源性最高，分别为 100.00%、99.86%、99.86%，与无乳支原体的同源性分别为 99.51%、99.38%、99.38%，三个分离株之间的同源性达到 99.93%。3 个分离株以牛支原体特异性引物扩增后测序结果与牛支原体的同源性分别为 99.81%、99.68%、99.80%，与无乳支原体的同源性分别为 82.45%、82.48%、82.48%。结果分析证实所分离菌株为牛支原体。同时采用 MEGA6.0 软件，使用 N-J 法对 3 个分离株的 16 SrRNA 序列与支原体代表菌株建立系统发育树，该系统发育树表明 3 个分离株与国际上的牛支原体代表菌株 PG45 亲缘关系最近，3 个分离株为牛支原体而不是无乳支原体，与特异性 PCR 扩增结果相同。将分离到的牛支原体菌株同牛支原体国际标准菌株和本实验室保存的 7 株牛支原体菌株进行亲缘性分析。同时采用 ELISA 诊断试剂盒对该地区 12 个规模化奶牛场的 230 份奶牛血清进行牛支原体血清学调查，结果显示牛群平均阳性率 26.52%，牛场阳性率高达 100%，表明牛支原体病在该地区呈流行趋势，亲缘性分析发现宁夏地区奶牛分离到的牛支原体无明显的时间性和地域性差别趋势。

**关键词**：牛支原体，分离与鉴定，流行病学调查，分子进化分析

* **通讯作者**：E-mail：heshenghu308@ 163. com

# 宁夏肉牛皮肤病原真菌的分离与鉴定

葛 松[1],蒋 万[2],何生虎[1*],余永涛[1],张蕾蕾[1],郭澍强[1],王 静[1]
([1]宁夏大学农学院,银川 750021;[2]宁夏大学实验农场,银川 750021)

**摘 要**:近年来,随着我国奶牛和肉牛产业的快速发展,牛真菌性皮肤病呈现出群发性的特点,发病率显著提高,人们对导致牛真菌性皮肤病的病原种类及其生物学特性缺乏了解,无法对该病进行有效的防控。本研究旨在从患真菌性皮肤病肉牛的皮屑、痂皮等病料样本中分离皮肤病原真菌并对其进行种属分类鉴定,为牛真菌性皮肤病的防治提供参考依据。对病料样本进行镜检观察并进行真菌的分离培养,根据菌株形态进行初步分类;通过真菌侵染ICR小鼠皮肤试验,根据其致病性筛选病原真菌;对病原真菌内转录间隔区(ITS)进行PCR扩增并测序,依据病原真菌ITS序列同源性比较及系统发育分析结果,结合菌株形态特点对病原真菌作出种属分类鉴定。结果显示,镜检发现皮屑病料中含有大量卵圆形小分生孢子,并且在毛发周围富集形成孢子壳,存在链状厚垣孢子,未见疥螨及虫卵等寄生虫;从病料样本中共分离到23株真菌,筛选出NXGY1、NXGY2等2株病原真菌,同源性比较与系统发育分析结果表明,这2株菌与疣状毛癣菌(*Trichophyton verrucosum*)同源性最高,遗传关系最近,并且这2株菌的菌丝及孢子形态与疣状毛癣菌也极为相似。结果提示,根据真菌形态特征及遗传进化分析结果,鉴定本研究分离的病原真菌NXGY1、NXGY2为疣状毛癣菌(*T. verrucosum*)。

**关键词**:肉牛,皮肤癣菌,系统发育分析,疣状毛癣菌

# Isolation and identification of dermatophytes from beef cattle in Ningxia

GE Song[1],JIANG Wan[2],HE Shenghu[1*],YU Yongtao[1],
ZHANG Leilei[1],GUO Shuqiang[1],WANG Jing[1]
[1]*College of Agronomy,Ningxia University,Yinchuan* 750021;[2]*Teaching Experimental Farm,*
*Ningxia University,Yinchuan* 750021

**Abstract**: In recent years,with the rapid development of dairy and beef cattle industry in our country,the cattle fungal skin disease presents the characteristic of clusters, and the incidence of which is getting significantly increased,and people are lack of understanding about the pathogenic species and their biological characteristics of the disease,so people are inability to make effective prevention and control of the disease. An experiment was carried out to isolate and identify the pathogenic fungi from the skin scrapings of beef cattle affected by dermatophytes in Ningxia, and provide reference for prevention and control of dermatophytes among the cattle population for people. The skin scrapings and hairs were observed under the light microscope and cultured to isolate fungi; the classification of each isolate was performed preliminarily according to the morphological characteristics; the experiment that the isolates infect the healthy ICR mice's skin was carried out to confirm whether the isolates do has pathogenicity and screening pathogenic fungi; the ITS sequences of the suspected pathogenic fungi were amplified and sequenced, and finally classified according to the homology comparison and phylogenetic analysis combine with the morphological characteristics. Under the light microscope,there's a lot oval microconidia in the skin scrapings and some of which are gathering around the hair to form like a shell, and there exists some chlamydospore in chains but none parasites like sarcoptes mites and their eggs in the skin scrapings; twenty-three fungi were isolated from the skin scrapings and hairs,and two of them NXGY1、NXGY2 were confirmed to be pathogenic fungi, and which were closely related to *Trichophyton verrucosum* according to the homology and phylogenetic analysis, and the morphologies of hypha and conidia in the two fungi were also highly similar to *T. verrucosum*. The results indicated that NXGY1、NXGY2 isolated in this research were *T. verrucosum*.

**Keyword**: Beef Cattle, Dermatophytes, Phylogenetic analysis, *Trichophyton verrucosum*

**通讯作者**: E-mail: heshenghu308@163. com

## 牛源胸腺嘧啶依赖型金黄色葡萄球菌小菌落突变株的生理生化特性分析

朱立力，王奇惠，曲伟杰*

（云南农业大学动物科学技术学院，昆明 650201）

**摘　要**：金黄色葡萄球菌在患乳房炎的奶牛乳汁中检出率最高，该属细菌在自然条件下可以发生变异，抵抗力较强，容易对抗生素产生耐药性。1957 年国外学者首次报道了金黄色葡萄球菌小菌落突变株的存在，在后来的陆续研究表明，小菌落是一种利于持续复发性感染的细菌致病形式。笔者对由某牛场患奶牛乳房炎奶样中分离出的一株疑似金黄色葡萄球菌小菌落突变株（SCVs）进行了鉴定、生理生化特性研究、药敏试验和补偿实验等，希望为由金黄色葡萄球菌 SCVs 引起的奶牛慢性乳房炎的预防和控制及其致病机制的研究奠定前期基础。本研究对由患慢性奶牛乳房炎奶样中分离出的一株疑似金黄色葡萄球菌 SCVs 进行耐盐能力实验、生化编码鉴定管试验、触酶试验和血浆凝固酶试验等生理生化特性研究；通过金黄色葡萄球菌相关保守基因片段（nuc 、nucA、16 SrDNA）多重 PCR 扩增鉴定；药敏试验；并进行甲萘醌、硫胺素、胸腺嘧啶和血红素等补偿实验。分离出一株黄色葡萄球菌 SCVs，该菌含有金黄色葡萄球菌菌种特异性基因 nuc 和 nucA。菌落形态主要表现为菌落细小、生长缓慢、溶血能力下降；凝固酶活性较亲本株下降；耐盐能力较亲本株降低；革兰染色为革兰阳性球菌，呈葡萄状排列。药敏试验中，对磺胺甲基异噁唑等磺胺类药物具有抗药性。补偿试验发现分离到的 SASCV 在含硫胺素、甲萘醌及血红素的培养基中依然生长不良，在含胸腺嘧啶培养基中生长良好，较原株大，比亲本株稍小，鉴定该金黄色葡萄球菌 SCVs 为胸腺嘧啶依赖型。成功分离鉴定出一株胸腺嘧啶依赖型金黄色葡萄球菌 SCVs，并对其生理生化特性进行了研究，为由金黄色葡萄球菌 SCVs 引起的奶牛慢性乳房炎的预防和控制及其致病机制的研究奠定前期基础。

**关键词**：金黄色葡萄球菌，小菌落突变株，分离鉴定，胸腺嘧啶依赖型

**基金项目**：国家自然科学基金资助（No. 31260629）

* **通讯作者**：E-mail：qwj1204@ sohu. com

## 浅析抵抗素与炎症的相关性

万涛梅，王　婵，苟丽萍，任　毅，李　浪，袁贵强，左之才*

（四川农业大学，雅安 625014）

**摘　要**：抵抗素是近年发现的一种新型小分子蛋白，不仅能诱导胰岛素抵抗，还能调节细胞因子的分泌与表达，行使趋化因子的功能，在某些自身免疫性疾病和慢性炎症疾病中发挥重要的作用。本文就抵抗素的受体、信号通路、与炎症标志物的关系等方面阐述其与炎症和免疫的相关性，为进一步探索抵抗素的功能奠定基础。

**关键词**：抵抗素，炎症，免疫，相关性

* **通讯方式**：E-mail：zzcjl@ 126. com

## 奇异变形杆菌和大肠杆菌混合感染导致奶牛关节脓肿和腹泻的报道

王旭荣,王国庆,张景艳,杨志强,孟嘉仁,李建喜*
(中国农业科学院兰州畜牧与兽药研究所,兰州 730050)

**摘 要**:为查明2012年底甘肃省某奶牛场奶牛关节脓肿和腹泻的病因,选择合适的治疗药物。经牛病毒性腹泻黏膜病毒(BVDV)和牛轮状病毒(BRV)检测、细菌分离鉴定、小鼠致病性试验、药敏试验及血清中钙磷含量测定等综合分析,确诊该病例是由致病性奇异变形杆菌和大肠杆菌混合感染引起的。这2种分离菌对青霉素、阿莫西林、庆大霉素、卡那霉素、四环素、氟哌酸及复方新诺明等常用药物严重耐药,均对氟苯尼考中度敏感。经氟苯尼考治疗和加强饲养管理后该病得到有效控制。

**关键词**:奇异变形杆菌,大肠杆菌,关节脓肿,腹泻,奶牛

* 通讯方式:E-mail:lzjianxil@163.com

## 犬尿石症 X 光和 B 超诊断对比研究

许楚楚,孙雨航,夏 成*,罗春海,郑家三
(黑龙江八一农垦大学动物科技学院,大庆 163319)

**摘 要**:对X光和B超对犬尿石症的诊断做对比。选取10例尿石症患犬病例,经临床检查确定3例膀胱结石,5例尿道结石和2例肾结石。通过分析X光和B超图像,对X光和B超两种诊断方法做对比。结果显示,对所有病例均拍摄腹部正位和侧位X射线片,在腹部平片上肾脏、输尿管、膀胱和尿道区出现高密度影像者诊断为阳性结石。对于高密度、体积较大的膀胱结石、尿道结石和肾脏结石,X线可以确定其位置及大小形态。膀胱结石B超检查表现为膀胱腔内有强回声点或团块,声影明显,尤其是使用高频探头时,结石会移向膀胱的最低处,在冲击式触诊或移动动物时结石位置会改变,如果同时伴有炎症变化时,结石可能会粘连在膀胱壁上,这些结石很难与膀胱壁的营养不良性钙化区分开,这些钙化的位置是固定的。膀胱结石有时也能看到膀胱的最低处有沉积,当被搅动后,沉积会向云一样移动,并重新沉积排列,表面呈现水平状。B超检查显示肾脏钙化会引起皮质回声增强,血钙过高性肾病的犬,皮质髓质结合部会出现强回声边界或缘。结果提示对于密度高、体积较大的膀胱结石、尿道结石,X线检查可确定位置及大小形态。B超扫查能判定结石的大小、数目、形状、部位和膀胱壁有无增厚等,显示X射线阴性结石,弥补了X线平片检查的不足,但B超检查对结石大小具有局限性。X线平片与B超实时扫描结合可以促进犬尿石症得到更好的诊断及治疗。

**关键词**:尿石症,犬,X光,B超,诊断

* 通讯方式:E-mail:xcwlxyf@sohu.com

## 妊娠期与泌乳期氟摄入对子代雄性小鼠学习记忆能力的影响

张玉良,曾 威,韩海军,牛瑞燕,孙子龙*
(山西农业大学,太谷 030801)

**摘 要**:众所周知,氟是一种全身性毒物,能透过血脑屏障进入脑组织,从而导致人或动物学习记忆能力的下降。然而,妊娠期与泌乳期的母源氟摄入对子代动物学习记忆能力的影响未见报道。本试验将24只怀孕母鼠随机分为四组,即对照组(供

给蒸馏水)、低氟组(含 25mg/L NaF)、中氟组(含 50mg/L NaF)、高氟组(含 100mg/L NaF),饮水摄氟直至分娩后 21d,从每组中随机选取雄性仔鼠 6 只,采用开场实验检测小鼠的自发活动及探究行为,八臂迷宫实验记录其工作记忆错误和参考记忆错误。结果表明,在开场试验中,低氟组、中氟组、高氟组小鼠的自发活动和探究行为与正常对照组相比没有显著差异;八臂迷宫实验中高氟组、中氟组的工作记忆错误次数和参考记忆错误次数都比正常对照组显著增加,低氟组没有显著性变化。结论:妊娠期与泌乳期氟摄入对子代雄性小鼠的自发性活动没有显著影响,而其学习记忆能力随着摄入氟化钠浓度的增加呈下降趋势。

**关键词**:氟摄入,子代小鼠,学习记忆,开场实验,八臂迷宫

* **通讯作者**:E-mail:sunzilong2000@ 163. com

## 日粮硒对小鼠肾脏硒沉积及抗氧化酶基因表达的影响

刘碧涛,张 琴,郑良焰,郭 凯,余文兰,刘正伟,郭翠丽,陈 叶,唐兆新*

(华南农业大学兽医学院,广州 510642)

**摘 要**:硒元素自 1817 年被发现以来,其在动物机体抗氧化、免疫力调节、代谢调节及癌症预防治疗等方面的功能都已得到证实。硒元素在体内主要以硒蛋白的形式发挥生物学作用,硫氧还蛋白还原酶(thioredoxin reductase,TrxR)作为含硒酶之一与硫氧还蛋白(thioredoxin,Trx)和还原型辅酶Ⅱ(NADPH)共同构成了硫氧还蛋白系统,在抗氧化系统中发挥重要作用。超氧化物歧化酶(superoxide dismutase,SOD)和过氧化氢酶(hydrogen peroxidase,CAT)分别在清除生物体内超氧阴离子自由基($O_2^-$)和分解过氧化氢($H_2O_2$)方面发挥着巨大作用,是机体抗氧化酶系统中重要组成部分。本试验以小鼠为试验对象,在日粮中添加不同剂量的无机硒($Na_2Se_2O_3$),探讨硒对小鼠生长性能、肾脏硒沉积和肾脏 TrxR2、SOD1、SOD2 和 CAT 基因表达的影响,旨在研究日粮硒对小鼠肾脏硒沉积及抗氧化酶基因表达的影响,为进一步研究硒对动物机体的生物学功能及肾脏抗氧化机制提供理论依据。

结果显示:肾脏硒沉积水平随日粮中硒含量增加而升高,Ⅰ组显著低于其他组($P<0.05$),饲喂到第 56 天,Ⅱ、Ⅲ、Ⅳ组组间差异均不显著($P>0.05$);Ⅰ组的硫氧还蛋白还原酶(TrxR2)mRNA 表达量显著低于Ⅱ、Ⅲ组($P<0.05$),Ⅱ组高于Ⅲ组,且差异显著($P<0.05$);Ⅰ组的超氧化物歧化酶 1(SOD1)和超氧化物歧化酶 2(SOD2)mRNA 表达量均显著低于Ⅱ、Ⅲ组($P<0.05$),Ⅱ、Ⅲ组间 SOD1 表达量差异不显著($P>0.05$),SOD2 表达量差异显著($P<0.05$),且Ⅱ组高于Ⅲ组;Ⅰ、Ⅱ组的过氧化氢酶(CAT)mRNA 表达量显著高于Ⅲ组($P<0.05$),但Ⅰ、Ⅱ组间差异不显著($P>0.05$)。

综上所述,补硒可提高肾脏组织硒沉积水平,且随着饲喂时间增加,肾脏硒沉积会在高硒添加组中趋于饱和;缺硒和补硒过量都会降低肾脏 TrxR2、SOD1、SOD2 和 CAT mRNA 的表达水平。

**关键词**:硒,肾脏硒沉积,抗氧化酶,基因表达

* **通讯方式**:E-mail:tangzhaoxin@ hotmail. com

## 肉鸡硫氧还蛋白原核表达载体的构建、蛋白纯化及其活性鉴定

余文兰[1],胡莲美[1],王朵朵[2],吴富旺[2],张晓战[1],郭凯[1],张浩[1],郭剑英[1],潘家强[1],唐兆新*[1]

(1. 华南农业大学 兽医学院,广州 510642;2. 中国科学院华南植物园,广州 510642)

**摘 要**:旨在克隆肉鸡硫氧还蛋白(Trx)基因,建立肉鸡硫氧还蛋白的原核表达体系,纯化具有生物活性的重组肉鸡硫氧还蛋白。根据 GenBank 中 Trx 的 cDNA 序列及原核表达载体 pET28a(+)中的多克隆位点设计引物,进行 Trx 的克隆和原核表达载体的构建;将重组载体在大肠杆菌中表达;应用镍柱亲和层析的方法对该融合蛋白进行纯化;采用胰岛素还原法对重组肉鸡

硫氧还蛋白进行活性鉴定。结果显示成功构建了 pET28a-Trx 重组质粒,该重组蛋白的最适诱导条件是 37℃下用 0.4 mmol/L IPTG 诱导 4 h。纯化的目的蛋白纯度可达到90%,浓度为 4 mg/ml。纯化的重组肉鸡硫氧还蛋白(GgTrx)具有较高的还原胰岛素二硫键的能力,并且呈现出浓度效应。结果提示已成功在大肠杆菌中表达了重组肉鸡硫氧还蛋白,纯化的目的蛋白纯度高,具有还原二硫键的活性,为将该蛋白应用于治疗氧化应激性疾病奠定了基础。

**关键词**:硫氧还蛋白,肉鸡,原核表达,蛋白纯化

**通讯作者**:E-mail:tangzx@ scau. edu. cn

## 山羊羔"猝死症"的诊治

达能太*,达布拉,敖日格乐

(内蒙古阿拉善动物中毒病防治研究所,阿拉善 750300)

**摘　要**:我国缺硒地带分布在北纬 21 ~57°和东经 97 ~135°,内蒙古阿拉善盟左旗地理坐标在北纬 37°24′ ~41°52′,东经 103°21′ ~106°51′,正好是在我国缺硒地带内。近些年来左旗羊缺硒病连年不断,尤其是干旱少雨,入春后羊只长期处于饥饿半饥饿状态,补饲牧草主要是玉米秸秆,饲喂牧草品种单一,致使硒等多种营养元素缺乏,给牧民的生产和生活造成极大的经济损失。

2014 年 6 月 9 日接到宗别立镇内蒙古阿拉善左旗新顺生态养殖有限责任公司的报告,该公司段某某的羊发生急性死亡,请求急速诊治,接到报告后我们迅速前去诊治,据调查,该公司共饲养羊 1600 只,其中绵羊 110 只,产羔 320 只,山羊羔 293 只,于6 月7 日开始发病,发病山羊羔110 只,发病率达37. 54% ,死亡50 只,死亡率达45. 45% 。畜主表述,发病羊主要表现食欲废绝,有的只采食玉米,不采食饲草,喜欢采食羊粪、骨头、土、石块、煤渣、布块等异物。经检查羊只体温 39. 5 ~42℃ ,心跳75 ~115 次/分,呼吸 70 ~80 次/分,一般 3 ~5 天内死亡。死羊尸体膘情良好,眼结膜苍白。今年已注射口蹄疫疫苗、羊痘疫苗、羊三联四防菌苗。曾用重症头孢、五号神针、牛羊怪病、热毒目赤、青霉素等药物治疗无效。

本次发生的疾病根据发病情况,临床症状、病理剖检和试验性治疗等综合分析,确诊为山羊羔缺硒症。

治疗上采取对全群羊,肌肉注射亚硒酸钠维生素 E 注射液(由重庆金帮动物药业有限公司,生产日期:201172,有效期:20136,批号:110701,每 10ml 含亚硒酸钠 10mg、维生素 E0. 5g)2ml,隔 10d 再注射 1 次。经过上述方法治疗彻底控制了病情。

**关键词** 山羊羔,猝死症,硒缺乏,诊治

**通讯作者**:E-mail:fdant08@ 126. com

## 上海规模牧场奶牛疾病流行状况及分析

张峥臻,张瑞华,张克春*

(上海市奶牛研究所)

**摘　要**:牛只淘汰是牛群更新、产量提高和效益增加的一个手段,现阶段上海地区奶牛场淘汰多为疾病引起的被动淘汰,严重制约了奶牛场经济效益的提高。因此分析奶牛场疾病流行状况,并分析深层次的原因刻不容缓。旨在分析上海地区奶牛疾病淘汰情况,了解引起该地区牛只被淘汰的主要疾病,为控制疾病淘汰寻找方向,同时使用实验室检测手段进一步的了解上海地区牛群各种病原在牛只疾病淘汰中的作用,为疾病的控制提供科学参考。本试验(1)通过奶牛场疾病淘汰记录来分析奶牛场成乳牛以及犊牛的发病情况。(2)使用 ELISA 方法对引起奶牛疾病的各种重要病毒性疾病进行血清学普查和抽查,并对个别典型病例进行筛查。(3)使用传统的细菌分离鉴定和药敏试验技术持续的对各牧场的乳房炎样品进行检测分析。(4)使用酮粉法和血清 BHBA 检测法调查新产牛群酮病(含亚临床酮病)发病规律。

结果显示,(1)从成乳牛疾病淘汰统计结果来看:泌乳系统疾病、肢蹄病和繁殖疾病是导致牛只被淘汰的主要疾病,分别占疾病淘汰总数的32.04%,23.97%和20.79%;新产牛(分娩后前3个月)是疾病淘汰发生的主要阶段,疾病淘汰数超过了总淘汰数的35%;新产牛因泌乳和肢蹄病被淘汰数分别占新产牛淘汰总数的37.11%和28.85%。(2)犊牛因呼吸道疾病和消化道疾病而被淘汰的数量分别占到了疾病总淘汰数的30.25%和27.94%,此外因肢蹄病被淘汰的犊牛也占到了犊牛总淘汰数的31.64%。另外通过对28头犊牛肺炎的血清检测发现,71.43%的牛只呼吸道合胞体(BRSV)血清抗体阳性,100%的牛只牛副流感3型(PI-3)血清抗体阳性,96.43%的牛只BCV血清抗体阳性,46.43%的牛只IBR血清抗体阳性,78.57%的牛只BVDV血清抗体阳性。(3)上海地区奶牛场牛传染性鼻气管炎(IBR)平均血清抗体阳性率为39.04%;牛病毒性腹泻(BVDV)血清抗原阳性率为0.80%;牛冠状病毒(BCV)血清抗体阳性率为97%;轮状病毒(BRV)血清抗体阳性率为97.10%;牛病毒性白血病(BLV)血清抗体阳性率为35.89%。(4)对上海地区奶牛临床性和隐性乳房炎样品的检测结果发现,凝固酶阴性葡萄球菌、金黄色葡萄球菌和无乳链球菌是主要致病菌,分别从21.78%,11.46%和33.72%的样品中分离到,此外还有7.62%,2.09%和3.84%的样品分离到酵母菌、大肠杆菌和克雷伯杆菌。分离到的金黄色葡萄球菌和凝固酶阴性葡萄球菌对头孢噻呋的敏感比例为100%和72.73%;分离到的大肠杆菌和其他肠杆菌科细菌对环丙沙星敏感率最高但分别只有51.35%和56.90%。

结果提示:(1)上海地区成泌牛的疾病淘汰主要病因是肢蹄、乳房炎和繁殖障碍,围产牛和新产牛管理不到位是新产牛被大量淘汰的根本原因,而繁殖障碍只是诸多原因累积到最后的结果。需要开展围产牛和新产牛的代谢病研究,如SARRA、脂肪肝、酮病发病原因等。(2)犊牛疾病淘汰主要病因是消化道、呼吸道和肢蹄疾病,病原微生物在犊牛疾病中有着重要作用。(3)上海地区牛群各种病毒性疾病感染状况不容乐观,各种病毒性疾病几乎都有流行,急需开展相关的诊断和防控研究。

**关键词**:奶牛疾病淘汰,奶牛乳房炎,奶牛病毒性疾病,酮病

**通讯作者**:E-mail:zhangkc68@126.com

# 肾型IBV感染对鸡临床病理学及相关基因表达的影响

邹跃龙,朱书梁,郭小权,张彩英,刘　平,曹华斌,胡国良

(江西农业大学动物科学技术学院,南昌 330045)

**摘　要**:家禽痛风是商品鸡中的常见多发病,遍布于世界各地,其发病率可达85%,死亡率高达30%,引起家禽痛风的原因有多种。在我国,肾型传染性支气管炎病毒(IBV)感染是引起鸡痛风的几种主要原因之一,给养禽业造成很大的损失。有关其发病机理国内外虽做过一些研究,但迄今仍不清楚。为探讨肾型IBV感染对鸡机体的影响,从而进一步阐明其诱发鸡痛风的机理,本论文将35日龄海兰褐蛋鸡240羽,随机分为对照组和病毒组,每组3个重复,每个重复40羽。试验第0天,对照组滴鼻双蒸水,0.2ml/羽,病毒组按鸡胚半数致死量测定结果($10^{-5}$/0.2ml)滴鼻攻毒,0.2ml/羽,试验期22 d,人工复制痛风病例,于试验第8天、第15天、第22天,每次每组随机抽取6羽,采血、分离血清,同时剖检、采集肾脏,分别应用临床病理学方法、实时荧光定量PCR方法,测定血清电解质、肾功能、抗氧化功能以及肾脏中XOD、Bak1、P53 mRNA的表达。结果表明,肾型IBV感染初期鸡血清Ca、P和K显著下降,感染后期血清Ca、P和K显著下降;肾型IBV感染可导致鸡肾脏肿大,呈花斑肾,血清UA、CR显著升高,血清SOD、T-AOC、GSH-Px活性显著下降,血清XOD活性、MDA含量显著升高;肾型IBV感染可导致鸡肾脏中XOD、Bak1和P53 mRNA表达显著升高。说明,肾型IBV感染会导致鸡机体血清电解质紊乱,抗氧化功能降低,肾功能受损,肾脏XOD、Bak1和P53 mRNA表达增强。

**关键词**:鸡,肾型IBV,XOD,凋亡基因

**项目来源**:国家自然科学基金资助项目(31260627),江西省青年科学家培养对象项目(20122BCB23022),江西省科技支撑计划(2010BNB00501)

**作者简介**:邹跃龙(1989.2-),男,硕士研究生

**通讯作者**:郭小权,(1976.6-),男,博士,教授,E-mail:xqguo20720@aliyun.com

# 饲料中 $AFB_1$、ZEA、DON 的污染情况分析

周　闯,吴文达,张海彬*

(南京农业大学动物医学院动物中毒病实验室,南京 210095)

**摘　要**:霉菌毒素(Mycotoxin)是由某些真菌生长成熟后产生的有毒次级代谢产物,它会对人畜造成损害。近年来饲料中霉菌毒素污染日益严重,严重阻碍了畜牧业的发展,因此不断地检测饲料中霉菌毒素含量是很有必要的。从 2012 到 2013 年 9 月份,通过对来自全国的 255 份饲料样品进行黄曲霉毒素 $B_1$(Aflatoxin$B_1$,$AFB_1$)、玉米赤霉烯酮(Zearalenone,ZEA)和脱氧雪腐镰刀菌烯醇(Deoxynivalenol,DON)含量的检测分析来了解目前我国饲料中这三种霉菌毒素的污染情况。本试验利用高效液相色谱法(HPLC)对饲料样品进行检测,共进行了 476 次检测,并对检测结果进行数据分析。结果表明:三种霉菌毒素污染情况仍然严重,总检出率高达 63.11%,总超标率高达 24.95%;一直令人重视的黄曲霉毒素 $B_1$,其污染情况有所减轻,但最大值达 488.86 μg/kg,高出国家标准近 25 倍;三种霉菌毒素中,呕吐毒素的污染情况最为严重,其检出率从 2012 年到现在,一直居高不下,均在 75% 以上,风险很高;与其他原料相比,饼粕类饲料原料霉菌毒素污染情况更为严重,污染率可达 72.73%,其中黄曲霉毒素的污染情况特别严重,其阳性平均值(109.88 μg/kg)和最大值(488.86 μg/kg)已经超出国家相关标准 5 和 20 倍,因此饼粕类饲料原料存在高度风险;检测两种以上毒素的饲料样品有 72 份,能同时检出两种以上毒素的样品占到 77.78%,造成不同毒素产生联合毒性的机率变大。此次调查结果为今后有针对性的防控饲料中霉菌毒素的污染提供一定理论依据。

**关键词**:霉菌毒素,饲料,原料,污染,HPLC

* **通讯方式**:E-mail:haibinzh@ njau. edu. cn

# 羧基化多壁碳纳米管对大鼠睾丸 p38 和 JNK 通路的影响

张建海[1*],郄明丽[1],罗广营[1],李志慧[1],王俊东[1*],郭良宏[2]

(1. 山西农业大学动科院,生态畜牧与环境兽医学山西省重点实验室,太谷 030801;

2. 中国科学院生态环境研究中心,环境化学与生态毒理学国家重点实验室,北京 100085)

**摘　要**:随着纳米技术的日新月异,各种纳米材料日益进入我们日常生活环境周围,也不断被应用于畜牧兽医生产实践,如纳米饲料、药物制剂及畜舍建筑等。这些材料是否对动物和人具有潜在危害,尤其是生殖毒害成为我们最关心的问题。因此,本研究以典型纳米材料多壁碳纳米管为对象,经 TEM、红外和拉曼光谱、ICP-MS 及 Zeta 电位等充分表征后,将分散性良好的不同剂量(2.5,5,10mg/kg · bt)羧基化多壁碳纳米管溶液连续给雄性大鼠灌胃 64d,观察小鼠体重、脏体比、精液质量和睾丸组织结构及超微结果变化,用 FCM 检测生精细胞周期和凋亡,后提取睾丸组织 RNA,采用定量 PCR 技术对 MAPKs 通路的 JNK 通路和 P38 通路的 26 个相关基因进行了分析。结果表明,纳米管长期染毒可引起大鼠精子质量和形态学发生显著变化,睾丸组织的 MAPK 不同等级的相关基因 *CASP*、*TRAF*2、*MEKK*1、*MLK*3、*MKK*4、*MKK*7、*JNK*1、*JNK*3、*EIK*-1、*JnuD*、*DAXX*、*P*38、*Sapla*、*MEF*2*C* 和 *P*53 表达显著下调,基因 *JNK*2 和 *GADD*153 表达显著上升,*TNE-α*、*IL-β*、*TGF-β*、*ASK*1、*MKK*3、*MKK*6、*c-Jun*、*ATF*-2 和 *MAX mRNA* 表达没有明显变化。说明羧基化修饰多壁碳纳米管可能通过引起睾丸组织 MAPKs 通路中相关基因的变化来影响雄性生殖功能。

**关键词**:羧基化多壁碳纳米管,睾丸,MAPKs 通路

* E-mail:wangjd53@ outlook. com;jianhaiz@ 163. com

## 脱氧雪腐镰刀菌烯醇抑制大鼠胃分泌及消化功能研究

王育伟，吴文达，张海彬*

（南京农业大学动物医学院，动物中毒病实验室，南京 210095）

**摘　要**：脱氧雪腐镰刀菌烯醇（deoxynivalenol，DON）是常见的真菌毒素之一，在小麦、燕麦和玉米等谷物及副产品污染率极高，对食品安全及人畜造成了严重威胁。国内外研究表明消化系统是最敏感的靶系统，也是研究热点之一。研究表明 DON 可导致肠病理损伤及吸收功能下降。胃功能正常与否直接影响吸收功能，尤其是胃液分泌功能与胃液消化吸收关系密切，而有关 DON 对胃分泌与消化功能的研究却很少。为了研究 DON 对胃分泌与消化能力的影响，本试验对 wistar 大鼠进行 DON 染毒，利用幽门结扎模型收集胃液。并对胃液分泌功能、消化能力和金属离子水平进行了测定。结果表明 1、5 和 25 mg/kg bw DON 染毒后胃液总体积分别下降了 25%、51% 和 61%。胃液酸度也呈现剂量依赖性下降，pH 值分别达到了 3.2、3.81 和 6.65，对照组只有 1.9。胃蛋白酶含量分别下降了 8%，18% 和 51%，阿利新蓝法对黏蛋白测定结果表明胃液黏蛋白显著下降。胃液体积、pH 值、粘蛋白含量、蛋白酶含量测定结果表明，DON 染毒抑制了大鼠胃液的分泌功能。本试验利用福林酚法对大鼠胃液蛋白酶活性进行了测定，结果表明染毒后胃液胃蛋白酶总活性分别下降了 27.97%、55.36% 和 67.82%。DON 染毒后胃内残渣湿重分别增加 2 倍、2.58 倍和 5.4 倍；干重分别增加了 43%、2.5 倍和 24.6 倍。以上结果从不同角度表明 DON 染毒抑制了大鼠胃液的消化功能。综上所述，本试验表明 DON 染毒抑制了大鼠胃液的分泌功能，同时胃液消化能力也明显降低，DON 也引起了大鼠胃内金属离子的紊乱。揭示出 DON 对消化系统功能的抑制与胃功能下降有关，其详细机理仍待进一步研究。

**关键词**：脱氧雪腐镰刀菌烯醇，大鼠，胃分泌，消化

* **通讯作者**：E-mail：haibinzh@ njau. edu. cn

## 武定鸡抗马立克氏病育种的研究

周　勇，白文顺[1]，张　健[1]，杨斯涵[1]，杨亮宇[2]*

（1. 云南农业大学动物科学技术学院；2. 云南农业大外语学院，昆明 650201）

**摘　要**：目前，有关武定鸡抗马立克氏病（MD）育种的研究尚未见报道，为明确 MHC-B 单倍型在武定鸡中分布情况以开展武定鸡抗 MD 品系的选育，研究采用对血细胞 DNA 进行微卫星 LEI0258 位点和微卫星 MCW0371 位点 PCR 扩增及测序的方法，对 20 个试验鸡群 B 单倍型进行分型，将试验鸡按不同鸡群分为公鸡组和母鸡组，并对发现的部分 B 单倍型纯合子试验鸡 F1 代 10 日龄武定雏鸡进行京-1 株马立克氏病毒攻毒试验，选出抗 MD 最强的试验鸡。

结果在试验鸡群中发现了 B1、B2/29、B5、B6、B10/24/26、B11、B13、B14、B15、B17、B21、B22 和 B27 单倍型，其中 B11 单倍型出现频率最高，达 37%。母鸡组中发现存在 B5/B5、B11/B11、B14/B14、B15/B15 和 B21/B21 单倍型纯合子试验鸡，分别为 1%、20%、2%、2% 和 4%；公鸡组中发现存在 B5/B5、B6/B6、B11/B11、B14/B14、B15/B15、B21/B21 单倍型纯合子试验鸡，分别为 5%、1%、14%、3%、1%、4%。测序发现 MD 抗性和易感 B 单倍型在 193 ~ 230 bp“TTCTTTCTTTCC”重复单元存在个数上的差异，并且易感 B 单倍型在 230 ~ 266 bp 的重复单元部分变为“GGATTTTGAGCC”、“AAAAAAATCACC”和“ACAAAATGAGCC”重复单元。攻毒试验表明 B5/B5、B15/B15、B11/B11、B14/B14 和 B21/B21 单倍型纯合子雏鸡对 MD 易感性依次减弱，发病率为 100%，100%，80%，40%，20%。综上所述，选育出了 MD 抗性最强的 B21/B21 单倍型纯合子试验公鸡和母鸡，它们不存在个体差异，具有高度的遗传稳定性，可作为抗 MD 品系进行培育。

**关键词**：武定鸡，马立克氏病，B 单倍型

**通讯作者**：E-mail：yangliangyu@ 163. com

# 小花棘豆对小鼠脑组织 NO-cGMP-Glu 途径的影响

王　帅[1,2],贾琦珍[2],陈根元[2]
(1.塔里木大学动物科学学院,阿拉尔 843300;2.新疆生产建设兵团塔里木畜牧科技重点实验室,阿拉尔 843300)

**摘　要**:小花棘豆(*Oxytropis glabra* DC)是西北地区危害草原畜牧业的主要毒草之一,动物大量摄食后会造成机体广泛性损伤,其中以神经系统损伤最为显著。前期研究表明,苦马豆素(Swainsonine。SW)是小花棘豆的主要毒性物质,可通过影响神经系统的自由基代谢而发挥毒性作用。一氧化氮(NO)是具有高度反应性的自由基,也是重要的细胞内信号转导分子。本研究通过研究小花棘豆中毒对小鼠不同脑区 NO、环磷鸟苷(cGMP)和游离谷氨酸(Glu)的影响,进一步探讨小花棘豆的中毒机理。将 40 只小鼠随机分为 4 组,即对照组和试验Ⅰ、Ⅱ、Ⅲ组,对照组仅饲喂全价日粮,3 个试验组分别按照每千克体重 1、5、10 g 的剂量饲喂小花棘豆,第 63 天小鼠出现典型小花棘豆中毒临床症状后全部捕杀,采集全脑,通过检测试剂盒测定小鼠各脑区 NO、cGMP 及 Glu 含量。试验Ⅰ组小鼠大脑、小脑、丘脑及海马的 NO、cGMP 及 Glu 含量均显著高于对照组($P < 0.05$);试验Ⅱ组和试验Ⅲ组小鼠大脑、小脑、丘脑及海马的 NO、cGMP 及 Glu 含量均极显著高于对照组($P < 0.01$),但 3 个试验组脑干中 3 种物质含量与对照差异均不显著($P > 0.05$)。同一试验组中小脑 3 种物质的含量变化较海马、大脑和丘脑明显。结果提示小花棘豆中毒可在一定程度上可影响小鼠脑组织中 NO、cGMP 及 Glu 含量,并表现出明显的剂量效应,小脑、海马、大脑和丘脑是小花棘豆毒性物质作用的靶区,通过影响这几个脑区信号转导而产生毒性作用。

**关键词**:小花棘豆,一氧化氮,环磷鸟苷,谷氨酸

**基金项目**:国家自然科学基金(30110497),新疆生产建设兵团塔里木畜牧科技重点实验室开放课题(HS201409)。

**作者简介**:王帅(1984 -),男,山西长治人,硕士,实验师,主要从事动物中毒病与毒理学方面的研究。E-mail:wangshuaidky@126.com

**通讯作者**:陈根元(1984 -),男,湖南岳阳人,硕士,助理研究员,主要从事动物基因工程方面的研究。

# 小花棘豆中毒对小鼠睾丸 α-甘露糖苷酶活性及基因表达量的影响

贾琦珍[1,2],王　帅[2],陈根元[2]
(1.塔里木大学生命科学学院,阿拉尔 843300;2.新疆生产建设兵团塔里木畜牧科技重点实验室,阿拉尔 843300)

**摘　要**:小花棘豆(*Oxytropis glabra* DC)是新疆分布最广泛、危害最严重的毒草,动物大量采食后会造成中毒。其中中毒动物的繁殖机能下降明显。国内外研究表明,小花棘豆主要毒性成分苦马豆素可通过抑制 α-甘露糖苷酶(AMA)的活性而发挥毒性作用。现有研究表明小花棘豆中毒动物睾丸出现细胞空泡变性和坏死等病理变化,但尚未有关于 AMA 在睾丸中分布和表达的报道。本研究通过探讨小花棘豆中毒对小鼠睾丸 AMA 活性及表达的影响,以期为小花棘豆中毒的防治提供基础。将 40 只小鼠随机分为 4 组,即对照组和试验Ⅰ、Ⅱ、Ⅲ组,对照组仅饲喂全价日粮,3 个试验组分别按照每千克体重 1、5、10 g 的剂量饲喂小花棘豆,第 63 天小鼠出现典型小花棘豆中毒临床症状后全部捕杀,采集睾丸,通过 ELISA 法和实时荧光定量 PCR 法检测小鼠睾丸中 AMA 的活性及表达。试验组及对照组 AMA 基因均有表达,但各试验组中转录表达水平不同,试验Ⅰ组小鼠睾丸 AMA 基因的相对表达量高于对照组,但差异不显著($P > 0.05$);试验Ⅱ组和试验Ⅲ组小鼠睾丸 AMA 基因的相对表达量均低于对照组,其中试验Ⅲ组显著低于对照组($P < 0.05$)。试验Ⅰ组和试验Ⅱ组小鼠睾丸 AMA 活性与对照组差异不显著($P > 0.05$),但试验Ⅲ组小鼠睾丸 AMA 活性显著低于对照组($P < 0.05$)。结果提示小花棘豆中毒在一定程度上可影响小鼠睾丸中 AMA 的活性及其基因的转录表达,其中低剂量可促进 AMA 表达,而高剂量对其表达产生抑制。

**关键词**:小花棘豆,睾丸,α-甘露糖苷酶

**基金项目**：国家自然科学基金(30110497)，新疆生产建设兵团塔里木畜牧科技重点实验室开放课题(HS201409)。

**作者简介**：贾琦珍(1984－)，女，新疆伊犁人，硕士，助理研究员，主要从事动物营养代谢病方面的研究。E-mail：xiaxue1984521@126.com

**通讯作者**：陈根元(1984－)，男，湖南岳阳人，硕士，助理研究员，主要从事动物基因工程方面的研究。

## 硒缺乏对雏鸡胰腺中硒蛋白 mRNA 表达的影响

赵　霞，姚海东，张子威，徐世文*

(东北农业大学动物医学学院 临床兽医内科教研室，哈尔滨 150030)

**摘　要**：硒作为机体所必需的营养性微量元素，在动物的生理过程中发挥着不可替代的作用。动物通过日粮来摄取机体所需的硒，硒进入体内以硒蛋白的形式发挥着各类生物学功能。目前已知的硒蛋白有25种，但目前对于硒蛋白的生物学功能和机制尚不清楚。硒缺乏可引起雏鸡胰腺的萎缩、纤维化，因此胰腺是硒缺乏的重要靶器官之一。本研究通过检测缺硒鸡胰腺中硒蛋白 mRNA 表达的水平，意在探讨缺硒导致鸡胰腺损伤的可能机制。本试验中，将1日龄雏鸡160只随机分为两组。缺硒组饲喂含硒量为0.02mg/kg的硒缺乏基础日粮；对照组饲喂含硒量为0.2mg/kg的全价基础日粮，饲喂期为55天。分别于15、25、35、45和55日龄，分别于两组各取15只，将鸡颈静脉放血处死，剖杀后采集胰腺组织。样本采集后，一部分立即进行组织匀浆，进行抗氧化指标的检测；剩余组织样本，保存于-80℃，用于实时定量 PCR 分析。结果显示，胰腺组织中23种硒蛋白 mRNA 表达水平呈不同程度下降，而抗氧化水平降低，胰腺发生氧化损伤。本研究表明，雏鸡硒缺乏会导致胰腺组织发生氧化应激，同时硒蛋白的含量降低，进而影响胰腺组织的正常功能，本研究结果将为硒缺乏引发的胰腺疾病的诊治提供理论依据。

**关键词**：硒蛋白，胰腺，鸡，缺硒

* E-mail：shiwenxu@neau.edu.cn

## 硒缺乏对肉鸡中性粒细胞中炎症因子及硒蛋白 mRNA 表达的影响

陈　晰，姚海东，张子威，徐世文*

(东北农业大学动物医学学院 临床兽医内科教研室，哈尔滨 150030)

**摘　要**：硒作为人和动物必需的微量元素，在机体的新陈代谢过程中起着重要的作用，其生物学功能主要是以硒蛋白的形式表现的。硒蛋白对动物机体具有抗氧化作用，参与免疫应答及甲状腺激素的分泌等生理过程。日粮硒的缺乏能够引起一系列缺硒性疾病的发生。硒缺乏可引起肉鸡中性粒细胞介导炎症反应及免疫应答反应，因此中性粒细胞是硒缺乏的重要靶细胞之一。本研究通过检测缺硒肉鸡中性粒细胞中炎症因子(*NF-κB*、*TNF-α*、*iNOS*、*COX-2*、*PTGES*)以及硒蛋白 mRNA 表达水平，意在探讨缺硒对肉鸡中性粒细胞损伤的可能机制。本试验中，将1日龄雏鸡50只随机分为两组。缺硒组饲喂含硒量为0.02mg/kg的硒缺乏基础日粮；对照组饲喂含硒量为0.2mg/kg的全价基础日粮，饲喂期为35天。在35日龄时，分别于两组各取15只，在鸡翅中静脉采血，利用试剂盒提取中性粒细胞，保存于-80℃，用于实时定量PCR分析。结果显示，中性粒细胞中5种炎症因子 mRNA 表达水平呈不同程度上升而23种硒蛋白的 mRNA 表达水平呈不同程度下降。相关性分析结果显示，在缺硒所诱导的中性粒细胞损伤的过程中，炎症因子的表达与硒蛋白的表达具有负相关性。本研究表明，缺硒能够引起鸡中性粒细胞炎症的发生，同时也导致硒蛋白水平的降低。结果提示缺硒所导致的中性粒细胞炎症损伤可能与硒蛋白的低表达有关。

**关键词**：炎症因子，中性粒细胞，硒蛋白，肉鸡，缺硒

* E-mail：shiwenxu@neau.edu.cn

## 硒缺乏对肉鸡肝脏凋亡及内质网应激的影响

姚琳琳,姚海东,张子威,徐世文*
(东北农业大学动物医学学院 临床兽医内科教研室, 哈尔滨 150030)

**摘　要**:硒是一种必需的微量元素,对人类健康起重要作用,人类和动物的许多疾病与硒缺乏有关,例如牲畜的白肌病,人的克山病,冠心病,癌症,阿尔茨海默氏病,糖尿病。肝脏是缺硒重要的靶器官之一。缺硒能够引起小鼠的肝坏死,猪的肝功能障碍,兔的肝脏坏死等。内质网应激被认为是细胞承受应激或损伤的初始反应。当内质网应激过强时,常常会诱导细胞凋亡的发生。而缺硒能否引起鸡肝脏的内质网应激及凋亡还很少有报道。本研究通过检测缺硒鸡模型中肝脏的凋亡水平,内质网发生的情况,探讨缺硒诱导鸡肝脏损伤的可能机制。本实验通过成功复制缺硒鸡模型,并利用实时定量 PCR 及 western blot 技术检测肝脏中凋亡基因 *Caspase*-3 及 *Bcl*-2,内质网相关基因 *ATF*4,*ATF*6,*GRP*78 和 *GRP*94 的表达水平。利用 TUNEL 法检测肝脏细胞凋亡的水平。结果表明,缺硒组鸡的肝脏中 *Caspase*-3 的 mRNA 和蛋白表达量都显著升高,*Bcl*-2 的表达量显著降低。HE 染色结果显示缺硒肝脏发生炎性细胞浸润。TUNEL 检测可见低硒组肝脏凋亡数量显著增加。同时缺硒引起了肝脏内质网相关基因表达的升高,提示缺硒诱导内质网应激的发生。结果表明,长时间饲喂缺硒日粮能够引起鸡肝脏内质网应激的发生,并会导致凋亡的发生。而在内质网应激发生的过程中,相关基因如 *ATF*4,*ATF*6,*GRP*78 和 *GRP*94 具有重要的作用。内质网应激在缺硒所引起的肝脏损伤中具有重要的作用。

**关键词**:硒,鸡肝脏,凋亡,内质网应激

* E-mail:shiwenxu@ neau. edu. cn

## 缺硒对雏鸡肠道热休克蛋白及抗氧化功能的影响

于　娇,姚海东,张子威,徐世文*
(东北农业大学动物医学学院 临床兽医内科教研室,哈尔滨 150030)

**摘　要**:硒是机体所必需的营养性微量元素,在动物的生理过程中发挥着不可替代的作用。硒进入体内以硒蛋白的形式发挥着各类生物学功能。目前已知的硒蛋白有 25 种,但目前对于硒蛋白的生物学功能尚不清楚。肠道是鸡摄取硒的主要器官,同时肠道黏膜也是机体中重要的防御器官。硒缺乏可引起雏鸡肠道内稳态失衡,引发肠道损伤,但硒缺乏对肠道黏膜免疫应答的影响及其机制仍不清楚。

本试验通过成功复制缺硒鸡模型,并利用光镜观察肠道组织病理学变化,通过 RT-PCR 技术以及 western blot 技术检测 3 个不同肠组织中 5 种热休克蛋白的 mRNA 和蛋白的表达水平,以及检测肠道中的抗氧化能力,研究硒缺乏对鸡肠组织黏膜免疫的影响,以及其可能发生的机制。结果表明:

1. 日粮中硒元素的缺乏可引起雏鸡肠道组织发生病理性损伤,肠道黏膜出血,镜检可观察到肠道黏膜完整性破坏,炎性细胞浸润。

2. 在缺硒初期可引起肠道中热休克蛋白的表达量显著升高,但随着试验的进行则呈显著的下降趋势。证明热休克蛋白功能的降低可能是缺硒引起肠道损伤的机制之一。

3. 硒缺乏可雏鸡引起肠道组织 CAT 活性、GPx 活性和 GSH 含量的下降,MDA 和 $H_2O_2$ 含量显著升高,表明硒可通过降低肠道的抗氧化能力而引起的肠道损伤。

**关键词**:硒,鸡,肠道,热休克蛋白,抗氧化

* E-mail:shiwenxu@ neau. edu. cn

# 硒缺乏所导致的肉鸡血管炎症损伤机理的研究

杜　强，姚海东，张子威，徐世文*

（东北农业大学动物医学学院 临床兽医内科教研室，哈尔滨 150030）

**摘　要**：硒是生物体所必需的微量元素，缺硒会导致多种疾病的发生。研究发现，硒对维持心血管系统的正常结构和功能起着重要的作用，而硒缺乏却是引起心血管疾病的重要因素。随着对硒生物学功能的不断研究，人们发现硒蛋白的表达水平受到硒的调控，且在缺硒诱导的疾病发生过程中也具有一定的作用。本试验通过复制缺硒鸡模型，探讨缺硒所诱导的心血管炎症损伤的可能机制。本试验中，将 1 日龄雏鸡 160 只随机分为两组。缺硒组饲喂含硒量为 0.02mg/kg 的硒缺乏基础日粮；对照组饲喂含硒量为 0.2mg/kg 的全价基础日粮，饲喂期为 55 天。分别于 15、25、35、45 和 55 日龄，分别于两组各取 15 只，将鸡颈静脉放血处死，剖杀后采集主动脉血管组织。通过 RT-PCR 技术以及 western blot 技术检测血管组织中炎症因子及硒蛋白的 mRNA 和蛋白的表达水平。结果显示，不同时间点的缺硒组鸡炎症因子的 mRNA 表达均显著升高，25 个硒蛋白 mRNA 的表达均有不同程度的下调。而蛋白的表达结果与 mRNA 水平基本一致。另外相关性分析结果显示，硒蛋白表达水平与炎症因子具有一定的负相关性。结果表明，缺硒能够导致鸡血管炎症的发生。而在此过程中，炎症因子如：*NF-κB*、*TNF-α*、*iNOS*、*COX-2*、*PTGES* 在缺硒所导致的炎症中可能具有一定的作用。另外缺硒所引起的硒蛋白降低与血管的损伤也有一定的关系。

**关键词**：缺硒，鸡，硒蛋白，炎症，血管

E-mail：shiwenxu@ neau. edu. cn

# 亚硒酸钠对鸡心肌细胞中硒蛋白 mRNA 表达的影响

赵文超，姚海东，张子威，徐世文*

（东北农业大学动物医学学院 临床兽医内科教研室，哈尔滨 150030）

**摘　要**：硒作为人及动物所必需的营养性微量元素，对人体及多种动物的生理功能发挥着不可替代的作用。在生物体内硒以硒蛋白的形式发挥着各类生物学功能，如调节代谢、拮抗有毒金属元素、清除自由基、抗氧化、延缓衰老、提高免疫力、防治癌症等等。有报道已指出，在不同动物体内及多种细胞中，硒对硒蛋白的表达具有一定的调节作用。而在体外实验中，硒对心肌细胞中硒蛋白表达的相关研究还未有报道。本实验通过检测亚硒酸钠对体外培养的鸡心肌细胞中硒蛋白 mRNA 表达的影响，探讨硒对硒蛋白的调节模式。本研究通过在体外培养鸡胚心肌细胞，并用不同浓度的亚硒酸钠（$10^{-9}$ mol/L，$10^{-8}$ mol/L，$10^{-7}$ mol/L）孵育，后用过氧化氢组处理细胞。采用实时荧光定量 PCR 法检测 25 种硒蛋白 mRNA 表达量。结果表明适量的亚硒酸钠（$10^{-8}$ mol/L）能够升高鸡心肌细胞 25 种硒蛋白的表达量。而过高浓度的硒对硒蛋白的表达有一定的抑制作用，相反过低的硒浓度对硒蛋白的表达作用不明显。同时亚硒酸钠的处理降低了细胞对过氧化氢的敏感性，表明硒的处理增加了心肌细胞的抗氧化能力。本研究表明，适量的亚硒酸钠在鸡心肌细胞内能够上调硒蛋白的表达，同时高表达的硒蛋白也降低了细胞对氧化应激的敏感性。

**关键词**：亚硒酸钠，心肌细胞，氧化应激，硒蛋白

* E-mail：shiwenxu@ neau. edu. cn

# 疫苗免疫应激对鸡脾淋巴细胞增殖和凋亡的影响

李荣芳,李晓文,文　琼,Froilan Bernard R. Matias,邬　静,易金娥,文利新*

(湖南农业大学动物医学院)

**摘　要**:养鸡生产特别是肉鸡饲养,疫苗接种是饲养管理中关键的一环,但在短时间内需要接种十几种疫苗,其疫苗免疫应激强度很大,临床实践中常常发生免疫应激事件,甚至是免疫麻痹,诱发家禽疾病。这与免疫功能抑制和细胞凋亡有着极为密切的关系。脾脏作为机体的重要免疫器官,其功能状态的正常与否,直接关系到到机体整体免疫情况的好坏,因此,检测脾淋巴细胞的凋亡与增殖情况是从客观上判断机体整体免疫功能好坏的一个重要手段。将 90 只 1 ~ 3 日龄 AA 肉鸡随机分为 5 组,在 11 日龄进行滴鼻免疫鸡新城疫Ⅳ系苗。A、B、C、D 组分别接种鸡新城疫Ⅳ系苗 16、8、4、2 羽份,E 组为空白组,生理盐水滴鼻。分别于免疫后 7d、14d、21d 颈部采血收集血清,用放免法测定 COR 激素水平;同时,无菌分离脾淋巴细胞,进行凋亡检测和增殖实验。凋亡检测用流式凋亡试剂盒于流式细胞仪上进行,增殖实验给予有丝分裂素 ConA 刺激细胞增殖。各剂量组与对照组比较,在免疫后 7d 和 14d,血清 COR 水平均极显著升高,免疫后 21d,A 组极显著升高,B、C 组显著性升高,D 组无差异性。说明本实验疫苗免疫应激模型造模成功。在免疫后 7d、14d、21d,各剂量组脾淋巴细胞都有不同程度的细胞凋亡现象发生,与对照组比较,除了免疫后 21d,C 组差异显著外,其他各组均差异极显著。在免疫后 7d、14d、21d,各剂量组对于 ConA 诱导的脾淋巴细胞增殖有不同程度的抑制作用,与对照组相比,除了免疫后 7d,A、D 组差异显著外,其他各组均差异极显著。疫苗免疫应激可以导致脾淋巴细胞凋亡,并且抑制 ConA 诱导的脾淋巴细胞增殖。

**关键词**:疫苗免疫应激,淋巴细胞增殖,淋巴细胞凋亡

**通讯作者**:E-mail:sfwlx8015@ sina. com

# 应用 iTRAQ-LCMS/MS 技术筛选硒缺乏雏鸡肠黏膜差异表达蛋白的研究

刘　哲,孙东波,王建发,武　瑞*

(黑龙江八一农垦大学动物科技学院,大庆 163319)

**摘　要**:为了研究硒缺乏对雏鸡肠道黏膜系统功能的影响,本试验以三黄蛋用雏鸡为试验动物,通过饲喂低硒饲料的方法构建了硒缺乏雏鸡动物模型,运用 iTRAQ-LCMS/MS 技术筛选和定量鉴定雏鸡肠黏膜差异表达蛋白,利用生物信息学方法对差异表达蛋白功能、所涉及信号通路及其在生物途径中的作用进行了分析。结果发现,共鉴定出 3 448 个蛋白质,包括 20 667 个肽,其中在 7d、21、35d 三个时间点鉴定出 29 个表达一致且差异显著的蛋白。在鉴定出的 29 个蛋白点中,以正常组为对照,其中 14 个表达上调,15 个表达下调,这 29 个差异蛋白的功能主要是参与催化及氧化还原酶活性、金属离子结合、蛋白及核甘酸结合。通过对差异性蛋白进行 GO 分析发现,在 3 448 个蛋白质中,2 379 个蛋白质(68.99%)在基因本体论(GO)类别生物过程有注释;2 610 个蛋白质(75.69%)注释分子功能类别;2082 个蛋白质(60.38%)注释为细胞组件类别。雏鸡肠黏膜差异性蛋白质主要富集于抗氧化活性、蛋白及核甘酸结合、金属离子结合和氧化应激反应。KEGG pathway 分析发现 PI3K-Akt-mTOR、NF-κB 及 MAPK 信号通路与硒缺乏雏鸡肠黏膜功能有关。相关差异蛋白的 western blot 及 qRT-PCR 验证工作正在进行过程中。本研究为从硒营养免疫学的角度解析其在动物肠道黏膜功能中的作用提供了研究依据。

* **通讯作者**:E-mail:fuhewu@ 126. com

# 肿瘤标志物在犬肿瘤疾病诊断中的应用

冯士彬，王希春，闫妮娜，吴金节*

（安徽农业大学动物科技学院，合肥 230036）

**摘　要**：肿瘤标志物（Tumor marker，TM）是指由肿瘤组织产生并可反映肿瘤自身存在的化学物质。TM的存在和量变可以提示肿瘤的性质，了解肿瘤的组织发生、细胞分化、细胞功能，以帮助肿瘤的诊断、分类、判断预后以及指导治疗。随着对肿瘤研究的不断深入，TM已经成为肿瘤早期诊断的研究热点。该方法常用于人类肿瘤的诊断、预后和疗效观察，但在动物肿瘤疾病的诊断中少见报道。为了探讨血清TM在犬肿瘤疾病的早期诊断、多种恶性肿瘤的诊断及肿瘤筛查中的应用价值，本试验选择7只健康犬作为对照组，以临床接诊的32例肿瘤患犬（其中乳腺及生殖器官肿瘤14例、皮肤及软组织肿瘤12例、其他肿瘤6例，经病理组织学检查确诊良性肿瘤18例，恶性肿瘤14例）为试验组，分别测定犬血清中癌胚抗原（CEA）、甲胎蛋白（AFP）、糖类抗原125（CA125）、糖类抗原15-3（CA15-3）和肿瘤特异性生长因子（TSGF）等五项TM的含量。结果显示，与良性肿瘤组相比，恶性肿瘤组血清中TSGF极显著升高（$P<0.01$）；恶性肿瘤组的5种TM的阳性检测率均高于良性肿瘤组；TM联合检测阳性率均高于单一标志物阳性检测率；与乳腺及生殖器官肿瘤组相比，皮下及软组织肿瘤组血清中CA125、CA15-3含量显著降低（$P<0.05$）。试验结果表明，TSGF可以作为恶性肿瘤早期诊断的特异性TM；采用多种TM联合检测可显著提高恶性肿瘤的检出率；血清TM具有一定的特异性，在不同组织的肿瘤表达中，存在较大的差异。试验证实了TM可以作为犬肿瘤疾病的早期诊断依据，有助于多种恶性肿瘤的诊断，并可用于肿瘤的筛查。

**关键词**：犬，肿瘤，肿瘤标志物，检测

**通讯作者**：E-mail：wjj@ahau.edu.cn

# 紫茎泽兰对生态环境的影响及其开发利用研究

廖　飞，胡延春*，何亚军，莫　全，陈伟红，胡　洋，邓俊良，左之才，王　娅

（四川农业大学动物医学院，温江 611130；动物疾病与环境公害四川省高校重点实验室，温江 611130；
四川农业大学动物疫病与人类健康四川省重点实验室，雅安 625000）

**摘　要**：调查紫茎泽兰对畜牧业发展及生态环境的影响，开发利用紫茎泽兰，变废为宝。采用实地调查、走访调查及采样调查等方法，弄清紫茎泽兰的分布、对畜牧业发展及生态环境的影响。提取主要毒素，研究其杀螨活性、抗肿瘤活性及主要毒素微生物降解菌的分离与鉴定。紫茎泽兰已在云南、四川、贵州、广西和西藏有广泛的分布，并以每年约30～60km的速度随西南风向东和向北扩散；四川省入侵紫茎泽兰已分布于凉山、甘孜、乐山、峨眉、自贡、雅安等地，凉山州经紫茎泽兰入侵天然草地三年后盖度达80%～95%以上，而牧草减少70.1%～79.36%，产草量仅为2 400～2 940kg/hm$^2$，天然草地失去放牧利用价值。经甲醇超声提取，用硅胶柱层析和大孔吸附树脂进行纯化和HPLC分析，其主要毒性成分为9-羰基-10，11-去氢泽兰酮（euptox A），体外杀螨活性优于阿维菌素类杀螨药物，对A549，Hela和Hep-2三株肿瘤细胞系的半数抑制浓度（$IC_{50}$）分别为369，401和427 μg/ml。分别从紫茎泽兰叶片、土壤、当地山羊瘤胃中分离到三株降解euptox A的细菌*Stenotrophomonas* sp. XC-07、*Klebsiella* sp. XC-08和*Pseudomonas* sp. XC-09，通过薄层色谱法和高效液相色谱法检测，3株菌在24h内对45mg/L浓度的euptox A降解率分别为：91.2%、94.3%和93.2%。紫茎泽兰对我国草原畜牧业发展及生态环境具有严重影响。其主要毒素为9-羰基-10，11-去氢泽兰酮（euptox A），具有较好的体外杀螨活性及抗肿瘤性，可以作为新药开发利用。紫茎泽兰可经降解菌去毒处理后作为饲料资源开发利用。

**关键词**：紫茎泽兰，生态影响，主要毒素，开发利用，微生物降解

通讯作者:胡延春(1975 - ),博士,教授。主要从事动物环境公害性疾病研究。E-mail:yanchunhu@ 126. com

# 乙二醛与甲基乙二醛对 BRL_3A 细胞的毒性机制及其作用机制

李盼盼,张天宇,穆颍颍,董 强
(西北农林科技大学 动物医学院,杨凌 712100)

**摘 要**:肥胖已成为全球性影响人类和动物健康的问题,肥胖已经成为一种全球流行病,并有加重的趋势。研究表明,人和动物肥胖最主要的原因是饮食和生活习惯的改变,如食用高脂,高糖,高能量饮食而又缺乏锻炼造成脂肪在体内累计。这些饮食中富含蔗糖、葡萄糖和果糖,其体内代谢产物为大量的活性羰基化合物(Reactive Carbonyl Species,RCS),如甲基乙二醛(Methylglyoxal,MGO)和乙二醛(Glyoxal,GO)。那么活性羰基化合物 MGO、GO 对肝细胞有着细胞毒性,但其具体的毒性机制以及 MGO 和 GO 毒性机制的区别尚未明确。本试验以羰基化合物 MGO 和 GO 为研究对象,以不同浓度的 MGO 或 GO 孵育大鼠肝细胞 BRL 3A 细胞,通过台盼蓝方法检测细胞通透性及完整性变化;通过 MTT 实验检测细胞活性;通过对氧化敏感的荧光染料 2,7-二氢二氯荧光素二乙酸酯(DCFH-DA)孵育检测细胞和培养液内的总 ROS(Reactive oxygen species)及细胞内 ROS 水平;通过 Rhodamine 123(Rh123)荧光染料孵育检测 MGO、GO 对线粒体膜电位的影响,为阐述 MGO、GO 对 BRL 3A 细胞的毒性机制提供理论依据。显示,400 ~ 2 000 μmol/L 的 MGO 可以显著降低 BRL 3A 细胞的活力,细胞死亡率成梯度依赖性增加;100 ~ 1 600 μmol/L 的 MGO 可以使细胞内的 ROS 显著增加($P < 0.05$),具有明显的梯度依赖性,且总 ROS 也显著增加($P < 0.05$);100 ~ 800 μmol/L 的 MGO 可以使线粒体膜电位显著升高($P < 0.05$),1 600 μmol/L 时使线粒体膜电位显著下降,具有浓度梯度依赖性。200 ~ 2 000 μmol/L 的 GO 可以使细胞通透性显著增加,线粒体损伤加剧,细胞死亡率增加;100 ~ 1 600 μmol/L 的 GO 细胞内 ROS 水平没有明显变化($P > 0.05$),总 ROS 水平显著增加($P < 0.05$),但没有 MGO 引起的 ROS 水平高;100 ~ 1 600 μmol/L 的 GO 使线粒体膜电位升高,但无浓度梯度依赖性。结果提示,MGO 可能通过使细胞内 ROS 增加,造成线粒体膜电位发生变化,破坏线粒体功能,最终引起细胞凋亡。GO 孵育使 BRL 3A 细胞总 ROS 含量显著增加,而细胞内 ROS 无明显变化,线粒体膜电位显著上升,其机制可能是 GO 孵育 12 h 后,细胞内产生的 ROS,自由扩散进入培养液,导致培养液 ROS 含量增加。而细胞内产生的 ROS,激活细胞内抗氧化酶,抑制细胞内 ROS 含量的上升,但其造成线粒体膜电位发生变化,破坏线粒体功能,最终诱导细胞凋亡。

**关键词**:MGO,GO,氧化应激,线粒体损伤

**作者简介**:李盼盼,女,硕士研究生,主要从事氧化损伤和生化分子毒理学的研究,E-mail:lpp0801@ 126. com
**通讯作者**:董 强,E-mail:ardour@ 126. com

# 柔嫩艾美耳球虫沉默信息调节因子 2 特性初步研究

杨斯涵[1,2],韩红玉[2],赵其平[2],朱顺海[2],黄 兵[2],董 辉[2*],杨亮宇[1*]
(1. 云南农业大学昆明 650201;2. 中国农业科学院上海兽医研究所上海 200241)

**摘 要**:球虫病是鸡场中最普遍、损失最严重的疾病之一,并呈世界性分布。鸡球虫有个种,其中柔嫩艾美耳球虫(*Eimeria tenella*)的分布最普遍、致病力最强。目前,鸡球虫病的防治主要依赖于抗球虫药物和疫苗。对抗球虫药物而言,随着药物的不断使用,耐药性的问题日益严重,需致力于找到新的药物靶标,来合成新型抗球虫药。沉默信息调节因子 2 (Silent Information Regulator 2,SIR2)是一类具有 $NAD^+$ 依赖型去乙酰化酶活性的组蛋白去乙酰化酶,其高度保守,参与染色质沉默,对细胞的存活、衰老、凋亡等的调节起到十分重要的作用。寄生虫的 SIR2 基因已有报道,并发现其在寄生虫的抗原变异、端粒

沉默和 DNA 修复等生理过程中发挥重要作用，且寄生虫与宿主之间的 SIR2 特性明显不同，是理想的抗寄生虫药物靶标，因此成为了研究的热点。本研究拟通过对柔嫩艾美耳球虫 SIR2（*Et*SIR2）基因的特性进行研究，从而了解 SIR2 的结构功能及在球虫体内所承担的重要生理学功能，为寻找潜在药物靶标筛选新药物奠定基础。通过克隆柔嫩艾美耳球虫 SIR2 基因，对其进行生物信息学分析；通过构建 *Et*SIR2 基因表达载体诱导表达重组蛋白，制备抗体，利用免疫荧光技术对 *Et*SIR2 蛋白在球虫不同发育阶段虫体中进行体内定位；利用定量 PCR 技术分析 *Et*SIR2 在球虫不同发育阶段虫体中 mRNA 转录水平的差异。成功克隆了 *Et*SIR2 基因，对 *Et*SIR2 蛋白进行亚细胞定位分析，结果显示其属于线粒体蛋白，不含跨膜结构域，不具有典型信号肽；通过免疫荧光技术发现 *Et*SIR2 蛋白定位于子孢子前端；*Et*SIR2 基因在未孢子化卵囊阶段 mRNA 转录水平明显高于其他三个阶段。*Et*SIR2 蛋白属于线粒体蛋白，表明其可能参与调控三羧酸循环，*Et*SIR2 基因在未孢子化卵囊阶段 mRNA 转录水平最高，表明其可能参与柔嫩艾美耳球虫孢子囊的形成过程。

**关键词**：球虫病，柔嫩艾美耳球虫，沉默信息调节因子 2，去乙酰化酶，药物靶标

**通讯作者**：E-mail：yangliangyu2004@163.com

# 285 例役用牛前胃弛缓病因和治疗分析

胡俊杰，华永丽，魏彦明

（甘肃农业大学动物医学院，兰州 730070）

**摘　要**：为了探讨前胃弛缓的病因、诊断及中西兽医结合的防治效果。对甘肃农业大学原武威家畜病院门诊病例中 285 例记录完整的前胃弛缓病例，从病因、诊断和治疗进行了回顾性分析。引起前胃弛缓的病因中以饲草因素为主，约占 45%，其次是管理因素、继发性因素和其他因素；发病季节以 4～5 月份和 9～10 月份发病较多；以中西兽医结合治疗为主，根据病因和病情的不同，应用自拟方健脾理气散、当归导滞汤、小柴胡汤、谷维素合剂和导胃洗胃及输液疗法等治疗原发性前胃弛缓，取得了较好的疗效。牛前胃弛缓以原发性为主，饲草料性质和饲养管理不当是主要病因，且发病与役用牛的使役密切相关。针对病情和病因制定治疗措施，采取中西兽医相结合能有效的治疗该病。

**关键词**：牛，前胃弛缓，病因，治疗

**基金项目**：公益性行业（农业）科研专项，牛重要脾胃病辨证施治关键技术研究与示范（201403051—09）

**通讯作者**：魏彦明，教授，从事中兽医学、兽医中药学和临床兽医学教学和研究工作，E-mail：weiym@gsau.edu.cn

**作者简介**：胡俊杰，讲师，博士，从事兽医临床诊断学和兽医内科学教学和研究工作. E-mail：lijianlijing111@163.com

# 输血疗法在犬病临床上的应用探讨

王江豪，李思远，秦建辉，陶田谷晟，肖　啸

（云南农业大学动物科学技术学院，昆明　650201）

**摘　要**：输血疗法是利用输入正常血液或血液成分制品进行补血、止血、解毒的一种治疗措施。对于治疗一些危重病例方面具有重要作用。输血作为一种治疗方法，在犬病临床上的应用越来越广泛。然而在有些情况下由于对输血疗法的认识不到位，导致不合理输血的情况时有发生。本文旨在通过对输血疗法的介绍，使科学合理的输血疗法得以推广。在犬病临床中运用输血疗法应当先明确以下问题：犬的血型，输血的适应症，供血犬的选择，血液如何采集和保存，如何进行交叉配血试验，还有输血的不良反应以及输血的并发症状，最后是输血中应当注意的事项。这些方面决定了我们对输血病例的治疗能否达到预期的效果，因此我们应当明确以下问题：是不是应该输血，是输注血液还是血液产品（如纯红细胞、富含血小板的血浆、新鲜冷冻的血浆、冷凝蛋白或血浆），是否我们有能力按照正规的操作流程来进行科学的输血。输血疗法在犬病临床中主要在以

下几个方面应用如:郭永久和李金龙分别报道过输血疗法在治疗犬急性失血病例中的应用;孙姝等报道一例牧羊犬产后子宫脱出手术及输血治疗;李志鹏等报道,1只博美犬极度贫血的输血治疗;李义春等报道,通过输血疗法治愈5只犬细小病毒病患犬;李鹏等曾治疗昆明某宠物市场的18只犬瘟热病犬,常规治疗用药3d后患犬病情没有明显好转,转用老龄健康犬或犬瘟热康复犬血进行输血治疗,结合常规治疗方案,成功治愈了17只患病犬。肖啸等报道对一只8岁的北京雌犬,5kg,诊断为肝细胞型黄疸。另外一只5岁的圣伯纳雌犬,50kg,诊断为传染性肝炎。对这两个例分别进行输血,3个月后回访,一切恢复正常。目前来说最主要应用的还是全血和红细胞,随着技术的不断进步,相信成分输血会应用的更好。

**关键词**:犬病,输血疗法,临床应用

**通讯作者**:肖　啸,E-mail:3116299366@126.com

# 三、中国畜牧兽医学会兽医内科与临床诊疗学分会历届终身成就奖获得者名单

**1. 第一届终身成就奖名单（2009 年，青岛）**

李毓义　教授（吉林大学动物医学学院）

刘应义　教授（吉林大学动物医学学院）

史　言　教授（东北农业大学动物医学学院）

金久善　教授（中国农业大学动物医学院）

李庆怀　教授（中国农业大学动物医学院）

王小龙　教授（南京农业大学动物医学院）

**2. 第二届终身成就奖名单（2011 年，南昌）**

崔治国　教授（内蒙古农业大学动物医学院）

曹光荣　教授（西北农林科技大学动物医学院）

张才骏　教授（青海大学农牧学院）

刘　鑫　教授（江西农业大学动物科学技术学院）

郭成裕　教授（云南农业大学动物科技学院）

袁　慧　教授（湖南农业大学动物医学院）

**3. 第三届终身成就奖名单（2012 年，兰州）**

张庆斌　教授（甘肃农业大学动物医学院）

邹康南　教授（南京农业大学动物医学院）

付有丰　教授（东北农业大学动物医学学院）

张志良　教授（甘肃农业大学动物医学院）

张一贤　教授（宁夏大学农学院）

王书林　教授（东北农业大学动物医学学院）

熊道焕　教授（华中农业大学动物医学院）

张德寿　教授（甘肃农业大学动物医学院）

崔中林　教授（西北农林科技大学动物医学院）

贺普霄　教授（西北农林科技大学动物医学院）

吴治礼　教授（江西农业大学动物科学技术学院）

白善义研究员（内蒙古鄂尔多斯市动物疫病预防控制中心）

**4. 第四届终身成就奖名单（2014 年，昆明）**

王　志　教授（中国农业大学动物医学院）

康世良　教授（东北农业大学动物医学学院）

石发庆　教授（东北农业大学动物医学学院）

# 中国畜牧兽医学会兽医内科与临床诊疗学分会第四届终身成就奖获得者

## 王志简历

王志，原名张铭新，男，辽宁省沈阳人，汉族，生于1924年3月。中国农业大学教授。1948年毕业于东北大学农学院畜牧兽医系。经历了伪满政府、国民党政府、中国共产党政府三个时期，新中国成立前，进入华北大学学习，后留该校农学院任教。1949年并入北京农业大学，先后任兽医系助教、讲师、副教授、教授；家畜内科学教研室副主任、主任，兽医院副院长、院长；兽医系党总支书记；并兼任中国畜牧兽医学会家畜内科学研究会副理事长兼秘书长。《中国农业百科全书·兽医卷》编委兼兽医内科学分支副主编，《动物毒物学》杂志副主编，《中国兽医杂志》编委等职。讲授过“家畜内科学”、“兽医临床诊断学”、“畜禽真菌毒素中毒病”等课程。

王志教授主要从事家畜内科学等方面的教学和研究工作，40余年致力于兽医临床工作，孜孜不倦，培养并影响了一批优秀的兽医学人才。取得“马属动物腹痛的研究”等项科研成果。编著有《家畜内科学》（部分章节）、《家畜皮肤病学》、《畜禽真菌毒素中毒病》、《简明日汉农业词典》（畜牧兽医部分）等著作；并发表论文多篇。1984年11月王志教授获得北京市高等教育局“从事教育工作三十年”表彰；1992年10月获得国务院政府特殊津贴；2013年12月22日，王志教授以90岁高龄获得《中国畜牧杂志》和《中国兽医杂志》颁发的“耕耘奖”，以表示其多年对杂志投稿、校稿工作所做出的贡献。

耄耋之年，王志教授有很多的人生感悟，值得我们后辈学习体悟。2013年12月24日，他在90岁生日的宴会上欣然做诗一首：

**髦耋之年感悟**

垂髫年代遭涂炭，弱冠人生沐党恩；
深感内疚遗恨是，双亲健在尽孝无；
平生虽怀宏图志，评定业绩属平常；
结发夫妻钻石婚，相濡以沫度残年；
伉俪抚养五子女，成家立业奔小康；
兄妹同根亲情重，同胞和睦重泰山；
父母生前应尽孝，长逝悔之已枉然；
生儿育女人丁旺，男女双全四世堂；
誓志不做违心事，身教言教子嗣传；
身患致命恶性病，任凭病逝别吓死；
人生七十古来稀，现今八十不惊奇；
老伴年岁八十八，我的生龄已九十；
阖家欢乐除夕夜，髦耋之人乐融融。

## 石发庆教授简介

黑龙江肇州人，生于1936年11月，1960年毕业于东北农学院并留校任教，先后任助教、讲师、副教授、教授、硕士研究生导师、博士研究生导师和黑龙江省重点学科兽医内科学学科带头人，是国务院特殊津贴获得者。曾任中国家畜内科学会常务理事、东北兽医内科研究会副理事长、黑龙江省兽医内科研究会副理事长。主持黑龙江省攻关项目2项，省自然科学基金1项，参加国家自然科学基金项目2项。研究成果获省科技进步二等奖1项、三等奖2项、农业部科技进步三等奖1项。发表第一作者和通讯作者论文100余篇。主编研究专著《禽病诊断》，参编研究专著2部。副主编中国面向21世纪课程教材《兽医内科学》，参编全国统编教材《兽医临床诊断学》和《兽医临床诊疗基础》。

## 康世良教授简介

1935年生，东北农业大学兽医62级。教授、博导。培养博士生4人、硕士生4人。曾任内科教研室主任、中共临床支部书记、省重点学科带头人、学术带头人、动物医学院学术委员会和学位评定委员会委员。全国动物毒物学会副理事长、中国家畜内科学会理事、东北地区兽医内科学会副理事长兼秘书长、黑龙江省畜牧兽医学会常务理事和兽医内科学会理事长、中国畜牧业协会犬业分会理事、黑龙江畜牧业协会专家顾问成员、犬业分会常务理事及技术总监、黑龙江农业科学咨询专家。《黑龙江畜牧兽医》杂志编委。主编《家畜内科学》（获校自编教材二等奖）、《经济动物病诊断》；副主编《家畜中毒学》（校表扬教材）；参编全国教材《家畜中毒学与毒物检验》（获内蒙古高校优秀教材三等奖）和《兽医学》；参编《实用兽医学》和《畜禽硒和维生素E缺乏症》；主审《奶牛疾病图谱》。长期从事动物中毒病、动物营养代谢病和微量元素营养强化蛋方面的研究。主持或负责国家自然科学基金课题2项、省级课题2项、自选和合作课题3项。发表论文90余篇（含通讯作者），被国外权威杂志（英）摘录4篇。获省畜牧科技进步二等奖1项（主持人），部级科技进步一、三等奖各1项（主要成员），省优秀教学成果一等奖1项（副主持人）。入编《世界名人录》、《世界科技名人录》、《中国当代科学和发明家大辞典》。

# 四、中国畜牧兽医学会兽医内科与临床诊疗学分会历届SCI收录论文奖励名单

## （一）第一届SCI收录论文奖励名单（2011年11月于南昌）

**1. SCI收录的论文特等奖获得者（3人）**

王俊东教授（山西农业大学副校长）

王　哲教授（吉林大学农学部）

黄克和教授（南京农业大学动物医学院）

**2. SCI收录的论文一等奖获得者（4人）**

刘国艳博士（上海交通大学农业与生物技术学院）

韩　博教授（中国农业大学动物医学院）

王小龙教授（南京农业大学动物医学院）

刘宗平教授（扬州大学兽医学院）

**3. SCI收录的论文二等奖获得者（5人）**

袁　慧教授（湖南农业大学动物医学院）

王九峰教授（中国农业大学动物医学院）

王建华教授（西北农林科技大学动物医学院）

徐世文教授（东北农业大学动物医学学院）

李艳飞教授（东北农业大学动物医学学院）

## （二）第二届SCI收录论文奖励名单（2012年8月于兰州）

**1. SCI收录的论文特等奖获得者（1人）**

黄克和教授（南京农业大学动物医学院）

**2. SCI收录的论文一等奖获得者（1人）**

王俊东教授（山西农业大学副校长）

**3. SCI收录的论文二等奖获得者（3人）**

王　哲教授（吉林大学农学部）

徐世文教授（东北农业大学动物医学学院）

王九峰教授（中国农业大学动物医学院）

**4. SCI收录的论文三等奖获得者（1人）**

韩　博教授（中国农业大学动物医学院）

## （三）第三届SCI收录论文奖励名单（2014年7月于昆明）

**1. SCI收录的论文IF大于5的一等奖获得者（4人）**

王俊东教授（山西农业大学）**（发表2篇）**

刘宗平教授（扬州大学）

刘建柱博士、副教授（山东农业大学）

张志刚博士、副教授（东北农业大学）

**2. SCI收录的论文IF在4和5之间的二等奖获得者（6人）**

王俊东教授（山西农业大学）**（发表2篇）**

张乃生教授（吉林大学农学部动物医学学院）

郭定宗教授（华中农业大学动物医学院）

徐世文教授（东北农业大学动物医学学院）

韩　博教授（中国农业大学动物医学院）

杨正涛博士、副教授（吉林大学农学部动物医学学院）**（发表2篇）**

### 第三届SCI收录论文获奖论文的题目

| 论　文 | IF | 学校 | 通讯作者 | 名次 |
|---|---|---|---|---|
| Meng K, Sun W, Zhao P, Zhang L, Cai D, Cheng Z, Guo H, **Liu J**, Yang D, Wang S, Chai T. Development of colloidal gold-based immunochromatographic assay for rapid detection of Mycoplasma suis in porcine plasma. **Biosensors and Bioelectronics.** 2014, 55: 396-399. | 5.437 | 山东农业大学 | 刘建柱 | 一等 |
| Z. Lu, S. Wang, Z. Sun, R. Niu, J. Wang. In vivo influence of sodium fluoride on sperm chemotaxis in male mice. **Arch Toxicol.** 2014, 88(2):533-9. | 5.215 | 山西农业大学 | 王俊东 | 一等 |
| F. Jia, Z. Sun, X. Yan, B. Zhou, **J. Wang**. Effect of pubertal nano-TiO2 exposure on testosterone synthesis and spermatogenesis in mice. **Arch Toxicol.** 2014, 88(3):781-8. | 5.215 | 山西农业大学 | 王俊东 | 一等 |

| | | | | |
|---|---|---|---|---|
| Zhu Jia-Qiao, Si Yang-Jun, Cheng Lai-Yang, Xu Bao-Zeng, Wang Qi-Wen, Zhang Xiao, Wang Heng, **Liu Zong-Ping**. Sodium fluoride disrupts DNA methylation of H19 and Peg3 imprinted genes during the early development of mouse embryo. **Arch Toxicol.** 2014, 88(2): 241-8. | 5.215 | 扬州大学 | 刘宗平 | 一等 |
| Yu ML, Xue JD, Li YJ, Zhang WQ, Ma DX, Liu L, **Zhang ZG**. Resveratrol protects against arsenic trioxide-induced nephrotoxicity by facilitating arsenic metabolism and decreasing oxidative stress. **Arch Toxicol.** 2013, 87: 1025-1035. | 5.215 | 东北农业大学 | 张志刚 | 一等 |
| Fu Y, Zhou E, Wei Z, Wang W, Wang T, Yang Z, Zhang N. Cyanidin-3-O-β-glucoside ameliorates lipopolysaccharide-induced acute lung injury by reducing TLR4 recruitment into lipid rafts. Biochem Pharmacol. 2014, 90:126–134 | 4.576 | 吉林大学 | 张乃生 | 二等 |
| Tian WX, Li JK, Qin P, Wang R, Ning GB, Qiao JG, Li HQ, Bi DR, Pan SY. **Guo DZ**.Screening of differentially expressed genes in the growth plate of broiler chickens with Tibial Dyschondroplasia by microarray analysis. **BMC Genomics**. 2013,14:276 | 4.397 | 华中农业大学 | 郭定宗 | 二等 |
| Fu Y, Wei Z, Zhou E, Zhang N, **Yang Z.** Cyanidin-3-O-β-glucoside inhibits lipopolysaccharide-induced inflammatory response in mouse mastitis model. **J Lipid Res.** 2014, 55(6):1111-1119. | 4.386 | 吉林大学 | 杨正涛 | 二等 |
| G. Luo, S. Wang, Z. Li, R. Wei, L. Zhang, H. Liu, C. Wang, R. Niu, J. Wang. Maternal Bisphenol A Diet Induces Anxiety-like Behavior in Female Juvenile with Neuroimmune Activation. **Toxicol Sci.** 2014, May 13. pii: kfu085. [Epub ahead of print] | 4.328 | 山西农业大学 | 王俊东 | 二等 |
| Fu Y, Zhou E, Wei Z, Liang D, Wang W, Wang T, Guo M, Zhang N, **Yang Z.** Glycyrrhizin inhibits the inflammatory response in mouse mammary epithelial cells and a mouse mastitis model. **FEBS J.** 2014, 281(11):2543-57. | 4.250 | 吉林大学 | 杨正涛 | 二等 |
| Hai-Dong Yao Qiong Wu, Zi-Wei Zhang, Jiu-Li Zhang, Shu Li, Jia-Qiang Huang, Fa-Zheng Ren, **Shi-Wen Xu**, Xiao-Long Wang, XinGen Lei. Gene Expression of Endoplasmic Reticulum Resident Selenoproteins Correlates with Apoptosis in Various Muscles of Se-Deficient Chicks. **The Journal of Nutrition.** 2013, 143(5): 613-619 | 4.196 | 东北农业大学 | 徐世文 | 二等 |
| Zahra Mohammad, Ferreri M, Alkasir R, Yin J, **Han B**, Su J. Isolation and Characterization of Small-Colony Variants of Ornithobacterium rhinotracheale. **Journal of Clinical Microbiology.** 51(10):3228-3236, 2013 | 4.068 | 中国农业大学 | 韩 博 | 二等 |
| C. Wang, R. Niu, Y. Zhu, H. Han, G. Luo, B. Zhou, **J. Wang.** Changes in memory and synaptic plasticity induced in male rats after maternal exposure to bisphenol A. **Toxicology**. 2014 May 10. pii: S0300-483X(14)00095-X. doi: 10.1016/j.tox.2014.05.001. [Epub ahead of print] | 4.017 | 山西农业大学 | 王俊东 | 二等 |

# 五、人才培养

截至目前，我国有13所学校拥有兽医内科诊断学科博士点（中国农业大学、吉林大学、东北农业大学、山西农业大学、内蒙古农业大学、西北农林科技大学、甘肃农业大学、南京农业大学、华中农业大学、湖南农业大学、华南农业大学、扬州大学和四川农业大学），50所学校拥有50多个硕士点。据统计2012年7月30日之前我国兽医内科诊断学培养已毕业的硕士研究生778名（实际超过1000人），培养已毕业的博士研究生214名，培养已出站的博士后12名。据统计，2012年8月1日至2014年6月30日我国兽医内科诊断学培养已毕业的硕士研究生381名，培养已毕业的博士研究生50名，培养已出站的博士后11名。

## 1.硕士研究生论文名单

我国兽医内科诊断学培养已毕业的硕士研究生381名（2012秋至2014年）

| 姓名 | 论文题目 | 毕业时间 | 单位 | 指导教师 |
|---|---|---|---|---|
| 班晓敏 | AFB_1、OTA、DON高效液相色谱检测方法的优化和应用 | 2012 | 南京农业大学 | 张海彬 |
| 王智群 | 犬瘟热病毒单克隆抗体夹心ELISA方法的建立与胶体金免疫层析试纸条的制备 | 2012 | 南京农业大学 | 张海彬 |
| 张广斌 | 复方吡喹酮片对犬的安全性试验及临床疗效的研究 | 2012 | 南京农业大学 | 张海彬 |
| 杨彦琼 | 新型复合吸附剂对黄曲霉毒素B_1的吸附脱毒研究 | 2012 | 南京农业大学 | 张海彬 |
| 马金荣 | 传染性法氏囊病病毒VP2与C3d串联基因重组表达质粒的构建及其免疫功能分析 | 2012 | 南京农业大学 | 张海彬 |
| 吴云飞 | 盐酸特比萘芬的安全性和抗真菌药物的敏感性及临床疗效试验 | 2012 | 南京农业大学 | 张海彬 |
| 张　鹏 | ZEN、T-2、FB_1高效液相色谱检测方法的优化和应用 | 2012 | 南京农业大学 | 张海彬 |
| 赵士侠 | 赭曲霉毒素A诱导BHK细胞凋亡及维生素C对其毒性干预的研究 | 2012 | 南京农业大学 | 张海彬 |
| 刘　宏 | 美洛昔康片对犬的临床安全性试验及对诱导型骨关节炎的疗效研究 | 2012 | 南京农业大学 | 张海彬 |
| 赖晓云 | 当前国内宠物医院管理现状的调研 | 2012 | 南京农业大学 | 张海彬 |
| 王志敏 | 盐酸贝那普利对犬的安全性试验及犬充血性心力衰竭的疗效研究 | 2013 | 南京农业大学 | 张海彬 |
| 顾舒舒 | 赭曲霉毒素A单链抗体的制备及初步应用 | 2013 | 南京农业大学 | 张海彬 |
| 孟玲玲 | NJA-1菌中DON降解酶基因在毕赤酵母中的表达及其对DON降解效果分析 | 2013 | 南京农业大学 | 张海彬 |
| 周　闯 | 饲料中DON的污染情况调查及DON制备工艺的研究 | 2014 | 南京农业大学 | 张海彬 |
| 唐龙琴 | 阿米卡星对犬的临床安全性试验与药代动力学研究 | 2014 | 南京农业大学 | 张海彬 |
| 孙一涵 | 阿米卡星对靶动物犬的药效学研究 | 2014 | 南京农业大学 | 张海彬 |
| 刘　堃 | 盐酸特比萘芬和伊曲康唑联合用药的安全性和临床疗效试验 | 2014 | 南京农业大学 | 张海彬 |
| 梁荣发 | 犬瘟热病毒冻干单克隆抗体的制备及临床试验 | 2014 | 南京农业大学 | 张海彬 |
| 李俊娴 | 酿酒酵母NDE1基因缺失突变株的构建及其生物学性质研究 | 2013 | 南京农业大学 | 黄克和 |
| 吕晨辉 | 富硒益生菌对高温条件下仔猪生产性能、抗氧化能力和肠道菌群的影响 | 2013 | 南京农业大学 | 黄克和 |
| 石　俊 | 富硒酵母生产工艺优化及酵母多糖抗PCV2作用的研究 | 2013 | 南京农业大学 | 黄克和 |
| 石秀丽 | 谷氨酰胺对PCV2在PK15细胞中复制的影响 | 2013 | 南京农业大学 | 黄克和 |
| 钟文婷 | 富硒乳酸菌对食源性草酸过多犬尿石症的预防作用 | 2013 | 南京农业大学 | 黄克和 |

| 姓名 | 论文题目 | 毕业时间 | 单位 | 指导教师 |
|---|---|---|---|---|
| 王　亮 | 亚临床酮病对围产期奶牛临床发病情况、生化指标及免疫功能动态变化的影响 | 2013 | 南京农业大学 | 张克春<br>黄克和 |
| 陈　敏 | 公猪异味控制疫苗对公猪生产性能、胴体品质的影响 | 2013 | 南京农业大学 | 黄克和 |
| 伏金笋 | 维生素E对东北虎高温热应激缓解作用的研究 | 2013 | 南京农业大学 | 黄克和 |
| 肖　华 | 一种新型微生态制剂在断奶仔猪上的应用研究 | 2013 | 南京农业大学 | 黄克和 |
| 车超平 | 不同硒源对黄曲霉毒素B1致鸡原代肝细胞损伤的缓解作用及其机理研究 | 2014 | 南京农业大学 | 黄克和 |
| 黄　钰 | 猪肾传代细胞过表达SelS细胞系的构建及其对OTA毒性的拮抗作用 | 2014 | 南京农业大学 | 黄克和 |
| 周媛丽 | 酿酒酵母GPD1基因超表达突变株的构建及其生物学性质研究 | 2014 | 南京农业大学 | 黄克和 |
| 李　盟 | EHEC O157:H7紧密素单克隆抗体的制备及夹心ELISA检测方法的建立 | 2014 | 南京农业大学 | 吴德华<br>黄克和 |
| 林　梅 | 上海地区牛传染性鼻气管炎感染情况调查及病毒分离鉴定 | 2014 | 南京农业大学 | 张克春<br>黄克和 |
| 李玫毅 | 警犬搜爆用TNT气味替代品的研制及其生物安全性评估 | 2014 | 南京农业大学 | 黄克和 |
| 王谷雨 | 犬新鲜血浆、HES40和LR溶液对失血性休克犬复苏效果的比较研究 | 2012 | 中国农业大学 | 夏兆飞 |
| 陈江楠 | 二烯丙基二硫醚对犬体外红细胞抗氧化系统和膜蛋白的影响 | 2012 | 中国农业大学 | 夏兆飞 |
| 毛军福 | 北京地区犬肥胖症流行病学调查 | 2012 | 中国农业大学 | 夏兆飞 |
| 宋璐莎 | 犬高脂血症病例分析：内分泌疾病对脂质和脂蛋白浓度影响 | 2013 | 中国农业大学 | 夏兆飞 |
| 陈艳云 | 北京地区犬尿源性大肠杆菌种系分群及其对氟喹诺酮类药物耐药机制初探 | 2013 | 中国农业大学 | 夏兆飞 |
| 葛冰倩 | 北京地区猫肥胖症流行病学调查 | 2013 | 中国农业大学 | 吕艳丽<br>夏兆飞 |
| 樊　君 | 54例犬急性胰腺炎临床病例调查分析 | 2013 | 中国农业大学 | 夏兆飞 |
| 高进东 | 北京地区犬泌尿系统大肠杆菌的分离鉴定及药敏试验 | 2013 | 中国农业大学 | 夏兆飞 |
| 孙玉祝 | 犬源大肠杆菌的耐药性调查以及其介导对β-内酰胺类抗生素耐药相关基因的检测 | 2014 | 中国农业大学 | 夏兆飞 |
| 陈珊珊 | 住院患犬营养状况与疾病转归和血清IGF-1水平关系的研究 | 2014 | 中国农业大学 | 夏兆飞 |
| 刘比一 | 不同储存时间的玉米对肉鸡血清抗氧化、血清生化及组织学的影响 | 2014 | 中国农业大学 | 夏兆飞 |
| 周少芳 | 北京地区犬猫糖尿病病例分析 | 2014 | 中国农业大学 | 夏兆飞 |
| 唐伟红 | 鸭呼肠孤病毒抗体检测及细胞传代致弱研究 | 2012 | 中国农业大学 | 韩　博<br>苏敬良 |
| 阳世勇 | 氟化钠诱导MC3T3-E1成骨细胞凋亡以及对其Bcl-2家族蛋白表达的影响 | 2012 | 中国农业大学 | 韩　博 |
| 罗立平 | 犬奇异变形杆菌分子流行病学及耐药特征初探 | 2012 | 中国农业大学 | 韩　博<br>苏敬良 |
| 董　微 | 血液滤过对犬急性肾衰竭治疗的生化指标研究（在职兽医硕士） | 2012 | 中国农业大学 | 韩　博<br>董悦农 |
| 刘爱民 | 上消化道内镜对犬猫食道异物和胃内异物的诊疗研究（在职兽医硕士） | 2012 | 中国农业大学 | 韩　博<br>董悦农 |
| 侯荣光 | 奶牛乳房炎源左氏无绿藻的分离鉴定及分子流行病学调查 | 2013 | 中国农业大学 | 韩　博 |
| 常瑞龙 | 庆大霉素对小鼠无绿藻性乳房炎的疗效评价 | 2013 | 中国农业大学 | 韩　博<br>苏敬良 |
| 杨巧玲 | 犬奇异变形杆菌毒力相关基因分析 | 2013 | 中国农业大学 | 韩　博 |
| 魏　鹏 | 北京西南地区奶牛和奶山羊乳房炎病原菌的分离鉴定及药敏试验（在职兽医硕士） | 2013 | 中国农业大学 | 韩　博 |
| 梁志强 | 犬猫疾病诊断专家系统的设计与应用研究（在职兽医硕士） | 2013 | 中国农业大学 | 韩　博 |
| 袁　晨 | 匹莫苯丹治疗犬慢性房室瓣膜病的临床观察（在职兽医硕士） | 2013 | 中国农业大学 | 韩　博 |
| 高金花 | 肝癌Li-050细胞模型的建立及厄洛替尼抗癌的效果研究（在职兽医硕士） | 2014 | 中国农业大学 | 韩　博 |

| 姓名 | 论文题目 | 毕业时间 | 单位 | 指导教师 |
|---|---|---|---|---|
| 王　平 | 北京某奶牛场奶牛隐性乳房炎的动态监测（在职兽医硕士） | 2014 | 中国农业大学 | 韩　博 |
| 崔燕滨 | 北京地区奶牛乳房炎致病菌分离鉴定与葡萄球菌属菌株耐药性分析（在职兽医硕士） | 2014 | 中国农业大学 | 韩　博 |
| 彭雨佳 | 血清和尿液胱抑素 C 对犬慢性肾衰的评估 | 2014 | 中国农业大学 | 韩　博 |
| 王淑红 | 犬奇异变形杆菌的耐药性分析及 PFGE 基因指纹图谱研究 | 2014 | 中国农业大学 | 韩　博 |
| 蔡正兴 | 使用腹泻模型评价鼠李糖乳杆菌对仔猪肠黏膜免疫应答的影响 | 2012 | 中国农业大学 | 王九峰 |
| 李小琼 | 不同剂量鼠李糖乳杆菌对大肠 杆菌性腹泻仔猪肠道黏膜免疫应答的影响 | 2012 | 中国农业大学 | 王九峰 |
| 张　荣 | 2,4,6-三硝基甲苯诱导的犬氧化应激与免疫反应研究 | 2012 | 中国农业大学 | 王九峰 |
| 朱　强 | 北京工作犬基地新引进犬弓形虫感染分子流行病学调查 | 2013 | 中国农业大学 | 王九峰 |
| 张　琼 | 充血性心力衰竭犬血清中 NT-proBNP 和 cTnI 的相关性研究 | 2013 | 中国农业大学 | 王九峰<br>汤小朋 |
| 杨金彩 | 鼠李糖乳杆菌对产肠毒素大肠杆菌感染猪肠上皮细胞保护机制的研究 | 2013 | 中国农业大学 | 王九峰 |
| 翟丛丛 | 芽孢杆菌对大肠杆菌感染仔猪肠道黏膜免疫应答的影响 | 2014 | 中国农业大学 | 王九峰 |
| 刘玉丽 | 微创主动脉弓缩窄致小鼠心衰模型的建立及高频超声检测 | 2014 | 中国农业大学 | 王九峰 |
| 崔丽丽 | 北京地区犬、猫弓形虫病血清学调查 | 2012 | 中国农业大学 | 于咏兰 |
| 魏晓宏 | 关于人接触动物与弓形虫感染关系的研究 | 2013 | 中国农业大学 | 于咏兰 |
| 张伟伟 | 北京地区犬细小病毒流行特性及基于 VP2 基因片段的基因型分析 | 2013 | 中国农业大学 | 于咏兰 |
| 李　鹏 | 酮病奶牛肝脏脂肪酸氧化代谢特征及其调控 | 2012 | 吉林大学 | 王　哲 |
| 陈　灰 | 脂联素激活 AMPK 信号通路调控奶牛肝细胞脂代谢的相关机制 | 2013 | 吉林大学 | 王　哲 |
| 张　良 | 黄曲霉毒素 B1 间接竞争酶免疫分析方法的研究 | 2013 | 吉林大学 | 王　哲 |
| 赵晨旭 | 人工诱导 SARA 山羊瘤胃微生物区系的变化及微生态制剂的调控作用 | 2013 | 吉林大学 | 王　哲 |
| 刘兆喜 | NEFA 和 BHBA 对酮病奶牛氧化应激状态的影响 | 2013 | 吉林大学 | 刘国文 |
| 杨文涛 | 腺病毒介导的 SREBP-1c 基因过表达对奶牛肝细胞脂肪沉积的影响 | 2013 | 吉林大学 | 李小兵 |
| 李德鹏 | 大黄素对 LPS 诱发的小鼠乳腺炎性损伤保护效果的研究 | 2013 | 吉林大学 | 张乃生 |
| 付云鹤 | 厚朴酚抗炎作用及机制研究 | 2013 | 吉林大学 | 杨正涛 |
| 付云云 | CSF 疫苗免疫猪对 PRRS 特异性免疫应答以及细胞表型分析 | 2013 | 吉林大学 | 高英杰 |
| 单　悦 | CSFV 免疫猪接种 PRRSV nsp2Δ1882-2241 弱毒疫苗后 IL-10、IL-12、TNF-α 的应答特点研究 | 2013 | 吉林大学 | 高英杰 |
| 宋玉祥 | 非酯化脂肪酸激活 p38MAPK 信号通路诱导体外培养的奶牛肝细胞凋亡 | 2014 | 吉林大学 | 王　哲 |
| 韩盈盈 | 金属 Cd2+免疫层析胶体金试纸条的研制 | 2014 | 吉林大学 | 王　哲 |
| 史晓霞 | NEFAs 和 BHBA 诱导氧化应激对奶牛肝细胞 NF-κB 信号通路的研究 | 2014 | 吉林大学 | 刘国文 |
| 丁红研 | 胰岛素、胰高血糖素通过 AMPK 信号通路调控奶牛肝脂代谢的机制 | 2014 | 吉林大学 | 李小兵 |
| 梁德洁 | 百里香酚对 LPS 诱发的乳腺炎小鼠的保护作用及机制研究 | 2014 | 吉林大学 | 张乃生 |
| 李丰阳 | 紫云英苷对小鼠乳腺炎的保护作用及其机制研究 | 2014 | 吉林大学 | 张乃生 |
| 孙　勇 | 栀子苷对 LPS 诱发的小鼠乳腺炎性损伤保护效果的研究 | 2014 | 吉林大学 | 张乃生 |
| 赵福义 | 某猪场猪链球菌感染病例诊断及防治研究 | 2014 | 吉林大学 | 张乃生 |
| 刘志成 | 棉酚抗炎活性及作用机制初步研究 | 2014 | 吉林大学 | 杨正涛 |
| 郭　敏 | PRRSV 多肽对 CSF、PRRS 免疫猪淋巴细胞亚群的影响 | 2014 | 吉林大学 | 高英杰 |
| 郎广平 | PRRS 特异性肽对 CSF、PRRS 免疫猪 IFN-γ、IL-10 分泌的影响 | 2014 | 吉林大学 | 高英杰 |
| 宋岩岩 | 苦马豆素致大鼠细胞凋亡与 a-甘露糖苷酶动态变化研究 | 2012 | 西北农林科技大学 | 赵宝玉 |
| 陈基萍 | 疯草内生真菌种属鉴定及次生代谢产物研究 | 2012 | 西北农林科技大学 | 赵宝玉 |
| 庞　龙 | 苦马豆素体外诱导大鼠神经元凋亡的研究 | 2012 | 西北农林科技大学 | 赵宝玉 |
| 孔祥雅 | 山羊溶酶体 α-甘露糖苷酶基因的克隆、表达及酶特性的研究 | 2012 | 西北农林科技大学 | 李勤凡 |
| 张丽慧 | A2A 腺苷受体拮抗剂嘧啶类衍生物的结合模式和结构特征 | 2012 | 西北农林科技大学 | 李勤凡 |
| 李艳红 | 节杆菌 HW08 降解 SW 酶及其特性的初步研究 | 2013 | 西北农林科技大学 | 王建华 |

| 姓名 | 论文题目 | 毕业时间 | 单位 | 指导教师 |
|---|---|---|---|---|
| 康丹菊 | 疯草内生真菌菌丝体中苦马豆素的分离与鉴定 | 2013 | 西北农林科技大学 | 王建华 |
| 李金城 | HPLC-ELSD 检测疯草中苦马豆素方法的建立 | 2013 | 西北农林科技大学 | 王建华 |
| 周启武 | 疯草内生真菌多样性及检测方法研究 | 2013 | 西北农林科技大学 | 赵宝玉 |
| 温伟利 | 野生型斜茎黄芪生物碱成分与营养成分分析及毒性评价 | 2013 | 西北农林科技大学 | 赵宝玉 |
| 张　樑 | 苦马豆素诱导新生 SD 大鼠大脑皮质神经细胞凋亡机制研究 | 2013 | 西北农林科技大学 | 赵宝玉 |
| 厉秀秀 | 黄连产小檗碱内生真菌的分离鉴定 | 2013 | 西北农林科技大学 | 李勤凡 |
| 杨　凯 | 产乌头碱内生真菌的分离、鉴定及固体培养基的优化 | 2013 | 西北农林科技大学 | 李勤凡 |
| 方玉鹏 | 鱼腥草产槲皮素内生真菌的分离鉴定 | 2013 | 西北农林科技大学 | 李勤凡 |
| 梁　洁 | 产萱草根素内生真菌的分离鉴定及其对小鼠的病理学观察 | 2013 | 西北农林科技大学 | 李勤凡 |
| 王文龙 | 甘肃棘豆提取物致大鼠毒性损伤及“棘防 E 号”保护作用研究 | 2014 | 西北农林科技大学 | 赵宝玉 |
| 王姗姗 | 苦马豆素致 BRL-3A 细胞毒性损伤及“棘放 E 号”和 $Fe2^{+}$ 保护效应研究 | 2014 | 西北农林科技大学 | 赵宝玉 |
| 杨晓雯 | 甘肃棘豆内生真菌种群分布特征及遗传多样性研究 | 2014 | 西北农林科技大学 | 赵宝玉 |
| 严杜建 | 新疆阿克苏天然草地毒草分布及危害调查与防控技术 | 2014 | 西北农林科技大学 | 赵宝玉 |
| 张江业 | 山羊溶酶体 AMA 的定点突变及突变产物对 SW 敏感性研究 | 2014 | 西北农林科技大学 | 李勤凡 |
| 吴　莹 | 山羊溶酶体 α-甘露糖苷酶基因的截短表达及表达产物特性的研究 | 2014 | 西北农林科技大学 | 李勤凡 |
| 陈　睿 | 鱼腥草内生真菌生物多样性研究 | 2014 | 西北农林科技大学 | 李勤凡 |
| 马正南 | 甘草中产芦丁内生真菌的分离与鉴定 | 2014 | 西北农林科技大学 | 李勤凡 |
| 杨立军 | 儿茶素对蛋白羧甲基赖氨酸形成的抑制作用及其机制 | 2014 | 西北农林科技大学 | 董　强 |
| 张天宇 | 儿茶素对甲基乙二醛致人血管内皮细胞毒性的抑制作用及其机制 | 2014 | 西北农林科技大学 | 董　强 |
| 邹　苗 | 石首保护区麋鹿致病性大肠杆菌分离研究与肠道寄生虫感染调查 | 2014 | 华中农业大学 | 郭定宗 |
| 回　琳 | 胰岛素样生长因子-1 对 Caco2 细胞斯钙素 1 表达的影响 | 2014 | 华中农业大学 | 郭定宗 |
| 马　锐 | 1,25(OH)2D3 对牛肾细胞 STC-1 的表达作用及其凋亡的影响 | 2014 | 华中农业大学 | 郭定宗 |
| 孙晓转 | 不同日龄仔猪胃肠道 Obestatin 蛋白含量及 GHRL 基因转录水平测定 | 2014 | 华中农业大学 | 郭定宗 |
| 刁加亮 | 犬瘟热患犬呼吸道组织中蛋白酶激活受体 2 的分布变化研究 | 2014 | 华中农业大学 | 郭定宗 |
| 夏　菲 | 日本乙型脑炎病毒 NS1 蛋白单克隆抗体的制备及鉴定 | 2014 | 华中农业大学 | 郭定宗 |
| 白莹盈 | STC-1 在新生奶牛胃肠道中的定位和定量表达 | 2013 | 华中农业大学 | 郭定宗 |
| 赵　靖 | ARC 蛋白对胰岛素诱发的肉鸡心肌细胞肥大 c-myc 基因表达的影响 | 2013 | 华中农业大学 | 郭定宗 |
| 董世起 | Obestatin 在仔猪不同阶段胃肠壁中的分布研究 | 2012 | 华中农业大学 | 郭定宗 |
| 陈　森 | 麋鹿主要组织、器官的组织学观察与 STC-1 在肾脏中的定位 | 2012 | 华中农业大学 | 郭定宗 |
| 王建林 | 山羊松果体细胞微囊化组织相容性研究 | 2012 | 华中农业大学 | 郭定宗 |
| 韩胜旗 | 纳米氧化铜和硫酸铜对猪 SIEC 细胞铜代谢酶影响的比较研究 | 2013 | 华中农业大学 | 周东海 |
| 邓　荣 | FF3 在断奶腹泻仔猪肠道中的分布表达及对肠黏膜上皮细胞的作用 | 2012 | 华中农业大学 | 周东海 |
| 雷颂苹 | 硫酸铜和纳米铜对断奶仔猪血清抗氧化能力的影响 | 2013 | 华中农业大学 | 周东海 |
| 王　进 | 犬瘟热、犬细小病毒病犬临床病理学研究 | 2012 | 华中农业大学 | 周东海 |
| 李朝阳 | 纳米铜对猪肾细胞线粒体凋亡途径的影响 | 2014 | 华中农业大学 | 周东海 |
| 姚人升 | 纳米氧化铜和硫酸铜对断奶仔猪几种免疫因子的影响 | 2014 | 华中农业大学 | 周东海 |
| 赵丽茹 | 缺氧对肉鸡肺血管重塑_ET_1 及其受体表达水平的影响 | 2012 | 华中农业大学 | 李家奎 |
| 刘孝东 | 低氧状态对肉鸡肺组织中内皮素_1 及其受体的基因表达的影响 | 2012 | 华中农业大学 | 李家奎 |
| 张可荣 | 依普菌素透皮剂的配制及应用 | 2013 | 华中农业大学 | 李家奎 |
| 王立秋 | 武汉某马场马的常见病调查及过敏性皮炎的防治试验 | 2013 | 华中农业大学 | 李家奎 |
| 刘梦媛 | 牦牛 BVDV 血清流行病学调查及 BVDV 疫苗和猪瘟疫苗对牦牛的免疫效果 | 2014 | 华中农业大学 | 李家奎 |
| 翟　佳 | 武汉地区家犬两种重要人畜共患病血清学调查 | 2014 | 华中农业大学 | 李家奎 |
| 吴　岳 | 吴岳;IFN-τ 调控奶牛子宫上皮细胞表达 MICB 及 BoLA-A 的研究 | 2012 | 华中农业大学 | 邓干臻 |
| 舒　蕾 | BoLA--A 在妊娠奶牛和新生牛中的差异表达研究 | 2012 | 华中农业大学 | 邓干臻 |
| 王　潇 | 孕酮、雌二醇调控奶牛滋养层细胞表达 IFN-τ 研究 | 2012 | 华中农业大学 | 邓干臻 |
| 何明月 | BoLA--I 外显子 2、3 在亲本和子代奶牛中同源性研究 | 2012 | 华中农业大学 | 邓干臻 |
| 王彩霞 | 带蒂大网膜移植治疗犬腹壁大面积损失价值评价 | 2012 | 华中农业大学 | 邓干臻 |

| 姓名 | 论文题目 | 毕业时间 | 单位 | 指导教师 |
|---|---|---|---|---|
| 肖　雅 | 奶牛 BoLA-I 重链在子宫上皮细胞的表达调控及其在主要组织细胞分布的研究 | 2013 | 华中农业大学 | 邓干臻 |
| 祝　晶 | 奶牛 MIC1 表达检测及其在主要组织细胞分布研究 | 2013 | 华中农业大学 | 邓干臻 |
| 许　显 | 我国执业兽医师职数测算及发展趋势预测 | 2013 | 华中农业大学 | 邓干臻 |
| 刘宏靖 | 奶牛 BoLAA 表达检测及在主要组织细胞分布研究 | 2103 | 华中农业大学 | 邓干臻 |
| 肖　禹 | 湖北十堰地区犬吉氏巴贝斯虫流行病学调查 | 2013 | 华中农业大学 | 邓干臻 |
| 王　超 | IFN-tau 对奶牛 BoLA-I 表达作用研究 | 2014 | 华中农业大学 | 邓干臻 |
| 李彬彬 | 缺氧调节奶牛子宫内膜上皮细胞 BoLA-A 和 MIC1 表达 | 2014 | 华中农业大学 | 邓干臻 |
| 刘相洋 | 主要生殖激素和 IFN-tau 调控奶牛子宫上皮细胞表达 GM-CSF 表达 | 2014 | 华中农业大学 | 邓干臻 |
| 陈启宏 | 主要生殖激素好 IFN-tau 调节奶牛子宫内膜上皮细胞 BoLA-A 和 MIC1 研究 | 2014 | 华中农业大学 | 邓干臻 |
| 吴小彦 | 非泼罗尼对犬感染跳蚤的驱杀效果试验 | 2014 | 华中农业大学 | 邓干臻 |
| 李广胜 | 氟对大鼠成骨细胞间隙通讯的影响 | 2012 | 山西农业大学 | 王俊东 |
| 边升太 | 几丁聚糖抗氟中毒小鼠骨骼损伤的机理研究 | 2012 | 山西农业大学 | 王俊东 |
| 卢绪秀 | 青春期壬基酚与铅暴露对小鼠焦虑和抑郁样行为的影响 | 2912 | 山西农业大学 | 王俊东 |
| 刘双玲 | 氟铅联合对发育期小鼠海马和皮质中蛋白表达影响的研究 | 2012 | 山西农业大学 | 王俊东 |
| 田玉虎 | 氟对巨噬细胞前炎症因子表达的影响及其调节途径的研究 | 2012 | 山西农业大学 | 王俊东 |
| 贾　芳 | 纳米二氧化钛对青春期小鼠睾酮生成的影响及其机制研究 | 2013 | 山西农业大学 | 王俊东 |
| 陆兆静 | 氟对小鼠精子趋化性的影响及其机制研究 | 2013 | 山西农业大学 | 王俊东 |
| 魏瑞芬 | 氟对小鼠睾丸免疫毒性的研究及精子双向电泳图谱的构建 | 2014 | 山西农业大学 | 王俊东 |
| 闫杭杭 | 氟对小鼠 Y 染色体微缺失基因的影响及其研究机制 | 2014 | 山西农业大学 | 王俊东 |
| 张雪莲 | RNA 干扰法对羊驼 MC1R 基因表达量及黑色素合成量的影响 | 2012 | 山西农业大学 | 庞全海 |
| 赵礼军 | IFN-γ 对羊驼皮肤黑素细胞增殖和黑色素合成的影响 | 2012 | 山西农业大学 | 庞全海 |
| 陈娟娟 | pcDNA3.1(+)-MC1R 转染羊驼毛囊干细胞对 MC1R、TYR 及 MITF 表达量的影响 | 2012 | 山西农业大学 | 庞全海 |
| 牛　峰 | ASIP-YY 对羊驼皮肤黑色素细胞增殖和黑色素生成的影响 | 2012 | 山西农业大学 | 庞全海 |
| 夏　玉 | Gnαq 在小鼠不同毛色皮肤中的表达研究 | 2013 | 山西农业大学 | 庞全海 |
| 吴崇晖 | 乙基麦芽酚络合物对小鼠的一般毒性研究 | 2013 | 山西农业大学 | 庞全海 |
| 张　超 | 酸性蛋白酶对脂多糖刺激肉鸡应激反应的抑制作用 | 2013 | 山西农业大学 | 庞全海 |
| 王　平 | 益生素对免疫应激肉仔鸡生长性能及血液指标的影响 | 2013 | 山西农业大学 | 庞全海 |
| 肖　凯 | 兔斯氏艾美尔球虫野外毒株的分离与纯化 | 2013 | 山西农业大学 | 庞全海 |
| 柴　娟 | MMP-2 对 ET-1 诱导的肉鸡肺动脉平滑肌细胞增殖的调控及其在 PHS 肉鸡肺血管重构中的作用 | 2012 | 浙江大学 | 谭　勋 |
| 李文文 | 围产期奶牛 PMN 细胞 Nod-1、Nod-2 的表达变化及与 PMN 细胞功能的关系 | 2013 | 浙江大学 | 谭　勋 |
| 毕师诚 | HGF 与 ROS 联合作用对肉鸡内皮祖细胞功能的影响 | 2014 | 浙江大学 | 谭　勋 |
| 李君君 | BMAP27/ BMAP28 牛源抗菌肽的克隆表达及抑菌活性分析 | 2014 | 浙江大学 | 谭　勋 |
| 王世涛 | OPG 调控破骨细胞 NF-κB 通路的研究 | 2012 | 扬州大学 | 刘宗平 |
| 甄建伟 | 玉米赤霉烯酮对睾丸间质细胞的毒性作用 | 2012 | 扬州大学 | 刘宗平 |
| 徐　卉 | 镉致大鼠大脑皮质神经元凋亡的线粒体途径 | 2013 | 扬州大学 | 刘宗平 |
| 王玲玲 | 镉对大鼠肝细胞 DNA 甲基化和自噬的影响 | 2013 | 扬州大学 | 刘宗平 |
| 斯洋军 | 氟对小鼠早期胚胎 DNA 甲基化的影响 | 2013 | 扬州大学 | 刘宗平 |
| 刘　青 | 玉米赤霉烯酮抑制离体小鼠 Leydig 细胞睾酮分泌的机理研究 | 2013 | 扬州大学 | 卞建春 |
| 胡飞飞 | 镉诱导神经细胞凋亡的 mTOR 通路及其 NAC 保护效应 | 2014 | 扬州大学 | 刘宗平 |
| 程来洋 | 自噬在小鼠卵母细胞老化与 XRCC1 在早期胚胎 DNA 去甲基化中的作用 | 2014 | 扬州大学 | 刘宗平 |
| 张家铭 | 奶牛围产期血清氧化应激与微量元素含量变化的研究 | 2014 | 扬州大学 | 刘宗平 |
| 王家晶 | 奶牛围产期血清脂肪代谢及肝脏功能相关指标变化的研究 | 2014 | 扬州大学 | 刘宗平 |
| 王亚军 | 玉米赤霉烯酮诱导大鼠 Leydig 细胞的凋亡及自噬保护作用 | 2014 | 扬州大学 | 卞建春 |
| 仝锡帅 | 1α, 25-$(OH)_2D_3$ 对破骨细胞形成与活化相关功能蛋白的影响 | 2014 | 扬州大学 | 顾建红 |
| 姚路连 | 奶牛围产期血清相关激素水平动态变化的研究 | 2014 | 扬州大学 | 刘学忠 |
| 张　丹 | 淀粉糖化酶抗体的研制及其双抗夹心 ELISA 法的初步建立与应用 | 2012 | 扬州大学 | 王捍东 |

| 姓名 | 论文题目 | 毕业时间 | 单位 | 指导教师 |
|---|---|---|---|---|
| 姜　威 | β-呋喃果糖苷酶单克隆抗体的研制及其在蜂蜜掺假 ELISA 中的应用 | 2012 | 扬州大学 | 王捍东 |
| 汪惠泽 | 盐酸克伦特罗单克隆抗体的制备及其 CiELISA 检测方法的建立 | 2012 | 扬州大学 | 王捍东 |
| 罗宏鹏 | 猪源沙门氏菌的分离鉴定和耐药性检测（在职兽医硕士） | 2012 | 扬州大学 | 王捍东 |
| 唐　静 | 西藏山南地区动物检疫现状和思考（在职兽医硕士） | 2012 | 扬州大学 | 王捍东 |
| 李双喜 | 邻苯二甲酸二丁酯单克隆抗体的制备及 CiELISA 检测方法的建立与应用 | 2013 | 扬州大学 | 王捍东 |
| 冯云飞 | 邻苯二甲酸二（2-乙基己）酯单克隆抗体的制备及 ELISA 检测方法的建立 | 2013 | 扬州大学 | 王捍东 |
| 刘云迎 | 孔雀石绿单克隆抗体的研制及 ELISA 检测方法的建立 | 2014 | 扬州大学 | 王捍东 |
| 蒋志惠 | 硒蛋白 W 对鸡肝脏细胞凋亡影响的研究 | 2013 | 东北农业大学 | 徐世文 |
| 刘　焘 | 阿特拉津和毒死蜱单一及联合暴露对鲤鱼热休克蛋白影响 | 2013 | 东北农业大学 | 徐世文 |
| 孟繁宇 | 小鹅瘟肠道损伤机制的研究与 VP3 基因重组腺病毒的构建 | 2013 | 东北农业大学 | 徐世文 |
| 王　超 | 阿特拉津和毒死蜱单一及联合暴露对鲤鱼组织 DNA 甲基化的影响 | 2013 | 东北农业大学 | 徐世文 |
| 王亮亮 | 阿特拉津和毒死蜱单一及联合暴露对鲤鱼脑组织自噬的影响 | 2013 | 东北农业大学 | 徐世文 |
| 姚海东 | 硒蛋白 W 抗氧化功能及其在鸡缺硒性骨骼肌细胞凋亡作用中的研究 | 2013 | 东北农业大学 | 徐世文 |
| 吴　琼 | 硒对鸡成肌细胞分化及骨骼肌炎症基因表达影响的研究 | 2013 | 东北农业大学 | 徐世文 |
| 张　博 | 城市污水中肠杆菌的抗生素耐药性和多种耐药基因的检测 | 2013 | 东北农业大学 | 徐世文 |
| 赵福庆 | 冷应激对雏鸡组织热休克蛋白和小肠免疫功能的影响 | 2013 | 东北农业大学 | 徐世文 |
| 关　博 | 哈尔滨市鸡饲料中 DON 和 ZEA 的调查及其对鸡脾淋巴细胞毒性的影响 | 2013 | 东北农业大学 | 徐世文 |
| 赵　妍 | 山奈酚对脂多糖诱导小鼠急性肺损伤的保护作用 | 2013 | 东北农业大学 | 林洪金 |
| 商思伟 | 复方青天葵颗粒剂的研制及其对断奶仔猪生长性能和血液指标的影响 | 2013 | 东北农业大学 | 林洪金 |
| 林石磊 | 缺硒对鸡甲状腺和肾脏功能的影响 | 2014 | 东北农业大学 | 徐世文 |
| 良　杨 | 低硒对鸡脂肪组织中脂肪代谢与硒蛋白表达的影响 | 2014 | 东北农业大学 | 徐世文 |
| 曹　也 | 阿维菌素对王鸽组织 DNA 总甲基化水平的影响 | 2014 | 东北农业大学 | 徐世文 |
| 陈德纯 | 阿特拉津和毒死蜱单一及联合暴露对鲤鱼免疫器官自噬的影响 | 2014 | 东北农业大学 | 徐世文 |
| 于　娇 | 缺硒对鸡肠道免疫功能的影响 | 2014 | 东北农业大学 | 徐世文 |
| 周红艳 | 某鸡场大肠杆菌耐药分析及预防效果 | 2014 | 东北农业大学 | 徐世文 |
| 王从武 | 硒缺乏对鸡肝脏内源性硫化氢及其相关生成酶的影响 | 2014 | 东北农业大学 | 林洪金 |
| 李　楠 | 镉经内质网应激途径诱导鸡肝细胞自噬的机制 | 2014 | 东北农业大学 | 孙　刚 |
| 王　楠 | 亚慢性铝暴露对雌性大鼠的生殖毒性 | 2012 | 东北农业大学 | 李艳飞 |
| 佘　玥 | DFP 对染铝大鼠脾淋巴细胞免疫功能的影响 | 2012 | 东北农业大学 | 李艳飞 |
| 韩彦飞 | 三氯化铝对体外培养大鼠成骨细胞的毒性作用 | 2013 | 东北农业大学 | 李艳飞 |
| 张继红 | 肾上腺素对体外培养染铝大鼠脾淋巴细胞免疫功能的影响 | 2013 | 东北农业大学 | 李艳飞 |
| 夏世亮 | 铝致大鼠淋巴细胞凋亡的线粒体/Caspase 依赖性途径机制 | 2013 | 东北农业大学 | 李艳飞 |
| 白崇生 | CRH、ACTH 和 GC 对体外培养染铝大鼠脾淋巴细胞免疫功能的影响 | 2013 | 东北农业大学 | 李艳飞 |
| 王　静 | 铝致大鼠淋巴细胞凋亡的 cAMP/PKA 信号传导机制 | 2014 | 东北农业大学 | 李艳飞 |
| 宋　淼 | 铝对大鼠成骨细胞矿化过程的影响 | 2014 | 东北农业大学 | 李艳飞 |
| 富　杨 | 铝对大鼠成骨细胞 Wnt/β-catenin 信号通路影响 | 2014 | 东北农业大学 | 李艳飞 |
| 王功臣 | 大蒜素对蛋雏肝脏结构与功能的影响 | 2014 | 东北农业大学 | 李艳飞 |
| 王子久 | 生育三烯酚对高脂小鼠肝脏保护功能及脂代谢机理的研究 | 2013 | 湖南农业大学 | 文利新 |
| 何沙沙 | 生育三烯酚对小鼠胆固醇代谢及其相关基因表达的研究 | 2013 | 湖南农业大学 | 文利新 |
| 李　锐 | 蚯蚓镇痛有效成分的镇痛机制初探 | 2013 | 湖南农业大学 | 李文平 |
| 韩　燃 | EHEC O157:H7 保护性抗原的筛选和鉴定 | 2013 | 湖南农业大学 | 李文平 |
| 廖鹏运 | 蚯蚓提取物对肉品的保鲜作用与机理研究 | 2013 | 湖南农业大学 | 李文平 |
| 文　琼 | 发芽糙米及其功效成分对大鼠和 LO2 细胞抗氧化以及脂代谢的影响 | 2014 | 湖南农业大学 | 文利新 |
| Froilan | Cytoprotective Effect of Oryzanol and γ-Aminobutric Acid(GABA) in | 2014 | 湖南农业大学 | 文利新 |

| 姓名 | 论文题目 | 毕业时间 | 单位 | 指导教师 |
|---|---|---|---|---|
| Bernard R.Matias | fatty Acid treated RAW 264.7 Macrophages | | | |
| 赖小龙 | 应用基因表达谱芯片筛选结肠癌差异表达基因 | 2014 | 湖南农业大学 | 李文平 |
| 常争艳 | 全外显子测序筛选及验证结肠癌相关基因突变 | 2014 | 湖南农业大学 | 李文平 |
| 戴梦南 | 生物处理畜禽粪便及对蚯蚓养殖效果的研究 | 2014 | 湖南农业大学 | 李文平 |
| 贺　琦 | 蚯蚓提取物对激素性股骨头坏死的防治作用研究 | 2014 | 湖南农业大学 | 李文平 |
| 孟　锐 | 镰刀菌毒素DON、ZEA及其联合染毒对小鼠肠道功能损伤的研究 | 2012 | 四川农业大学 | 邓俊良 |
| 朱于斌 | 脱氧雪腐镰刀菌烯醇、玉米赤霉烯酮及其联合染毒对小鼠肝肾损伤机制的研究 | 2012 | 四川农业大学 | 邓俊良 |
| 陈仓良 | 复方中药“猪康散”对仔猪体液免疫功能及细胞因子水平影响的研究 | 2012 | 四川农业大学 | 邓俊良 |
| 刘晋生 | 转基因植酸酶玉米的营养成分及其对猪的营养价值评定 | 2012 | 四川农业大学 | 邓俊良 |
| 李　洋 | 镰刀菌毒素ZEA、DON及其联合染毒对小鼠脑损伤机制的研究 | 2013 | 四川农业大学 | 邓俊良 |
| 郑健宇 | 复合抗菌肽”态康利保”对断奶仔猪生长性能及血液生理生化指标的影响(兽医硕士) | 2013 | 四川农业大学 | 邓俊良 |
| 胥世洪 | 复合抗菌肽”态康利保”对断奶仔猪细胞免疫功能的影响(兽医硕士) | 2014 | 四川农业大学 | 邓俊良 |
| 周　锐 | 镰刀菌毒素ZEA、DON及联合作用对小白鼠体液免疫及抗氧化功能影响的研究 | 2014 | 四川农业大学 | 邓俊良 |
| 张　超 | DON、ZEA及其联合染毒对小鼠脾脏细胞免疫及T细胞亚群的影响研究 | 2014 | 四川农业大学 | 邓俊良 |
| 石　凯 | 隐孢子虫棒状体cgd8_2530蛋白sushi片段亚单位疫苗和核酸疫苗的制备及其免疫保护性研究 | 2014 | 四川农业大学 | 邓俊良 |
| 赵春蕊 | “主动免疫增强剂”与疫苗协同对比格犬免疫功能的影响 | 2012 | 四川农业大学 | 胡延春 |
| 魏继涛 | 黄曲霉毒素B1单克隆抗体的制备及鉴定 | 2012 | 四川农业大学 | 胡延春 |
| 谢　鹏 | 畜禽用多种微量元素注射液的处方、工艺与稳定性研究 | 2013 | 四川农业大学 | 胡延春 |
| 廖　飞 | 紫茎泽兰主要毒素euptox A的提取及其生物活性初步研究 | 2014 | 四川农业大学 | 胡延春 |
| 谭　辉 | 牛瘤胃液中玉米赤霉烯酮降解菌的分离、鉴定及降解特性研究 | 2014 | 四川农业大学 | 胡延春 |
| 罗丽娟 | 主要微量元素肠溶微囊的制备及其特性的研究 | 2014 | 四川农业大学 | 胡延春 |
| 代有兵 | 猪肉产品溯源系统的构建及初步应用 | 2014 | 四川农业大学 | 胡延春 |
| 刘曦 | 黄帚橐吾提取物抑菌活性初步研究 | 2014 | 四川农业大学 | 胡延春 |
| 金仕强 | 10株新城疫病毒（弱毒）的遗传进化分析 | 2012 | 四川农业大学 | 左之才 |
| 白　欣 | 不同药物处理对深II度烫伤犬细胞因子影响的研究 | 2013 | 四川农业大学 | 左之才 |
| 张　乐 | 胸腺五肽对AA肉鸡生产性能及抗氧化功能的影响研究 | 2013 | 四川农业大学 | 左之才 |
| 吴　鹏 | 竹鼠大肠杆菌病的病原特性及其综合防治研究 | 2013 | 四川农业大学 | 左之才 |
| 甘　梦 | 自拟中药复方对骨髓抑制小鼠造血及免疫功能的影响研究 | 2013 | 四川农业大学 | 左之才 |
| 夏祥昕 | 钙、钾注射液及钙钾混合液对犬心脏功能的影响研究 | 2013 | 四川农业大学 | 左之才 |
| 赵志贤 | 早期断奶应激对仔猪肠道屏障损伤作用的研究 | 2013 | 四川农业大学 | 左之才 |
| 谢懋英 | 肉鸡产业链中肠炎沙门菌的分离鉴定、耐药性分析及PFGE分型 | 2014 | 四川农业大学 | 左之才 |
| 钟　洁 | 重组蛋白Cfr的单克隆抗体制备及初步应用 | 2014 | 四川农业大学 | 左之才 |
| 赵晓东 | 冷季全舍饲对麦洼牦牛生长性能、血液生化和血清矿物元素的影响 | 2014 | 四川农业大学 | 左之才 |
| 姚慧敏 | 犬疾病专用口服补液盐配方的筛选及应用 | 2014 | 四川农业大学 | 左之才 |
| 王　健 | 锰缺乏对肉仔鸡胫骨生长板、软骨细胞发育相关因子的影响 | 2012 | 山东农业大学 | 王振勇 |
| 刘　然 | 锰缺乏对肉仔鸡胫骨形态学及OPG/RANKL信号传导通路的影响 | 2012 | 山东农业大学 | 王振勇 |
| 王照军 | 锰缺乏对肉仔鸡骨骼发育相关血清学指标的影响 | 2012 | 山东农业大学 | 王振勇 |
| 周　川 | 牛磺鹅去氧胆酸抗小鼠肺纤维化作用的研究 | 2012 | 山东农业大学 | 刘建柱 |
| 刘永夏 | 应用重组MSG1蛋白检测猪附红细胞体抗体的ELISA方法的建立及抗MSG1基因单克隆抗体的制备 | 2012 | 山东农业大学 | 刘建柱 |
| 杨　溢 | 抗猪流感病毒单味中药的体外筛选 | 2013 | 山东农业大学 | 王振勇 |
| 仲　峰 | 转基因紫花苜蓿对家兔的慢性毒性实验 | 2013 | 山东农业大学 | 王振勇 |

| 姓名 | 论文题目 | 毕业时间 | 单位 | 指导教师 |
|---|---|---|---|---|
| 王　珍 | 抗犬流感病毒单味中草药体外筛选 | 2013 | 山东农业大学 | 王振勇 |
| 刘明超 | 三氯化铬对鸡胚成纤维细胞活性的影响 | 2013 | 山东农业大学 | 刘建柱 |
| 白艳飞 | 吡啶甲酸铬对鸡胚成纤维细胞活性的氧化损伤作用及机制研究 | 2013 | 山东农业大学 | 刘建柱 |
| 禚雯超 | J亚群ALV的分离鉴定及三种方法动态比较分析其在DF-1细胞中的增殖 | 2013 | 山东农业大学 | 刘建柱 |
| 刘　刚 | 线粒体膜通透性转换在铅诱导大鼠肾小管上皮细胞凋亡中的作用 | 2014 | 山东农业大学 | 王振勇 |
| 牛茂源 | Lpin1基因的多态性及其分析 | 2014 | 山东农业大学 | 王振勇 |
| 樊瑞锋 | 奶牛胰岛素抵抗与脂肪肝发病关系的研究 | 2014 | 山东农业大学 | 王振勇 |
| 李忠鹏 | 氧化应激在奶牛跛行发病中的作用及其机制探讨 | 2014 | 山东农业大学 | 王振勇 |
| 张立梅 | 脂质体包被ALV-GP85蛋白的疫苗制备及其免疫效果研究 | 2014 | 山东农业大学 | 刘建柱 |
| 才冬杰 | 禽白血病J亚群GP85基因乳酸杆菌表达及其表达产物对动物机体的影响 | 2014 | 山东农业大学 | 刘建柱 |
| 李　展 | 中国部分地区犬肠道寄生虫感染情况的调查 | 2014 | 山东农业大学 | 刘建柱 |
| 孙志阔 | 大豆抗原蛋白对仔猪的致敏性和免疫原性的研究 | 2012 | 安徽农业大学 | 吴金节 |
| 石念进 | 中药复方多糖的提取方法优化及对雏鸡免疫功能的影响 | 2012 | 安徽农业大学 | 吴金节 |
| 刘芳芳 | 3种离子对种猪生殖机能的作用及其机理研究 | 2012 | 安徽农业大学 | 吴金节 |
| 闫妮娜 | 合肥地区犬肿瘤流行病学调查及临床诊断技术的研究 | 2013 | 安徽农业大学 | 吴金节 |
| 汪　洋 | 中药复方透皮贴剂有效成分的提取及测定方法的研究 | 2013 | 安徽农业大学 | 吴金节 |
| 钟　刚 | 中药添加剂和包被γ-氨基丁酸对奶牛热应激的作用机理研究 | 2013 | 安徽农业大学 | 吴金节 |
| 徐同铜 | 包被γ-氨基丁酸对奶牛热应激的防治效果及机理的研究 | 2013 | 安徽农业大学 | 李锦春 |
| 罗　莹 | 中药提取物对奶牛乳腺上皮细胞增殖及蛋白含量的影响 | 2014 | 安徽农业大学 | 吴金节 |
| 马良友 | 大豆抗原蛋白对断奶仔猪血清中细胞因子的影响及肠道致敏作用 | 2014 | 安徽农业大学 | 吴金节 |
| 周冠宇 | 奶牛乳房炎中药复方灌注液有效成分测定方法的研究 | 2014 | 安徽农业大学 | 吴金节 |
| 寇亚楠 | 大豆抗原蛋白对仔猪肠道的致敏作用及ELISA检测方法的建立 | 2014 | 安徽农业大学 | 吴金节 |
| 文中涛 | p-ASAP对SD大鼠的促生长作用及其机理研究 | 2014 | 安徽农业大学 | 吴金节 |
| 李复辉 | FB1 Ci-ELISA检测方法的建立及初步应用 | 2014 | 安徽农业大学 | 王希春 |
| 杨雪艳 | 合肥地区犬皮肤病的调查及鉴别诊断流程的建立与应用 | 2014 | 安徽农业大学 | 李锦春 |
| 张　茜 | 纳米对氨基苯胂酸对SD大鼠毒性作用的研究 | 2014 | 安徽农业大学 | 李锦春 |
| 孙喜安 | 绵羊肺炎支原体内蒙古地方株的分离鉴定及交叉反应性的测定 | 2012 | 内蒙古农业大学 | 韩　敏 |
| 王　荣 | 新西兰引进娟珊牛、荷斯坦奶牛与中国荷斯坦奶牛的抗病性比较 | 2012 | 内蒙古农业大学 | 韩　敏 |
| 张　利 | 环介导等温扩增技术（LAMP）与荧光定量PCR检测布鲁菌方法的建立及对比试验 | 2012 | 内蒙古农业大学 | 莫　内 |
| 邵　贺 | 丙泊酚麻醉对犬生理机能影响的研究 | 2012 | 内蒙古农业大学 | 莫　内 |
| 陈　圆 | H9N2亚型禽流感病毒鸭感染模型的建立 | 2013 | 内蒙古农业大学 | 韩　敏 |
| 张建军 | 小花棘豆生物碱对小白鼠免疫功能及α-甘露糖苷酶表达的影响 | 2013 | 内蒙古农业大学 | 韩　敏 |
| 李晓菲 | 雌激素对奶牛输卵管平滑肌PGE2和PGF2α受体表达的影响 | 2014 | 内蒙古农业大学 | 莫　内 |
| 胡广胜 | 兰州地区奶牛子宫内膜炎病原菌分离鉴定及抗生素耐药性研究 | 2013 | 甘肃农业大学 | 李宏胜 |
| 肖　敏 | 牛源金黄色葡萄球菌基因分型、毒力基因和耐药基因研究 | 2014 | 甘肃农业大学 | 李宏胜 |
| 罗秀刚 | 河西绒山羊DRB3、DQB1基因外显子2多态性及其连锁分析 | 2012 | 甘肃农业大学 | 马小军 |
| 岳　燕 | 河西绒山羊MHC-DQB2基因第2外显子和第3外显子多态性及其与流产性状关联分析 | 2012 | 甘肃农业大学 | 马小军 |
| 李富强 | 合作猪血液生理生化及免疫指标的测定和IFN-γ序列分析 | 2013 | 甘肃农业大学 | 马小军 |
| 姜力飞 | 合作猪SLA-DQB、Mx1基因多态性分析 | 2013 | 甘肃农业大学 | 马小军 |
| 张国华 | 合作猪SLA-DQA第2、4外显子多态性分析 | 2014 | 甘肃农业大学 | 马小军 |
| 吕伟丽 | 甘肃地区小尾寒羊TLR4、TLR9基因多态性与呼吸系统疾病的相关性分析 | 2014 | 甘肃农业大学 | 马小军 |
| 王恩丽 | 猪嵴病毒CH441株和猪博卡病毒CH437株全基因组克隆及序列分析 | 2014 | 甘肃农业大学 | 马小军 |
| 曹　志 | 治疗奶牛乳房炎中草药复方透皮软膏的研究 | 2012 | 宁夏大学 | 何生虎 |
| 安泓霏 | DHI报告对奶牛场的管理应用研究 | 2012 | 宁夏大学 | 何生虎 |
| 邵　倩 | 牛支原体分离鉴定、药敏试验及耐喹诺酮类药基因的监测 | 2012 | 宁夏大学 | 何生虎 |
| 郭　磊 | 防治奶牛子宫内膜炎复方黄连泡腾片的制备与研究 | 2013 | 宁夏大学 | 何生虎 |

| 姓名 | 论文题目 | 毕业时间 | 单位 | 指导教师 |
|---|---|---|---|---|
| 蒋魏娟 | 治疗奶牛乳房炎中草药涂膜剂的研究 | 2013 | 宁夏大学 | 何生虎 |
| 吴学荣 | 中草药泡腾片对肉鸡后期增长保健作用的研究 | 2013 | 宁夏大学 | 何生虎 |
| 孙文龙 | 绵羊 GPR54、LIN28、LIN28b 和 SCD 基因多态性及生物信息学分析 | 2014 | 宁夏大学 | 何生虎 |
| 张蕾蕾 | 疯草内生真菌-Undifilum Oxytiopis 合成苦马豆素的研究 | 2014 | 宁夏大学 | 何生虎 |
| 杨萌萌 | 防治牛前胃疾病中药复方分散片的制备与研究 | 2014 | 宁夏大学 | 何生虎 |
| 潘美娟 | 燕麦草、羊草及其组合 TMR 日粮对奶牛瘤胃消化代谢的影响 | 2012.6 | 南京农业大学 | 张克春<br>黄克和 |
| 王　亮 | 亚临床酮病对围产期奶牛临床发病情况、生化指标及免疫功能动态变化的影响 | 2012.12 | 南京农业大学 | 张克春<br>黄克和 |
| 高　潮 | 奶牛隐性乳房炎病原微生物区系分析及金黄色葡萄球菌基因分型 | 2013.6 | 南京农业大学 | 张克春<br>黄克和 |
| 林　梅 | 上海地区牛传染性鼻气管炎感染情况调查及病毒分离鉴定 | 2014.5 | 南京农业大学 | 张克春<br>黄克和 |
| 鲁俊杰 | 复方中药乳剂对奶牛子宫内膜炎防治效果研究 | 2012 | 黑龙江八一农垦大学 | 武　瑞 |
| 王跃强 | 弱激光治疗对 LPS 诱导大鼠乳腺炎粘附分子的影响 | 2013 | 黑龙江八一农垦大学 | 武　瑞 |
| 郭婷婷 | 防治奶牛乳房炎中药成膜剂研究 | 2013 | 黑龙江八一农垦大学 | 武　瑞 |
| 王宏刚 | 复方氨基酸螯合钙口服液临床效果研究 | 2013 | 黑龙江八一农垦大学 | 武　瑞 |
| 黄显烨 | 生化黄腐酸治疗奶牛隐性乳房炎应用研究 | 2013 | 黑龙江八一农垦大学 | 武　瑞 |
| 宋国希 | 奶牛血清钙检测试剂盒研制与应用 | 2014 | 黑龙江八一农垦大学 | 武　瑞 |
| 张国岩 | 血清钙检测试剂盒的应用及围产期奶牛低钙血症的调查 | 2014 | 黑龙江八一农垦大学 | 武　瑞 |
| 孙玲伟 | 基于 1H NMR 和 GC/MS 技术的奶牛酮病血浆代谢组学分析 | 2014 | 黑龙江八一农垦大学 | 张洪友 |
| 包　凯 | 基于孕酮单抗的奶牛妊娠诊断胶体金试纸条研制与临床应用 | 2014 | 黑龙江八一农垦大学 | 张洪友 |
| 王　博 | 奶牛围产期酮病和乳热的早期预警体系的研究 | 2014 | 黑龙江八一农垦大学 | 夏　成 |
| 段　宇 | 孕酮传感器检测奶牛妊娠和繁殖障碍疾病的效果 | 2014 | 黑龙江八一农垦大学 | 夏　成 |
| 遥　远 | 黑龙江省集约化牛场奶牛Ⅱ型酮病的调查研究 | 2014 | 黑龙江八一农垦大学 | 夏　成 |
| 刘健男 | 黑龙江省泌乳奶牛主要维生素和微量元素状况的调查研究 | 2014 | 黑龙江八一农垦大学 | 夏　成 |
| 舒　适 | 奶牛乳热血浆差异表达蛋白的分离鉴定及其生物信息学分析 | 2013 | 黑龙江八一农垦大学 | 夏　成 |
| 孙照磊 | 奶牛酮病的发病调查及酮病与胰岛素抵抗的关系 | 2013 | 黑龙江八一农垦大学 | 夏　成 |
| 许　文 | 不同钼水平对山羊肾脏线粒体和红细胞膜自由基代谢及组织金属含量的影响 | 2012 | 江西农业大学 | 胡国良 |
| 田　山 | 高钙日粮和肾型传染性支气管炎病毒致鸡痛风的临床病理学比较研究 | 2012 | 江西农业大学 | 胡国良 |
| 国　鑫 | 肾型传支病毒致鸡痛风过程中肾损伤及机理的研究 | 2012 | 江西农业大学 | 郭小权 |
| 熊关越 | 南昌地区家养犬血液常规及其生化指标的测定与探讨 | 2012 | 江西农业大学 | 胡国良 |
| 庄　煜 | 钼镉联合胁迫对山羊肝线粒体自由基代谢及肝部分基因 mRNA 表达的影响 | 2013 | 江西农业大学 | 胡国良 |

| 姓名 | 论文题目 | 毕业时间 | 单位 | 指导教师 |
|---|---|---|---|---|
| 彭成诚 | 镉钼联合胁迫对山羊血清自由基代谢及肝凋亡相关基因表达量的影响 | 2013 | 江西农业大学 | 胡国良 |
| 顾小龙 | 钼镉联合胁迫对山羊红细胞膜损伤和脾脏相关基因表达的影响 | 2013 | 江西农业大学 | 胡国良 |
| 杨建珍 | 钼镉联合胁迫对山羊肾线粒体自由基代谢及凋亡相关基因表达的影响 | 2013 | 江西农业大学 | 胡国良 |
| 金丽珠 | 不同钼水平对山羊瘤胃内环境及组织凋亡相关基因 mRNA 表达的影响 | 2013 | 江西农业大学 | 胡国良 |
| 赵　海 | 肾型 IBV 致鸡痛风的临床病理学与中药防治研究 | 2013 | 江西农业大学 | 胡国良 |
| 匡　俊 | 复方中药对肾型 IBV 诱发鸡痛风的防治作用 | 2013 | 江西农业大学 | 郭小权 |
| 肖莉春 | 29 种中药营养价值及对耐药大肠杆菌抑菌活性研究 | 2013 | 江西农业大学 | 郭小权 |
| 黄名钱 | 猪源大肠杆菌四环素耐药基因检测及大肠杆菌抑制剂筛选 | 2013 | 江西农业大学 | 郭小权 |
| 彭梦华 | 钼镉联合胁迫对山羊组织器官微量元素含量的影响 | 2013 | 江西农业大学 | 胡国良 |
| 谢晓鹏 | 中药方剂对肾型传染性支气管炎病毒致痛风鸡的生产性能和血常规的影响 | 2013 | 江西农业大学 | 郭小权 |
| 庄智明 | 钼镉联合胁迫对山羊消化系统微量元素含量的影响 | 2013 | 江西农业大学 | 胡国良 |
| 夏安琪 | 江西不同地方品种鸡 apoA-Ⅰ、apoB 基因多态性与其脂肪代谢的关联性分析 | 2014 | 江西农业大学 | 胡国良 |
| 乔煜婷 | 水体中甲肝病毒富集检测方法的研究 | 2014 | 江西农业大学 | 胡国良 |
| 陈　花 | 钼镉联合诱导鸭睾丸损伤及对相关基因表达的影响 | 2014 | 江西农业大学 | 张彩英 |
| 饶　丹 | 猪瘟野毒株的分离鉴定及与疫苗株的 HRM 鉴别方法的建立 | 2014 | 江西农业大学 | 郭小权 |
| 邹跃龙 | 肾型 IBV 感染对鸡临床病理学及相关基因表达的影响 | 2014 | 江西农业大学 | 郭小权 |
| 朱书梁 | 高钙和高钙高蛋白日粮对鸡临床病理学和相关基因表达的影响 | 2014 | 江西农业大学 | 郭小权 |
| 李山威 | 钼镉联合诱导对鸭肾功能和 CP、MT-1 mRNA 表达量的影响 | 2014 | 江西农业大学 | 郭小权 |
| 龚　婷 | 钼镉联合诱导对鸭肝功能和 CP、MT-1 mRNA 表达的影响 | 2014 | 江西农业大学 | 曹华斌 |
| 苏强 | 三聚氰胺对肉鸡的急性、蓄积和亚慢行毒性及三聚氰胺对肉鸡血液生化指标的影响 | 2012 | 青岛农业大学 | 朱连勤 |
| 贾姗姗 | 硫酸化壳聚糖硒的合成及不同硒源对鸡肝细胞抗氧化功能的影响 | 2012 | 青岛农业大学 | 朱连勤 |
| 齐娟 | EGM 合成工艺优化、结构表征及其体外吸附霉菌毒素的能力和对淋巴细胞的保护作用 | 2012 | 青岛农业大学 | 朱连勤 |
| 张娇娇 | 纳米氧化铜的稳定性及对肝细胞生长代谢的影响 | 2012 | 青岛农业大学 | 朱连勤 |
| 柳永振 | 玉米赤霉烯酮降解菌的分离与鉴定 | 2013 | 青岛农业大学 | 朱连勤 |
| 张婷 | 脱氧雪腐镰刀烯醇对小鼠免疫功能的影响 | 2013 | 青岛农业大学 | 朱连勤 |
| 张静 | 呕吐毒素、玉米赤霉烯酮和烟曲霉毒素 B1 互作对昆明小鼠毒性的研究 | 2013 | 青岛农业大学 | 朱连勤 |
| 王学鑫 | 复方霉菌毒素吸附剂的研制及对饲喂霉变饲料雏鸡的保护作用 | 2013 | 青岛农业大学 | 朱连勤 |
| 朱祖贤 | 饲料中霉菌毒素在肉鸡混合感染发生的诱导作用及防控措施的研究 | 2014 | 青岛农业大学 | 朱连勤 |
| 崔晓妮 | 子宫内膜炎患牛子宫中 MMPs 的表达及其调控机制 | 2013 | 青岛农业大学 | 曹荣峰 |
| 陶爽 | 奶牛子宫内膜组织 β-防御素与炎症因子的关系 | 2013 | 青岛农业大学 | 曹荣峰 |

## 2.博士研究生论文名单

我国兽医内科诊断学培养已毕业的博士研究生 50 名（2012 秋至 2014 年）

| 姓名 | 论文题目 | 毕业时间 | 单位 | 指导教师 |
|---|---|---|---|---|
| 吴文达 | B 型单端孢霉烯族毒素诱导拒食和呕吐的机理研究 | 2012 | 南京农业大学 | 张海彬 |
| 王　莹 | 伏马菌素 B1 和赭曲霉毒素 A 噬菌体单链抗体库的构建及重组抗体蛋白的原核表达 | 2012 | 南京农业大学 | 张海彬 |
| 王育伟 | 脱氧雪腐镰刀菌烯醇对大鼠胃毒性损伤及机理研究 | 2014 | 南京农业大学 | 张海彬 |

| 姓名 | 论文题目 | 毕业时间 | 单位 | 指导教师 |
| --- | --- | --- | --- | --- |
| 任　喆 | 大叶贯众对小鼠增强免疫和抗氧化作用的研究 | 2014 | 南京农业大学 | 张海彬 |
| 谢伟东 | 犬糖尿病的病因与临床诊疗研究 | 2013 | 南京农业大学 | 黄克和 |
| 吴　聪 | 肠道共生菌感染基因敲除小鼠后黏膜免疫的变化及中药成分的改善作用 | 2014 | 南京农业大学 | 黄克和 |
| 叶耿坪 | 富甘油酵母菌制剂对围产期奶牛能量负平衡的影响及机理研究 | 2014 | 南京农业大学 | 黄克和 |
| 刘　瑾 | 产甘油酵母培养物缓解奶牛热应激的作用及机理研究 | 2014 | 南京农业大学 | 黄克和 |
| Fahmida Parveen | 黄曲霉毒素 B1 诱导的氧化应激和细胞损伤及硒和β胡萝卜素对其的改善作用 | 2014 | 南京农业大学 | 黄克和 |
| 刘永旺 | 不同硒源对乙二醇诱导的犬草酸钙尿结石形成的影响与机理研究 | 2014 | 南京农业大学 | 黄克和 |
| 李　爽 | 鸭坦布苏病毒分离鉴定及感染性 cDNA 克隆的构建 | 2013 | 中国农业大学 | 韩　博<br>苏敬良 |
| Mohammad Zahra | 鼻气管鸟杆菌小菌落的分离鉴定 | 2014 | 中国农业大学 | 韩　博 |
| 翁晓刚 | 北京地区牛病毒性腹泻病毒分子流行病学调查及其持续性感染对牛 IFN-α/β 反应的影响 | 2014 | 中国农业大学 | 王九峰 |
| 李心慰 | 乙酸、非酯化脂肪酸、生长激素和催乳素调控奶牛肝细胞脂代谢的信号机制 | 2013 | 吉林大学 | 王　哲 |
| 王建国 | 围产期健康奶牛与酮病、亚临床低钙血症病牛血液代谢谱的比较与分析 | 2013 | 吉林大学 | 王　哲 |
| 刘　博 | 小鼠巨噬细胞 TLR2、TLR4 及 RP105 在金黄色葡萄球菌感染中的天然免疫应答机制 | 2013 | 吉林大学 | 张乃生 |
| 刘春杰 | 阿魏酸钠抗炎分子机制及对奶牛子宫内膜炎疗效初步观察 | 2013 | 吉林大学 | 张乃生 |
| 赵立香 | 吉林市奶牛子宫内膜炎致病菌调查及 TLR4 封闭肽作用机制研究 | 2013 | 吉林大学 | 张乃生 |
| 刘　磊 | 仔猪补充甘油三酯 DHA 或磷脂 DHA 效能的比较及初步分析 | 2014 | 吉林大学 | 王　哲 |
| 李　玉 | 奶牛酮病氧化应激致肝细胞凋亡的相关信号转导机制研究 | 2014 | 吉林大学 | 王　哲 |
| 杨　威 | 绵羊脂肪肝超声诊断方法的研究 | 2014 | 吉林大学 | 王　哲 |
| 郭梦尧 | 黄芩苷对小鼠金黄色葡萄球菌性乳腺炎的作用及机制研究 | 2014 | 吉林大学 | 张乃生 |
| 张　雯 | 硒对奶牛乳腺炎抗炎作用和相关信号转导通路调节机制的研究 | 2014 | 吉林大学 | 张乃生 |
| 高瑞峰 | 绿原酸抗乳腺炎作用及机制研究 | 2014 | 吉林大学 | 张乃生 |
| 杨国栋 | 疯草内生真菌合成苦马豆素的研究 | 2012 | 西北农林科技大学 | 王建华 |
| 吴黎明 | 斯钙素-1 在初生奶牛胃肠道的表达及其对氧化应激和新型钙离子通道的影响研究 | 2014 | 华中农业大学 | 郭定宗 |
| 高建峰 | 低氧/低硒对肉鸡肺脏内皮素-1 和其受体基因表达的影响 | 2014 | 华中农业大学 | 李家奎 |
| 马艳琴 | 氟与砷致动脉粥样硬化性血管损伤及其分子机制 | 2012 | 山西农业大学 | 王俊东 |
| 张向杰 | TFHL 和 TSG 对肉鸡脂质代谢的影响及其作用机制 | 2012 | 山西农业大学 | 王俊东 |
| 杨忠丽 | Studies on susceptibility genes and functional variants for smoking dependence by using molecular biology and bioinformatics approaches | 2013 | 山西农业大学 | 王俊东 |
| 梁占学 | 氟致兔肝损伤及蛋白质和钙干预性研究 | 2013 | 山西农业大学 | 王俊东 |
| 孙好学 | 缺硒对感染禽流感 H9N2 亚型病毒小鼠致病作用的研究 | 2013 | 山西农业大学 | 王俊东 |
| 李志刚 | 胎儿期补充胆碱诱导成年期小鼠抗抑郁和抗焦虑行为的表观机制研究 | 2014 | 山西农业大学 | 王俊东 |
| 白　峰 | 氟对大鼠骨组织 MSCs 成脂成骨分化的影响 | 2014 | 山西农业大学 | 王俊东 |
| 罗广营 | 双酚 A 诱导行为障碍的潜在神经生物学机制 | 2014 | 山西农业大学 | 王俊东 |
| 王　冲 | 胚胎期双酚 A 暴露致雄性仔鼠学习记忆降低的分子机制研究 | 2014 | 山西农业大学 | 王俊东 |
| 孙耀贵 | 柴术抗激颗粒对早期断奶仔猪抗应激作用及其机制研究 | 2014 | 山西农业大学 | 王俊东 |
| 李勇军 | SPZ-DVD 混悬液在鸡体内的药动学、靶动物安全性及残留消除研究（兽医博士） | 2012 | 扬州大学 | 刘宗平 |
| 付应霄 | OPG 对破骨细胞活性的影响及其信号转导机制 | 2013 | 扬州大学 | 刘宗平 |
| 张义冉 | 镉致大鼠肝 BRL 3A 细胞的毒性机理及 NAC 的保护作用 | 2013 | 扬州大学 | 刘宗平 |
| 张　敏 | 流产布氏杆菌脂多糖突变株ΔrfbE 和ΔrfbD 的构建及其生物学特性研究 | 2013 | 扬州大学 | 刘宗平 |
| 陈淑芳 | 浙江宁波地区猪弓形体病的流行病学及防治研究（兽医博士） | 2013 | 扬州大学 | 刘宗平 |

| 姓名 | 论文题目 | 毕业时间 | 单位 | 指导教师 |
|---|---|---|---|---|
| 杨开红 | 奶牛围产期血液生理生化指标动态变化的研究（兽医博士） | 2013 | 扬州大学 | 刘宗平 |
| 王　怡 | 镉对大鼠成骨细胞的毒性机理及NAC的保护效应 | 2013 | 扬州大学 | 刘宗平 |
| 宋瑞龙 | OPG对破骨细胞骨架的影响和分子机理 | 2014 | 扬州大学 | 刘宗平 |
| 江辰阳 | 镉诱导神经细胞线粒体凋亡的机制及NAC的保护作用 | 2014 | 扬州大学 | 刘宗平 |
| 王琪文 | 自噬在镉致神经细胞毒性损伤中的作用及调控机理 | 2014 | 扬州大学 | 刘宗平 |
| 于　东 | 硒蛋白W对鸡免疫机能的影响 | 2013 | 东北农业大学 | 徐世文 |
| 盛鹏飞 | 硒蛋白W与低硒致鸡脑损伤的相关性研究 | 2014 | 东北农业大学 | 徐世文 |
| 高　晨 | 纳米氧化铜在Caco-2细胞模型及鸡肠道中吸收转运机制的研究 | 2014 | 吉林大学 | 朱连勤 |

## 3.博士后出站人员人单

我国兽医内科诊断学培养已出站的博士后11人（2012—2014年）

| 作者 | 题目 | 出站时间 | 进站学校 | 合作导师 |
|---|---|---|---|---|
| 王　亨 | 大肠杆菌急性感染奶牛子宫内膜的先天性免疫机制研究 | 2012 | 扬州大学 | 刘宗平 |
| 龙　淼 | 新型产甘油益生菌饲料添加剂的研制及其在奶牛生产上的应用 | 2013 | 南京农业大学 | 黄克和 |
| 郑博文 | 奶牛日粮颗粒度及其评价方法的比较研究 | 2013 | 中国农业大学 | 韩　博 |
| 王国卿 | 奶牛子宫内膜炎的发病机制研究 | 2013.6 | 吉林大学 | 张乃生 |
| 马海利 | 氟对鲤鱼鳃组织的影响及其机制研究 | 2013 | 山西农业大学 | 王俊东 |
| 朱家桥 | 氟对生殖系统DNA甲基化的影响与自噬在卵母细胞老化中的作用 | 2013 | 扬州大学 | 刘宗平 |
| 赵春超 | 蓬子菜化学成分及其抗血栓活性研究 | 2013 | 扬州大学 | 刘宗平 |
| 崔世全 | TLR4及其相关因子在断奶仔猪肠道中表达规律的研究 | 2013 | 东北农业大学 | 徐世文 |
| 易金娥 | 活性肽对过氧化氢致Caco-2细胞氧化应激的调控作用及机理研究 | 2013 | University of Alberta | Jianping Wu |
| 曹荣峰 | 子宫内膜炎的免疫防御机制研究 | 2014.6 | 吉林大学 | 张乃生 |
| 曹瑾玲 | 氟影响动物体液免疫的分子机制 | 2014 | 山西农业大学 | 王俊东 |

## 4.获国家、省级优秀博士、硕士论文名单

| 姓名 | 论文题目 | 毕业时间 | 单位 | 指导教师 |
|---|---|---|---|---|
| 陈兴祥 | PCV2感染、氧化应激与硒的相互作用关系及其机理研究 | 2012 | 南京农业大学 | 黄克和 |
| 吴文达 | B型单端孢霉烯族毒素诱导拒食和呕吐的机理研究 | 2012 | 南京农业大学 | 张海彬 |
| 王彩霞 | 全国兽医专业学位研究生教育指导委员会优秀硕士论文 | 2013 | 华中农业大学 | 邓干臻 |

# 六、成果展示

从改革开放到2012年7月30日，中国畜牧兽医学会家畜内科学分会代表获得国家科技进步三等奖2项，省部级科技成果三等以上奖励134项，授权发明、实用新型专利77项，制定标准（国家、行业和省级地方标准）9项，发表SCI收录的论文383篇，其中2011年10月1日至2012年7月30日，发表SCI收录的论文98篇。2012年8月1日至2014年6月30日，中国畜牧兽医学会家畜内科学分会代表获得省部级科技成果三等以上奖励5项，授权发明、实用新型专利49项，发表SCI收录的论文250篇。

## 1. 获奖成果

（国家级和省部级）11项（2012—2014年）

| 获奖成果名称 | 主持人（或主要参加人） | 获奖单位 | 奖励类别 | 等级 | 获奖日期（年） |
|---|---|---|---|---|---|
| 新型多功能生物饲料添加剂的创制与应用 | 黄克和 等 | 南京农业大学 | 教育部技术发明奖 | 二等奖 | 2012 |
| 富硒益生菌促进动物生长及免疫研究与应用 | 张健騂，黄克和 等 | 南京农业大学 | 广东省科技成果奖 | 三等奖 | 2012 |
| 谷物重要真菌毒素检测与安全控制关键技术研究 | 张海彬(2) | 南京农业大学 | 江苏省科学技术奖 | 三等奖 | 2012.12 |
| “环境兽医学”特色化课程建设与应用 | 王俊东(1) | 山西农业大学 | 山西省高等学校教学成果 | 一等奖 | 2013.01 |
| 反刍动物几种重要群发性营养代谢病防控技术的研究与应用 | 刘宗平，卞建春，刘学忠，马小军，曲亚玲，袁燕，顾建红 | 扬州大学、甘肃农业大学 | 农业部中华农业科技奖 | 二等奖 | 2013.11 |
| 霍英东教育基金会高等院校青年教师奖 | 张志刚 | 东北农业大学 | 第十四届教育部霍英东教育基金会高等院校青年教师奖 | 三等奖 | 2014.3 |
| 烟村低胆固醇、无抗安全猪肉的研发 | 文利新，屠迪，李荣芳，邬静，易金娥，谢岚 | 湖南农业大学 | 湖南省教职工科技创新奖 | 一等奖 | 2013.7 |
| 奶牛乳房炎综合防治技术研究与应用 | 武瑞(1) | 黑龙江八一农垦大学 | 黑龙江省政府科学技术进步奖 | 二等奖 | 2012.12 |
| 防治奶牛乳房炎系列产品研制与应用 | 武瑞(1) | 黑龙江八一农垦大学 | 北京大北农集团科技奖 | 促进奖 | 2013.11 |
| 兽医内科学精品课程国家资源共享课 | 郭定宗，周东海 | 华中农业大学 | | | 2013 |
| 黑龙江省青年科技奖 | 徐闯(1) | 黑龙江八一农垦大学 | 黑龙江省奖励 | | 2013 |

## 2. 授权专利

授权的发明、实用新型专利 49 项（2012—2014 年）

| 单 位 | 发明人或设计人 | 发 明 名 称 | 专 利 号 | 分 类 | 授权日期 |
|---|---|---|---|---|---|
| 南京农业大学 | 黄克和，朱永兴，刘瑾，叶耿坪，李灵恩，刘永杰 | 产甘油酿酒酵母NAU-ZH-GY1及其应用 | ZL 201110057770.6 | 发明专利 | 2012.8 |
| 南京农业大学 | 黄克和，赵治平，杨家军，张克春，卫程武 | 一种富含有机硒和有机锌的复合益生菌制剂 | ZL 200810103540.7 | 发明专利 | 2012.2 |
| 南京农业大学 | 张海彬，张爱华，王希春，何成华，王莹 | 杂交瘤细胞株D6D | ZL 201010204412.9 | 实用新型 | 2012.03 |
| 南京农业大学 | 余祖功，李晶，张海彬，张军忍，江善祥 | 家畜用利福昔明阴道栓剂及其制备方法 | ZL201110425645.6 | 实用新型 | 2012.06 |
| 南京农业大学 | 何成华，樊彦红，皇超英，孟玲玲，杨彦琼，张海彬 | 一种DON 降解酶的编码基因和应用 | ZL201210359882.1 | 实用新型 | 2012.12 |
| 吉林大学 | 张志刚，刘国文，李小兵，王哲 | 一种奶牛亚临床酮病诊断试纸 | ZL2200710056031.9 | 发明专利 | 2012.01 |
| 吉林大学 | 张燚，刘国文，李小兵，王哲 | 一种Pb2+抗原和相应单克隆抗体及其制备方法 | ZL201010216757.6 | 发明专利 | 2012.07 |
| 浙江大学 | 谭勋，丁守强，潘韬 | 纳米金免疫电极的制备方法 | ZL200910153183.X | 发明专利 | 2012.12.5 |
| 东北农业大学 | 李艳飞 | 一种治疗羊腹泻的组合物及其制备方法 | ZL201010032434.1 | 发明专利 | 2012.5 |
| 东北农业大学 | 李艳飞 | 一种治疗犬瘟热的中药的组合物及其制备方法 | ZL200910217453.9 | 发明专利 | 2012.3 |
| 湖南农业大学长沙绿叶生物科技有限公司 | 文利新，邬静，王国强，等 | 猪蓝耳病疫苗的一种免疫增强调节剂及应用 | ZL200810143916.7 | 发明专利 | 2012 |
| 上海市奶牛研究所 | 张克春 | 大米草为原料的奶牛后备牛 TMR 发酵饲料及其制备方法 | ZL201010578656.3 | 实用新型 | 2012.11.4 |
| 上海市奶牛研究所 | 张克春 | 水葫芦为原料的奶牛后备牛 TMR 发酵饲料及其制备方法 | ZL201010578322.6 | 实用新型 | 2012.12.26 |
| 上海市奶牛研究所 | 张克春 | 一种保护奶牛安全度过围产后期的奶牛灌服液 | ZL201010149700.9 | 实用新型 | 2012.7.4 |
| 宁夏大学 | 何生虎，曹志 | 治疗奶牛乳房炎的中药透皮软膏及其制备方法 | ZL 2010 1 0574717.9 | 发明专利 | 2012-02-22 |
| 黑龙江八一农垦大学 | 郭婷婷，林云成，武瑞，贺显晶 | 防止软膏剂、凝胶剂、成膜剂药物交叉感染的涂抹瓶 | ZL201220183457.7 | 实用新型 | 2012.11.07 |
| 南京农业大学 | 黄克和，甘芳，陈兴祥，任飞，吕晨辉，赵如茜，潘翠玲，周红，叶耿坪，石秀丽 | 一种抗猪热应激的富硒复合菌饲料添加剂及其应用 | ZL 201210333510.1 | 发明专利 | 2013.7 |
| 南京农业大学 | 黄克和，任志华，蒋鲁岩，陈兴祥 | 草酸降解菌NJODE1及其应用 | ZL 201110169374.2 | 发明专利 | 2013.9 |
| 南京农业大学 | 黄克和，任志华 | 草酸降解菌NJODL1及其应用 | ZL 201110169482.X | 发明专利 | 2013.9 |

| 单　　位 | 发明人或设计人 | 发　明　名　称 | 专 利 号 | 分　类 | 授权日期 |
|---|---|---|---|---|---|
| 吉林大学 | 孔涛，刘国文，李小兵，王哲 | 重金属$Cu2^+$完全抗原及其制备方法 | ZL200910217866.7 | 发明专利 | 2013. 05 |
| 西北农林科技大学 | 王建华，杨国栋，王妍，李金成，耿果霞 | 一种检测疯草内生真菌中苦马豆素含量的方法 | CN102590378A | 发明专利 | 2013.09 |
| 西北农林科技大学 | 李勤凡，杨凯，梁洁，孔祥雅，夏爽 | 产乌头碱的枝孢菌XJ-AC03及其应用 | ZL201210105591.X | 发明专利 | 2013.11 |
| 山西农业大学 | 张建海，王俊东，梁琛，李宏全，王金明，牛瑞燕，孙子龙，罗广营 | 一种简易二氧化硫气体动式染毒实验装置 | ZL 2011 1 0058955.9 | 发明专利 | 2013.04 |
| 浙江大学 | 谭勋，丁守强，潘韬 | 检测牛奶中结合珠蛋白含量的免疫传感器及检测方法 | ZL200910153182.5 | 发明专利 | 2013.3.27 |
| 湖南农业大学 | 邬静，文利新，屠迪，陈宇科，袁莉芸，肖娟，丁慧昕 | 一种减少肥育猪体内铜蓄积的复方丹参动物保健品及应用 | ZL201210280870.X | 发明专利 | 2013 |
| 长沙树人牧业科技有限公司长沙绿叶生物科技有限公司 | 文利新，袁慧，尹恒，谢岚，文舸，李逢慧，谭琼 | 预防猪PSE肉的功能性饲料和应用 | ZL200910310261.2 | 发明专利 | 2013 |
| 湖南烟村生态农牧科技股份有限公司 | 屠迪，陈瑶，杨阳，邬静，刘成国，文利新 | 一种冰鲜成熟猪肉的加工工艺 | ZL201110320205 | 发明专利 | 2013 |
| 湖南农业大学长沙绿叶生物科技有限公司 | 许道军，文利新，周正 | 一种用于畜舍及动物体表微生态改良的益生菌组合菌剂及应用方法 | ZL201110267214 | 发明专利 | 2013 |
| 湖南烟村生态农牧科技股份有限公司 | 刘建丰，文利新 | 一种富含γ-氨基丁酸的保健米粉及其生产方法 | ZL201010555882 | 发明专利 | 2013 |
| 山东农业大学 | 刘建柱，孟凯 | 一种快速检测猪全血中猪附红细胞体的胶体金试纸条 | ZL2013203780834. | 实用新型 | 2013.11 |
| 山东农业大学 | 刘建柱，孟凯 | 一种快速检测猪血清中猪附红细胞体的胶体金试纸条 | ZL201320378081.X. | 实用新型 | 2014.01 |
| 中国农业科学院兰州畜牧与兽药研究所 | 李宏胜，罗金印，杨峰，王旭荣，李新圃，陈灵然，张世栋 | 一种无菌脱纤绵羊全血采集装置 | ZL 201320366280.9 | 实用新型 | 2013.11 |
| 中国农业科学院兰州畜牧与兽药研究所 | 李宏胜，罗金印，李新圃，杨峰，王旭荣，陈灵然，尚立宏 | 奶牛乳房炎和子宫内膜炎样品采集储运管 | ZL201320285644.0 | 实用新型 | 2013.11 |
| 中国农业科学院兰州畜牧与兽药研究所 | 杨峰，王旭荣，李宏胜，罗金印，李新圃，张世栋 | 一种耐高温耐高压试管斜面培养基洗菌棒 | ZL201320270662.1 | 实用新型 | 2013.10 |
| 武威市金绿源农业科技发展有限公司 | 雷华，徐庚全，谢占玲，王明奎，王成泰 | 生物环保发酵床养猪发酵菌剂及其制备方法 | ZL 2012 1 0044903.0 | 实用新型 | 2013.06 |
| 宁夏大学 | 何生虎，蒋魏娟，郭磊，付少刚，李勇 | 一种治疗奶牛乳房炎的中药涂膜剂制备方法 | ZL 2012 1 0243002.4 | 发明专利 | 2013.12.04 |
| 宁夏大学 | 何生虎，郭磊，蒋魏娟、付少刚，李勇 | 一种治疗奶牛子宫内膜炎的中药复方泡腾片制备方法 | ZL 2012 1 0243004.3 | 发明专利 | 2013. 09.25 |
| 黑龙江八一农垦大学 | 武瑞，王建发，贺显晶，张鹏宇，王爽，孙东波 | 新型瓶式兽用灌药器 | ZL201220539846.9 | 实用新型 | 2013.03.13 |

| 单　位 | 发明人或设计人 | 发　明　名　称 | 专 利 号 | 分　类 | 授权日期 |
|---|---|---|---|---|---|
| 黑龙江八一农垦大学 | 武瑞，杨利民，吴国军，贺显晶，孙东波，王建发 | 奶牛电动洗胃机 | ZL201220458767.5 | 实用新型 | 2013.05.08 |
| 上海市奶牛研究所 | 张克春 | 水稻秸秆为原料的奶牛后备牛 TMR 发酵饲料及其制备 | ZL201010578138.1 | 实用新型 | 2013.01.02 |
| 上海市奶牛研究所 | 张克春 | 麦秸秆为原料的奶牛后备牛 TMR 发酵饲料及其制备方法 | ZL201010578162.5 | 实用新型 | 2013.01.02 |
| 上海市奶牛研究所 | 张克春 | 应用 TMR 日粮高产奶牛精料补充料 | ZL201010150435.6 | 实用新型 | 2013.02.06 |
| 上海市奶牛研究所 | 张克春 | 大豆秸秆为原料的奶牛后备牛 TMR 发酵饲料及其制备 | ZL201010578643.6 | 实用新型 | 2013.03.20 |
| 上海市奶牛研究所 | 张克春 | 一种 14-23 月龄后备牛发酵 TMR 日粮及其制备方法 | ZL201110086346.4 | 实用新型 | 2013.03.20 |
| 上海市奶牛研究所 | 张克春 | 一种采用牛粪发酵菌处理牛粪制备活性炭的方法 | ZL201110071341.4 | 实用新型 | 2013.06.26 |
| 上海市奶牛研究所 | 张克春 | 一种专用于奶牛发酵 TMR 的预混料及其制备工艺 | ZL201010273520.1 | 实用新型 | 2013.09.11 |
| 上海市奶牛研究所 | 张克春 | 一种盐酸聚六甲基胍碘药浴液及其制备方法 | ZL201210322610.4 | 实用新型 | 2014.03.26 |
| 宁夏大学 | 何生虎，杨萌萌 | 一种防治牛前胃疾病的中药及其复方口服片剂的制备方法 | ZL 2013 1 0244217.2 | 发明专利 | 2014.06.30 |
| 四川农业大学 | 左之才 | 治疗各种顽固性皮肤病的兽用纯中药酊剂 | ZL 2012 1 0016531.0 | 发明专利 | 2014.02 |

## 3. SCI 收录论文

SCI 收录的论文 250 篇（2012.8.1—2014. 6.30）

截至 2014 年 6 月 30 日，中国畜牧兽医学会家畜内科学分会的代表发表 SCI 收录的论文 633 篇，其中 2011 年 10 月 1 日之前发表 285 篇，2011 年 10 月 1 日至 2012 年 7 月 30 日发表 SCI 收录的论文 98 篇，2012 年 8 月 1 日至 2014 年 6 月 30 日发表 SCI 收录的论文 250 篇。

| 作者单位 | 题目 | 作者简称或全称 | 通讯作者及单位 | 杂志名称 | 卷期页码 |
|---|---|---|---|---|---|
| 南京农业大学 | A mitochondria-mediated apoptotic pathway induced by deoxynivalenol in human colon cancer cells | Ma Y, Zhang A, Shi Z, He C, Ding J, Wang X, Ma J, Zhang HB. | Haibin Zhang 南京农业大学 | Toxicol In Vitro **(SCI IF=2.650)** | 2012,26(3):414-420 |
| 南京农业大学 | **Characterization and comparison of fumonisinb(1)-protein conjugates by six methods** | Wang Y, He CH, Zheng H, Zhang HB | Haibin Zhang 南京农业大学 | Int J Mol Sci**(SCI IF=2.464)** | 2012,13(1):84-96 |

| 作者单位 | 题目 | 作者简称或全称 | 通讯作者及单位 | 杂志名称 | 卷期页码 |
| --- | --- | --- | --- | --- | --- |
| 南京农业大学 | T-2 toxin, zearalenone and fumonisin B1 in feedstuffs from China | Wang Y, Liu S, Zheng H, He C, Zhang HB | Haibin Zhang 南京农业大学 | Food Addit Contam Part B Surveill **(SCI IF=0.831)** | 2013,6(2):116-122 |
| 南京农业大学 | Inhibitory effects of DON on gastric secretion in rats | Wang Y, Wu W, Wang X, He C, Yue H, Ren Z, Zhang HB | Haibin Zhang 南京农业大学 | J Food Protect**(SCI IF=1.832)** | 2014, Accepted |
| 南京农业大学 | Feeding glycerol-enriched yeast culture improves performance, energy status, and heat shock protein gene expression of lactating Holstein cows under heat stress. | Liu J, Ye G, Zhou Y, Liu Y, Zhao L, Liu Y, Chen X, Huang D, Liao SF, Huang K. | Kehe Huang 南京农业大学 | J Anim Sci**(SCI IF=2.093)** | 2014, 92(6):2494-502 |
| 南京农业大学 | Selenium-Enriched Probiotics Improve Antioxidant Status, Immune Function, and Selenoprotein Gene Expression of Piglets Raised under High Ambient Temperature | Gan F, Chen X, Liao SF, Lv C, Ren F, Ye G, Pan C, Huang D, Shi J, Shi X, Zhou H, Huang K. | Kehe Huang 南京农业大学 | J Agric Food Chem**(SCI IF=2.906)** | 2014, 62(20):4502-8 |
| 南京农业大学 | Preparation of glycerol-enriched yeast culture and its effect on blood metabolites and ruminal fermentation in goats | Ye G, Zhu Y, Liu J, Chen X, Huang K | Kehe Huang 南京农业大学 | PLoS One**(SCI IF=3.730)** | 2014, 9(4):e94410 |
| 南京农业大学 | Protective effect of selenomethionine on aflatoxin B1-induced oxidative stress in MDCK cells | Parveen F, Nizamani ZA, Gan F, Chen X, Shi X, Kumbhar S, Zeb A, Huang K | Kehe Huang 南京农业大学 | Biol Trace Elem Res**(SCI IF=1.307)** | 2014, 157(3):266-74. |
| 南京农业大学 | Construction and immunogenicity of DNA vaccines encoding fusion protein of porcine IFN- λ 1 and GP5 gene of porcine reproductive and respiratory syndrome virus | Du L, Li B, He K, Zhang H, Huang K, Xiao S | Kehe Huang 南京农业大学 | Biomed Res Int. **(SCI IF= 2.880)** | 2013:318698. doi: 10.1155/2013/318698 |
| 南京农业大学 | Interaction of porcine circovirus type 2 replication with intracellular redox status in vitro | Chen X, Ren F, Hesketh J, Shi X, Li J, Gan F, Hu Z, Huang K | Kehe Huang 南京农业大学 | Redox Rep**(SCI IF= 1.662)** | 2013, 18(5):186-92 |
| 南京农业大学 | Effects of selenium-enriched probiotics on heat shock protein mRNA levels in piglet under heat stress conditions | Gan F, Ren F, Chen X, Lv C, Pan C, Ye G, Shi J, Shi X, Zhou H, Shituleni SA, Huang K | Kehe Huang 南京农业大学 | J Agric Food Chem**(SCI IF=2.906)** | 2013, 61(10):2385-91 |
| 南京农业大学 | Effects of the Chinese medicine matrine on experimental C. parvum infection in BALB/c mice and MDBK cells | Chen F, Huang K | Kehe Huang 南京农业大学 | Parasitol Res**(SCI IF=2.852)** | 2012, 111(4):1827-32 |

| 作者单位 | 题目 | 作者简称或全称 | 通讯作者及单位 | 杂志名称 | 卷期页码 |
|---|---|---|---|---|---|
| 南京农业大学 | Rescue of recombinant peste des petits ruminants virus: creation of a GFP-expressing virus and application in rapid virus neutralization test | Hu Q, Chen W, Huang K, Baron MD, Bu Z | Kehe Huang 南京农业大学 | Vet Res**(SCI IF=3.426)** | 2012, 43(1):48 |
| 中国农业大学 | Prevalence and risk factors for canine obesity surveyed in veterinary practices in Beijing, China | Mao, Junfu; Xia, Zhaofei; Chen, Jiangnan; Yu, Jinhai | Zhaofei Xia 中国农业大学 | Preventive Veterinary Medicine**(SCI IF=2.277)** | 2012,26(3):414-420 |
| 中国农业大学 | **Seroepidemiology of canine brucellosis in Beijing, China** | Xiang, Fu; Xia, Zhaofei; Wu, Qingming; Chen, Yanyun; Yu, Jinhai; Wan, Jianqing | Zhaofei Xia 中国农业大学 | Turkish J Vet & Ani Sci **(SCI IF=0.479)** | 2012,13(1):84-96 |
| 中国农业大学 | Xylitol poisoning of dogs is associated with increased glycogenolysis, coagulopathy, and oxidative stress | Xia, Zhaofei; Cai, Liyuan; He, Yuying; Wan, Jianqing; Yu, Jinhai | Zhaofei Xia 中国农业大学 | Toxicol Environ Chemistry**(SCI IF=Null)** | 2013,6(2):116-122 |
| 中国农业大学 | Characteristics of Staphylococcus aureus Small Colony Variant and Its Parent Strain Isolated from Chronic Mastitis at a Dairy Farm in Beijing, China | Alkasir R, Liu X, Zahra M, Ferreri M, Su J, Han B. | Bo Han 中国农业大学 | Microbiology Drug Resistance **(SCI IF= 2.364)** | 2013,19(2):138-45, |
| 中国农业大学 | Treatment with Gentamicin on a Murine Model of Protothecal Mastitis | Chang R, Yang Q, Liu G, Liu Y, Zheng B, Su J, Han B. | Bo Han, Jingliang Su 中国农业大学 | Mycopathologia(**SCI IF= 1.489)** | 2013, 175:241–248, |
| 中国农业大学 | Isolation and Characterization of Small-Colony Variants of Ornithobacterium rhinotracheale | Zahra Mohammad, Ferreri M, Alkasir R, Yin J, Han B, Su J. | Bo Han, Jingliang Su 中国农业大学 | Journal of Clinical Microbiology**(SCI IF=4.096)** | 2013, 51(10):3228-3236 |
| 中国农业大学 | Phylogenetic group, virulence factors and antimicrobial resistance of Escherichia coli associated with bovine mastitis. | Liu Y, Liu G, Liu W, Liu Y, Ali T, Chen W, Yin J, Han B. | Bo Han 中国农业大学 | Res Microbiol. **(SCI IF=2.889)** | 2014, 165: 273-277 |
| 中国农业大学 | Biofilm formation and biofilm-associated genes assay of Staphylococcus aureus isolated from bovine subclinical mastitis in China | Jian-zhong He, An-qi Wang, Gang Liu, Jian Gao, Tariq Ali and Bo Han | Bo Han 中国农业大学 | Pak Vet J**(SCI IF=1.365)** | 2014, Accepted: May 09 |
| 中国农业大学 | Dose-Dependent Effects of Lactobacillus rhamnosus on Serum Interleukin-17 Production and Intestinal T-Cell Responses in Pigs Challenged with Escherichia coli | Zhu YH, Li XQ, Zhang W, Zhou D, Liu HY, Wang JF | Wang JF 中国农业大学 | Appl Environ Microbiol. **(SCI IF=3.678)** | 2014, 80(5):1787-98 |
| 中国农业大学 | Risks associated with high-dose Lactobacillus rhamnosus in an | Li XQ, Zhu YH, Zhang HF, Yue Y, | Wang JF 中国农业大学 | PLoS One**(SCI IF=3.730)** | 2012;7(7):e40666 |

| 作者单位 | 题目 | 作者简称或全称 | 通讯作者及单位 | 杂志名称 | 卷期页码 |
|---|---|---|---|---|---|
| | Escherichia coli model of piglet diarrhoea: intestinal microbiota and immune imbalances. | Cai ZX, Lu QP, Zhang L, Weng XG, Zhang FJ, Zhou D, Yang JC,Wang JF. | | | |
| 吉林大学 | SREBP-1c overexpression induces triglycerides accumulation through increasing lipid synthesis and decreasing lipid oxidation and VLDL assembly in bovine hepatocytes | Xinwei Li, Yu Li，Wentao Yang, Chong Xiao, Shixin Fu, Qinghua Deng, Hongyan Ding, Zhe Wang, Guowen Liu. | Guowen Liu, Xiaobing Li 吉林大学 | The Journal of Steroid Biochemistry and Molecular Biology**(SCI IF=3.984)** | 2014, 143:174–182 |
| 吉林大学 | Non-esterified fatty acids activate the ROS–p38–p53/Nrf2 signaling pathway to induce bovine hepatocyte apoptosis in vitro | Yuxiang Song, Xinwei Li, Yu Li, Xiaobing Li, Guowen Liu, Zhe Wang | Xiaobing Li, Guowen Liu 吉林大学 | Apoptosis **(SCI IF=3.949)** | 2014, 19(6):984-97 |
| 吉林大学 | SREBP-1c Gene Silencing can Decrease Lipid Deposits in Bovine Hepatocytes Cultured in Vitro | Qinghua Deng, Xinwei Li, Shixin Fu, Jianguo Wang, Lei Liu, Xue Yuan, Zhe Wang, Guowen Liu, Xiaobing Li | Guowen Liu, Xiaobing Li 吉林大学 | Cell Physiol Biochem **(SCI IF=3.415)** | 2014;33:1568-1578 |
| 吉林大学 | β-Hydroxybutyrate Activates the NF-κB Signaling Pathway to Promote the Expression of Pro-Inflammatory Factors in Calf Hepatocytes | Xiaoxia Shi, Xinwei Li, Guowen Liu, Xiaobing Li,Wang Zhe | Xiaobing Li, Guowen Liu 吉林大学 | Cell Physiol Biochem **(SCIIF=3. 415)** | 2014;33:920-932 |
| 吉林大学 | Effects of nonesterified fatty acids on the synthesis and assembly<br>of very low density lipoprotein in bovine hepatocytes in vitro | Lei Liu, Xinwei Li, Guowen Liu, Xiaobing Li,Wang Zhe | Xiaoyu Yang, Guowen Liu 吉林大学 | J. Dairy Sci. **(SCI IF= 2.566)** | 2014;97 :1328–1335 |
| 吉林大学 | Adiponectin activates the AMPK signaling pathway to regulate lipidmetabolism in bovine hepatocytes | Hui Chen, Liang Zhang, Xinwei Li | Hui Zhang 吉林农业科技学院，Guowen Liu 吉林大学 | The Journal Steroid Biochemistry and Molecular Biology**(SCIIF= 3.984)** | 2013; 138: 445–454 |
| 吉林大学 | Acetic Acid Activates the AMP-Activated Protein Kinase Signaling Pathway to Regulate Lipid Metabolism in Bovine Hepatocytes | Xinwei Li, Hui Chen, Yuan Guan, Xiaobing Li | Guowen Liu, Zhe Wang 吉林大学 | PLoS ONE**(SCI IF= 3.73)** | 2013, 8(7): e67880 |
| 吉林大学 | Non-Esterified Fatty Acids Activate the AMP-Activated Protein Kinase Signaling Pathway to Regulate Lipid Metabolism in Bovine Hepatocytes | Xinwei Li, Xiaobing Li, Hui Chen, Liancheng Lei | Guowen Liu, Zhe Wang 吉林大学 | Cell Biochem Biophys**(SCI IF= 1.912)** | 2013-013-9629-1 |

| 作者单位 | 题目 | 作者简称或全称 | 通讯作者及单位 | 杂志名称 | 卷期页码 |
|---|---|---|---|---|---|
| 吉林大学 | Effect of leptin on the gluconeogenesis in calf hepatocytes cultured in vitro | Li Y, Li X, Song Y, Shi X | Wang Z. 吉林大学 | Cell Biol Int.**(SCI IF= 1.64)** | 2013, 37(12):1350-3 |
| 吉林大学 | Evaluation of the Difference of L-selectin, Tumor Necrosis Factor-α and Sialic Acid Concentration in Dairy Cows with Subclinical Ketosis and without Subclinical Ketosis. | Z. G. Zhang, J. D. Xue, R. F. Gao, J.Y. Liu | Z. Wang 吉林大学 | Pak Vet J **(SCI IF= 1.365)** | 2013, 33(2): 225-22 |
| 吉林大学 | Effects of High Zinc Levels on the Lipid Synthesis in Rat Hepatocytes. | Xinwei Li, Yuan Guan, Xiaoxia Shi, Hongyan Ding | Guowen Liu 吉林大学 | Biol Trace Elem Res**(SCI IF= 1.307)** | 2013, 154(1):97-102 |
| 吉林大学 | Preparation of Novel Monoclonal Antibodies Against Chelated Cadmium Ions | Tao Kong, Xue-Qin Hao, Xiao-Bing Li, Guo-Wen Liu | Xiaobing Li, Guowen Liu 吉林大学 | Biol Trace Elem Res **(SCI IF= 1.307)** | 2013, 152(1):117–124 |
| 吉林大学 | Effects of Strontium on Collagen Content and Expression of Related Genes in Rat Chondrocytes Cultured | Jianguo Wang, Xiaoyan Zhu, Lei Liu, Xiaoxia Shi | Zhe Wang, Guowen Liu. 吉林大学 | Biol Trace Elem Res **(SCI IF= 1.307)** | 2013, 153(1-3):212-9 |
| 吉林大学 | Effects of non-esterified fatty acids on the gluconeogenesis in bovine hepatocytes | Xinwei Li, Xiaobing Li, Ge Bai, Hui Chen, | Guowen Liu, Zhe Wang 吉林大学 | Mol Cell Biochem**(SCI IF= 2.057)** | 2012; 359(1):385-388. |
| 吉林大学 | Detection of subclinical ketosis in dairy cows. | Zhang Z, G Liu, H Wang, X Li | Z Wang 吉林大学 | Pak Vet J**(SCI IF= 1.365)** | 2012, 32(2): 156-160. |
| 吉林大学 | High Lever Dietary Copper Promote Ghrelin Gene expression in the fundic gland of growing pigs | Wenyan Yang, Jianguo Wang, Xiaoyan Zhu, Yunhang Gao | Lianyu Yang 吉林农业大学，Guowen Liu 吉林大学 | Biol Trace Elem Res**(SCI IF= 1.307)** | 2012, 150: 154–157 |
| 吉林大学 | P-cymene protects mice against lipopolysaccharide-induced acute lung injury by inhibiting inflammatory cell activation. | Guanghong Xie , Na Chen, Lanan Wassy Soromou | Haihua Feng, Guowen Liu 吉林大学 | Molecules**(SCI IF= 2.428)** | 2012, 17, 8159-8173 |
| 吉林大学 | Increase of Fatty Acid Oxidation and VLDL Assembly and Secretion Overexpression of PTEN in Cultured Hepatocytes of Newborn Calf. | Shixin Fu, Qinghua Deng, Wengtao Yang, Hongyan Ding | Xiaobin Li, Guowen Liu 吉林大学 | Cell Physiol Biochem**(SCI IF= 1.912)** | 2012;30:1005-1013. |
| 吉林大学 | Effect of deleting acetic acid-producing key enzyme gene of Selenomonas ruminantium on the ruminal fermentation in vitro. | Miao LONG, Xiaoyang PANG, Xia QIN | Guowen Liu 吉林大学 | African J Microbiol Res**(SCI IF= 0.539)** | 2012, 13: 6476-6482 |
| 吉林大学 | Removal of Zearalenone by Strains of Lactobacillus sp Isolated from Rumen in vitro. | Long, M; Li, P; Zhang, WK; Li, XB | Liu, GW 吉林大学 | J Ani Vet Advan**(SCI IF= 0.365)** | 2012, 11(14): 2417-2422 |
| 黑龙江八一农垦大学 | Concentrations of Plasma Metabolites, Hormones, and mRNA Abundance of Adipose Leptin and Hormone-Sensitive Lipase in Ketotic and | Xia, C; Wang, Z; Xu, C; Zhang, HY. | Wang, Z 吉林大学 | Vet Internal med**(SCI IF= 2.064)** | 2012, 26(2): 415-417 |

| 作者单位 | 题目 | 作者简称或全称 | 通讯作者及单位 | 杂志名称 | 卷期页码 |
|---|---|---|---|---|---|
| | Nonketotic Dairy Cows. | | | | |
| 吉林大学 | An updated method for the isolation and culture of primary calf hepatocytes | Zhi-Gang Zhang, Xiao-Bing Li, Li Gao | Xiaobin Li, Zhe Wang 吉林大学 | Vet J**(SCI IF=2.239)** | 2012, 191(3): 323-326. |
| 吉林大学 | Genome shuffling of Megasphaera elsdenii for improving acid-tolerance and propionate production | Miao Long, Peng Li, Wen Kui Zhang, Yi Zhang | Guowen Liu 吉林大学 | African J Microbiol Res**(SCI IF=0.539)** | 2012, 6(18): 4041-4047 |
| 吉林大学 | Dietary Selenium Deficiency Exacerbates Lipopolysaccharide-Induced Inflammatory Response in Mouse Mastitis Models. | Wei Z, Yao M, Li Y, He X, Yang Z. | Yang Zhengtao 吉林大学 | Inflammation**(SCI IF=2.457)** | 2014 May 21 |
| 吉林大学 | Cyanidin-3-O-β-glucoside ameliorates lipopolysaccharide-induced acute lung injury by reducing TLR4 recruitment into lipid rafts. | Fu Y, Zhou E, Wei Z, Wang W, Wang T, Yang Z, Zhang N. | Yang Zhengtao, Zhang Naisheng 吉林大学 | Biochem Pharmacol **(SCI IF=4.576)** | 2014, 90:126–134 |
| 吉林大学 | Thymol attenuates allergic airway Inflammation in ovalbumin (OVA)-induced mouse asthma. | Zhou E, Fu Y, Wei Z, Yu Y, Zhang X, Yang Z. | Xichen Zhang, Zhengtao Yang 吉林大学 | Fitoterapia**(SCI IF=2.231)** | 2014, 96:131–137 |
| 吉林大学 | Geniposide Plays an Anti-inflammatory Role via Regulating TLR4 and Downstream Signaling Pathways in Lipopolysaccharide-Induced Mastitis in Mice. | Song X, Zhang W, Wang T, Jiang H, Zhang Z, Fu Y, Yang Z, Cao Y, Zhang N. | Zhang Naisheng 吉林大学 | Inflammation**(SCI IF=2.457)** | 2014, Apr 27 |
| 吉林大学 | Cyanidin-3-O-β-glucoside inhibits lipopolysaccharide-induced inflammatory response in mouse mastitis model. | Fu Y, Wei Z, Zhou E, Zhang N, Yang Z. | Yang Zhengtao 吉林大学 | J Lipid Res**(SCI IF=4.386)** | 2014, 55(6):1111-1119. |
| 吉林大学 | Effects of niacin on Staphylococcus aureus internalization into bovine mammary epithelial cells by modulating NF-κB activation. | Wei Z, Fu Y, Zhou E, Tian Y, Yao M, Li Y, Yang Z, Cao Y. | Yongguo Cao 吉林大学 | Microb Pathog**(SCI IF=1.974)** | 2014, 71-72C:62-67 |
| 吉林大学 | Glycyrrhizin inhibits the inflammatory response in mouse mammary epithelial cells and a mouse mastitis model. | Fu Y, Zhou E, Wei Z, Liang D, Wang W, Wang T, Guo M, Zhang N, Yang Z. | Yang Zhengtao 吉林大学 | FEBS J**(SCI IF=4.25)** | 2014, 281(11):2543-57. |
| 吉林大学 | Thymol inhibits Staphylococcus aureus internalization into bovine mammary epithelial cells by inhibiting NF-κB activation. | Wei Z, Zhou E, Guo C, Fu Y, Yu Y, Li Y, Yao M, Zhang N, Yang Z. | Yang Zhengtao 吉林大学 | Microb Pathog**(SCI IF=1.974)** | 2014, 71-72C:15-19. |
| 吉林大学 | Curcumin attenuates | Fu Y, Gao R, Cao | Zhang | Int | 2014, |

| 作者单位 | 题目 | 作者简称或全称 | 通讯作者及单位 | 杂志名称 | 卷期页码 |
|---|---|---|---|---|---|
| | inflammatory responses by suppressing TLR4-mediated NF-κB signaling pathway in lipopolysaccharide-induced mastitis in mice | Y, Guo M, Wei Z, Zhou E, Li Y, Yao M, Yang Z, Zhang N. | Naisheng 吉林大学 | Immunopharmacol **(SCI IF=2.417)** | 20(1):54-8. |
| 吉林大学 | Glycyrrhizin inhibits lipopolysaccharide-induced inflammatory response by reducing TLR4 recruitment into lipid rafts in RAW264.7 cells. | Fu Y, Zhou E, Wei Z, Song X, Liu Z, Wang T, Wang W, Zhang N, Liu G, Yang Z. | Liu Guowen, Yang Zhengtao 吉林大学 | Biochim Biophys Acta**(SCI IF=3.848)** | 2014, 1840(6):1755-64. |
| 吉林大学 | Baicalin inhibits Staphylococcus aureus-induced apoptosis by regulating TLR2 and TLR2-related apoptotic factors in the mouse mammary glands. | Guo M, Cao Y, Wang T, Song X, Liu Z, Zhou E, Deng X, Zhang N, Yang Z. | Naisheng Zhang, Zhengtao Yang 吉林大学 | Eur J Pharmacol**(SCI IF=2.592)** | 2014, 723:481-8. |
| 吉林大学 | Selenium inhibits LPS-induced pro-inflammatory gene expression by modulating MAPK and NF-κB signaling pathways in mouse mammary epithelial cells in primary culture. | Zhang W, Zhang R, Wang T, Jiang H, Guo M, Zhou E, Sun Y, Yang Z, Xu S, Cao Y, Zhang N. | Yongguo Cao, Naisheng Zhang 吉林大学 | Inflammation**(SCI IF=2.457)** | 2014, 37(2):478-85 |
| 吉林大学 | Thymol inhibits LPS-stimulated inflammatory response via down-regulation of NF-κB and MAPK signaling pathways in mouse mammary epithelial cells. | Liang D, Li F, Fu Y, Cao Y, Song X, Wang T, Wang W, Guo M, Zhou E, Li D, Yang Z, Zhang N. | Zhengtao Yang, Naisheng Zhang 吉林大学 | Inflammation**(SCI IF=2.457)** | 2014, 37(1):214-22 |
| 吉林大学 | Farrerol regulates antimicrobial peptide expression and reduces Staphylococcus aureus internalization into bovine mammary epithelial cells. | Yang Z, Fu Y, Liu B, Zhou E, Liu Z, Song X, Li D, Zhang N. | Zhang Naisheng 吉林大学 | Microb Pathog**(SCI IF=1.974)** | 2013, 65:1-6 |
| 吉林大学 | Staphylococcus aureus and Escherichia coli elicit different innate immune responses from bovine mammary epithelial cells. | Fu Y, Zhou E, Liu Z, Li F, Liang D, Liu B, Song X, Zhao F, Fen X, Li D, Cao Y, Zhang X, Zhang N, Yang Z | Yang Zhengtao 吉林大学 | Vet Immunol Immunopathol**(SCI IF=1.877)** | 2013, 155(4):245-52. |
| 吉林大学 | Astragalin suppresses inflammatory responses via down-regulation of NF-κB signaling pathway in lipopolysaccharide-induced mastitis in a murine model. | Li F, Liang D, Yang Z, Wang T, Wang W, Song X, Guo M, Zhou E, Li D, Cao Y, Zhang N. | Yongguo Cao, Naisheng Zhang 吉林大学 | Int Immunopharmacol **(SCI IF=2.417)** | 2013, 17(2):478-82 |
| 吉林大学 | Shikonin exerts anti-inflammatory effects in a murine model of lipopolysaccharide-induced acute lung injury by inhibiting | Liang D, Sun Y, Shen Y, Li F, Song X, Zhou E, Zhao F, Liu Z, Fu Y, Guo M, Zhang | Yongguo Cao 吉林大学 | Int Immunopharmacol **(SCI IF=2.417)** | 2013, 16(4):475-80 |

| 作者单位 | 题目 | 作者简称或全称 | 通讯作者及单位 | 杂志名称 | 卷期页码 |
|---|---|---|---|---|---|
| | the nuclear factor-kappaB signaling pathway. | N, Yang Z, Cao Y. | | | |
| 吉林大学 | Baicalin plays an anti-inflammatory role through reducing nuclear factor-κB and p38 phosphorylation in S. aureus-induced mastitis. | Guo M, Zhang N, Li D, Liang D, Liu Z, Li F, Fu Y, Cao Y, Deng X, Yang Z. | Zhengtao Yang 吉林大学 | Int Immunopharmacol **(SCI IF=2.417)** | 2013, 16(2):125-30 |
| 吉林大学 | Emodin ameliorates lipopolysaccharide-induced mastitis in mice by inhibiting activation of NF-κB and MAPKs signal pathways. | Li D, Zhang N, Cao Y, Zhang W, Su G, Sun Y, Liu Z, Li F, Liang D, Liu B, Guo M, Fu Y, Zhang X, Yang Z. | Zhengtao Yang 吉林大学 | Eur J Pharmacol**(SCI IF=2.592)** | 2013, 705(1-3):79-85 |
| 吉林大学 | RP105 involved in activation of mouse macrophages via TLR2 and TLR4 signaling.<br>Protective effect of gossypol on lipopolysaccharide-induced acute lung injury in mice. | Liu B, Zhang N, Liu Z, Fu Y, Feng S, Wang S, Cao Y, Li D, Liang D, Li F, Song X, Yang Z. | Yang Zhengtao 吉林大学 | Mol Cell Biochem **(SCI IF=2.329)** | 2013, 378(1-2):183-93. |
| 吉林大学 | Protective effect of gossypol on lipopolysaccharide-induced acute lung injury in mice. | Liu Z, Yang Z, Fu Y, Li F, Liang D, Zhou E, Song X, Zhang W, Zhang X, Cao Y, Zhang N. | Zhengtao Yang 吉林大学 | Inflamm Res**(SCI IF=1.964)** | 2013, 62(5):499-506 |
| 吉林大学 | Lipopolysaccharide increases Toll-like receptor 4 and downstream Toll-like receptor signaling molecules expression in bovine endometrial epithelial cells. | Fu Y, Liu B, Feng X, Liu Z, Liang D, Li F, Li D, Cao Y, Feng S, Zhang X, Zhang N, Yang Z. | Zhengtao Yang 吉林大学 | Vet Immunol Immunopathol**(SCI IF=1.877)** | 2013, 151(1-2):20-27 |
| 吉林大学 | Magnolol inhibits lipopolysaccharide-induced inflammatory response by interfering with TLR4 mediated NF-κB and MAPKs signaling pathways. | Fu Y, Liu B, Zhang N, Liu Z, Liang D, Li F, Cao Y, Feng X, Zhang X, Yang Z. | Xichen Zhang, Zhengtao Yang 吉林大学 | J Ethnopharmacol. **(SCI IF=2.755)** | 2013, 145(1):193-199 |
| 吉林大学 | Salidroside attenuates inflammatory responses by suppressing nuclear factor-κB and mitogen activated protein kinases activation in lipopolysaccharide-induced mastitis in mice. | Li D, Fu Y, Zhang W, Su G, Liu B, Guo M, Li F, Liang D, Liu Z, Zhang X, Cao Y, Zhang N, Yang Z. | Zhengtao Yang 吉林大学 | Inflamm Res **(SCI IF=1.964)** | 2013, 62(1):9-15. |
| 吉林大学 | Geniposide, from Gardenia jasminoides Ellis, inhibits the inflammatory response in the primary mouse macrophages and mouse models. | Fu Y, Liu B, Liu J, Liu Z, Liang D, Li F, Li D, Cao Y, Zhang X, Zhang N, Yang Z. | Zhengtao Yang 吉林大学 | Int Immunopharmacol **(SCI IF=2.417)** | 2012, 14(4):792-798 |
| 吉林大学 | Stevioside Plays an Anti-inflammatory Role by | Wang T, Guo M, Song X, Zhang Z, | LianqinZhu 青岛农业大 | Inflammation**(SCI IF=2.457)** | 2014, May 25 |

| 作者单位 | 题目 | 作者简称或全称 | 通讯作者及单位 | 杂志名称 | 卷期页码 |
|---|---|---|---|---|---|
| | Regulating the NF-κB and MAPK Pathways in S. aureus-infected Mouse Mammary Glands. | Jiang H, Wang W, Fu Y, Cao Y, Zhu L, Zhang N. | 学 Naisheng Zhang 吉林大学 | | |
| 吉林大学 | Protective Effects of Kaempferol on Lipopolysaccharide-Induced Mastitis in Mice. | Cao R, Fu K, Lv X, Li W, Zhang N. | Naisheng Zhang 吉林大学 | Inflammation(**SCI IF=2.457)** | 2014, April18 |
| 吉林大学 | Protective effect of taraxasterol on acute lung injury induced by lipopolysaccharide in mice. | San Z, Fu Y, Li W, Zhou E, Li Y, Song X, Wang T, Tian Y, Wei Z, Yao M, Cao Y, Zhang N. | Naisheng Zhang 吉林大学 | Int Immunopharmacol **(SCI IF=2.417)** | 2014, 19(2):342-350 |
| 吉林大学 | Endometrial inflammation and abnormal expression of extracellular matrix proteins induced by Mycoplasma bovis in dairy cows. | Guo M, Wang G, Lv T, Song X, Wang T, Xie G, Cao Y, Zhang N, Cao R. | Naisheng Zhang, Rongfeng Cao 吉林大学 | Theriogenology.(**SCI IF=2.082)** | 2014, 81(5):669-674. |
| 吉林大学 | Dietary selenium influences calcium release and activation of MLCK in uterine smooth muscle of rats. | Guo M, Lv T, Liu F, Yan H, Wei T, Cai H, Tian W, Zhang N, Wang Z, Xie G. | Guanghong Xie 吉林大学 | Biol Trace Elem Res(**SCI IF=1.307)** | 2013, 154(1):127-133 |
| 沈阳农业大学<br>吉林大学 | High insulin concentrations inhibit fatty acid oxidation-related gene expression in calf hepatocytes cultured in vitro | P. Li, C. C. Wu, M. Long, Y. Zhang, X. B. Li, J. B. He, Z. Wang, G. W. Liu. | X. B. Li, G. W. Liu 吉林大学 | J. Diary Sci.(**SCI IF=2.566**) | 2013, 96: 3840–3844. |
| 沈阳农业大学<br>吉林大学 | Effect of non-esterified fatty acids on fatty acid metabolism-related genes in Calf Hepatocytes Cultured in Vitro | Peng Li, Yiming Liu, Yi Zhang, Miao Long, Yang Guo, Zhe Wang, Xinwei Li, Cai Zhang, Xiaobing Li, Jianbin He and Guowen Liu | Guowen Liu, Xiaobing Li, Jianbin He 吉林大学 | Cellular Physiology and Biochemistry(**SCI IF=3.415**) | 2013, 32:1509-1516 |
| 沈阳农业大学 | Effect of different yeasts on selenomonas ruminantium utilizing lactate in vitro | Miao Long, Jing Li, Xia Qin, Yiming Li1, Peng Li, Guo-Wen Liu* and Xiaobing Li | Guowen Liu and Xiaobing Li 吉林大学 | Indian J. Anim. Res.,(**SCI IF=0.031**) | 2013, 47 (2) : 126-131 |
| 沈阳农业大学<br>吉林大学 | Effects of the acid-tolerant engineered bacterial strain Megasphaera elsdenii H6F32 on ruminal pH and the lactic acid concentration of simulated rumen acidosis in vitro | M. Long, W.J. Feng, P. Li, Y. Zhang, R.X. He, L.H. Yu, J.B. He, W.Y. Jing, Y.M. Li, Z. Wang, G.W. Liu | Guowen Liu 吉林大学 | Research in Veterinary Science(**SCI IF=1.774**) | 2014, 96: 28–29 |
| 沈阳农业大学<br>南京农业大学 | Genome Shuffling of Saccharomyces cerevisiae for | Long Miao, He Runxia, Dong | Kehe Huang | Journal of Pure and Applied | 2014, 8(2): 1217-1223 |

| 作者单位 | 题目 | 作者简称或全称 | 通讯作者及单位 | 杂志名称 | 卷期页码 |
|---|---|---|---|---|---|
| | Improving Thermotolerance and Glycerol Production | Shuang, Jing Wenying and Huang Kehe | 南京农业大学 | Microbiology(**SCI IF=0.054)** | |
| 东北农业大学 | Cold stress induces antioxidants and Hsps in chicken immune organs | Fu Qing Zhao, ZiWei Zhang, Jian Ping Qu, Hai Dong Yao, Ming Li, Shu Li, ShiWen Xu | Shu Li, Shi-Wen Xu 东北农业大学 | Cell Stress and Chaperones (**SCI IF=2.484**) | 2014 Jan 4. [Epub ahead of print] |
| 东北农业大学 | Effects of cold stress on mRNA expression of immunoglobulin and cytokine in the small intestine of broilers | Fu-qing Zhao, Zi-wei Zhang, Hai-dong Yao, Liang-liang Wang, Tao Liu, Xian-yi Yu, Shu Li, Shi-Wen Xu | Shi-Wen Xu 东北农业大学 | Research in Veterinary Science (**SCI IF=1.774**) | 2013, 95(1): 146-155 |
| 东北农业大学 | The role of heat shock proteins in inflammatory injury induced by cold stress in chicken hearts | Fu-Qing Zhao, Zi-Wei Zhang, ChaoWang, Bo Zhang, Hai-Dong Yao, Shu Li, Shi-Wen Xu | Shi-Wen Xu 东北农业大学 | Cell Stress and Chaperones (**SCI IF=2.484**) | 2013, 18(6): 773-783 |
| 东北农业大学 | Effects of atrazine and chlorpyrifos on cytochrome P450 in common carp liver | Houjuan Xing, Ziwei Zhang, Haidong Yao, Tao Liu, Liangliang Wang, Shiwen Xu, Shu Li | Shi-Wen Xu 东北农业大学 | Chemosphere (**SCI IF=3.137**) | 2014, 104: 244–250 |
| 东北农业大学 | Antioxidant response, CYP450 system, and histopathological changes in the liver of nitrobenzene-treated drakes | Houjuan Xing, Haibo Wang, Gang Sun, Hongda Wu, Junfeng Zhang, Mingwei Xing, Shiwen Xu | Shi-Wen Xu 东北农业大学 | Research in Veterinary Science (**SCI IF=1.774**) | 2013, 95(3): 1088-93 |
| 东北农业大学 | Effects of atrazine and chlorpyrifos on the mRNA levels of HSP70 and HSC70 in the liver, brain, kidney and gill of common carp (Cyprinus carpio L.) | Houjuan Xing, Shu Li, Xu Wanga, Xuejiao Gao, Shiwen Xu, Xiaolong Wang | Shi-Wen Xu 东北农业大学 | Chemosphere (**SCI IF=3.137**) | 2013, 90: 910–916 |
| 东北农业大学 | Gene Expression of Endoplasmic Reticulum Resident Selenoproteins Correlates with Apoptosis in Various Muscles of Se-Deficient Chicks | Hai-Dong Yao Qiong Wu, Zi-Wei Zhang, Jiu-Li Zhang, Shu Li, Jia-Qiang Huang, Fa-Zheng Ren, Shi-Wen Xu, Xiao-Long Wang, XinGenLei | Shi-Wen Xu 东北农业大学 | The Journal of Nutrition.(**SCI IF=4.196**) | 2013, 143(5): 613-619 |
| 东北农业大学 | SelenoproteinW serves as an antioxidant in chicken | Hai-Dong Yao, Qiong Wu, Zi-Wei | Shi-Wen Xu 东北农业大 | Biochimica et Biophysica Acta | 2013, 1830: 3112–3120 |

| 作者单位 | 题目 | 作者简称或全称 | 通讯作者及单位 | 杂志名称 | 卷期页码 |
|---|---|---|---|---|---|
| | myoblasts | Zhang, Shu Li, Xiao-Long Wang, Xin-Gen Lei, Shi-Wen Xu | 学 | **(SCI IF=3.848)** | |
| 东北农业大学 | Effects of oxidative stress on apoptosis in manganese-induced testicular toxicity in cocks | Xiao-fei Liu, Li-ming Zhang, Hua-nan Guan, Zi-wei Zhang, Shi-wen Xu | Shi-Wen Xu 东北农业大学 | Food and Chemical Toxicology **(SCI IF=3.01)** | 2013, 60: 168–176 |
| 东北农业大学 | Effects of manganese-toxicity on immune-related organs of cocks | Xiao-fei Liu, Zhi-peng Li, Feng Tie, Ning Liu, Zi-wei Zhang, Shi-wen Xu | Shi-Wen Xu 东北农业大学 | Chemosphere **(SCI IF=3.137)** | 2013, 90(7): 2085-2100 |
| 东北农业大学 | Effect of selenium on selenoprotein expression in the adipose tissue of chickens | Liang Y, Lin SL, Wang CW, Yao HD, Zhang ZW, Xu SW | Ziwei Zhang, Shi-Wen Xu 东北农业大学 | Biol Trace Elem Res **(SCI IF=1.307)** | 2014, 160: 41-48 |
| 东北农业大学 | Effect of oxygen free radicals and nitric oxide on apoptosis of immune organ induced by selenium deficiency in chickens | Zi-wei Zhang, Jiu-li Zhang, Yu-hong Zhang, Qiao-hong Wang, Shu Li, Xiao-long Wang, Shi-wen Xu | Shi-Wen Xu 东北农业大学 | Biometals **(SCI IF=3.284)** | 2013, 26: 355–365 |
| 东北农业大学 | Effects of Dietary Selenium Deficiency or Excess on Gene Expression of Selenoprotein N in ChickenMuscle Tissues | Jiu-li Zhang, Zi-Wei Zhang, An-Shan Shan, Shi-wen Xu | Shi-Wen Xu 东北农业大学 | Biol Trace Elem Res **(SCI IF=1.307)** | 2014, 157: 234–241 |
| 东北农业大学 | Antioxidative role of selenoprotein W in oxidant-induced chicken splenic lymphocyte death | Dong Yu, Shu Li, Zi-wei Zhang, Hai-dong Yao, Shi-wen Xu | Shu Li, Shi-Wen Xu 东北农业大学 | Biometals **(SCI IF=3.284)** | 2014, 27(2):277-91 |
| 东北农业大学 | The effect of Se-deficient diet on gene expression of inflammatory cytokines in chicken brain | Peng-fei Sheng, Yue Jiang , Zi-wei Zhang, Jiu-li Zhang, , Shu Li, Zi-qun Zhang, Shi-wen Xu | Shi-Wen Xu 东北农业大学 | Biometals **(SCI IF=3.284)** | 2014, 27(1):33-43 |
| 东北农业大学 | Cadmium supplement triggers endoplasmic reticulum stress response and cytotoxicity in primary chicken hepatocytes | Cheng-Cheng Shao, Nan Li, Zi-Wei Zhang, Jian Su, Shu Li, Jin-Long Li, Shi-Wen Xu | Shi-Wen Xu 东北农业大学 | Ecotoxicology and Environmental Safety **(SCI IF=2.203)** | 2014, 106: 109–114 |
| 东北农业大学 | Analysis of the structure of bacteria communities and detection of resistance genes of quinolones from pharmaceutical wastewater | Bo Zhang, Zi-wei Zhang, Fan-Yu Meng, Qiong Wu, Shi-Wen Xu, Xiao-Long Wang | Shi-Wen Xu 东北农业大学 Xiao-Long Wang 东北林业大学 | Ann Microbiol **(SCI IF=1.549)** | 2014, 64(1): 23-29 |
| 东北农业大学 | Effect of atrazine and | Tao Liu, Ziwei | Shi-Wen Xu | Pesticide | 2013, 107: |

| 作者单位 | 题目 | 作者简称或全称 | 通讯作者及单位 | 杂志名称 | 卷期页码 |
|---|---|---|---|---|---|
| | chlorpyrifos exposure on heat shock protein response in the brain of common carp (Cyprinus carpio L.) | Zhang, Dechun Chen, Liangliang Wang, Haidong Yao, Fuqing Zhao, Houjuan Xing, Shiwen Xu | 东北农业大学 | Biochemistry and Physiology **(SCI IF=2.111)** | 277–283 |
| 东北农业大学 | Cadmium induced hepatotoxicity in chickens (Gallus domesticus) and ameliorative effect by selenium | Jin-Long Li, Cheng-Yu Jiang, Shu Li, Shi-Wen Xu | Shi-Wen Xu 东北农业大学 | Ecotoxicology and Environmental Safety **(SCI IF=2.203)** | 2013, 96: 103–109 |
| 东北农业大学 | Cadmium supplement triggers endoplasmic reticulum stress response and cytotoxicity in primary chicken hepatocytes | Shao CC, Li N, Zhang ZW, Su J1, Li S, Li JL, Xu SW | Shiwen Xu and Jinlong Li 东北农业大学 | Ecotoxicol Environ Saf **(SCI IF=2.203)** | 2014, 106: 109–114 |
| 东北农业大学 | Influence of inflammatory pathway markers on oxidative stress induced by cold stress in intestine of quails | Jing Fu, Chun-peng Liu, Zi-wei Zhang, Ming-wei Xing, Shi-wen Xu | Shi-Wen Xu 东北农业大学 | Res Vet Sci **(SCI IF=1.774)** | 2013, 95: 495–501 |
| 东北农业大学 | Effects of atrazine and chlorpyrifos on the production of nitric oxide and expression of inducible nitric oxide synthase in the brain of common carp (Cyprinus carpio L.) | Liang-Liang Wang, Tao Liu, Chao Wang, Fu-Qing Zhao, Zi-Wei Zhang, Hai-Dong Yao,Hou-Juan Xing, Shi-Wen Xu | Shi-Wen Xu 东北农业大学 | Ecotoxicology and Environmental Safety **(SCI IF=2.203)** | 2013, 93: 7–12 |
| 东北农业大学 | Cadmium induced hepatotoxicity in chickens (Gallus domesticus) and ameliorative effect by selenium | Li JL, Jiang CY, Li S, Xu SW. | Shi-wen Xu 东北农业大学 | Ecotoxicol Environ Saf **(SCI IF=2.203)** | 2013, 96:103-109 |
| 东北农业大学 | The Oxidative Damage and Disbalance of Calcium Homeostasis in Brain of Chicken Induced by Selenium Deficiency | Shi-Wen Xu, Hai-Dong Yao, Jian Zhang, Zi-Wei Zhang, Jin-Tao Wang, Jiu-Li Zhang, Zhi-Hui Jiang | Shi-Wen Xu 东北农业大学 | Biol Trace Elem Res **(SCI IF=1.307)** | 2013, 151: 225–233 |
| 东北农业大学 | Possible Correlation between Selenoprotein Wand Myogenic Regulatory Factors in Chicken Embryonic Myoblasts | Qiong Wu, Hai-Dong Yao, Zi-Wei Zhang, Bo Zhang, Fan-Yu Meng, Shi-Wen Xu, Xiao-Long Wang | Shi-Wen Xu 东北农业大学 | Biol Trace Elem Res **(SCI IF=1.307)** | 2012, 150: 166–172 |
| 东北农业大学 | The disruption of mitochondrial metabolism and ion homeostasis in chicken hearts exposed to manganese | Jing-Jun Shao, Hai-Dong Yao, Zi-Wei Zhang, Shu Li, Shi-Wen Xu | Shi-Wen Xu 东北农业大学 | Toxicology Letters **(SCI IF=3.145)** | 2012, 214: 99–108 |

| 作者单位 | 题目 | 作者简称或全称 | 通讯作者及单位 | 杂志名称 | 卷期页码 |
|---|---|---|---|---|---|
| 东北农业大学 | Reducing lipid peroxidation for improving colour stability of beef and lamb: on-farm considerations | Yanfei Li, Shimin Liu | Shimin Liu 西澳大学 | J Sci Food Agric **(SCI IF=1.759)** | 2012, 92: 719–726 |
| 吉林大学 | Aluminum Induces Osteoblast Apoptos is Through the Oxidative Stress-Media ted JNK Signaling Pathway | Xinwei Li, Yanfei Han, Yuan Guan | 李艳飞 东北农业大学 | Biol Trace Elem Res**(IF=1.307)** | 2012, 150:502–508 |
| 吉林特产所 | Suppressive effect of accumulated aluminum trichloride on the hepatic microsomal cytochrome P450 enzyme system in rats | Yanzhu Zhu, Yanfei Han, Hansong Zhao | 李艳飞 东北农业大学 | Food and Chemical Toxicology**(SCI IF=3.01)** | 2013,51: 210–214 |
| 吉林特产所 | Impact of aluminum exposure on the immune system: a mini review | Y Z Zhu, D W Liu, Y F Li, Z.Y. Liu | 李艳飞 东北农业大学 | Environ Toxicol Pharmacol **(IF=2.005)** | 2013,35:82-87 |
| 东北农业大学 | Effects of Norepinephrine on Immune Functions of Cultured Splenic Lymphocytes Exposed to Aluminum Trichloride. | Ji-Hong Zhang, Chong-Wei Hu | 李艳飞 东北农业大学 | Biol Trace Elem Res(**SCI IF=1.307)** | 2013,154:275-280 |
| 福建农林大学 | Effects of Al on the splenic immune function and NE in rats. | Chongwei Hu, Jing Li, Yanzhu Zhu | 李艳飞 东北农业大学 | Food and Chemical Toxicology**(SCI IF=3.01)** | 2013,62:194-198 |
| 东北农业大学 | Effects of Sub-chronic Aluminum Chloride Exposure on Ovary of Rats | Fu, Y, Jia FB, Wang J，Song M | 李艳飞 东北农业大学 | Life Sciences**(IF=2.55)** | 2014,100:61-66. |
| 东北农业大学 | Effects of sub-chronic aluminum chloride on spermatogenesis and testicular enzymatic activity in male rats | Y.Z. Zhu, H. Sun, Yang Fu | 李艳飞 东北农业大学 | Life Sciences**(IF=2.55)** | 2014, 102: 36–40 |
| 吉林特产所 | Immunotoxicity of Aluminum. | Zhu YZ, Li YF, Miao LG | 李艳飞 东北农业大学 | Chemosphere**(IF=3.137)** | 2014,104(6): 1–6 |
| 东北农业大学 | Resveratrol protects against arsenic trioxide-induced nephrotoxicity by facilitating arsenic metabolism and decreasing oxidative stress | Yu ML, Xue JD, Li YJ, Zhang WQ, Ma DX, Liu L, Zhang ZG | Zhigang Zhang 东北农业大学 | Arch Toxicol**(IF=5.215)** | 2013,87, 1025-1035 |
| 东北农业大学 | Resveratrol attenuates hepatotoxicity of rats exposed to arsenic trioxide | Zhang WQ, Xue JD, Ge M, Y, Zhang ZG | Zhigang Zhang 东北农业大学 | Food Chem Toxicol**(IF=3.01)** | 2013, 51, 87-92 |
| 东北农业大学 | Attenuation of arsenic retention by resveratrol in lung of arsenic trioxide-exposed rats | Zhang WQ, Yao CY, Ge M, Xue JD, Ma DX, Liu Y, Liu JY, Zhang ZG | Zhigang Zhang 东北农业大学 | Environ Toxicol Phar**(IF=2.005)** | 2013,36, 35-39 |
| 东北农业大学 | Evaluation of the difference of and sialic acid concentration in dairy cows L-selectin, tumor necrosis factor-α with subclinical ketosis and without | Zhang ZG, Xue JD, Gao RF, Liu JY, Wang JG, Yao CY, Liu Y, Li XW, Li XB, Liu GW, | Zhe Wang 吉林大学 | Pak Vet J**(IF=1.365)** | 2013,33, 25-228 |

| 作者单位 | 题目 | 作者简称或全称 | 通讯作者及单位 | 杂志名称 | 卷期页码 |
|---|---|---|---|---|---|
| | subclinical ketosis | Wang Z | | | |
| 东北农业大学 | Attenuation of arsenic retention by resveratrol in lung of arsenic trioxide-exposed rats | Zhang WQ, Yao CY, Ge M, Xue JD, Ma DX, Liu Y, Liu JY, Zhang ZG | Zhigang Zhang 东北农业大学 | Environ Toxicol Phar**(IF=2.005)** | 2013, 36, 35-39 |
| 东北农业大学 | The protective role of resveratrol against arsenic trioxide-induced cardiotoxicity | Zhang WQ, Guo CM, Gao RF, Ge M, Zhu YZ, Zhang ZG | Zhigang Zhang 东北农业大学 | Evid Based Complement Alternat Med**(IF=1.722)** | 2013,407839 |
| 东北农业大学 | Nephroprotective effect of astaxanthin against trivalent inorganic arsenic-induced renal injury in wistar rats | Wang XN, Zhao HY, Shao YL, Wang P, Wei YR, Zhang WQ, Jiang J, Chen Y, Zhang ZG | Zhigang Zhang 东北农业大学 | Nutrition research and practice**(IF=0.973)** | 2014,1, 46-53 |
| 东北农业大学 | Effects on liver hydrogen peroxide metabolism induced by dietary selenium deficiency or excess in chickens | Xu JX, Cao CY, Sun YC, Wang LL, Li N, Xu SW, Li JL. | Jinlong Li 东北农业大学 | Biol Trace Elem Res**(SCI IF=1.307)** | 2014, 5,online |
| 西北农林科技大学 | Screening Genes Related to Breast Blister (keel cyst) in Chicken by Delta Differential Display.Asian | Jianzhou Shi, Guirong Sun, Yadong Tian, Ruili Han, Guoxi Li,Yanqun Huang, Jianhua Wang and Xiangtao Kang | Wang jian hua 西北农林科技大学 | J Ani Vet Adv**(Sci IF ＝ 0.365)** | 2012. 7(10): 989-997 |
| 西北农林科技大学 | A Comparison of Different Biochemical Parameters in Blood Serum of Healthy and Breast Blister Chickens. | Jianzhou Shi, Yadong Tian, Jianhua Wang and Xiangtao Kang | Wang jian hua 西北农林科技大学 | J Ani Vet Adv**(Sci IF ＝ 0.365)** | 2012, 11(13):2313-2315 |
| 西北农林科技大学 | Swainsonine accumulation by endophytic Undifilum fungi in liquid media and determined by means of a modified enzymatic assay | Guo-Dong Yang, Dan-Ju Kang, Yan-Hong Li, Jin-Cheng Li, Yan Wang, Xiang-Ya Kong, Qin-Fan Li,Jian-Hua Wan | Wang jian hua 西北农林科技大学 | J Ani Vet Adv **(Sci IF＝0.365)** | 2012, 11(21): 3876-3881 |
| 西北农林科技大学 | Protein extraction methods for the two-dimensional gel electrophoresis analysis of the slow growing fungus Undifilum oxytropis. African | Haili Li, Jianna Wang, Jianhua Wang, Guoxia Geng, Haocai Ju1 and Rebecca Creamer. | Rebecca Creamer美国新墨西哥州立大学 | Journal of Microbiology Research **(Sci IF ＝0.528)** | 2012, 6(4) : 757-763 |
| 西北农林科技大学 | Potential Degradation of Swainsonine by Intracellular Enzymes of Arthrobacter sp. HW08 | Yan Wang, Yanhong Li, Yanchun Hu, Jincheng Li, Guodong Yang, Danju Kang, Haili Li and Jianhua Wang | Wang jian hua 西北农林科技大学 | Toxins**(Sci IF＝2.129)** | 2013,5(11):2161-2171 |

| 作者单位 | 题目 | 作者简称或全称 | 通讯作者及单位 | 杂志名称 | 卷期页码 |
|---|---|---|---|---|---|
| 西北农林科技大学 | Optimization and validation of HPLC-ELSD method for determination of swainsonine in Chinese locoweeds, | Li Jin-cheng, Yang Guodong, Kang Danju, Li Yanhong, Wng Yan, Geng Guoxia, Wang Jianhua | Wang jian hua 西北农林科技大学 | Asian Journal of Chemistry**(IF=0.253)** | 2013, 25(17): 9635-9657 |
| 西北农林科技大学 | Oxidative stress in farmed minks with self-biting behavior. | Defa Sun, Jianhua Wang, Xiurong Xu | Xiurong Xu 西北农林科技大学 | J Vet Behavior: Clin Appl Res **(IF=0.786)** | 2013,8（1）: 51-57 |
| 西北农林科技大学 | Swainsonine as a lysosomal toxin affects dopaminergic neurons | Qinfan Li, Yingzi Wang, Rudolf Moldzio, Weimin Lin, Wolf-Dieter Rausch | Wolf-Dieter Rausch 维也纳兽医大学 | J Neural Transm **(SCI IF=2.730)** | 2012, 119: 1483-1490 |
| 西北农林科技大学 | Structural determinants of imidacloprid-based nicotinic acetylcholine receptor inhibitors identified using 3D-QSAR, docking and molecular dynamics | Qinfan Li, Xiangya Kong, Zheng tao Xiao, Lihui Zhang, Fang fang Wang, Hong Zhang, Yan Li, Yonghua Wang | Yonghua Wang 西北农林科技大学 | J Mol Model **(SCI IF=1.797)** | 2012, 18: 2279–2289 |
| 西北农林科技大学 | Identification of a New Locoweed (Oxytropis serioopetala) and Its Clinical and Pathological Features in Poisoned Rabbits | Qin-Fan LI, Cai-Ju HAO, Yong-Ping XU, Jie LIANG, Kai YANG, Zhong-Hua CUI | Yong-Ping XU 西北农林科技大学 | J. Vet. Med. Sci **(SCI IF= 0.851)** | 2012,74(8): 989–993 |
| 西北农林科技大学 | Cladosporium cladosporioides XJ-AC03, an aconitine-producing endophytic fungus isolated from Aconitum leucostomum | Kai Yang, Jie Liang, Qinfan Li, Xiangya Kong, Rui Chen, Yimin Jin. | Qinfan Li 西北农林科技大学 | World J Microb Biot **(SCI IF=1.262)** | 2013, 29(5): 933- 938 |
| 西北农林科技大学 | Mechanism of the plant cytochrome P450 for herbicide resistance: a modelling study | Qinfan Li, Yupeng Fang, Xiuxiu Li, Hong Zhang, Mengmeng Liu, Huibin Yang, Zhuo Kang, Yan Li, Yonghua Wang. | Yonghua Wang 西北农林科技大学 | J Enzyme Inhibition and Medicinal Chemistry **(SCI, IF=1.495)** | 2013, 28(6): 1182-1191 |
| 西北农林科技大学 | Molecular characterization of Capra hircus lysosomal α-mannosidase and potential mutant site for the therapy of locoweed poisoning | Kong Xiangya, Zhang Jiangye, Wu Ying, Li Jianfei , Li Qinfan | Li Qinfan 西北农林科技大学 | Acta Biochim Pol **(SCI, IF=1.185)** | 2014, 61(1): 77-84. |
| 西北农林科技大学 | Effect of Swainsonine in Oxytropis kansuensis on Golgi α□Mannosidase II Expression in the Brain Tissues of Sprague−Dawley Rats | H. Lu, S. S. Wang, W. l. Wang, L. Zhang, B. Y. Zhao | Bao-yu Zhao 西北农林科技大学 | J Agri Food Chemistry**(SCI IF=2.906)** | 2014, dx.doi.org/10.1021/jf501299d |
| 西北农林科技 | Characterization of Locoweeds | Hao Lu, Dandan | Bao-yu Zhao | The Rangeland | 2014, 36: |

| 作者单位 | 题目 | 作者简称或全称 | 通讯作者及单位 | 杂志名称 | 卷期页码 |
|---|---|---|---|---|---|
| 大学 | and Their Effect on Livestock Production in the Western Rangelands of China | Cao, Feng Ma, ShanShan Wang, Xiaowen Yang, Wenlong Wang, QiWu Zhou, Bao Yu Zhao | 西北农林科技大学 | Journal**(SCI IF=1.276)** | 121-131 |
| 西北农林科技大学 | The Study of the Oxytropis kansuensis-Induced Apoptotic Pathway in the Cerebrum of SD Rats | Hao Lu, Liang Zhang, Shan Shan Wang, Wen long Wang, Bao Yu Zhao | Bao-yu Zhao 西北农林科技大学 | *BMC* Vet Res**(SCI IF=1.861)** | 2013, 9: 217 |
| 西北农林科技大学 | Damage and control of major poisonous plants in the western grasslands of China | Hao Lu, Shan Shan Wang, Qi Wu Zhou, Yi Nan Zhao and Bao Yu Zhao | Bao-yu Zhao 西北农林科技大学 | The Rangeland Journal**(SCI IF=1.276)** | 2012, 34: 329-339 |
| 西北农林科技大学 | Isolation and Identification of Swainsonine from Oxytropis Glabra and its Pathological Lesions to SD Rats | H. Lu, S. S. Wang, B.Y. Zhao | Bao-yu Zhao 西北农林科技大学 | Asian J Anim Vet Adv**(SCI IF=0.869)** | 2012, 7(9): 822-831 |
| 西北农林科技大学 | Isolation and Identification of Swainsonine-Producing Fungi Found in Locoweeds and Their Rhizosphere Soil | Hao Lu, Ji Ping Chen, Wei Lu, Yao Ma, BaoYu Zhao and Jin Yi Wang. | Bao-yu Zhao 西北农林科技大学 | Afr J Microbiol Res**(SCI IF=0.539)** | 2012, 6(23): 4959-4969 |
| 西北农林科技大学 | Pathogenesis and preventive treatment for animal disease due to locoweed poisoning. | Wuchenchen,Wang wenlong, Wang shanshan,Yang xiaowen, Lu hao, Zhao baoyu | Bao-yu Zhao 西北农林科技大学 | Environ. Toxicol. Pharmacol.**(SCI IF=2.005)** | 2014, 4:355-369 |
| 西北农林科技大学 | Prevention of Calcium-Phosphor-Plural Gel in Milk Fever of Dairy Cows | Wu chenchen, Guo xi, Zhao baoyu | Bao-yu Zhao 西北农林科技大学 | Acta Scientiae Veterinariae**(SCI IF=0.273)** | 2014,3:11-13 |
| 西北农林科技大学；吉林大学 | Changes in Serum Copper and Zinc Levels in Peripartum Healthy and Subclinically Hypocalcemic Dairy Cows | Jianguo Wang, Xiaoyan Zhu, Zhe Wang, Xiaobing Li, Baoyu Zhao, Guowen Liu | Bao-yu Zhao 西北农林科技大学；Guo-wen Liu 吉林大学 | Trace Element Research**(SCI IF=1.307)** | 2014, DOI 10.1007/s12011-014-9997-4 |
| 湖南农业大学 | Influence of Betulinic Acid on Lymphocyte Subsets and Humoral Immune Response in Mice | Yi JE, Lis M, Szczypka M, Obmińska-Mrukowicz B. | **Obmińska-Mrukowicz B, Wrocław University of Environmental and Life Sciences** | **Pol. J. Vet. Sci**(SCI IF=1.307) | **2012, 15(2): 305-313.** |
| 湖南农业大学 | Gossypol acetic acid induces apoptosis in RAW264.7 cell via a caspase-dependent mitochondrial signaling pathway | Sijun Deng, Hui Yuan, Jine Yi, Yin Lu, Qiang Wei, Chengzhi Guo, Jing Wu, Liyun Yuan, Zuping He | Liyun Yuan 湖南农业大学 | J. Vet. Sci. **(SCI IF=0.926)** | 2013.14.3.281 |
| 湖南农业大学 | **T-2 toxin exposure induces** | Wu J1, Tu D,Yuan | Wen Lixin | Environ Toxicol | 2013, |

| 作者单位 | 题目 | 作者简称或全称 | 通讯作者及单位 | 杂志名称 | 卷期页码 |
|---|---|---|---|---|---|
| | **apoptosis in rat ovarian granulose cells through oxidative stress** | LY, Yuan H, Wen LX. | 湖南农业大学 | Pharmacol. **(SCI IF=2.005)** | 36(2):493-500 |
| 湖南农业大学 | Arsanilic acid causes apoptosis and oxidative stress in rat kidney epithelial cells (NRK-52e cells) by the activation of the caspase-9 and -3 signaling pathway | Yin Lu, Hui Yuan, Sijun Deng, Qiang Wei, Chengzhi Guo, Jine Yi, Jing Wu, Rongfang Li, Lixin Wen, Zuping He, Liyun Yuan | Liyun Yuan 湖南农业大学 | Drug and Chemical Toxicology**(SCI IF=1.293)** | 2014, 37(1): 55 |
| 湖南农业大学 | Hypocholesterolemic and anti-oxidative properties of germinated brown rice (GBR) in hypercholesterolemia-induced rats. | Matias,F. B.; Wen Qiong; Wen LiXin; Li RongFang; Tu Di; He ShaSha; Wang ZiJiu; Huang HaiBin;Wu Jing | Wu Jing 湖南农业大学 | J Microbiol Biotechnol Food Sci**(SCI IF=1.399)** | 2014, 3(4): 295-298 |
| 湖南农业大学 | **Differential gene expression of the key signalling pathway in para-carcinoma, carcinoma and relapse human pancreatic cancer.** | Chang ZY, Sun R, Ma YS, Fu D, Lai XL, Li YS, Wang XH, Zhang XP, Lv ZW, Cong XL, Li WP. | Li Wenping 湖南农业大学 | Cell Biochem Funct. **(SCI IF=1.854)** | 2014; 32(3):258-67 |
| 湖南农业大学 | Betulinic acid protects against alcohol-induced liver damage by improving antioxidant system in mice | Jine Yi, Wei Xia, Jianping Wu, Liyun Yuan, Jing Wu, Di Tu, Jun Fang, Zhuliang Tan | **Jine Yi 湖南农业大学** | **J Vet Sci**(SCI IF=0.926) | **2014, 15(1), 141-148** |
| 浙江大学 | Down-regulation of NOD1 in neutrophils of periparturient dairy cows | Tan X, Li W, Guo J, Zhou J. | 谭勋 浙江大学 | Vet Immunol Immunop **(IF=1.877)** | 2012, 150:133-139 |
| 浙江大学 | Involvement of matrix metalloproteinase-2 in the medial hypertrophy of pulmonary arterioles in broilers with pulmonary arterial hypertension | Tan X, Chai J, Bi S, Li J, Li W, Zhou J. | 谭勋 浙江大学 | The Vet J **(IF=2.42)** | 2012, 193:420-425. |
| 浙江大学 | Development of an immunosensor assay for the detection of haptoglobin in mastitic milk. | Tan X, Ding S, Li J, Li W, Zhou J. | 谭勋 浙江大学 | Vet Clin Pathol **（IF= 1.29）** | 2012, 41:575-581 |
| 浙江大学 | Cost effective and time efficient measurement of CD4, CD8, major histocompatibility complex class II, and macrophage antigen expression in the lungs of chickens. | Fletcher OJ, Tan X, Cortes L, Gimeno I. | Fletcher OJ 北卡州立大学 | Vet Immunol Immunop **(IF=1.877)** | 2012, 146: 225-236 |
| 浙江大学 | Differential characteristics and in vitro angiogenesis of bone | Shah QA, Tan X, Bi S, Liu X, Hu S. | 谭勋 浙江大学 | Cell proliferation **（IF=2.27）** | 2014, Epub ahead of print |

| 作者单位 | 题目 | 作者简称或全称 | 通讯作者及单位 | 杂志名称 | 卷期页码 |
|---|---|---|---|---|---|
| | marrow- and peripheral blood-derived endothelial progenitor cells: evidence from avian species | | | | |
| 四川农业大学 | Effects of the Fusarium toxin zearalenone(ZEA) and/or deoxynivalenol(DON) on the serum IgA,IgG and IgM levels | Z.H.Ren,R.Zhou,J.L.Deng,Z.C.Zuo, X.Peng,Y.C.Wang,Y.Wang,S.M.Yu, L.H.Shen,H.M.Cui, J.Fang | J.L.Deng 四川农业大学 | Food and Agricultural Immunology **(SCI IF=0.73)** | On:16 December 2013 |
| 四川农业大学 | Effects of zearalenone on calcium homeostasis of splenic lymphocytes of chickens in vitro | Wang YC, Deng JL, Xu SW, Peng X, Zuo ZC, Cui HM, Wang Y, Ren ZH | J.L.Deng 四川农业大学 | Poultry Science **(SCI IF=1.782)** | 2012, 91(8): 1956-1963 |
| 四川农业大学 | Effects of Zearalenone on IL-2,IL-6,and INF-γ mRNA Levels in the Splenic Lymphocytes of Chickens | Y. C. Wang ,J. L. Deng, S.W. Xu,X. Peng,Z. C. Zuo,H.M. Cui,Y. Wang,andZ.H.Ren | J.L.Deng 四川农业大学 | The Scientific World Journal**(SCI IF=1.73)** | 2012:567327. doi: 10.1100/2012/567327 |
| 四川农业大学 | Isolation and Identification of Bacteria Capable of Degrading Euptox A from Eupatorium adenophorum Spreng | Fei Liao, Yunfei Wang, Yue Huang, Quan Mo, Hui Tan, Yanchun Hu | Yanchun Hu 四川农业大学 | Toxicon **(SCI IF=2.924)** | 2014,77: 87-92 |
| 四川农业大学 | Acaricidal Activity of 9-oxo-10,11-dehydroageraphorone Extracted from Eupatorium adenophorum | Fei Liao, Yanchun Hu, Hui Tan, Lei Wu, Yunfei Wang, Yue Huang, Quan Mo, Yahui Wei. | Yanchun Hu 四川农业大学 | Experimental Parasitology**(IF=2.154)** | 2014,140: 8-11 |
| 四川农业大学 | The Antitumor Activity in Vitro by 9-Oxo-10, 11-dehydroageraphorone Extracted from Eupatorium adenophorum. | Fei Liao, Yanchun Hu, Lei Wu, Hui Tan, Quan Mo, Biao Luo, Yajun He, Junliang Deng, Yahui Wei. | Yanchun Hu 四川农业大学 | Asian Journal of Chemistry. **(SCI IF=0.251)** | 2014,Accepted( MS No.16696/2013) |
| 四川农业大学 | Transcriptional Profiling of Swine Lung Tissue after Experimental Infection with Actinobacillus pleuropneumoniae | Zhicai Zuo, Hengmin Cui, Mingzhou Li, etc | Hengmin Cui 四川农业大学 | International Journal of Molecular Sciences（**SCI IF=2.464**） | 2013, 14, 10626-10660 |
| 四川农业大学 | Transcriptional Profiling of Hilar Nodes from Pigs after Experimental Infection with Actinobacillus Pleuropneumonia | Shumin Yu, Zhicai Zuo, Hengmin Cui, etc | Hengmin Cui 四川农业大学 | International Journal of Molecular Sciences（**SCI IF=2.464**） | 2013, 14, 23516-23532 |
| 四川农业大学 | Improved Establishment of Embryonic Stem (ES) Cell Lines from the Chinese Kunming Mice by Hybridization with 129 Mice | Shumin Yu, Xingrong Yan, Zhicai Zuo, etc | Zhicai Zuo 四川农业大学 | Int J Mol Sci（**SCI IF=2.464**） | 2014, 15, 3389-3402 |
| 四川农业大学 | Prokaryotic Expression and | Zhicai Zuo, Shan | Wanzhu Guo | Journal of | 2012, 11 (7): |

| 作者单位 | 题目 | 作者简称或全称 | 通讯作者及单位 | 杂志名称 | 卷期页码 |
|---|---|---|---|---|---|
| | Sequence Analysis of Porcine BCL10 Gene | Liao, Xinqiao Qi, etc | 四川农业大学 | Animal and Veterinary Advances(**SCI IF=0.365**） | 1046-1051 |
| 四川农业大学 | Isolation and the Analysis of 16S rDNA Sequence of Swine Bordetella bronchiseptica | Zuo Zhicai, Fu Tongchao, Wan Hongping | Zhu Lin 四川农业大学 | Journal of Animal and Veterinary Advances(**SCI IF=0.365**） | 2012, 11 (17) : 3156-3159 |
| 四川农业大学 | Development and Application of a Polymerase Chain Reaction to Early Detect Haemophilus parasuis | Zuo Zhicai, Hongbo Dai, Lei Chen | Wanzhu Guo 四川农业大学 | Journal of Animal and Veterinary Advances（**SCI IF=0.365**） | 2013, 12 (2): 140-145 |
| 四川农业大学 | Cytokine and Chemokine Microarray Profiles in Lung and Hilar Nodes from Pigs after Experimental Infection with Actinobacillus Pleuropneumoniae | Zhicai Zuo, Hengmin Cui, Xi Peng | Wanzhu Guo 四川农业大学 | Journal of Animal and Veterinary Advances（**SCI IF=0.365**） | 2012, 11 (24):4603-4610 |
| 山东农业大学 | Effects of manganese deficiency on serum hormones and biochemical markers of bone metabolism in chicks | Zhaojun W,Lin W, Zhenyong W, Jian W, Ran L. | 王振勇 山东农业大学 | J Bone Miner Metab(**SCI IF=2.38)** | 2013, 31(3):285-92. |
| 山东农业大学 | Effects of manganese deficiency on chondrocyte development in tibia growth plate of Arbor Acres chicks. | Wang J, Wang ZY, Wang ZJ, Liu R, Liu SQ, Wang L. | 王振勇 山东农业大学 | J Bone Miner Metab(**SCI IF=2.38)** | 2014, Feb 28. [Epub ahead of print] |
| 山东农业大学 | Protective effects of quercetin on cadmium-induced cytotoxicity in primary cultures of rat proximal tubular cells. | Wang L, Lin SQ, He YL, Liu G, Wang ZY. | 王振勇 山东农业大学 | Biomed Environ Sci.(**SCI IF=1.345)** | 2013, Apr;26(4):258-67 |
| 山东农业大学 | Protective effects of puerarin on experimental chronic lead nephrotoxicity in immature female rats. | Wang L, Lin S, Li Z, Yang D, Wang Z. | 王振勇 山东农业大学 | Hum Exp Toxicol(**IF=1.453)** | 2013, Feb;32(2):172-85 |
| 山东农业大学 | Detection of Anaplasma marginale in Hyalomma asiaticum ticks by PCR assay | Zhang L, Wang Y, Cai D, He G, Cheng Z, Liu J, Meng K, Yang D, Wang S. | 刘建柱 山东农业大学 | Parasitology research(**IF=2.852)** | 2013, 112(7): 2697-2702. |
| 山东农业大学 | The effects of taurochenodeoxycholic acid in preventing pulmonary fibrosis in mice | Zhou C, Shi Y, Li J, Zhang W, Wang Y, Liu Y, Liu J | 刘建柱 山东农业大学 | Pak J pharmaceutical sci (**IF=0.947)** | 2013, 26(4):761-5 |
| 山东农业大学 | Scutellaria polysaccharide inhibits the infectivity of Newcastle disease virus to chicken embryo fibroblast | Xiaona Z, Jianzhu L. | 刘建柱 山东农业大学 | Journal of the Science of Food and Agriculture (**IF=1.759)** | 2014, 94(4):779-84. |
| 山东农业大学 | PCR-based detection of Theileria annulata in Hyalomma asiaticum ticks in northwestern | Meng K, Li Z, Wang Y, Jing Z, Zhao X, Liu J, Cai | 刘建柱 山东农业大学 | Ticks and tick-borne diseases(**IF=2.35** | 2014, 5(2): 105-106 |

| 作者单位 | 题目 | 作者简称或全称 | 通讯作者及单位 | 杂志名称 | 卷期页码 |
|---|---|---|---|---|---|
| | China | D, Zhang L, Yang D, Wang S | | **3)** | |
| 山东农业大学 | Effects of chromium picolinate on the viability of chick embryo fibroblast | Bai Y, Zhao X, Qi C, Wang L, Cheng Z, Liu M, Liu J, Yang D, Wang S, Chai T | 刘建柱 山东农业大学 | Human & experimental toxicology**(IF=1.453)** | 2014, 33(4):403-13 |
| 山东农业大学 | Prevalence of fur mites in canine dermatologic disease in Henan, Hebei, Heilongjiang Provinces and Xinjiang Uygur Autonomous Region, China | Cai D, Q Zhang, L Zhang, H Zhang, Z Fu, G He, G Liu and J Liu | 刘建柱 山东农业大学 | International Journal of Veterinary Science**(IF=2.175)** | 2014, 3(1): 29-32 |
| 山东农业大学 | Development of colloidal gold-based immunochromatographic assay for rapid detection of Mycoplasma suis in porcine plasma | Meng K, Sun W, Zhao P, Zhang L, Cai D, Cheng Z, Guo H, Liu J, Yang D, Wang S, Chai T. | 刘建柱 山东农业大学 | Biosensors and Bioelectronics**(IF= 5.437)** | 2014, 55: 396-399. |
| 山东农业大学 | Liposomes containing recombinant gp85 protein vaccine against ALV-J in chickens | Zhang L, Cai D, Zhao X, Cheng Z, Guo H, Qi C, Liu J, Xu R, Zhao P, Cui Z. | 刘建柱 山东农业大学 | Vaccine**(IF= 3.492)** | 2014, 32(21): 2452-2456. |
| 山东农业大学 | Prevalence and risk factors of Giardia doudenalis in dogs from China | Yang D, Zhang Q, Zhang L, Dong H, Jing Z, Li Z, Liu J | 刘建柱 山东农业大学 | Int J Environl health res **(IF=1.203)** | 2014, (ahead-of-print): 1-7. |
| 山西农业大学 | Decreased percentages of CD4+CD25+ regulatory T cells and Foxp3 expression in the spleen of female mice exposed to fluoride. | G.H. Zhang, B.R. Zhou, T.L. Han, M. Wang, X.P. Du, Q. Li, J.D. Wang | Jundong Wang 山西农业大学 | Fluoride **(SCI IF=0.8)** | 2012, 45(4):317-324. |
| 山西农业大学 | Determination of allelic expression of SNP rs1880676 in choline acetyltransferase gene in HeLa cells. | Z. Yang, C. Lin, S. Wang, C. Seneviratne, J.D. Wang, M.D. Li | Jundong Wang 山西农业大学 | Neurosci Lett. **(SCI IF=2.03)** | 2013; 555:215-219. |
| 山西农业大学 | Effect of Dietary Yeast Chromium and L-Carnitine on Lipid Metabolism of Sheep. | B. Zhou, H. Wang, G. Luo, R. Niu, J. Wang | Jundong Wang 山西农业大学 | Biol Trace Elem Res **(SCI IF=1.31)** | 2013, 155(2):221-227. |
| 山西农业大学 | Pubertal exposure to Bisphenol A increases anxiety-like behavior and decreases acetylcholinesterase activity of hippocampus in adult male mice. | G. Luo, R. Wei, R. Niu, C. Wang, J. Wang | Jundong Wang 山西农业大学 | Food Chem Toxicol. **(SCI IF=3.01)** | 2013, 60:177-180 |
| 山西农业大学 | Tissue distributions of fluoride and its toxicity in the gills of a freshwater teleost, Cyprinus carpio. | J. Cao, J. Chen, J. Wang, X. Wu, Y. Li, L. Xie | Jundong Wang 山西农业大学 | Aquat Toxicol **(SCI IF=3.73)** | 2013, 130-131:68-76. |
| 山西农业大学 | Effects of fluoride on liver apoptosis and Bcl-2, Bax protein expression in freshwater | J. Cao, J. Chen, J. Wang, R. Jia, W. Xue, Y. Luo, X. | Jundong Wang 山西农业大 | Chemosphere **(SCI IF=3.14)** | 2013, 91(8):1203-1212. |

| 作者单位 | 题目 | 作者简称或全称 | 通讯作者及单位 | 杂志名称 | 卷期页码 |
|---|---|---|---|---|---|
| | teleost, Cyprinus carpio. | Gan | 学 | | |
| 河南科技学院 | Effects of high fluoride and low iodine on thyroid function in offspring rats. | Y.M. Ge, H.M. Ning, X.L. Gu, M. Yin, X.F. Yang, Y.H. Qi, J.D. Wang | Jundong Wang 山西农业大学 | J Integr Agric **(SCI IF=0.53)** | 2013, 12(3): 502-508. |
| 山西农业大学 | Effects of fluoride on bacterial growth and its gene/protein expression. | H. Ma, X. Wu, M. Yang, J. Wang, J. Wang, J. Wang | Jundong Wang 山西农业大学 | Chemosphere **(SCI IF=3.14)** | 2014, 100:190-193. |
| 山西农业大学 | In vivo influence of sodium fluoride on sperm chemotaxis in male mice. | Z. Lu, S. Wang, Z. Sun, R. Niu, J. Wang | Jundong Wang 山西农业大学 | Arch Toxicol **(SCI IF=5.22)** | 2014, 88(2):533-9 |
| 山西农业大学 | Effect of pubertal nano-TiO2 exposure on testosterone synthesis and spermatogenesis in mice. | F. Jia, Z. Sun, X. Yan, B. Zhou, J. Wang | Jundong Wang 山西农业大学 | Arch Toxicol **(SCI IF=5.22)** | 2014, 88(3):781-8 |
| 山西农业大学 | Altered sperm chromatin structure in mice exposed to sodium fluoride through drinking water. | Z.L. Sun, R.Y. Niu, B. Wang, J.D. Wang | Jundong Wang 山西农业大学 | Environ Toxicol **(SCI IF=2.71)** | 2014, 29(6):690-696. |
| 山西农业大学 | Effects of sodium fluoride on MAPKs signaling pathway in the gills of a freshwater teleost, Cyprinus carpio. | J. Cao, J. Chen, J. Wang, P. Klerks, L. Xie | Jundong Wang 山西农业大学 | Aquat Toxicol **(SCI IF=3.73)** | 2014, 152C:164-172. |
| 山西农业大学 | Maternal Bisphenol A Diet Induces Anxiety-like Behavior in Female Juvenile with Neuroimmune Activation. | G. Luo, S. Wang, Z. Li, R. Wei, L. Zhang, H. Liu, C. Wang, R. Niu, J. Wang | Jundong Wang 山西农业大学 | Toxicol Sci **(SCI IF=4.33)** | 2014 May 13. pii: kfu085. [Epub ahead of print] |
| 山西农业大学 | Changes in memory and synaptic plasticity induced in male rats after maternal exposure to bisphenol A. | C. Wang, R. Niu, Y. Zhu, H. Han, G. Luo, B. Zhou, J. Wang | Jundong Wang 山西农业大学 | Toxicology **(SCI IF=4.02)** | 2014 May 10. pii: S0300-483X(14)00095-X. doi: 10.1016/j.tox.2014.05.001. [Epub ahead of print] |
| 黑龙江八一农垦大学 | Shotgun Proteomic Analysis of Plasma from Dairy Cattle Suffering from Footrot: Characterization of Potential Disease-Associated Factors | Sun D, Zhang H, Guo D, Sun A, Wang H. | Dongbo Sun 黑龙江八一农垦大学 | PLoS ONE **(SCI IF=3.73)** | 2013, 8(2): e55973. |
| 黑龙江八一农垦大学 | Identification of a 43-kDa outer membrane protein of Fusobacterium necrophorum that exhibits similarity with pore-forming proteins of other Fusobacterium species | Sun D, Zhang H, Lv S, Wang H, Guo D. | Dongbo Sun 黑龙江八一农垦大学 | Research in Veterinary Sciences **(SCI IF=1.77)** | 2013, 95: 27-33 |
| 黑龙江八一农垦大学 | Proteomic Analysis of Membrane Proteins of Vero | Guo D, Zhu Q, Zhang H, Sun D. | Dongbo Sun 黑龙江八一 | DNA and Cell Biology | 2014, 33(1): 20-28 |

| 作者单位 | 题目 | 作者简称或全称 | 通讯作者及单位 | 杂志名称 | 卷期页码 |
|---|---|---|---|---|---|
| | Cells: Exploration of Potential Proteins Responsible for Virus Entry | | 农垦大学 | **(SCI IF=2.34)** | |
| 黑龙江八一农垦大学 | Screening and antiviral analysis of phages that display peptides with an affinity to subunit C of porcine aminopeptidase | Guo D, Zhu Q, Feng L, Sun D. | Dongbo Sun 黑龙江八一农垦大学 | Monoclon Antib Immunodiagn Immunother **(SCI IF=0.33)** | 2013,32(5): 326-329 |
| 黑龙江八一农垦大学 | Expression and Purification of the scFv from Hybridoma Cells Secreting a Monoclonal Antibody Against S Protein of PEDV | QH Zhu, DH Guo, L Feng, DB Sun | Dongbo Sun 黑龙江八一农垦大学 | Monoclon Antib Immunodiagn Immunother **(SCI IF=0.33)** | 2013, 32(1): 41-46. |
| 黑龙江八一农垦大学 | 1H NMR-based Plasma Metabolic Profiling of Dairy Cows with Clinical and Subclinical Ketosis | Lingwei Sun, Hongyou Zhang, Ling Wu, Shi Shu, Cheng Xia, Chung Xu, Jiasan, Zheng | Hongyou Zhang 黑龙江八一农垦大学 | J. Dairy Sci **(SCI IF=2.566)** | 2014, 97(3): 1552-1562 |
| 黑龙江八一农垦大学 | Plasma metabolomic profiling of dairy cows affected with ketosis using gas chromatography/mass spectrometry | Hongyou Zhang, Ling Wu, Chuang Xu, Cheng Xia*, Lingwei Sun and Shi Shu. | Cheng Xia 黑龙江八一农垦大学 | BMC Vet Res**(SCI IF=1.86)** | 2013, 26(9):186 |
| 黑龙江八一农垦大学 | Detection of progesterone in bovine milk using an electrochemical immunosensor | Hongyou Zhang, Xiayan Du,Qian Liu,Cheng Xia, Lingwei Sun | Cheng Xia 黑龙江八一农垦大学 | Society of Dairy Technology **(SCI .IF=1.107)** | 2013, 66(4): 461-467 |
| 黑龙江八一农垦大学 | Plasma Proteomics Analysis of Dairy Cows with Milk Fever Using SELDI-TOF-MS | Shi Shu, Cheng Xia, Hongyou Zhang, Zhaolei Sun, Jiannan Liu and Bo Wang | Cheng Xia 黑龙江八一农垦大学 | Asian J Anim Vet Adv**(SCI .IF=0.869)** | 2014,9(1):1-12 |
| 黑龙江八一农垦大学 | Effects of Anionic Salts on Hypocalcaemia and Ca Homeostasis in Periparturient Dairy Cows in China | Cheng Xia, Shi Shu, Bo Wang, Chuang Xu*, Hong-You. Zhang, Ling Wu, and Jia-San Zheng | Chuang Xu 黑龙江八一农垦大学 | J anim vet adv**(SCI IF=0.39)** | 2013,12(13):1193-1197 |
| 黑龙江八一农垦大学 | Proteomic analysis of plasma from cows affected with milk fever using two-dimensional differential in-gel electrophoresis and mass spectrometry | C Xia, H Y Zhang, L Wu, C Xu, J S Zheng, Y J Yan, L J Yang, S Shu | Hongyou Zhang 黑龙江八一农垦大学 | Res Vet Sci **(SCI .IF=1.64)** | 2012,93(2):857-861 |
| 黑龙江八一农垦大学 | Concentrations of plasma metabolites, hormones, and mRNA abundance of adipose leptin and hormone-sensitive lipase in ketotic and nonketotic dairy cows. | Cheng. Xia, Zhe. Wang, C. Xu, and H.Y. | Wang Zhe 吉林大学 | J Vet Intern Med**(SCI .IF=1.992)** | 2012, 26(2): 415-417. |
| 安徽农业大学 | Effects of β-conglycinin on | Xichun Wang, | Wu Jinjie | Archives of | 2014, 68(3): |

| 作者单位 | 题目 | 作者简称或全称 | 通讯作者及单位 | 杂志名称 | 卷期页码 |
|---|---|---|---|---|---|
| | growth performance, immunoglobulins and intestinal mucosal morphology in piglets | Fangfang Geng, Jinjie Wu, Yanan Kou, Shuliang Xu, Zhikuo Sun, Shibin Feng, Liangyou Ma and Ying Luo | 安徽农业大学 | Animal Nutrition(**SCI IF=1.095**) | 186-195 |
| 安徽农业大学 | Characterization and Comparison of Ochratoxin A-Ovalbumin (OTA-OVA) Conjugation by Three Methods | WANG Xi-chun, BAO Ming, WU Jin-jie, LUO Ying, MA Liang-you, WANG Ying, ZHANG Aihua, HE Cheng-hua and ZHANG Hai-bin | Wu Jinjie 安徽农业大学 | Journal of Integrative Agriculture(**SCI IF=0.53**) | 2014, 13(5): 1130-1136 |
| 扬州大学 | Oxidative Stress and Apoptotic Changes of Rat Cerebral Cortical Neurons Exposed to Cadmium in Vitro | YUAN Yan, BIAN Jian Chun, LIU Xue Zhong, ZHANG Ying, SUN Ya, and LIU Zong Ping | Zongping Liu 扬州大学 | Biomedical and Environmental Sciences (**SCI IF=1.154**) | 2012, 25(2): 172-181 |
| 扬州大学 | Protective Effect of Naringenin Against Lead-Induced Oxidative Stress in Rats | Wang Jicang, Yang Zijun, Lin Lin, Zhao Zhanqin, Liu Zongping, Liu Xuezhong | Zongping Liu/Liu Xuezhong 扬州大学 | Biol Trace Element Res (**SCI IF=1.307**) | 2012, 146: 354–359 |
| 扬州大学 | Cadmium-Induced Apoptosis in Primary Rat Cerebral Cortical Neurons Culture Is Mediated by a Calcium Signaling Pathway | Yan Yuan., Chen-yang Jiang., Hui Xu, Ya Sun, Fei-fei Hu, Jian-chun Bian, Xue-zhong Liu, Jian-hong Gu, Zong-ping Liu | Zongping Liu 扬州大学 | PLOS ONE (**SCI IF=3.73**) | 2013, 8(5): e64330 |
| 扬州大学 | Oxidative Stress and Mitogen-Activated Protein Kinase Pathways Involved in Cadmium-Induced BRL 3A Cell Apoptosis | Zhang Yiran, Jiang Chenyang, Wang Jiajing, Yuan Yan, Gu Jianhong, Bian Jianchun, Liu Xuezhong, Liu Zongping | Zongping Liu 扬州大学 | Oxidative Medicine and Cellular Longevity (**SCI IF=3.393**) | 2013, 1-12 |
| 扬州大学 | Osteoprotegerin influences the bone resorption activity of osteoclasts | Fu Ying-Xiao, Gu Jian-Hong, Zhang Yi-Ran, Tong Xi-Shuai, Zhao Hong-Yan, Yuan Yan, Liu Xue-Zhong, Bian | Zongping Liu 扬州大学 | Int J Mol Med(**SCI IF=1.957**) | 2013,31(6): 1411-1417 |

| 作者单位 | 题目 | 作者简称或全称 | 通讯作者及单位 | 杂志名称 | 卷期页码 |
|---|---|---|---|---|---|
| | | Jian-Chun, Liu Zong-Ping | | | |
| 扬州大学 | Influence of osteoprotegerin on differentiation, activation, and apoptosis of Gaoyou duck embryo osteoclasts in vitro | Fu Ying-Xiao, Gu Jian-Hong, Zhang Yi-Ran, Tong Xi-Shuai, Zhao Hong-Yan, Yuan Yan, Liu Xue-Zhong, Bian Jian-Chun, Liu Zong-Ping | Zongping Liu 扬州大学 | Poultry Science **(SCI IF=1.156)** | 2013, 92 (6): 1613-1620 |
| 扬州大学 | Induction of cytoprotective autophagy in PC-12 cells by cadmium | Wang Qiwen, Zhu Jiaqiao, Zhang Kangbao, Jiang Chenyang, Wang Yi, Yuan Yan, Bian Jianchun, Liu Xuezhong, Gu Jianhong, Liu Zongping | Zongping Liu 扬州大学 | Biochem Biophys Res Commun **(SCI IF=2.406)** | 2013,438: 186-192 |
| 扬州大学 | Inhibitory effects of osteoprotegerin on osteoclast formation and function under serum-free conditions | Ying-Xiao Fu, Jian-Hong Gu, Yi-Ran Zhang, Xi-Shuai Tong, Hong-Yan Zhao, Yan Yuan, Xue-Zhong Liu, Jian-Chun Bian, Zong-Ping Liu | Zongping Liu 扬州大学 | Journal of Veterinary Science **(SCI IF=0.926)** | 2013, 14(4): 405-412 |
| 扬州大学 | Inactivation of the ABC transporter ATPase gene in Brucella abortus strain 2308 attenuated the virulence of the bacteria | Zhang Min, Han Xiangan, Liu Haiwen, Tian Mingxing, Ding Chan, Song Jun, Sun Xiaoqing, Liu Zongping, Yu Shengqing | Yu Shengqing 上海兽医研究所 | Veterinary Microbiology **(SCI IF=3.127)** | 2013,164 (3-4): 322-329 |
| 扬州大学 | Sodium fluoride disrupts DNA methylation of H19 and Peg3 imprinted genes during the early development of mouse embryo | Zhu Jia-Qiao, Si Yang-Jun, Cheng Lai-Yang, Xu Bao-Zeng, Wang Qi-Wen, Zhang Xiao, Wang Heng, Liu Zong-Ping | Zongping Liu 扬州大学 | Arch Toxicol **(SCI IF=5.215)** | 2014, 88(2): 241-8 |
| 扬州大学 | Calcium–calmodulin signaling elicits mitochondrial dysfunction and the release of cytochrome c during cadmium-induced apoptosis in primary osteoblasts | Liu Wei, Zhao Hongyan, Wang Yi, Jiang Chenyang, Xia Pengpeng, Gu Jianhong, Liu Xuezhong, Bian Jianchun, Yuan | Zongping Liu 扬州大学 | Toxicology Letters **(SCI IF=3.145)** | 2014,224 (1): 1-6 |

| 作者单位 | 题目 | 作者简称或全称 | 通讯作者及单位 | 杂志名称 | 卷期页码 |
|---|---|---|---|---|---|
| | | Yan, Liu Zongping | | | |
| 扬州大学 | 1α, 25-(OH)2D3 regulation of matrix metalloproteinase-9 protein expression during osteoclast differentiation | Gu JH, Tong XS, Liu XZ, Bian JC, Yuan Y, Liu ZP | Zongping Liu 扬州大学 | J Vet Sci **(SCI IF=0.926)** | 2014, 15(1): 133-140 |
| 扬州大学 | Cadmium Induces PC12 Cells Apoptosis via an Extracellular Signal-Regulated Kinase and c-Jun N-Terminal Kinase-Mediated Mitochondrial Apoptotic Pathway | Jiang Chenyang, Yuan Yan, Hu Feifei, Wang Qiwen, Zhang Kangbao, Wang Yi, Gu Jianhong, Liu Xuezhong, Bian Jianchun, Liu Zongping | Zongping Liu 扬州大学 | Biol Trance Elem Res **(SCI IF=1.307)** | 2014, 158(2):249-58 |
| 扬州大学 | Cadmium induces the differentiation of duck embryonic bone marrow cells into osteoclasts in vitro | Yi Wang, Ying-Xiao Fu,Jian-Hong Gu, Yan Yuan, Xue-Zhong Liu, Jian-Chun Bian, Zong-Ping Liu | Zongping Liu 扬州大学 | The Veterinary Journal **(SCI IF=2.239)** | 2014, 200:181-185 |
| 扬州大学 | Effect of Cadmium on Rat Leydig Cell Testosterone Production and DNA Integrity in vitro | Liu Qing, Gu Jian Hong, Yuan Yan, Liu Xue Zhong, Wang Ya Jun, Wang Han Dong, Liu Zong Ping, Wang Zong Yuan, Bian Jian Chun | Jianchun Bian 扬州大学 | Biomed Environ Sci **(SCI IF=1.154)** | 2013, 26(9): 769-773 |
| 扬州大学 | Zearalenone induces apoptosis and cytoprotective autophagy inprimary Leydig cells | Wang Yajun, Zheng Wanglong, Bian Xiaojiao, Yuan Yan, Gu Jianhong, Liu Xuezhong, Liu Zongping, Bian Jianchun | Jianchun Bian 扬州大学 | Toxicology Letters **(SCI IF=3.145)** | 2014, 226: 182-191 |
| 扬州大学 | Zearalenone inhibits testosterone biosynthesis in mouse Leydig cells via the crosstalk of estrogen receptor signaling and orphan nuclear receptor Nur77 expression | Liu Qing, Wang Yajun, Gu Jianhong, Yuan Yan, Liu Xuezhong, Zheng Wanglong, Huang Qinyi, Liu Zongping, Bian Jianchun | Jianchun Bian 扬州大学 | Toxicology in Vitro **(SCI IF=2.65)** | 2014, 28: 647-656 |
| 陕西师范大学 华中农业大学 | Mussel-inspired polydopamine coating as a versatile platform for synthesizing polystyrene/Ag nanocomposite particles with enhanced antibacterial activities | Ying Cong,Tian Xia, Miao Zou, Zhenni Li, Bo Peng, Dingzong Guo and Ziwei | Bo Peng 荷兰 Dingzong Guo 华中农业大学 Ziwei Deng 陕西师 | Journal of Materials Chemistry B **(IF=6.101)** | 2014, (2): 3450–3461 |

| 作者单位 | 题目 | 作者简称或全称 | 通讯作者及单位 | 杂志名称 | 卷期页码 |
| --- | --- | --- | --- | --- | --- |
| | | Deng | 范大学 | | |
| 华中农业大学 | Screening of differentially expressed genes in the growth plate of broiler chickens with Tibial Dyschondroplasia by microarray analysis | Tian WX, Li JK, Qin P, Wang R, Ning GB, Qiao JG, Li HQ, Bi DR, Pan SY | Guo DZ 华中农农业大学 | BMC Genomics(SCI **IF=4.397**) | 2013, 14:276 |
| 华中农业大学 | Exogenous ARC down-regulates caspase-3 expression and inhibits apoptosis of broiler chicken cardiomyocytes exposed to hydrogen peroxide | Wu, Liming 吴黎明 | Guo DZ 华中农农业大学 | Vet J(**SCI IF=2.424**) | 2013, 42(1): 32-37 |
| 华中农业大学 | Effects of ceftiofur sodium liposomes on free radical formation in mice | SR Liu, DH Zhou, SJ Yang and DZ Guo | Guo DZ 华中农业大学 | Pak Vet J(**SCI IF=1.365**) | 2012, 32(4): 543-546 |
| 华中农业大学 | Expression and Localization of Stanniocalcin-1 in Bovine Osteoblasts | Dong-yang Liu, Shi-jin Yang, Zhao-fang Xi, Liming-Wu, Sen Chen, Shi-qi Dong, Jian-lin Wang and Ding-zong Guo | Guo DZ 华中农业大学 | Pak Vet J(**SCI IF=1.365**) | 2012,32(2): 242-246 |
| 华中农业大学 | ROS Induce Cardiomyocyte Apoptosis in Ascitic Broiler Chickens | Zhaofang Xi, Shijin Yang, Dongyang Liu, Liming Wu, Xiaodong Liu, Jing Zhao and Dingzong Guo | Guo DZ 华中农业大学 | Pak Vet J(**SCI IF=1.365**) | 2012, 32(4): 613-617 |
| 华中农业大学 | IGF-1 Protects Myocardial Cells from ROS Stress-Induced Apoptosis via Up-Regulating ARC | Zhaofang Xi, Haibao Zhu, Dongyang Liu, Liming Wu and DingZong Guo | Guo DZ 华中农业大学 | J Ani Vet Adv(**SCI IF=0.39**) | 2012, 11(11): 1901-1906 |
| 华中农业大学 | Expression of endothelin-1 and its receptors in the lungs of broiler chickens exposed to high altitude hypoxia | Jianfeng Gao, Liru Zhao, Ding Zhang, Muhammad Shahzad, Ding Zhang, Guoquan Liu, Bo Hou, Jiakui Li | 李家奎 华中农业大学 | Avian pathol **(SCI IF=1.729）** | 2013, 42(5):416-419 |
| 华中农业大学 | Cultivation and characterization of pulmonary microvascular endothelial cells from chick embryos | Jianfeng Gao, Ding Zhang, Muhammad Shahzad, Kerong Zhang, Liru Zhao, Jiakui Li | 李家奎 华中农业大学 | Pak Vet J **(SCI IF=1.365）** | 2013, 33(3): 300-303 |
| 华中农业大学 | Effects of Selenium Supplementation on Expression of Endothelin-1 and its | Jianfeng Gao, Ding Zhang, Kerong Zhang, | 李家奎 华中农业大学 | Biol Trace Elem Res **(SCI IF=1.307)** | 2012, 150 (1-3): 173-177 |

| 作者单位 | 题目 | 作者简称或全称 | 通讯作者及单位 | 杂志名称 | 卷期页码 |
| --- | --- | --- | --- | --- | --- |
| | Receptors in Pulmonary Microvascular Endothelial Cells from Chick Embryos | Mengyuan Liu, Zhaoqing Han, Jiakui Li | | | |
| 华中农业大学 | Protective effects of Herpetospermum caudigerum extract against liver injury induced by carbon tetrachloride in mice | Jianfeng Gao, Xiong Jiang, Li Teng, Muhammad Shahzad, Ding Zhang, Bo Hou, Zhaoqing Han, Liqiu Wang, Kerong Zhang, Jiakui Li | Gao, Jianfeng 华中农业大学 | J Invest Med **(SCI IF=1.964)** | 2013, 61 (4), S5-S5 Meeting Abstract |
| 华中农业大学 | Seroprevalence of bovine viral diarrhea infection in Yaks (Bos grunniens) on the Qinghai-Tibetan Plateau of China | Jianfeng Gao, Mengyuan Liu, Xianrong Meng, Zhaoqing Han, Ding Zhang, Bo Hou, Kerong Zhang, Suolang SIzhu, Jiakui Li | 李家奎 华中农业大学 | Trop Anim Health Prod **(SCI IF=1.115)** | 2012, 45 (3):791-793 |
| 华中农业大学 | Effects of Chronic Cadmium Poisoning on Zn, Cu, Fe, Ca and Metallothionein in Liver and Kidney of Rats | Ding Zhang, Jianfeng Gao, Kerong Zhang, Xiaodong Liu, Jiakui Li | 李家奎 华中农业大学 | Biol Trace Elem Res **(SCI IF=1.307)** | 2012, 149 (1) :57–63 |
| 华中农业大学 | Seroprevalence of bovine tuberculosis infection in Yaks (Bos grunniens) on the Qinghai-Tibetan Plateau of China | Zhaoqing Han, Jianfeng Gao, Mengyuan Liu, Kerong Zhang, Ding Zhang, Jiakui Li | 李家奎 华中农业大学 | Trop Anim Health Prod **(SCI IF=1.115)** | 2012, 45: 1277-1279 |
| 华中农业大学 | Seroprevalence of Brucella infection in yaks (Bos grunniens) in Tibet, China | Jianfeng Gao, Kerong Zhang, Muhammad Shahzad, Ding Zhang, Zhaoqing Han, Jiakui Li | 李家奎 华中农业大学 | Cattle Pract **(SCI IF=0.194)** | 2013, 21: 109-111 |
| 华中农业大学 | Estimation and Forecast of the Possible Position Numbers for Licensed Veterinarian in China | Xian Xu, Jinhua Xiao and Ganzhen Deng | Ganzhen Deng, 华中农业大学 | Asian J Ani Vet Adv **(SCI IF=0.87)** | 2014, 9(1): 47-55 |
| 华中农业大学 | Non-classical Major Histocompatibility Complex Class Makes a Crucial Contribution to Reproduction in the Dairy Cow | Lei SHU, Xiuli PENG, Shen ZHANG, Ganzhen DENG, Yue WU, Mingyue HE, Beibei LI, Chengye LI, Kechun | Ganzhen Deng 华中农业大学 | J Reprod Develop **(SCI IF=1.76)** | 2012, 58(5): 569-575 |
| 华中农业大学 | Value of the Pedicle Omentum Transfer for the Healing of Large Skin Wound in Dogs | CaiXia Wang, Chengye Li, Ganzhen Deng, Xian Xu, Lei Shu, | Ganzhen Deng 华中农业大学 | Int J Applied Research in Vet Med **(SCI IF=1.38)** | 2013,10(4):300-304 |

| 作者单位 | 题目 | 作者简称或全称 | 通讯作者及单位 | 杂志名称 | 卷期页码 |
|---|---|---|---|---|---|
| | | Xiangyang Liu, Qihong Chen | | | |
| 北京农学院 | Baicalin Induces IFN-α/βand IFN-γExpressions in Cultured Mouse Pulmonary Microvascular Endothelial Cells | Hu Ge, Xue Jiu-zhou, Liu Jing, Zhang Tao, Dong Hong, Duan Hui-qin, Yang Zuo-jun, Ren Xiao-ming, Mu Xiang | Xiaoming Ren<br>北京农学院 | Journal of Integrative Agriculture (**SCI IF=2.38**) | 2012,11(4):646-654 |
| 北京农学院 | Lactic Acid Reduces LPS-induced TNF-α dnd IL-6 mRNA Levels Through Decreasing IκBα Phosphorylation | G. Xu, J. Shu, X. Ren | Xiaoming Ren<br>北京农学院 | Journal of Integrative Agriculture (**SCI IF=0.53**) | 2013,12(6):1073-1078 |

## 4. 国家、省部级人才培养计划获得者名单

**指本会会员纳入国家和省部级人才培养计划的人员**

李金龙，黑龙江省普通高等学校新世纪优秀人才培养计划，2012

李金龙，教育部新世纪优秀人才培养计划，2013

张志刚，黑龙江省新世纪优秀人才支持计划，2013

孙东波，黑龙江省高校新世纪优秀人才项目获得者，2012

徐　闯，黑龙江省高校新世纪优秀人才项目获得者，2012

郭小权，江西省青年科学家（井冈之星）培养对象，2012

邓俊良，四川省学术和技术带头人

胡延春，四川省学术和技术带头人后备人选

左之才，四川省学术和技术带头人后备人选

# 七、教材与专著

从收集的材料中可知，2012 年 7 月 30 日以前，中国畜牧兽医学会家畜内科学分会的代表主编兽医内科学与兽医诊断学的教材 55 部，主编兽医内科学与诊断学方面的专著 102 部，主译兽医内科学与动物诊断学相关的著作 17 部，编写兽医内科学丛书 18 本。2012 年 8 月 1 日至 2014 年 6 月 30 日，中国畜牧兽医学会家畜内科学分会的代表主编兽医内科学与兽医诊断学的教材 11 部，主编兽医内科学与诊断学方面的专著 19 部，主译兽医内科学与动物诊断学相关的著作 3 部。

## 1.教材

主编兽医内科学与兽医诊断学的教材 11 部。

| 主编 | 教材名称 | 出版社 | 出版时间 |
|---|---|---|---|
| 黄克和， 王小龙 | 兽医临床病理学 | 中国农业出版社 | 2012 |
| 王建华 | 兽医内科学实习指导 | 中国农业出版社 | 2012 |
| 李勤凡 | 兽医临床诊断学实习指导 | 中国农业出版社 | 2012 |
| 庞全海 | 兽医大意 | 中国农业大学出版社 | 2012 |
| 徐世文 | 兽医临床诊疗基础（第三版） | 中国农业出版社 | 2012 |
| 郭定宗 | 兽医实验室诊断指南 | 中国农业出版社 | 2013.6 |
| 左之才 | 小动物疾病诊疗技术 | 河南科学技术出版社 | 2013 |
| 刘建柱 | 动物临床诊断学（十二五规划） | 中国林业出版社 | 2013 |
| 孙东波 | 兽医临床诊断学 | 东北林业大学出版社 | 2013.7 |
| 刘建柱 | 特种经济动物疾病防治学（十二五规划） | 中国农业大学出版社 | 2014 |
| 庞全海 | 兽医内科学 | 中国林业出版社 | 2014 |

## 2.专著

主编兽医内科学与诊断学方面的专著 19 部。

| 主编 | 书名 | 出版社 | 出版时间 |
|---|---|---|---|
| 王俊东 | 环境与健康 | 人民教育出版社 | 2012 |
| 徐世文，郭东华 | 奶牛病防治技术 | 中国农业出版社 | 2012 |
| 王宗元主编，刘宗平副主编 | 动物临床症状鉴别诊断 | 中国农业出版社 | 2013 |

| 主编 | 书名 | 出版社 | 出版时间 |
|---|---|---|---|
| 夏 成，张洪友，徐 闯 | 奶牛酮病与脂肪肝综合征 | 中国农业出版社 | 2013 |
| 李金龙，高 利 | 肉犬安全生产技术指南 | 中国农业出版社 | 2013 |
| 余树民、邓俊良 | 执业兽医考试通关宝典（临床兽医学部分） | 化学工业出版社 | 2013 |
| 左之才，邓俊良 | 全国执业兽医师资格考试通关宝典（临床兽医学部分） | 化学工业出版社 | 2013 |
| 任晓明 | 猪病临床快速诊疗指南 | 中国农业出版社 | 2013 |
| 刘建柱 | 奶牛场技术管理要点和常见疾病防治 | 中国农业出版社 | 2013 |
| 徐庚全，岳海宁 | 执业兽医资格考试单元强化自测与详解（兽医临床） | 中国农业出版社 | 2013 |
| 徐庚全，岳海宁 | 执业兽医资格考试单元强化自测与详解（兽医临床） | 中国农业出版社 | 2014 |
| 王九峰 | 小动物内科学 | 中国农业出版社 | 2013 |
| 刘建柱 | 宠物医师临床急救手册 | 中国农业出版社 | 2014 |
| 刘建柱 | 养殖场处方药速查 | 中国农业出版社 | 2014 |
| 刘建柱 | 宠物处方药速查 | 中国农业出版社 | 2014 |
| 郭东华，孙东波 | 奶牛变形蹄与腐蹄病 | 东北林业大学出版社 | 2014 |
| 叶俊华 | 犬病诊疗技术（第二版） | 中国农业出版社 | 2014 |
| 吴心华 | 肉羊肥育与疾病防治 | 北京金盾出版社 | 2014.5 |
| 吴心华 | 奶牛健康养殖技术 | 宁夏阳光出版社 | 2014.4 |

## 3. 译著

主译兽医内科学与动物诊断学相关的著作 3 部。

| 主译 | 书名 | 出版社 | 出版时间 |
|---|---|---|---|
| 夏兆飞，张海彬 | 小动物内科学 | 中国农业大学出版社 | 2012 |
| 任晓明 | 动物医院临床诊疗技术 | 中国农业科学技术出版社 | 2014.1 |
| 任晓明 | 动物医院临床技术指南 | 中国农业科学技术出版社 | 2014.6 |

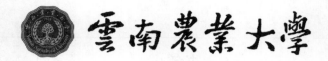

# 雲南農業大學

## 动物科学技术学院简介

学院创建于1950年，前身是国立云南大学农学院畜牧系，1995年成立云南农业大学动物科学技术学院。

办学64年来，学院秉承“开学养正、耕读至诚”的大学精神，立足云南，植根边疆，服务三农，彰显特色，精心培育边疆民族人才，创建特色鲜明的教学研究型学院。

目前学院共有5个本科专业，其中动物科学和动物医学2个专业为云南省省级重点专业；草业和畜牧学2个一级学科博士授权点；畜牧学、兽医学、草学3个一级学科硕士学位授权点；动物遗传育种与繁殖、动物营养与饲料科、特种经济动物养殖、草业科学、预防兽医学、基础兽医学、临床兽医学7个二级学科硕士学位授权点；农业推广硕士和兽医专业硕士学位授予权。

学院现有专任教师115人，其中正高职22人，副高职41人，其中具有博士学位的教师51人，博士生导师8人。学院柔性引进中国工程院院士1人，长江学者1人，国家杰出青年基金获得者1人。学院长期依托云南独具的畜禽品种和饲草饲料资源优势，开展了大量富有成效的科学研究，获得了一批有影响的重大科技成果，并产生了显著的社会经济效益。

近五年来，主持项目及获得的科研成果在数量、层次和水平上有了大幅提升。共承担国家“973”前期专项、“863”计划、国家科技支撑计划、国家自然科学基金、农业科技成果转化资金等国家级项目50余项，主持云南省科技攻关项目、自然科学基金项目等省部级项目60余项，其他各类项目100余项。获省、部级科技成果奖17项（其中云南省科技进步一等奖2项，云南省科技进步二等奖3项，云南省自然科学一等奖1项，云南省自然科学二等奖2项，云南省科技进步三等奖7项，全国农牧渔业丰收二等奖1项、三等奖1项）。

学院科研团队成功培育出世界上第一个大型哺乳动物近交系——版纳微型猪近交系，历时28年，繁育25世代，近交系数达99.47%；成功培育出我国第一个以纯地方猪种选育的专门化品系配套而成的猪配套系——滇撒猪配套系，取得巨大的经济及社会效益。成功获得孤雌生殖克隆猪，该成果不仅是理论上的一个重大突破，而且对研究猪孤雌生殖的机理、生殖发育、遗传疾病以及品种选育等方面都有重要的意义，该科研成果成功入选“2013年云南十大科技进步”。此外，发表SCI文章100余篇，以第一作者及通讯作者发表的单篇最高影响因子为14.4（Genome Research）。

◀ 云南省动物营养与饲料重点实验室

▼ 云南省版纳微型猪近交系重点实验室

# 致　谢

2014 年 7 月 19—21 日，中国畜牧兽医学会兽医内科与临床诊疗学分会 2014 年学术研讨会暨小动物临床技术交流会将在云南昆明召开，为了本次盛会的顺利召开，更为了将兽医内科与临床诊疗学研究成果广泛交流与传播，很多单位、专家、学者都积极参与、鼎力支持，并将两年来已有的研究成果集结成册，付梓出版。

尤其得益于美国动物蛋白及油脂提炼协会（National Renderers Association，NRA）、爱得士缅因生物制品贸易（上海）有限公司、北京世纪元亨动物防疫技术有限公司、北京中拓奕腾科技有限公司、百奥明饲料添加剂（上海）有限公司、长沙绿叶生物科技（集团）有限公司、香港岑氏实业集团、北京东方联鸣科技发展有限公司、北京中联动保科技有限责任公司、无锡祥生医学影像有限责任公司、云南省养犬服务协会、昆明市爱犬协会、云南龙升医药科技有限公司、云南马帮贡茶茶业有限公司、云南农业大学动物科学技术学院等单位的积极参与。

本次会议的承办方云南农业大学动物科学技术学院，为本次会议的圆满召开，进行了细心周详的准备工作。

在此，谨代表中国畜牧兽医学会兽医内科与临床诊疗学分会对这些参与者的工作，致以诚挚的谢意！

中国畜牧兽医学会兽医内科与临床诊疗学分会

2014 年 6 月

**图书在版编目（CIP）数据**

兽医内科与临床诊疗学研究：2012—2014／中国畜牧兽医学会兽医内科与临床诊疗学分会组编. —北京：中国农业科学技术出版社，2014.7

ISBN 978-7-5116-1752-1

Ⅰ.①兽… Ⅱ.①中… Ⅲ.①兽医学—内科学—文集 ②动物疾病—诊疗—文集 Ⅳ.①S85-53

中国版本图书馆 CIP 数据核字（2014）第 145364 号

**责任编辑** 朱 绯
**责任校对** 贾晓红

**出　　版** 中国农业科学技术出版社
北京市中关村南大街 12 号　邮编：100081
**电　　话** （010）82106626（编辑室）　（010）82109702（发行部）　（010）82109709（读者服务部）
**传　　真** （010）82106626
**网　　址** http：//www. castp. cn
**经　　销** 各地新华书店
**印　　刷** 北京富泰印刷有限责任公司
**开　　本** 880 mm×1230 mm　1/16
**印　　张** 25.75
**字　　数** 758 千字
**版　　次** 2014 年 7 月第 1 版　2014 年 7 月第 1 次印刷
**定　　价** 80.00 元